Methods in Enzymology

Volume 362

RECOGNITION OF CARBOHYDRATES IN BIOLOGICAL SYSTEMS

Part A

General Procedures

METHODS IN ENZYMOLOGY

Methods in Enzymology

Volume 362

Recognition of Carbohydrates in Biological Systems

Part A

General Procedures

EDITED BY

Yuan C. Lee

Reiko T. Lee

BIOLOGY DEPARTMENT
THE JOHNS HOPKINS UNIVERSITY
BALTIMORE, MARYLAND

ACADEMIC PRESS

An imprint of Elsevier

Amsterdam Boston Heidelberg London New York Oxford
Paris San Diego San Francisco Singapore Sydney Tokyo

This book is printed on acid-free paper. ♾

Academic Press
An Elsevier Imprint.
525 B Street, Suite 1900, San Diego, California 92101-4495, USA
http://www.academicpress.com

Academic Press
84 Theobald's Road, London WC1X 8RR, UK
http://www.academicpress.com

International Standard Book Number: 0-12-182265-6

PRINTED IN THE UNITED STATES OF AMERICA
03 04 05 06 07 08 9 8 7 6 5 4 3 2 1

Table of Contents

Section I. Preparative Methods

Section II. General Techniques

Contributors to Volume 362

Article numbers are in parentheses and following the names of contributors.
Affiliations listed are current.

BADRULHISAM ADBUL-RAHMAN (512), *Biology Department, Johns Hopkins University, 3400 North Charles Street, Baltimore, Maryland 21218*

SABINE ANDRE (288, 417), *Institute of Physiological Chemistry, Ludwig-Maximilians University, Veterinarstrasse 13, Munich D-80539, Germany*

YOICHIRO ARATA (353), *Department of Biological Chemistry, Teikyo University, Sagamiko, Kanagawa 199-0195, Japan*

KIRAN BACHHAWAT-SIKDER (312), *Center for Biomedical Inventions, University of Texas South Western Medical Center, Dallas, Texas 75390*

MYUNG-GI BAEK (240), *Department of Chemistry, University of Ottawa, Centre for Research in Biopharmaceuticals, Ottawa, Ontario, Canada K1N 6N5*

CAROLYN R. BERTOZZI (250), *Department of Chemistry, University of California, Berkeley, California 94720*

M. JACK BORROK (301), *Departments of Chemistry and Biochemistry, University of Wisconsin, 1101 University Avenue, Madison, Wisconsin 53706*

C. FRED BREWER (455), *Departments of Molecular Pharmacology, and Microbiology and Immunology, Albert Einstein College of Medicine, 1300 Morris Park Avenue, Bronx, New York 10461*

MARIAN C. BRYAN (218), *Department of Chemistry, The Scripps Research Institute, 10550 North Torrey Pines Road, La Jolla, California 92307*

DAVID R. BUNDLE (86), *Department of Chemistry, University of Alberta, Edmonton, Alberta, Canada T6G 2G2*

CHRISTOPHER W. CAIRO (301), *Departments of Chemistry and Biochemistry, University of Wisconsin, 1101 University Avenue, Madison, Wisconsin 53706*

WENGANG CHAI (160), *MRC Glycosciences laboratory, Imperial College London, Northwick Park Hospital Campus, Harrow, Middlesex HA1 3UJ, United Kingdom*

XI CHEN (106), *Neose Technologies, Inc., 102 Witmer Road, Horsham, Pennsylvania 19044*

ANATOLY CHERNYAK (125, 140), *Laboratory of Medicinal Chemistry, Section on Carbohydrates, National Institutes of Health, NIDDK, 8 Center Drive, Building 8, Bethesda, Maryland 20892*

TRINE CHRISTENSEN (486), *Department of Chemistry, Duke University, Durham, North Carolina 27708*

TARUN K. DAM (455, 567), *Departments of Molecular Pharmacology, and Microbiology and Immunology, Albert Einstein College of Medicine, 1300 Morris Park Avenue, Bronx, New York 10461*

SANJOY K. DAS (3, 17), *Department of Chemistry, University of Ottawa, Centre for Research in Biopharmaceuticals, Ottawa, Ontario, Canada K1N 6N5*

ROMYR DOMINIQUE (3), *Department of Chemistry, University of Ottawa, Centre for Research in Biopharmaceuticals, Ottawa, Ontario, Canada K1N 6N5*

KURT DRICKAMER (560), *Department of Biochemistry, Glycobiology Institute, University of Oxford, South Parks Road, Oxford 0X1 3QU, United Kingdom*

DANIELLE H. DUBE (250), *Department of Chemistry, University of California, Berkeley, California 94720*

ERKANG FAN (209), *Departments of Biochemistry and Biological Structure, University of Washington, Biomolecular Structure Center, Seattle, Washington 98195*

JIAN-QIANG FAN (64), *Department of Human Genetics, Mount Sinai School of Medicine, One Gustave L. Levy Place, Box 1498, New York, New York 10029*

TEN FEIZI (160), *MRC Glycosciences laboratory, Imperial College London, Northwick Park Hospital Campus, Harrow, Middlesex HA1 3UJ, United Kingdom*

JURGEN R. FISCHER (288), *Immunology-Molecular Biology Laboratory, Thoraxklinik Heidelberg gGmbHm, Amalienstrasse 5, Heidelberg D-69126, Germany*

MIRNA FLOGEL (29), *Department of Biochemistry and Molecular Biology, University of Zagreb, A. Kovacica 1, Zagreb, 1000 Croatia*

HANS-JOACHIM GABIUS (288, 417), *Institute of Physiological Chemistry, Ludwig-Maximilians University, Veterinarstrasse 13, Munich D-80539, Germany*

CHRISTINE GALUSTIAN (140), *MRC Glycosciences laboratory, Imperial College London, Northwick Park Hospital Campus, Harrow, Middlesex HA1 3UJ, United Kingdom*

ZHONGHONG GAN (17), *Department of Chemistry, University of Ottawa, Centre for Research in Biopharmaceuticals, Ottawa, Ontario, Canada K1N 6N5*

JASON E. GESTWICKI (301), *Departments of Chemistry and Biochemistry, University of Wisconsin, 1101 University Avenue, Madison, Wisconsin 53706*

JOCELYN R. GRUNWELL (250), *Department of Chemistry, University of California, Berkeley, California 94720*

KATSUJI HANEDA (74), *The Noguchi Institute, 1-8-1 Kaga, Itabashi-ku, Tokyo 173-0003, Japan*

HOWARD C. HANG (250), *Department of Chemistry, University of California, Berkeley, California 94720*

OLE HINDSGAUL (369), *Department of Chemistry, University of Alberta, Edmonton, Alberta, Canada T6G 2G2*

JUN HIRABAYASHI (353), *Department of Biological Chemistry, Teikyo University, Sagamiko, Kanagawa 199-0195, Japan*

ANDREAS HOEFLICH (288), *Institute of Molecular Animal Breeding, Ludwig-Maximilians University, Amalienstrasse 5, Heidelberg D-69126, Germany*

SUSUMU HONDA (434), *Faculty of Pharmaceutical Sciences, Kinki University, 3-4-1 Kowakee, Higashi-Osaka 577-8502, Japan*

TOSHIYUKI INAZU (74), *The Noguchi Institute, 1-8-1 Kaga, Itabashi-ku, Tokyo 173-0003, Japan*

SADAKO INOUE (543), *Institute of Biological Chemistry, Academia Sinica, Taipei 11529, Taiwan*

YASUO INOUE (543), *Institute of Biological Chemistry, Academia Sinica, Taipei 11529, Taiwan*

JESUS JIMENEZ-BARBERO (417), *Institute of Physiological Chemistry, Ludwig-Maximilians University, Veterinarstrasse 13, Munich D-80539, Germany*

LEKH RAJ JUNEJA (44), *Central Research Laboratories, Taiyo Kagaku Co., Ltd, Yokkaichi 510-0825, Japan*

BARBRO KAHL-KNUTSSON (504), *Department of Bioorganic Chemistry, Lund University, Lund SE-22100, Sweden*

YASUHIRO KAJIHARA (44), *Graduate School of Integrated Science, Yokohama City University, Yokohama 236-0027, Japan*

HERBERT KALTNER (288, 417), *Institute of Physiological Chemistry, Ludwig-Maximilians University, Veterinarstrasse 13, Munich D-80539, Germany*

MILI KAPOOR (312, 567), *Molecular Biophysics Unit, Indian Institute of Science, Bangalore 560012, India*

ALEX KARAVANOV (125), *Ciphergen Biosystems, Inc., Palo Alto, California 94306-4636*

KEN-ICHI KASAI (353, 376), *Department of Biological Chemistry, Teikyo University, Sagamiko, Kanagawa 199-0195, Japan*

LAURA L. KIESSLING (301), *Departments of Chemistry and Biochemistry, University of Wisconsin, 1101 University Avenue, Madison, Wisconsin 53706*

KAZUAKI KAKEHI (512), *Faculty of Pharmaceutical Sciences, Kinki University, 3-4-1 Kowakee, Higashi-Osaka 577-8502, Japan*

PAVEL I. KITOV (86), *Department of Chemistry, University of Alberta, Edmonton, Alberta, Canada T6G 2G2*

ELENA N. KITOVA (376), *Department of Chemistry, University of Alberta, Edmonton, Alberta, Canada T6G 2G2*

JOHN S. KLASSEN (376), *Department of Chemistry, University of Alberta, Edmonton, Alberta, Canada T6G 2G2*

PAVOL KOVAC (125, 140), *Laboratory of Medicinal Chemistry, Section on Carbohydrates, National Institutes of Health, NIDDK, 8 Center Drive, Building 8, Bethesda, Maryland 20892*

PRZEMYSLAW KOWAL (106), *Department of Chemistry, Wayne State University, Detroit, Michigan 48202*

HARALD LAHM (288), *Immunology-Molecular Biology Laboratory, Thoraxklinik Heidelberg gGmbHm, Amalienstrasse 5, Heidelberg D-69126, Germany*

GORDAN LAUC (29), *Department of Biochemistry and Molecular Biology, University of Zagreb, A. Kovacica 1, Zagreb, 1000 Croatia*

ALEXANDER M. LAWSON (160), *MRC Glycosciences laboratory, Imperial College London, Northwick Park Hospital Campus, Harrow, Middlesex HA1 3UJ, United Kingdom*

REIKO T. LEE (38, 330), *Biology Department, Johns Hopkins University, 3400 North Charles Street, Baltimore, Maryland 21218*

YUAN C. LEE (29, 38, 330), *Biology Department, Johns Hopkins University, 3400 North Charles Street, Baltimore, Maryland 21218*

HAKON LEFFLER (504), *Department of Bioorganic Chemistry, Lund University, Lund SE-22100, Sweden*

FU-SEN LIANG (340), *Department of Chemistry, The Scripps Research Institute, 10550 North Torrey Pines Road, La Jolla, California 92307*

BINCAN LIU (17, 226), *Department of Chemistry, University of Ottawa, Centre for Research in Biopharmaceuticals, Ottawa, Ontario, Canada K1N 6N5*

ZIYE LIU (106), *Department of Chemistry, Wayne State University, Detroit, Michigan 48202*

YUQUAN LU (106), *Department of Chemistry, Wayne State University, Detroit, Michigan 48202*

SARAH J. LUCHANSKY (250), *Department of Chemistry, University of California, Berkeley, California 94720*

SANDRA MISQUITH (567), *Department of Chemistry, St. Joseph's College, Bangalore 560025, India*

MAMORU MIZUNO (74), *The Noguchi Institute, 1-8-1 Kaga, Itabashi-ku, Tokyo 173-0003, Japan*

KENJI MONDE (274), *Laboratory of Bio-Macromolecular Chemistry, Division of Biological Sciences, Hokkaido University, N10 W8, Kita-ku, Sapporo 060-0810, Japan*

JOE NAHRA (17), *Department of Chemistry, University of Ottawa, Centre for Research in Biopharmaceuticals, Ottawa, Ontario, Canada K1N 6N5*

AYUMI NATSUME (196), *Department of Advanced Biosciences, Ochanomizu University, 2-1-1 Otsuka, Bunkyo-ku, Tokyo 112-8610, Japan*

KENICHI NIKURA (274), *Sapporo Laboratory for Glycocluster Project, Japan Bioindustry Association, N10 W8, Kita-ku, Sapporo 060-0810, Japan*

ULF J. NILSSON (504), *Department of Bioorganic Chemistry, Lund University, Lund SE-22100, Sweden*

SHIN-ICHIRO NISHIMURA (274), *Laboratory of Bio-Macromolecular Chemistry, Division of Biological Sciences, Hokkaido University, N10 W8, Kita-ku, Sapporo 060-0810, Japan*

YASUO ODA (512), *Faculty of Pharmaceutical Sciences, Kinki University, 3-4-1 Kowakee, Higashi-Osaka 577-8502, Japan*

HARUKO OGAWA (196), *Department of Advanced Biosciences, Ochanomizu University, 2-1-1 Otsuka, Bunkyo-ku, Tokyo 112-8610, Japan*

MONICA M. PALCIC (369), *Department of Chemistry, University of Alberta, Edmonton, Alberta, Canada T6G 2G2*

EUGENIA PASZKIEWICZ (86), *Department of Chemistry, University of Alberta, Edmonton, Alberta, Canada T6G 2G2*

ANITA RAMDAS PATIL (567), *Department of Medicinal Chemistry, University of Kansas, Lawrence, Kansas 66045*

EMMANUAL POIROT (140), *Laboratory of Medicinal Chemistry, Section on Carbohydrates, National Institutes of Health, NIDDK, 8 Center Drive, Building 8, Bethesda, Maryland 20892*

XIANGPING QIAN (369), *Department of Chemistry, University of Alberta, Edmonton, Alberta, Canada T6G 2G2*

BRIAN REMPEL (369), *Department of Chemistry, University of Alberta, Edmonton, Alberta, Canada T6G 2G2*

RENE ROY (3, 17, 226, 240), *Department of Chemistry, University of Ottawa, Centre for Research in Biopharmaceuticals, Ottawa, Ontario, Canada K1N 6N5*

REIKO SADAMOTO (274), *Sapporo Laboratory for Glycocluster Project, Japan Bioindustry Association, N10 W8, Kita-ku, Sapporo 060-0810, Japan*

RINA SAKSENA (125, 140), *Laboratory of Medicinal Chemistry, Section on Carbohydrates, National Institutes of Health, NIDDK, 8 Center Drive, Building 8, Bethesda, Maryland 20892*

KEN SASAKI (44), *Central Research Laboratories, Taiyo Kagaku Co., Ltd, Yokkaichi 510-0825, Japan*

ELIANA SAXON (250), *Department of Chemistry, University of California, Berkeley, California 94720*

BRAD SCHMOR (17), *Department of Chemistry, University of Ottawa, Centre for Research in Biopharmaceuticals, Ottawa, Ontario, Canada K1N 6N5*

JUN SHAO (106), *Department of Chemistry, Wayne State University, Detroit, Michigan 48202*

SHILPI SHARMA (312), *Molecular Biophysics Unit, Indian Institute of Science, Bangalore 560012, India*

VIVEK SHARMA (567), *Centre for Structural Biology, Institute of Biosciences and Technology, 2121 West Holcombe Boulevard, Houston, Texas 77030*

JEANE SHI (523), *Bioscience Division, Los Alamos National Laboratory, Los Alamos, New Mexico 87545*

KIYOHITO SHIMURA (398), *Department of Biological Chemistry, Teikyo University, Sagamiko, Kanagawa 199-0195, Japan*

YASURO SHINOHARA (330), *Tokyo R & D, Amersham Biosciences, 3-25-1, Hyakunincho, Shinjuku-ku, Tokyo 169-0073, Japan*

HANS-CHRISTIAN SIEBERT (417), *Institute of Physiological Chemistry, Ludwig-Maximilians University, Veterinarstrasse 13, Munich D-80539, Germany*

XUEDONG SONG (523), *Corporate Emerging Technologies, Kimberly-Clark Corporation, 1400 Holcomb Bridge Road, Roswell, Georgia 30076*

BERNARD SORDAT (288), *Experimental Pathology Unit, Swiss Institute for Experimental Cancer Research, Chemin des Boveresses 155, Espalinges CH-1066, Switzerland*

PERNILLA SORME (504), *Department of Bioorganic Chemistry, Lund University, Lund SE-22100, Sweden*

HIROYUKI SOTA (330), *Tokyo R & D, Amersham Biosciences, 3-25-1, Hyakunincho, Shinjuku-ku, Tokyo 169-0073, Japan*

MARK S. STOLL (160), *MRC Glycosciences laboratory, Imperial College London, Northwick Park Hospital Campus, Harrow, Middlesex HA1 3UJ, United Kingdom*

SANDRA SUPRAHA (29), *Department of Biochemistry and Molecular Biology, University of Zagreb, A. Kovacica 1, Zagreb, 1000 Croatia*

AVADHESHA SUROLIA (312, 567), *Molecular Biophysics Unit, Indian Institute of Science, Bangalore 560012, India*

YASUHIRO SUZUKI (44), *Graduate School of Integrated Science, Yokohama City University, Yokohama 236-0027, Japan*

BASIL I. SWANSON (523), *Bioscience Division, Los Alamos National Laboratory, Los Alamos, New Mexico 87545*

ATSUSHI TAGA (434), *Faculty of Pharmaceutical Sciences, Kinki University, 3-4-1 Kowakee, Higashi-Osaka 577-8502, Japan*

KAORU TAKEGAWA (64), *Department of Life Sciences, Kagawa University, Mikicho, Kita-gun, Kagawa 761-0795, Japan*

MAUREEN E. TAYLOR (560), *Department of Biochemistry, Glycobiology Institute, University of Oxford, South Parks Road, Oxford 0X1 3QU, United Kingdom*

CELESTINE J. THOMAS (312), *Center for Biomedical Inventions, University of Texas South Western Medical Center, Dallas, Texas 75390*

ERIC J. TOONE (486), *Department of Chemistry, Duke University, Durham, North Carolina 27708*

M. CORAZON TRONO (3), *Department of Chemistry, University of Ottawa, Centre for Research in Biopharmaceuticals, Ottawa, Ontario, Canada K1N 6N5*

HARUKO UEDA (196), *Department of Advanced Biosciences, Ochanomizu University, 2-1-1 Otsuka, Bunkyo-ku, Tokyo 112-8610, Japan*

WARREN W. WAKARCHUK (86), *Institute of Biological Sciences, National Research Council of Canada, Ottawa, Ontario, Canada K1A 0R6*

PENG G. WANG (106), *Department of Chemistry, Wayne State University, Detroit, Michigan 48202*

WEIJIE WANG (376), *Department of Chemistry, University of Alberta, Edmonton, Alberta, Canada T6G 2G2*

ULF WELLMAR (504), *Lundonia Biotech AB, IDEON Research Park, Lund SE-22370, Sweden*

ECKHARD WOLF (288), *Institute of Molecular Animal Breeding, Ludwig-Maximilians University, Amalienstrasse 5, Heidelberg D-69126, Germany*

CHI-HUEY WONG (218, 340), *Department of Chemistry, The Scripps Research Institute, 10550 North Torrey Pines Road, La Jolla, California 92307*

KENJI YAMAMOTO (74), *Graduate School of Biostudies, Kyoto University, Kita-Shirakawa, Sakyo-ku, Kyoto 606-8502, Japan*

CHONG YU (250), *Department of Chemistry, University of California, Berkeley, California 94720*

BOYAN ZHANG (369), *Department of Chemistry, University of Alberta, Edmonton, Alberta, Canada T6G 2G2*

JIANBO ZHANG (106), *Department of Chemistry, Wayne State University, Detroit, Michigan 48202*

ZHONGSHENG ZHANG (209), *Departments of Biochemistry and Biological Structure, University of Washington, Biomolecular Structure Center, Seattle, Washington 98195*

Preface

After genomics and proteomics, glycomics is often said to be the next frontier. Like genomics and proteomics, glycomics includes both the determination of structures and the deciphering of functions and correlations between them. Structural determination of carbohydrates is orders of magnitude more difficult than those of nucleotides and peptides because of the inherent microheterogeneity associated with glycan structures as well as the innate complexity of the structures—anomeric configurations, positional isomers, and various branching patterns. Happily, there has been a great deal of progress in this area of glycomics, although much more is needed for the full maturation of this field.

Accompanying the progress in structural determination, the role of carbohydrates in biological systems is also beginning to be better understood. The progress in this area owes a great deal to newer technologies for measuring biological interactions involving carbohydrates ("recognition of Carbohydrates"). The areas in which carbohydrates play a role are indeed very broad, covering the clearance of hormone and other materials from circulation, cell–cell and cell–matrix adhesion, fertilization, cell migration and trafficking, parasitic invasion, bacterial and viral attachment, and even apoptosis. We felt it was time to gather many diverse methodologies and concepts in a single volume which would cover wide areas of carbohydrate recognition. We received such an overwhelming response from potential contributors that our original plan of a single volume of Methods in Enzymology developed into two volumes. Volume 362, Part A: General Procedures covers Preparative Methods and General Techniques. Volume 363, Part B: Specific Applications includes sections on Carbohydrate-Binding Proteins, Glycoproteins and Glycolipids, Polysaccharides, and Enzymes and Cells. The division and categorization are rather arbitrary, but we feel reasonable.

The scope of these volumes is by no means comprehensive, since it would be impossible to include all interesting areas of carbohydrate biology in two volumes. But they do present a sampling of established as well as newer methodology and findings. We received overwhelming positive support and cooperation from the invited authors, and express our sincere gratitude to all of them, for without their conscientious undertaking and timely submission of the articles, these volumes would not have been possible.

Yuan C. Lee
Reiko T. Lee

METHODS IN ENZYMOLOGY

VOLUME XXXV. Lipids (Part B)
Edited by JOHN M. LOWENSTEIN

VOLUME XXXVI. Hormone Action (Part A: Steroid Hormones)
Edited by BERT W. O'MALLEY AND JOEL G. HARDMAN

VOLUME XXXVII. Hormone Action (Part B: Peptide Hormones)
Edited by BERT W. O'MALLEY AND JOEL G. HARDMAN

VOLUME XXXVIII. Hormone Action (Part C: Cyclic Nucleotides)
Edited by JOEL G. HARDMAN AND BERT W. O'MALLEY

VOLUME XXXIX. Hormone Action (Part D: Isolated Cells, Tissues, and Organ Systems)
Edited by JOEL G. HARDMAN AND BERT W. O'MALLEY

VOLUME XL. Hormone Action (Part E: Nuclear Structure and Function)
Edited by BERT W. O'MALLEY AND JOEL G. HARDMAN

VOLUME XLI. Carbohydrate Metabolism (Part B)
Edited by W. A. WOOD

VOLUME XLII. Carbohydrate Metabolism (Part C)
Edited by W. A. WOOD

VOLUME XLIII. Antibiotics
Edited by JOHN H. HASH

VOLUME XLIV. Immobilized Enzymes
Edited by KLAUS MOSBACH

VOLUME XLV. Proteolytic Enzymes (Part B)
Edited by LASZLO LORAND

VOLUME XLVI. Affinity Labeling
Edited by WILLIAM B. JAKOBY AND MEIR WILCHEK

VOLUME XLVII. Enzyme Structure (Part E)
Edited by C. H. W. HIRS AND SERGE N. TIMASHEFF

VOLUME XLVIII. Enzyme Structure (Part F)
Edited by C. H. W. HIRS AND SERGE N. TIMASHEFF

VOLUME XLIX. Enzyme Structure (Part G)
Edited by C. H. W. HIRS AND SERGE N. TIMASHEFF

VOLUME L. Complex Carbohydrates (Part C)
Edited by VICTOR GINSBURG

VOLUME LI. Purine and Pyrimidine Nucleotide Metabolism
Edited by PATRICIA A. HOFFEE AND MARY ELLEN JONES

VOLUME LII. Biomembranes (Part C: Biological Oxidations)
Edited by SIDNEY FLEISCHER AND LESTER PACKER

VOLUME LIII. Biomembranes (Part D: Biological Oxidations)
Edited by SIDNEY FLEISCHER AND LESTER PACKER

VOLUME LIV. Biomembranes (Part E: Biological Oxidations)
Edited by SIDNEY FLEISCHER AND LESTER PACKER

VOLUME LV. Biomembranes (Part F: Bioenergetics)
Edited by SIDNEY FLEISCHER AND LESTER PACKER

VOLUME LVI. Biomembranes (Part G: Bioenergetics)
Edited by SIDNEY FLEISCHER AND LESTER PACKER

VOLUME LVII. Bioluminescence and Chemiluminescence
Edited by MARLENE A. DELUCA

VOLUME LVIII. Cell Culture
Edited by WILLIAM B. JAKOBY AND IRA PASTAN

VOLUME LIX. Nucleic Acids and Protein Synthesis (Part G)
Edited by KIVIE MOLDAVE AND LAWRENCE GROSSMAN

VOLUME LX. Nucleic Acids and Protein Synthesis (Part H)
Edited by KIVIE MOLDAVE AND LAWRENCE GROSSMAN

VOLUME 61. Enzyme Structure (Part H)
Edited by C. H. W. HIRS AND SERGE N. TIMASHEFF

VOLUME 62. Vitamins and Coenzymes (Part D)
Edited by DONALD B. MCCORMICK AND LEMUEL D. WRIGHT

VOLUME 63. Enzyme Kinetics and Mechanism (Part A: Initial Rate and Inhibitor Methods)
Edited by DANIEL L. PURICH

VOLUME 64. Enzyme Kinetics and Mechanism (Part B: Isotopic Probes and Complex Enzyme Systems)
Edited by DANIEL L. PURICH

VOLUME 65. Nucleic Acids (Part I)
Edited by LAWRENCE GROSSMAN AND KIVIE MOLDAVE

VOLUME 66. Vitamins and Coenzymes (Part E)
Edited by DONALD B. MCCORMICK AND LEMUEL D. WRIGHT

VOLUME 67. Vitamins and Coenzymes (Part F)
Edited by DONALD B. MCCORMICK AND LEMUEL D. WRIGHT

VOLUME 68. Recombinant DNA
Edited by RAY WU

VOLUME 69. Photosynthesis and Nitrogen Fixation (Part C)
Edited by ANTHONY SAN PIETRO

VOLUME 70. Immunochemical Techniques (Part A)
Edited by HELEN VAN VUNAKIS AND JOHN J. LANGONE

VOLUME 71. Lipids (Part C)
Edited by JOHN M. LOWENSTEIN

Volume 107. Posttranslational Modifications (Part B)
Edited by Finn Wold and Kivie Moldave

Volume 108. Immunochemical Techniques (Part G: Separation and Characterization of Lymphoid Cells)
Edited by Giovanni Di Sabato, John J. Langone, and Helen Van Vunakis

Volume 109. Hormone Action (Part I: Peptide Hormones)
Edited by Lutz Birnbaumer and Bert W. O'Malley

Volume 110. Steroids and Isoprenoids (Part A)
Edited by John H. Law and Hans C. Rilling

Volume 111. Steroids and Isoprenoids (Part B)
Edited by John H. Law and Hans C. Rilling

Volume 112. Drug and Enzyme Targeting (Part A)
Edited by Kenneth J. Widder and Ralph Green

Volume 113. Glutamate, Glutamine, Glutathione, and Related Compounds
Edited by Alton Meister

Volume 114. Diffraction Methods for Biological Macromolecules (Part A)
Edited by Harold W. Wyckoff, C. H. W. Hirs, and Serge N. Timasheff

Volume 115. Diffraction Methods for Biological Macromolecules (Part B)
Edited by Harold W. Wyckoff, C. H. W. Hirs, and Serge N. Timasheff

Volume 116. Immunochemical Techniques (Part H: Effectors and Mediators of Lymphoid Cell Functions)
Edited by Giovanni Di Sabato, John J. Langone, and Helen Van Vunakis

Volume 117. Enzyme Structure (Part J)
Edited by C. H. W. Hirs and Serge N. Timasheff

Volume 118. Plant Molecular Biology
Edited by Arthur Weissbach and Herbert Weissbach

Volume 119. Interferons (Part C)
Edited by Sidney Pestka

Volume 120. Cumulative Subject Index Volumes 81–94, 96–101

Volume 121. Immunochemical Techniques (Part I: Hybridoma Technology and Monoclonal Antibodies)
Edited by John J. Langone and Helen Van Vunakis

Volume 122. Vitamins and Coenzymes (Part G)
Edited by Frank Chytil and Donald B. McCormick

Volume 123. Vitamins and Coenzymes (Part H)
Edited by Frank Chytil and Donald B. McCormick

Volume 124. Hormone Action (Part J: Neuroendocrine Peptides)
Edited by P. Michael Conn

VOLUME 140. Cumulative Subject Index Volumes 102–119, 121–134

VOLUME 141. Cellular Regulators (Part B: Calcium and Lipids)
Edited by P. MICHAEL CONN AND ANTHONY R. MEANS

VOLUME 142. Metabolism of Aromatic Amino Acids and Amines
Edited by SEYMOUR KAUFMAN

VOLUME 143. Sulfur and Sulfur Amino Acids
Edited by WILLIAM B. JAKOBY AND OWEN GRIFFITH

VOLUME 144. Structural and Contractile Proteins (Part D: Extracellular Matrix)
Edited by LEON W. CUNNINGHAM

VOLUME 145. Structural and Contractile Proteins (Part E: Extracellular Matrix)
Edited by LEON W. CUNNINGHAM

VOLUME 146. Peptide Growth Factors (Part A)
Edited by DAVID BARNES AND DAVID A. SIRBASKU

VOLUME 147. Peptide Growth Factors (Part B)
Edited by DAVID BARNES AND DAVID A. SIRBASKU

VOLUME 148. Plant Cell Membranes
Edited by LESTER PACKER AND ROLAND DOUCE

VOLUME 149. Drug and Enzyme Targeting (Part B)
Edited by RALPH GREEN AND KENNETH J. WIDDER

VOLUME 150. Immunochemical Techniques (Part K: *In Vitro* Models of B and T Cell Functions and Lymphoid Cell Receptors)
Edited by GIOVANNI DI SABATO

VOLUME 151. Molecular Genetics of Mammalian Cells
Edited by MICHAEL M. GOTTESMAN

VOLUME 152. Guide to Molecular Cloning Techniques
Edited by SHELBY L. BERGER AND ALAN R. KIMMEL

VOLUME 153. Recombinant DNA (Part D)
Edited by RAY WU AND LAWRENCE GROSSMAN

VOLUME 154. Recombinant DNA (Part E)
Edited by RAY WU AND LAWRENCE GROSSMAN

VOLUME 155. Recombinant DNA (Part F)
Edited by RAY WU

VOLUME 156. Biomembranes (Part P: ATP-Driven Pumps and Related Transport: The Na, K-Pump)
Edited by SIDNEY FLEISCHER AND BECCA FLEISCHER

VOLUME 157. Biomembranes (Part Q: ATP-Driven Pumps and Related Transport: Calcium, Proton, and Potassium Pumps)
Edited by SIDNEY FLEISCHER AND BECCA FLEISCHER

VOLUME 158. Metalloproteins (Part A)
Edited by JAMES F. RIORDAN AND BERT L. VALLEE

VOLUME 159. Initiation and Termination of Cyclic Nucleotide Action
Edited by JACKIE D. CORBIN AND ROGER A. JOHNSON

VOLUME 160. Biomass (Part A: Cellulose and Hemicellulose)
Edited by WILLIS A. WOOD AND SCOTT T. KELLOGG

VOLUME 161. Biomass (Part B: Lignin, Pectin, and Chitin)
Edited by WILLIS A. WOOD AND SCOTT T. KELLOGG

VOLUME 162. Immunochemical Techniques (Part L: Chemotaxis and Inflammation)
Edited by GIOVANNI DI SABATO

VOLUME 163. Immunochemical Techniques (Part M: Chemotaxis and Inflammation)
Edited by GIOVANNI DI SABATO

VOLUME 164. Ribosomes
Edited by HARRY F. NOLLER, JR., AND KIVIE MOLDAVE

VOLUME 165. Microbial Toxins: Tools for Enzymology
Edited by SIDNEY HARSHMAN

VOLUME 166. Branched-Chain Amino Acids
Edited by ROBERT HARRIS AND JOHN R. SOKATCH

VOLUME 167. Cyanobacteria
Edited by LESTER PACKER AND ALEXANDER N. GLAZER

VOLUME 168. Hormone Action (Part K: Neuroendocrine Peptides)
Edited by P. MICHAEL CONN

VOLUME 169. Platelets: Receptors, Adhesion, Secretion (Part A)
Edited by JACEK HAWIGER

VOLUME 170. Nucleosomes
Edited by PAUL M. WASSARMAN AND ROGER D. KORNBERG

VOLUME 171. Biomembranes (Part R: Transport Theory: Cells and Model Membranes)
Edited by SIDNEY FLEISCHER AND BECCA FLEISCHER

VOLUME 172. Biomembranes (Part S: Transport: Membrane Isolation and Characterization)
Edited by SIDNEY FLEISCHER AND BECCA FLEISCHER

VOLUME 173. Biomembranes [Part T: Cellular and Subcellular Transport: Eukaryotic (Nonepithelial) Cells]
Edited by SIDNEY FLEISCHER AND BECCA FLEISCHER

VOLUME 174. Biomembranes [Part U: Cellular and Subcellular Transport: Eukaryotic (Nonepithelial) Cells]
Edited by SIDNEY FLEISCHER AND BECCA FLEISCHER

VOLUME 193. Mass Spectrometry
Edited by JAMES A. MCCLOSKEY

VOLUME 194. Guide to Yeast Genetics and Molecular Biology
Edited by CHRISTINE GUTHRIE AND GERALD R. FINK

VOLUME 195. Adenylyl Cyclase, G Proteins, and Guanylyl Cyclase
Edited by ROGER A. JOHNSON AND JACKIE D. CORBIN

VOLUME 196. Molecular Motors and the Cytoskeleton
Edited by RICHARD B. VALLEE

VOLUME 197. Phospholipases
Edited by EDWARD A. DENNIS

VOLUME 198. Peptide Growth Factors (Part C)
Edited by DAVID BARNES, J. P. MATHER, AND GORDON H. SATO

VOLUME 199. Cumulative Subject Index Volumes 168–174, 176–194

VOLUME 200. Protein Phosphorylation (Part A: Protein Kinases: Assays, Purification, Antibodies, Functional Analysis, Cloning, and Expression)
Edited by TONY HUNTER AND BARTHOLOMEW M. SEFTON

VOLUME 201. Protein Phosphorylation (Part B: Analysis of Protein Phosphorylation, Protein Kinase Inhibitors, and Protein Phosphatases)
Edited by TONY HUNTER AND BARTHOLOMEW M. SEFTON

VOLUME 202. Molecular Design and Modeling: Concepts and Applications (Part A: Proteins, Peptides, and Enzymes)
Edited by JOHN J. LANGONE

VOLUME 203. Molecular Design and Modeling: Concepts and Applications (Part B: Antibodies and Antigens, Nucleic Acids, Polysaccharides, and Drugs)
Edited by JOHN J. LANGONE

VOLUME 204. Bacterial Genetic Systems
Edited by JEFFREY H. MILLER

VOLUME 205. Metallobiochemistry (Part B: Metallothionein and Related Molecules)
Edited by JAMES F. RIORDAN AND BERT L. VALLEE

VOLUME 206. Cytochrome P450
Edited by MICHAEL R. WATERMAN AND ERIC F. JOHNSON

VOLUME 207. Ion Channels
Edited by BERNARDO RUDY AND LINDA E. IVERSON

VOLUME 208. Protein–DNA Interactions
Edited by ROBERT T. SAUER

VOLUME 209. Phospholipid Biosynthesis
Edited by EDWARD A. DENNIS AND DENNIS E. VANCE

VOLUME 244. Proteolytic Enzymes: Serine and Cysteine Peptidases
Edited by ALAN J. BARRETT

VOLUME 245. Extracellular Matrix Components
Edited by E. RUOSLAHTI AND E. ENGVALL

VOLUME 246. Biochemical Spectroscopy
Edited by KENNETH SAUER

VOLUME 247. Neoglycoconjugates (Part B: Biomedical Applications)
Edited by Y. C. LEE AND REIKO T. LEE

VOLUME 248. Proteolytic Enzymes: Aspartic and Metallo Peptidases
Edited by ALAN J. BARRETT

VOLUME 249. Enzyme Kinetics and Mechanism (Part D: Developments in Enzyme Dynamics)
Edited by DANIEL L. PURICH

VOLUME 250. Lipid Modifications of Proteins
Edited by PATRICK J. CASEY AND JANICE E. BUSS

VOLUME 251. Biothiols (Part A: Monothiols and Dithiols, Protein Thiols, and Thiyl Radicals)
Edited by LESTER PACKER

VOLUME 252. Biothiols (Part B: Glutathione and Thioredoxin; Thiols in Signal Transduction and Gene Regulation)
Edited by LESTER PACKER

VOLUME 253. Adhesion of Microbial Pathogens
Edited by RON J. DOYLE AND ITZHAK OFEK

VOLUME 254. Oncogene Techniques
Edited by PETER K. VOGT AND INDER M. VERMA

VOLUME 255. Small GTPases and Their Regulators (Part A: Ras Family)
Edited by W. E. BALCH, CHANNING J. DER, AND ALAN HALL

VOLUME 256. Small GTPases and Their Regulators (Part B: Rho Family)
Edited by W. E. BALCH, CHANNING J. DER, AND ALAN HALL

VOLUME 257. Small GTPases and Their Regulators (Part C: Proteins Involved in Transport)
Edited by W. E. BALCH, CHANNING J. DER, AND ALAN HALL

VOLUME 258. Redox-Active Amino Acids in Biology
Edited by JUDITH P. KLINMAN

VOLUME 259. Energetics of Biological Macromolecules
Edited by MICHAEL L. JOHNSON AND GARY K. ACKERS

VOLUME 260. Mitochondrial Biogenesis and Genetics (Part A)
Edited by GIUSEPPE M. ATTARDI AND ANNE CHOMYN

VOLUME 261. Nuclear Magnetic Resonance and Nucleic Acids
Edited by THOMAS L. JAMES

Volume 300. Oxidants and Antioxidants (Part B)
Edited by Lester Packer

Volume 301. Nitric Oxide: Biological and Antioxidant Activities (Part C)
Edited by Lester Packer

Volume 302. Green Fluorescent Protein
Edited by P. Michael Conn

Volume 303. cDNA Preparation and Display
Edited by Sherman M. Weissman

Volume 304. Chromatin
Edited by Paul M. Wassarman and Alan P. Wolffe

Volume 305. Bioluminescence and Chemiluminescence (Part C)
Edited by Thomas O. Baldwin and Miriam M. Ziegler

Volume 306. Expression of Recombinant Genes in Eukaryotic Systems
Edited by Joseph C. Glorioso and Martin C. Schmidt

Volume 307. Confocal Microscopy
Edited by P. Michael Conn

Volume 308. Enzyme Kinetics and Mechanism (Part E: Energetics of Enzyme Catalysis)
Edited by Daniel L. Purich and Vern L. Schramm

Volume 309. Amyloid, Prions, and Other Protein Aggregates
Edited by Ronald Wetzel

Volume 310. Biofilms
Edited by Ron J. Doyle

Volume 311. Sphingolipid Metabolism and Cell Signaling (Part A)
Edited by Alfred H. Merrill, Jr., and Yusuf A. Hannun

Volume 312. Sphingolipid Metabolism and Cell Signaling (Part B)
Edited by Alfred H. Merrill, Jr., and Yusuf A. Hannun

Volume 313. Antisense Technology (Part A: General Methods, Methods of Delivery, and RNA Studies)
Edited by M. Ian Phillips

Volume 314. Antisense Technology (Part B: Applications)
Edited by M. Ian Phillips

Volume 315. Vertebrate Phototransduction and the Visual Cycle (Part A)
Edited by Krzysztof Palczewski

Volume 316. Vertebrate Phototransduction and the Visual Cycle (Part B)
Edited by Krzysztof Palczewski

Volume 317. RNA–Ligand Interactions (Part A: Structural Biology Methods)
Edited by Daniel W. Celander and John N. Abelson

VOLUME 336. Microbial Growth in Biofilms (Part A: Developmental and Molecular Biological Aspects)
Edited by RON J. DOYLE

VOLUME 337. Microbial Growth in Biofilms (Part B: Special Environments and Physicochemical Aspects)
Edited by RON J. DOYLE

VOLUME 338. Nuclear Magnetic Resonance of Biological Macromolecules (Part A)
Edited by THOMAS L. JAMES, VOLKER DÖTSCH, AND ULI SCHMITZ

VOLUME 339. Nuclear Magnetic Resonance of Biological Macromolecules (Part B)
Edited by THOMAS L. JAMES, VOLKER DÖTSCH, AND ULI SCHMITZ

VOLUME 340. Drug–Nucleic Acid Interactions
Edited by JONATHAN B. CHAIRES AND MICHAEL J. WARING

VOLUME 341. Ribonucleases (Part A)
Edited by ALLEN W. NICHOLSON

VOLUME 342. Ribonucleases (Part B)
Edited by ALLEN W. NICHOLSON

VOLUME 343. G Protein Pathways (Part A: Receptors)
Edited by RAVI IYENGAR AND JOHN D. HILDEBRANDT

VOLUME 344. G Protein Pathways (Part B: G Proteins and Their Regulators)
Edited by RAVI IYENGAR AND JOHN D. HILDEBRANDT

VOLUME 345. G Protein Pathways (Part C: Effector Mechanisms)
Edited by RAVI IYENGAR AND JOHN D. HILDEBRANDT

VOLUME 346. Gene Therapy Methods
Edited by M. IAN PHILLIPS

VOLUME 347. Protein Sensors and Reactive Oxygen Species (Part A: Selenoproteins and Thioredoxin)
Edited by HELMUT SIES AND LESTER PACKER

VOLUME 348. Protein Sensors and Reactive Oxygen Species (Part B: Thiol Enzymes and Proteins)
Edited by HELMUT SIES AND LESTER PACKER

VOLUME 349. Superoxide Dismutase
Edited by LESTER PACKER

VOLUME 350. Guide to Yeast Genetics and Molecular and Cell Biology (Part B)
Edited by CHRISTINE GUTHRIE AND GERALD R. FINK

VOLUME 351. Guide to Yeast Genetics and Molecular and Cell Biology (Part C)
Edited by CHRISTINE GUTHRIE AND GERALD R. FINK

VOLUME 352. Redox Cell Biology and Genetics (Part A)
Edited by CHANDAN K. SEN AND LESTER PACKER

Section I

Preparative Methods

[1] Transition Metal-Catalyzed Cross-Coupling Reactions toward the Synthesis of α-D-Mannopyranoside Clusters

By SANJOY K. DAS, M. CORAZON TRONO, and RENÉ ROY

Introduction

Cell surface carbohydrates are well-recognized markers that can mediate "specific" carbohydrate–protein interactions leading to a wide range of cell adhesions. However, it is increasingly surprising to find that the key binding interactions are limited to a handful of carbohydrate ligands of limited "complexity" and yet lead to a vast range of different biological consequences. One emerging postulate is that the diversity may originate more from the varied topography of these simple glycoconjugates than from their intrinsic molecular identities.[1] As mentioned in the companion article related to galactosides (see [16] in this volume) the same trends seem to hold for α-D-mannopyranosides, in which both macrophages[2] and mannose-binding proteins[3,4] from the innate immune system play analogous but subtly different roles wherein the inherent structures of mannoside clusters are key factors for specificity.[5] Moreover, this trend also prevails with other animal and plant lectins having α-D-mannoside-binding requirements.[6]

In ongoing activities directed at the design of small glycoclusters of defined topologies,[7] we were interested in using transition metal-catalyzed reactions that could readily provide carbohydrate assembly together with imparting conformational restrictions.[8] To this end, we describe herein the useful manipulations of palladium-catalyzed cross-coupling reactions leading to "sugar rods" having aryl–alkyne linkages, together with ruthenium carbenoid- or dicobalt octacarbonyl-catalyzed cyclotrimerization of

[1] Y. C. Lee and R. T. Lee, *Acc. Chem. Res.* **28,** 321 (1995).
[2] P. D. Stahl, *Curr. Opin. Immunol.* **4,** 49 (1992).
[3] J. Epstein, Q. Eichbaum, S. Sheriff, and R. A. B. Ezekowitz, *Curr. Opin. Immunol.* **8,** 29 (1996).
[4] T. Kawasaki, *Biochim. Biophys. Acta* **1473,** 186 (1999).
[5] M. S. Quesenberry, R. T. Lee, and Y. C. Lee, *Biochemistry* **36,** 2724 (1997).
[6] H. Lis and N. Sharon, *Chem. Rev.* **98,** 637 (1998).
[7] R. Roy, *Top. Curr. Chem.* **187,** 241 (1997).
[8] R. Roy, S. K. Das, F. Hernández-Meteo, F. Santoyo-González, and Z. Gan, *Synthesis* **7,** 1049 (2001).

0076-6879/03 $35.00

mono- and disubstituted alkyne-containing carbohydrate moieties.[9] Thus, in the last two steps of operation, dimannoside clusters interconnected by an acetylene bridge are transformed in one step into hexamers built around a benzene nucleus. The relative cross-linking abilities of these compounds are evaluated with the phytohemagglutinin concanavalin A from *Canavalia ensiformis*.

Discussions

The required 4-iodophenyl (**2**) and prop-2-ynyl (**3**) α-D-mannopyranoside residues are prepared by a standard procedure involving the treatment of mannose pentaacetate (**1**) with either 4-iodophenol or propargyl alcohol under Lewis acid-catalyzed glycosidation [BF_3–$(C_2H_5)_2O$] in dichloromethane (Scheme 1).[8,10] Because of the anomeric effect combined with the 2-acetoxy group anchimeric participation, the axially substituted α-mannosides **2** and **3** are obtained in 54 and 74% yields, respectively. A first unsymmetrical and semirigid dimannoside cluster, compound **4**, is then assembled from compounds **2** and **3** using the palladium-catalyzed cross-coupling Sonogashira reaction[11] with a catalytic amount of tetrakis

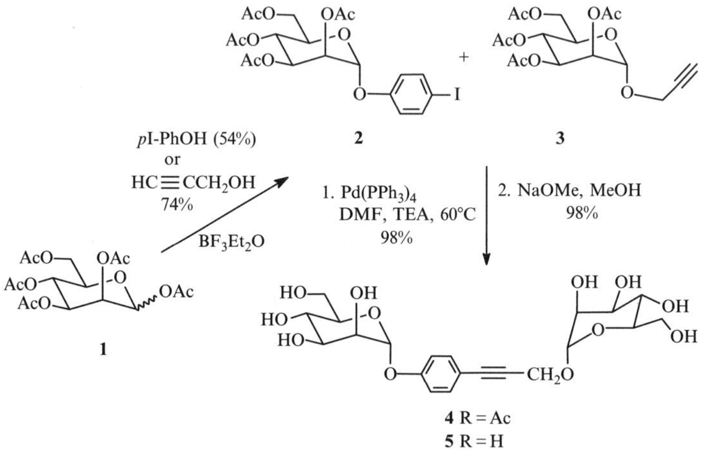

pI-PhOH (54%) or HC≡CCH2OH 74%, BF3Et2O
1. Pd(PPh3)4, DMF, TEA, 60°C, 98%
2. NaOMe, MeOH, 98%
4 R = Ac
5 R = H

SCHEME 1

[9] R. Roy, S. Das, R. Dominique, M. C. Trono, F. Hernández-Meteo, and F. Santoyo-González, *Pure Appl. Chem.* **71,** 565 (1999).

[10] H. B. Mereyala and S. R. Gurrala, *Carbohydr. Res.* **307,** 351 (1998).

[11] K. Sonogashira, *in* "Metal-Catalyzed Cross-Coupling Reactions" (F. Diederich and P. J. Stang, eds.), p. 203. Wiley-VCH, Weinheim, Germany, 1998.

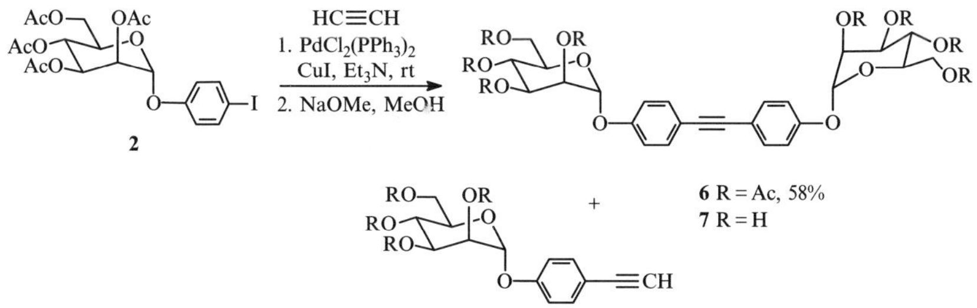

2

$HC \equiv CH$
1. $PdCl_2(PPh_3)_2$, CuI, Et_3N, rt
2. NaOMe, MeOH

6 R = Ac, 58%
7 R = H

8 R = Ac, 27%
9 R = H

SCHEME 2

(triphenylphosphine)palladium(0) [$Pd(PPh_3)_4$] [dimethylformamide–triethylamine (DMF, TEA), 1:1 (v/v), 60° in almost quantitative yield (98%). The dimer **4** is then deprotected under Zemplén conditions ($NaOCH_3$, CH_3OH) to afford water-soluble compound **5**.

The next symmetrical dimer, **6**, is prepared by a single one-step double cross-coupling reaction of 4-iodophenyl mannoside **2** onto acetylene gas, using dichlorobis(triphenylphosphine)palladium(II) [$Pd(Cl_2)(PPh_3)_2$], copper(I) iodide, and triphenylphosphine [$(C_2H_5)_3N$, 60°C, 5 h). The procedure provides dimer **6** in 58% yield, inevitably accompanied by monomer **8** in 27% yield (Scheme 2). Attempts to optimize the conditions with other palladium catalysts, pressure, temperature, or solvent have not improved the formation of the required dimer **6**. Finally, compounds **6** and **8** were de-O-acetylated under Zemplén conditions to give compounds **7** and **9**, which were only moderately soluble in water.

Trimers **11** and **12** are readily prepared as an inseparable mixture of regioisomers by a classic [2 + 2 + 2] cycloaddition[12] using dicobalt octacarbonyl [$Co_2(CO)_8$] under refluxing conditions in dioxane (Scheme 3). The nonsymmetrical 1,2,4-tris[(mannosyloxy)methyl]benzene derivative **11** is obtained as the major isomer in a 9:1 ratio together with the 1,3,5-symmetrical isomer **12** as the minor product. The reaction can also be carried out under milder conditions, using the olefin metathesis ruthenium carbenoid catalyst **10** (Grubbs's catalyst), which affords the same regioisomeric mixture of trimer **11** and **12** at room temperature (CH_2Cl_2, 12 h) in a slightly higher yield (75%).[13] This unprecedented application of the Grubbs's catalyst expands the usefulness of this type of catalyst[14,15] into

[12] R. J. Kaufman and R. S. Sidhu, *J. Org. Chem.* **47,** 4941 (1982).
[13] S. K. Das and R. Roy, *Tetrahedron Lett.* **40,** 4015 (1999).
[14] R. H. Grubbs and S. Chang, *Tetrahedron* **54,** 4413 (1998).
[15] A. Fürstner, *Angew. Chem. Int. Ed.* **39,** 3012 (2000).

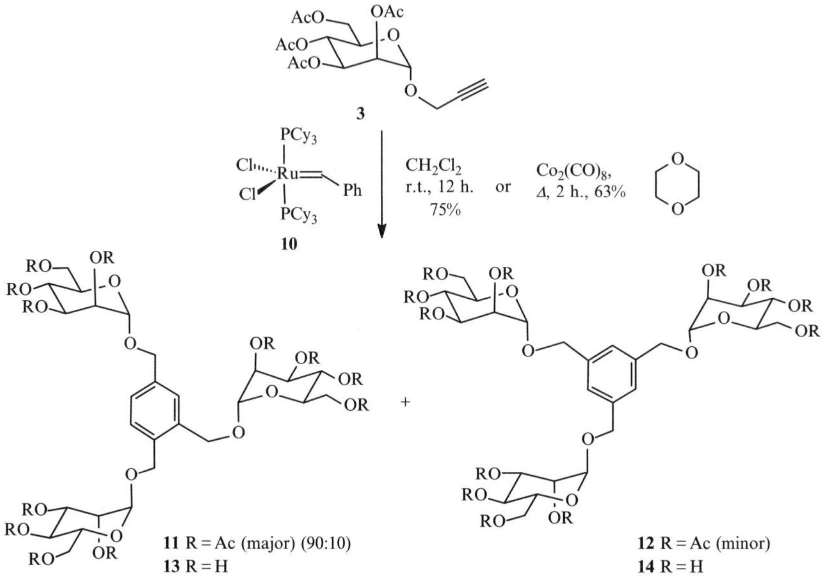

SCHEME 3

carbohydrate chemistry.[16] The mixture of isomers is next treated as described above for **4** ($NaOCH_3$, CH_3OH) to give water-soluble compounds **13** and **14**.

A related fully substituted hexameric benzenoid cluster, **17**, is prepared by an analogous strategy in which the known symmetrical 1,4-bis(mannosyloxy)but-2-yne derivative **16**[12] is treated with the cobalt catalyst mentioned above (Scheme 4). To this end, mannose pentaacetate **1** is glycosidated with but-2-yne-1,4-diol **15** under boron trifluoride diethyl dietherate [BF_3–$(C_2H_5)_2O$] in dichloromethane to afford dimer **16** in 65% yield. When dimer **16** is transformed in a one-step [2 + 2 + 2] cycloaddition process, the expected hexamer **17** is obtained in 61% yield. Interestingly, Grubbs's catalyst **10** is ineffective in accomplishing this transformation. In fact, catalyst **10** provides the cycloadditions only with terminal alkynes.[13] After Zemplén deprotection, water-soluble hexamer **18** is obtained in 94% yield. It is noteworthy that **18** is analogous to trimers **11/12** but represents a more crowded cluster.

Finally, the same cycloaddition operation is performed on the other symmetrical dimer, **6**, having added benzene rings tethered by an acetylene bridge. Thus, treatment of **6** with dicobalt octacarbonyl in refluxing

[16] R. Roy and S. K. Das, *Chem. Commun.* 519 (2000).

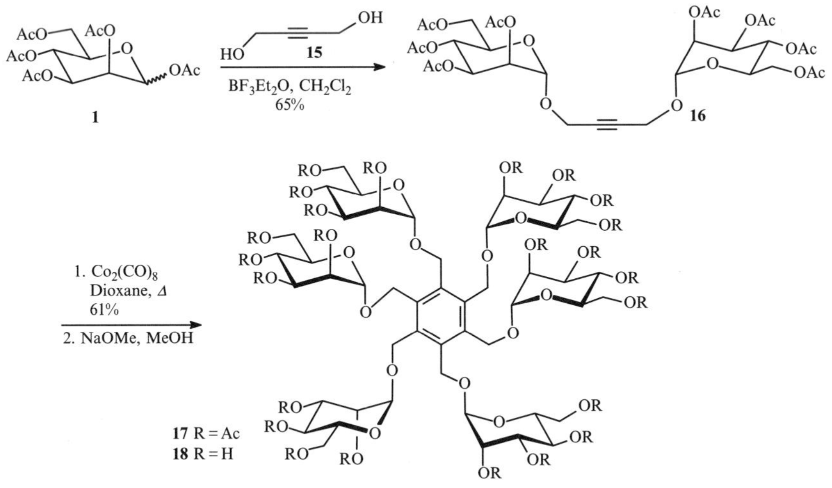

SCHEME 4

dioxane affords "molecular asterisk" **19** in 42% yield (Scheme 5). The symmetry of **19** and its deprotected analog **20** ($NaOCH_3$, CH_3OH) is clearly visible from their decoupled ^{13}C nuclear magnetic resonance (NMR) spectra, which show only one signal for each of the benzenoid and anomeric carbons. This family of compounds is known to potentially exist as enantiomeric mixtures depending on the rotational energy of activation and on the "tilt" angle between the external benzene rings and the inner one. So far, we have not yet obtained evidence for this situation. Hexakismannoside **20** differs from its lower congener **18** in that it is less water soluble, but represents more accessible mannoside residues because the interresidue distances are longer.

Cross-Linking Properties

The relative cross-linking abilities of these mannoside clusters toward concanavalin A are evaluated in a microtiter plate version of a turbidimetric analysis that is somewhat related to a kinetic nephelometry assay. Therefore, compounds **7**, **13/14**, **18**, and **20** are allowed to form insoluble cross-linked complexes with concanavalin A (Con A). The reactions are monitored as a function of time and the results are shown in Fig. 1. The time course to reach maximum turbidity is plotted (not shown) and the data are expressed as a percentage of their maximum optical density

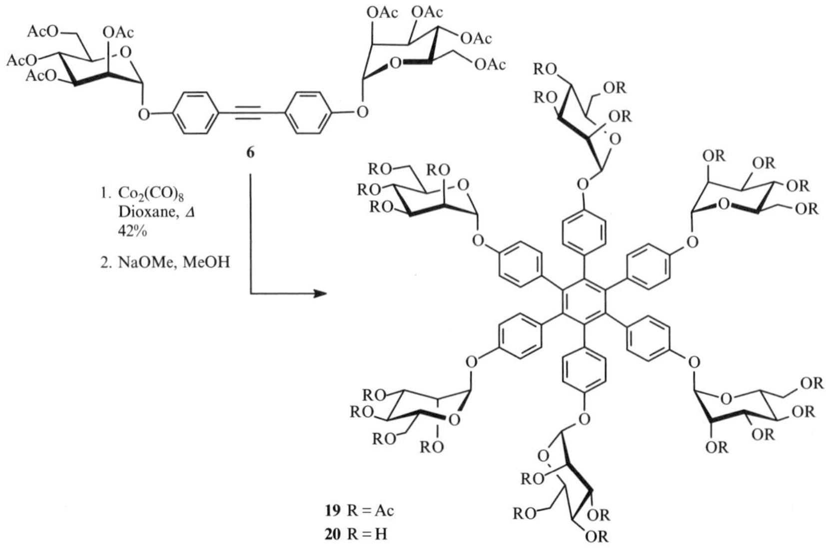

6

1. $Co_2(CO)_8$
Dioxane, Δ
42%

2. NaOMe, MeOH

19 R = Ac
20 R = H

SCHEME 5

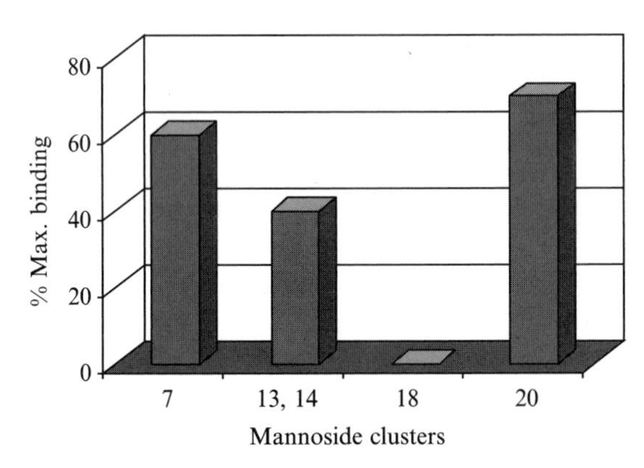

FIG. 1. Relative cross-linking abilities of mannoside clusters **7** (dimer), **13/14** (trimer), **18** (hexamer), and **20** (hexamer) toward the phytohemagglutinin Con A from *Canavalia ensiformis*. Maximum binding in each case was observed after 2 h. The measurements were done after 5 min.

(OD) after 5 min (middle of the slope). Hexamer **20** is the fastest to form the cross-linked complex, followed by dimer **7** and trimers **13/14**. Hexamer **18**, having the shortest intermannoside distances, fails to form such complexes.

Experimental Methods

Materials

^1H and ^{13}C NMR spectra are recorded at 500 or 200 MHz and at 125 or 50 MHz, respectively, with tetramethylsilane (0.00 ppm) as an internal reference. Thin-layer chromatography (TLC) is performed with silica gel 60 F_{254} aluminum sheets purchased from E. Merck (Darmstadt, Germany). Reagents used for developing plates include ceric sulfate (1%, w/v) and ammonium sulfate (2.5%, w/v) in 10% (v/v) aqueous sulfuric acid, iodine, dilute aqueous potassium permanganate, and ultraviolet (UV) light. TLC plates are heated to approximately 150° when necessary. Purifications are performed by gravity or flash chromatography on silica gel 60 (230–400 mesh; E. Merck). Solvents are evaporated under reduced pressure in a Buchi rotary evaporator connected to a water aspirator. All chemicals used in experiments are of reagent grade. Solvents are purified according to a published procedure.[8]

*4-Iodophenyl 2,3,4,6-tetra-*O*-acetyl-α-D-mannopyranoside* (**2**)

To a solution of penta-*O*-acetyl-α,β-D-mannopyranose (1 g) and 4-iodophenol (1 g) in dry dichloromethane (20 ml) is added BF_3–etherate (0.5 ml). The reaction mixture is kept at room temperature and the course of the reaction is monitored by TLC [ethyl acetate–hexane, 1:1 (v/v)] until complete disappearance of the starting material (18 h). Dichloromethane (75 ml) is added and the solution is washed with Na_2CO_3 aqueous saturated solution (twice, 50 ml each), 0.5 *N* NaOH solution (twice, 50 ml each), water (50 ml), 5% HCl solution (twice, 50 ml each), and water (twice, 50 ml each). After drying and evaporation the resulting crude product is crystallized from ether–hexane, giving compound **2** (0.955 g, 54%): m.p. 127–129°; $[\alpha]_D^{22} = +65°$ ($c = 1$ in chloroform); infrared (IR) (film) 1751, 1483, 1386, 1224 cm^{-1}; ^1H NMR (300 MHz, $CDCl_3$): δ 7.57, 6.85 (2 d, 4 H, $J = 9.0$ Hz, C_6H_4), 5.50 (dd, 1 H, $J = 10.1$ and 3.5 Hz, H-3), 5.46 (d, 1 H, $J = 1.9$ Hz, H-1), 5.40 (dd, 1 H, $J = 3.5$ and 1.9 Hz, H-2), 5.33 (t, 1 H, $J = 10.0$ Hz, H-4), 4.24 (dd, 1 H, $J = 12.4$, 5.5 Hz, H-6), 4.06–3.99 (m, 2 H, H-5, 6′), 2.17, 2.03, 2.00 (3 s, 12 H, 4 Ac); ^{13}C NMR (75 MHz, $CDCl_3$): δ 170.5, 170.0, 169.7 (CO), 155.4, 138.5, 118.8, 85.8 (C_6H_4), 95.8 (C-1), 69.4, 69.3, 68.8, 65.8 (C-2, 3, 4, 5), 62.1 (C-6), 20-9, 20.7 (CH_3CO). Mass spectrometry (MS) (fast atom bombardment, FAB): *m/z*: calculated for $C_{20}H_{23}IO_{10}$, 550.04; found, 551.2269 [M + 1].

*2-Propynyl 2,3,4,6-tetra-O-acetyl-α-D-mannopyranoside (**3**)*

A solution of 1,2,3,4,6-penta-*O*-acetyl-α,β-D-mannopyranose (5 g), freshly distilled propargyl alcohol (3 ml), and BF_3–etherate (3 ml) in CH_2Cl_2 (50 ml) is kept at room temperature for 2.5 days. Dichloromethane (50 ml) is added and the resulting solution is washed with 20% Na_2CO_3 aqueous solution (150 ml) and water (100 ml). The organic phase is dried and evaporated to obtain a crude product that is acetylated with acetic anhydride–pyridine [Ac_2O–Py, 10:10 ml (v/v)]. Conventional workup gives a crude product that is purified to afford the title compound **3**. Compound **3** is purified by silica gel column chromatography using ethyl acetate and hexane (1:1, v/v), which is isolated as a white solid, m.p. 100°; $[\alpha]_D^{22} = +56°$ ($c = 2$ in chloroform); ^{1}H NMR (500 MHz, $CDCl_3$): δ 5.31 (dd, 1 H, $J = 3.4$, 10.0 Hz, H-3), 5.26 (t, 1 H, $J = 10.0$ Hz, H-4), 5.23 (dd, 1 H, $J = 3.4$, 1.7 Hz, H-2), 4.99 (d, 1H, $J = 1.7$ Hz, H-1), 4.24 (dd, 1 H, $J = 5.2$, 12.2 Hz, H-6b), 4.24 (d, 2H, $J = 2.4$, H-1′), 4.07 (dd, 1 H, $J =$ 2.5 Hz, H-6a), 4.00 (ddd, 1H, H-5), 2.44 (t, 1 H, $J = 2.4$ Hz, H-2′), 2.12, 2.07, 2.01, 1.96 (4s, 12H, OAc); ^{13}C NMR (125 MHz, $CDCl_3$): δ = 170.5, 169.8, 169.7, 169.6 (CO), 96.2 (C-1), 77.9 (C-2′), 75.0 (C-3′), 70.6, 69.3, 68.9, 67.9 (C-2, 3, 4, 5), 62.3 (C-6), 54.9 (C-1′), 20.8, 20.7, 20.6, 20.6 (C-Me); MS (FAB): *m/z*: calculated for $C_{17}H_{22}O_{10}$: 386.36. Found: 387.1264 [M + 1].

*Coupling Reaction between 4-Iodophenyl 2,3,4,6-tetra-O-acetyl-α-D-mannopyranoside (**2**) and Prop-2 ynyl 2,3,4,6-tetra-O-acetyl-α-D-mannopyranoside (**3**) (**4**)*

To a degassed solution of the 4-iodophenylmannoside **2** (1 mmol) and the prop-2-ynyl mannoside **3** (1.1 mmol) in DMF–TEA [8:8 ml (v/v)] is added $Pd(PPh_3)_4$ (0.1 mmol). The solution is then heated at 60° under a nitrogen atmosphere for 3–4 h. The TEA is removed by evaporation under vacuum. Ether–toluene (100–50 ml) is then added to the residue and the solution is washed with 5% HCl (50 ml), $NaHCO_3$ saturated aqueous solution (50 ml), and water (50 ml). The organic phase is dried (Na_2SO_4) and evaporated, yielding a crude product that is purified by column chromatography [ethyl acetate–hexane, 1:1 (v/v)] to provide compound **4** as a solid: m.p. 68–70°; $[\alpha]_D^{22} = +88°$ ($c = 1$ in chloroform); IR (KBr): 2351, 1772, 1741, 1604, 1509 cm^{-1}; ^{1}H NMR (500 MHz, $CDCl_3$): δ 7.37, 7.02 (2 d, 4 H, $J = 8.8$ Hz, C_6H_4), 5.51 (dd, 1 H, $J = 10.0$ and 3.5 Hz, H-3a), 5.50 (d, 1 H, $J = 1.7$ Hz, H-1a), 5.40 (dd, 1 H, $J = 3.5$ and 1.8 Hz, H-2a), 5.35 (dd, 1 H, $J = 9.9$ and 3.4 Hz, H-3b), 5.32 (t, 1 H, $J = 9.8$ Hz, H-4a or 4b), 5.29 (t, 1 H, $J = 9.9$ Hz, H-4b or 4a), 5.28 (dd, 1 H, $J = 3.4$, 1.7 Hz, H-2b), 5.07 (d, 1 H, $J = 1.6$ Hz, H-1b), 4.45 (AB system, 2 H, $J = 15.8$ Hz,

$J = 18.2$ Hz, $CH_2C{\equiv}CH$), 4.27 (dd, 1 H, $J = 12.3$, 5.0 Hz, H-6), 4.24 (dd, 1 H, $J = 12.3$ and 5.3 Hz, H-6), 4.09 (dd, 1 H, $J = 12.3$ and 3.4 Hz, H-6′), 4.05–4.08 (m, 3 H, H-5a, 5b, 6′), 2.17, 2.13, 2.07, 2.02, 2.01, 2.00, 2.00, 1.97 (8 s, 24 H, 8 Ac); ^{13}C NMR (175 MHz, $CDCl_3$): δ 170.6, 170.5, 169.9, 169.8, 169.7 (CO), 155.8, 133.4, 116.5 (C_6H_4), 96.1 (C-1b), 95.6 (C-1a), 86.5, 82.6 (C≡C), 69.5, 69.3, 69.2, 69.0, 68.9, 68.7 (C-2a, 2b, 3a, 3b, 5a, 5b), 66.1, 65.9 (C-4a, 4b) 62.3, 62.1 (C-6a, 6b), 55.7 ($CH_2C{\equiv}C$), 20.8, 20.1, 20.7, 20.6 (CH_3CO); MS (FAB): *m/z*: 847 for $[M+K]^+$; $C_{37}H_{44}O_{20}$: calculated: C 54.95, H 5.48. Found: C 54.97, H 5.46.

*Deacetylation of Compound **4***

A solution of compound **4** (50 mg, 0.06 mmol) in CH_3ONa–methanol (0.2 *M*, 10 ml) is stirred at ambient temperature for 6 h. The resulting solution is neutralized [Dowex-50W (H^+)], and the resin is filtered and washed with CH_3OH. Volatiles are evaporated from the combined filtrates and the residue is dissolved in water (2 ml) and lyophilized to obtain pure compound **5** (29 mg, 100%) as a foam; $[\alpha]_D^{22} = -2°$ ($c = 1.0$ in water); ^1H NMR (500 MHz, D_2O): δ 7.54, 7.20 (2d, 4 H, $J = 8.9$ Hz, C_6H_4), 5.70 (d, 1 H, $J = 1.6$ Hz, H-1a), 5.16 (d, 1 H, $J = 1.6$ Hz, H-1b), 4.60 (Abq, 2 H, $J = 16.0$ Hz, $CH_2C{\equiv}CH$), 4.23 (dd, 1 H, $J = 3.5$, 1.6 Hz, H-2a), 4.10 (dd, 1 H, $J = 3.5$, 9.6 Hz, H-3a), 4.03 (dd, 1 H, $J = 1.6$, 3.4 Hz, H-2b), 3.90–3.70 (m, 9 H, H-4a, H-5a, H-6, H-3b, H-4b, H-5b, H-6′); MS (FAB): *m/z*: 511.6312 for $[M+K]^+$.

*Bis-para-(2,3,4,6-tetra-O-acetyl-α-D-mannopyranosyloxy) diphenylacetylene (**6**)*

Nitrogen gas is bubbled through a mixture of 4-iodophenyl 2,3,4,6-tetra-*O*-acetyl-α-D-mannopyranoside (**2**) (1.0 g, 1.81 mmol) in 30 ml of triethylamine. Dichlorobis(triphenylphosphine)palladium(II) [$PdCl_2(PPh_3)_2$] (0.014 g, 0.019 mmol) is added and the mixture is stirred under a stream of N_2 for 5 min. Copper(I) iodide (CuI) (0.025 g, 0.13 mmol) and triphenylphosphine (PPh_3) (0.012 g, 0.047 mmol) are added and stirring is continued under N_2 for 5 min. The reaction mixture is heated to 60° with acetylene gas bubbling through for 5 h, after which time the reaction is judged complete by TLC [hexane–ethyl acetate, 1:1 (v/v)]. The solvent is removed under reduced pressure, the residue is dissolved in CH_2Cl_2, and washed two times with H_2O. The organic extracts are dried over $NaSO_4$, filtered, and concentrated under reduced pressure. The concentrate is purified by silica gel column chromatography with hexane–ethyl acetate (1:1, v/v) as eluent. Pure compound **6** is obtained as a yellow crystalline solid (0.4598 g, 58% yield); m.p. 89°; $[\alpha]_D^{23} = 117.5°$ ($c = 1.2$, $CHCl_3$); ^1H NMR ($CDCl_3$, 500 MHz): δ (ppm) 7.40 (C_6H_4, dd,

$J = 2.1$ Hz, $J = 6.8$ Hz, 4H), 7.01 (C_6H_4, dd, $J = 2.1$ Hz, $J = 6.8$ Hz, 4H), 5.48 (H-3, dd, $J_{2,3} = 3.5$ Hz, $J_{3,4} = 9.9$ Hz, 2H), 5.49 (H-1, d, $J = 1.8$ Hz, 2H), 5.39 (H-2, dd, $J_{1,2} = 1.7$ Hz, $J_{2,3} = 3.5$ Hz, 2H), 5.30 (H-4, dd, $J = 10.0$ Hz, $J = 10.0$ Hz, 2H), 4.21 (H-6a, dd, $J_{5,6a} = 6.0$ Hz, $J_{6a,6b} =$ 12.7 Hz, 2H), 4.03–4.00 (H-5, H-6b, m, 4H), 2.14, 2.00, 1.98, 1.97 ($COCH_3$, 4 s, 24H); ^{13}C NMR ($CDCl_3$, 125.7 MHz): δ (ppm) 170.4, 169.8, 169.8, 169.6 (C=O), 155.3 (C-ipso–C_6H_4), 132.9, 116.5 (C_6H_4), 95.6 (C-1), 88.1 (C), 69.2 (C-5), 69.2 (C-3), 68.7 (C-2), 65.8 (C-4), 62.0 (C-6), 20.7, 20.6 ($COCH_3$); high-resolution mass spectrometry (HRMS) FAB $[M]^+$ (*m/z*) calculated for $C_{42}H_{46}O_{20}$: 870.26. Found: 870.32. Analysis: calculated for $C_{42}H_{46}O_{20}$: C 57.93, H 5.32. Found: C 57.32, H 5.31.

*Bis-para-(α-D-mannopyranosyloxy)diphenylacetylene (**7**)*

Compound **7** is obtained by deacetylation of bis-*para*-[(2,3,4,6-tetra-*O*-acetyl-α-D-mannopyranosyl)diphenylacetylene (**6**) (0.11 g, 0.12 mmol), using the procedure described above for compound **4** (97% yield, 0.067 g, yellow solid); m.p. 240–242°; $[\alpha]_D^{23} = +114.0°$ ($c = 1.0$, DMF); ^1H NMR (DMSO-d_6, 500 MHz): δ (ppm) 7.44 (C_6H_4, d, $J = 8.7$ Hz, 4H), 7.09 (C_6H_4, d, $J = 8.8$ Hz, 4H), 5.41 (H-1, d, $J = 1.2$ Hz, 2H), 5.04 (C2-OH, d, $J = 4.3$ Hz, 2H), 4.82 (C4-OH, d, $J = 5.7$ Hz, 2H), 4.74 (C3-OH, d, $J = 6.0$ Hz, 2H), 4.44 (C6-OH, t, $J = 5.9$ Hz, 2H), 3.82 (H-2, bs, 2H), 3.66 (H- 3, m, 2H), 3.58 (H-6a, m, 2H), 3.36 (H-4, m, 2H), 3.47 (H-5, H-6b, m, 4H); ^{13}C NMR (DMSO-d_6, 125.7 MHz): δ (ppm) 156.4 (C-ipso–C_6H_4), 132.7 (C_6H_4), 117.0 (C_6H_4), 115.9 (C-ipso-), 98.7 (C-1), 88.1 (C), 75.1 (C-4), 70.6 (C-3), 70.0 (C-2), 66.7 (C-5), 61.0 (C-6); FAB MS $[M + K]^+$ *m/z* (relalitive intensity %) calculated for $C_{26}H_{30}O_{12}K$: 573.14. Found: 573.16 (2.3). $[M]^+$ calculated for $C_{26}H_{30}O_{12}$: 534.17. Found: 534.17 (0.6).

*4-Ethynylphenyl 2,3,4,6-tetra-*O*-acetyl-α-D-mannopyranoside (**8**)*

Further elution of the by-product obtained from the reaction described above yields 4-ethynylphenyl 2,3,4,6-tetra-*O*-acetyl-α-D-mannopyranoside (**8**) (0.2232 g, 27% yield, yellow solid); m.p. 93–95°; $[\alpha]_D^{23} = +33.8°$ ($c = 1.3$, $CHCl_3$); IR (neat): 3277, 2960, 2108, 1751 cm^{-1}; ^1H NMR ($CDCl_3$, 500 MHz): δ (ppm) 7.41 (C_6H_4, d, $J = 8.9$ Hz, 2H), 7.01 (C_6H_4, d, $J = 8$ Hz, 2H), 5.48 (H-3, dd, $J_{2,3} = 3.5$ Hz, $J_{3,4} = 10.1$ Hz, 1H), 5.50 (H-1, d, $J = 1.4$ Hz, 1H), 5.40 (H-2, dd, $J_{1,2} = 2.0$ Hz, $J_{2,3} = 3.6$ Hz, 1H), 5.32 (H-4, dd, $J = 10.1$, $J = 10.1$ Hz, 1H), 4.23 (H-6a, dd, $J_{5-6a} = 5.7$ Hz, $J_{6a-6b} = 12.4$ Hz, 1H), 4.03 (H-5, H-6b, m, 2H), 3.00 (CCH, s, 1H), 2.16, 2.02, 2.00, 1.99 ($COCH_3$, 4 s, 12H); ^{13}C NMR ($CDCl_3$, 125.7 MHz): δ (ppm) 170.4, 169.9, 169.8, 169.6

(C=O), 155.7 (C-ipso), 133.6, 116.4 (Ar), 95.6 (C-1), 82.9, 76.6 (C≡), 69.3 (C-5), 69.2 (C-2), 68.7 (C-3), 65.8 (C-4), 62.0 (C-6), 20.8, 20.6, 20.6 (CO*C*H$_3$); FAB-MS $[M+K]^+$ *m/z* (relative intensity %): calculated for $C_{22}H_{24}O_{10}K$: 487.10. Found: 486.70 (23.2).

*1,2,4-/1,3,5-Tris-{[(2,3,4,6-tetra-O-acetyl-α-D-mannopyranosyl)oxy] methyl}benzene (**11, 12**)*

Method A with Grubbs's Catalyst. To a solution of compound **3** (100 mg, 0.259 mmol) in dry CH_2Cl_2 (1 ml) is added Grubbs's catalyst **10** (11 mg, 15 mol%). After stirring for 12 h at room temperature, the solvent is evaporated to dryness and the crude mixture is purified by silica gel column chromatography [ethyl acetate–hexane, 3:2 (v/v)] to provide a mixture of **11** and **12** (75 mg, 75% yield) (1,2,4 and 1,3,5-regioisomers; 90:10).

Method B with $Co_2(CO)_8$. A solution of prop-2 ynyl-2,3,4,6-tetra-*O*-acetyl-α-D-mannopyranoside (**3**) (0.51 g, 1.3 mmol) in 5 ml of freshly distilled 1,4-dioxane is refluxed under a stream of N_2 gas. After 5 min of reflux, dicobalt octacarbonyl [$Co_2(CO)_8$] (0.045 g, 0.13 mmol) is added. The reaction mixture is refluxed for a further 2 h, after which time the reaction is judged complete by TLC on silica gel [ethyl acetate–toluene, 6.4:3.2 (v/v)]. The solvent is evaporated under reduced pressure and the concentrate is purified by column chromatography on silica gel with hexane–ethyl acetate (1:2, v/v) as eluent. The cyclotrimerized mixture of inseparable regioisomers is obtained as white crystalline powder (0.0.32 g, 63% yield); m.p. 74°; $[\alpha]_D^{23} = +56.0°$ ($c = 1.5$, $CHCl_3$); ^{1}H NMR ($CDCl_3$, 500 MHz): δ (ppm) 7.37 (H$_a$-C_6H_4, d, 1H, $J = 7.8$ Hz, 1H), 7.3 (H$_b$-C_6H_4, dd, 1H, $J = 1.2$ Hz, $J = 7.5$ Hz, 1H), 7.28 (H$_c$-C_6H_4, s, 1H), 7.23 (C_6H_4–1,3,5-isomer, s), 5.33 (H-3, m, 3H), 5.28 (H-4, m, 3H), 5.20 (H-2, m, 3H), 4.83 (H-1, m, 3H), 4.27 (H-6a, m, 3H), 4.07 (H-6b, m, 3H), 4.00 (H-5, m, 3H), 4.62 (CH_2Ar, m, 6H), 2.10 2.09, 2.08, 2.00, 1.99, 1.94, 1.93 ($COCH_3$, overlapping singlets, 36 H); ^{13}C NMR ($CDCl_3$, 125.7 MHz): δ (ppm) 170.6, 170.5, 169.8, 169.7, 169.7, 169.6, 169.6, 169.5 (C=O), 137.1 (C_6H_4–1,3,5-isomer), 136.7, 135.0, 134.8, 130.0, 129.3, 128.1 (C_6H_4–1,2,4-isomer), 127.4 (C_6H_4–1,3,5-isomer), 96.9, 96.8, 96.7 (C-1), 69.5, 69.4, 69.4 (C-2), 69.2, 69.0, 69.0 (C-3), 68.9 (*C*H$_2$Ar), 68.8, 68.7 (C-5), 67.4, 66.8 (*C*H$_2$Ar), 66.1, 66.0, 66.0 (C-4), 62.4, 62.3 (C-6), 21.0, 20.8, 20.7, 20.6, 20.5 (CO*C*H$_3$); FAB-MS $[M+K]^+$ *m/z* (relative intensity, %): calculated for $C_{51}H_{66}O_{30}K$: 1197.33. Found: 1197.42 (0.4).

1,2,4-/1,3,5-Tris-{[(α-D-mannopyranosyl)oxy]methyl}benzene (**13, 14**)

The title regioisomers are obtained by deacetylation of a mixture of 1,2,4-/1,3,5-tris{[(2,3,4,6-tetra-*O*-acetyl-α-D-mannopyranosyl)oxy]methyl}-benzene (**1**, **12**) (0.098 g, 0.082 mmol), using the procedure described above for compound **4** (91% yield, 0.049 g, yellow solid); m.p. 75°; $[\alpha]_D^{23} = +118.9°$ ($c = 0.9$, CH_3OH); ^1H NMR (D_2O, 500 MHz): δ (ppm) 7.56 (H_a-C_6H_4/H_c-C_6H_4, bd, $J = 7.9$ Hz, 2H), 7.50 (H_b-C_6H_4, d, $J = 7.8$ Hz, 1H), 7.48 (C_6H_4–1,3,5-isomer), 5.03 (H-1, s, 3H), 4.92 (CH_2, d, $J = 12.0$ Hz, 1H), 4.91 (CH_2, d, $J = 11.9$ Hz, 1H), 4.85 (CH_2, s, 1H), 4.78 (CH_2, d, $J = 11.9$ Hz, 1H), 4.77 (CH_2, d, $J = 12.0$ Hz, 1H), 4.68 (CH_2, d, $J = 11.8$ Hz, 1H), 4.02 (H-2, m, 3H), 3.89 (H-6a, m, 3H), 3.86 (H-3, dd, $J_{2,3} = 3.4$ Hz, $J_{3,4} = 9.2$ Hz, 3H), 3.82 (H-6b, m, 3H), 3.74 (H-4, dd, $J = 9.7$ Hz, $J = 9.6$ Hz, 3H), 3.62 (H-5, m, 3H); ^{13}C NMR (D_2O, 125.7 MHz): δ (ppm) 139.9, 137.3 (C-aryl–1,3,5-isomer), 135.3, 134.9 (C-ipso), 129.9, 129.3, 128.2 (C-aryl), 99.2, 99.1, 99.9 (C-1), 72.6, 72.6 (C-5), 70.1 (C-3), 69.6 (C-2), 68.4, 66.4, 66.30 (CH_2), 66.2 (C-4), 60.3 (C-6); FAB MS $[M + K]^+$ m/z (relative intensity %) calculated for $C_{27}H_{42}O_{18}K$: 693.20. Found: 693.26 (0.4).

1,4-Bis-(2,3,4,6-tetra-O-acetyl-α-D-mannopyranosyl)but-2-yne (**16**)

Trimethylsilyltrifluoromethane sulfonate [$CF_3SO_3Si(CH_3)_3$] (0.25 ml, 1.4 mmol) is added via a syringe to a solution of mannose pentaacetate **1** (1.0 g, 2.7 mmol) and but-2-yne-1,4-diol (0.12 g, 1.4 mmol) in dry CH_2Cl_2 (50 ml) cooled to 0° and under a stream of N2. The solution is allowed to come to room temperature and stirred for a further 6 h, after which time the reaction is judged complete by TLC [hexane–ethyl acetate, 1:1 (v/v)]. The reaction mixture is washed with saturated $NaHCO_3$ (twice) and H_2O (twice). The organic layer is dried over $NaSO_4$ and filtered, and the solvent is removed under vacuum. The crude product is purified by column chromatography on silica gel, using gradient elution [hexane–ethyl acetate, 1:1 (v/v); then hexane–ethyl acetate, 1:2 (v/v)]. The compound is obtained as a white crystalline solid (0.2 g, 27% yield). When the same procedure is used in the presence of boron trifluoride diethyl etherate, the yield of **16** is raised to 65%: m.p. 52–53°; $[\alpha]_D^{23} = +68.9°$ ($c = 0.9$, $CHCl_3$); ^1H NMR ($CDCl_3$, 500 MHz): δ (ppm) 5.30 (H-3, dd, $J_{2,3} = 3.9$ Hz, $J_{3,4} = 10.0$ Hz, 2H), 5.27 (H-4, dd, $J = 9.8$ Hz, $J = 10.8$ Hz, 2H), 5.23 (H-2, dd, $J_{1,2} = 1.8$ Hz, Hz, $J_{2,3} = 3.2$ Hz, 2H), 4.95 (H-1, d, $J = 1.6$ Hz, 2H), 4.26 (H-6a, $H_2CC{\equiv}$, m, 6H), 4.07 (H-6b, m, 2H), 3.96 (H-5, m, 2H), 2.12, 2.12, 2.06, 1.95 ($COCH_3$, 4 s, 24H); ^{13}C NMR ($CDCl_3$, 125.7 MHz,): δ (ppm) 170.5, 169.8, 169.7, 169.6 (C=O), 95.1 (C-1), 81.6 (C≡C), 69.3 (C-2), 68.9 (C-3, C-5), 68.9 (C-3, C-5), 66.0 (C-4), 62.3 (C-6), 54.9 ($H_2CC{\equiv}$), 20.7, 20.6, 20.6, 20.5

(COCH_3); HRMS (FAB) $[M+1]^+$ *m/z* calculated for $C_{32}H_{42}O_{20}$: 747.23473. Found: 747.2221.

*1,2,3,4,5,6-Hexakis{[(2,3,4,6-tetra-O-acetyl-α-D-mannopyranosyl)oxy] methyl}benzene (**17**)*

A solution of 1,4-bis(2,3,4,6-tetra-*O*-α-D-mannopyranosyl)but-2-yne (**16**) (0.59 g, 0.79 mmole) in 5 ml of freshly distilled 1,4-dioxane is refluxed under a stream of N_2 gas. After 5 min of reflux, dicobalt octacarbonyl [$Co_2(CO)_8$] (0.040 g, 0.12 mmol) is added. The reaction mixture is refluxed for a further 20 h, after which time the reaction is judged complete by TLC on silica gel [ethyl acetate–toluene, 10:1 (v/v)]. The solvent is evaporated under reduced pressure and the concentrate is purified by column chromatography on silica gel with $CHCl_3$–ethyl acetate (1:10, v/v) as eluent. The cyclotrimerized product is obtained as a white crystalline powder (0.36 g, 61% yield); m.p. 102–104°; $[\alpha]_D^{23} = +87.0°$ ($c = 1.0$, $CHCl_3$); IR (neat) 2940, 1750, 1435, 1371, 1226, 1083, 1046 cm^{-1}; ^1H NMR ($CDCl_3$, 500 MHz): δ (ppm) 5.30 (H-4, dd, $J = 9.9$, $J = 9.9$ Hz, 6H), 5.25 (H-3, dd, $J_{2,3} = 3.4$ Hz, $J_{3,4} = 10.1$ Hz, 6H), 5.19 (H-2, dd, $J_{1,2} = 1.7$ Hz, $J_{2,3} = 3.4$ Hz, 6H), 5.05 (Ph-CH_2, d, $J = 12$ Hz, 6H), 4.94 (H-1, d, $J = 1.4$ Hz, 6H), 4.88 (Ph-CH_2, d, $J = 12$ Hz, 6H), 4.38 (H-6a, m, $J_{5,6a} = 3.9$ Hz, $J_{6a,6b} = 12.5$ Hz, 6H), 4.11 (H-6b, dd, $J_{5,6a} = 2.6$ Hz, $J_{6a,6b} = 12.5$ Hz, 6H), 4.05 (H-5, m, 6H), 2.11, 2.08, 1.99, 1.91 ($COCH_3$, 4 s, 72H); ^{13}C NMR ($CDCl_3$, 125.7 MHz): δ (ppm) 170.6, 169.9, 169.7, 169.6 (C=O), 137.7 (C_6H_4), 97.6 (C-1), 69.4, 69.3 (C-2, C-5), 69.0 (C-3), 65.9 (C-4), 63.7 (Ph-CH_2), 62.0 (C-6) 20.8, 20.6, 20.6, 20.6 (COCH_3); HRMS (FAB) $[M+K]^+$ calculated for $C_{96}H_{126}O_{60}K$: 2279.6513. Found: 2279.6592 *m/z*; Analysis: calculated for $C_{96}H_{126}O_{60}$: C 51.47, H 5.67. Found: C 51.37, H 5.79.

*Hexakis{[(α-D-mannopyranosyl)oxy]hexamethyl}benzene (**18**)*

The title compound is obtained by the procedure described above for compound **4** (94% yield, 0.0454 g, white solid); m.p. 166–167°; $[\alpha]_D^{23} = +73.3°$ ($c = 0.9$, CH_3OH); ^1H NMR (D_2O, 500 MHz): δ (ppm) 5.18 (Ph-CH_2, d, $J = 12.5$, 6H), 5.02 (PhCH_2, d, $J = 12.1$ Hz, 6H), 5.03 (H-1, s, 6H), 3.96 (H-2, bs, 6H), 3.80 (H-3/H-6, m, 12H), 3.74 (H-4, dd, $J = 9.74$ Hz, $J = 9.7$ Hz, 6H), 3.57 (H-5, m, 6H); ^{13}C NMR (D_2O, 125.7 MHz): δ (ppm) 137.4 (C-ipso-aryl), 99.6 (C-1), 73.0 (C-5), 70.0 (C-3), 69.7 (C-2), 66.0 (C-4), 63.1 (Ph-CH_2), 60.1 (C-6).

Hexakis(2,3,4,6-tetra-O-acetyl-α-D-mannopyranosyl) hexaphenylbenzene ***(19)***

A solution of bis-*para*-(2,3,4,6-tetra-*O*-acetyl-α-D-mannopyranosyl)-diphenyl-acetylene (**16**) (0.35 g, 0.40 mmol) in 5 ml of freshly distilled 1,4-dioxane is refluxed under a stream of N_2 gas. After 5 min of reflux, dicobalt octacarbonyl [$Co_2(CO)_8$] (0.019 g, 0.056 mmol) is added. The reaction mixture is refluxed for a further 24 h, after which time the reaction is judged complete by TLC on silica gel [ethyl acetate–toluene, 6.4:1.5 (v/v)]. The solvent is evaporated under reduced pressure and the concentrate is purified by column chromatography on silica gel with hexane–ethyl acetate (1:4, v/v) as eluent. The cyclotrimerized product is obtained as a yellow crystalline powder (0.14 g, 42% yield); m.p. 85°; $[\alpha]_D^{23} = +0.0°$ ($c = 1.1$, $CHCl_3$); IR 2940, 1750, 1435, 1371, 1226 cm^{-1}; ^1H NMR ($CDCl_3$, 500 MHz): δ (ppm) 6.62 (C_6H_4, d $J = 8.749$ Hz, 12H), 6.58 (C_6H_4, d, $J = 8.466$ Hz, 12H), 5.40 (H-3), dd, $J_{2,3} = 3.7$ Hz, $J_{3,4} = 10.0$ Hz, 6H), 5.26 (H-2, H-4, dd, $J = 9.7$ Hz, $J = 10.7$ Hz, 12H), 5.24 (H-1, s, 6H), 4.20 (H-6a, dd, $J_{5,6a} = 5.2$ Hz, $J_{6a,6b} = 12.5$ Hz, 6H), 3.93 (H-5, H-6b, m, 12H), 2.10, 2.00, 1.99, 1.98, 1.94, 1.94 ($COCH_3$, overlapping singlets, 72H); ^{13}C NMR ($CDCl_3$, 125.7 MHz): δ (ppm) 170.3, 169.8, 169.7, 169.7, (C=O), 153.3, 139.9, 135.3, 132.3, 115.2 (C_6H_4), 96.0 (C-1), 69.4, 69.2, 69.0, 68.8 (C-2, C-3, C-5), 65.8 (C-4), 20.9, 20.7, 20.6 ($COCH_3$); FAB-MS $[M + K]^+$ m/z (relative intensity %) calculated for $C_{126}H_{138}O_{60}$: 2649.74. Found 2650.70 (0.1).

Hexakis(α-D-mannopyranosyl)hexaphenylbenzene ***(20)***

The compound is obtained by deacetylation of hexakis(2,3,4,6-tetra-*O*-acetyl-α-D-mannopyranosyl)hexaphenylbenzene (**19**) (0.14 g, 0.056 mmol), using general Zemplén conditions ($NaOCH_3$, CH_3OH). The solution turns cloudy after stirring overnight in sodium methoxide (10 drops, 1 *M* in methanol). Dimethyl sulfoxide (30 ml) is added and the solution turns clear. The final product is obtained as a brown sticky solid after evaporation and freeze–drying (0.089 g, 100% yield); m.p. 187°; $[\alpha]_D^{23} = +0.9°$ ($c = 1.1$, DMSO); ^1H NMR (DMSO-d_6, 500 MHz): δ (ppm) 6.67 (C_6H_4, d, $J = 7.8$ Hz, 12H), 6.54 (C_6H_4, d, $J = 8.0$ Hz, 12H), 5.09 (H-1, s, 6H), 3.70 (H-2, s, 6H), 3.56 (H-3, d, $J = 7.5$ Hz, 6H), 3.54 (H-6b, d, $J = 10.3$ Hz, 6H), 3.48 (H-4, H-6a, m, 12H), 3.19 (H-5, bs, 6H); ^{13}C NMR (DMSO-d_6, 125.7 MHz): δ (ppm) 153.8, 140.0, 134.4 (C-ipso), 132.1 (C_6H_4), 115.5 (C_6H_4), 98.9 (C-1), 74.5 (C-5), 70.8 (C-3), 70.2 (C-2), 66.5 (C-4), 60.6 (C-6).

Turbidimetric Analysis

Equivalent amounts of the compounds (corresponding to 9.1 μmol of monosaccharide residue) are delivered separately into microtiter wells. The lectin from *Canavalia ensiformis*, concanavalin A (Con A; Sigma-Aldrich, st. Louis, MO) (80 μl of a stock solution of 1.0 mg in 100 ml of phosphate-buffered saline [PBS]), is delivered via a microsyringe in each well. PBS solution is added to each well so that the total volume in each well is 120 μl. The blank consists of a solution of 80 μl of Con A solution and 40 μl of PBS. The optical density (OD) at 490 nm is monitored for 2 h at room temperature. Each test is done in triplicate. The turbidity is measured on a Dynatech MR (Burlington, MA) 600 microplate reader at regular time intervals.

[2] Ruthenium Carbenoids as Catalysts for Olefin Metathesis of ω-Alkenyl Glycosides

By ROMYR DOMINIQUE, SANJOY K. DAS, BINGCAN LIU, JOE NAHRA, BRAD SCHMOR, ZHONGHONG GAN, and RENÉ ROY

Introduction

Olefin metathesis is a powerful synthetic process that translates, in its most simplified version, into the combination of two alkenes to generate a new double bond between the reacting partners with the evolution of ethylene gas when dealing with terminal alkenes.[1–3] The procedure can lead to self- and cross-metathesis, to ring-closing (RCM), and to ring-opening metathesis polymerization (ROMP) (Scheme 1). The most widely used catalysts for this process are transition metal carbenoids **1–3**. Although Schrock's molybdenum catalyst **1** is also efficient, it requires more stringent operating conditions and, consequently, Grubbs's ruthenium catalysts **2** and **3** are used more often, particularly in carbohydrate chemistry where functional group tolerance plays a critical factor.[4] Moreover, the most recent, air-stable, and more versatile carbenoid **3** has been shown to provide access to better chemo- and stereoselective transformations.

[1] R. R. Schrock, *Acc. Chem. Res.* **12,** 98 (1979).
[2] R. H. Grubbs and S. Chang, *Tetrahedron* **54,** 4413 (1998).
[3] A. Fürstner, *Angew. Chem. Int. Ed.* **39,** 3012 (2000).
[4] R. Roy and S. K. Das, *Chem. Commun.* 519 (2000).

0076-6879/03 $35.00

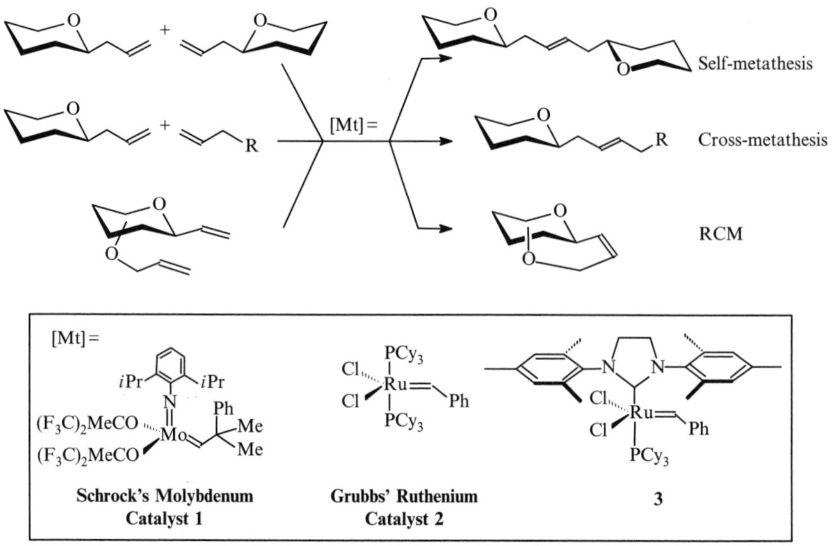

SCHEME 1. General olefin metathesis processes using transition metal carbenoid species. Mt, Metal.

Carbohydrate derivatives with allyl ethers as well as with *O*- and *C*-linked allyl glycosides have been known for a long time and several of these derivatives are well-established precursors in synthetic carbohydrate chemistry. Obviously, olefin metathesis in its various forms represents a particularly appealing chemical process for the design of small carbohydrate clusters (e.g., dimers), glycomimetics, conformationally restrained oligosaccharides, and perhaps, more importantly, functionalized aglycones with pendant functionalities suitable for further neoglycoconjugate syntheses. We report here some of the above-cited applications in which the terminal alkene groups have been introduced in key positions around the ring.

Olefin Self-Metathesis

Olefin self-metathesis of *O*- or *C*-linked allyl, pentenyl, and vinyl glycosides is a straightforward process leading to homodimers having either *trans* or *cis* geometries (Scheme 2, compounds **4–21**; Table I[5–9]).

[5] Z. Gan and R. Roy, *Tetrahedron* **56,** 1423 (2000).
[6] S. K. Das, R. Dominique, C. Smith, J. Nahra, and R. Roy, *Carbohydr. Lett.* **3,** 361 (1999).
[7] R. Dominique, B. Liu, S. K. Das, and R. Roy, *Synthesis* 862 (2000).
[8] R. Roy, R. Dominique, and S. K. Das, *J. Org. Chem.* **64,** 5408 (1999).
[9] B. Liu, S. K. Das, and R. Roy, *Org. Lett.* **4,** 2723 (2002).

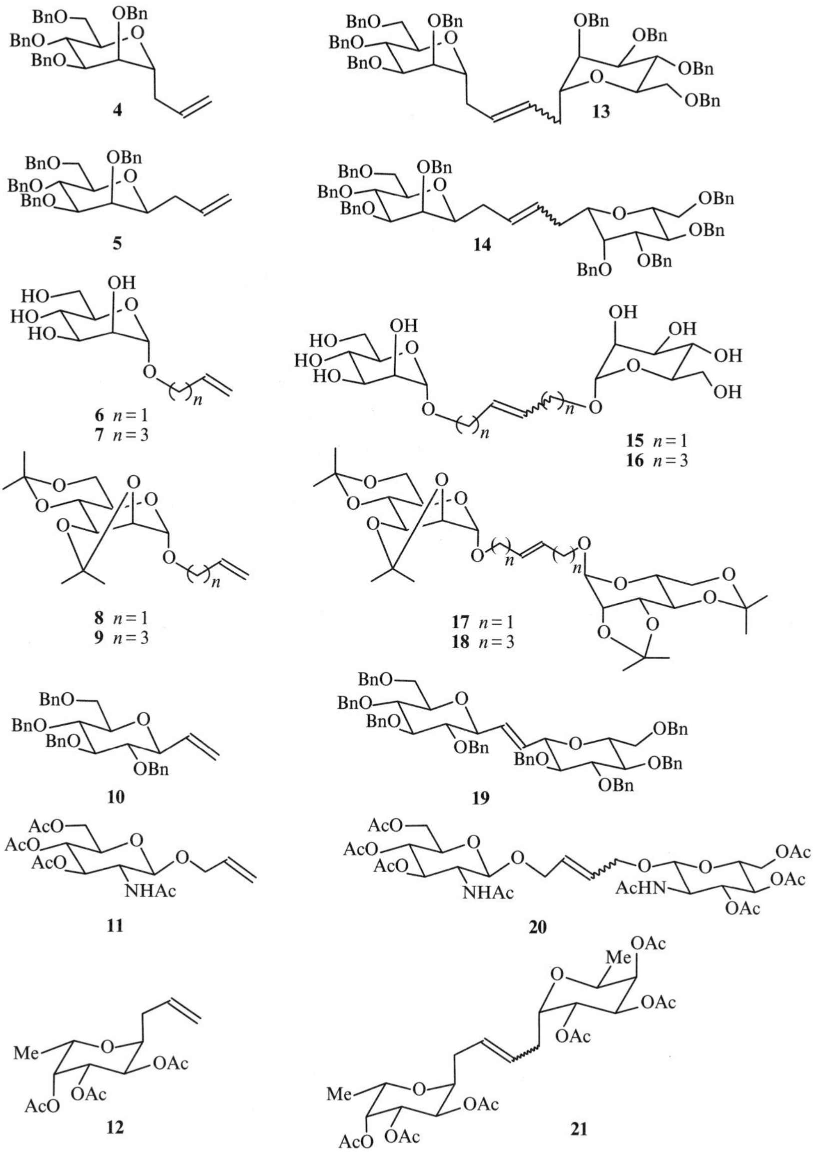

SCHEME 2

TABLE I
OLEFIN SELF-METATHESIS OF ALKENYL *O*- AND *C*-GLYCOPYRANOSIDES[a]

Entry	Alkene	Product	Reaction conditions	Reaction time (h)	Yield (%)	*E/Z* ratio
1	**4**	**13**	CH_2Cl_2, Δ	8	83	1:1
2	**5**	**14**	CH_2Cl_2, Δ	8	70	1.4:1
3	**6**	**15**	CH_2Cl_2:CH_3OH (3:1)[b]	8	60	1.7:1
4	**6**	**15**	CH_2Cl_2:CH_3OH (3:1), 40°	18	34	1.4:1
5	**7**	**16**	CH_2Cl_2:CH_3OH (3:1)[b]	8	67	1.3:1
6	**7**	**16**	CH_2Cl_2:CH_3OH (3:1), 40°	16	42	1.1:1
7	**8**	**17**	CH_2Cl_2, Δ		85	5:1
8	**9**	**18**	CH_2Cl_2, Δ		85	5:1
9	**10**	**19**	CH_2Cl_2[b]	24	NR	—
10	**10**	**19**	CH_2Cl_2, Δ	16	8	100:0
11	**10**	**19**	$ClCH_2CH_2Cl$, 70°	16	24	100:0
12	**11**	**20**	CH_2Cl_2, Δ	8	66	5:2
13	**12**	**21**	CH_2Cl_2, Δ	8	95	3:1
14	**24**	**25**	CH_2Cl_2, Δ	4	91	100:0
15	**26**	**29**	CH_2Cl_2, Δ	6	82	7:1
16	**27**	**30**	CH_2Cl_2, Δ	2	88	3:1
17	**28**	**31**	CH_2Cl_2, Δ	24	26	2.5:1
18	**32**	**33**	CH_2Cl_2, Δ	24	37	100:0
19	**34**	**35**	CH_2Cl_2, Δ	2	78	1.7:1

Abbreviation: NR, Not recorded.
[a]From Refs. 5–9.
[b]At room temperature.

Unfortunately, separation of the two stereoisomers is sometimes difficult and requires high-performance liquid chromatography (HPLC) separation. Alternatively, hydrogenation of the double bonds generates single aliphatic spacers that are otherwise generally obtained in much lower yields and anomeric stereoselectivities when double glycosidation reactions are utilized with the corresponding diols. As can be seen from dimerization of vinyl glycoside **10** (Table I, entries 9–11), steric factors may drastically reduce the overall efficacy of the reaction with, however, increased stereoselectivity. Interestingly, even with catalyst **2**, the olefin metathesis reaction allows for a wide range of functional group tolerance. Even unprotected allyl glycosides **6** and **7** afford the desired products **15** and **16**, albeit in lower yields, when the reactions are performed under refluxing conditions. In this latter case, performing the reaction at room temperature has a beneficial effect on the yield (compare entries 5 and 6 in Table I). The usual acetates, benzyl ethers, and acetal protecting groups are all compatible with the reaction conditions.

The olefin self-metathesis is not restricted to aliphatic alkenes, because even styryl (Table I, entries 14 and 18) and allylbenzene derivatives (Table I, entry 19) provide good to moderate yields of the corresponding glycosyl stilbene analogs **25**, **33**, and **35**, respectively (Scheme 3, compounds **22–35**). *para*-Vinylphenyl-*O*-glycosides such as **24** and **32** are readily accessible under phase transfer catalysis (PTC). The synthesis of the sialic acid-containing dimers is also noteworthy (see compounds **29**, **30**, **33**, and **35** in Scheme 3). More important is the finding that *S*-allyl glycoside **28** also provides the expected dimer **31** (26%; Table I, entry 17). The lower yield in this case may be attributed to reduction of the catalyst turnover, because the remaining starting material **28** can be recovered intact and reentered into the catalytic cycle with fresh catalyst. The α-pentenyl sialoside **27** (Table I, entry 16) affords a better yield of dimer **30** than the corresponding allyl sialoside **26** (entry 15), although with a decrease in *E/Z* stereoselectivity due to steric decompression (3:1 for **30** versus 7:1 for **29**). Once deprotected, some of the above-described homodimers represent useful bivalent ligands for protein cross-linking studies.

Olefin Cross-Metathesis

Olefin cross-metathesis is conceptually more complicated than self-metathesis simply because each of the intervening partners can, in principle, participate in the catalytic process leading to each homodimer, together with the desired cross-products. Two easy solutions to this problem have been undertaken. In the most simple cases, a slight excess of one of the olefins can be used. This protocol is favored when one of the alkenes is commercially available or is readily accessible. A conceptually different approach can be used in which one of the reacting alkenes can be anchored to a solid phase, thus preventing self-metathesis of the immobilized alkene from occurring.

With the advent of the more reactive and more selective ruthenium alkylidene catalyst **3**, somewhat "controlled" cross-metathesis reactions are possible. The usually random [2 + 2] cycloadditions/cycloreversions catalytic cycles involving alkenes, carbenes, and metallo-cyclobutane intermediates can be partly controlled with catalyst **3**, which has been shown to be more chemoselective toward electron-poor olefins. For instance, allyl chloride (Table II, entries 1–3; Scheme 4, compounds **36–50**) and *tert*-butyl acrylate (Table II, entries 6 and 7) are unreactive with the original Grubbs's catalyst **2**, although they have been found to react efficiently with catalyst **3**. It also appears that the new ruthenium catalyst **3** is slightly better

SCHEME 3. *(continued)*

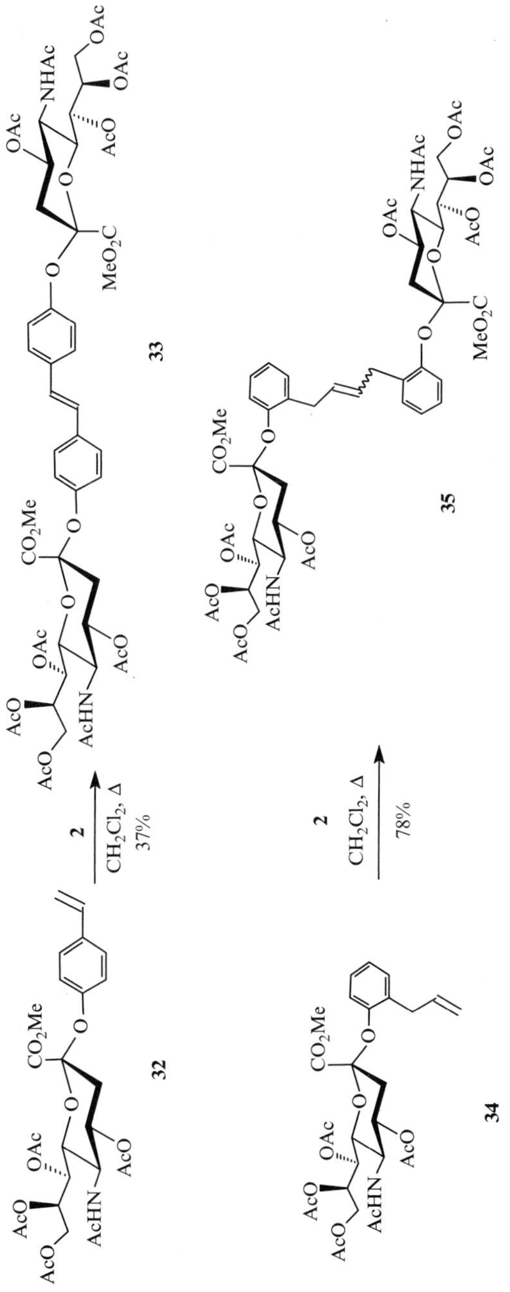

SCHEME 3

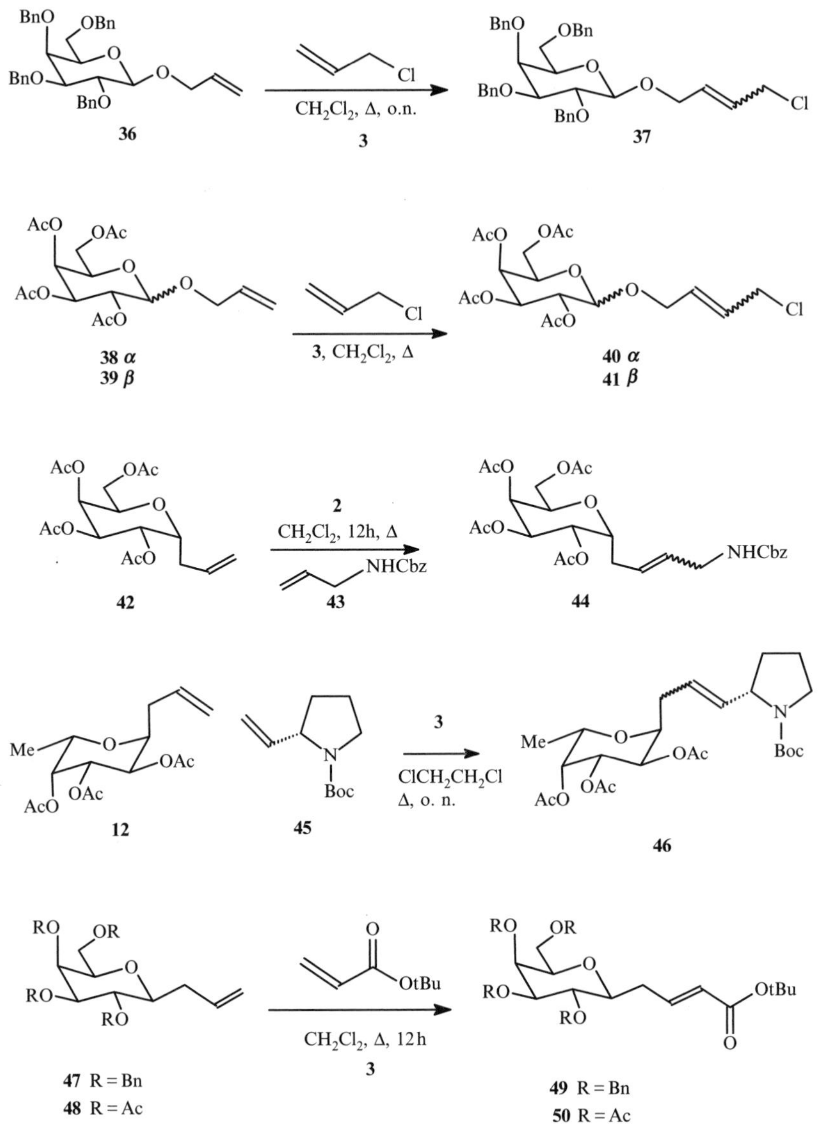

SCHEME 4

at overcoming steric hindrance problems because *N*-Boc-2-vinylpyrrolidine (compound **45**; Table II, entry 5) reacts smoothly to provide cross-product **46** (64%) containing the fucoside residue (**12**) whereas catalyst **2** fails to afford the desired cross-metathesis.

TABLE II
OLEFIN CROSS-METATHESIS OF ALKENYL *O*- AND *C*-GLYCOPYRANOSIDES WITH CATALYST **3**

Entry	Compound	Alkene	Product	Reaction conditions	Time (h)	Yield (%)	*E/Z* ratio
1	**36**	$CH_2 = CHCH_2Cl$	**37**	CH_2Cl_2, Δ	6	75	>20:1
2	**38**	$CH_2 = CHCH_2Cl$	**40**	CH_2Cl_2, Δ	6	65	17:1
3	**39**	$CH_2 = CHCH_2Cl$	**41**	CH_2Cl_2, Δ	6	55	15:1
4	**42**	$CH_2 = CHCH_2NHCbz$	**44**	CH_2Cl_2, Δ	12	45[a]	>20:1
5	**12**	**45**	**46**	$ClCH_2CH_2Cl$, Δ	16	64[b]	3.3:1
6	**47**	$CH_2 = CHC(O)Ot$-Bu	**49**	CH_2Cl_2, Δ	12	74	95:5
7	**48**	$CH_2 = CHC(O)Ot$-Bu	**50**	CH_2Cl_2, Δ	12	91	97:3
8	**51**	**8**	**52a**	CH_2Cl_2, Δ	13	48[a]	5:1
9	**42**	**54**	**55**	CH_2Cl_2, Δ	6	75[a]	100:0
10	**42**	**56**	**57**	CH_2Cl_2, Δ	6	67	4:1

[a]Catalyst **2** is used.
[b]Homodimer **21** (22%) is also formed. No trace of **45** homodimer is detected.

More complex reactions can also be undertaken by cross-metathesis. Scheme 5 (compounds **51–57**) illustrates the double cross-metathesis reactions between diene **51** and allyl mannoside **8** (4 equivalents), using catalyst **2**. The reaction proceeds smoothly to provide tetramer **52a** (fully protected intermediate) in 48% yield. In this particular case, the reaction conditions must be fine-tuned because both the homodimer of **8** (**17**) and the competitive ring-closing metathesis (RCM) product **53** are inevitably formed during the process (Scheme 5). The cross-metathesis reaction has also been extended to C-linked glycoside-containing aryl functionalities (**55**, 50%; *E/Z*, 100:0) using 4-acetoxystyrene **54** and to the synthesis of pseudodisaccharide as seen with the coupling of *C*-allyl galactoside **42** and 6-*O*-allyl isopropylidene galactoside **56**, which affords cross-dimer **57** in 67% yield (*E*/*Z*, 4:1) (entries 9 and 10, Table II).

Typical Procedures

Reagents

Catalyst **2** [$Cl_2Ru(PCy_3)_2$ CHPh; bis(tricyclohexylphosphine)benzylideneruthenium(IV)dichloride, Grubbs's catalyst] and catalyst **3** [$Cl_2Ru(PCy_3)(=CHPh)(IMes_2)$; tricyclohexylphosphine[1,3-bis(2,4,6-trimethylphenyl)-4,5-dihydroimidazol-2-ylidene]benzylidene]ruthenium(IV) dichloride] are obtained from Strem Chemicals (Newburyport, MA).

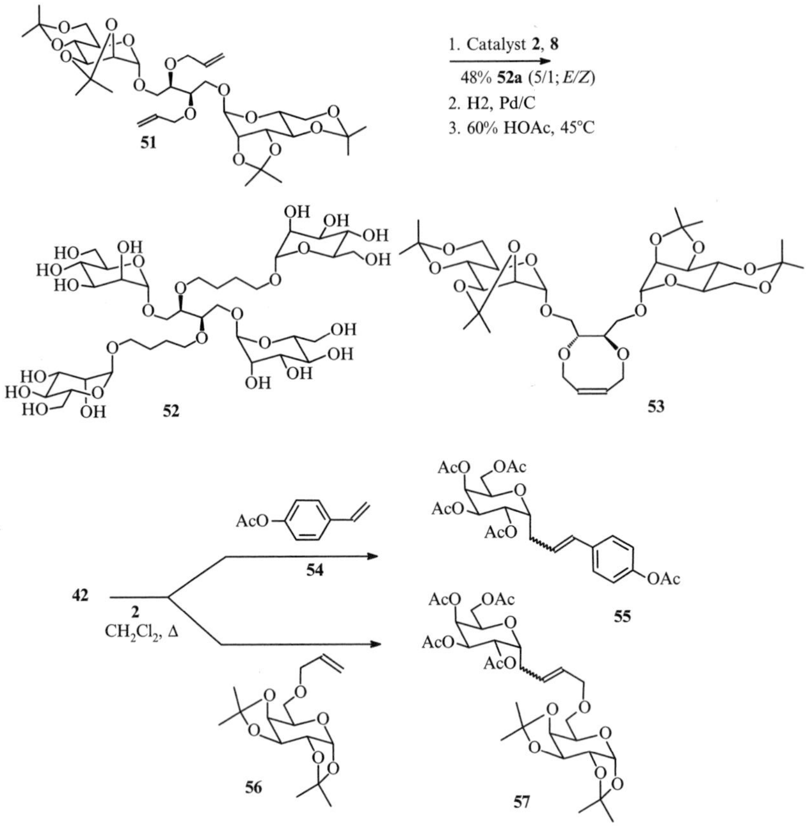

1. Catalyst **2**, **8**
48% **52a** (5/1; *E/Z*)
2. H2, Pd/C
3. 60% HOAc, 45°C

51 **52** **53**

42 **2** CH_2Cl_2, Δ **54** **55** **56** **57**

SCHEME 5

These reagents are best conserved in dry boxes under a nitrogen atmosphere at room temperature. *O*-Allyl glycosides are prepared by standard procedures using allyl alcohol and a Lewis acid.[10,11] *C*-Allyl glycosides are synthesized from peracetylated or perbenzylated sugars, allyl trimethylsilane, and a Lewis acid[12–15] or by the lactone procedure of Lewis *et al.*[16]

[10] J. Gigg, R. Gigg, S. Payne, and R. Conant, *Carbohydr. Res.* **141,** 91 (1985).
[11] B. Fraser Reid, U. E. Udolong, Z. Wu, H. Otosson, J. R. Meritt, C. S. Rao, and R. Madsen, *Synlett* 927 (1992).
[12] A. P. Kozikowski and K. L. Sorgi, *Tetrahedron Lett.* 2281 (1982).
[13] A. Giannis and K. Sandhoff, *Tetrahedron Lett.* 1479 (1985).
[14] D. Horton and T. Miyake, *Carbohydr. Res.* **184,** 221 (1988).
[15] A. Hosomi, Y. Sakata, and H. Sakurai, *Carbohydr. Res.* **171,** 223 (1987).
[16] M. D. Lewis, J. K. Cha, and Y. Kishi, *J. Am. Chem. Soc.* **104,** 4976 (1982).

Self-Metathesis

*(E,Z)-1,4-Di-C-(2,3,4,6-tetra-O-benzyl-α-D-mannopyranosyl)but-2-ene (**13**).* To a solution of 3-(2,3,4,6-tetra-*O*-benzyl-α-D-mannopyranosyl)-propene (**4**)[15] (200 mg, 0.355 mmol) in CH_2Cl_2 (3 ml) and under nitrogen atmosphere is added ruthenium catalyst **2** (15 mg, 10 mol%). The purple reaction mixture is heated under reflux and the solution turns black over 1 h. The reaction mixture is stirred at this temperature for another 5 h. The solution is then concentrated under reduced pressure and the residue is purified by silica gel column chromatography using ethyl acetate–hexane (1:2, v/v) as eluent to give syrupy compound **13**[17] (162 mg, 83%) as a *cis/trans* mixture. ^{1}H nuclear magnetic resonance (NMR) (500 MHz, $CDCl_3$): δ (ppm) 7.35–7.19 (m, 40 H, aromatic), 5.46 (t, 2H, $J = 4.4$ Hz, H-2′ *cis*), 5.37 (t, 2H, $J = 3.8$ Hz, H-2′ *trans*), 4.72–4.49 (m, 16H, 4× CH_2Ph), 4.02 (m, 2H, H-1), 3.88 (t, 2H, $J = 6.7$ Hz, H-4 *trans*), 3.87 (t, 2H, $J = 6.7$ Hz, H-4 *cis*), 3.82–3.70 (m, 8H, H-3, H-5, H-6a, H-6b), 3.62 (dd, 2H, $J = 3.3$ and 4.6 Hz, H-2′ *trans*), 3.61 (dd, 2H, $J = 3.1$ and 4.5 Hz, H-2′ *cis*), 2.30 (m, 4H, H-1′ *cis*), 2.30–2.10 (m, 4H, H-1′ *trans*); ^{13}C NMR (125 MHz): δ (ppm) 138.5–127.4 (aromatic), 75.2 (C-3), 74.9 (C-2), 74.9 (C-4), 73.8, 73.3, 71.5, 71.5 (CH_2Ph), 73.6 (C-5), 72.1 (C-1), 69.2 (C-6), 33.4 (C-1′ *trans*), and 28.4 (C-1′ *cis*): Analysis: calculated for $C_{72}H_{76}O_{10}$: C 78.51, H 6.95: Found: C 78.17, H 7.00.

Cross-Metathesis

*N-(tert-Butoxycarbonyl)-2(S)-[3-(2,3,4-tri-O-acetyl-α-L-fucopyranosyl)] propen-1-ylpyrrolidine (**46**).* 3-(2,3,4-Tri-*O*-acetyl-α-L-fucopyranosyl)-propene (**12**)[18] (51.4 mg, 164 μmol) along with (*S*)-*N*-(*tert*-butoxycarbonyl)-2-vinylpyrrolidine (**45**)[19] (64.5 mg, 327 μmol, 2.0 equivalents) are dissolved in 1.4 ml of dry 1,2-dichloroethane, under a nitrogen atmosphere. Catalyst **3** (6.9 mg, 5 mol%) is then added to the solution, which is stirred for 30 min at room temperature, and then brought to reflux for 16 h. The solution is then evaporated, and the residue is chromatographed on a silica gel column [hexane–ethyl acetate, 3:1 (v/v)] to give the *cis* isomer (**46***Z*) (11.9 mg, 15%) and the *trans* isomer (**46***E*) (39.0 mg, 49%) as thick oils with a brown color, which can be removed by chromatographing a second

[17] R. Dominique and R. Roy, *Tetrahedron Lett.* **43,** 395 (2002).

[18] T. Uchiyama, T. J. Woltering, W. Wong, C.-C. Lin, T. Kajimoto, M. Takebayashi, G. Wetz-Schmidt, T. Asakura, M. Noda, and C.-H. Wong, *Bioorg. Med. Chem.* **4,** 1149 (1996).

[19] Compound **45** was made from (L)-prolinal and triphenylmethylidenephosphorane under conventional Wittig conditions. The physical data of **45** compared well with those published in: C. Serino, N. Stehle, Y. S. Park, S. Florio, and P. Beak, *J. Org. Chem.* **64,** 1160 (1999).

time. The samples are stable, showing no decomposition over several months at room temperature. The expected homodimer **21** is also obtained in 22% yield and no pyrrolidine homodimer of **45** is detected. Thin-layer chromatography (TLC) (1:1 hexane–ethyl acetate): R_f: 0.67 (**3**), 0.62 (**45**), 0.53 (**12**), 0.32 (**46***Z*), 0.28 (**46***E*).

Trans-N-(tert-Butoxycarbonyl)-2(S)-[3-(2,3,4-tri-O-acetyl-α-L-fucopyranosyl)]propen-1-ylpyrrolidine (***46E***). ^{1}H NMR (500 MHz, $CDCl_3$): δ (ppm) 5.38 (2H, m, C=CH × 2), 5.27 (1H, dd, $J_{1,2} = 10.0$ Hz, $J_{2,3} = 5.6$ Hz, H-2), 5.24 (1H, dd, $J_{3,4} = 3.4$ Hz, $J_{4,5} = 2.0$ Hz, H-4), 5.17 (1H, dd, $J_{2,3} = 10.0$ Hz, $J_{3,4} = 3.4$ Hz, H-3), 4.25 (1H, m, 2-pyrrolidine-H), 4.18 (1H, m, H-1), 3.93 (1H, dq, $J_{5,6} = 6.4$ Hz, $J_{4,5} = 1.8$ Hz, H-6), 3.33 (2H, dd, $J = 3.7$, 7.0, 5-pyrrolidine-H), 2.47 (1H, m, H-1′), 2.17 (1H, m, H-1′), 2.12, 2.03, 1.98 (3 × 3H, s, $COCH_3$), 1.95 (1H, m, 3-pyrrolidine-H), 1.80 (2H, m, 4-pyrrolidine-H), 1.64 (1H, m, 3-pyrrolidine-H), 1.41 (9H, s, *tert*-butyl-H), 1.10 (3H, d, $J = 6.4$ Hz, H-6). ^{13}C NMR (125 MHz): δ (ppm) 170.5, 170.1, 169.9 (3 × CO), 154.5 (NCO), 133.6, 124.8 (C=C × 2), 79.0 [C-$(CH_3)_3$], 72.2 (C-1), 70.7 (C-4), 68.6 (C-2), 68.3 (C-3), 65.6 (C-5), 58.2, (2-pyrrolidine), 46.2 (5-pyrrolidine), 32.0 (3-pyrrolidine), 29.7 (4-pyrrolidine), 28.8 (C-1′), 28.5 [C-$(CH_3)_3$], 20.8, 20.7, 20.6, ($COCH_3$) 15.9 (C-6). MS: (+FAB-HRMS, *m/z*): ($C_{24}H_{38}NO_9^+$) calculated: 484.2547: Found: 484.2559.

Comments

Some of the homodimers described in this chapter show improved binding affinities with plant lectins in comparison with their corresponding monomers. For instance, the fully hydrogenated derivative of the α-mannoside **13** shows K_a values of $5.3 \times 10^{-4}\ M^{-1}$ and $10.6 \times 10^{-4}\ M^{-1}$ with concanavalin A and *Dioclea grandiflora* lectins, respectively (methyl-α-mannoside, $1.2 \times 10^{-4}\ M^{-1}$).[20] There is also strong evidence that these simple dimers form cross-linked lattices with tetrameric lectins.

[20] T. K. Dam, R. Roy, S. K. Das, S. Oscarson, and C. F. Brewer, *J. Biol. Chem.* **275,** 14223 (2000).

[3] Digoxin Derivatives as Tools for Glycobiology

By GORDAN LAUC, SANDRA ŠUPRAHA, YUAN C. LEE, and MIRNA FLÖGEL

Studies have demonstrated that nearly all proteins are glycosylated, and that glycan structures have important structural, functional, and modulatory roles. One of the main mechanisms by which glycans convey their information is through interactions with specific lectin receptors. As the level of knowledge about the structure and function of glycoconjugates increases, it is becoming increasingly important to be able to analyze functional properties of their receptors. Lectins are involved in numerous vital physiological processes. For example, selectins mediate lymphocyte adhesion to inflamed endothelium,[1] and mannose-binding lectin (MBP) functions as a part of the host defense system.[2] Lectin activity is also being discovered in an expanding number of previously known proteins, but there are still whole families of lectins, for example, galectins,[3] whose role remains elusive even after extensive research.[4]

One of the major problems in the study of lectins is the lack of adequate methods to analyze their activity. Currently available methods are mainly based on either affinity chromatography or the use of radiolabeled neoglycoconjugates and neoglycoproteins.[5] However, with a few rare exceptions, most of these methods require that lectins be purified before their properties can be analyzed.

Because carbohydrates are relatively difficult to detect, they frequently need to be linked to an appropriate tag. Biotin is a widely used nonradioactive tag, whose main advantage is the availability of excellent binding proteins such as avidin, streptavidin, and extravidin, which bind biotin with great specificity and avidity. In addition, the multimeric nature of avidin–biotin interactions results in clustering of reporter enzyme molecules and increased sensitivity. However, there is a significant drawback in using biotin. Biotin is a prosthetic group that is covalently linked to a number of ubiquitously expressed endogenous enzymes (e.g., pyruvate carboxylase). Because the endogenous biotin complex cannot be differentiated from biotin that is used as a label, all methods that use biotin-containing

[1] L. A. Lasky, *Science* **258,** 964 (1992).
[2] D. L. Jack, N. J. Klein, and M. W. Turner, *Immunol. Rev.* **180,** 86 (2001).
[3] G. Lauc and M. Flögel, *Croat. Chim. Acta* **69,** 339 (1996).
[4] H. Leffler, *Results Probl. Cell Differ.* **33,** 57 (2001).
[5] Y. C. Lee and R. T. Lee (Eds.), *Methods Enzymol.* **247** (1994).

0076-6879/03 $35.00

conjugates are prone to false-positive results. As is clearly shown in Fig. 1, many tissues have several proteins that contain covalently linked biotin and will produce false positives in any assay system that uses biotin as a label.

An alternative label that can alleviate the shortcoming of biotin is digoxigenin. Digoxigenin is a derivative of the cardiac glycoside digoxin (Fig. 2), which has been used as a drug for centuries. Highly specific and strong binding monoclonal and polyclonal antibodies have been developed

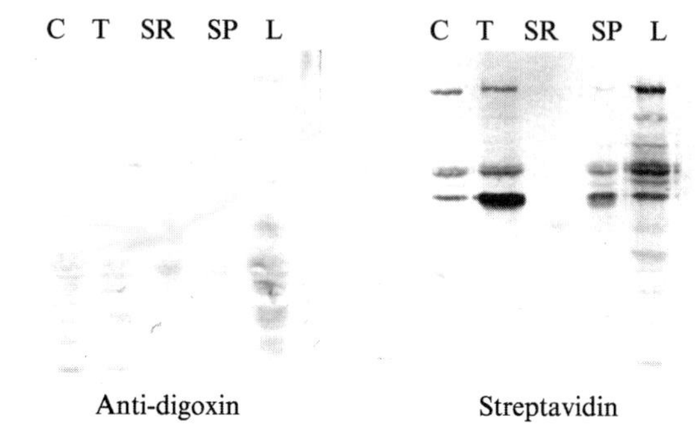

FIG. 1. False-positive bands when biotin is used as a label. Homogenates (10 μg of protein per lane) of rat cortex (C), thymus (T), serum (SR), spleen (SP), and liver (L) were separated by reducing sodium dodecyl sulfate (SDS)–polyacrylamide gel electrophoresis, blotted onto polyvinyledene difloride (PVDF) membrane, and incubated with anti-digoxin antibody labeled with alkaline phosphatase (diluted 1:5000: Sigma) or with streptavidin–biotinylated alkaline phosphatase complex (diluted 1:2000; Amersham Biosciences, Piscataway, NJ). 5-Bromo-4-chloro-3-indolyl phosphate (0.02 mg/ml) in 50 m*M* Tris-HCl (pH 9.5), 100 m*M* NaCl, 5 m*M* $MgCl_2$ was used as a substrate, and the color was intensified by addition of nitro blue tetrazolium (0.04 mg/ml).

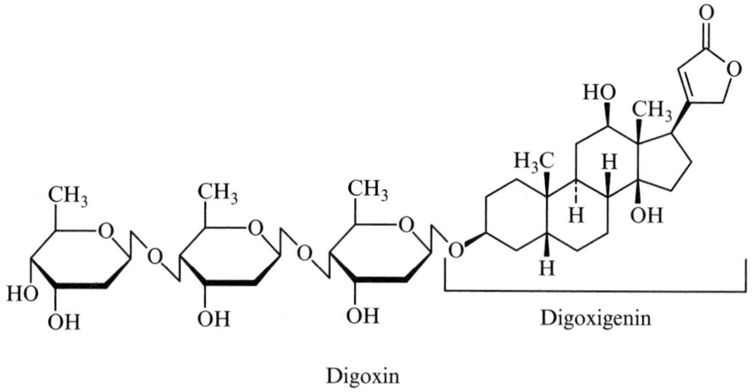

FIG. 2. Structures of digoxin and its deglycosylated derivative digoxigenin.

as an antidote to treat digoxin overdose.[6,7] When labeled with an appropriate enzyme these antibodies can be used to detect both digoxin and digoxigenin. Digoxigenin was first introduced to glycobiology by Boehringer Mannheim (now Roche Molecular Biochemicals, Indianapolis, IN) to label lectins.[8] It proved to be an excellent label that eliminated the problem of false positives due to the presence of endogenous biotin, but this line of products has been significantly downsized and individual digoxigenin-labeled lectins can no longer be obtained commercially.

To exploit commercially available anti-digoxin antibodies and their derivatives, we decided to use digoxin as a label for glycobiological studies. Digoxin-labeled compounds can be used interchangeably with digoxigenin-labeled compounds, but they also offer additional advantage such as easy cleavability of the label, when necessary.

Digoxin-Labeled Lectins

Plant lectins are a simple and versatile tool to analyze glycoconjugates. In a way, they are analogous to antibodies and can be used in a variety of immunoassay-like techniques including enzyme-linked lectin assay (ELLA), lectin Western blot, and lectin affinity chromatography. However, in contrast to antibodies, lectins do not share common structural characteristics, and thus there are no universal secondary antibodies to lectins. To be used in immunoassay-like techniques lectins must first be labeled with a tag, frequently biotin or digoxigenin. Because both biotin and digoxigenin have serious drawbacks as mentioned above, we decided to use digoxin as a lectin tag.

Because of the high acid sensitivity of the glycosidic bond between digitoxoses, and the alkaline sensitivity of its lactone ring, handling of digoxin requires special care. We have developed a method of activation of digoxin using cyanogen bromide,[9] enabling the use of digoxin as a label.

Because lectins vary significantly in the availability of free amino groups, solubility, and stability, it is necessary to optimize labeling conditions for each individual lectin. Labeling is generally performed in a mixture of sodium carbonate or triethylamine–carbonate buffer (pH 8.5–9.5) and a water-miscible organic solvent such as ethanol, methanol, or

[6] S. Tyutyulkova, M. Stamenova, V. Tsvetkova, I. Kehayov, and S. Kyurkchiev, *Immunobiology* **188,** 113 (1993).

[7] P. D. Jeffrey, J. F. Schildbach, C. Y. Chang, P. H. Kussie, M. N. Margolies, and S. Sheriff, *J. Mol. Biol.* **248,** 344 (1995).

[8] A. Haselbeck, E. Schickaneder, H. von der Eltz, and W. Hosel, *Anal. Biochem.* **191,** 25 (1990).

[9] G. Lauc, R. T. Lee, J. Dumic, and Y. C. Lee, *Glycobiology* **10,** 357 (2000).

acetonitrile. The addition of an organic solvent (10–30%) is necessary to keep activated digoxin in solution. Depending on the lectin being labeled, activated digoxin is added in 5- to 20-fold molar excess and incubated overnight. Because too many attached digoxin molecules can decrease lectin activity, the exact ratio of digoxin to lectin concentrations must be determined experimentally for each lectin. After labeling, the CNBr-activated digoxin that did not hydrolyze is inactivated by addition of an equimolar amount of amino groups (glycine, ethanolamine, etc.) and removed by gel filtration or dialysis. Lectins labeled in this way can be applied in any application analogous to other commercially obtained labeled lectins.

Labeling of Concanavalin A with Digoxin

Pure concanavalin A (2 mg; Sigma, St. Louis, MO) is dissolved in 1 ml of 50 m*M* sodium carbonate buffer, pH 9.5, and dialyzed twice against 1 liter of the same buffer at 4°. CNBr-activated digoxin (2 μmol; BioMed Laboratories, Newcastle upon Tyne, UK) is dissolved in 200 μl of methanol and rapidly mixed with concanavalin A (Con A) solution. After overnight incubation at room temperature, 2 μmol of glycine is added to inactivate any remaining CNBr-activated digoxin. The reaction mixture is dialyzed against three 1-liter changes of 10 m*M* sodium phosphate buffer, pH 7.5, at 4° and freeze dried.

Digoxin-Labeled Glycans

One of the main functional roles of carbohydrates in glycoconjugates is their interaction with specific lectin receptors. Labeled glycans are usually employed as various neoglycoconjugates and neoglycoproteins,[10,11] but in contrast to the widely used labeled lectins, the use of labeled glycans to detect lectin activity is still limited.

Using the same activation procedure as described for labeling lectins, we have developed a method to label carbohydrates with digoxin. Free oligosaccharide is first reductively aminated with hydrazine with a 10- to 50-fold molar excess of hydrazine. It is essential to convert all the carbohydrate to hydrazones because any remaining free material would act as a competitive inhibitor in the final labeling reaction. The reaction is usually performed in a volatile buffer (e.g., triethylamine/CO_2 buffer, pH 8.5) in the presence of a suitable reducing agent such as pyridine borane at slightly

[10] R. T. Lee, and Y. C. Lee *in* "Glycoproteins II" (J. Montruil, J. F. G. Vliegentharat, and H. Schachter, eds.), pp. 601–620. Elsevier, Amsterdam, The Netherlands, 1997.

[11] R. T. Lee, and Y. C. Lee *in* "Glycosciences: Status and Perspective" (H.-J. Gabius and S. Gabius, eds.), pp. 55–77. Chapman & Hall, Weinheim, Germany, 1997.

elevated temperature (42°). Any other diamine or dihydrazide can be used instead of hydrazine, but hydrazine is convenient because it can be nearly quantitatively removed by repeated evaporation from toluene.

Hydrazone derivatives of glycans are then incubated with a 5- to 10-fold molar surplus of CNBr-activated digoxin in a mixture of a suitable buffer (phosphate, carbonate, triethylamine) and an organic solvent (tetrahydrofuran, acetonitrile). This reaction should also be allowed to proceed as close to completion as possible because remaining free oligosaccharides will again be inhibitory to future labeling. Activated digoxin decomposes quickly at elevated pH, and thus repeated addition of "fresh" activated digoxin is a better choice than a large excess of digoxin at the beginning of the reaction. After labeling, any unreacted or hydrolyzed digoxin can be easily removed by simple extraction with toluene or chloroform.

Once prepared, labeled glycans can be used as antibodies in any enzyme-linked immunosorbent assay (ELISA) or histochemical assay. However, there is a significant difference between labeled antibodies and labeled glycans. Antibodies assay only the existence of a specific structure, whereas labeled glycans measure the activity of lectins. This has both an advantage and a disadvantage: the advantage is that the actual binding activity and not just the presence of lectins is analyzed; the disadvantage is that the method is incompatible with any denaturing techniques, such as sodium dodecyl sulfate–polyacrylamide gel electrophoresis (SDS–PAGE).

Labeling of Oligosaccharides from Pigeon Ovalbumin with Digoxin

Oligosaccharides are purified from pigeon ovalbumin as described previously.[12] Free oligosaccharide (1 μmol) is reductively aminated with hydrazine in a total reaction volume of 35 μl of 0.1 *M* triethylamine/CO_2 buffer, pH 8.5 A large excess of hydrazine (20 μmol) is added to avoid any significant dimerization. Pyridine borane complex is used as a reducing agent in a final concentration of 10%. After 24 h at 42°, the remaining pyridine borane is extracted with two 0.5-ml volumes of ether. Borate is removed by two evaporations from 20 μl of 10 m*M* acetic acid, and hydrazine is removed by repeated evaporation from toluene (10 evaporations).

Hydrazone derivatives of pigeon ovalbumin oligosaccharides are dissolved in 50 μl of 0.1 *M* triethylamine/CO_2 buffer, pH 8.5, and rapidly mixed with CNBr-activated digoxin (5 μmol; BioMed Laboratories) dissolved in 50 μl of tetrahydrofuran. After overnight incubation at room

[12] N. Suzuki, K. H. Khoo, H. C. Chen, J. R. Johnson, and Y. C. Lee, *J. Biol. Chem.* **276,** 23221 (2001).

temperature, the reaction is stopped by evaporation under reduced pressure. The product is dissolved in 100 μl of water and all nonreacted digoxin is removed by two extractions with 500 μl of chloroform.

Glycoprobes

Glycoprobes are complex neoglycoconjugates based on the patented Bioprobe technology[13] we have been developing for a number of years.[9,14,15] They are composed of the following: a carbohydrate ligand, a digoxin tag, and a photoreactive cross-linker. A schematic representation of glycoprobe application is shown in Fig. 3. Glycoprobes are incubated in the dark with the sample to be analyzed to allow the formation of

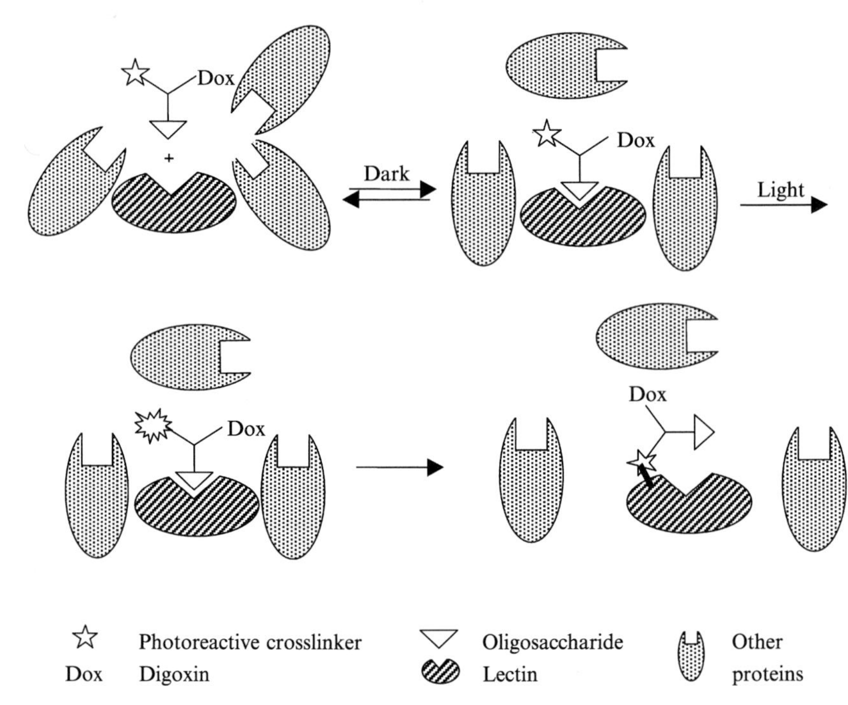

FIG. 3. Schematic representation of glycoprobe application. For detailed explanation see text.

[13] Bioprobes are protected by patent GB 2361699 A, which is the property of BioMed Reagents Ltd., Newcastle upon Tyne, UK.

[14] G. Lauc, M. Flögel, and W. E. G. Müller, *Z. Naturforsch. C* **49,** 843 (1994).

[15] G. Lauc, K. Bariši, T. Žani, and M. Flögel, *Eur. J. Clin. Chem. Clin. Biochem.* **33,** 933 (1995).

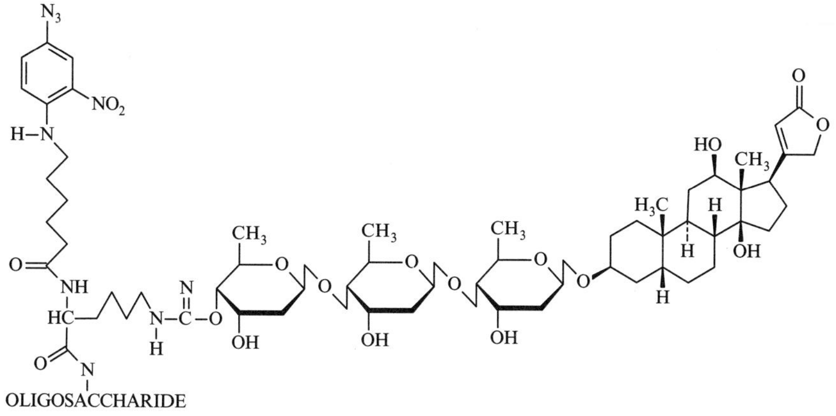

FIG. 4. General structure of a glycoprobe.

noncovalent complexes between an oligosaccharide structure and any lectin in the sample. After UV illumination, the azido group in the photoreactive cross-linker becomes a highly reactive nitrene group that nonselectively forms a covalent bond with adjacent structures[16] on their lectin receptor. Although these covalent bonds are randomly formed, under carefully optimized conditions, carbohydrates that form weak noncovalent complexes with their receptors, can be firmly linked to these structures. The end products are lectins that are covalently tagged with digoxin what enables their subsequent identification and/or isolation using specific anti-digoxin antibodies.

The general structure of a glycoprobe is shown in Fig. 4. *De novo* synthesis of a glycoprobe is relatively complex and probably too demanding for the average biochemical laboratory, but complete glycoprobes and activated bioprobes are now available commercially. In contrast to their synthesis, the application of glycoprobes is simple.

Oligomannose Glycoprobe

Glycoprobes containing oligomannose-type oligosaccharides are prepared by incorporating Man_9-$GlcNAc_2$-Asn glycopeptides from soybean agglutinin. Activated bioprobes[17] (200 nmol) are dissolved in 0.5 ml of

[16] H. Bayley, and J. R. Knowles, *Methods Enzymol.* **44,** 69 (1977).

[17] We have prepared activated bioprobes as described previously [G. Lauc, R. T. Lee, J. Dumiae, and Y. C. Lee, *Glycobiology* **10,** 357 (2000)], but they can also be purchased from Biomed Labs, Newcastle upon Tyne, UK.

distilled dimethylformamide (DMF) and added to 100 nmol of dry Man_9-$GlcNAc_2$-Asn glycopeptides prepared from soybean agglutinin as described previously.[18] Man_9-$GlcNAc_2$-Asn glycopeptides are poorly soluble in DMF and do not fully dissolve immediately, but as the reaction progresses, all glycopeptides are dissolved and react with the activated bioprobe. The reaction mixture is evaporated and dissolved in a mixture of 100 μl of chloroform and 100 μl of water. After extraction, all reaction product remains in the water phase while unreacted and hydrolyzed bioprobes are extracted into chloroform. Piperidine (25 μl) is added to the water phase and incubated for 10 min at room temperature to deprotect the α-amino group of lysine. Free fluorenylmethanol and piperidine are removed by toluene extraction and oligomannose-containing glycoprobe that remains in the water phase is evaporated to dryness under reduced pressure. A photoreactive cross-linker is incorporated and all further reactions are performed in darkness. 6-(4-Azido-2-nitrophenylamino)hexanoic acid *N*-hydroxysuccinimide ester (150 nmol; Sigma) is incubated with approximately 100 nmol of oligomannose glycoprobe in 100 μl of 50% tetrahydrofuran, 50% 0.1 *M* triethylamine/CO_2 buffer, pH 8.5, for 8 h at room temperature. Water (100 μl) and chloroform (200 μl) are added to the

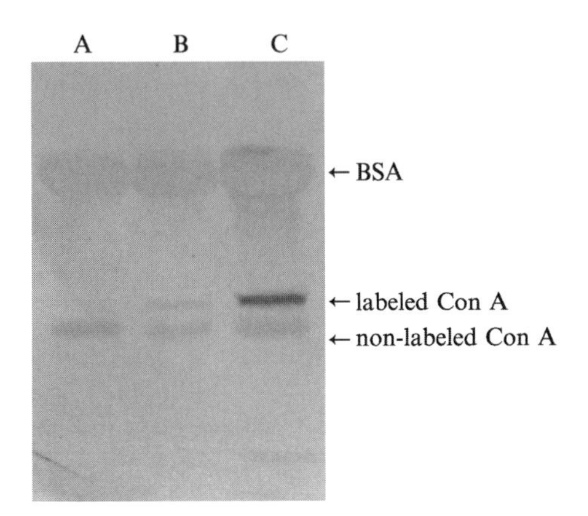

FIG. 5. Glycoprobe assay. Concanavalin A (5 μg) and bovine serum albumin (50 μg) are incubated in the presence of 0, 0.01, and 0.1 nmol (lanes A, B, and C, respectively) of oligomannose-glycoprobe for 60 min in the dark. After 10 min of illumination under a UV lamp, the sample is separated by 10% SDS–PAGE, transferred onto PVDF membrane, and developed with anti-digoxin antibodies labeled with alkaline phosphatase. It is clearly visible that even in the presence of a 10-fold excess of BSA, glycoprobe specifically labels only Con A.

[18] J. G. Fan, A. Kondo, I. Kato, and Y. C. Lee, *Anal. Biochem.* **219,** 224 (1994).

mixture and, after vigorous shaking, the layers are separated. The water phase containing final photoaffinity glycoprobe and *N*-hydroxysuccinimide is collected, while the chloroform phase containing other reactants is discharged. The photoaffinity glycoprobes are divided into small portions (about 20 nmol), evaporated, and stored dry at -20°.

To demonstrate specificity of glycoprobe labeling, mannose-binding Con A lectin is labeled in the presence of a 10-fold excess of bovine serum albumin (BSA) (Fig. 5). After separating the two proteins by SDS–PAGE and blotting, labeled proteins are detected with antibodies against digoxin. Practically all the glycoprobe that is cross-linked to proteins in the sample (and consequently all the digoxin) is linked to Con A, and there is virtually no detectable reaction with BSA.

Conclusions

Immense versatility of carbohydrate structures and the unraveling complexity of their interactions with lectin receptors require continuous development of novel analytical methods. Here we have presented a line of techniques that introduce digoxin as a novel label in glycobiology. Because they surpass some of the problems in the currently used methods, it is hoped that they will prove to be a useful addition to the existing array of glycobiological techniques. In combination with specific antibodies, glycoprobes might also allow the study of the regulation of lectin activity. This is still an undeveloped field, but it has been reported that phosphorylation may, in some cases, modify lectin activity.[19]

Acknowledgments

Work in the authors' laboratories is supported by NIH FIRCA grant 5 R03 TW01477 and by the Croatian Ministry for Science and Technology (grant TP-01/0006-01).

[19] N. Mazurek, J. Conklin, J. C. Byrd, A. Raz, and R. S. Bresalier, *J. Biol. Chem.* **275,** 36311 (2000).

[4] Synthesis of Peptide-Based Trivalent Scaffold for Preparation of Cluster Glycosides

By REIKO T. LEE and YUAN C. LEE

Introduction

The presence of two or more sugar residues that are the recognition target of lectins on one ligand molecule can sometimes bring about a large enhancement in binding affinity.[1] For instance, for the mammalian hepatic lectin, also known as asialoglycoprotein receptor (ASGP-R), compounds containing two and three β-D-galactopyranosides or 2-acetamido-2-deoxy-β-D-galactoparanosides proved to be excellent inhibitors of galactose-terminated glycoproteins, which are thought to be natural ligands of ASGP-R.[2] We have prepared a number of high-affinity, di- and trivalent inhibitors, using an amino acid- or peptide-based scaffold to provide branching points: aspartic acid having two carboxylic acid groups as a divalent scaffold and γ-L-glutamyl-L-glutamic acid (EE) having three carboxylic acid groups as a trivalent scaffold. ω-Amino-terminated glycosides are attached to these scaffolds to produce, respectively, di- and trivalent glycosides conveniently. Advantages of peptide-based scaffolds are as follows: (1) the existence of well-established synthetic peptide chemistry, and (2) the easy availability of ω-aminoglycosides and glycosylamines via synthesis or from natural sources.

One of the trivalent scaffolds we used, γ-glutamylglutamic acid, was commercially available, but was expensive and its solubility was not high. For this reason, we developed an alternative scaffold, β-L-aspartyl-L-aspartic acid (DD), which can be prepared by an easy synthetic scheme without involving any chromatographic separations.[3] We have used the new scaffold for the synthesis of a trivalent mammalian hepatic lectin inhibitor, and found it to be as effective as the derivatives based on γ-glutamylglutamic acid.[3]

[1] R. T. Lee and Y. C. Lee, *Glycoconjug. J.* **17,** 543 (2000).

[2] R. T. Lee, *in* "Liver Diseases: Targeted Diagnosis and Therapy Using Specific Receptors and Ligands" (G. Y. Wu and C. H. Wu, eds.), pp 65–86. Marcel Dekker, New York, 1991.

[3] R. T. Lee and Y. C. Lee *Bioconjug. Chem.* **8,** 762 (1997).

0076-6879/03 $35.00

Materials and Methods

Materials

N-Benzyloxycarbonyl-L-aspartic acid (Z-D), L-aspartic acid dibenzyl ester *p*-toluene sulfonate, *N*-methylmorpholine (NMM), methyl chloroformate, α-L-aspartyl-L-aspartic acid, 1-hydroxybenzotriazole (1-OH-Bt), and dicyclohexylcarbodiimide (DCC) are obtained from Sigma (St. Louis, MO). The *p*-nitrophenyl ester of *N*-benzyloxycarbonyl-L-tyrosine (Z-Y PNP ester) is from Research Organics (Cleveland, OH). The preparation of 6-aminohexyl-2-acetamido-2-deoxy-β-D-galactopyranoside (β-ah-GalNAc) has been reported.[4] The aglycon can be extended by one or more glycine residues by the 1-OH-Bt/DCC coupling method.[5,6] Freshly isolated rat hepatocytes are kindly provided by the Center for Alternatives to Animal Testing *In Vitro* Toxicology Laboratory (Johns Hopkins University, Baltimore, MD).

Methods

Thin-layer chromatography (TLC) is done on a silica gel F_{254} layer coated on an aluminum sheet (E. Merck, Darmstadt, Germany). After the chromatography, the plates are dried and compounds are visualized by inspection under an ultraviolet (UV) lamp (aromatic groups), by spraying with 15% (v/v) sulfuric acid in 50% (v/v) ethanol followed by heating on a hot plate (for sugars and certain peptide backbones), or by spraying with 0.5% (w/v) ninhydrin in 95% (v/v) ethanol and heating briefly (for amino groups). Amino acids and galactosamine (GalN) are analyzed according to the Waters PicoTag automated method (Waters Chromatography Division, Millipore, Milford, MA) with the following modification: in order to separate GalN from amino acids, the initial eluting condition is held for 2 min before gradient is initiated. Nuclear magnetic resonance (NMR) spectra are recorded on a Bruker BioSpin (Billerica, MA) AMX 300 spectrometer, and molecular weight is determined by FAB-MS (VG Instruments, Manchester, UK), using nitrobenzene or glycerol as matrix. Desialylated α_1-acid glycoprotein (ASOR),[7] 10 μg, is iodinated with 0.5 mCi of carrier-free $Na^{125}I$ by the chloramine-T method.[8] The

[4] R. T. Lee, T. C. Wong, and Y. C. Lee, *J. Carbohydr. Chem.* **5,** 343 (1986).
[5] T. B. Johnson and J. K. Coward, *J. Org. Chem.* **52,** 1771 (1987).
[6] R. T. Lee, Y. Ichikawa, T. Kawasaki, K. Drickamer, and Y. C. Lee, *Arch. Biochem. Biophys.* **299,** 129 (1992).
[7] M. J. Kranz, N. A. Holtzman, C. P. Stowell, and Y. C. Lee, *Biochemistry* **15,** 3963 (1976).
[8] F. C. Greenwood, N. M. Hunter, and J. S. Glover, *Biochem. J.* **89,** 114 (1963).

affinity of test compounds is estimated by inhibition assay.[9] Briefly, hepatocytes are incubated with [^{125}I]ASOR and an inhibitor at various concentrations in a pH 7.5 medium at 2° for 2 h with end-over-end tumbling of the incubation tubes. The hepatocyte-bound radioactivity is separated from the bulk radioactivity by placing an aliquot of the suspension on an oil layer (a mixture of silicone oil and mineral oil having a density higher than that of the aqueous medium) in a slim microcentrifuge tube (0.4 ml), and centrifuging it in a microcentrifuge for 40 to 60 s. Cells in the aqueous layer break through the water–oil interface and pellet at the bottom of the tube. The tip of the microcentrifuge tube containing pelleted cells is clipped off and counted in a γ counter (Minaxi; Packard, Rockford, IL). From the plot of percentage of inhibition (y axis) versus inhibitor concentration in logarithmic scale (x axis), the concentration of the inhibitor that causes 50% reduction in the bound amount of radiolabeled ligand (IC_{50}) is obtained.

Experimental Procedures

Synthesis of β-L-*Aspartyl*-L-aspartic acid (β-*DD*, **2**)

The synthetic procedure is shown in Scheme 1. A solution of Z-D (1.34 g, 5 mmol) in dimethylformamide (DMF, 20 ml) is cooled in a dry ice–ethanol bath, and methyl chloroformate (0.5 ml, 6.4 mmol) and NMM (1.5 ml, 13.65 mmol) are added while being stirred. After 20 min, a solution of L-aspartyldibenzyl ester *p*-toluene sulfonate (2.43 g, 5 mmol) and NMM (0.55 ml, 5 mmol) in DMF (10 ml) is added and the mixture is slowly brought to room temperature. After standing overnight, the mixture is filtered and the filtrate is evaporated. On addition of water (30 ml) to the residue, the desired product, **1**, separates as a white precipitate, which is broken up into small pieces with a spatula, and the suspension is stirred overnight at room temperature. The solid is filtered and dried to yield 2.64 g of **1** (80% yield as mono-NMM salt). The product is purified by repeated crystallization from absolute ethanol. TLC was done in ethyl acetate–acetone (50:1, v/v): R_f 0.41, mp 126–127°. ^{1}H NMR in $CDCl_3$ shows the presence of two well-separated methine hydrogens of aspartate and the correct number of aromatic hydrogens. The linkage between two aspartic acid residues is determined to be β after deprotection of **1** (see below).

All the benzyl protecting groups of **1** are removed by hydrogenolysis of **1** (0.85 g, 1.5 mmol) in 90% (v/v) acetic acid (15 ml) in an H_2 atmosphere

[9] Y. C. Lee, R. R. Townsend, M. R. Hardy, J. Lonngren, J. Arnarp, M. Haraldsson, and H. Lonn, *J. Biol. Chem.* **258,** 199 (1983).

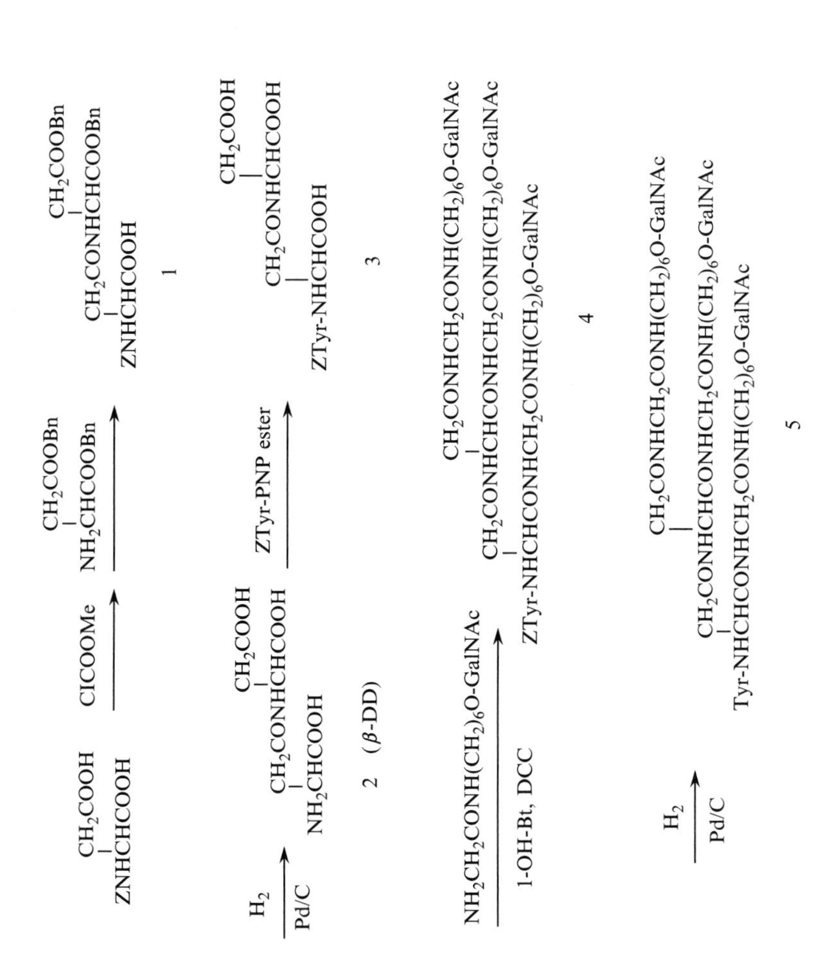

ZNHCHCOOH CH2COOH
ClCOOMe
NH2CHCOOBn CH2COOBn
ZNHCHCOOH CH2CONHCHCOOBn CH2COOBn
1
H2 Pd/C
NH2CHCOOH CH2CONHCHCOOH CH2COOH
2 (β-DD)
ZTyr-PNP ester
ZTyr-NHCHCOOH CH2CONHCHCOOH CH2COOH
3
NH2CH2CONH(CH2)6O-GalNAc
1-OH-Bt, DCC
ZTyr-NHCHCONHCH2CONH(CH2)6O-GalNAc
CH2CONHCHCONHCH2CONH(CH2)6O-GalNAc
CH2CONHCH2CONH(CH2)6O-GalNAc
4
H2 Pd/C
Tyr-NHCHCONHCH2CONH(CH2)6O-GalNAc
CH2CONHCHCONHCH2CONH(CH2)6O-GalNAc
CH2CONHCH2CONH(CH2)6O-GalNAc
5

Scheme 1

generated in a Brown hydrogenator,[10] using 10% palladium on carbon (100 mg) as catalyst. When the uptake of H_2 ceases (~5 h), the suspension is filtered and the filtrate is evaporated to a syrup. Stirring of this syrup in 20 ml of dry DMF produces crystalline **2** in quantitative yield. Crystals are washed with DMF and ether, and dried (mp 152–153°). TLC in ethyl acetate–acetic acid–water (3:2:1, v/v/v) indicates it to be pure. The product **2** is determined to be β-L-aspartyl-L-aspartic acid (β-DD) by comparing its ^{1}H NMR spectrum with that of the authentic isomer, α-L-aspartyl-L-aspartic acid. For detailed characterization, see Ref. 3.

Tyrosine-Containing Trivalent Scaffold

Because β-DD contains both carboxylic acid and amino groups, it is difficult to attach ω-aminoglycoside directly to it. It is also advantageous to have a tyrosine residue present in the scaffold, as this will allow a ready radioiodination for biochemical and biological studies, as well as make it convenient for quantification of the compound by measuring the phenolic group chemically or spectrophotometrically.

p-Nitrophenyl ester of Z-Y (633 mg, 1.45 mmol) and triethylamine (TEA, 0.5 ml, 3.6 mmol) are added to a solution of β-DD (300 mg, 0.86 mmol) in 10 ml of dimethyl sulfoxide (DMSO), and the solution is left at room temperature overnight. The solvent is removed by evaporation, the syrupy residue is dissolved in ethyl acetate (~2 ml), and ether (20 ml) is added while the solution is being stirred. After standing overnight at room temperature, the supernatant is decanted off to yield an amorphous solid, which is dissolved in 95% ethanol. After filtering off undissolved material, the ethanolic solution is evaporated to yield **3** (Z-YDD, 450 mg, 0.7 mmol as mono-TEA salt). TLC in ethyl acetate–2-propanol–water (4:2:1, v/v/v) shows the presence of one major spot that is UV absorbing and chars weakly with the sulfuric acid spray. For other characterizations, see Ref. 3. Because compound **3** as obtained is difficult to dissolve in most of the commonly used solvents, it is used in the next reaction without further purification.

Trivalent GalNAc Ligand, YDD(G-ah-GalNAc)$_3$

Earlier binding studies from our laboratory[11] have shown that, for inhibition of the mammalian hepatic lectin system, the optimal aglycon length for aspartic acid-based divalent inhibitors is provided by the 6-*N*-(glycyl)aminohexyl (G-ah) group. Because GalNAc is a considerably

[10] C. A. Brown and H. C. Brown, *J. Org. Chem.* **31,** 3989 (1966).
[11] R. T. Lee and Y. C. Lee, *Glycoconjug. J.* **4,** 317 (1987).

better inhibitor than Gal,[12] G-ah-GalNAc [$NH_2CH_2CONH(CH_2)_6O$-Gal-NAc] is attached to the aspartate-based trivalent scaffold. Formation of the amide linkage is accomplished by the 1-OH-Bt method.[5] Z-YDD (120 mg, 0.185 mmol) and G-ah-GalNAc (250 mg, 0.66 mmol) are dissolved in DMSO (3 ml) and diluted with DMF (2 ml). While this solution is being stirred, 1-OH-Bt (90 mg, 0.66 mmol), DCC (157 mg, 0.76 mmol), and NMM (24 μl, 0.22 mmol) are added, and the mixture is stirred for 48 h at room temperature. The precipitate (*N,N′*-dicyclohexylurea) is filtered off, and a large volume of toluene is added to the filtrate. After standing overnight at room temperature, the liquid is decanted off, and the amorphous solid is washed once with ether and air dried. TLC of this solid in ethyl acetate–acetic acid–water (3:2:1, v/v/v) shows by charring the presence of one major product, Z-YDD(G-ah-GalNAc)$_3$ (**4**) (R_f 0.06), and a small amount of remaining G-ah-GalNAc (R_f 0.17). Z-YDD(G-ah-GalNAc)$_3$ is crystallized out from 50% ethanol. The remaining **4** in the mother liquor is fractionated on a column of Sephadex G-15 (2.5 × 140 cm) with 0.1 *M* acetic acid as eluant. The combined yield of **4** is 218 mg (72.4%). The D_2O-exchanged **4** shows the correct ratios of aromatic hydrogen, methine hydrogen, anomeric hydrogen, and methyl hydrogen of GalNAc. Fast atom bombardment-mass spectroscopy (FAB-MS) gives the correct molecular ion peak ([M + Na]: 1645.8).

The N-protecting group of **4** is removed by hydrogenolysis, primarily for the purpose of improving the solubility of the trivalent ligand in aqueous media. Hydrogenolysis is carried out as described previously in 60% (v/v) acetic acid to produce **5** in quantitative yield. The product hardly moves in TLC with ethyl acetate–acetic acid–water (3:2:1, v/v/v) (R_f 0.01). Amino acid analysis and FAB-MS indicate the product to be the correct compound.

Inhibitory Potency of YDD(G-ah-GalNAc)$_3$

We have reported that an analogous trivalent GalNAc inhibitor, YEE(ah-GalNAc)$_3$, is a potent inhibitor of mammalian hepatic lectin[11] that can be used effectively as a carrier for the delivery of antisense mucleotides to liver.[13] An inhibition binding assay of [^{125}I]ASOR to isolated rat hepatocytes using both YEE(ah-GalNAc)$_3$ and YDD(G-ah-GalNAc)$_3$ as inhibitors shows both compounds to be equally potent, with an IC_{50} of 8 n*M*.

[12] D. T. Connolly, R. R. Townsend, K. Kawaguchi, W. R. Bell, and Y. C. Lee, *J. Biol. Chem.* **257,** 939 (1982).

[13] J. J. Hangeland, J. E. Flesher, S. F. Deamond, Y. C. Lee, P. O. P., Ts'o, and J. J. Frost, *Antisense Nucleic Acid Drug Dev.* **7,** 141 (1997).

[5] Chemoenzymatic Synthesis of Diverse Asparagine-Linked Oligosaccharides

By YASUHIRO KAJIHARA, YASUHIRO SUZUKI, KEN SASAKI, and LEKH RAJ JUNEJA

Introduction

Most oligosaccharides on proteins are divided into *O*-glycans and *N*-glycans. To investigate the roles of the oligosaccharides, chemical and chemoenzymatic syntheses of these oligosaccharides have been performed.[1–3] Technologies for the synthesis of large oligosaccharides have advanced considerably; however, chemical synthesis is time consuming because of repetitive protection/deprotection steps. *N*-Glycans in glycoproteins show structural diversity in their oligosaccharides, called glycoforms. To study why glycoproteins show different glycoforms, *N*-glycans having diverse structures must be prepared and are used for bioassay. Semisynthetic methods have also been developed.[4–7] Using these approaches[4–6] oligosaccharides from natural sources such as commercially available glycoproteins can yield large quantities of pure oligosaccharide.

We have used asparagine-linked biantennary complex-type sialylundecasaccharide **1**[8] (see Scheme 1) obtained from egg yolk to prepare more than 20 kinds of a pure asparagine-linked oligosaccharide (Asn-oligosaccharide), using branch-specific glycosidase digestion. In short, monosialyloligosaccharides are prepared from **2** by acid hydrolysis of NeuAc, and subsequent exoglycosidase digestion (β-galactosidase, *N*-acetyl-β-D-glucosaminidase, and α-D-mannosidase) of the individual asialo branch affords corresponding diverse oligosaccharides.

[1] K. Toshima and K. Tatsuta, *Chem. Rev.* **93,** 1503 (1993).
[2] C. H. Wong, R. L. Halcomb, Y. Ichikawa, and T. Kajimoto, *Angew. Chem. Int. Ed.* **34,** 521 (1995).
[3] C. Unverzagt, *Carbohydr. Res.* **305,** 423 (1998).
[4] T. Tamura, M. S. Wadhwa, and K. G. Rice, *Anal. Biochem.* **216,** 335 (1994).
[5] K. G. Rice, P. Wu, L. Brand, and Y. C. Lee, *Biochemistry* **32,** 7264 (1993).
[6] E. Meinjohanns, M. Meldal, H. Paulsen, R. A. Dwek, and K. J. Bock, *J. Chem. Soc. Perkin Trans.* **1,** 549 (1998).
[7] C. H. Lin, M. Shimazaki, C. H. Wong, M. Koketsu, L. R. Juneja, and M. Kim, *Bioorg. Med. Chem.* **3,** 1625 (1995).
[8] A. Seko, M. Koketsu, M. Nishizono, Y. Enoki, H. R. Ibrahim, L. R. Juneja, M. Kim, and T. Yamamoto, *Biochim. Biophys. Acta* **1335,** 23 (1997).

0076-6879/03 $35.00

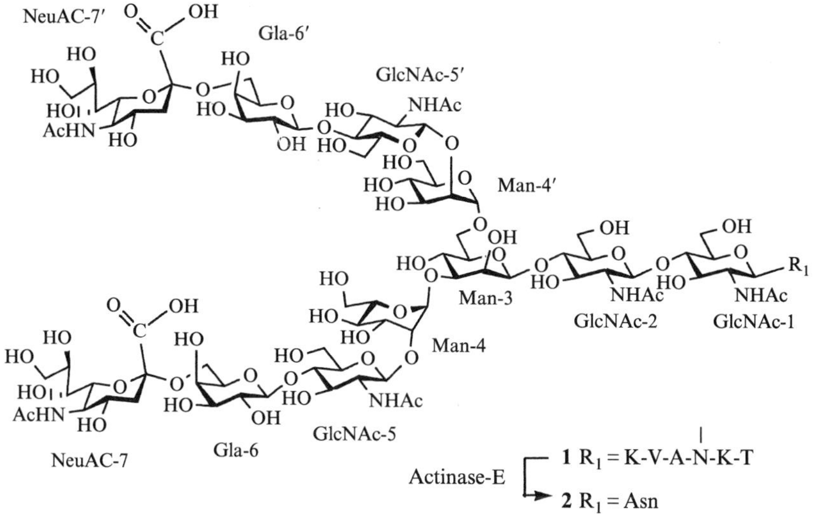

Actinase-E

1 R_1 = K-V-A-N-K-T

2 R_1 = Asn

SCHEME 1

Synthesis of Asparagine-Linked Oligosaccharide Derivatives

Asn-oligosaccharide **2** is prepared by protease (actinase E) digestion of **1**. To release one of the two NeuAc residues from **2**, acid hydrolysis of Asn-oligosaccharide **2** by 40 m*M* HCl solution is performed. This reaction affords four kinds of Asn-oligosaccharide, **2**, **3**, **4**, and **5**, along with **6** and **7** as contaminants. To obtain pure monosialyloligosaccharides **3** and **4**, purification by high-performance liquid chromatography (HPLC) (ODS column) was considered. However, because these oligosaccharides are hydrophilic, purification could not be attained in milligram scale. Therefore, oligosaccharides **2–7** are protected by a hydrophobic protecting group such as 9-fluorenylmethyl group (Fmoc) in order to increase their hydrophobicity (Scheme 2; compounds **2–13**). As expected, this increases interaction between Fmoc-oligosaccharides **8–13** and the ODS column and affords each Asn-oligosaccharide (Fig. 1A) except **9** and **10**. The mixture of **9** and **10** is further protected by forming a benzyl ester of NeuAc. This treatment enables us to purify monosialyloligosaccharides **14** and **15** (Scheme 3; compounds **14–17**).

Galactosidase digestion of **9** and **10** affords nonasaccharides **12** and **16**. The limit of this purification is about 20 mg in one purification step by HPLC [ODS; inner diameter (ϕ) 20 × 250 mm]. Repeat chromatography on the same column yields 200 mg of each monosialyloligosaccharide, **12** and **16** (Scheme 3 and Fig. 1B). Each monosialyloligosaccharide, **12** and

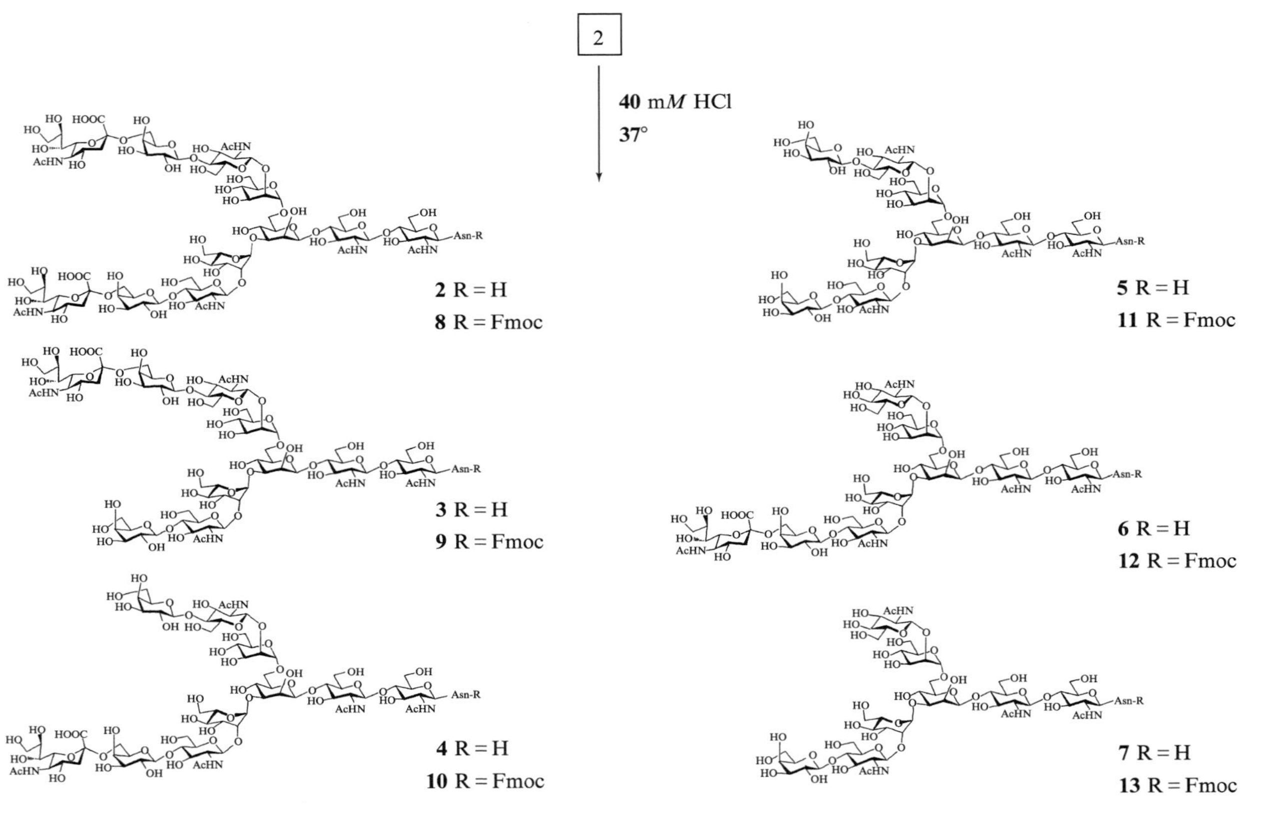
2
40 mM HCl
37°
2 R = H
8 R = Fmoc
3 R = H
9 R = Fmoc
4 R = H
10 R = Fmoc
5 R = H
11 R = Fmoc
6 R = H
12 R = Fmoc
7 R = H
13 R = Fmoc

SCHEME 2

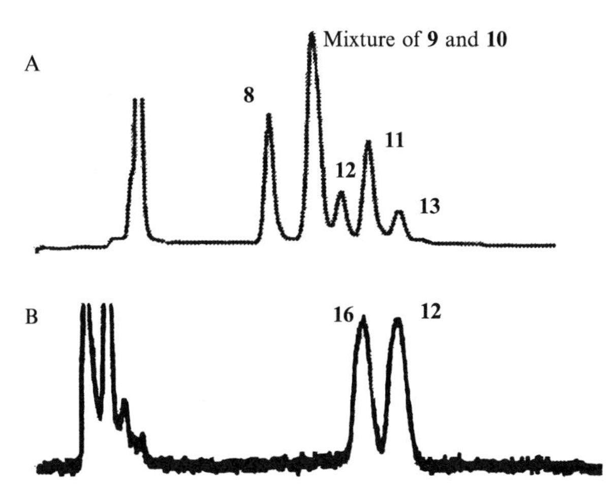

FIG. 1. HPLC profile of (A) acid hydrolysis reaction and (B) purification of **12** and **16** after galactosidase digestion of **9** and **10**.

16, is separately treated by exoglycosidase digestion as shown in Schemes 4 (compounds **18–27**) and 5 (compounds **28–37**). Each exoglycosidase digestion can be performed on a 100-mg scale and yield ranges from 70 to 90% (isolated yield). In addition, simultaneous treatment of several exoglycosidase digestions can also be performed as shown in Scheme 6 (compounds **38** and **39**). Asn-oligosaccharides, shown in Fig. 2 (compounds **40–51**), can also be prepared from **12** by the same strategy shown in Schemes 2–5. The structure of these oligosaccharides thus obtained can be determined by use of a reporter group[9] and high-resolution mass spectroscopy. The nuclear magnetic resonance (NMR) data for these oligosaccharides identifies the product. Removal of the *N*-9-fluorenylmethyl group is performed with morpholine in dimethylformaside (DMF) solution in moderate yield. Using this strategy, more than 24 kinds of Asn-oligosaccharides are obtained as shown in Schemes 2–5 and Fig. 2.

Procedures

General Methods

Crude sialylglycopeptide **1** is prepared from egg yolk by the reported method[8] of extraction with aqueous phenol and simple gel permeation

[9] J. F. G. Vliegenthart, L. Dorland, and H. Halbeek, *Adv. Carbohydr. Chem. Biochem.* **41,** 209 (1983).

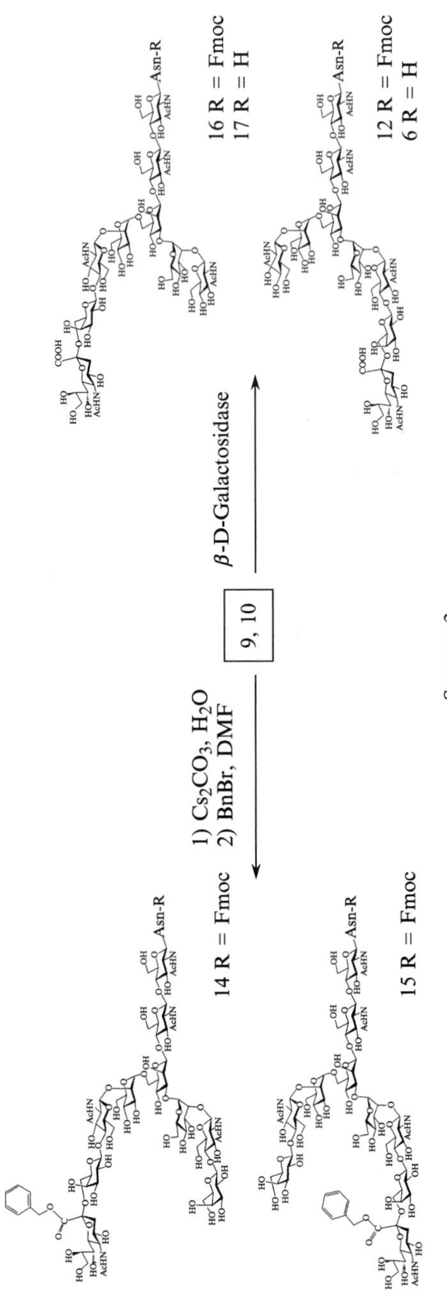
16 R = Fmoc
17 R = H
12 R = Fmoc
6 R = H
β-D-Galactosidase
9, 10
1) Cs2CO3, H2O
2) BnBr, DMF
14 R = Fmoc
15 R = Fmoc
Asn-R

SCHEME 3

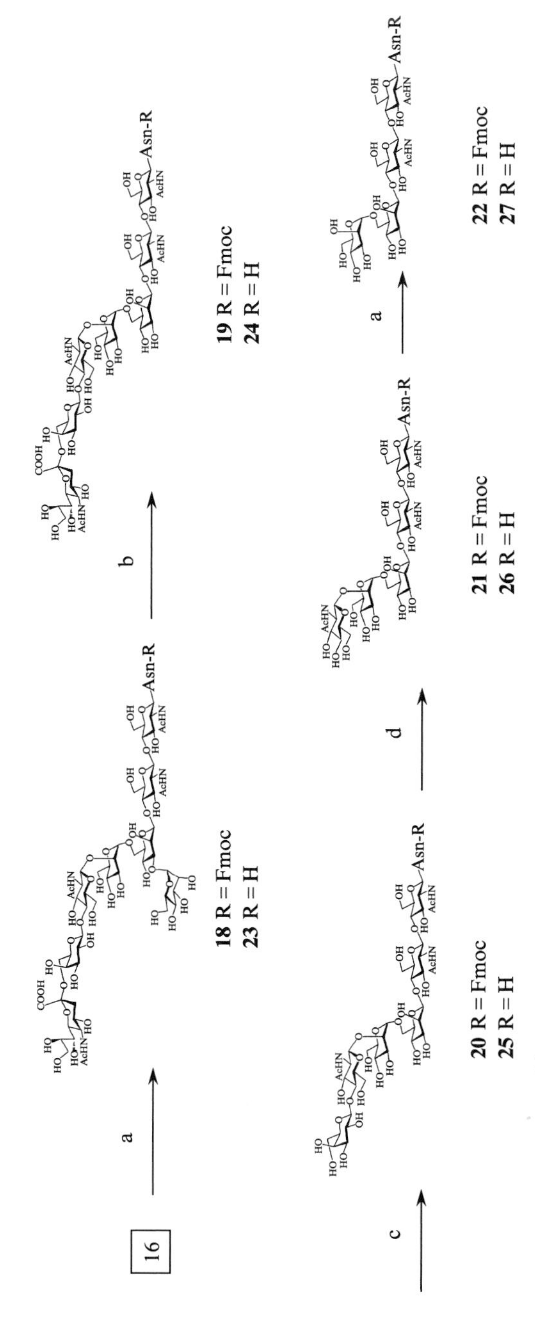

SCHEME 4. (a) *N*-Acetyl-β-D-glucosaminidase; (b) α-D-mannosidase; (c) α-D-neuraminidase; (d) β-D-galactosidase.

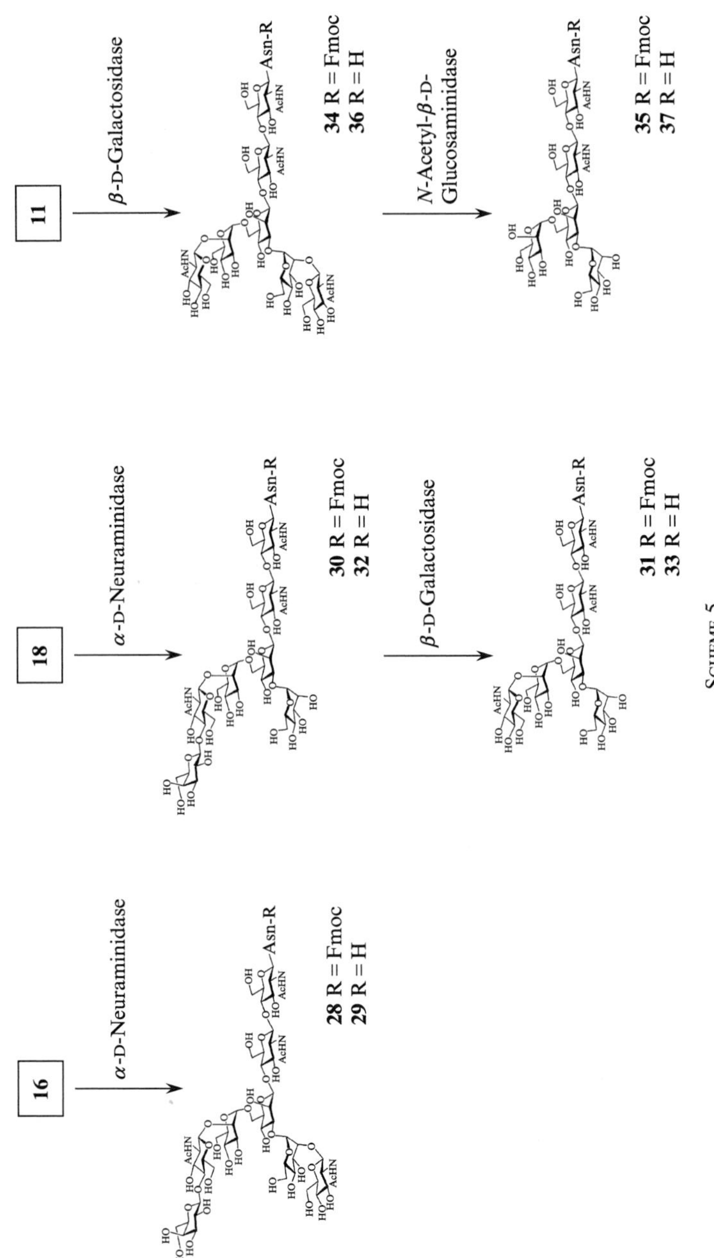
16
α-D-Neuraminidase
28 R = Fmoc
29 R = H
18
α-D-Neuraminidase
30 R = Fmoc
32 R = H
β-D-Galactosidase
31 R = Fmoc
33 R = H
11
β-D-Galactosidase
34 R = Fmoc
36 R = H
N-Acetyl-β-D-
Glucosaminidase
35 R = Fmoc
37 R = H
Asn-R

SCHEME 5

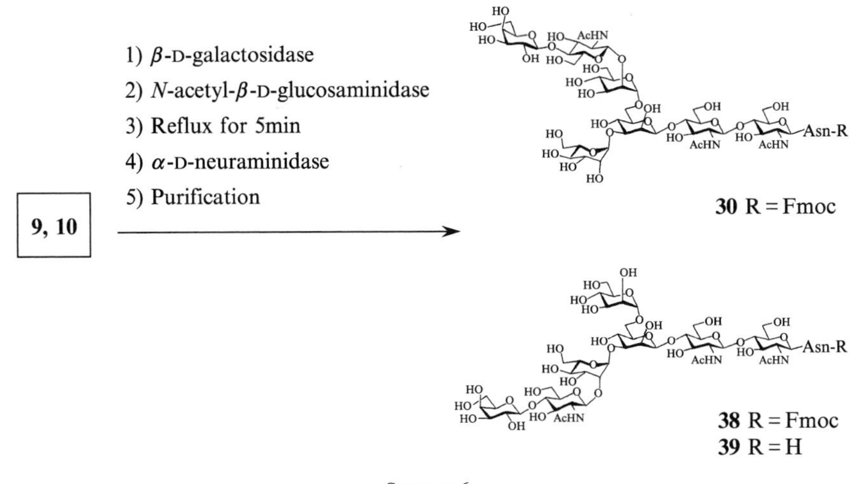

SCHEME 6

(Sephadex G-25) and is about 50% pure. Actinase E is purchased from Kaken Pharmaceutical (Osaka, Japan). Jack bean β-D-galactosidase is purchased from Seikagakukogyo (Japan). Jack beans *N*-acetyl-β-D-glucosaminidase, jack bean β-D-mannosidase, and *Vibrio cholerae* α-D-neuraminidase are purchased from Sigma (St. Louis, MO). NMR spectra are measured with a Bruker BioSpin (Billerica, MA) Avance 400 [30°, internal standard HOD = 4.718 ppm; external (or internal) standard acetone = 2.225 ppm][9] instrument.

*Synthesis of Compound **2***

To a solution of crude sialylglycopeptide **1** (809 mg) and NaN_3 in Tris-HCl buffer (50 m*M*; with 10 m*M* $CaCl_2$, pH 7.5; 32 ml) is added actinase E (263 mg, the same Tris-HCl buffer, 8 ml) and this mixture is incubated at 37° for 60 h. During incubation, the pH is kept at pH 7.5. After 60 h, additional actinase E (25 mg) is added and the mixture is incubated for a further 55 h. This reaction is monitored by thin-layer chromatography (TLC) [1 *M* ammonium acetatc–2-propanol, 1:1 (v/v)]. After the reaction, the mixture is lyophilized and purified by gel permeation (Sephadex G-25; ϕ 2.5 × 100 cm; H_2O) to afford Asn-sialyloligosaccharide **2**[10] (301 mg).

^{1}H NMR δ 5.13 (s, 1H, Man4-H-1), 5.07 (d, 1H, $J = 9.5$ Hz, GlcNAc1-H-1), 4.95 (s, 1H, Man4′-H-1), 4.77 (s, 1H, Man3-H-1), 4.60 (m,

[10] M. Mizuno, K. Haneda, R. Iguchi, I. Muramoto, T. Kawakami, S. Aimoto, K. Yamamoto, and T. Inazu, *J. Am. Chem. Soc.* **121,** 284 (1999).

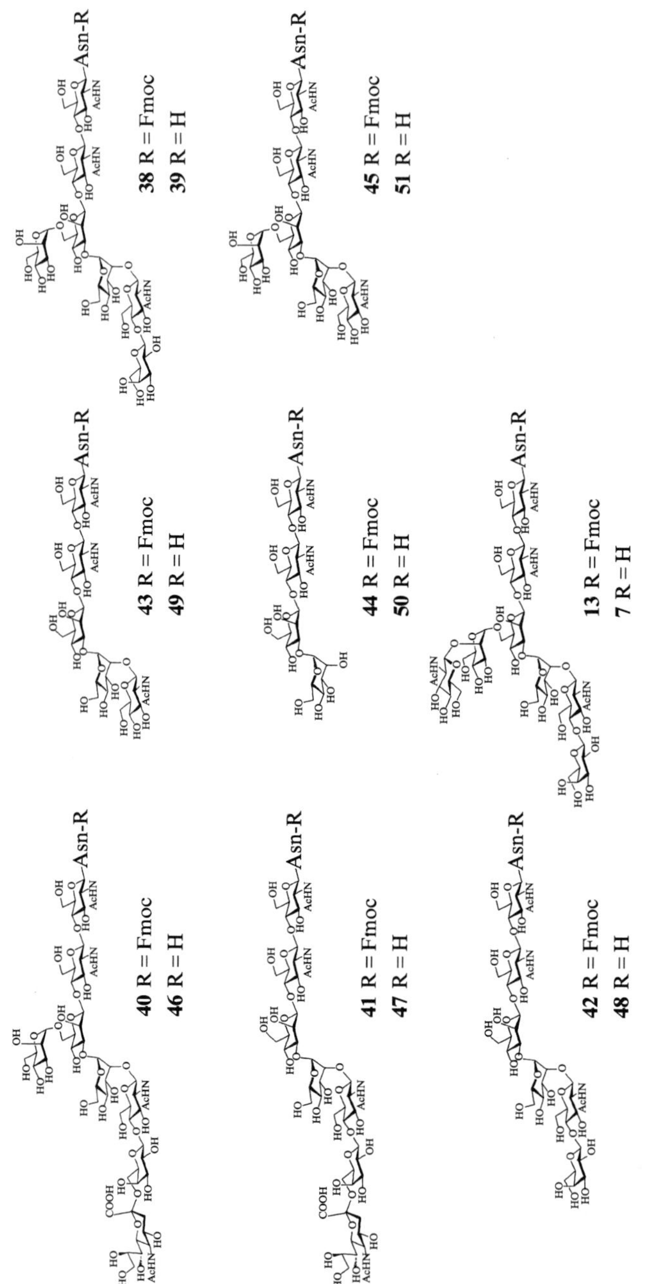

FIG. 2. Structure of oligosaccharide derivatives.

3H, GlcNAc2,5,5′-H-1), 4.44 (d, 2H, $J=8.0$ Hz, Gal6,6′-H-1), 4.25 (bs, 1H, Man3-H-2), 4.20 (bd, 1H, Man4-H-2), 4.12 (bd, 1H, Man4′-H-2), 2.67 (bdd, 2H, $J=4.6$ Hz, 12.4 Hz, NeuAc7,7′-H-3eq), 2.16 (s, 3H, Ac), 2.15 (s, 6H, Ac $\times$ 2), 2.02 (s, 6H, Ac $\times$ 2), 1.71 (dd, 2H, $J=12.4$ Hz, 12.4 Hz, NeuAc7,7′-H-3ax).

*Asialononasaccharide **5***

To a solution of Asn-disialyloligosaccharide **2** (98 mg, 42 μmol) in water (5.8 ml) is added HCl solution (2 *M*, 100 μl) and this mixture is stirred at 80°. After 2 h, the mixture is cooled to 4° and neutralized with aqueous $NaHCO_3$. This solution is then lyophilized. Purification of the residue by gel-permeation column (Sephadex G-25, ϕ 2.5 $\times$ 100 cm; water) affords asialooligosaccharide **5**[6] (67 mg, 91%).

^{1}H NMR δ 5.12 (s, 1H, Man4-H-1), 5.07 (d, 1H, $J=9.7$ Hz, GlcNAcl-H-1), 4.92 (s, 1H, Man4′-H-1), 4.76 (s, 1H, Man3-H-1), 4.62 (d, 1H, $J=8.0$ Hz, GlcNAc2-H-1), 4.58 (d, 2H, $J=7.8$ Hz, GlcNAc5,5′-H-1), 4.47 (d, 2H, $J=7.9$ Hz, Gal6,6′-H-1), 4.24 (bd, 1H, Man3-H-2), 4.19 (bdd, 1H, $J=3.2$ Hz, 1.4 Hz, Man4-H-2), 4.12 (bdd, 1H, J = 3.2 Hz, 1.4 Hz, Man4′-H-2), 2.08 (s, 3H, Ac), 2.05 (s, 6H, Ac $\times$ 2), 2.01 (s, 3H, Ac).

*Fmoc-oligosaccharides **8**, **9**, **10**, and **11***

To a solution of Asn-oligosaccharide **2**, (1.07 g, 456 μmol) in water (11.4 ml) is added HCl solution (80 m*M*, 11.4 ml) and this mixture is stirred at 37°. After 6 h, the mixture is cooled to 4° and neutralized with aqueous $NaHCO_3$, and lyophilized. Purification of the residue by gel-permeation column (Sephadex G-25, ϕ 2.5 $\times$ 100 cm; water) affords a mixture (778 mg) of disialo substrate **2**, monosialyloligosaccharides **3** and **4**, and asialooligosaccharide **5**, along with **6**, and **7**. This mixture is then used for the next protecting reaction. To a solution of this mixture (778 mg) in H_2O–acetone (3.8 ml–5.7 ml) is added $NaHCO_3$ (162 mg, 1.93 mmol) and 9-fluorenylmethyl-*N*-succimidylcarbonate (432 mg, 1.28 mmol), and the mixture is stirred at room temperature. After 2 h, the mixture is evaporated to remove acetone and desalted on an ODS column (ϕ 20 $\times$ 250 mm; eluted with H_2O, 100 ml, and then with 25% CH_3CN, 200 ml) to yield mixture of Fmoc-oligosaccharides, **8–13** (681 mg). This mixture is further purified by HPLC on an ODS column [YMC packed column D-ODS-5 S-5 120A, ϕ 20 $\times$ 250 mm; 50 m*M* ammonium acetate: CH_3CN, 82:18 (v/v); 3.5 ml/min; monitoring at 215 nm) to obtain disialyloligosaccharide **8** (retention time, 86 min), a mixture of monosialyloligosaccharides **9** and **10** (100 min), and asialooligosaccharide **11** (127 min). In addition, monosialylnonasaccharide **12** (112 min) and asialooctasaccharide **13** (139 min) are also obtained. This purification is performed repeatedly with about 30 mg of mixture each time (total, about 500 mg). The individual oligosaccharides thus obtained are combined, lyophilized, and then desalted using HPLC (ODS column, ϕ 5 × 150 mm; eluted with H_2O, 100 ml, and then with 25% CH_3CN, 200 ml) to yield pure oligosaccharide **8** (148 mg, 13%), a mixture of **9** and **10** (249 mg, 24%), **11** (101 mg, 11%), **12** (68 mg, 7%), and **13** (35 mg, 4%).

*Disialooligosaccharide **8***

^{1}H NMR δ 7.92 (d, 2H, $J = 7.5$ Hz, Fmoc), 7.72 (d, 2H, $J = 7.5$ Hz, Fmoc), 7.51 (dd, 2H, $J = 7.5$ Hz, Fmoc), 7.44 (dd, 2H, $J = 7.5$ Hz, Fmoc), 5.14 (s, 1H, Man4-H-1), 5.00 (d, 1H, J = 9.4 Hz, GlcNAc1-H-1), 4.95 (s, 1H, Man4′-H-1), 4.77 (s, 1H, Man3-H-1), 4.60 (m, GlcNAc2,5,5′-H-1), 4.45 (d, 2H, $J = 7.8$ Hz, Gal6,6′-H-1), 4.35 (1H, Fmoc), 4.25 (bd, 1H, Man3-H-2), 4.20 (bdd, 1H, Man4-H-2), 4.11 (bd, 1H, Man4′-H-2), 2.73 (bdd, 1H, $J = 1.7$ Hz, Asn-βCH), 2.52 (bdd, 1H, Asn-βCH), 2.67 (dd, 2H, $J = 4.8$ Hz, 12.5 Hz, NeuAc7,7′-H-3eq), 2.07 (s, 9H, Ac × 3), 2.03 (s, 6H, Ac × 2), 1.89 (s, 3H, Ac), 1.72 (dd, 2H, $J = 12.1$ Hz, 12.1 Hz, NeuAc7,7′-H-3ax).

*Asialooligosaccharide **11***

^{1}H NMR δ 7.91 (d, 2H, $J = 7.5$ Hz, Fmoc), 7.70 (d, 2H, $J = 7.5$ Hz, Fmoc), 7.50 (dd, 2H, $J = 7.5$ Hz, Fmoc), 7.43 (dd, 2H, $J = 7.5$ Hz, Fmoc), 5.12 (s, 1H, Man4-H-1), 5.00 (d, 1H, $J = 9.4$ Hz, GlcNAc1-H-1), 4.93 (s, 1H, Man4′-H-1), 4.76 (s, 1H, Man3-H-1), 4.58 (d, 3H, GlcNAc2,5,5′-H-1), 4.47 (d, 2H, $J = 7.9$ Hz, Gal6,6′-H-1), 4.33 (1H, Fmoc), 4.24 (bd, 1H, $J = 2.1$ Hz, Man3-H-2), 4.19 (bd, 1H, $J = 2.7$HZ, Man4-H-2), 4.11 (bd, 1H, $J = 3.0$ Hz, Man4′-H-2), 2.72 (bdd, 1H, $J = 1.7$ Hz, 15.4 Hz, Asn-βCH), 2.52 (bdd, 1H, $J = 9.2$ Hz, 15.4 Hz, Asn-βCH), 2.07 (s, 3H, Ac), 2.05 (s, 6H, Ac × 2), 1.89 (s, 3H, Ac).

*Monosialyloligosaccharide **12***

^{1}H NMR δ 7.92 (d, 2H, $J = 7.5$ Hz, Fmoc), 7.71 (d, 2H, $J = 7.5$ Hz, Fmoc), 7.51 (dd, 2H, $J = 7.5$ Hz, Fmoc), 7.43 (dd, 2H, $J = 7.5$ Hz, Fmoc), 5.14 (s, 1H, Man4-H-1), 5.00 (d, 1H, GlcNAc1-H-1), 4.92 (s, 1H, Man4′-H-1), 4.77 (s, 1H, Man3-H-1), 4.61 (d, 1H, $J = 8.0$ Hz GlcNAc2-H-1), 4.56 (d, 2H, GlcNAc5,5′-H-1), 4.45 (d, 1H, $J = 7.8$ Hz, Gal6-H-1), 4.35 (bt, 1H, Fmoc), 4.25 (bd, 1H, Man3-H-2), 4.20 (bd, 1H, Man4-H-2), 4.11 (bd, 1H, Man4′-H-2), 2.72 (bdd, 1H, Asn-βCH), 2.67 (dd, 1H, $J = 4.3$ Hz, 12.4 Hz, NeuAc7-H-3eq), 2.54 (bdd, 1H, $J = 9.2$ Hz, 15.9 Hz, Asn-βCH),

2.07 (s, 6H, Ac × 2), 2.05, 2.03, 1.89 (each s, each 3H, Ac), 1.72 (dd, 1H, $J = 12.1$ Hz, 12.1 Hz, NeuAc7-H-3ax).

*Asialooctasaccharide **13***

^{1}H NMR δ 7.92 (d, 2H, $J = 7.5$ Hz, Fmoc), 7.72 (d, 2H, $J = 7.5$ Hz, Fmoc), 7.51 (dd, 2H, $J = 7.5$ Hz, Fmoc), 7.44 (dd, 2H, $J = 7.5$ Hz, Fmoc), 5.12 (s, 1H, Man4-H-1), 5.00 (d, 1H, $J = 9.5$ Hz, GlcNAc1-H-1), 4.92 (s, 1H, Man4′-H-1), 4.76 (s, 1H, Man3-H-1), 4.57 (bd, 1H, GlcNAc2-H-1), 4.55 (d, 2H, $J = 8.9$ Hz, GlcNAc5,5′-H-1), 4.46 (dd, $J = 7.8$ Hz, Gal6-H-1), 4.35 (t, 1H, Fmoc), 4.24 (bd, 1H, Man3-H-2), 4.19 (bd, 1H, Man4-H-2), 4.11 (bd, 1H, Man4′-H-2), 2.72 (1H, bdd, $J = 15.4$ Hz, Asn-βCH), 2.51 (bdd, 1H, $J = 9.6$ Hz, 13.8 Hz, Asn-βCH), 2.06 (s, 3H, Ac), 2.05 (s, 6H, Ac × 2), 1.89 (s, 3H, Ac).

*Benzyl Ester Monosialyloligosaccharides **14** and **15***

A solution of a mixture of Fmoc-monosialyldecasaccharides **9** and **10** (5.0 mg) in cold H_2O (1 ml, 4°) is passed through to a column [ϕ 0.5 cm × 5 cm containing Dowex 50W × 8(H^+) resin], and the column is washed with 10 ml of cold water. The eluant and the washing are pooled and lyophilized. This residue is dissolved in H_2O (0.22 ml) and neutralized (pH 7) by stepwise addition of a solution of Cs_2CO_3 (2.5 mg/ml) while monitoring with a pH meter, and lyophilized. The residue is dissolved in dry DMF (0.43 ml) and is mixed with a solution of benzyl promide (BnBr, 6.6 μl) in DMF (20 μl) and stirred at room temperature under an argon atmosphere. After 48 h, diethyl ether (5 ml) is added and the precipitate formed is collected. Purification of the precipitated material by HPLC [ODS column, ϕ 20 × 250 mm; ammonium acetate–CH_3CN, 78:22 (v/v)] affords monobenzylsialyloligosaccharides **14** (91 min) and **15** (88 min). Desalting of each product by HPLC (ODS column, ϕ 5 × 150 mm; H_2O, 50 ml, and then 25% CH_3CN, 100 ml) affords pure monobenzylsialyloligosaccharides, **14** (1.6 mg) and **15** (1.8 mg).

*Compound **14***

^{1}H NMR δ 7.91 (d, 2H, $J = 7.5$ Hz, Fmoc), 7.71 (d, 2H, $J = 7.5$ Hz, Fmoc), 7.53–7.41 (m, 9H, Fmoc, -CH_2-Ph), 5.38 (d, 1H, $J = 12.1$ Hz, -CH_2-Ph), 5.31 (d, 1H, $J = 12.1$ Hz, -CH_2-Ph), 5.12 (s, 1H, Man4-H-1), 4.99 (d, 1H, $J = 9.8$ Hz, GlcNAc1-H-1), 4.93 (s, 1H, Man4′-H-1), 4.76 (s, 1H, Man3-H-1), 4.58 (m, 3H, GlcNAc2,5,5′-H-1), 4.46 (1H, d, $J = 7.8$ Hz, Gal6-H-1), 4.33 (d, 1H, $J = 7.8$ Hz, Gal6′-H-1), 4.24 (bs, 1H, Man3-H-2), 4.19 (bs, 1H, Man4-H-2), 4.11 (bs, 1H, Man4′-H-2), 2.72 (bd, 1H, Asn-βCH), 2.68

(dd, 1H, $J = 4.8$ Hz, 13.0 Hz, NeuAc7-H-3eq), 2.52 (bdd, 1H, $J = 9.7$ Hz, 14.1 Hz, Asn-βCH), 2.06, 2.05, 2.04, 2.02, 1.89 (each s, each 3H, Ac), 1.84 (dd, 1H, $J = 13.0$ Hz, 13.0 Hz, NeuAc7-H-3ax).

*Compound **15***

^{1}H NMR δ 7.91 (d, 2H, $J = 7.5$ Hz, Fmoc), 7.71 (d, 2H, $J = 7.5$ Hz, Fmoc), 7.53–7.41 (m, 9H, Fmoc, -CH_2-Ph), 5.38 (d, 1H, $J = 12.1$ Hz, -CH_2-Ph), 5.31 (d, 1H, $J = 12.1$ Hz, -CH_2-Ph), 5.12 (s, 1H, Man4-H-1), 4.99 (d, 1H, $J = 9.5$ Hz, GlcNAc1-H-1), 4.92 (s, 1H, Man4′-H-1), 4.76 (s, 1H, Man3-H-1), 4.58 (m, 3H, GlcNAc2,5,5′-H-1), 4.47 (d, 1H, $J = 7.9$ Hz, Gal6′-H-1), 4.33 (d, 1H, $J = 7.9$ Hz, Gal6-H-1), 4.24 (bs, 1H, Man3-H-2), 4.19 (bs, 1H, Man4-H-2), 4.11 (bs, 1H, Man4′-H-2), 2.72 (bd, 1H, Asn-βCH), 2.68 (dd, 1H, $J = 4.6$ Hz, 12.7 Hz, NeuAc7-H-3eq), 2.52 (dd, 1H, J = 8.7 Hz, 15.0 Hz, Asn-βCH), 2.06, 2.05, 2.04, 2.02, 1.89 (each s, each 3H, Ac) 1.84 (dd, 1H, $J = 12.7$ Hz, 12.7 Hz, NeuAc7-H3ax).

*Galactosidase Digestion of Monosialyldecasaccharides **9** and **10***

To a mixture of monosialyldecasaccharides **9** and **10** (135 mg, 59.4 μmol) in HEPES buffer (50 m*M*, pH 6.0, 5.6 ml) containing bovine serum albumin (1.0 mg) is added β-galactosidase (390 mU in 50 m*M* HEPES buffer, 100 μl, pH 6.0) and this mixture is incubated at 37° for 19 h, and lyophilized. Purification of the residue by HPLC [YMC packed column D-ODS-5 S-5 120A, ϕ 20 × 250 mm; 50 m*M* ammonium acetate–CH_3CN, 82:18 (v/v); 3.5 ml/min] affords monosialyloligosaccharides **16** (170 min) and **12** (182 min). This purification is performed repeatedly with ~10 to 20-mg portions of the reaction mixture (total, ~120 mg). Desalting of each product by HPLC (ODS column; ϕ 5 × 150 mm; H_2O, 100 ml and then 25% CH_3CN solution, 200 ml) yields pure oligosaccharides **16** (51 mg, 41%) and **12** (60 mg, 48%).

*Monosialylnonasaccharide **16***

^{1}H NMR δ 7.92 (d, 2H, Fmoc), 7.71 (d, 2H, $J = 7.5$ Hz, Fmoc), 7.51 (dd, 2H, $J = 7.5$ Hz, Fmoc), 7.44 (dd, 2H, $J = 7.5$ Hz, Fmoc), 5.12 (s, 1H, Man4-H-1), 5.00 (d, 1H, $J = 9.1$ Hz, GlcNAcl-H-1), 4.94 (s, 1H, Man4′-H-1), 4.77 (s, 1H, Man3-H-1), 4.60 (b, 1H, GlcNAc2-H-1), 4.55 (d, 2H, $J = 8.4$ Hz, GlcNAc5,5′-H-1), 4.45 (d, 1H, $J = 8.0$ Hz, Gal6′-H-1), 4.35 (t, 1H, Fmoc), 4.25 (bd, 1H, Man3-H-2), 4.19 (bd, 1H, $J = 1.8$ Hz, Man4-H-2), 4.11 (bd, 1H, $J = 1.9$ Hz, Man4′-H-2), 2.72 (bd, 1H, Asn-βCH), 2.67 (dd, 1H, $J = 4.7$ Hz, 12.6 Hz, NeuAc7′-H-3eq), 2.52 (bdd, 1H, $J = 9.7$ Hz, 14.8 Hz, Asn-βCH), 2.06 (s, 6H, Ac × 2), 2.05, 2.03, 1.89 (each s, each 3H, Ac), 1.79 (dd, 1H, $J = 11.9$ Hz, 11.9 Hz, NeuAc7′-H-3ax).

General Purification Method for Exoglycosidase Digestion

After exoglycosidase digestion, the mixture is lyophilized, Fractionation by HPLC [YMC packed column D-ODS-5 S-5 120A, ϕ 20×250 mm; eluted with 50 m*M* ammonium acetate–CH_3CN, 3:2 (v/v); 4.0 ml/min; monitoring at 215 nm] affords oligosaccharide. The fraction containing the desired product is pooled and lyophilized. Removal of ammonium acetate is accomplished by HPLC (ODS column; ϕ 5×150 mm; H_2O, 100 ml, and then 25% CH_3CN, 200 ml, monitoring at 215 nm), affording a pure oligosaccharide.

General Procedure for N-Acetyl-β-D-glucosaminidase Digestion

To a solution of substrate (48 μmol) in HEPES buffer (50 n*M*, pH 6.0, 1.8 ml) containing bovine serum albumin (1 mg) is added *N*-acetyl-β-D-glucosaminidase (1.6 U in 50 m*M* HEPES buffer, 100 μl, pH 6.0), with incubation at 37°. After completion of the digestion (~1 day) as monitored by HPLC [ODS, ϕ 5×150 mm; 50 m*M* ammonium acetate–CH_3CN, 8:2 (v/v) monitoring at 215 nm], the mixture is lyophilized.

General Procedure for β-D-Mannosidase Digestion

To a solution of substrate (24 μmol) in HEPES buffer (50 m*M*, pH 6.0, 0.9 ml) containing bovine serum albumin (1 mg) is added β-D-mannosidase (0.5 U in 50 m*M* HEPES buffer, 50 μl, pH 6.0) and this solution is incubated at 37°. After completion of the digestion (~1 day) as monitored by HPLC [ODS, ϕ 5×150 mm, 50 m*M* ammonium acetate–CH_3CN, 8:2 (v/v) monitoring at 215 nm], the mixture is lyophilized.

General Procedure for α-D-Neuraminidase Digestion

To a solution of substrate (18.4 μmol) in HEPES buffer (50 m*M*, pH 6.0, 0.72 ml) containing bovine serum albumin (1 mg) is added α-D-neuraminidase (134 mU in 50 m*M* HEPES buffer, 50 μl, pH 6.0) and this solution is incubated at 37°. After completion of the digestion (~1 day) as monitored by HPLC [ODS, ϕ 5×150 mm, 50 m*M* ammonium acetate–CH_3CN, 8:2 (v/v) monitoring 215 nm], the mixture is lyophilized.

General Procedure for β-D-Galactosidase Digestion

To a solution of substrate (6.2 μmol) in HEPES buffer (50 m*M*, pH 6.0, 0.6 ml) containing bovine serum albumin (1 mg) is added β-D-galactosidase (180 mU in 50 m*M* HEPES buffer, 50 μl, pH 6.0) and this solution is incubated at 37°. After completion of the digestion (~30 h) as monitored by HPLC [ODS, ϕ 5 × 150 mm, 50 mM ammonium acetate–CH_3CN, 8:2 (v/v) monitoring at 215 nm], the mixture is lyophilized.

*Compound **18***

^{1}H NMR δ 5.11 (s, 1H, Man4-H-1), 5.00 (d, 1H, $J=9.7$ Hz, GlcNAc1-H-1), 4.95 (s, 1H, Man4′-H-1), 4.77 (s, 1H, Man3-H-1), 4.45 (d, 1H, $J=7.9$ Hz, Gal6′-H-1), 4.25 (bs, 1H, Man3-H-2), 4.12 (bdd, 1H, Man4′-H-2), 4.07 (bdd, 1H, Man4-H-2), 2.68 (dd, 2H, $J=4.8$ Hz, 12.4 Hz, NeuAc7′-H-3eq), 1.72 (dd, 1H, $J=12.4$ Hz, 12.4 Hz, Neu Ac7′-H-3ax).

*Compound **19***

^{1}H NMR δ 5.00 (d, 1H, $J=9.4$ Hz, GlcNAc1-H-1), 4.94 (s, 1H, Man4′-H-1), 4.77 (s, 1H, Man3-H-1), 4.64–4.53 (1H, GlcNAc2-H-1), 4.64–4.53 (1H, GlcNAc5′-H-1), 4.45 (d, 1H, $J=8.0$ Hz, Gal6′-H-1), 4.28 (b, 1H, Man3-H-2), 4.11 (b, 1H, Man4′-H-2), 2.68 (dd, 1H, $J=4.6$ Hz, 12.5 Hz, NeuAc7′-H-3eq), 1.72 (dd, 1H, $J=12.5$ Hz, 12.5 Hz, NeuAc7′-H-3ax).

*Compound **40***

^{1}H NMR δ 5.14 (s, 1H, Man4-H-1), 5.00 (d, 1H, J = 9.4 Hz, GlcNAc1-H-1), 4.92 (s, 1H, Man4′-H-1), 4.78 (s, 1H, Man3-H-1), 4.44 (d, 1H, $J=8.0$ Hz, Gal6-H-1), 4.35 (t, 1H, Fmoc), 4.25 (bd, 1H, Man3-H-2), 4.20 (bdd, 1H, Man4-H-2), 2.67 (dd, 1H, $J=4.6$ Hz, 12.2 Hz, NeuAc7-H-3eq), 1.72 (dd, 1H, $J=12.2$ Hz, 12.2 Hz, NeuAc7-H-3ax).

*Compound **41***

^{1}H NMR δ 5.14 (s, 1H, Man4-H-1), 5.00 (d, 1H, $J=9.4$ Hz, GlcNAc1-H-1), 4.78 (s, 1H, Man3-H-1), 4.45 (d, 1H, $J=8.0$ Hz, Gal6-H-1), 4.23 (bd, 1H, Man3-H-2), 4.20 (bd, 1H, Man4-H-2), 2.67 (dd, 1H, $J=4.8$ Hz, 12.4 Hz, NeuAc7-H-3eq), 1.72 (dd, 1H, $J=12.4$ Hz, 12.4 Hz, NeuAc7-H-3ax).

*Compound **28***

^{1}H NMR δ 5.12 (s, 1H, Man4-H-1), 4.99 (d, 1H, $J=9.5$ Hz, GlcNAc1-H-1), 4.92 (s, 1H, Man4′-H-1), 4.76 (s, 1H, Man3-H-1), 4.58 (d, 1H, $J=8.0$ Hz, GlcNAc2-H-1), 4.55 (d, 1H, $J=8.4$ Hz, GlcNAc5,5′-H-1), 4.47 (d, 1H, $J=7.8$ Hz, Gal6′-H-1), 4.24 (bd, 1H, $J=1.9$ Hz Man3-H-2), 4.18 (bdd, 1H, $J=1.4$ Hz, 3.3 Hz, Man4-H-2), 4.11 (bdd, 1H, $J=1.4$ Hz, 3.5 Hz, Man4′-H-2).

*Compound **30***

^{1}H NMR δ 5.10 (s, 1H, Man4-H-1), 4.99 (d, 1H, $J=9.5$ Hz, GlcNAc1-H-1), 4.92 (s, 1H, Man4′-H-1), 4.76 (s, 1H, Man3-H-1) 4.58 (d, 2H,

GlcNAc2,5′-H-1), 4.47 (d, 1H, $J=8.0$ Hz, Gal6′-H-1), 4.24 (bd, 1H, $J=1.9$ Hz, Man3-H-2), 4.11 (bs, 1H, Man4′-H-2), 4.07 (bs, 1H, Man4-H-2).

Compound ***31***

^{1}H NMR δ 5.10 (s, 1H, Man4-H-1), 4.99 (d, 1H, $J=9.5$ Hz, GlcNAc1-H-1), 4.91 (s, 1H, Man4′-H-1), 4.76 (s, 1H, Man3-H-1), 4.55 (d, 2H, GlcNAc2,5′-H-1), 4.24 (bs, 1H, Man3-H-2), 4.10 (bs, 1H, Man4′-H-2), 4.06 (bs, 1H, $J=1.3$ Hz, Man4-H-2).

Compound ***38***

^{1}H NMR δ 5.12 (s, 1H, Man4-H-1), 4.99 (d, 1H, $J=9.4$ Hz, GlcNAc1-H-1), 4.91 (s, 1H, Man4′-H-1), 4.77 (s, 1H, Man3-H-1), 4.57 (bd, 2H, GlcNAc2,5-H-1), 4.46 (d, 1H, $J=7.5$ Hz, Gal6-H-1), 4.24 (bs, 1H, Man3-H-2), 4.19 (bs, 1H, Man4-H-2).

Compound ***45***

^{1}H NMR δ 5.11 (s, 1H, Man4-H-1), 4.99 (d, 1H, $J=9.4$ Hz, GlcNAc1-H-1), 4.91 (s, 1H, Man4′-H-1), 4.76 (s, 1H, Man3-H-1), 4.55 (d, 2H, GlcNAc2,5-H-1), 4.24 (bs, 1H, Man3-H-2), 4.18 (bs, 1H, Man4-H-2), 3.97 (dd, 1H, $J=1.8$Hz, 3.3 Hz, Man4′-H-2).

Compound ***20***

^{1}H NMR δ 5.00 (d, 1H, $J=9.9$ Hz, GlcNAc1-H-1), 4.92 (s, 1H, Man4′-H-1), 4.75 (s, 1H, Man3-H-1), 4.58 (d, 2H, $J=7.5$ Hz, GlcNAc2,5′-H-1), 4.47 (d, 1H, $J=7.8$ Hz, Gal6′-H-1), 4.10 (bd, 1H, Man4′-H-2), 4.07 (bs, 1H, Man3-H-2).

Compound ***21***

^{1}H NMR δ 4.99 (d, 1H, $J=9.7$ Hz, GlcNAc1-H-1), 4.91 (s, 1H, Man4′-H-1), 4.55 (d, 1H, $J=8.1$ Hz, GlcNAc2,5′-H-1), 4.09 (bd, 1H, Man4′-H-2), 4.07 (s, 1H, Man3-H-2).

Compound ***22***

^{1}H NMR δ 5.00 (d, 1H, $J=9.7$ Hz, GlcNAc1-H-1), 4.91 (d, 1H, $J=1.6$ Hz, Man4′-H-1), 4.76 (s, 1H, Man3-H-1), 4.58 (d, 1H, $J=7.8$ Hz, GlcNAc2-H-1), 4.07 (d, 1H, $J=2.7$ Hz, Man3-H-2), 3.97 (dd, 1H, $J=1.6$ Hz, 3.7 Hz, Man4′-H-2).

Compound ***42***

^{1}H NMR δ 5.12 (s, 1H, Man4-H-1), 4.99 (d, 1H, $J = 9.5$ Hz, GlcNAc1-H-1), 4.77 (s, 1H, Man3-H-1), 4.57 (d, 2H, $J = 7.2$ Hz, GlcNAc2-H-1), 4.46 (d, 1H, $J = 7.8$ Hz, Gal6-H-1), 4.22 (bd, 1H, $J = 2.7$ Hz, Man3-H-2), 4.19 (b, 1H, Man4-H-2).

Compound ***43***

^{1}H NMR δ 5.12 (s, 1H, Man4-H-1), 4.99 (d, 1H, $J = 9.7$ Hz, GlcNAc1-H-1), 4.76 (s, 1H, Man3-H-1), 4.55 (d, 2H, $J = 8.6$ Hz, GlcNAc2,5-H-1), 4.22 (d, 1H, $J = 2.2$ Hz, Man3-H-2), 4.18 (bdd, 1H, $J = 1.3$ Hz, 3.3 Hz, Man4-H-2).

Compound ***44***

^{1}H NMR δ 5.11 (s, 1H, Man4-H-1), 4.99 (d, 1H, $J = 9.7$ Hz, GlcNAc1-H-1), 4.77 (s, 1H, Man3-H-1), 4.57 (d, 1H, $J = 6.5$ Hz, GlcNAc-H-1), 4.22 (d, 1H, $J = 3.0$ Hz, Man3-H-2), 4.07 (bdd, 1H, $J = 2.1$ Hz, Man4-H-2).

Compound ***34***

^{1}H NMR δ 5.11 (s, 1H, Man4-H-1), 4.99 (1H, d, $J = 9.9$ Hz, GlcNAc1-H-1), 4.91 (s, 1H, Man4′-H-1), 4.76 (s, 1H, Man3-H-1), 4.55 (d, 3H, $J = 8.6$ Hz, GlcNAc2,5,5′-H-1), 4.24 (s, 1H, Man3-H-2), 4.18 (s, 1H, Man4-H-2), 4.10 (s, 1H, Man4′-H-2).

Compound ***35***

^{1}H NMR δ 5.10 (s, 1H, Man4-H-1), 4.99 (d, 1H, $J = 9.7$ Hz, GlcNAc1-H-1), 4.91 (bd, 1H, $J = 1.6$ Hz, Man4′-H-1), 4.77 (s, 1H, Man3-H-1), 4.24 (bs, 1H, Man3-H-2), 4.06 (dd, 1H, $J = 1.6$ Hz, 3.2 Hz, Man4-H-2), 3.97 (dd, 1H, $J = 1.6$ Hz, 3.5 Hz, Man4′-H-2).

General Procedure for De-N-9-fluorenylmethyl Group

To a solution of Fmoc substrate (1 μmol) in DMF (240 μl) is added morpholine (160 μl) and this mixture is stirred at room temperature. After finishing the reaction (~7–10 h), to this mixture is added ether (4.0 ml) and the precipitate is then collected. Purification of the residue by HPLC (ODS column; ϕ 5 × 150 mm; H_2O, 100 ml, and then 25% CH_3CN, 200 ml, with monitoring at 215 nm) affords Asn-oligosaccharide.

Compound ***17***

^{1}H NMR δ 5.13 (s, 1H, Man4-H-1), 5.07 (d, 1H, $J = 9.9$ Hz, GlcNAc1-H-1), 4.95 (s, 1H, Man4′-H-1), 4.78 (s, 1H, Man3-H-1), 4.62 (2H, GlcNAc2,5′-H-1), 4.56 (d, 1H, $J = 8.1$ Hz, GlcNAc5-H-1), 4.52

(d, 1H, $J = 7.8$ Hz, Gal6′-H-1), 4.25 (bs, 1H, Man3-H-2), 4.19 (bs, 1H, Man4-H-2), 4.12 (bs, 1H, Man4′-H-2), 2.68 (dd, 1H, $J = 4.6$ Hz, 12.4Hz, NeuAc7′-H-3eq), 1.72 (dd, 1H, $J = 12.1$ Hz, 12.1 Hz, NeuAc7′-H-3ax).

Compound **23**

^{1}H NMR δ 5.11 (s, 1H, Man4-H-1), 5.08 (d, 1H, J = 9.7 Hz, GlcNAc1-H-1), 4.95 (s, 1H, Man4′-H-1), 4.78 (s, 1H, Man3-H-1), 4.62 (d, 2H, GlcNAc2,5′-H-1), 4.45 (d, 1H, $J = 7.6$ Hz, Gal6′-H-1), 4.26 (bd, 1H, Man3-H-2), 4.12 (bd, 1H, Man4′-H-2), 4.08 (bdd, 1H, $J = 1.6$ Hz, 3.3 Hz, Man4-H-2), 2.68 (dd, 1H, $J = 4.1$ Hz, 12.1 Hz, NeuAc7′-H-3eq), 1.72 (dd, 1H, $J = 12.1$ Hz, 12.1 Hz, NeuAc7′-H-3ax).

Compound ***24***

^{1}H NMR δ 5.07 (d, 1H, $J = 9.4$ Hz, GlcNAc1-H-1), 4.94 (s, 1H, Man4′-H-1), 4.76 (s, 1H, Man3-H-1), 4.61 (d, 1H, GlcNAc2-H-1), 4.59 (d, 1H, GlcNAc5′-H-1), 4.44 (d, 1H, $J = 7.8$ Hz, Gal6′-H-1), 4.10 (bs, 1H, Man4′-H-2), 4.07 (1H, Man3-H-2), 2.67 (dd, 1H, $J = 4.6$ Hz, 12.2 Hz, NeuAc7′-H-3eq), 1.71 (2H, dd, $J = 12.2$ Hz, 12.2 Hz, NeuAc7′-H-3ax).

Compound ***6***

^{1}H NMR δ 5.13 (s, 1H, Man4-H-1), 5.06 (d, 1H, $J = 9.9$ Hz, GlcNAc1-H-1), 4.91 (s, 1H, Man4′-H-1), 4.77 (s, 1H, Man3-H-1), 4.61 (d, 1H, $J = 8.0$ Hz, GlcNAc2-H-1), 4.60 (d, 1H, $J = 8.0$ Hz, GlcNAc5-H-1), 4.55 (d, 1H, $J = 8.4$ Hz, GlcNAc5′-H-1), 4.44 (d, 1H, $J = 8.0$ Hz, Gal6-H-1), 4.24 (bd, 1H, Man3-H-2), 4.19 (bdd, 1H, $J = 1.3$ Hz, 3.2 Hz, Man4-H-2), 4.10 (bdd, 1H, $J = 1.4$ Hz, 3.2 Hz, Man4′-H-2), 2.66 (dd, 1H, $J = 4.6$ Hz, 12.4 Hz, NeuAc7-H-3eq), 1.71 (dd, 1H, $J = 12.4$ Hz, 12.4 Hz, NeuAc7-H-3ax).

Compound ***46***

^{1}H NMR δ 5.12 (s, 1H, Man4-H-1), 5.06 (d, 1H, $J = 9.5$ Hz, GlcNAc1-H-1), 4.91 (s, 1H, Man4′-H-1), 4.77 (s, 1H, Man3-H-1), 4.61 (d, 1H, GlcNAc2-H-1), 4.59 (d, 1H, GlcNAc5-H-1), 4.43 (d, 1H, $J = 8.0$ Hz, Gal6-H-1), 4.24 (bd, 1H, Man3-H-2), 4.18 (bdd, 1H, Man4-H-2), 2.66 (dd, 1H, $J = 4.6$ Hz, 12.6 Hz, NeuAc7-H-3eq), 1.70 (dd, 1H, $J = 12.6$ Hz, 12.6 Hz, NeuAc7-H-3ax).

Compound ***47***

^{1}H NMR δ 5.14 (s, 1H, Man4-H-1), 5.07 (d, 1H, $J = 9.4$ Hz, GlcNAc1-H-1), 4.78 (s, 1H, Man3-H-1), 4.61 (d, 1H, GlcNAc2-H-1), 4.60 (d, 1H, GlcNAc5-H-1), 4.44 (d, 1H, $J = 8.0$ Hz, Gal6-H-1), 4.23 (d, 1H, $J = 3.0$ Hz, Man3-H-2), 4.19 (bdd, 1H, $J = 1.3$ Hz, 2.9 Hz, Man4-H-2), 2.67 (dd, 1H, $J = 4.6$ Hz, 12.7 Hz, NeuAc7-H-3eq), 1.71 (dd, 1H, $J = 12.7$ Hz, 12.7 Hz, NeuAc7-H-3ax).

Compound ***29***

^{1}H NMR δ 5.11 (s, 1H, Man4-H-1), 5.06 (d, 1H, $J = 9.5$ Hz, GlcNAc1-H-1), 4.92 (s, 1H, Man4′-H-1), 4.75 (s, 1H, Man3-H-1), 4.61 (d, 1H, $J = 7.5$ Hz, GlcNAc2-H-1), 4.57, 4.55 (each d, each 1H, $J = 7.5$ Hz, GlcNAc5,5′-H-1), 4.46 (d, 1H, $J = 7.3$ Hz, Gal6′-H-1), 4.23 (bs, 1H, Man3-H-2), 4.18 (bs, 1H, Man4-H-2), 4.10 (bs, 1H, Man4′-H-2).

Compound ***7***

^{1}H NMR δ 5.11 (s, 1H, Man4-H-1), 5.06 (d, 1H, $J = 9.5$ Hz, GlcNAc1-H-1), 4.91 (s, 1H, Man4′-H-1), 4.76 (s, 1H, Man3-H-1), 4.60 (d, 1H, GlcNAc2-H-1), 4.57, 4.55 (each d, each 1H, GlcNAc5,5′-H-1), 4.46 (d, 1H, $J = 7.8$ Hz, Gal6-H-1), 4.28 (s, 1H, Man3-H-2), 4.18 (s, 1H, Man4-H-2), 4.10 (s, 1H, Man4′-H-2).

Compound ***32***

^{1}H NMR δ 5.10 (s, 1H, Man4-H-1), 5.07 (d, 1H, $J = 9.4$ Hz, GlcNAc1-H-1), 4.92 (s, 1H, Man4′-H-1), 4.76 (s, 1H, Man3-H-1), 4.61 (d, 1H, $J = 7.8$ Hz, GlcNAc2-H-1), 4.57 (d, 1H, $J = 7.8$ Hz, GlcNAc5′-H-1), 4.47 (d, 1H, $J = 7.8$ Hz, Gal6′-H-1), 4.24 (d, 1H, $J = 2.3$ Hz, Man3-H-2), 4.10 (bd, 1H, Man4′-H2), 4.06 (bd, 1H, Man4-H-2).

Compound ***33***

^{1}H NMR δ 5.09 (s, 1H, Man4-H-1), 5.06 (d, 1H, $J = 9.8$ Hz, GlcNAc1-H-1), 4.91 (s, 1H, Man4′-H-1), 4.76 (s, 1H, Man3-H-1), 4.61 (d, 1H, GlcNAc2-H-1), 4.54 (d, 1H, GlcNAc5′-H-1), 4.24 (s, 1H, Man3-H-2), 4.10 (bd, 1H, Man4′-H2), 4.06 (bs, 1H, Man4-H-2).

Compound ***39***

^{1}H NMR δ 5.11 (s, 1H, Man4-H-1), 5.07 (d, 1H, $J = 9.5$ Hz, GlcNAc1-H-1), 4.91 (s, 1H, Man4′-H-1), 4.77 (s, 1H, Man3-H-1), 4.61 (d, 1H,

GlcNAc2-H-1), 4.57 (d, 1H, GlcNAc5-H-1), 4.46 (d, 1H, Gal6-H-1), 4.24 (s, 1H, Man3-H-2), 4.18 (bs, 1H, Man4-H-2).

*Compound **51***

^{1}H NMR δ 5.11 (s, 1H, Man4-H-1), 5.06 (d, 1H, $J=9.9$ Hz, GlcNAc1-H-1), 4.91 (s, 1H, Man4′-H-1), 4.77 (s, 1H, Man3-H-1), 4.60 (d, 1H, $J=7.9$ Hz, GlcNAc2-H-1), 4.54 (d, 1H, $J=7.9$ Hz, GlcNAc5-H-1), 4.24 (s, 1H, Man3-H-2), 4.18 (dd, 1H, $J=1.6$ Hz, 1.6 Hz, Man4-H-2), 3.96 (1H, dd, $J=1.6$ Hz, 1.6 Hz, Man4′-H-2).

*Compound **25***

^{1}H NMR δ 5.07 (d, 1H, $J=9.7$ Hz, GlcNAc1-H-1), 4.92 (s, 1H, Man4′-H-1), 4.75 (s, 1H, Man3-H-1), 4.62 (d, 1H, GlcNAc2-H-1), 4.58 (d, 1H, GlcNAc5′-H-1), 4.09 (s, 1H, Man4′-H-2), 4.08 (s, 1H, Man3-H-2).

*Compound **26***

^{1}H NMR δ 5.07 (d, 1H, $J=9.5$ Hz, GlcNAc1-H-1), 4.91 (s, 1H, Man4′-H-1), 4.76 (s, 1H, Man3-H-1), 4.62 (d, 1H, GlcNAc2-H-1), 4.55 (d, 1H, GlcNAc5′-H-1), 4.10 (bs, 1H, Man4′-H2), 4.07 (s, 1H, Man3-H-2).

*Compound **27***

^{1}H NMR δ 5.07 (d, 1H, $J=9.5$ Hz, GlcNAc1-H-1), 4.91 (s, 1H, Man4′-H-1), 4.76 (s, 1H, Man3-H-1), 4.62 (d, 1H, $J=7.8$ Hz, GlcNAc2-H-1), 4.08 (d, 1H, $J=2.9$ Hz, Man3-H-2), 3.97(bs, 1H, Man4′-H-2).

*Compound **48***

^{1}H NMR δ 5.11 (s, 1H, Man4-H-1), 5.06 (d, 1H, J = 9.8 Hz, GlcNAc1-H-1), 4.77 (s, 1H, Man3-H-1), 4.61 (d, 1H, GlcNAc2-H-1), 4.57 (d, 1H, GlcNAc5-H-1), 4.46 (d, 1H, $J=7.8$ Hz, Gal6-H-1), 4.22 (bs, 1H, Man3-H-2), 4.18 (bs, 1H, Man4-H-2).

*Compound **49***

^{1}H NMR δ 5.12 (s, 1H, Man4-H-1), 5.07 (d, 1H, J = 9.7 Hz, GlcNAc1-H-1), 4.77 (s, 1H, Man3-H-1), 4.61 (d, 1H, GlcNAc2-H-1), 4.54 (d, 1H, GlcNAc5-H-1), 4.22 (d, 1H, $J=2.5$ Hz, Man3-H-2), 4.18 (dd, 1H, $J=1.4$ Hz, 3.0 Hz, Man4-H-2).

Compound ***50***

^{1}H NMR δ 5.10 (s, 1H, Man4-H-1), 5.06 (d, 1H, J = 9.5 Hz, GlcNAc1-H-1), 4.77 (s, 1H, Man3-H-1), 4.61 (d, 1H, J = 7.3 Hz, GlcNAc2-H-1), 4.22 (d, 1H, J = 2.4 Hz, Man3-H-2), 4.07 (dd, 1H, J = 1.6 Hz, 3.0 Hz, Man4-H-2).

Compound ***36***

^{1}H NMR δ 5.11 (s, 1H, Man4-H-1), 5.07 (d, 1H, J = 10.0 Hz, GlcNAc1-H-1), 4.91 (s, 1H, Man4′-H-1), 4.77 (s, 1H, Man3-H-1), 4.61 (d, 1H, J = 6.8 Hz, GlcNAc2-H-1), 4.55 (d, 2H, GlcNAc5,5′-H-1), 4.24 (bs, 1H, Man3-H-2), 4.18 (bs, 1H, Man4-H-2), 4.10 (bs, 1H, Man4′-H-2).

Compound ***37***

^{1}H NMR δ 5.10 (s, 1H, Man4-H-1), 5.07 (d, 1H, J = 9.7 Hz, GlcNAc1-H-1), 4.91 (s, 1H, Man4′-H-1), 4.78 (s, 1H, Man3-H-1), 4.61 (d, 1H, J = 8.0 Hz, GlcNAc2-H-1), 4.25 (bs, 1H, Man3-H-2), 4.06 (bs, 1H, Man4-H-2), 3.97 (bs, 1H, Man4′-H-2).

[6] Enzymatic Synthesis of Neoglycoconjugates by Transglycosylation with Endo-β-N-acetylglucosaminidase A

By KAORU TAKEGAWA and JIAN-QIANG FAN

Introduction

The oligosaccharide moieties of glycoproteins have been shown to play important roles in biological processes such as cellular recognition, lectin binding, viral infection, and substrate–receptor recognition.[1] In general, naturally occurring glycoproteins have a high degree of heterogeneity in their oligosaccharide moiety, and it is difficult to clarify the biological roles of individual oligosaccharides. Therefore, to effectively elucidate the significance and function of each oligosaccharide, efficient methods for the synthesis and construction of a neoglycoprotein with a single type of carbohydrate moiety are required. Several strategies for the attachment of naturally occurring N-linked oligosaccharides to proteins have been described. These include the chemical addition of N-linked oligosaccharides to

[1] A. Varki, *Glycobiology* **3,** 97 (1993).

0076-6879/03 $35.00

proteins by reductive amination,[2] solid-phase synthesis using glycoamino acids as building blocks,[3] and enzymatic synthesis using transglutaminase.[4] These methods are useful for attaching N-linked oligosaccharides to proteins; however, no effective method is currently available for the conversion of heterologous N-linked oligosaccharides to homogeneous N-linked oligosaccharides in glycoproteins.

Endo-β-*N*-acetylglucosaminidase (EC 3.2.1.96) hydrolyzes the glycosidic bond in the *N,N′*-diacetylchitobiose moiety of N-linked oligosaccharides of various glycoproteins. The enzyme is one of the most well-studied endoglycosidases used for the structural analysis of glycoproteins and has been used for isolation of both N-linked oligosaccharides and partially deglycosylated proteins without damaging the glycoproteins.[5,6] *Arthrobacter protophormiae*, a gram-positive bacterium, produces endo-β-*N*-acetylglucosaminidase (Endo-A) when the cells are grown in medium containing ovalbumin, and the purification and properties of the enzyme have been reported.[7] During studies on enzymatic kinetics of the enzyme, we found that Endo-A is capable of transferring N-linked oligosaccharides to a suitable acceptor by transglycosylation activity of the enzyme.[8] We have reported that Endo-A can transfer high mannose-type oligosaccharides to several acceptor substrates: monosaccharides, disaccharides, glycosides, and glycoproteins. Thus, the transglycosylation activity of Endo-A is useful for the remodeling of glycoproteins or the synthesis of neoglycoconjugates.

This article describes the enzymatic synthesis of neoglycoconjugates and neoglycoproteins using the transglycosylation catalyzed by Endo-A.

Synthesis of Neoglycoconjugates: Assay Methods

Principle

Scheme 1 illustrates the hydrolytic and transglycosylation acitivities of Endo-A. After forming the enzyme–substrate complex, the enzyme will transfer the donor oligosaccharide to either water (leading to hydrolysis)

[2] Y. C. Lee and R. T. Lee, *in* "Glycoconjugates" (H. J. Allen and E. C. Kisalius, eds.), pp. 121–165. Marcel Dekker, New York, 1992.

[3] M. Mendal, *Glycoconj. J.* **11,** 59 (1994).

[4] S. C. B. Yan, *Methods Enzymol.* **138,** 413 (1987).

[5] A. Kobata, *Anal. Biochem.* **100,** 1 (1979).

[6] K. Yamamoto, *J. Biochem. (Tokyo)* **116,** 229 (1994).

[7] K. Takegawa, M. Nakoshi, S. Iwahara, K. Yamamoto, and T. Tochikura, *Appl. Environ. Microbiol.* **55,** 3107 (1989).

[8] K. Takegawa, S. Yamaguchi, A. Kondo, H. Iwamoto, M. Nakoshi, I. Kato, and S. Iwahara, *Biochem. Int.* **24,** 849 (1991).

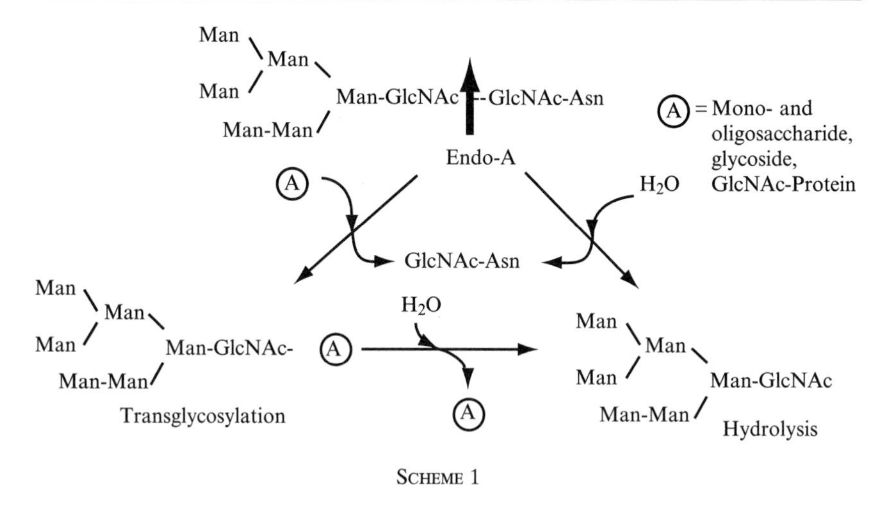

SCHEME 1

or an acceptor (leading to transglycosylation). Because the transglycosylation product can be the substrate of the same enzyme, the process repeats until all substrates including the transglycosylation products are transferred to water (complete hydrolysis).

Materials

$Man_6GlcNAc_2Asn$ (Man, mannose; GlcNAc, *N*-acetylglucosamine) is prepared from ovalbumin glycopeptides by the method of Huang *et al.*, using Dowex 50W-X2 column chromatography.[9] Recombinant Endo-A is purified from transformed *Escherichia coli* (XL1-Blue) cells carrying Endo-A plasmid as described.[10] Dansylated derivative of $Man_6GlcNAc_2Asn$ is prepared by the method of Gray,[11] followed by paper chromatography to remove dansylsulfonic acid. Endo-A activity is assayed with $Man_6GlcNAc_2Asn$-dansyl as the substrate, and 1 unit is defined as the amount yielding 1 μmol of GlcNAc-Asn-dansyl/min at 37° as described.[7] The other reagents used are commercially available products.

Procedure

The reaction mixture for transglycosylation contains $Man_6GlcNAc_2Asn$ (1.3 mg), 40 μl of 1 *M* ammonium acetate buffer (pH 6.0), 340 μl of 0.1–0.5 *M* acceptor substrate, and 10 mU of Endo-A solution having a total

[9] C. C. Huang, H. E. Mayer, Jr., and R. Montgomery, *Carbohydr. Res.* **13,** 127 (1970).

[10] K. Takegawa, K. Yamabe, K. Fujita, M. Tabuchi, M. Mita, H. Izu, A. Watanabe, Y. Asada, M. Sano, A. Kondo, I. Kato, and S. Iwahara, *Arch. Biochem. Biophys.* **388,** 22 (1997).

[11] W. R. Gray, *Methods Enzymol.* **11,** 139 (1967).

volume of 680 μl. After incubation at 37° for 10 min, the enzyme reaction is terminated by boiling for 3 min in a water bath, and the mixture is evaporated with a SpeedVac (Savant, Holbrook, NY), using a vacuum pump. An aliquot of the reconstituted sample is analyzed by high-performance anion-exchange chromatography (HPAEC)[12] or thin-layer chromatography (TLC).[10]

Analytical Methods

An HPAEC equipped with a pulsed amperometric detector (PAD-II) and a Dionex (Sunnyvale, CA) CarboPac PA-1 column is used for analysis of the Endo-A reaction as described.[13] The separation of the transglycosylation products is accomplished by elution with 100 m*M* sodium hydroxide with a linear gradient of sodium acetate (0–80 m*M*) developed in 24 min at a rate of 1.0 ml/min. The quantification of transglycosylation product is by subtraction of the hydrolysis product and remaining donor substrate $Man_6GlcNAc_2Asn$ from the initial donor substrates. Quantification is based on the assumption that the PAD response is approximately equal on a molar basis. TLC is performed on silica gel 60 plates (Merck, Darmstadt, Germany). For the separation of the oligosaccharides, the following solvent system is used: *n*-propanol–acetic acid–water, 3:3:1.2 (v/v/v). The orcinol–H_2SO_4 reagent is used for the detection of the oligosaccharides.[14]

Oligosaccharide Transfer

The reaction time course of Endo-A in aqueous solution has been examined (Fig. 1A).[8,13] The transglycosylation product is formed predominantly in the initial stage of the reaction. As the reaction proceeds, the product is rapidly hydrolyzed by the same enzyme; the reaction conditions should be carefully attained to obtain optimal transglycosylation activity.

We found that the hydrolytic activity of Endo-A can be suppressed and the transglycosylation activity can be enhanced (Table I) in reaction media containing organic solvents such as acetone, dimethyl sulfoxide (DMSO), and *N,N'*-dimethylformamide.[13] Endo-A exerts exclusive transglycosylation activity in a reaction mixture containing acetone, and the hydrolysis product is undetectable under these conditions (Fig. 1B). The effective use of acetone in the suppression of hydrolysis is not common, and the

[12] J.-Q. Fan, L. H. Huynh, B. B. Reinhold, V. N. Reinhold, K. Takegawa, S. Iwahara, A. Kondo, I. Kato, and Y. C. Lee, *Glycoconj. J.* **13,** 643 (1996).
[13] J.-Q. Fan, K. Takegawa, S. Iwahara, A. Kondo, I. Kato, C. Abeygunawardana, and Y. C. Lee, *J. Biol. Chem.* **270,** 17723 (1995).
[14] E. W. Holmes and J. S. O'Brien, *Anal. Biochem.* **93,** 167 (1979).

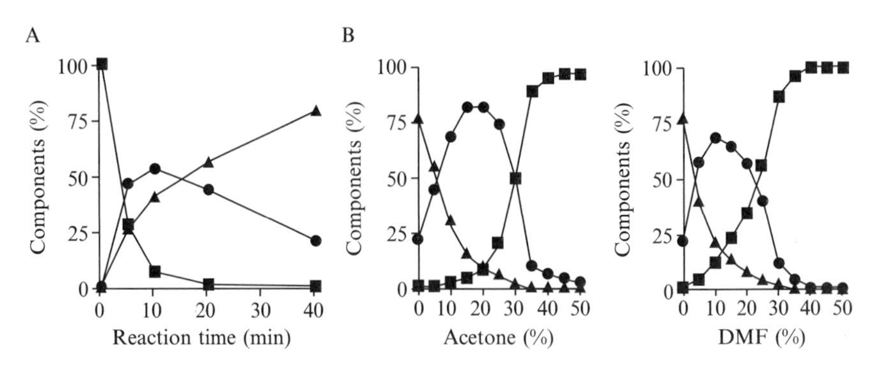

FIG. 1. (A) Time course of Endo-A action in aqueous solution. (B) Effect of organic solvents on the transglycosylation activity of Endo-A. The enzymatic reaction was carried out with 15 nmol of $Man_6GlcNAc_2Asn$, 10 μmol of GlcNAc, and 2 mU of Endo-A in 20 μl of 25 mM ammonium acetate buffer (pH 6.0) at 37° containing different amounts of acetone (*left*) and *N,N'*-dimethylformamide (DMF) (*right*). The hydrolysis and transglycosylation products were quantified with an HPAEC-PAD system. (●) Transglycosylation product; (▲) hydrolysis product; (■) remaining substrate.

mechanism is not well known. However, these hydrophilic organic solvents may decrease the water content (hydrolytic acceptor) in the enzyme reaction mixture. A reaction mixture containing 30% acetone (v/v) or 20% dimethyl sulfoxide (v/v), which achieves the maximum transglycosylation yield, is recommended (Fig. 1B).[13]

The acceptor specificity of Endo-A transglycosylation toward various monosaccharides and disaccharides has been examined.[15,16] The enzyme requires the 3- and 4-OHs of a sugar residue to be equatorial as an acceptor, in a normal conformation, but 1-, 2-, and 6-OHs are relatively unimportant. For example, Endo-A can transfer oligosaccharides to GlcNAc, Glc, Man, and 2-deoxy-Glc, all possessing equatorial 3-OH and 4-OH. In contrast, galactose and 3-deoxy-Glc are not acceptors for the transglycosylation activity. The enzyme also transfers N-linked oligosaccharides more efficiently to β-linked disaccharides (cellobiose, gentiobiose, sophorose, and labinaribiose) than to α-linked disaccharides (isomaltose, maltose, nigerose, and kojibiose) as acceptor substrates. These results indicate that Endo-A recognizes not only nonreducing monosaccharides but also the linkage of disaccharides, and β-linked glycosides are suitable for the synthesis of transglycosylation products.

[15] K. Takegawa, S. Yamaguchi, A. Kondo, I. Kato, and S. Iwahara, *Biochem. Int.* **25,** 829 (1991).

[16] K. Fujita, T. Miyamura, M. Sano, I. Kato, and K. Takegawa, *J. Biosci. Bioeng.* **93,** 614 (2002).

TABLE I
TRANSGLYCOSYLATION ACTIVITY OF Endo-A IN ORGANIC SOLVENTS[a]

Solvent	Remaining substrate	Yield (%)	
		Hydrolysis	Transglycosylation
None (H_2O)	2.0	66.6	32.1
Acetone	8.1	0.8	89.4
Acetonitrile	99.8	0.0	1.4
Dioxane	85.6	0.4	12.6
Dimethylformamide	71.7	0.0	27.5
Dimethyl sulfoxide	24.8	1.5	74.6
Ethanol	47.1	0.9	55.8
Methanol	2.5	5.9	74.9
Tetrahydrofuran	98.6	0.0	0.0

[a]The enzymatic reaction was carried out with 3 nmol of $Man_9GlcNAc_2Asn$, 10 mmol of GlcNAc, and 2 mU of Endo-A in 25 m*M* ammonium acetate buffer (pH 6.0) containing 30% (v/v) organic solvent. After incubation at 37° for 15 min, the hydrolysis and transglycosylation products were quantified with an HPAEC-PAD system.

Synthesis of Neoglycoproteins

A method for the conversion of heterogeneous to homogeneous N-linked sugar chains in glycoproteins, utilizing the transglycosylation activity of Endo-A, is illustrated in Scheme 2.

Enzymatic Synthesis of Neoribonuclease B

(See Ref. 18.) Bovine pancreatic ribonuclease B (RNase B) is purchased (type III-B; Sigma, St. Louis, MO), and further purified by concanavalin A (Con A)–agarose column chromatography. The RNase B is partially deglycosylated by endo-β-*N*-acetylglucosaminidase to prepare GlcNAc-RNase as an acceptor of the transglycosylation reaction. The reaction mixture and conditions for transglycosylation are as follows. One milligram of $Man_6GlcNAc_2$-Asn is incubated with 10 mU of Endo-A in 50 m*M* ammonium acetate buffer (pH 6.0) in the presence of GlcNAc-RNase (3.0 mg), having a total volume of 50 μl. After incubation of the mixture for 10 min at 37°, the reaction is terminated by boiling for 3 min, and the mixture is then placed on a Con A–agarose column at 4°. The unreacted GlcNAc-RNase is passed through the Con A column, and the newly synthesized glycosylated RNase is recovered from the column by eluting with methyl-α-mannoside. The molecular weight of the bound RNase is

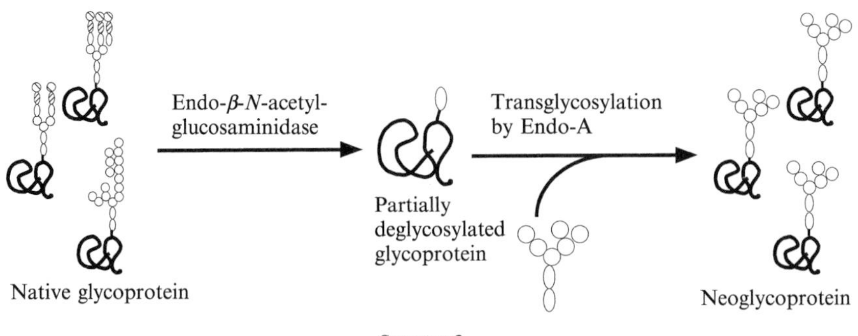

SCHEME 2

increased, and it migrates to the same position as native RNase B by sodium dodecyl sulfate–polyacrylamide gel electrophoresis (SDS–PAGE) analysis (Fig. 2), indicating that the $Man_6GlcNAc$ has been transferred to the GlcNAc-RNase. RNase B contains one N-linked sugar chain in a heterogeneous form.[19] Our result indicates that RNase B with a homogeneous N-linked sugar chain (i.e., $Man_6GlcNAc_2$) can be obtained by the transglycosylation of Endo-A.

For the synthesis of biologically active neoglycoproteins, we attempted to express Endo-A as a fusion protein linked to glutathione *S*-transferase (GST).[21] The fusion protein is purified to homogeneity with a glutathione–Sepharose 4B column and shows transglycosylation activity without removing the GST moiety. The GST–Endo-A immobilized on glutathione–Sepharose retains its transglycosylation activity. We have examined the transglycosylation reaction using the immobilized Endo-A. After examination of the transglycosylation reaction, the reaction is terminated by centrifugation in order to remove the immobilized GST–Endo-A. We have reported that the immobilized enzyme could transfer $Man_6GlcNAc_2$ *en bloc* to partially deglycosylated RNase B without damaging its enzyme activity. The immobilized GST–Endo-A should be useful for synthesizing biologically active and glycosylation-reformed enzymes that contain homogeneous N-linked sugar chains.

Applications of Neoglycoconjugates

Novel oligosaccharides, $Man_6GlcNAcMan$ and $Man_6GlcNAcGlc_2$, have been synthesized by the transglycosylation of Endo-A.[15] Using

[19] C.-J. Liang, K. Yamashita, and A. Kobata, *J. Biochem. (Tokyo)* **88,** 51 (1980).

[21] K. Fujita, N. Tanaka, S. Sano, I. Kato, Y. Asada, and K. Takegawa, *Biochem. Biophys. Res. Commun.* **267,** 134 (2000).

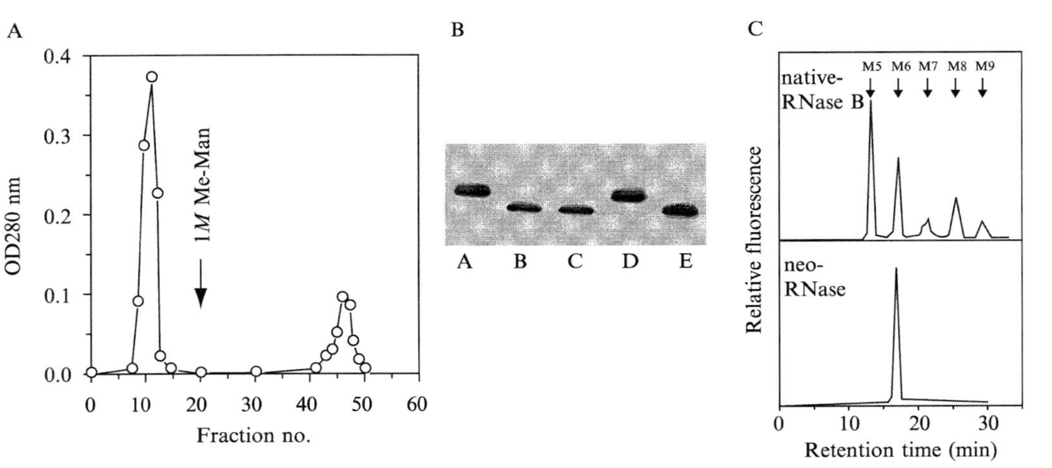

FIG. 2. (A) Con A-Sepharose column chromatography and (B) SDS–PAGE of the reaction products produced by Endo-A. $Man_6GlcNAc$ was incubated with Endo-A in the presence of GlcNAc-RNase. The reaction was eluted through a Con A–Sepharose column, and bound RNase was then eluted with 1 *M* methyl-α-D-mannoside. Each protein peak was subjected to SDS–PAGE. Lane A, Native RNase B; lane B, GlcNAc-RNase by endo-β-*N*-acetylglucosaminidase digestion; lane C, unbound RNase; lane D, bound RNase; lane E, bound RNase digested by endo-β-*N*-acetylglucosaminidase. (C) Comparison of the N-linked oligosaccharides in native and neo-RNases. The N-linked sugar chains were released from native and neo-RNases by hydrazinolysis, and derivatized with 2-aminopyridine.[20] A portion of each pyridylamino-oligosaccharide was analyzed by HPLC with a TaKaRa PALPAK N column. The arrows indicating the elution positions of standard PA oligosaccharides M5–M9 correspond to $Man_{5-9}GlcNAc_2$-PA, respectively.

[20] S. Hase, *Methods Mol. Biol.* **14,** 69 (1993).

these oligosaccharides, the susceptibility of these oligosaccharides to endo-β-*N*-acetylglucosaminidases was examined. Endo-β-*N*-acetylglucosaminidases degraded these oligosaccharides, $Man_6GlcNAcMan$ and $Man_6GlcNAcGlc_2$, to Man and Glc_2. These results show that endo-β-*N*-acetylglucosaminidases degrade not only the GlcNAc-GlcNAc linkage but also GlcNAc-Man and GlcNAc-Glc_2 linkages, and that the enzyme recognizes only one GlcNAc residue of the diacetylchitobiose moiety in the N-linked oligosaccharides.

The transglycosylation activity of Endo-A was used for the enzymatic synthesis of the novel oligosaccharide, Man_6GlcNAc-*p*-nitrophenyl-α-D-glucose (Man_6GlcNAc-Glc-pNP).[22] Various endo-β-*N*-acetylglucosaminidases hydrolyzed this oligosaccharide, and released Man_6GlcNAc and Glc-pNP. We have established a new procedure for the colorimetric detection of endo-β-*N*-acetylglucosaminidase activity using Man_6GlcNAc-pNP, which is as simple as that for other exoglycosidase assays with pNP-glycosides as substrates.[22] Endo-A was also used for the synthesis of Man_6GlcNAc-5-bromo-4-chloro-3-indolyl-β-glucoside (Man_6GlcNAc-Glc-β-X).[23] The *E. coli* strains coexpressing Endo-A and β-glucosidase formed blue colonies in the presence of Man_6GlcNAc-Glc-β-X. This compound can be useful for detecting endo-β-*N*-acetylglucosaminidase-producing bacterial colonies on a culture plate.

An unnatural product, Man_9GlcNAcGlc-pentapeptide, was synthesized by transglycosylation of Endo-A using a synthetic pentapeptide containing a Glc-Asn linkage,[24] and was tested as an inhibitor for glycoamidases. The product was not hydrolyzed by glycoamidases from plant and bacterial sources, but it inhibited both enzymes in the micromolar range.[24] An unusual neoglycopeptide containing glutamine-linked oligosaccharide was also synthesized.[25]

Formation of glycopolymers is a convenient way to provide a glycoside clustering effect.[26] A novel oligosaccharide, $Man_9GlcNAc\text{-}O\text{-}(CH_2)_3\text{-}NHCOCH{=}CH_2$, was synthesized by the Endo-A tranglycosylation. The neoglycopolymer having this oligosaccharide as side chains was shown to be a more potent inhibitor than soybean agglutinin, which contains a

[22] K. Takegawa, K. Fujita, J.-Q. Fan, M. Tabuchi, N. Tanaka, A. Kondo, H. Iwamoto, I. Kato, Y. C. Lee, and S. Iwahara, *Anal. Biochem.* **257,** 218 (1998).

[23] K. Fujita, Y. Asada, K. Yamamoto, and K. Takegawa, *J. Biosci. Bioeng.* **90,** 462 (2000).

[24] I. L. Deras, K. Takegawa, A. Kondo, I. Kato, and Y. C. Lee, *Bioorg. Med. Chem. Lett.* **8,** 1763 (1998).

[25] K. Haneda, T. Inazu, M. Mizuno, R. Iguchi, H. Tanabe, K. Fujimori, K. Yamamoto, H. Kumagai, K. Tsumori, and E. Munekata, *Biochem. Biophys. Acta* **1526,** 242 (2001).

[26] Y. C. Lee, *Biochem. Soc. Trans.* **21,** 460 (1993).

single $Man_9GlcNAc_2$ chain, for rat mannose binding protein–carbohydrate recognition domains.[17]

Comments

Transglycosylation of Endo-A is an attractive method for remodeling glycoproteins or synthesizing neoglycoconjugates, because N-linked sugar chains can be transferred to a glycoside acceptor *en bloc*. Moreover, Endo-A has broad ranges of pH and organic solvent stabilities, and does not require nucleotide sugars for the reactions. Because of the substrate specificity of the enzyme, however, the conversion of the N-linked oligosaccharides by Endo-A is limited to high mannose-type oligosaccharides. Yamamoto and co-workers reported that endo-β-*N*-acetylglucosaminidase from *Mucor hiemalis* (Endo-M) could transfer sialo and asialo complex-type oligosaccharides to various glycopeptides such as peptide T, substance P, and eel calcitonin derivatives.[27–30] Therefore, a combination of Endo-A and Endo-M will enable us to design more complex neoglycoproteins with different N-linked oligosaccharide moieties.

Our data showed that the GlcNAc-peptides derived from ovalbumin (10 amino acids) and eel calcitonin (32 amino acids) are better acceptors than bovine RNase B (GlcNAc-protein) for the transglycosylation reaction of Endo-A.[18,31] For example, Endo-A can readily deglycosylate RNase B but shows less oligosaccharide-transferring activity toward the native form of partially deglycosylated RNase B.[18,32] This may be considered a result of the accessibility of glycosylation sites to Endo-A, based on the stereo structure of the protein.

We have cloned and sequenced Endo-A, and Endo-A has no sequence similarity with Endo-H from *Streptomyces plicatus*, which does not have

[17] J.-Q. Fan, M. S. Quesenberry, K. Takegawa, S. Iwahara, A. Kondo, I. Kato, and Y. C. Lee, *J. Biol. Chem.* **270,** 17730 (1995).

[18] K. Takegawa, M. Tabuchi, S. Yamaguchi, A. Kondo, I. Kato, and S. Iwahara, *J. Biol. Chem.* **270,** 3094 (1995).

[27] K. Yamamoto, S. Kadowaki, J. Watanabe, and H. Kumagai, *Biochem. Biophys. Res. Commun.* **203,** 244 (1994).

[28] K. Yamamoto, K. Fujimori, K. Haneda, M. Mizuno, T. Inazu, and H. Kumagai, *Carbohydr. Res.* **305,** 415 (1997).

[29] K. Haneda, T. Inazu, M. Mizuno, R. Iguchi, H. Tanabe, K. Fujimori, K. Yamamoto, H. Kumagai, K. Tsumori, and E. Munekata, *Biochim. Biophys. Acta* **1526,** 242 (2001).

[30] K. Haneda, T. Inazu, M. Mizuno, R. Iguchi, K. Yamamoto, H. Kumagai, S. Aimoto, H. Suzuki, and T. Noda, *Bioorg. Med. Chem. Lett.* **8,** 1303 (1998).

[31] K. Yamamoto, K. Haneda, R. Iguchi, T. Inazu, M. Mizuno, K. Takegawa, A. Kondo, and I. Kato, *J. Biosci. Bioeng.* **87,** 175 (1999).

[32] K. Fujita and K. Takegawa, *Biochem. Biophys. Res. Commun.* **282,** 678 (2001).

transglycosylation activity.[10] Interestingly, Endo-A homologs can also be found in multiple eukaryotic organisms, such as worm (*Caenorhabditis elegans*), plant (*Arabidopsis thaliana*), fly (*Drosophila melanogaster*), and human (*Homo sapiens*).[33] Thus, Endo-A seems to belong to a family that has a well-conserved structure from prokaryotes to eukaryotic multicellular organisms. The transglycosylation activity of endo-β-*N*-acetylglucosaminidases might positively participate in various cellular functions in eukaryotic cells.

Acknowledgments

We thank Drs. Yuan C. Lee (Johns Hopkins University), Shojiro Iwahara (Kagawa University), Kenji Yamamoto (Kyoto University), Kiyotaka Fujita (Kyoto University), Kondo Akihiro (Takara Shuzo), Mitsuaki Tabuchi (Yamaguchi University), Katsuji Haneda (Noguchi Institute), and Naotaka Tanaka (Kagawa University), as well as all of our collaborators.

[33] K. Fujita and K. Takegawa, *Biochem. Biophys. Res. Commun.* **283,** 680 (2001).

[7] Chemoenzymatic Synthesis of Neoglycopeptides Using Endo β-*N*-acetylglucosaminidase from *Mucor hiemalis*

By Katsuji Haneda, Toshiyuki Inazu, Mamoru Mizuno, and Kenji Yamamoto

Introduction

Endo-β-*N*-acetylglucosaminidase (EC 3.2.1.96) is an enzyme that hydrolytically cleaves the *N,N'*-diacetylchitobiose moiety of the asparagine (N)-linked oligosaccharide of various glycoproteins and releases intact oligosaccharides. This enzyme is a unique endoglycosidase that leaves one *N*-acetyl-D-glucosamine (GlcNAc) residue on the protein moiety.

Several microbial endo-β-*N*-acetylglucosaminidases show the transglycosylation activity. Trimble *et al.* reported the transglycosylation reaction by an endo-β-*N*-acetylglucosaminidase of *Flavobacterium meningosepticum* (Endo-F).[1] Endo-β-*N*-acetylglucosaminidases of *Arthrobacter protophormiae* (Endo-A) and of *Mucor hiemalis* (Endo-M) also show the

[1] R. B. Trimble, P. H. Atkinson, A. L. Tarentino, T. H. Plummer, Jr., F. Maley, and K. B. Tomer, *J. Biol. Chem.* **261,** 12000 (1986).

0076-6879/03 $35.00

X-GlcNAc-GlcNAc-Asn-peptide + GlcNAc-Asn-R (Acceptor) ⟶ X-GlcNAc-GlcNAc-Asn-R + GlcNAc-Asn-peptide

(X: Oligosaccharide; R: Peptide or peptide derivative)

SCHEME 1

transglycosylation activity.[2,3] Endo-A acts on a high mannose-type oligosaccharide, whereas Endo-M acts on complex-type, high mannose-type, and hybrid-type oligosaccharides[4] and can transfer the oligosaccharides from glycopeptides to suitable acceptors with a GlcNAc residue during hydrolysis of the glycopeptides (Scheme 1). The transglycosylation reaction by endo-β-*N*-acetylglucosaminidase may be a useful tool for neoglycoconjugate synthesis.

We have developed a chemoenzymatic synthetic method for a glycopeptide combined with the chemical synthesis of a peptide containing GlcNAc as a glycosylation tag and the transglycosylation catalyzed by Endo-M.[5] Scheme 2 shows the principal procedure for this method including the following four steps.

1. The first step is the new synthetic method for N-protected Asn(GlcNAc)-OH with the reaction of the corresponding glycosylazide and N-protected L-aspartic acid in the presence of a tertiary phosphine.[6–8]

2. The second step is the chemical synthesis of a peptide containing GlcNAc (GlcNAc-peptide). The GlcNAc-peptide is synthesized using N-protected Asn(GlcNAc)-OH by the usual coupling method without protection of sugar hydroxyl functions. To prevent O-acylation of the sugar moiety, the dimethylphosphinothioic mixed anhydride (Mpt-MA) method, which is superior to other methods, is used. The Mpt-MAs of amino acid derivatives have high reactivity and chemoselectivity with the amino

[2] K. Takegawa, S. Yamaguchi, A. Kondo, H. Iwamoto, M. Nakoshi, I. Kato, and S. Iwahara, *Biochem. Int.* **24,** 849 (1991).

[3] K. Yamamoto, S. Kadowaki, J. Watanabe, and K. Kumagai, *Biochem. Biophys. Res. Commun.* **203,** 244 (1994).

[4] K. Yamamoto, S. Kadowaki, M. Fujisaki, H. Kumagai, and T. Tochikura, *Biosci. Biotech. Biochem.* **58,** 72 (1994).

[5] K. Haneda, T. Inazu, K. Yamamoto, H. Kumagai, Y. Nakahara, and A. Kobata, *Carbohydr. Res.* **292,** 61 (1996).

[6] T. Inazu and K. Kobayashi, *Synlett* 869 (1993).

[7] M. Mizuno, I. Muramoto, K. Kobayashi, H. Yaginuma, and T. Inazu, *Synthesis* **1999,** 162 (1999).

[8] M. Mizuno, K. Haneda, R. Iguchi, I. Muramoto, T. Kawakami, S. Aimoto, K. Yamamoto, and T. Inazu, *J. Am. Chem. Soc.* **121,** 284 (1999).

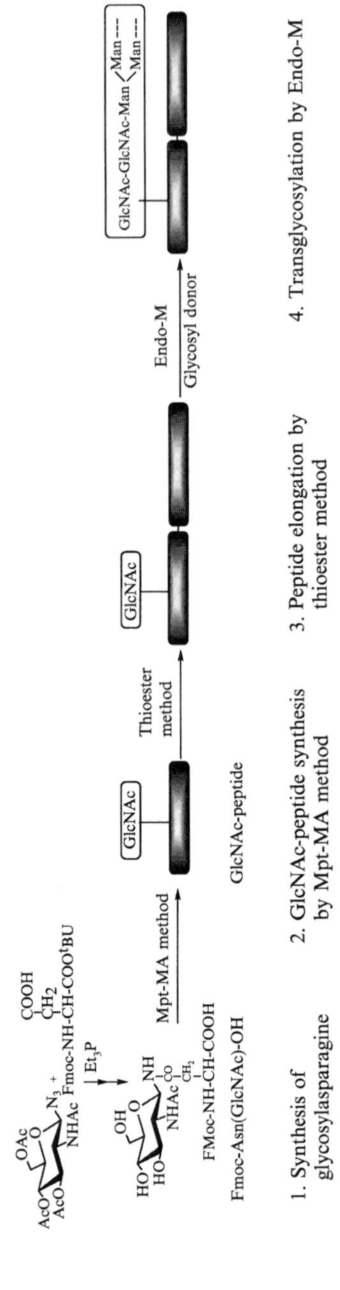

SCHEME 2. Chemoenzymatic synthesis of complex glycopeptide by the transglycosylation reaction catalyzed by Endo-M.

function to hydroxyl functions.[9,10] Because of the low coupling efficiency of H-Asn(GlcNAc)-peptide–resin,[9] it is necessary to monitor the coupling reaction by a method such as the Kaiser test.[11]

3. The third step is peptide elongation by the thioester segment condensation method.[12,13] This method requires only protection of the side chains of Lys and Cys.

4. The final step is an enzymatic transglycosylation reaction catalyzed by Endo-M. The reaction mixture contains a natural glycosyl donor, GlcNAc-peptide as a glycoside acceptor, and Endo-M. The oligosaccharide block cleaved from the glycosyl donor is transferred to the GlcNAc moiety of GlcNAc-peptide. As a result, a complex glycopeptide having a natural N-linked oligosaccharide attached to the targeted site of the peptide is prepared.

For the transglycosylation reaction, both native Endo-M of *Mucor hiemalis*[14] and a recombinant Endo-M[15] are available. The recombinant Endo-M is more suitable, because this preparation is free of protease activity. Endo-M transfers a complex-type oligosaccharide most efficiently,[5] although it is the poorest substrate for hydrolysis.[4] As a glycosyl donor for a disialo biantennary complex-type oligosaccharide, a sialylglycohexapeptide (SGP), H-Lys-Val-Ala-Asn[(NeuAc-Gal-GlcNAc-Man)$_2$-Man-GlcNAc$_2$]-Lys-Thr-OH, prepared from hen's egg yolk,[16] is available. Use of excess glycosyl donor promotes the transglycosylation reaction dose dependently. The reaction must be terminated at the maximum of the transglycosylation yield, because the transglycosylation product is gradually hydrolyzed to GlcNAc-peptide during the longer incubation.

As a practical example, the synthesis of an eel calcitonin (eCT) derivative having an N-linked oligosaccharide[8] is shown (Scheme 3).

Endo-M strictly recognizes the equatorial 4-OH of the D-glycopyranoside moiety of the acceptor substrate (GlcNAc, D-glucose, D-mannose, etc.) and transglycosylates an oligosaccharide block onto it. On the other hand, the recognition specificity of Endo-M for the amino acid

[9] T. Inazu, M. Mizuno, Y. Kohda, K. Kobayashi, and H. Yaginuma, "Peptide Chem. 1995", p. 61. Protein Res. Found., Osaka, Japan, 1996.

[10] T. Inazu, H. Hosokawa, and M. Amemiya, *Bull. Chem., Soc. Jpn.* **61,** 4467 (1988).

[11] E. Kaiser, R. L. Colescott, C. D. Bossinger, and P. I. Cook, *Anal. Biochem.* **34,** 595 (1970).

[12] H. Hojo and S. Aimoto, *Bull. Chem. Soc. Jpn.* **64,** 111 (1991); **65,** 3055 (1992).

[13] T. Kawakami, S. Kogure, and S. Aimoto, *Bull. Chem. Soc. Jpn.* **69,** 3331 (1996).

[14] S. Kadowaki, K. Yamamoto, M. Fujisaki, and T. Tochikura, *J. Biochem.* **110,** 17 (1991).

[15] K. Kobayashi, M. Takeuchi, A. Iwamatsu, K. Yamamoto, H. Kumagai, and S. Yoshida, *JP*11332568 (1999).

[16] A. Seko, M. Koketsu, M. Nishizono, Y. Enoki, H. R. Ibrahim, L. R. Juneja, M. Kim, and T. Yamamoto, *Biochim. Biophys. Acta* **1335,** 23 (1997).

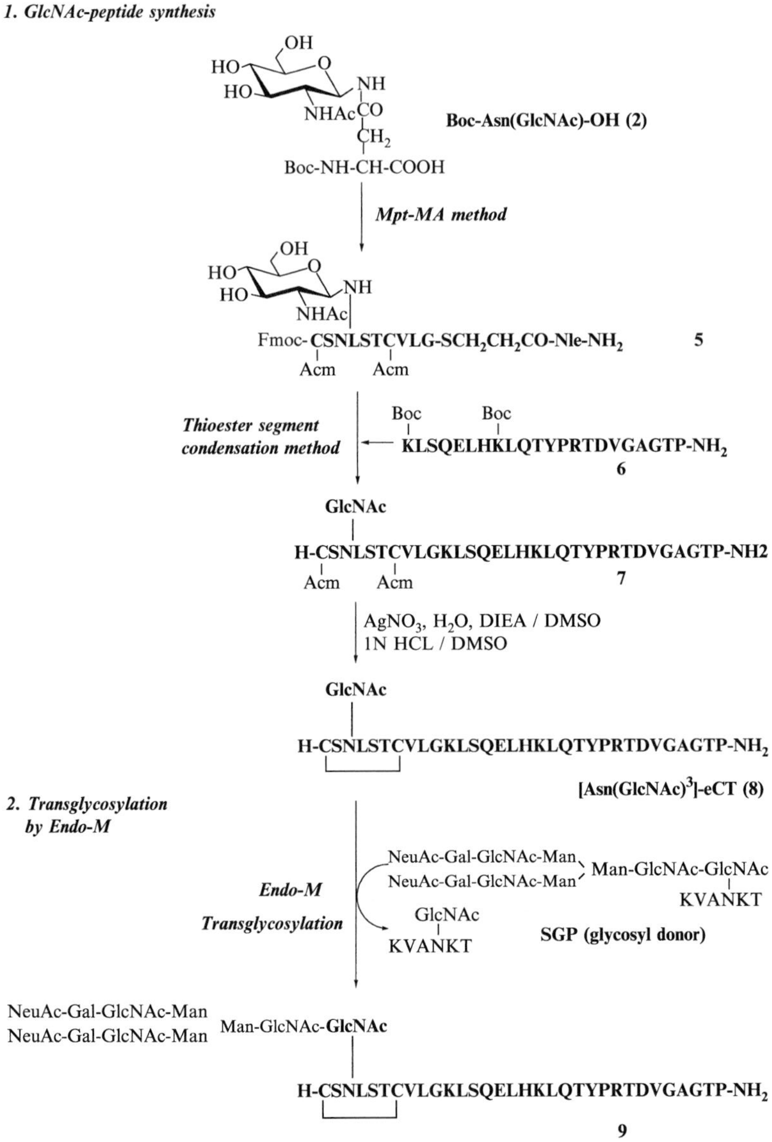

1. GlcNAc-peptide synthesis

HO, HO, OH, O, NH, NHAc, CO, CH_2

Boc-NH-CH-COOH

Boc-Asn(GlcNAc)-OH (2)

Mpt-MA method

HO, HO, OH, O, NH, NHAc

Fmoc-**CSNLSTCVLG-SCH$_2$CH$_2$CO-Nle-NH$_2$** **5**

Acm Acm

Thioester segment condensation method

Boc Boc

KLSQELHKLQTYPRTDVGAGTP-NH$_2$

6

GlcNAc

H-CSNLSTCVLGKLSQELHKLQTYPRTDVGAGTP-NH2

Acm Acm **7**

$AgNO_3$, H_2O, DIEA / DMSO

1N HCL / DMSO

GlcNAc

H-CSNLSTCVLGKLSQELHKLQTYPRTDVGAGTP-NH$_2$

[Asn(GlcNAc)^3]-eCT (8)

2. Transglycosylation by Endo-M

Endo-M

Transglycosylation

NeuAc-Gal-GlcNAc-Man

NeuAc-Gal-GlcNAc-Man

Man-GlcNAc-GlcNAc

KVANKT

GlcNAc

KVANKT

SGP (glycosyl donor)

NeuAc-Gal-GlcNAc-Man

NeuAc-Gal-GlcNAc-Man

Man-GlcNAc-**GlcNAc**

H-CSNLSTCVLGKLSQELHKLQTYPRTDVGAGTP-NH$_2$

9

SCHEME 3. Synthetic scheme of glycosylated eel calcitonin derivatives.

residue bound to GlcNAc is, however, less stringent. Therefore, we could synthesize several neoglycoconjugates, such as the substance P (H-Arg-Pro-Lys-Pro-Gln^5-Gln^6-Phe-Phe-Gly-Leu-Met-NH_2) derivative having a glutamine-linked oligosaccharide to which an oligosaccharide is introduced into the Gln residue,[17] glycopeptides having plural N-linked oligosaccharides,[18] a natural protein modified by the addition of an N-linked oligosaccharide,[19] and β-cyclodextrin (βCD) analogs having natural N-linked oligosaccharides.[20]

Experimental Methods

Synthesis of a Building Block: Boc-Asn(GlcNAc)-OH or Fmoc-Asn(GlcNAc)-OH

*Boc-Asn(GlcNAc)-OH, N^ω-(2-Acetamido-2-deoxy-β-D-glucopyranosyl)-N^α-(tert-butyloxycarbonyl)-L-asparagine (**2**).* 2-Acetamido-3,4,6-tri-*O*-benzyl-2-deoxy-β-D-glucopyranosyl azide (5.17 g, 10 mmol) and N^α-*tert*-butyloxycarbonyl-L-aspartic acid α-benzyl ester (Boc-Asp-OBzl) (3.88 g, 12 mmol) are dissolved in CH_2Cl_2 (80 ml) at -78° under an argon atmosphere. Next, tributylphosphine (Bu_3P) (2.5 ml, 10 mmol) is added to this solution. After being stirred for 22 h, the reaction mixture is diluted with chloroform and washed successively with aqueous $NaHCO_3$, water, and brine, and dried over anhydrous Na_2SO_4, and the solvent is evaporated. The resulting precipitate is filtered and washed with ether three times. Compound **1**, Boc-Asn[GlcNAc$(OBzl)_3$]-OBzl (5.11 g, 6.4 mmol), is obtained as a white powder in 54% yield. Compound **1** (3.32 g, 4.2 mmol) is dissolved in CH_2Cl_2 (40 ml)–methanol (100 ml) and hydrogenated over 10% Pd/C under atmospheric pressure. After 3.5 h, the mixture is filtered and the filtrate is evaporated *in vacuo*. The residue is washed with ether on a glass filter, and Boc-Asn(GlcNAc)-OH (**2**) (1.88 g, 4.2 mmol) is obtained as a white powder in quantitative yield. The structure is assigned by nuclear magnetic resonance (NMR) analysis.[8]

*Fmoc-Asn(GlcNAc)-OH, N^ω-(2-Acetamido-2-deoxy-β-D-glucopyranosyl)-N^α-(9-fluorenylmethyloxycarbonyl)-L-asparagine (**4**).* Triethylphosphine (Et_3P) (5.00 g, 42 mmol) is added to a CH_2Cl_2 solution of

[17] K. Haneda, T. Inazu, M. Mizuno, R. Iguchi, H. Tanabe, K. Fujimori, K. Yamamoto, H. Kumagai, K. Tsumori, and E. Munekata, *Biochim. Biophys. Acta* **1526,** 242 (2001).

[18] K. Haneda, M. Takeuchi, T. Inazu, K. Toma, M. Tagashira, K. Kobayashi, K. Yamamoto, and K. Takegawa, "Peptide Science 2001," p. 89. Japanese Peptide Society, Osaka, Japan, 2002.

[19] R. Suzuki, H. Ohmae, K. Kiroku, and T. Inazu, *Glycoconj. J.* **16,** S166 (1999).

[20] K. Matsuda, T. Inazu, K. Haneda, M. Mizuno, T. Yamanoi, K. Hattori, K. Yamamoto, and H. Kumagai, *Bioorg. Med. Chem. Lett.* **7,** 2353 (1997).

2-acetamido-3,4,6-tri-*O*-acetyl-2-deoxy-β-D-glucopyranosylazide (18.94 g, 51 mmol) and N^{α}-(9-fluorenylmethyloxycarbonyl)-L-aspartic acid α-*tert*-butyl ester (Fmoc-Asp-OBu^t) (17.42 g, 42 mmol) under argon at $-78°$. After being stirred overnight, the precipitate forms in the mixture, because of poor product solubility. The resulting precipitate is filtered and washed with CH_2Cl_2 and ether to yield compound **3**, Fmoc-Asn[GlcNAc(OAc)$_3$]-OBu^t (24.97 g, 34 mmol), as a white powder (80% yield). The filtrate is concentrated and the residue is dissolved in CH_2Cl_2 and washed with 5% citric acid, water, and brine, and dried (anhydrous Na_2SO_4) and evaporated. The residue is washed with ether, to yield more of compound **3** in 20% yield (7.06 g), resulting in the total quantitative yield.[7] When tributylphosphine (Bu_3P) is used instead of Et_3P, compound **3** is obtained in 76% yield.[7] The *O*-acetyl groups of compound **3** are deprotected by a catalytic amount of sodium methoxide in methanol to give Fmoc-Asn(Glc-NAc)-OBu^t. Deprotection of *tert*-butyl ester by trifluoroacetic acid (TFA) treatment gives the desired compound Fmoc-Asn(GlcNAc)-OH (**4**) in 76% yield (from **3**) as colorless crystals.[9]

GlcNAc-Peptide Synthesis

A Typical Procedure. GlcNAc-peptide is synthesized using Boc-Asn(GlcNAc)-OH or Fmoc-Asn(GlcNAc)-OH as a building block by the Mpt-MA method with solid-phase synthesis depending on the Boc or Fmoc strategy. The coupling of an Asn(GlcNAc) and coupling to an Asn(GlcNAc) are performed with great care, monitoring the coupling reaction with the Kaiser test.[9,11] For the chain elongation of GlcNAc-peptide, the thioester segment condensation method[12,13] is used. The GlcNAc-peptide is purified by preparative reversed-phase high-performance liquid chromatography (RP-HPLC). The characterization of the GlcNAc-peptide is performed by matrix-assisted laser desorption ionization time-of-flight mass spectrometry (MALDI-TOF MS) and amino acid analysis.

A practical example is the synthesis of an eel calcitonin (eCT) derivative containing GlcNAc. A hypocalcemic peptide hormone eCT is composed of 32 amino acids. Here, we artificially introduce an N-linked oligosaccharide into the Asn^3 site of eCT compound **8**. The synthetic scheme is shown in Scheme 3.

Boc-Gly-S-CH$_2$CH$_2$CO-Nle-MBHA–Resin. A mixture of Boc-norleucine (Boc-Nle-OH) (0.80 g, 3.46 mmol), 1.0 *M* 1-hydroxybenzotriazole (HOBt) in *N*-methylpyrrolidone (NMP) (2.8 ml), and 1.0 *M* *N,N*-dicyclohexylcarbodiimide (DCC) in NMP (2.8 ml) is stirred for 1 h. The solution is added to a neutralized 4-methylbenzhydrylamine (MBHA) resin (2.4 g; NH_2, 1.85 mmol), and this mixture is shaken for 3 h. The resulting resin is washed with NMP, 50% methanol in CH_2Cl_2 and CH_2Cl_2, and then treated with 50% TFA in CH_2Cl_2, washed with CH_2Cl_2, and treated with 5% *N,N*-diisopropylethylamine (DIEA) and CH_2Cl_2 and 5% DIEA in NMP. Trityl (Trt)-S-CH_2CH_2COOH (0.90 g, 2.71 mmol), 1.0 *M* HOBt in NMP (2.7 ml), and 1.0 *M* DCC in NMP (2.7 ml) are mixed for 30 min, and the resulting solution is added to the resin. The suspension is shaken for 4 h. The resin is washed with NMP, 50% methanol in CH_2Cl_2 and then CH_2Cl_2, and finally dried *in vacuo* to give the Trt-S-CH_2CH_2CO-Nle-MBHA–resin (2.74 g of Nle: 0.52 mmol g^{-1}). An aliquot of the resin (0.79 g) is treated with 5% 1,2-ethanedithiol in TFA, and then washed with CH_2Cl_2, treated with 5% DIEA in NMP, and washed with NMP. A mixture of Boc-Gly-OH (0.32 g, 1.81 mmol), 1.0 *M* HOBt in NMP (1.8 ml), and 1.0 *M* DCC in NMP (1.7 ml) is stirred for 1 h. This solution is added to the resin, and the suspension is shaken for 16 h. The resin is washed with NMP, 50% methanol in CH_2Cl_2 and then CH_2Cl_2, and finally dried *in vacuo* to give the Boc-Gly-S-CH_2CH_2CO-Nle-MBHA–resin (0.75 g; Nle, 0.55 mmol g^{-1}).

*Fmoc-[Cys(Acm)^{1,7},Asn(GlcNAc)^{3}]-eCT(1–10)-S-CH_2CH_2CO-Nle-NH_2 (**5**).* Peptide chain elongation of Boc-Gly-S-CH_2CH_2CO-Nle-MBHA–resin (0.75 g, 0.41 mmol) is carried out with a model 430 peptide synthesizer (Applied Biosystems, Foster City, CA) to give 1.40 g of a protected peptide resin of the sequence eCT(4–10), Boc-Leu-Ser(Bzl)-Thr(Bzl)-Cys(Acm)-Val-Leu-Gly-S-CH_2CH_2CO-Nle-NH–resin (where Acm is acetamidomethyl S-protecting group). Further, 3-Asn(GlcNAc), 2-Ser(Bzl), and 1-Cys(Acm) residues are coupled manually by double coupling of the corresponding Mpt-MA. The Mpt-MAs of Boc-Asn(GlcNAc)-OH, Boc-Ser(Bzl)-OH, and Fmoc-Cys(Acm)-OH are prepared by mixing with 1.23 mmol of Boc-Asn(GlcNAc)-OH, Boc-Ser(Bzl)-OH, and Fmoc-Cys(Acm)-OH and 1 *M* dimethylphosphinothioyl chloride (Mpt-Cl) in DMF (1.2 ml) and 2 *M* DIEA in NMP (0.62 ml) and NMP (1.0 ml), and after 30 min, 2 *M* DIEA in NMP (0.62 ml) is added to the solution of Mpt-MA. The reaction mixture of Mpt-MA is added to the protected peptide resin, and the coupling time is 1 h. Completion of these coupling reactions are determined by the Kaiser test. An aliquot of the glycopeptide resin (400 mg) is treated with HF (9.0 ml) and anisole (1.0 ml) at 0° for 90 min. After evaporation of the HF, ether (10 ml) is added to the mixture and the precipitate obtained is washed with ether and dissolved in TFA (10 ml). The TFA solution is filtered through a glass filter to remove the resin and poured into cold ether (300 ml). The resulting precipitate is isolated by decantation, and crude glycopeptide (156 mg) is obtained. After purification by RP-HPLC (20–40% acetonitrile in 0.1%

TFA over 40 min), Fmoc-[Cys(Acm),^{1,7}Asn(GlcNAc)^{3}]-eCT(1–10)-S-CH_2CH_2CO-Nle-NH_2 (**5**) is obtained (47 mg, 49 μmol, 12% based on Nle in the starting resin). MALDI-TOF MS: found *m/z* $[M+Na]^+$ 1787.4, calculated for $C_{77}H_{118}N_{16}O_{25}S_3$ $[M+Na]^+$ 1787.1.

[Lys(Boc)^{11,18}]-eCT(11–32) (***6***). By starting from MBHA–resin (1.00 g, 0.42 mmol; NH_2, 0.41 mmol g^{-1}), the chain elongation reaction is carried out and the peptide resin (2.29 g) is obtained. An aliquot of this peptide resin (500 mg) is treated with HF (8.5 ml), anisole (0.75 ml), and 1,4-butanedithiol (0.75 ml) at 0° for 90 min. After evaporation of the HF, ether (10 ml) is added to the mixture and the obtained precipitate is washed with ether and then extracted with aqueous acetonitrile containing 5% acetic acid (20 ml) to give the crude peptide (178 mg) after freeze–drying. This is purified by RP-HPLC (20–60% acetonitrile in 0.1% TFA over 40 min) to yield Fmoc-eCT(11–32) (96 mg, 30 μmol, 34% based on the amino group in the starting resin) after freeze–drying. Boc-*O*-succinimide (16 mg, 76 μmol) and DIEA (60 μl) are added to a solution of this peptide (26 mg, 7.6 μmol) in dimethyl sulfoxide (DMSO, 2.0 ml), and the resulting solution is stirred for 7 h. Ether is added to the reaction mixture, and the resulting precipitate is washed with ether and then dissolved in DMSO (2.0 ml). To the solution is added piperidine (0.1 ml), followed by stirring for 90 min. Ether is added to the reaction mixture, and the resulting precipitation is washed with ether. After purification by RP-HPLC (20–65% acetonitrile in 0.1% TFA over 30 min), [Lys(-Boc)^{11,18}]-eCT(11–32) (**6**) is obtained (11 mg, 4.3 μmol, 56%). Overall yield based on the amino group in the starting resin is 19%. MALDI-TOF MS: found *m/z* $[M+H]^+$ 2639.7, calculated for $C_{117}H_{192}N_{32}O_{37}$ $[M+H]^+$ 2640.0.

Fmoc-[Cys(Acm)^{1,7},Asn(GlcNAc)^{3}]-eCT (***7***). A solution of $AgNO_3$ (0.9 mg, 5.4 μmol), 3,4-dihydro-3-hydroxy-4-oxo-1,2,3-benzotriazine (HOOBt) (8.8 mg, 54 μmol), and DIEA (6.3 μl, 36 μmol) in DMSO (0.20 ml) is stirred for 1 h, and to the mixture is added a solution of the glycopeptide thioester segment **5** (5.8 mg, 1.8 μmol) and the peptide segment **6** (4.8 mg, 1.8 μmol) in DMSO (0.30 ml). After 16 h of stirring, dithiothreitol (DTT, 2.5 mg, 16 μmol) is added to the reaction mixture to terminate the coupling reaction. A partially protected glycopeptide Fmoc-[Cys(Acm)^{1,7},Asn(GlcNAc)^{3},Lys(Boc)^{11,18}]-eCT (4.3 mg, 1.0 μmol, 55%) is isolated by RP-HPLC (30–70% acetonitrile in 0.1% TFA over 40 min). This partially protected glycopeptide (4.0 mg, 0.77 μmol) is treated with TFA containing 5% 1,4-butanedithiol (0.30 ml) at room temperature for 90 min. min. Water (7.0 ml) is added to the reaction mixture, and the solution is washed with ether. After freeze–drying, crude glycopeptide is obtained. This crude glycopeptide is treated with 5% piperidine in DMSO

(0.40 ml) at room temperature for 2.5 h. Acetic acid (50 μl) is added to the solution to quench the piperidine. After purification by RP-HPLC (20–60% acetonitrile in 0.1% TFA over 40 min), [Cys(Acm)^{1,7},Asn(GlcNAc)^{3}]-eCT (**7**) is obtained (3.5 mg, 0.64 μmol, 83%). MALDI-TOF MS: found *m/z* $[M+H]^+$ 3764.8, calculated for $C_{160}H_{266}N_{46}O_{54}S_2$ $[M+H]^+$ 3763.3.

*[Asn(GlcNAc)^{3}]-eCT (**8**).* Glycopeptide **7** (3.2 mg, 0.59 μmol) is dissolved in H_2O (0.30 ml) to which is added a solution of $AgNO_3$ (0.5 mg, 3.0 μmol) and DIEA (1.5 ml, 8.9 μmol) in DMSO (100 μl). After 2 h of stirring, a mixture of 1 *M* hydrochloric acid and DMSO [1:1 (v/v), 3.0 ml] is added to the solution, and the reaction mixture is stirred for 20 h. After purification by RP-HPLC (20–50% acetonitrile in 0.1% TFA over 40 min), [Asn(GlcNAc)^{3}]-eCT (**8**) is obtained (1.8 mg, 0.4 μmol, 67%). MALDI-TOF MS: found *m/z* $[M+H]^+$ 3618.9, calculated for $C_{154}H_{254}N_{44}O_{52}S_2$ $[M+H]^+$ 3619.1.

Transglycosylation reaction by Endo-M

Preparation of native Endo-M of Mucor hiemalis. Spores of *Mucor hiemalis* are inoculated into the medium containing yeast extract (25 g/liter; Difco, Detroit, MI), peptone (5 g/liter; Difco), D-glucose (1 g/liter), and glycerol (5 g/liter) at pH 6.5, and cultivated at 28° for 5 days. The enzyme (0.93 mU/ml; hydrolytic activity is determined with dansylasialo biantennary complex-type glycosylasparagine as substrate at 37°, pH 6.0) is obtained in the filtrate of the culture broth. The enzyme in the filtrate is purified[14] by the following procedure: CM cellulose treatment, ammonium sulfate (70%) precipitation, DEAE-Sepharose FF column chromatography [20 m*M* potassium phosphate buffer (pH 7.0), 0–0.5 *M* NaCl linear gradient], and hydroxyapatite (Gigapite; Seikagaku-kogyo, Tokyo, Japan) column chromatography (20–800 m*M* potassium phosphate buffer, pH 7.0, linear gradient). The active fraction of the final step is concentrated and desalted by an ultracentrifuge (Centricon Plus-20, PL-30; Millipore, Bedford, MA). The partially purified Endo-M is free from other glycosidase activities but contains some protease activity.

Preparation of Recombinant Endo-M. The recombinant *Candida boidinii* protease-deficient strain (*pep4 prb1*) (donated by Kirin Brewery, Tokyo, Japan) grown in medium containing methanol is disrupted by a French press. The recombinant Endo-M is purified from the cell-free extract by Q-Sepharose chromatography [10 m*M* potassium phosphate buffer (pH 7.0), 0–0.3 *M* NaCl gradient], phenyl-Sepharose chromatography [10 m*M* acetate buffer (pH 5.5), 1–0 *M* $(NH_4)_2SO_4$ gradient], and Sephacryl S300 gel filtration [20 m*M* potassium phosphate buffer (pH 7.0) containing 0.15 *M* NaCl]. The purified recombinant Endo-M is electrophoretically pure and free from protease activity.

Preparation of Sialylglycohexapeptide, H-Lys-Val-Ala-Asn[(NeuAc-Gal-GlcNAc-Man)$_2$-Man-GlcNAc$_2$]-Lys-Thr-OH. A sialylglycohexapeptide SGP as a glycosyl donor is prepared by the method described in the literature.[16] Using the 1000 ml of egg yolk from 60 commercial hen's eggs, 280 mg of SGP is obtained as a white powder. MALDI-TOF MS: found *m/z* $[M - H]^-$ 2864.8, calculated for $C_{112}H_{189}N_{15}O_{70}$ $[M - H]^-$ 2864.8.

Typical Transglycosylation Procedure by Endo-M

The typical reaction mixture for transglycosylation of a disialo complex-type oligosaccharide by Endo-M is composed of 10 m*M* GlcNAc-peptide (glycoside acceptor), 100 m*M* SGP (glycosyl donor), and Endo-M (60–120 mU/ml) in 60 m*M* potassium phosphate buffer, pH 6.25, and it is incubated at 37° for 20 to 30 min. The reaction is started by adding the enzyme solution into the preincubated substrate in the buffer solution, and is terminated by adding an equal volume of 0.5% cold TFA solution. The reaction mixture is analyzed by RP-HPLC eluted with a linear increase in aqueous acetonitrile using 0.1% TFA in water (buffer A) and 0.1% TFA in acetonitrile (buffer B). The transglycosylation product monitored at 214 nm is eluted as a single peak several minutes earlier than the remaining glycoside acceptor. The transglycosylation yield (mol%) is calculated from the ratio (%) of the peak area of the transglycosylation product to the initial peak of the acceptor added, which is based on the assumption that the molar absorption coefficient of the transglycosylation product at 214 nm is the same as that of the acceptor.

As a practical example, the synthesis of an eCT derivative containing a disialo biantennary complex-type oligosaccharide at Asn^3 using compound **8** as a glycoside acceptor and SGP as a glycosyl donor is shown in Fig. 1.

Preparation of Glycosylated eCT

[Asn(GlcNAc)^3]-eCT (**8**) (22.1 mg, 6.1 μmol) and SGP (172 mg, 60 μmol) are placed in a 2.0-ml Eppendorf tube, and dissolved with 360 μl of 0.1 *M* potassium phosphate buffer, pH 6.25. After preincubation at 37° for 5 min, the reaction is started by adding 240 μl of Endo-M solution (final concentration, 60 mU/ml), and the reaction mixture (600 μl) is incubated at 37° for 30 min with gentle shaking (120 rpm). The reaction is terminated by addition of an equal volume of 0.5% cold TFA solution. Ten microliters of the reaction mixture which is further diluted 10 times with 0.1% cold TFA solution is applied to the analytical RP-HPLC using an i.d. 4.6 mm × 250 mm ODS column (Mightysil RP-18; Kanto Chemical,

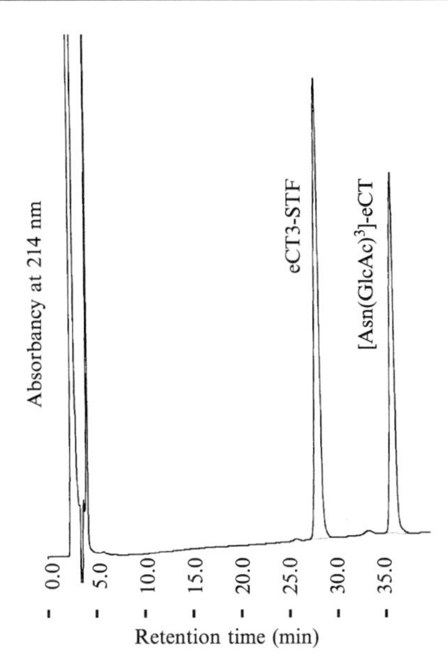

FIG. 1. HPLC profile of the transglycosylation reaction mixture to [Asn(GlcNAc)^3]-eCT.

Tokyo, Japan). Elution is carried out with a linear increase of 27.5 to 37.5% acetonitrile concentration in 0.1% TFA over 40 min at a flow rate of 0.8 ml/min. A single peak of the transglycosylation product is found at 28.2 min in front of the peak of the remaining acceptor at 36.0 min by monitoring at 214 nm. The yield of the transglycosylation product is 57% based on the acceptor used (Fig. 1). The transglycosylation product is purified by preparative RP-HPLC (i.d. 20 mm × 250 mm column), and freeze–dried. The transglycosylation product (16.7 mg, 2.96 μmol, 49%) is obtained as a white powder, and the remaining acceptor **8** (9.4 mg) is also recovered. The transglycosylation product is identified as an eCT derivative containing a disialo biantennary complex-type oligosaccharide, [Asn{(NeuAc-Gal-GlcNAc-Man)$_2$-Man-GlcNAc$_2$}^{3}]-eCT (eCT3-STF) (**9**). MALDI-TOF MS: found *m/z* $[M+H]^+$ 5623.8, calculated for $C_{230}H_{378}N_{49}O_{108}S_2$ $[M+H]^+$ 5621.9.

[8] Preparative-Scale Chemoenzymatic Synthesis of Large Carbohydrate Assemblies Using $\alpha(1\rightarrow4)$-Galactosyltransferase/UDP-4′-Gal-Epimerase Fusion Protein

By PAVEL I. KITOV, EUGENIA PASZKIEWICZ, WARREN W. WAKARCHUK, and DAVID R. BUNDLE

Introduction

The synthesis of a single oligosaccharide in nature requires a well-orchestrated interplay among a multitude of enzymes.[1] High specificity and efficiency are achieved either by compartmentalization of reactive species or by kinetic control. In the laboratory, oligosaccharides can also be synthesized with a set of enzymes acting in concert in a reaction mixture. Because of the high cost of most sugar nucleotides it is often convenient either to generate glycosyl donors from less expensive precursors[2] or to interconvert sugar nucleotides[3] *in situ*, using multienzyme systems. Recombinant technology has led to the creation of fused proteins, composed of enzymes with compatible intrinsic activities that are merged into a single fusion protein, and has reduced the number of laborious steps for enzyme purification.[4] It has also been demonstrated that the close proximity of epimerase and transferase in a fusion protein confers an increase in the reaction rate of up to 300% compared with equimolar mixtures of the individual enzymes.[5] The α-galactosyltransferase we have used for the construction of this fusion protein has been demonstrated to express extremely well as a recombinant protein, and to have a remarkable range of acceptor

[1] R. G. Spiro, *Glycobiology* **12,** 43R (2002); B. J. Mengeling and S. J. Turco, *Curr. Opin. Struct. Biol.* **8,** 572 (1998).

[2] M. Gilbert, R. Bayer, A. M. Cunningham, S. DeFrees, Y. Gao, D. C. Watson, M. N. Young, and W. W. Wakarchuk, *Nat. Biotechnol.* **16,** 769 (1998); Y. Ichikawa, R. Wang, and C. H. Wong, *Methods Enzymol.* **247,** 107 (1994).

[3] Y. Ichikawa, G. J. Shen, and C. H. Wong, *J. Am. Chem. Soc.* **113,** 4698 (1991); A. Zervosen and L. Elling, *J. Am. Chem. Soc.* **118,** 1836 (1996).

[4] J. Fang, X. Chen, W. Zhang, A. Janczuk, and P. G. Wang, *Carbohydr. Res.* **329,** 873 (2000).

[5] X. Chen, Z. Liu, J. Q. Wang, J. W. Fang, H. N. Fan, and P. G. Wang, *J. Biol. Chem.* **275,** 31594 (2000).

0076-6879/03 $35.00

and donor tolerance.[6] This enzyme, then, represents an example of the ideal glycosyltransferase for chemoenzymatic synthesis on a preparative scale.

Here we report an application of a fusion protein with two enzyme activities that consist of UDP-4′-Gal-epimerase and $\alpha(1\rightarrow4)$-galactosyltransferase to the synthesis of clustered carbohydrate derivatives containing large numbers of clustered oligosaccharides. Clusters of this type, which we termed "starfish," have been shown to have unprecedented activity in neutralizing bacterial toxins.[7]

The chemoenzymatic approach to glycoclusters **1** and **2**, each containing 10 copies of $\alpha(1\rightarrow4)$-galactopyranosyl-lactose (P^k-trisaccharide), involves assembly of a cluster containing a simple sugar (lactose in our case) followed by its enzymatic elaboration (Scheme 1). During the process, a C-4′ epimerase enzyme converts UDP-Glc to UDP-Gal, which is then utilized by the $\alpha(1\rightarrow4)$-galactosyltransferase as a glycosyl donor in order to create a galabiose, α-Gal-$(1\rightarrow4)$-β-Gal linkage.

Pendant oligosaccharide ligands can be attached to the scaffold via the glycosidic or nonglycosidic position (cf. compounds **1** and **2**). Each cluster consists of bivalent arms brought together into a pentavalent superstructure. This design reflects the topology of the homopentameric carbohydrate-binding domains of Shiga-like toxin types 1 and 2. Two P^k-trisaccharide ligands terminating each arm target two distinct binding sites per subunit.

Methods

Chemical Assembly of Glycoclusters

The glycosidically linked cluster **1** is synthesized as follows (Scheme 2). The allyl aglycone in lactose derivative **3** is extended by ultraviolet (UV)-initiated cysteamine addition to yield the carbohydrate building block **4**. Temporary *tert*-butoxycarbonyl protection of the terminal amino group is used to facilitate purification. The bidentant fork element **7** is obtained in three steps by selective N-acylation of diethanolamine with 10-undecenoyl chloride followed by UV-promoted addition of cysteamine to the terminal double bond of the acyl group of compound **5**, followed by the activation of hydroxyl groups in compound **6** as 4-nitrophenyl carbonates. Coupling of

[6] W. W. Wakarchuk, A. Cunningham, D. C. Watson, and N. M. Young, *Protein Eng.* **11,** 295 (1998); F. Yan, M. Gilbert, W. Wakarchuk, J.-R. Brisson, and D. M. Whitfield, *Org. Lett.* **3,** 3265 (2001); B. Lougheed, H. D. Ly, W. W. Wakarchuk, and S. G. Withers, *J. Biol. Chem.* **274,** 37717 (1999).

[7] P. I. Kitov, J. M. Sadowska, G. Mulvey, G. D. Armstrong, H. Ling, N. S. Pannu, R. J. Read, and D. R. Bundle, *Nature* **403,** 669 (2000).

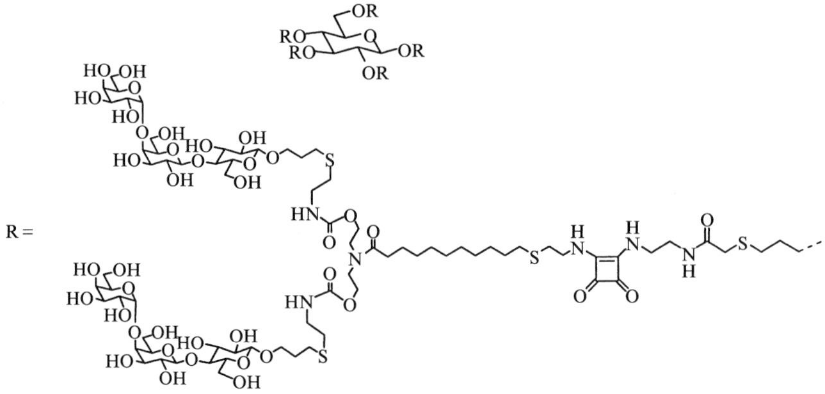

R =

1

R =

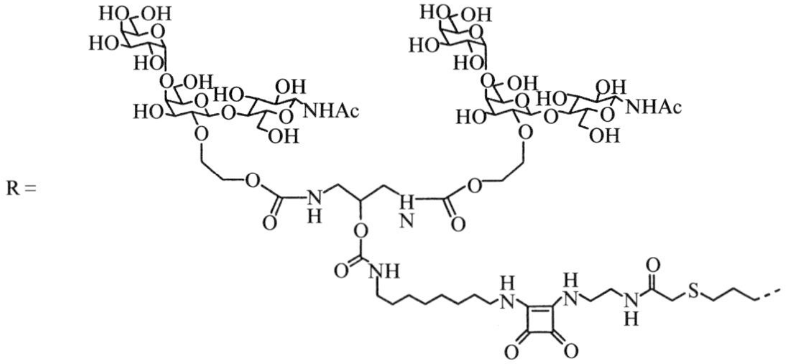

2

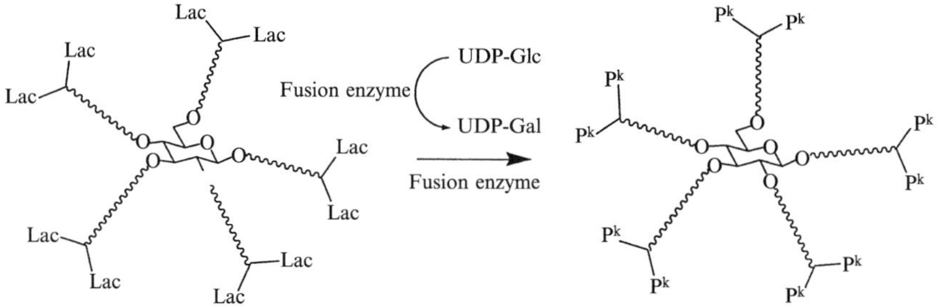

SCHEME 1

compounds **7** with two equivalents of compound **4a** gives the protected bivalent lactose derivative **8**. Sequential O- and N-deprotection of which, followed by activation of the amino group as the half-ester derivative of squaric acid, affords the "carbohydrate arm" building block **9**.

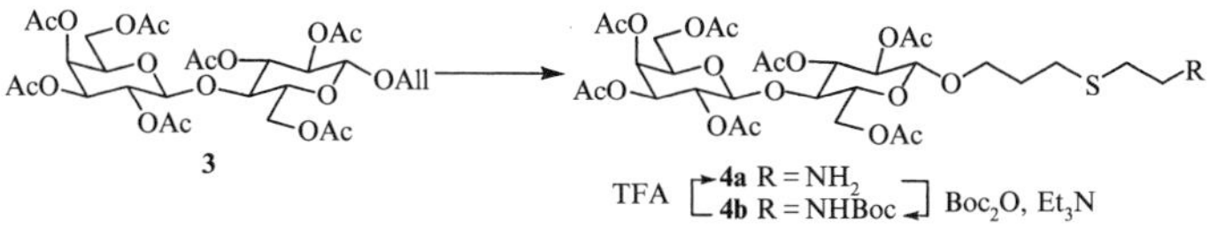

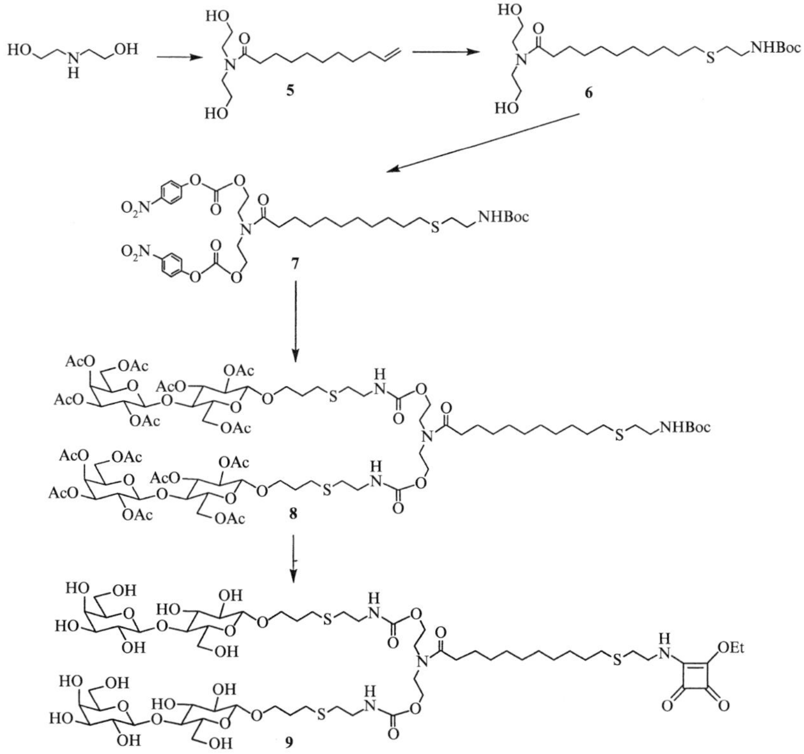

SCHEME 2

Condensation of compound **9** with the previously described[8] penta-amine derivative of glucose, **10**, gives the lactose cluster **11** (Scheme 3).

Synthesis of the nonglycosidically linked cluster **2** starts with the preparation of a suitably derivatized lactose derivative (Scheme 4). Allylation

[8] P. I. Kitov and D. R. Bundle, *J. Am. Chem. Soc.,* submitted.

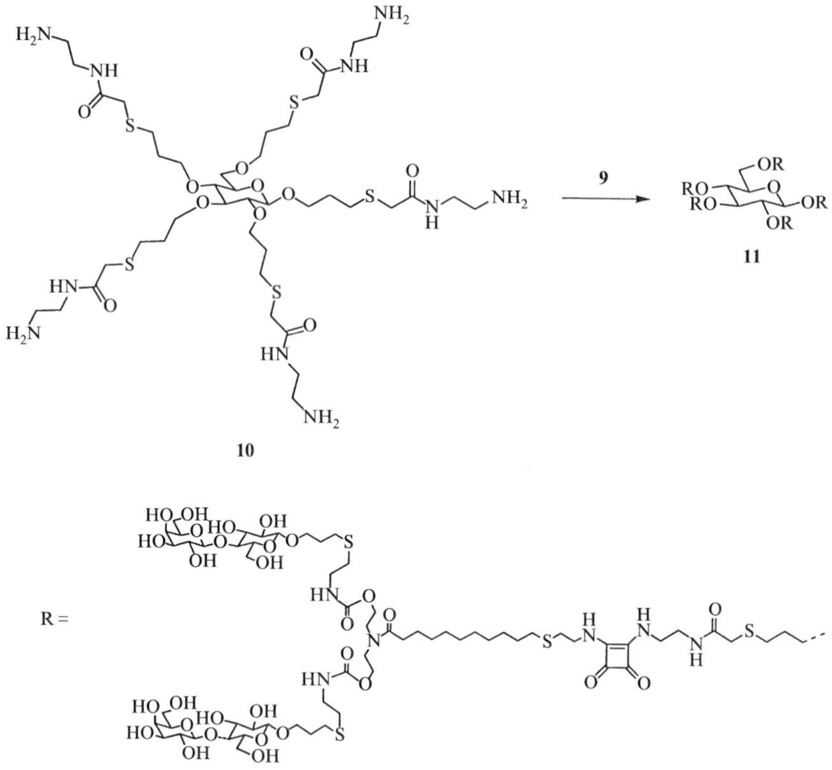

SCHEME 3

of the known per isopropylidene derivative of lactose, 12[9] followed by acidic hydrolysis gives 2-*O*′-allyllactose derivative **13**, which is then converted into the lactosylacetamide **14**. This reaction is accomplished by incubation of reducing sugar in saturated aqueous ammonia bicarbonate at 40° to form the glycosylamine according to Likhosherstov *et al.*[10] followed by the removal of the excess salt by lyophilization. The resulting glycosylamine is acetylated without purification to give compound **14**.

Oxidation of the allylic double bond, followed by reduction of the resulting aldehyde, gives hydroxyl compound **15**, which is activated as its *p*-nitrophenyl carbonate **16**. Reaction with 1,3-diamino-2-hydroxypropane

[9] P. L. Barili, G. Catalani, F. D'Andrea, F. D. Rensis, and P. Falcini, *Carbohydr. Res.* **298,** 75 (1997).

[10] L. M. Likhosherstov, O. S. Novikova, V. A. Derevitskaja, and N. K. Kochetkov, *Carbohydr. Res.* **146,** C1 (1986).

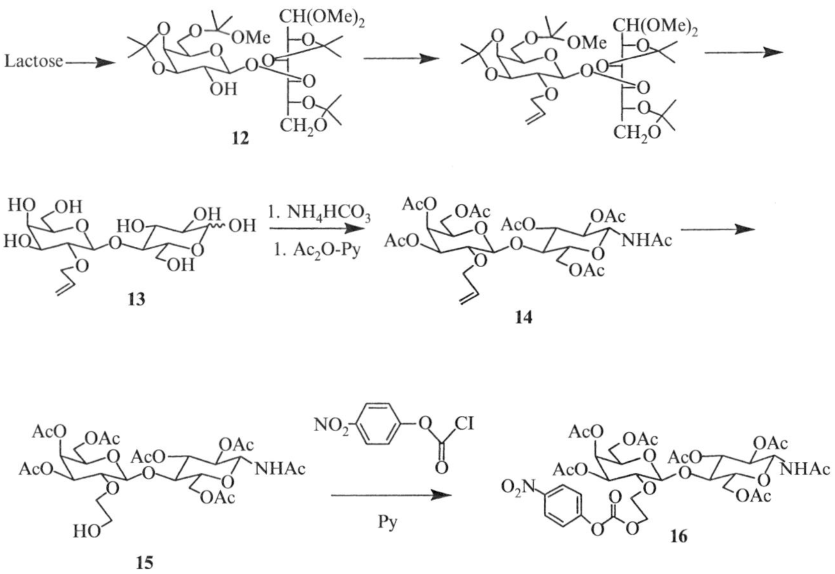

Lactose → 12 → ; 13 —1. NH_4HCO_3 / 1. Ac_2O-Py→ 14 → ; 15 —$NO_2C_6H_4OC(O)Cl$, Py→ 16

SCHEME 4

gives dimer **17** (Scheme 5), which after activation as the *p*-nitrophenyl carbonate **18** and condensation with 1,8-diaminooctane gives the lactose dimer with an extension arm terminated by an amine. The amine is activated as squaric acid semiester derivative **19**, which reacts with pentaamine **10** to give the lactose cluster **20** (Scheme 6).

Enzyme Assay

The activity of each enzyme of the fusion protein can be measured separately. The galactosyl transferase activity can be evaluated by a radiodiochemical assay,[11] high-performance liquid chromatography (HPLC), or capillary electrophoresis.[12]

The Gal-4′-epimerase activity for the interconversion between UDP-Galand UDP-Glc can be monitored directly by ^{1}H nuclear magnetic resonance (NMR) (Fig. 1a) because the H-1 resonances for UDP-Gal and UDP-Glc appear as well-separated multiplets at magnetic fields of 300 MHz or higher. Although in preparative experiments the epimerase converts UDP-Glc to UDP-Gal, which is then utilized by the galactosyl-

[11] M. M. Palcic, M. Pierce, and O. Hindsgaul, *Methods Enzymol.* **247,** 215 (1994).

[12] W. W. Wakarchuk and A. Cunningham, *in* "Capillary Electrophoresis of Oligosaccharides and Complex Glycoconjugates." Humana Press, Totowa, NJ, 2002.

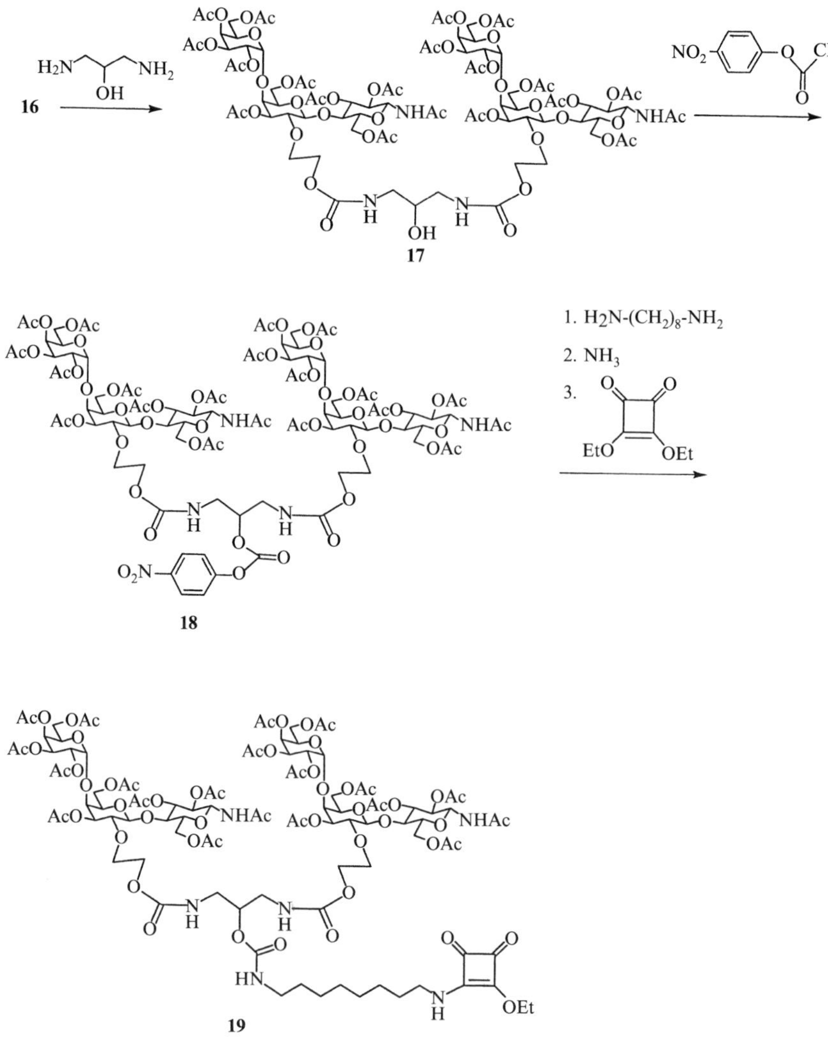

SCHEME 5

transferase, for the measurements of epimerase activity it is more convenient to use UDP-Gal as the reactant and regard UDP-Glc as the product. The reason for this is that, at equilibrium, UDP-Glc predominates over UDP-Gal (at 37°, the ratio of [Glc]/[Gal] is 2.85), therefore larger changes in the composition of the reaction mixture can be observed when UDP-Gal is chosen as a reactant.

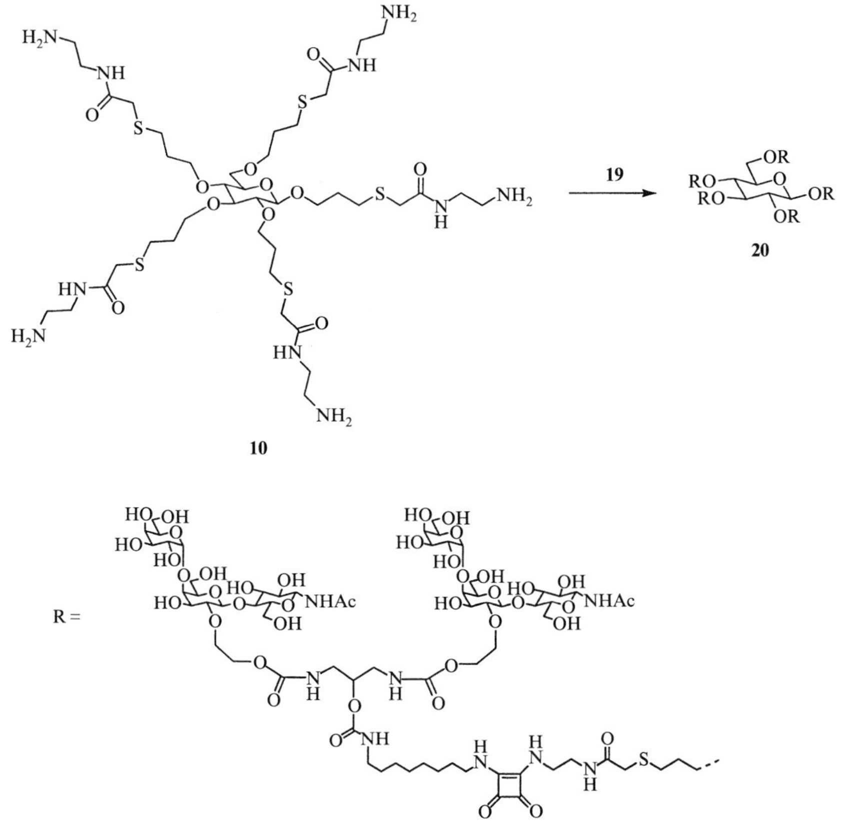

SCHEME 6

The kinetic model for equilibration of the two epimers is as follows:

$$\text{UDP-Gal} \underset{k_2}{\overset{k_1}{\rightleftharpoons}} \text{UDP-Glc}$$

$$\frac{d[\text{Glc}]}{dt} = k_2[\text{Glc}] - k_1[\text{Gal}] \tag{1}$$

$$K = \frac{k_1}{k_2} \tag{2}$$

$$[\text{Gal}]_0 = [\text{Glc}] + [\text{Gal}] \tag{3}$$

where [Glc] and [Gal] are concentrations of the respective monosaccharide nucleotides, $[\text{Gal}]_0$ is the initial concentration of UDP-Gal, k_1 and k_2 are rates of the direct and reverse reactions, and K is the equilibrium constant determined as the [Glc]:[Gal] ratio at $t \rightarrow \infty$.

By combining Eqs. (1)–(3) and solving the resulting differential equation we obtain the relationship between the molar fraction of UDP-Glc and the time of the reaction (t) with pseudo first-order rate constant (k_2) of the reverse reaction:

$$\frac{[\text{Glc}]}{[\text{Gal}]_0} = \frac{[\text{Glc}]}{[\text{Glc}] + [\text{Gal}]} = \frac{K - Ke^{(1+K)K_2 t}}{1 + K} \quad (4)$$

This pseudo first-order constant k_2 is proportional to the activity of the epimerase (A) and inversely proportional to the initial (or total) concentration of UDP-Gal (or Glc). Therefore, if the UDP-sugar concentration values are expressed as micromoles per milliliter the specific activity of the epimerase with respect to conversion of UDP-Glc to UDP-Gal is $A = k_2$ [UDP] DF in units per milliliter, where DF is the dilution factor.

A typical kinetic experiment is shown in Fig. 1. The reaction is conducted in an NMR tube in D_2O solution containing initially UDP-Gal (8.4 mM), NAD (0.84 mM), sodium HEPES buffer (200 mM, pH 8) at 37°. The data are fitted into Eq. (4), where K is assigned a value of 2.85.

NMR Assay of Transferase Activity

The NMR assay of transferase activity is conducted in a final volume of 100 μl at 37° for 15–60 min, containing the acceptor **21** (Scheme 7, 80 mM) and UDP-Gal (80 mM) in sodium HEPES buffer (200 mM, pH 8), $MnCl_2$ (1.25 mM), dithiothreitol (DTT, 5 mM) and bovine serum albumin (1%). Reactions are initiated by addition of the galactosyltransferase (10 μl) diluted in HEPES buffer, pH 7, and activated with DTT before use. In all cases, reactions are limited to a transfer of less than 20% of the UDP-Gal. The reactions are terminated by passage through a Pasteur pipette containing ion-retaining resin Dowex MR-3 (Sigma, St. Louis, MQ). The eluate is concentrated, the residue is dissolved in 1 ml of D_2O, and transfer of galactose is measured by integration of the signals for anomeric protons (see Fig. 2).

Enzymatic Processing of Glycoclusters

Compared with enzymatic reactions with monovalent substrates, in which the product can, in principle, be separated from unreacted starting material, the only acceptable outcome from enzymatic processing of a multivalent substrate is 100% conversion. Unfortunately, several factors may prevent a glycosyltransferase from driving the reaction to completion

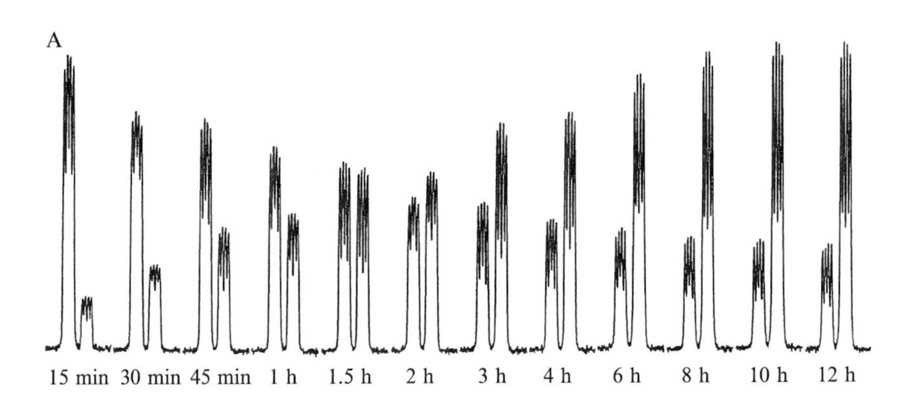

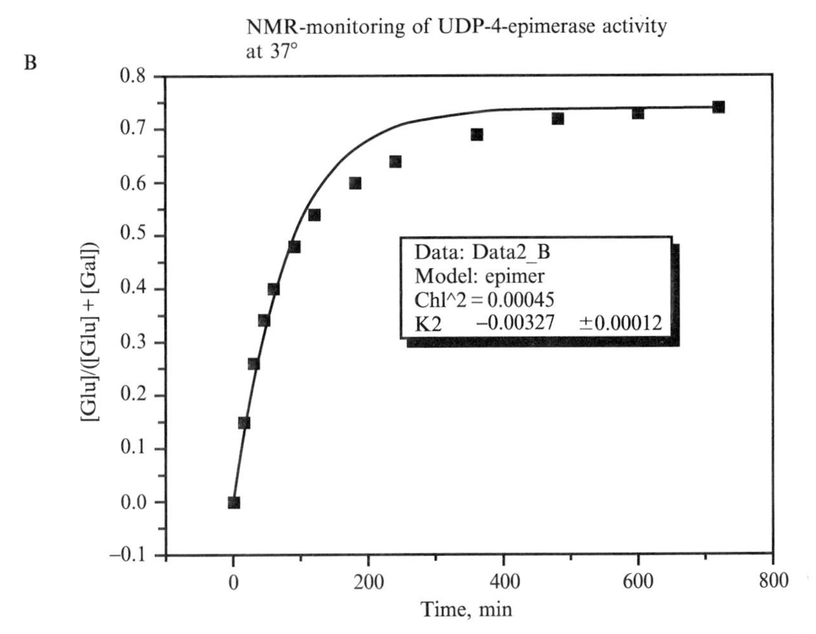

FIG. 1. NMR monitoring of UDP-4′-epimerase activity of the fusion protein at 37°. (A) Decrease in the signal for the galactose H-1 resonance with reciprocal growth of glucose H-1 resonance. (B) The time course of epimerization. Fitting was performed according to Eq. (4), using Microcal Origin 5.0 software.

in a single passage. These are as follows: poisoning of the enzyme by the product, a nucleotide monophosphate, precipitation of metal cofactor Mn^{2+} as a phosphate, change in pH, degradation of sugar nucleotide, and loss of enzyme activity during the course of reaction.

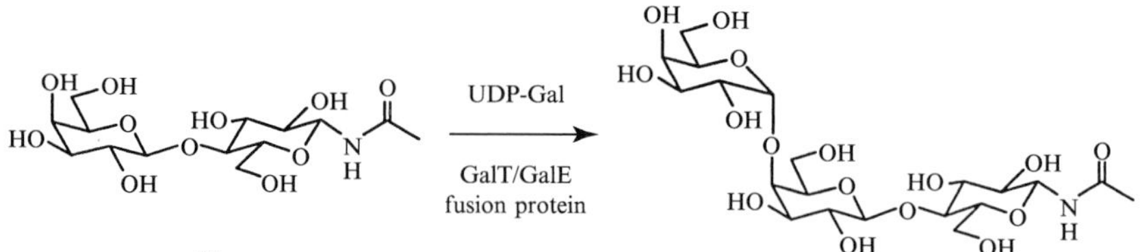

21

UDP-Gal

GalT/GalE fusion protein

22

SCHEME 7

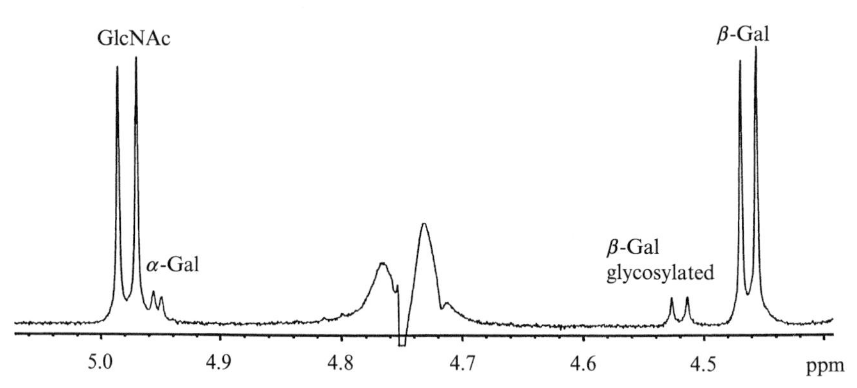

FIG. 2. Integration of proton resonances in the anomeric area of ^{1}H NMR spectrum of the mixture permits evaluation of degree of glycosylation. Doublet at 4.98 is Glc H-1 of **21** and **22**, doublet at 4.95 is α-Gal H-1 of **22**, doublet at 4.52 is β-Gal H-1 of **22**, doublet at 4.52 is β-Gal H-1 of **21**.

We have found that the best results can be achieved by removing low molecular weight products from the reaction mixture followed by a second treatment of the partially processed glycocluster. Thus, first processing of cluster **11** containing glycosidically linked lactose gives a product with more than 95% α-Gal incorporation, whereas substrate **20** with nonglycosidically linked lactose originally gives a product with 85% incorporation. This may be attributed in part to the reduced activities of glycosylacetamide derivatives, as has been previously reported for a different galactosyltransferase.[13] The second processing step converts both products into target P^k-trisaccharide clusters **1** and **2**. The 100% conversion has been ascertained by NMR and electrospray Fourier transform ion cyclotron resonance mass spectroscopy (ES FTICR MS).

[13] Y. Nishida, H. Tamakoshi, Y. Kitagawa, K. Kobayashi, and J. Thiem, *Angew. Chem. Int. Ed.* **39,** 2000 (2000).

Experimental

General Methods

Optical rotations are measured with a PerkinElmer (Korwalk, CT) 241 polarimeter in a 10-cm cell at ambient temperature ($22 \pm 2°$). Analytical thin layer chromatography (TLC) is performed on silica gel 60-F_{254} (Merck, Darmstadt, Germany) with detection by quenching of fluorescence and/or by charring with 10% H_2SO_4 in ethanol solution followed by heating at 180°. Column chromatography is performed on silica gel 60 (40–60 μm; Merck). Reversed-phase separations are conducted on Waters (Milford, CT) Delta 600 HPLC system, using a Beckman (Fulleston, CA) C_{18} semi-preparative column. ^{1}H NMR spectra are recorded at 500 and 600 MHz (Unity; Varian, Palo Atto, CA) in $CDCl_3$ (referenced to residual $CHCl_3$ at δ_H 7.24 ppm), CD_3OD (referenced to residual CD_2HOD at δ_H 3.3 ppm), ppm), or D_2O (referenced to external acetone at δ_H 2.225 ppm). All commercial reagents are used as supplied; solvents are distilled from appropriate desiccants before use.[14] Ultrafiltration is performed with cellulose membranes (molecular weight cutoff, 3000 and 10,000) in a 50-ml Amicon (Danvers, MA) chamber.

Freshly prepared fusion enzyme sample used in this work has 15 U/ml of epimerase activity and 4.2 U/ml of transferase activity with respect to 6-(5-fluorescein carboxamido)hexanoylaminophenyllactose (capillary electrophoresis assay)[Oa] and 3.5 U/ml with respect to *N*-acetyl-lactosamine (NMR assay).

6-tert-Butoxycarbamoyl-4-thiahexyl-2,3,6-tri-O-acetyl-4-O-(2,3,4,6-tetra-O-acetyl-β-D-galactopyranosyl)-β-D-glucopyranoside (**4b**)

A suspension of peracetylated allyllactoside **3**[15] (7.76 g, 11.47 mmol) and cysteamine hydrochloride (2.66 g, 2 equivalents, 23 mmol) in methanol (15 ml) and dichloromethane (DCM, 12 ml) is irradiated for 24 h and then di-*tert*-butyl dicarbonate (Boc_2O, 10.4 g, 46 mmol) is added, followed by triethylamine (6.4 ml, 46 m*M*). After 30 min the mixture is concentrated, taken up in DCM, washed with brine, and concentrated. Chromatography of the residue on silica gel in hexane–ethyl acetate (70:30–40:60, v/v) gives compound **4b** (8.7 g, 89%), $[\alpha]_D = -8.5°$ ($c = 0.91$, $CHCl_3$); ^{1}H NMR ($CDCl_3$): δ 5.32 (dd, 1 H, $J_{3',4'} = 3.51$ Hz, $J_{4',5'} = 1.07$ Hz, H-4′), 5.17 (t, 1 H, $J_{2,3} \approx J_{3,4} = 9.3$ Hz, H-3), 5.04 (dd, 1 H, $J_{1',2'} = 7.9$ Hz,

[14] D. D. Perrin, W. L. F. Armarego, and D. R. Perrin, "Purification of Laboratory Chemicals." Pergamon Press, Oxford, 1980.

[15] R. P. Evstigneeva, A. V. Lyubeshkin, M. V. Anikin, Y. L. Sebyakin, M. S. Barkhudaryan, *et al. Dokl. Chem.* (*Engl. Transl.*) **330,** 119 (1993).

$J_{2',3'} = 10.4$ Hz, H-2′), 4.94 (dd, 1 H, H-3′), 4.91–4.84 (m, 1 H, NH), 4.86 (dd, 1 H, $J_{1,2} = 7.94$ Hz, H-2), 4.47 (dd, 1 H, $J_{5,6a} = 2.14$ Hz, $J_{6a,6b} = 12.07$ Hz, H-6a), 4.45 (d, 1 H, H-1), 4.44 (d, 1 H, H-1), 4.12–4.03 (m, 3 H, H-6b, H-6′a, H-6′b), 3.90–3.82 (m, 2 H, H-5′, OCH_2), 3.76 (t, 1 H, $J_{4,5}$, 9.4 Hz, H-4), 3.60–3.54 (m, 2 H, H-5, OCH_2), 3.26 (dd, 2 H, $J = 4.4$ Hz, $J = 11.4$ Hz, NCH_2), 2.59 (t, 2 H, $J = 6.58$ Hz, SCH_2), 2.54 (t, 2 H, $J = 7.10$ Hz, SCH_2), 2.12, 2.10, 2.04, 2.02 × 3, 1.94 (21 H, OAc), 1.88–1.74 (m, 2 H, CH_2), 1.42 (s, 9 H, *t*-Bu). Electrospray MS: *m/z* 876.29332 (calculated for $[M+Na]^+$ $C_{36}H_{55}NO_{20}NaS$, *m/z* 876.29304).

*N,N-Di(2-hydroxyethyl)-10-undecenamide (**5**)*

To a solution of diethanolamine (2 g, 19 mmol) in DCM (40 ml), dry pyridine (10 ml) and 10-undecenoyl chloride (4 ml, 1 equivalent) are added at room temperature. After 30 min the mixture is concentrated and dissolved in methanol (100 ml) and NaOH (~1.5 g) is added. After the reaction is complete (~1 h), the mixture is neutralized with HCl solution, extracted with DCM, and concentrated and the residue is chromatographed on silica gel with DCM–methanol (5–8%) to give compound **5** (3 g, 58%), ^{1}H NMR ($CDCl_3$): δ 5.82–5.74 (m, 1 H, allyl), 4.98–4.88 (m, 2 H, allyl), 3.82 (t, 2 H, $J = 5.0$ Hz, OCH_2), 3.75 (t, 2 H, $J = 5.2$ Hz, OCH_2), 3.52 (t, 2 H, NCH_2), 3.47 (t, 2 H, NCH_2), 3.39 (broad s, 2 H, OH), 2.36 (t, 2 H, $J = 7.7$ Hz, CH_2CO), 2.03 (m, 2 H, allyl), 1.62 (m, 2 H, CH_2), 1.36–1.24 (m, 10 H, CH_2).

*N,N-Di(2-hydroxyethyl)-14-tert-butoxycarbamoyl-12-thiatetradecenamide (**6**)*

A suspension of compound **5** (3 g, 11 mmol) and cysteamine hydrochloride (1.63 g, 1.3 equivalents) in methanol (10 ml) and DCM (10 ml) is irradiated for 16 h and then Boc_2O (3.75 g) and triethylamine (1.5 g) are added. The mixture is refluxed for 15 min and then concentrated, dissolved in DCM, and washed with water. Chromatography on silica gel with DCM–methanol (5–10%) gives compound **6** (3.9 g, 79%). ^{1}H NMR ($CDCl_3$): δ 3.96–3.78 (m, 4 H, OCH_2), 3.58–3.48 (m, 4 H, NCH_2), 3.27 (t, 2 H, $J = 6.5$ Hz, Hz, NCH_2), 3.08 (broad s, 2 H, OH), 2.61 (t, 2 H, SCH_2), 2.49 (t, 2 H, $J = 7.40$ Hz, SCH_2), 2.42 (t, 2 H, $J = 7.7$ Hz, CH_2), 1.65–1.60 (m, 2 H, CH_2), 1.57–1.52 (m, 2 H, CH_2), 1.42 (s, 9 H, *t*-Bu), 1.36–1.22 (m, 12 H, CH_2).

*N,N-Di[2-(4-nitrophenoxycarbonyloxy)ethyl]-14-tert-butoxycarbamoyl-12-thiatetradecenamide (**7**)*

To a solution of compound **6** (3.34 g, 7.44 mmol) and 4-nitrophenyl chloroformate (4.9 g, 2.4 equivalents) in dry DCM (60 ml) is added dry pyridine (2.8 ml). After 0.5 h the mixture is diluted with DCM, washed with brine, and then concentrated. Chromatography of the residue on silica gel with hexane–ethyl acetate (2:1–1:1) gives compound **7** (3.8 g, 65%). ^{1}H NMR ($CDCl_3$): δ 8.30–8.22 [m, 2 H, aromatic (arom.)], 7.38–7.30 (m, 2 H, arom.), 4.86 (broad s, 1 H, NH), 4.46–4.41 (m, 4 H, CH_2), 3.78–3.70 (m, 4 H, NCH_2), 3.28 (broad s, 2 H, CH_2NBoc), 2.61 (t, 2 H, $J = 6.5$ Hz, SCH_2), 2.48 (t, 2 H, CH_2), 2.39 (t, 2 H, $J = 7.5$ Hz, CH_2), 1.67–1.62 (m, 2 H, CH_2), 1.55–1.50 (m, 2 H, CH_2), 1.42 (s, 9 H, *t*-Bu), 1.34–1.20 (m, 12 H, CH_2).

*N,N-Di{11-[2,3,6-tri-O-acetyl-4-O-(2,3,4,6-tetra-O-acetyl-β-D-galactopyranosyl)-β-D-glucopyranosyloxyl-5-aza-3-oxa-4-oxo-8-thiaundecyl}-14-tert-butoxycarbamoyl-12-thiatetradecenamide (**8**)*

Compound **4b** (11.1 g, 13 mmol) is dissolved in triflutroacetate (TFA) (20 ml). After 30 min at room temperature the mixture is concentrated, coevaporated with toluene (twice, 30 ml each time). The residue is dissolved in DCM, neutralized with triethylamine (~4 ml), and concentrated. To a solution of the residue in DCM (30 ml), triethylamine (2 g) is added followed by a solution of compound **7** (4.6 g, 5.9 mmol) in DCM (20 ml). The mixture is stirred for 3.5 h at room temperature and then concentrated, coevaporated with toluene, and chromatographed on silica gel with hexane–acetone (70:30–40:60) to give compound **8** (9.73 g, 82%). $[\alpha]_D$ $-10°$ ($c = 0.96$, $CHCl_3$); ^{1}H NMR ($CDCl_3$): δ 5.42 (broad s, 1 H, NH), 5.36 (broad s, 1 H, NH), 5.32 (d, 2 H, $J_{3',4'} = 3.5$ Hz, H-4′), 5.16 (t, 2 H, $J_{2,3} \approx J_{3,4} = 9.3$ Hz, H-3), 5.08 (dd, 2 H, $J_{1',2'} = 7.9$ Hz, $J_{2',3'} = 10.3$ Hz, H-2′), 4.94 (dd, 2 H, H-3′), 4.88 (broad s, 1 H, NH), 4.84 (t, 2 H, $J_{1,2} = 8.0$ Hz, H-2), 4.50–4.42 (m, 6 H, H-1, H-1′, H-6a), 4.22–4.18 (m, 4 H, OCH_2), 4.14–4.04 (m, 6 H, H-6b, H-6′ a, H-6′b), 3.90–3.84 (m, 4 H, H-5′, OCH_2), 3.78 (td, 2 H, $J = 3.2$ Hz, H-4), 3.60–3.54 (m, 8 H, H-5, OCH_2, NCH_2), 3.34–3.26 (m, 6 H, NCH_2), 2.62–2.58 (m, 6 H, SCH_2), 2.52 (t, 4 H, $J = 7$ Hz, SCH_2), 2.48 (t, 2 H, $J = 7.4$ Hz, SCH_2), 2.32 (t, 2 H, $J = 7.5$ Hz, CH_2CO), 2.12, 2.10, 2.04, 2.02, 1.94 (5 s, 42 H, OAc), 1.86–1.74 (m, 4 H, CH_2), 1.62–1.58 (m, 2 H, CH_2), 1.57–1.52 (q, 2 H, $J = 7.5$ Hz, Hz, CH_2), 1.42 (s, 9 H, *t*-Bu), 1.36–1.24 (m, 12 H, CH_2).

*N,N-Di [11-(4-O-β-D-galactopyranosyl-β-D-glucopyranosyloxy)-5-aza-3-oxa-4-oxo-8-thiaundecyl]-14-(4-ethoxy-2,3-dioxo-3-cyclo butenylamino)-12-thiatetradecenamide (**9**)*

Compound **8** (7 g, 3.48 mmol) is dissolved in dry methanol (60 ml) and 1 *M* MeONa (10 ml) is added. The mixture is stirred for 18 h and then Dowex (H^+) is added, the resin is filtered off, and the supernatant is concentrated. The residue is dissolved in TFA (15 ml), stirred for 30 min, and concentrated, coevaporated with water, and neutralized by $NaHCO_3$ solution. The pH of the solution (total volume, 45 ml) is adjusted to pH 8 and a solution of 3,4-diethoxy-3-cyclobuten-1,2-dione (1.18 g, 1.03 ml, 2 equivalents) in ethanol (5 ml) is added. After 1 h the mixture is applied to a C_{18} column and chromatographed with water–acetonitrile (10–40%) to give compound **9** (3.7 g, 73%). ^{1}H NMR (CD_3OD): δ 4.78–4.68 (m, 2 H, CH_2), 4.36 (d, 2 H, $J_{1',2'}=7.7$ Hz, H-1′), 4.29 (d, 2 H, $J_{1,2}=7.8$ Hz, H-1), 4.21 (t, 2 H, $J=5.5$ Hz, OCH_2), 4.17 (t, 2 H, $J=5.3$ Hz, OCH_2), 3.98–3.94 (m, 2 H, OCH_2), 3.89 (dd, 2 H, $J_{5,6a}=2.6$ Hz, $J_{6a,6b}=12.1$ Hz, H-6a), 3.84 (dd, 2 H, $J_{5,6b}=4.2$ Hz, H-6b), 3.81 (dd, 2 H, $J_{4',5'}=0.9$ Hz, $J_{3',4'}=3.3$ Hz, H-4′), 3.77 (dd, 2 H, $J_{5',6'a}=7.4$ Hz, $J_{6'a,6'b}=11.4$ Hz, H-6′a), 3.77–3.74 (m, 1 H, NCH_2), 3.69 (dd, 2 H, $J_{5',6'b}=4.7$ Hz, H-6′b), 3.68–3.63 (m, 4 H, NCH_2, OCH_2), 3.62–3.57 (m, 5 H, H-5′, OCH_2, NCH_2), 3.56 (t, 2 H, $J_{3,4} \approx J_{4,5}=9.4$ Hz, H-4), 3.54 (dd, 2 H, $J_{1',2'}=7.7$ Hz, $J_{2',3'}=9.7$ Hz, H-2′), 3.52 (t, 2 H, $J_{2,3}=8.8$ Hz, H-3), 3.47 (dd, 2 H, H-3′), 3.42–3.38 (m, 2 H, H-5), 3.28–3.26 (m, 4 H, NCH_2), 3.24 (dd, 2 H, H-2), 2.72 (t, 2H, $J=7.0$ Hz, SCH_2), 2.66(t, 4 H, $J=7.2$ Hz, SCH_2), 2.63–2.60 (m, 4 H, SCH_2), 2.56 (t, 2 H, $J=7.4$ Hz, SCH_2), 2.40 (t, 2 H, $J=7.5$ Hz, CH_2), 1.90–1.84 (m, 4 H, CH_2), 1.62–1.54 (m, 4 H, CH_2), 1.45 (dd, 3 H, CH_3), 1.42–1.30 (m, 12 H, CH_2).

*Glycocluster **11***

Pentamine derivative acetic acid salt **10** (334 mg, 0.247 mmol) and squaric acid derivative **9** (2.5 g, 1.7 mmol) are dissolved in water (50 ml) and the pH is adjusted to pH 9 by addition of a concentrated solution of Na_2CO_3. The mixture is stirred for 3 days at room temperature and then dialyzed through cellulose membrane (Amicon; molecular weight cutoff, 10,000). During dialysis the temperature is maintained at 65° by a water bath in order to prevent gel formation. The product is collected and lyophilized to give compound **11** (1.65 g, 83%). ^{1}H NMR (DMSO-d_6–D_2O = 19:1): δ 4.18 (d, 10 H, $J_{1',2'}=7.5$ Hz, H-1′), 4.16 (d, 10 H, $J_{1,2}=7.9$ Hz, H-1), 4.06–4.04 (m, 10 H, OCH_2), 3.98–3.96 (m, 10 H, OCH_2), 3.80–3.76 (m, 10 H, OCH_2), 3.71 (dd, $J_{5,6a}=5.4$ Hz, $J_{6a,6b}=11$ Hz, H-6a), 3.67–3.62 (broad s, 10 H, NCH_2), 3.60 (dd, 10 H, $J=4.3$ Hz, $J=3.4$ Hz, H-4′), 3.58 (dd, 10 H, $J_{5,6b}=3.9$ Hz, H-6b), 3.53–3.40 (m, 80 H, H-5′, H-6′a, H-6′b, NCH_2, OCH_2), 3.32–3.24 (m, 60 H, H-3, H-4, H-5, H-2′, H-3′, NCH_2), 3.12–3.06 (m, 30 H, NCH_2, SCH_2), 2.98 (t, 10 H, $J_{2,3}=8.3$ Hz, H-2), 2.65 (t, 10 H, $J=6.7$ Hz, SCH_2), 2.59–2.48 (m, 60 H, SCH_2), 2.25 (t, 10 H, $J=7.3$ Hz, CH_2), 1.77–1.69 (m, 30 H, CH_2), 1.50–1.46 (m, 10 H, CH_2), 1.46–1.40 (m, 10 H, CH_2), 1.32–1.26 (m, 10 H, CH_2), 1.21 (broad s, 50 H, CH_2).

*Glycocluster **1***

Lactose cluster **12** (1.1 g, 0.137 mmol) is dissolved in water (27 ml) and then sodium HEPES buffer [8 ml, 1 *M*, 10 m*M* $MnCl_2$, bovine serum albumin (BSA, 0.8 mg/ml), pH 8], DTT solution (2 ml, 100 m*M*), and alkaline phosphatase (0.5 ml) are added. To the mixture UDP-Glc (1.2 g, 19.6 mmol) is added followed by crude fusion enzyme (2.5 ml). The reaction is incubated at 37° for 24 h and then dialyzed, using cellulose membrane (molecular weight cutoff, 3000). The product is collected, concentrated, and lyophilized from water. ^1H NMR indicates >95% conversion. The procedure is repeated for this product, using 20% of original amounts of all reagents followed by purification on a preparative reversed-phase HPLC column (C_{18}) to give P^k-trisaccharide cluster **1**. Yield, 1.27 g (96%). Electrospray FTICR MS and ^1H NMR indicate 100% conversion. ^1H NMR [DMSO-d_6–D_2O, 19:1 (v/v)]; δ 4.76 (d, 10 H, $J_{1'',2''} = 3.8$ Hz, H-1″), 4.23 (d, 10 H, $J_{1',2'} = 7.7$ Hz, H-1′), 4.16 (d, 10 H, $J_{1,2} = 7.9$ Hz, H-1), 4.07–4.04 (m, 20 H, H-5″, OCH_2), 3.97 (t, 10 H, $J = 5.5$ Hz, OCH_2), 3.81–3.77 (m, 10 H, OCH_2), 3.76 (d, 10 H, $J_{3',4'} = 3.1$ Hz, H-4′), 3.72–3.70 (m, 20 H, H-4″, H-6a), 3.67–3.50 (m, 100 H, H-6b, H-5′, H-6′a, H-6′b, H-2″, H-3″, OCH_2, NCH_2), 3.49–3.40 (m, 30 H, H-6″a, H-6″b, NCH_2), 3.37 (dd, 10 H, $J_{2',3'} = 10.0$ Hz, H-3′), 3.32–3.20 (m, 60 H, H-3, H-4, H-5, H-2′, NCH_2), 3.14–3.06 (m, 30 H, SCH_2, NCH_2), 2.98 (t, 10 H, $J_{2,3} = 8.3$ Hz, H-2), 2.65 (t, 10 H, $J = 6.7$ Hz, SCH_2), 2.58–2.50 (m, 60 H, SCH_2), 2.25 (t, 10 H, $J = 7.3$ Hz, CH_2), 1.76–1.70 (m, 30 H, CH_2), 1.50–1.42 (m, 20 H, CH_2), 1.32–1.26 (m, 10 H, CH_2), 1.22 (broad s, 50 H, CH_2).

*1-Acetamido-1-deoxy-2,3,6-tri-O-acetyl-4-O-(3,4,6-tri-O-acetyl-2-O-allyl-β-D-galactopyranosyl)-β-D-glucopyranose **(14)***

A suspenstion of compound **12**[16] (6.81 g, 11.7 mmol), NaH (95%, 0.885 g, 35.2 mmol), and allyl bromide (2.03 ml, 23.4 mmol) in dry tetrahydrofuran (THF, 100 ml) is refluxed for 3 h until alkylation is complete. Methanol is then added and the mixture is neutralized by acetic acid. The mixture is filtered through Celite and the filtrate is concentrated. The residue is dissolved in aqueous 60% acetic acid, refluxed for 2 h, and concentrated. The residue is crystallized from ethanol and the crystals are collected and dried in a desiccator over P_2O_5 to give 2′-*O*-allyllactose **13** (3.26 g, 72%). A solution of 2′-*O*-allyllactose (1 g, 2.61 mmol) in water

[16] P. L. Barili, G. Catalani, F. D' Andrea, F. D. Rensis, and P. Falcini, *Carbohydr. Res.* **298,** 75 (1997).

(5 ml) is saturated with an excess of NH_4HCO_3 for 2 days at 45°, and new portions of NH_4HCO_3 are added as needed: The mixture is concentrated and coevaporated with water three times. The residue is dissolved in pyridine–acetic anhydride (Py–Ac_2O) mixture (1:1, 40 ml), stirred for 3 h at 50°, and then concentrated and coevaporated with toluene. Chromatography on silica gel in hexane–ethyl acetate (1:1–0:1, v/v) gives compound **14** (1.25 g, 71%). $[\alpha]_D + 2.9°$ ($c = 0.9$; $CHCl_3$); ^1H NMR ($CDCl_3$): δ 6.10 (d, 1 H, $J_{1,NH} = 9.4$ Hz, NH), 5.78 (m, 1 H, All), 5.32–5.08 (m, 5 H, H-1, H-3, H-4′, All), 4.86–4.77 (m, 2 H, H-2, H-3′), 4.50 (d, 1 H, $J_{6'a,6'b} = 12.2$ Hz, H-6′a), 4.30–3.99 (m, 6 H, H-1′, H-6′b, H-6a, H-6b, All), 3.80–3.72 (m, 3 H, H-4, H-5, H-5′), 3.40 (dd, 1 H, $J_{1',2'} = 7.8$ Hz, $J_{2',3'} = 10.1$ Hz, H-2′), 2.10, 2.07, 2.04 × 3, 1.98, 1.96 (21 H, Ac).

*1-Acetamido-1-deoxy-2,3, 6-tri-O-acetyl-4-O-[3,4,6-tri-O-acetyl-2-O-(2-hydroxyethyl)-β-D-galactopyranosyl]-β-D-glucopyranose **(15)***

A solution of allyl derivative **14** (32.9 g; 28.04 g, 41.44 mmol) in acetone (150 ml) is oxidized with *N*-methylmorpholine *N*-oxide (2 equivalents, 11.4 g; 10 g) in the presence of OsO_4 (5 ml; 10-mg/ml solution in *tert*-butyl alcohol) overnight at 50°. The mixture is diluted with 150 ml of water and $NaIO_4$ (2 equivalents, 17.7 g) is added. When oxidation is complete as judged from TLC [toluene–acetone, 65:35 (v/v)] as excess of $NaBH_4$ (10 g) is added in 1 g. The pH was maintained in the range of pH 5–7 by addition of acetic acid. The mixture is diluted with brine, extracted with DCM (three times), and concentrated. Chromatography of the residue on silica gel with toluene–acetone (75:35–60:40, v/v) gives compound **15** (25.5 g, 76%). $[\alpha]_D + 16.5°$ ($c = 0.3$; $CHCl_3$); ^1H NMR ($CDCl_3$): δ 6.14 (d, 1 H, $J_{1,NH} = 9.25$ Hz, NH), 5.33–5.28 (m, 2 H, H-3, H-4′), 5.20 (t, 1 H, $J_{1,2} = 9.3$ Hz, H-1), 4.88–4.26 (m, 2, H, H-2, H-3′), 4.49 (d, 1 H, $J_{6'a,\ 6'b} = 11.7$ Hz, H-6′a), 4.31–4.26 (m, 2 H, H-1′, H-6′b), 4.14–4.00 (m, 2 H, CH_2), 3.82–3.60 (m, 7 H, H-4, H-5, H-5′, H-6a, H-6b, CH_2), 3.45 (dd, 1 H, $J_{1',2'} = 7.7$ Hz, $J_{2',3'} = 10.1$ Hz, H-2′), 2.11, 2.10, 2.05 × 2, 2.04, 2.01, 1.97 (21 H, Ac).

*1-Acetamido-1-deoxy-2,3,6-tri-O-acetyl-4-O-(3,4,6-tri-O-acetyl-2-O-[2-(p-nitrophenyloxycarbonyloxy)ethyl]-β-D-galactopyranosyl)-β-D-glucopyranose **(16)***

To a solution of compound **15** (8.6 g, 12.65 mmol) and 4-nitrophenyl chloroformate (3.06 g, 1.2 equivalents) in dry DCM (50 ml), pyridine (2 ml) is added. After 20 min at room temperature the reaction is quenched with methanol, diluted with DCM, and washed with brine. Chromatography of the residue on silica gel, using a stepped gradient of pentane–ethyl acetate (1:1–2:8, v/v) gives compound **16** (7.2 g, 67%).

$[\alpha]_D + 13.2^\circ$ ($c = 0.8$; $CHCl_3$); ^1H NMR ($CDCl_3$): δ 8.29–8.26 (m, 2 H, arom.), 7.43–7.40 (m, 2 H, arom.), 6.12 (d, 1 H, $J_{1,NH} = 9.3$ Hz, NH), 5.31 (dd, 1 H, $J_{3',4'} = 3.5$ Hz, $J_{4',5'} = 0.55$ Hz, H-4′), 5.28 (t, 1 H, $J_{2,3} \approx J_{3,4} = 9.5$ Hz, H-3), 5.18 (t, 1 H, $J_{1,2} = \sim 9.3$ Hz, H-1), 4.88 (dd, 1 H, $J_{2',3'} = 10.1$ Hz, H-3′), 4.82 (t, 1 H, $J_{1,2} \approx J_{2,3} = \sim 9.7$ Hz, H-2), 4.56 (dd, 1 H, $J_{5',6'a} = 1.8$ Hz, $J_{6'a,6'b} = 12.3$ Hz, H-6′a), 4.35 (ddd, 1 H, $^2J = 11.9$ Hz, $^3J = 2.7$ Hz, $^3J = 6.0$ Hz, CH_2), 4.31 (d, 1 H, $J_{1',2'} = 7.7$ Hz, H-1′), 4.32–4.28 (m, 1 H, CH_2), 4.26 (dd, 1 H, $J_{5',6'} = 4.4$ Hz, H-6′b), 4.10 (dd, 1 H, $J_{5,6a} = 6.4$ Hz, $J_{6a,6b} = 11.2$ Hz, H-6a), 4.04 (dd, 1 H, $J_{5,6b} = 7.3$ Hz, H-6b), 3.97 (ddd, 1 H, $^2J = 11.5$ Hz, $^3J = 6.4$ Hz, $^3J = 2.5$ Hz, CH_2), 3.85 (ddd, 1 H, $^2J = 11.5$ Hz, $^3J = 6.0$ Hz, $^3J = 2.7$ Hz, CH_2), 3.82–3.75 (m, 3 H, H-4, H-5, H-5′), 3.41 (dd, 1 H, H-2′), 2.11, 2.05, 2.04, 2.04, 2.00, 1.96 (21 H, Ac).

*Bis{[1-acetamido-1-deoxy-2,3,6-tri-O-acetyl-4-O-(3,4,6-tri-O-acetyl-2-O-β-D-galactopyranosyl)-β-D-glucopyranose-2-yloxy]ethyl}oxycarbamoyl-2-hydroxypropane (**17**)*

A solution of compound **16** (300 mg, 0.355 mmol) and 1,3-diamino-2-hydroxypropane (16 mg, 0.209 mmol) in methyl cyanide (10 ml) in the presence on triethylamine (~2 equivalents) is stirred at 40° for 1 h. The mixture is concentrated and chromatographed on silica gel in DCM–methanol (3.4–5%) to give compound **17** (235 mg, 88%). $[\alpha]_D +15.7^\circ$ ($c = 0.3$; $CHCl_3$); ^1H NMR ($CDCl_3$): δ 6.37 and 6.36 (two d, 2 H, $J_{NH,1} = 9.16$ Hz, NH), 5.92–5.80 (m, 2 H, NH), 5.30–5.25 (m, 4 H, H-3, H-4′), 5.15 (t, 2 H, H-1), 4.77–4.70 (m, 4 H, H-2, H-3′), 4.49 (t, 2 H, $J_{6'a,6'b} = 9.76$ Hz, H-6′a), 4.40–4.13 (m, 6 H, H-1′, H-6′b, CH_2), 4.10–3.95 (m, 7 H, H-6a, H-6b, CH_2, CH), 3.85–3.70 (m, 7 H, H-4, H-5, H-5′, CH_2), 2.11, 2.10, 2.04, 2.03, 2.02, 1.98, 1.98 (m, 7 × 3 H, Ac).

*Bis{[1-acetamido-1-deoxy-2,3,6-tri-O-acetyl-4-O-(3,4,6-tri-O-acetyl-2-O-β-D-galactopyranosyl)-β-D-glucopyranose-2-yloxy]ethyl}oxycarbamoyl-2-(4-nitrophenyloxy)carbonyloxypropane (**18**)*

To a solution of compound **17** (12.9 g, 8.59 mmol) and 4-nitrophenyl chloroformate (1.3 equivalents, 2.26 g) in dry DCM (50 ml) pyridine (0.9 ml, 1.3 equivalents) is added. The mixture is stirred for 1 h at room temperature and then concentrated and coevaporated with toluene. Chromatography of the residue on silica gel with toluene–acetone (1:1, v/v) gives compound **18** (13.85 g, 97%). $[\alpha]_D + 15.8^\circ$ ($c = 0.5$; $CHCl_3$). The ^1H NMR spectrum in $CDCl_3$ is a complex superposition of two rotamers; selected signals: δ 8.23 (d, 2 H, $J = 9.0$ Hz, arom.), 7.48 (d, 2 H, arom.), 6.53 and 6.40 (2d, 1 H, $J_{NH,1} = 9.5$ Hz, NH), 5.43 and 5.18 (2t, 2 H, $J_{1,NH} = J_{1,2} = 9.5$ Hz, H-1).

*Bis{[1-acetamido-1-deoxy-4-O-(2-O-β-D-galactopyranosyl)-β-D-glucopyranose-2-yloxy]ethyl}carbamoyloxy-2-(R,S)-[8-(4-ethoxy-2,3-dioxo-3-cyclobutenylamino)octyl]carbamoyloxypropane (**19**)*

A solution of compound **18** (2 g, 1.2 mmol) in methyl cyanide (30 ml) is added dropwise to a solution of 1,8-diaminooctane (0.865 g, 6 mmol) in methyl cyanide (50 ml). After 1 h aqueons NH_3 (~5 ml) is added gradually with gentle heating. The mixture is stirred for 2 h and then concentrated. The residue is chromatographed on Bio-gel G-10 with water–10% methanol–0.01% TFA and concentrated to give a colorless oil. To a solution of the residue in methanol (150 ml) and water (50 ml), 3,4-diethoxy-3-cyclobutene-1,2-dione (410 mg, 2.4 mmol) and triathylamine (250 mg, 2.4 mmol) are added. After 3 h the mixture is concentrated. The residue is chromatographed on a reversed-phase (C_{18}) column with water–methanol (9:1–1:1, v/v) to give compound **19** (1.31 g, 84%). ^1H NMR (D_2O): δ 4.97 and 4.96 (two d, 2 H, $J_{1,2}=9.15$ Hz, H-1), 4.84 (broad s, 2 H, CH_2), 4.49 (d, 2 H, $J_{1',2'}=6.23$ Hz, H-1′), 4.26–4.18 (m, 4 H, CH_2), 4.07–4.00 (m, 2 H, CH_2), 3.99–3.90 (m, 4 H, H-6a, H-4′, CH_2), 3.80–3.10 (m, 16 H, H-3, H-4, H-5, H-6b, H-3′, H-5′, H-6′a, H-6′b), 3.50 (t, 2 H, $J=6.8$ Hz, CH_2), 3.45–3.25 (m, 8 H, H-2, H-2′, CH_2), 3.08 (t, 2 H, $J=6.6$ Hz, CH_2), 2.08 (s, 6 H, Ac), 1.65–1.60 (m, 2 H, CH_2), 1.50–1.43 (m, 5 H, CH_2, CH_3), 1.40–1.28 (m, 8 H, CH_2).

*Glycocluster **20***

A solution of the Boc derivative of pentaamine **10** (120 mg, 77.3 μmol or 0.386 μmol per NH_2) in TFA (0.5 ml) is stirred for 1 h and then concentrated and coevaporated with water. The residue is dissolved in water (1 ml) neutralized with $NaHCO_3$ and compound **19** (648 mg, 1.3 equivalents) is added. The pH is adjusted to pH9 with 1 *M* NaOH solution. The mixture is stirred overnight and then chromatographed on C_{18} (0–50% methanol) to give compound **20** (538 mg, 95%). ^1H NMR [DMSO-d_6–D_2O, 19:1 (v/v)]: δ 4.71 and 4.70 (2d, 10 H, $J_{1,2}=9.1$ Hz, H-1), 4.58–4.52 (m, 5 H, CH), 4.26 and 4.25 (2d, $J_{1',2'}=7.7$ Hz, H-1′), 4.08–3.97 (m, 20 H, OCH_2), 3.76–3.70 (m, 30 H, H-6a, OCH_2), 3.60 (d, 10 H, $J_{3',4'}=3.4$ Hz, H-4′), 3.54–3.41 (m, 70 H, H-5, H-6b, H-5′, H-6′a, H-6′b, OCH_2, NCH_2), 3.39 (dd, 10 H, $J_{2',3'}=9.6$ Hz, H-3′), 3.33 (t, 10 H, $J_{2,3}\approx J_{3,4}=8.4$ Hz, H-3), 3.28–3.21 (m, 30 H, H-4, NCH_2, SCH_2), 3.18–3.13 (m, 20 H, H-2′, NCH_2), 3.11 (t, 10 H, H-2), 3.08–3.02 (m, 20 H, NCH_2), 2.92 (t, 10 H, $J=6.9$ Hz, NCH_2), 2.58–2.53 (m, 10 H, SCH_2), 1.83 (s, 30 H, NAc), 1.77–1.68 (m, 10 H, CH_2), 1.52–1.46 (m, 10 H, CH_2), 1.38–1.32 (m, 10 H, CH_2), 1.28–1.17 (m, 40 H, CH_2).

*Glycocluster **2***

Lactose cluster **20** (214 mg, 0.029 mmol) is dissolved in hot water (5.7 ml) and then cooled to room temperature. HEPES buffer [0.86 ml, 1.6 *M*, 10 m*M* $MnCl_2$, bovine serum albumin (BSA, 0.8 mg/ml), pH 8], DTT solution (0.35 ml, 100 m*M*), and alkaline phosphatase (40 μl) are added. To the mixture UDP-Glc (200 mg, 0.328 mmol) is added, followed by the activated enzyme. Enzyme $\alpha(1 \rightarrow 4)$-GalT/UDP-4′-epimerase is activated by combining water (1.6 ml), buffer (0.32 ml), DTT (0.128 ml), and enzyme (0.506 ml) and incubating the mixture for 0.5 h at room temperature. The reaction is incubated at 37° for 2 days and then chromatographed on C_{18} (0–50% methanol). The product is collected, concentrated, and lyophilized from water. ES FTICR MS and ^{1}H NMR indicate 85% conversion. The procedure is repeated for the product, using half the quantity of all reagents to give P^k-trisaccharide cluster **2** (224 mg, 88%). ES FTICR MS indicates 100% conversion. ^{1}H NMR [DMSO-d_6–D_2O, 19:1 (v/v)]: δ 4.77 (d, 10 H, $J_{1'',2''} = 3.84$ Hz, H-1″), 4.72 and 4.71 (2 d, 10 H, $J_{1,2} = 9.2$ Hz, Hz, H-1, *R,S*-rotamers), 4.58–4.51 (m, 5 H, OCH), 4.31 and 4.30 (2 d, 10 H, $J_{1',2'} = 7.24$, H-1′, *R,S*-rotamers), 4.06–3.96 (m, 30 H, H-5″, OCH_2), 3.83–3.70 (m, 50 H, H-6a, OCH_2, H-4′, H-4″), 3.65–3.62 (m, 20 H, H-2″, H-6′a), 3.60–3.53 (m, 40 H, H-6b, H-5′, H-6′b, H-3″), 3.52–3.44 (m, 60 H, H-5, H-3′, H-6″ a, H-6″b, OCH_2, NCH_2), 3.34 (t, 10 H, $J_{2,3} \approx J_{3,4} = 8.4$ Hz, H-3), 3.28–3.20 (m, 30 H, H-4, NCH_2, SCH_2), 3.18–3.01 (m, 50 H, H-2, H-2′, NCH_2), 2.92–2.89 (m, 10 H, NCH_2), 2.60–2.52 (m, 10 H, SCH_2), 1.84 (s, 30 H, NAc), 1.77–1.68 (m, 10 H, CH_2), 1.52–1.46 (m, 10 H, CH_2), 1.38–1.32 (m, 10 H, CH_2), 1.25 and 1.22 (2 s, 40 H, CH_2).

Acknowledgments

We express our appreciation to Ms. M. J. Schur for production of recombinant cells for enzyme extraction; Ms. J. Sadowska for enzyme purification; and the Canadian Bacterial Diseases Network (CBDN) for financial support.

[9] Synthesis of Galactose-Containing Oligosaccharides through Superbeads and Superbug Approaches: Substrate Recognition along Different Biosynthetic Pathways

By JIANBO ZHANG, XI CHEN, JUN SHAO, ZIYE LIU, PRZEMYSLAW KOWAL, YUQUAN LU, and PENG G. WANG

Introduction

Galactosides as carbohydrate receptors play critical roles in biological recognition events.[1] α-Galactosyl (α-Gal) epitopes and globotriose are two representative galactosyloligosaccharides with therapeutic significance. α-Gal epitopes (Galα1,3Galβ1,4GlcOR, Fig. 1) are oligosaccharides abundantly expressed on the surface of pig vascular endothelial cells (1×10^7 epitopes per cell).[2] Their interactions with preexisting natural antibodies (anti-Gal) in human are believed to be the main cause of the hyperacute rejection (HAR) in pig-to-human organ xenotransplantation.[3] Neutralization of anti-Gal with free α-Gal and its derivatives is a promising treatment to overcome HAR.[4] Globotriose (Galα1,4Galβ1,4Glc; Fig. 2) is the carbohydrate moiety of a functional cell surface receptor, globotriaosylceramide (Gb_3; Galα1,4Galβ1,4GlcβOCer), which can be recognized by Shiga toxin (Stx)-producing *Escherichia coli* (STEC), including

[1] (a) A. Kobata, *Acc. Chem Res.* **26,** 319 (1993); (b) C.-H. Wong, R. L. Halcomb, Y. Ichikawa, and T. Kajimoto, *Angew. Chem. Int. Ed. Engl.* **34,** 412 (1995); (c) R. A. Dwek, *Chem. Rev.* **96,** 683 (1996); (d) A. Varki, R. D. Cummings, J. Esko, H. Freeze, and G. Hart (Eds.), "Essentials of Glycobiology." Cold Spring Harbor Laboratory Press, Cold Spring Harbor, NY, 1999.

[2] (a) D. K. C. Cooper, *Clin. Trans.* **6,** 178 (1992); (b) D. K. C. Cooper, A. H. Good, E. Koren, R. Oriol, A. J. Malcolm, R. M. Ippolito, F. A. Neethling, Y. Ye, E. Romano, and N. Zuhdi, *Transplant. Immunol.* **1,** 198 (1993); (c) U. Galili, *Sci. Med.* **5,** 28 (1998).

[3] (a) U. Galili, *Blood Cells* **14,** 205 (1988); (b) U. Galili, *Immunol. Ser.* **55,** 355 (1992); (c) U. Galili, *Immunol. Today* **14,** 480 (1993); (d) U. Galili, *Springer Semin. Immunopathol.* **15,** 155 (1993); (e) D. K. C. Cooper, E. Koren, and R. Oriol, *Immunol. Rev.* **141,** 31 (1994); (f) M. S. Sandrin, H. A. Vaughan and I. F. C. McKenzie, *Transplant. Rev.* **8,** 134 (1994); (g) B. E. Samuelsson, L. Rydberg, M. E. Breimer, A. Backer, M. Gustavsson, J. Holgersson, E. Karlsson, A.-E. Uyterwaal, T. Cairns, and K. Welsh, *Immunol. Rev.* **141,** 151 (1994); (h) R. Hamadeh, U. Galili, P. Zhou, and J. Griffiss, *Clin. Diagn. Lab. Immunol.* **2,** 125 (1995); (i) X. Chen, P. R. Andreana, and P. G. Wang, *Curr. Opin. Chem. Biol.* **3,** 650 (1999).

[4] U. Galili, *Biochimie* **83,** 557 (2001).

0076-6879/03 $35.00

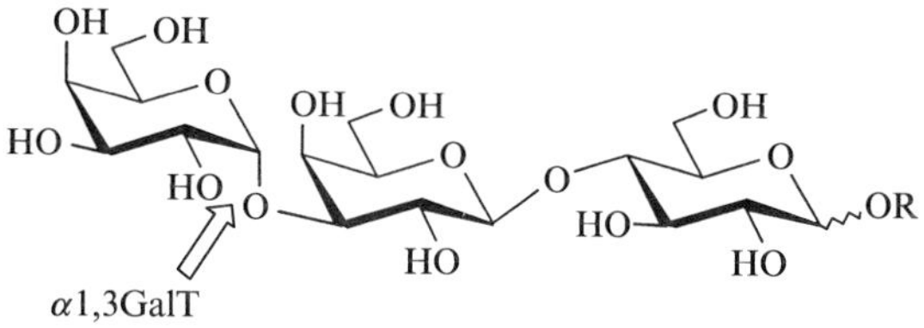

FIG. 1. Structure of α-Gal (Galα1,3Galβ1,4Glc) epitope and the linkage created by α1,3-galactosyltransferase (α1,3GalT).

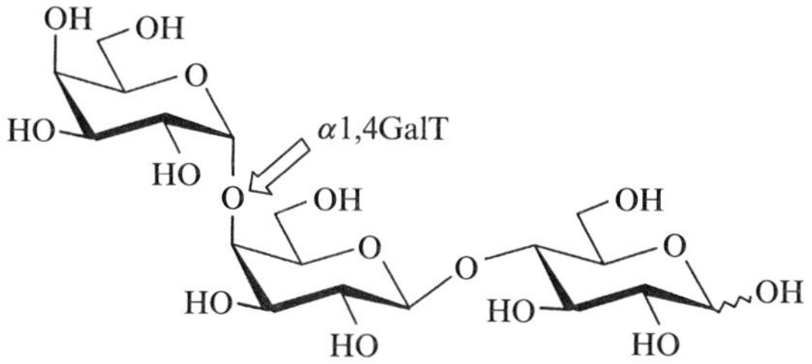

FIG. 2. Structure of globotriose (Galα1,4Galβ1,4Glc) and the linkage created by α1,4GalT.

O157:H7.[5] Research on Stx has revealed that Stx B subunit pentamer binds to the globotriose on human cells in the first stage of infection. This specific binding facilitates toxin entry into the host cell, resulting in cessation of protein synthesis, ultimately causing various secondary complications such as septic shock, multiple organ failure, and mortality.[6] In addition, globotriose found in the lipooligosaccharides of the bacterial pathogens *Neisseria meningitidis* immunotype L1 and *Neisseria gonorrhoeae* participates in the invasion of these pathogens into mammalian cells.[7]

[5] (a) C. A. Lingwood, *Biochim. Biophys. Acta* **1455,** 375 (1999); (b) J. C. Paton and A. W. Paton, *Clin. Microbiol. Rev.* **11,** 450 (1998); M. A. Karmali, M. Petric, C. Lim, P. C. Fleming, G. S. Arbus, and H. Lior, *J. Infect. Dis.* **151,** 775 (1985).

[6] (a) M. P. Jackson, *Microb. Pathog.* **8,** 235 (1990); (b) J. P. Nataro, and J. B. Kaper, *Clin. Microbiol. Rev.* **11,** 142 (1998); (c) J. C. Paton and A. W. Paton, *Clin. Microbiol. Rev.* **11,** 450 (1998).

[7] (a) R. J. P. M. Scholten, B. Kuipers, H. A. Valkenburg, J. Kankert, W. D. Zollinger, and J. T. Poolman, *J. Med. Microbiol.* **41,** 236 (1994); (b) J. Kihlberg and G. Magnusson, *Pure Appl. Chem.* **68,** 2119 (1996).

Therefore, globotriose and its derivatives have been used as effective Stx neutralizers[8] or inhibitors to prevent pathogen invasion.[9]

It is clear that further studies on preventing HAR and protecting humans from pathogen attack require easy access to substantial amounts of α-Gal and globotriose as well as their analogs.[10] Chemical synthesis of these compounds suffers from relatively low conversion due to multiple protection–deprotection steps and tedious purification at each stage of the synthesis.[11] Leloir glycosyltransferases, on the other hand, are highly regio- and stereoselective for the glycosidic linkage formation, thereby allowing the straightforward synthesis of glycoconjugates.[12] However, the scale-up application of glycosyltransferases is limited by the high cost of associated sugar nucleotide donors, especially the unnatural sugar nucleotides. One approach to overcome this drawback is to imitate the natural glycosylation pathway by which the sugar nucleotides are recycled during the reaction. Another advantage of sugar nucleotide recycling is that the low concentration of sugar nucleotide prevents its inhibitory effect on the glycosyltransferase-catalyzed reaction and, finally, increases the synthetic efficiency and yield.[13]

Sugar nucleotide regeneration is a complicated process involving interactions between multiple enzymes along the biosynthetic pathway. Besides their natural acceptor substrates, glycosyltransferases normally can recognize a variety of acceptor derivatives and produce glycoside derivatives in high yields.[14] Moreover, the enzymes involved in the generation of sugar nucleotides can also tolerate donor derivatives. Therefore, a wider range of

[8] K. Nishikawa, K. Matsuoka, E. Kita, N. Okabe, M. Mizuguchi, K. Kino, S. Miyazawa, C. Yamasaki, J. Aoki, S. Takashima, Y. Yamakawa, M. Nishijima, D. Terunuma, H. Kuzuhara, and Y. Natori, *Proc. Natl. Acad. Sci. USA* **99,** 7669 (2002).

[9] (a) J. J. Donnelly and R. Rappulo, *Nat. Med.* **6,** 257 (2000); (b) G. Mulvey, P. I. Kitov, P. Marcato, D. R. Bundle, and G. D. Armstrong, *Biochimie* **83,** 841 (2001).

[10] P. Sears and C.-H. Wong, *Angew. Chem. Int. Ed. Engl.* **38,** 2300 (1999).

[11] (a) K. Koike, M. Sugimoto, S. Sato, Y. Ito, Y. Nakahara, T. Ogawa, *Carbohydr. Res.* **163,** 189–208 (1987); (b) K. C. Nicolaou, T. Caufield, H. Kataoka, T. Kumazawa, *J. Am. Chem. Soc.* **110,** 7910–7912 (1998); (c) D. Qui, R. R. Schmidt, *Liebigs Ann.* 217–224 (1992); (d) W. Zhang, J.-Q. Wang, J. Li, L.-B. Yu, P. G. Wang, *J. Carb. Chem.* **18,** 1009–1017; 1999.

[12] (a) M. M. Palcic, *Curr. Opin. Biotechnol.* **10,** 616 (1999); (b) P. Sears and C.-H. Wong, *Science* **291,** 2344 (2001); (c) X. Chen, P. Kowal, and P. G. Wang, *Curr. Opin Drug Discov. Dev.* **3,** 756 (2000); (d) J. Zhang, B. Wu, Z. Liu, P. Kowal, X. Chen, J. Shao, and P. G. Wang, *Curr. Org. Chem.* **5,** 1169 (2001).

[13] (a) C.-H. Wong, S. L. Haynie, and G. M. Whitesides, *J. Org. Chem.* **47,** 5418 (1982); (b) M. Gilbert, R. Bayer, A.-M. Cunningham, S. DeFrees, Y. Gao, D. C. Watson, N. M. Young, and W. W. Wakarchuk, *Nat. Biotechnol.* **16,** 769 (1998).

[14] (a) J. Fang, J. Li, X. Chen, Y. Zhang, J. Wang, Z. Guo, K. Brew, and P. G. Wang, *J. Am. Chem. Soc.* **120,** 6635 (1998); (b) J. Zhang, P. Kowal, J. Fang, P. Andreana, and P. G. Wang, *Carbohydr. Res.* **337,** 969 (2002).

glycosides would be generated, starting from donor derivatives and acceptor derivatives. To provide convenient tools for glycobiologists and carbohydrate chemists, we have developed "superbeads" and "superbug" technologies and successfully applied them as independent toolkits for the synthesis of galactose-containing oligosaccharides.[15]

UDP-Gal Superbeads

An immobilized enzyme system has the advantages of ease of product separation, increased stability, reusability, and improved kinetics.[16] Superbeads, which are essentially agarose resin-immobilized enzymes along the biosynthetic pathway for glycoconjugate synthesis, inherit these advantages. This technology for sugar nucleotide regeneration involves the following steps: (1) cloning and overexpression of individual N-terminal hexahistidine (His_6)-tagged enzymes along the sugar nucleotide biosynthetic pathway, (2) stepwise testing of the activity of the recombinant enzymes, (3) coimmobilizing onto nickel-nitrilotriacetate (Ni^{2+}-NTA) beads, and (4) combining the superbeads with glycosyltransferases (either on the beads or in solution) for glycoside synthesis (Scheme 1).

Construction of in Vitro *Biosynthetic Pathway of UDP-Gal*

The biosynthetic pathway for galactoside synthesis involves at least one galactosyltransferase and four sugar nucleotide regeneration enzymes (Scheme 2): Galactokinase (GalK, EC 2.7.1.6) phosphorylates galactose to galactose 1-phosphate (Gal-1-P) and consumes one molecule of phosphoenolpyruvate (PEP). Galactose-1-phosphate uridylyltransferase (GalPUT, EC 2.7.7.10) and glucose-1-phosphate uridylyltransferase (GalU, EC 2.7.7.9) then work together to produce UDP-Gal from Gal-1-P and UTP through intermediates glucose 1-phosphate (Glc-1-P) and UDP-Glc. Pyruvate kinase (PykF, EC 2.7.1.40) is responsible for the formation of UTP from UDP, the by-product of the galactosylation, with the consumption of the second molecule of PEP.

[15] (a) X. Chen, J. Fang, J. Zhang, Z. Liu, J. Shao, P. Kowal, P. Andreana, and P. G. Wang, *J. Am. Chem. Soc.* **123,** 2081 (2001); (b) X. Chen, J. Zhang, P. Kowal, Z. Liu, P. R. Andreana, Y. Lu, and P. G. Wang, *J. Am Chem. Soc.* **123,** 8866 (2001); (c) X. Chen, Z. Liu, J. Zhang, W. Zheng, P. Kowal, and P. G. Wang, *ChemBioChem* **3,** 47 (2002).

[16] (a) N. L. S. Clair and M. A. Navia, *J. Am. Chem. Soc.* **114,** 7314 (1992); (b) J. Thiem and T. Wiemann, *Synthesis-Stuttgart* 141 (1992); (c) L. Bulow and K. Mosbach, *Trends Biotechnol.* **9,** 226 (1991); (d) L. Bulow, *Eur. J. Biochem.* **163,** 443 (1987); (e) C. Suge and C. Merienne, *Carbohydr. Res.* **151,** 147 (1986).

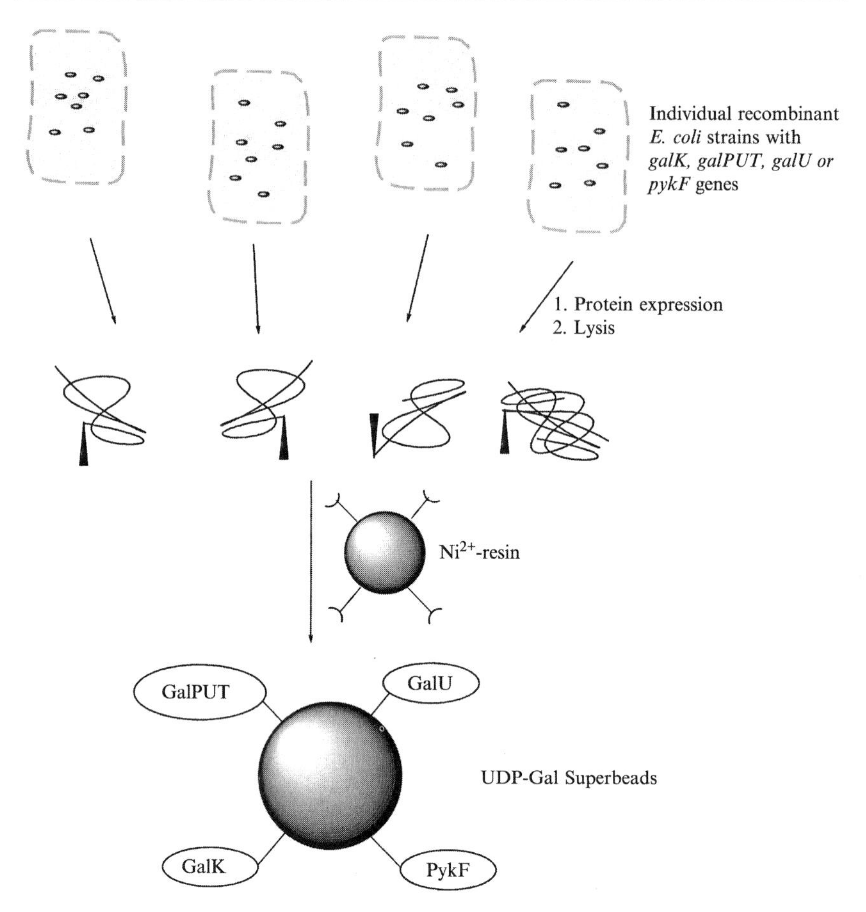

SCHEME 1. Preparation of UDP-Gal regeneration superbeads.

To construct the UDP-Gal regeneration cycle, genes for these four enzymes are individually amplified from the *E. coli* K-12 genome, cloned into the pET15b vector, and expressed in *E. coli* BL21 (DE3) with N-terminal His_6 tags.[15a,c,17]

The feasibility of UDP-Gal regeneration was tested by the combined activity of these biosynthetic enzymes and a recombinant α-1,3-galactosyltransferase (α1,3GalT). In the radioactivity assays, ^{3}H-labeled galactose (*Gal) is transferred to the acceptor substrate, $LacO(CH_2)_7CH_3$ (Lac-grease), via a stepwise combination of the action of several enzymes

[17] Z. Liu, J. Zhang, X. Chen, and P. G. Wang, *ChemBioChem* **3,** 348 (2002).

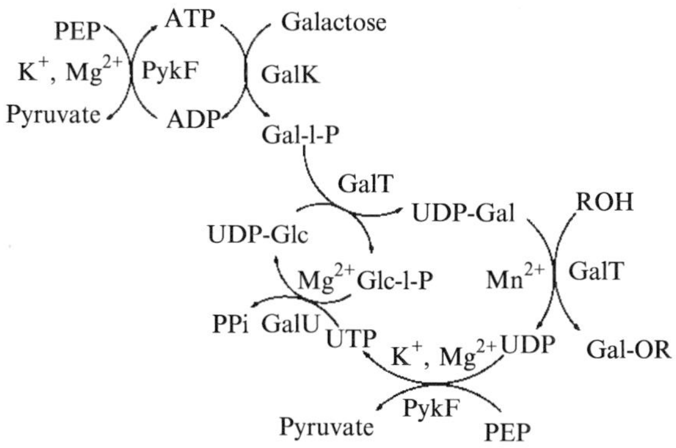

PEP ATP Galactose
K^+, Mg^{2+} PykF GalK
Pyruvate ADP Gal-l-P
GalT
UDP-Glc UDP-Gal ROH
Mg^{2+} Glc-l-P Mn^{2+} GalT
PPi GalU UTP K^+, Mg^{2+} UDP Gal-OR
PykF
Pyruvate PEP

2 PEP + Gal + ROH ⟶ 2 Pyruvate + PPi + Gal-OR

SCHEME 2. Biosynthetic pathway for galactosides with regeneration of UDP-Gal.

along the pathway. After the reactions, *Gal was removed from the reaction mixture and the radio-labeled product *Galα1, 3LacO$(CH_2)_7CH_3$ having hydrophobic moieties would bind to the Sep-Pak C_{18} cartridge and be eluted with methanol for radioactive counting. The second method of testing the regeneration cycle is to quantify the formation of α-Gal product by high-performance liquid chromatography (HPLC) with lactose as an acceptor. The high efficiency of each enzyme in the UDP-Gal regeneration cycle is demonstrated in stepwise experiments (steps 1–5 in Table I). Higher yields in this assay might represent the preference of the transferase for lactose (Lac) over Lac-grease.

Construction of UDP-Gal Superbeads

UDP-Gal regeneration beads (superbeads) are obtained by incubating Ni^{2+}-NTA resin with a mixture of the cell lysate consisting of the same number of activity units of His_6-tagged recombinant GalK, GalPUT, GalU, and PykF (Scheme 1). When combined with one or more galactosyltransferase(s), either on the beads or in solution, these UDP-Gal regeneration beads can catalyze the synthesis of a wider range of galactosides.

Synthesis of Galactose-Containing Oligosaccharides with UDP-Gal Superbeads

By using UDP-Gal regeneration beads in combination with β1,4-galactosyltransferases, LacNAc is synthesized from N-acetylglucosamine (GlcNAc) and galactose (Table II, entry 1). Furthermore, α-Gal and

TABLE I

HPLC Analysis for Production of Galα1,3LacOH with Purified Recombinant Enzymes of Biosynthetic Pathway[a]

Step	Enzyme(s)	Starting material	Product (%)
1	α1,3GalT	UDP-Gal + Lac	95
2	GalPUT + α1,3GalT	Gal-1-P + UDP-Glc + Lac	95
3	GalK + GalPUT + α1,3GalT	ATP + Gal + UDP-Glc + Lac	95
4	GalK + GalPUT + α1,3GalT + GalU	ATP + Gal + UTP + Glc-1-P (cat.) + Lac	90
5	GalK + GalPUT + α1,3GalT + GalU + PykF	ATP + Gal + PEP + UDP (cat.) + Glc-1-P (cat.) + Lac	90

[a]See text for abbreviations.

TABLE II

Syntheses of Galactoside Analogs with UDP-Gal Regeneration Superbeads

Entry	Gal derivative	GalT	Acceptor	Product		Yield (%)
1	Gal	α1,3GalT			1	92
2	Gal	α1,3GalT			2	85
3	Gal	α1,4GalT			3	86
4	Gal[a]	α1,3GalT β1,4GalT			4	76
5	2-Deoxy-Gal	α1,3GalT			5	69
6	[1-^{13}C]Gal	α1,3GalT			6	83
7	[1-^{13}C]Gal[a]	α1,3GalT β1,4GalT			7	95

[a] Two equivalents; others are 1 equivalent.

globotriose derivatives are synthesized from other oligosaccharide acceptors such as β-benzyllactoside (LacβOBn) and β-methyllactoside (LacβOMe) (Table II, entries 2, 3, 5, and 6). In addition, the UDP-Gal regeneration beads can be used in combination with multiple galactosyltransferases. For example, the combination of α1,3GalT and β1,4GalT with the superbeads can generate Galα1,3GAlβ1,4GlcNAc sequence starting from

GlcNAc and two equivalents of galactose (Table II, entries 4 and 7). Another powerful synthetic potential of the superbeads is that they can also synthesize unnatural sugar nucleotides and oligosaccharide derivatives with modifications at the nonreducing end of the sugar residues. When 2-deoxygalactose or 1-^{13}C-labeled galactose is used as starting monosaccharide instead of galactose, a novel 2″-deoxy-α-Gal epitope or a 1-^{13}C-labeled α-Gal is generated (Table II, entries 5–7). These labeled α-Gal derivatives are important tools to probe the interaction of α-Gal epitopes with anti-Gal antibodies. From these results, we conclude that the enzymes in the UDP-Gal regeneration cycle can accept different galactose derivatives as well as natural galactose.

Acceptor specificity studies on β1,4GalT, α1,3GalT, and LgtC have shown that these galactosyltransferases can utilize a broad spectrum of acceptors.[14,18] However, few data are available on the donor specificity of these glycosyltransferases because of the limited availability of the sugar nucleotide derivatives. Our superbeads technology is a tool to analyze the donor substrate specificity for a certain glycosyltransferase provided that a sugar nucleotide can be processed by the regeneration enzymes. Superbeads, therefore, can be used for the study of both the acceptor and substrate specificities.

UDP-Gal Superbug

Because the expression and purification of multiple enzymes individually are laborious and the purification process may result in decreases in enzymatic activities, whole cell reactions without isolating enzymes have been developed for carbohydrate synthesis.[19] After demonstration of the UDP-Gal synthetic pathway and the successful syntheses with recombinant enzymes on superbeads, we focused on using whole *E. coli* cells as biocatalysts. In this approach, genes encoding all related proteins in the biosynthetic pathway of galactoside are cloned in tandem into a single vector and transformed into the *E. coli* expression host. In contrast to

[18] (a) C. H. Hokke, A. Zervosen, L. Elling, D. H. Joziasse, and D. H. van Den Eihnden, *Glycoconj. J.* **13,** 687 (1996); (b) J. W. Fang, X. Chen, W. Zhang, A. Janczuk, and P. G. Wang, *Carbohydr. Res.* **329,** 873 (2000); (c) X. Chen, W. Zhang, J.-Q. Wang, J. W. Fang, and P. G. Wang, *Biotech. Prog.* **16,** 595 (2000); (d) K. Sujino, T. Uchiyama, O. Hindsgaul, N. O. L. Seto, W. Wakarchuk, and M. M. Palcic, *J. Am. Chem. Soc.* **122,** 1261 (2000).

[19] (a) S. Koizumi, T. Endo, K. Tabata, and A. Ozaki, *Nat. Biotechnol.* **16,** 847 (1998); (b) T. Endo, S. Koizumi, K. Tabata, S. Kakita, and A. Ozaki, *Carbohydr. Res.* **316,** 179 (1999); (c) K. Tabata, S. Koizumi, T. Endo, and A. Ozaki, *Biotech. Lett.* **22,** 479 (2000); (d) T. Endo, S. Koizumi, K. Tabata, and A. Ozaki, *Appl. Microbiol. Biotechnol.* **53,** 257 (2000); (e) T. Endo, S. Koizumi, K. Tabata, S. Kakita, and A. Ozaki, *Carbohydr. Res.* **330,** 439 (2001).

superbeads, superbug contains the necessary galactosyltransferases to accomplish the full biosynthetic cycle. Such a bacterial strain capable of simultaneous galactoside production and UDP-Gal regeneration has been named "UDP-Gal superbug" in our laboratory.[15b,c]

CKTUF Superbug: Synthesis of Globotriose and Its Derivatives

The superbug for globotrioside synthesis is constructed according to the biosynthetic pathway shown in Scheme 3B. To prevent the degradation of acceptor by β-galactosidase activity in the host strain of the pET system,

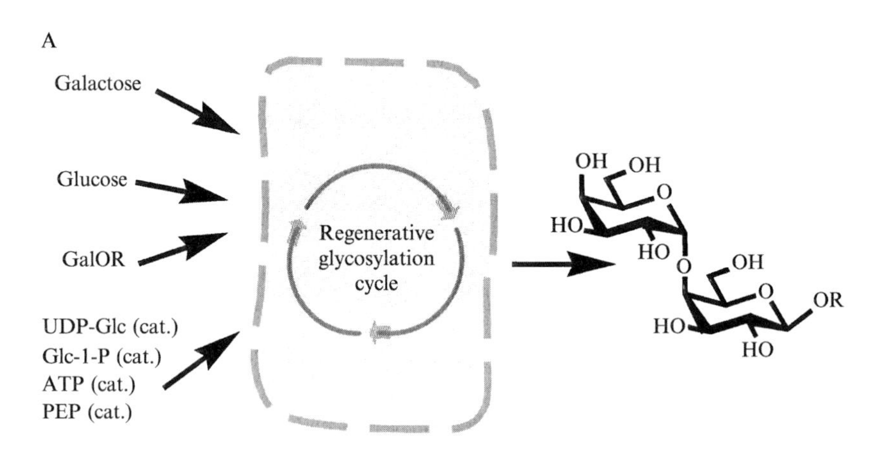

B

PEP, K$^+$, Mg^{2+}, Pyruvate, PykF, ATP, ADP, Galactose, GalK, Gal-l-P, GalT, UDP-Gal, UDP-Glc, Mg^{2+} Glc-l-P, PPi GalU, UTP, K$^+$, Mg^{2+}, UDP, Pyruvate, PykF, PEP, Galβ1,4GlcOH, Mn^{2+}, LgtC, Galα1,4Galβ1,4GlcOH

SCHEME 3. Synthesis of globotriose derivatives with superbug CKTUF. (A) Schematic presentation of the synthesis and (B) biosynthetic pathway of globotriose synthesis.

and to eliminate the need of isopropyl-1-thio-β-D-galactopyranoside (IPTG), β-galactosidase negative strain *E. coli* NM522 or DH5α and the heat-inducible pLDR20 vector are best suited for superbug application. The multienzyme plasmid construction is accomplished with the subsequent insertion of *galU*, *lgtC*, *pykF*, and *galPUT* + *galK* genes, respectively, with the corresponding ribosomal binding sites (rbs) and N-terminal His_6 tags into the pLDR20 plasmid. Each gene (except the *galK* gene, which has a natural rbs in the coding sequence of the upstream *galPUT* gene in the *gal* operon) is preceded by a Shine–Dalgarno sequence for ribosome binding to assure adequate translation. Because *galK* and *galPUT* exist in the same *gal* operon and close to each other, they are cloned together into the pET15b vector and then into the pLDR20 vector (Scheme 4). The plasmid pLDR20-CKTUF was then transformed into the NM522 strain to form globotriose-producing cells. Sodium dodecyl sulfate–polyacrylamide gel electrophoresis (SDS–PAGE) of this superbug indicates that all five enzymes are expressed. The activities of these enzymes are further confirmed by the synthesis of globotriose and its derivatives.

It should be mentioned that our whole cell synthesis of globotriose oligosaccharides is actually a two-step procedure, which is distinct from the commonly employed fermentative processes.[20] The first step involves the growth of the recombinant *E. coli* NM522 cells and the subsequent expression of the enzymes. In the second step the harvested cells are permeabilized by surfactant treatment and a multiple freeze–thaw procedure. They are then employed as catalysts in the reaction mixture that contains glucose, galactose, acceptor (lactose or its derivative), and catalytic amounts of PEP, ATP, Glc-1-P, and UDP-Glc. This two-step process avoids the possible inhibition of cell growth by substrates and product; moreover, it allows the use of high cell concentrations (i.e., high catalyst concentrations) and facile manipulation of substrate concentrations. The permeabilization allows for better transfer of substrates and sugar products into and out of the cells.[18c]

The CKTUF superbug has been compared with purified recombinant enzymes for capacity to synthesize globotriose and its derivatives[14b] (Table III). Good acceptors for purified LgtC are accepted well by the CKTUF superbug. For most substrates shown, the purified enzyme reactions give higher yields, probably because of the higher concentration of sugar nucleotide in the reaction mixtures. The most notable exception to this

[20] (a) R. Patnaik and J. Liao, *Appl. Environ. Microbiol.* **60,** 3903 (1994); (b) J. T. Kealey, L. Liu, D. V. Santi, M. C. Betlack, and P. J. Barr, *Proc. Natl. Acad. Sci. USA* **95,** 505 (1998); (c) K. Li and J. W. Frost, *J. Am. Chem. Soc.* **120,** 10545 (1998).

TABLE III

SYNTHESES OF GLOBOTRIOSE DERIVATIVES WITH RECOMBINANT ENZYMES AND WHOLE CELLS[a]

Entry	Acceptor	Product		Yield (%)	
				Whole cells	Enzyme
1	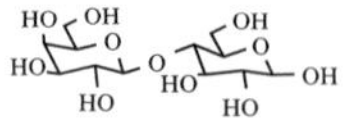	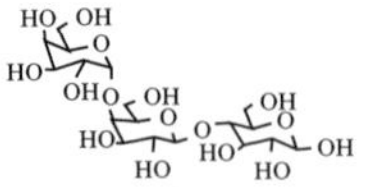	**8**	75	92
2	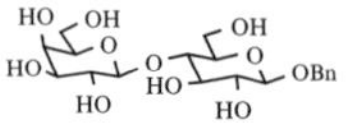	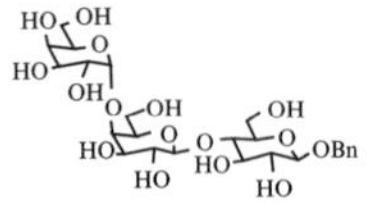	**3**	85	66
3	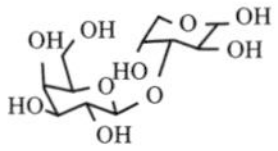	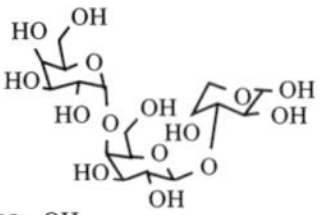	**9**	60	77
4	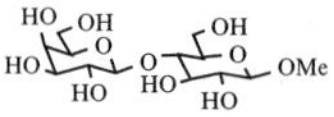	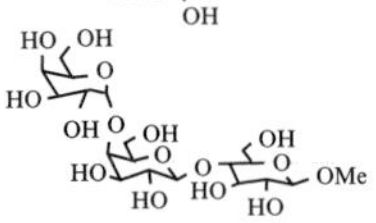	**10**	50	84
5	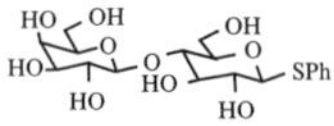	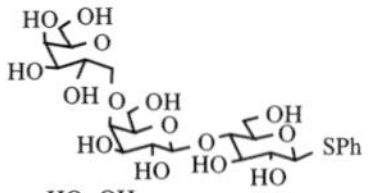	**11**	50	81
6	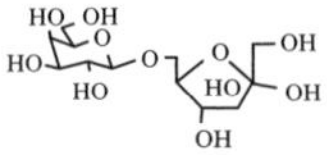	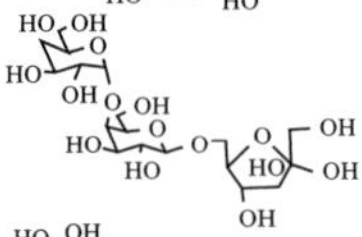	**12**	45	45
7	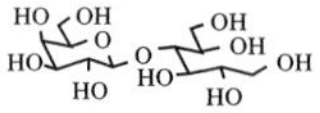	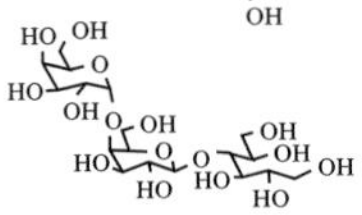	**13**	20	10
8	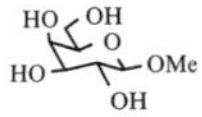	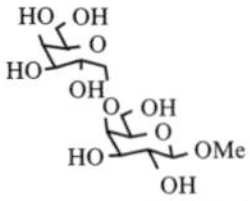	**14**	10	5

[a] See Ref. 14b for detailed reaction conditions for globotriose synthesis with purified enzymes. The whole cell-catalyzed reactions are described in the experimental section.

is benzyllactoside (LacOBn) (Table III, entry 2). This acceptor proved better in the whole cell-catalyzed reaction. In this case, the cell-confined LgtC transferase may prefer benzyllactose, as it is a better mimic of the natural lactosylceramide substrate.[21]

[21] W. W. Wakarchuk, A. Cunningham, D. C. Watson, and N. M. Young, *Protein Eng.* **11,** 295 (1998).

αKTUF Superbug: Synthesis of α-Gal Epitope

Given the efficiency of the superbug system, and the increasing availability of recombinant glycosyltransferases, various superbug strains can be conveniently constructed for specific glycosyl linkages. For example, once the *lgtC* gene is replaced by the gene encoding α1,3Gal T, the new superbug pLDR20-αKTUF is generated to synthesize α-Gal epitope. Similar to the globotriose superbug, plasmid pLDR20-αKTUF is obtained by sequential insertion of the genes encoding the five enzymes involved in the synthesis of α-Gal epitope into the pLDR20 vector. The resulting plasmid is transformed into NM522 to obtain the α-Gal superbug.[15c]

Another feature of this superbug reaction is that deoxygalactose can be utilized by superbug as a substitute for galactose to generate deoxy-α-Gal. The substrate tolerance of the superbug is being investigated further.

Conclusion

The superbeads and superbug are versatile tools that have shown utility in synthesis of galactoside analogs. At present, we are constructing other sugar–nucleotide regeneration systems to synthesize diverse glycoconjugates and unnatural derivatives. The success of superbeads and the superbug technology offers the promise of easily synthesizing diversified oligosaccharides and glycoconjugates with uncommon or even unnatural sugar residues to meet increasing biochemical demands.

On the other hand, the substrate recognition between substrates and enzymes has a dual function. With the superbeads and superbug approaches, it holds promise for the screening of possible inhibitors in the biosynthetic pathway, if the substrates cannot be adopted as suitable donor or acceptors.

General Methods and Procedures

Materials and Chemical Reagents

The polymerase chain reaction (PCR) purification kit, QIAEX II gel extraction kit, QIAamp tissue kit, and DNA miniprep spin kit are from Qiagen (Santa Clarita, CA). All restriction enzymes, *Taq* DNA polymerase, 1-kb DNA ladder, and T4 DNA ligase are obtained from Promega (Madison, WI). Sodium chloride and ScintiVerse BD are from Fisher Scientific (Pittsburgh, PA). IPTG, ampicillin, [6-^{3}H]galactose, Dowex 1X8 anion-exchange resin, PEP, ATP, UDP-Glc, Glc-1-P, HEPES hemisodium salt, $MgCl_2$, $MnCl_2$, KCl, β-D-lactose, β-methyllactoside, lactulose, lactitol, β-methyllactoside, *3-O-β*-D-galactopyranosyl-D-arabinose, and bovine

β1,4-GalT are obtained from Sigma (St. Louis, MO). β-Benzyllactoside and phenyl-1-thio-β-D-lactoside have been synthesized previously.[22] All other chemicals are obtained in reagent grade from commercially available sources.

Bacterial Strains and Plasmids

Plasmid vector pLDR20 (ATCC 87205) is purchased from American Tissue Culture Collection (ATCC, manassas, VA). Plasmid vector pET15b and *E. coli* competent strain BL21 (DE3) [F^- *ompT* $hsdS_B$ $(r_B^- m_B^-)$ *gal dcm* (DE3)] are from Novagen (Madison, WI). Plasmids pET15b-*α1,3GalT*, pET15b-*lgtC*, pET15b-*galK*, pET15b-*galPUT*, pET15b-*galU*, and pET15b-*pykF* are constructed as described previously.[14,15c,17] *Escherichia coli* competent strain DH5α (*lacZΔM15 hsdR recA*) is from Gibco–BRL Life Technology (Rockville, MD). *Escherichia coli* competent strain NM522 {*supE thi-1 Δ(lac-proAB) Δ(mcrB-hsdSM)5*$(r_K^- m_K^+)$ *[F proAB* $lacI^q$*ZΔM15]*} is from Stratagene (La Jolla, CA). Chromosomal DNA of *Neisseria meningitidis* MC58(L3) is a kind gift from M. Gilbert (Institute for Biological Sciences, National Research Council of Canada, Ottawa, ON, Canada).

HPLC Analysis for α-Gal Production

(See Table I.) The assays are carried out at room temperature (24°) for 3 days in a final volume of 250 ml in 2-[4-(2-hydroxyethyl)-l-piperazinyl] ethanesulfonic acid (HEPES) buffer (100 m*M*, pH 7.0) containing starting materials and enzyme solutions (20 mU for each enzyme). The reaction is stopped by heating in boiling water for 10 min to precipitate the enzymes. After centrifugation for 20 min, the supernatant is analyzed by HPLC [Microsorb, 100-Å pore size amino column with CH_3CN–H_2O (65:35, v/v) as the eluent] with a refractive index (RI) detector.

Starting materials: step 1, UDP-Gal (50 m*M*), lactose (50 m*M*), $MnCl_2$ (10 m*M*); step 2, Gal-1-P (50 m*M*), UDP-Glc (50 m*M*), lactose (50 m*M*), $MnCl_2$ (10 m*M*); step 3, ATP (50 m*M*), Gal (50 m*M*), UDP-Glc (50 m*M*), lactose (50 m*M*), $MnCl_2$ (10 m*M*); step 4, ATP (5 m*M*), Gal (5 m*M*), UTP (5 m*M*), Glc-1-P (1 m*M*), lactose (5 m*M*), $MgCl_2$ (10 m*M*), $MnCl_2$ (10 m*M*); step 5, ATP (5 m*M*), Gal (5 m*M*), PEP (5 m*M*), Glc-1-P (1 m*M*), UDP (1 m*M*), lactose (5 m*M*), $MgCl_2$ (10 m*M*), KCl (100 m*M*), $MnCl_2$ (10 m*M*).

[22] (a) W. Zhang, J. Wang, J. Li, L. Yu, and P. G. Wang, *J. Carbohydr. Chem.* **18,** 1009 (1999); (b) J. Fang, Ph.D. thesis. Wayne State University, Detroit, MI, 2000.

Synthesis of Galactose-Containing Oligosaccharides with UDP-Gal Superbeads

(See Table II; and the supplementary materials in Ref. 15a for detailed reaction conditions.) Superbeads are prepared by incubation of Ni^{2+}-NTA agarose with cell lysate of α1,3GalT, β1,4GalT, or LgtC (5 ml, 5 U), washed with Tris-HCl buffer (20 m*M*, pH 8.0) containing 0.5 *M* NaCl, and added to a mixture of starting materials in HEPES buffer (100 m*M*, pH 7.5). The reaction is carried out in a rotor mixer for 4 days at room temperature (24), when thin-layer chromatographic analysis [2-propanol–NH_4OH–H_2O, 7:3:2 (v/v/v)] indicates that the reaction is complete. After the reaction, the superbeads are separated from the reaction mixture by centrifugation and washed for another batch of reaction. The mixture is passed through Dowex 1-X8 (Cl^-) anion-exchange resin and purified by gel-permeation chromatography (Sephadex G-15; Sigma) with water as the mobile phase. The product-containing fractions are pooled and lyophilized.

The starting materials are acceptor (0.24 mmol), ATP (13.2 mg, 24 μmol), PEP (91.2 mg, 0.48 mmol), UDP (10 mg, 24 μmol), Glc-1-P (7.3 mg, 24 μmol), Gal (54 mg, 0.3 mmol), $MgCl_2$ (10 m*M*), $MnCl_2$ (10 m*M*), and KCl (100 m*M*); the total volume is 30 ml. For entry 1 (Table II), the superbeads are added directly to a reaction mixture containing 5 U of commercial available β1,4GalT. For entries 2, 3, 5, and 6 (Table II), superbeads (5 ml, containing 7.5 U of individual enzymes) are obtained by incubation with 16 ml of cell lysate mixture of GalK, GalPUT, GalU, and PykF with a volume ratio of 4:1:1:2. For entries 4 and 7 (Table II), the reaction mixture contains 2 equivalents of Gal or its derivatives and 4 equivalents of PEP. For the gram-scale synthesis of Galα1,3LacOBn: superbeads (40 ml) are obtained by incubation with 120 ml of cell lysate mixture (GalK, GalPUT, GalU, and PykF at a volume ratio of 4:1:1:2) and subsequent incubation with cell lysate of α1,3GalT (40 ml, 40 U). Starting materials are LacOBn (1 g, 2.4 mmol), ATP (132 mg, 240 μmol), PEP (912 mg, 4.8 mmol), UDP (100 mg, 240 μmol), Glc-1-P (73 mg, 240 μmol), μmol), Gal (540 mg, 3 mmol), $MgCl_2$ (10 m*M*), $MnCl_2$ (10 m*M*), and KCl (100 m*M*), and the total volume is 250 ml.

Construction of Superbug CKTUF-NM522

By using the preconstructed plasmids PET15b-*galK*, PET15b-*galPUT*, pET15b-*galU*, pET15b-*lgtC*, and pET15b-*pykF* as the PCR templates, the genes of the enzymes involved in the biosynthetic pathway of globotriose are subcloned one by one into the pLDR20 vector with the ribosomal binding site and the His_6 tag-encoding sequence preceding each gene (Scheme 4) to form the final plasmid pLDR20-CKTUF. This pLDR20 vector contains an ampicillin resistance gene, a p_R promoter, and a *c*I857 repressor gene. Because *galK* and *galPUT* exist in the same *gal* operon and close to each other, they are cloned together into the pET15b vector and then into the pLDR20 vector. Briefly, the stepwise construction of plasmid pLDR20-CKTUF is as follows: first, the *galU* gene is amplified by PCR from pET15b-*galU* with primers *galU*-F′ (5′-CCGGATATCCCGCGGGTCGACAATAATTTTGTTTAACTTTAAGAAGG-3′) and *galU*-R′ (5′-GCATCGATGGTCTAGAGGATCCTTACCITAATGCCCATCTC-3′), which introduces the *Eco*RV, *Sac*II, and *Sal*I, or *Xba*I and *Cla*I restriction sites, respectively. The PCR product is digested with *Eco*RV and *Cla*I and inserted into the multiple-cloning site of the pLDR20 vector previously cut with the same enzymes. Successful cloning is verified by restriction mapping and the expression of GalU is confirmed by SDS–PAGE. Second, primers *lgtC*-F′ (5′-GGATCCATATGACTAGT GATATCAATAATTTTGTTTAACTTTAAGAAGG-3′) and *lgtC*-R′ (5′-TCCCCGCGGTCATCAGTGCGGGACGGCAAGTTTGCC-3′) are used to amplify the *lgtC* gene from plasmid PET15b-*lgtC*. The PCR product is digested and inserted into the *Eco*RV and *Sac*II restriction sites of the plasmid pLDR20-U to form plasmid pLDR20-CU. Third, the pET15b-*pykF* plasmid is digested with *Xba*I and *Cla*I, and the smaller fragment containing the *pykF* gene, ribosomal binding site, sequence for the N-terminal His_6 tag, and T7 terminator is purified and inserted into the *Xba*I and *Cla*I restriction sites of the plasmid pLDR20-CU to form plasmid pLDR20-CUF. Finally, the *galK* and *galPUT* genes are amplified from plasmid pET15b-*galKT* (constructed by inserting the gene sequence encoding both *galK* and *galPUT* into a pET15b vector between the *Nde*I and *Bam*HI restriction sites) by using the primers *galKT*-F′ (5′-TCCCCGCGGCCCGGGAATAATTTTGTTTAACTTTAAGAAGG-3′) and *galKt*-R′ (5′-CGCGTCGACTCAGCACTGTCCTGCTCCTTG-3′). The PCR product is digested and inserted into the *Sac*II and *Sal*I restriction sites of plasmid pLDR20-CUF to form plasmid pLDR20-CKTUF. This final plasmid pLDR20-CKTUF, harboring five genes, is subsequently transformed into competent NM522 cells to make NM522 (pLDR20-CKTUF) globotriose-producing cells (Scheme 4).

Synthesis of Globotriose and Its Derivatives with Superbug CKTUF-NM522

(See Table III.) Globotriose-producing superbug NM522 (pLDR20-CKTUF) is grown in 4-liter shake flasks. The expression of the target genes in the superbug is initiated by increasing the temperature from 30 to 40°.

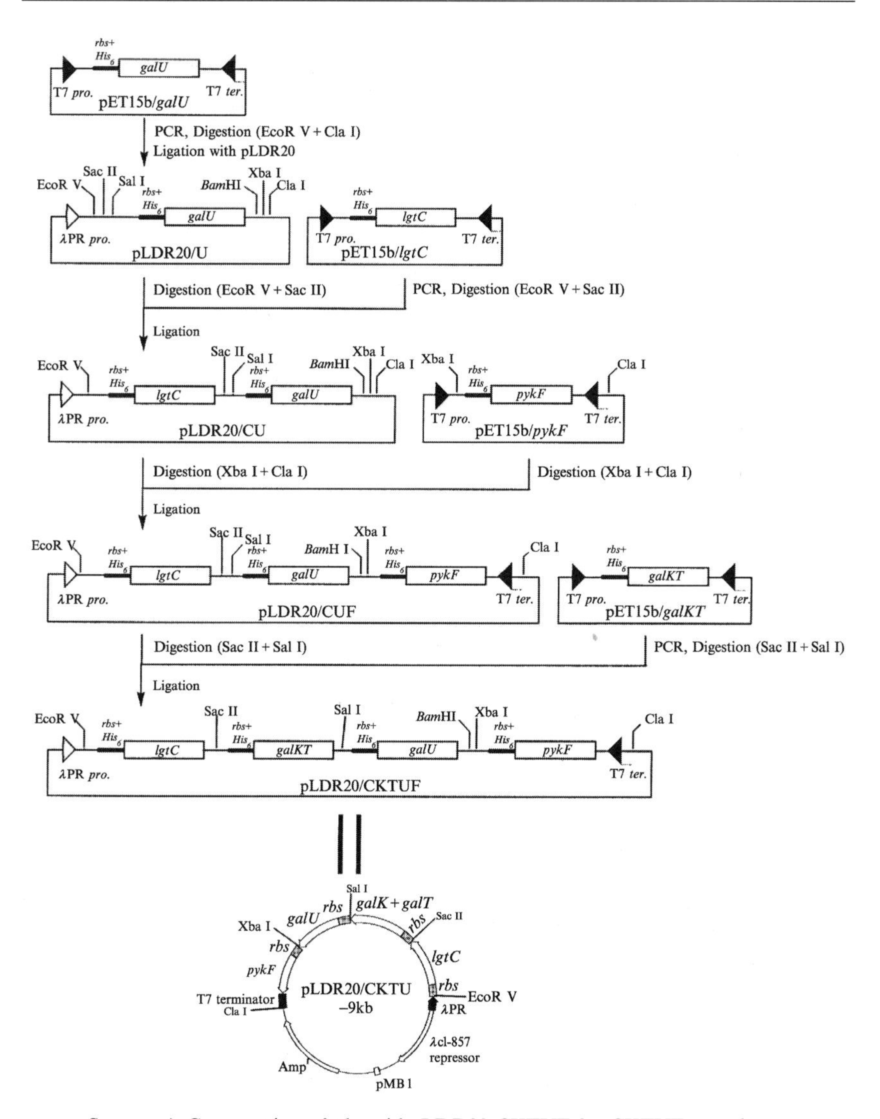

SCHEME 4. Construction of plasmid pLDR20-CKTUF for CKTUF superbug.

After being shaken at 40° for 3–3.5 h, the cells are separated from the medium by centrifugation (5000 *g* for 20 min at 4°) and suspended in 100 ml of Tris-HCl buffer (20 m*M*, pH 8.5) containing 1% Triton X-100. For better results, the cell suspension is freeze–thawed twice before being applied in the reaction. For small-scale analysis, the reaction is performed with 0.14 g (wet weight) of cells in a 1-ml reaction volume containing Gal (50 m*M*), Lac (25 m*M*), Glc (50 m*M*), PEP (5 m*M*), Glc-1-P (2 m*M*), UDP-Glc (2 m*M*), ATP (2 m*M*), $MgCl_2$ (10 m*M*), KCl (100 m*M*), $MnCl_2$ (10 m*M*), and HEPES (50 m*M*, pH 7.4). The reaction is carried out at room temperature and formation of the trisaccharide product is monitored by HPLC with a Microsorb 100-A pore size amino column using CH_3CN–H_2O (65:35, v/v) as eluent. To optimize the conditions, multiple 1-ml reactions are set up with different starting material compositions.

Large-scale production consists of two steps: culturing the superbug cells and producing trisaccharide as catalyzed by the cells. NM522 (pLDR20-CKTUF) cells are first grown at 30° in a 10-liter fermentor, and then enzyme expression is induced by increasing the temperature to 40° for 3 h. The cells are separated from the medium by centrifugation. The cell pellet (65 g, wet weight) is stored at −20° and freeze–thawed twice before use in the reaction. Gram-scale synthesis is performed with a variety of galactose or lactose derivatives as acceptors for the LgtC. For a typical synthesis reaction, to a 250-ml flask is added acceptor (2.92 mmol), Gal (1.05 g, 5.84 mmol), Glc (1.05 g, 5.84 mmol), PEP (111 mg, 0.584 mmol), ATP (129 mg, 0.234 mmol), UDP-Glc (143 mg, 0.234 mmol), Glc-1-P (72 mg, 0.234 mmol), and 12 ml of each of the following stock solutions: HEPES buffer (0.5 *M*, pH 7.4), $MnCl_2$ (0.1 *M*), $MgCl_2$ (0.1 *M*), and KCl (1 *M*). Superbug cells [12 g in 72 ml of Tris-HCl buffer (20 m*M*, pH 8.5) containing 1% Triton X-100, obtained from a 2-liter bacterial culture] are then added to bring the total reaction mixture volume to 120 ml. The reaction is agitated with a magnetic stirrer at room temperature (24°) for 36 h. The reaction is monitored by thin-layer chromatograpy (TLC) [2-proponal–H_2O–NH_4OH (7:3:2, v/v/v)] and HPLC. After 36 h, the reaction is stopped by heating the flask to 100° for 10 min. Insoluble components are sedimented by centrifugation at 5000 *g* for 20 min and the pellet is washed twice with 50 ml of deionized water. The combined supernatants are subsequently passed through an anion-exchange column and a cation-exchange column. The concentrated eluate is loaded onto a Sephadex G-15 gel-filtration column (120 × 4 cm) with water as the mobile phase. The desired fractions are pooled and lyophilized to give the derivatives of globotriose. The following compounds have been prepared according to this method.

*α-D-Galactopyranosyl-(1 → 4)-β-D-galactopyranosyl-(1 → 4)-D-glucopyranose (**8**).* (1.10 g, 75%) ^{1}H NMR (500 MHz, D_2O): δ 5.06 (d, $J=3.6$ Hz, 0.4 H), 4.78 (d, $J=4.1$ Hz, 1 H), 4.50 (d, $J=8.1$ Hz, 0.6 H), 4.34 (d, $J=7.6$ Hz, 1 H), 4.19 (t, $J=6.6$ Hz, 1 H), 3.87 (m, 2 H), 3.39–3.82 (m, 14.4 H), 3.11 (t, $J=8.6$ Hz, 0.6 H); ^{13}C NMR (125 MHz, D_2O): δ 103.41, 103.37, 100.46, 95.86, 91.94, 78.82, 78.71, 77.51, 75.58, 74.99, 74.56, 74.04, 72.30, 71.59, 71.35, 71.06, 70.96, 70.30, 69.28, 69.08, 68.71, 60.65, 60.53, 60.18, 60.06; mass spectrometry (MS) [fast atom bombardment (FAB)] 527 (M + Na$^+$); high-resolution MS (HRMS): calculated for $C_{18}H_{32}O_{16}Na$ (M + Na$^+$) 527.1588; found 527.1581.

*Benzyl-α-D-galactopyranosyl-(1 → 4)-β-D-galactopyranosyl-(1 → 4)-β-D-glucopyranoside (**3**).* (1.47 g, 85%) ^{1}H NMR (500 MHz, D_2O): δ 7.36–7.43 (m, 5 H), 4.89 (d, $J=4.1$ Hz, 1 H; d, $J=11.4$ Hz, 1 H), 4.71 (d, $J=11.4$ Hz, 1 H), 4.51 (d, $J=8.1$ Hz, 1 H), 4.46 (d, $J=8.1$ Hz, 1 H), 4.31 (t, $J=6.5$ Hz, 1 H), 3.48–3.98 (m, 16 H), 3.29 (t, $J=8.9$ Hz, 1 H); ^{13}C NMR (125 MHz, D_2O): δ 136.7, 129.02, 128.96, 128.7, 103.5, 101.2, 100.5, 78.8, 77.6, 75.7, 75.1, 74.7, 73.2, 72.4, 72.3, 71.7, 71.1, 71.0, 69.4, 69.1, 68.8, 62.7, 60.7, 60.6, 60.3; MS (FAB) 617 (M + Na$^+$); HRMS: calculated for $C_{25}H_{38}O_{16}Na$ (M + Na$^+$) 617.2058, found 617.2042.

*Methyl-α-D-galactopyranosyl-(1→4)-β-D-galactopyranosyl-(1→4)-β-D-glucopyranoside (**9**).* (756 mg, 50%) ^{1}H NMR (500 MHz, D_2O): δ4.79 (d, $J=4.1$ Hz, 1 H), 4.35 (d, $J=8.1$ Hz, 1 H), 4.25 (d, $J=8.1$ Hz, 1 H), 4.20 (t, $J=6.6$ Hz, 1 H), 3.40–3.89 (m, 16 H), 3.42 (s, 3 H), 3.14 (t, $J=8.6$ Hz, 1 H). ^{13}C NMR (125 M Hz, D_2O): δ 103.42, 103.18, 100.46, 78.78, 77.49, 75.58, 74.96, 74.60, 73.02, 72.30, 71.05, 70.96, 69.27, 69.08, 68.71, 60.64, 60.52, 60.15, 57.35. MS (FAB) 541 (M + Na$^+$); HRMS: calculated for $C_{19}H_{34}O_{16}Na$ (M + Na$^+$) 541.1745, found 541.1736.

*Phenyl-α-D-galactopyranosyl-(1→4)-β-D-galactopyranosyl-(1→4)-β-D-1-thioglucopyranoside (**10**).* (870 mg, 50%) ^{1}H NMR (500 MHz, D_2O): δ 7.24–7.41 (m, 5 H), 4.76 (d, $J=3.0$ Hz, 1 H), 4.32 (d, $J=7.6$ Hz, 1 H), 4.17 (t, $J=6.1$ Hz, 1 H), 3.85 (m, 2 H), 3.40–3.81 (m, 15 H), 3.22 (t, $J=9.1$ Hz, 1 H). ^{13}C NMR (125 MHz, D_2O): δ 132.00, 131.77, 129.45, 128.26, 103.33, 100.42, 87.24, 78.88, 78.29, 77.45, 75.95, 75.55, 72.25, 71.66, 71.00, 70.92, 69.25, 69.04, 68.68, 62.58, 60.60, 60.50, 60.17. MS (FAB) 619 (M + Na$^+$); HRMS: calculated for $C_{24}H_{36}O_{15}SNa$ (M + Na$^+$) 619.1673, found 619.1698.

*α-D-Galactopyranosyl-(1→4)-β-D-galactopyranosyl-(1→3)-β-D-arabinofuranose (**11**).* (830 mg, 60%) ^{1}H NMR (500 MHz, D_2O): δ5.11 (d, $J=3.6$ Hz, 0.4 H), 4.79 (d, $J=3.0$ Hz, 1 H), 4.44 (t, $J=8.1$ Hz, 1 H), 4.38 (d, $J=7.6$ Hz, 0.6 H), 4.21 (t, $J=6.6$ Hz, 1 H), 3.46–4.06 (m, 16 H). ^{13}C NMR (125 MHz, D_2O): δ101.34, 100.99, 100.40, 96.87, 92.61, 79.70, 77.48, 76.69, 75.55, 72.37, 70.94, 70.85, 70.37, 69.27, 69.09, 68.73, 67.12, 66.38,

66.02, 60.64. MS (FAB) 497 ($M + Na^+$); HRMS: calculated for $C_{17}H_{30}O_{15}Na$ ($M + Na^+$) 497.1482, found 497.1480.

*α-D-Galactopyranosyl-(1→4)-β-D-galactopyranosyl-(1→6)-β-D-fructofuranose (**12**).* (662 mg, 45%) ^{1}H NMR (500 MHz, D_2O): δ4.77 (d, $J = 4.1$ Hz, 0.6 H), 4.74 (d, $J = 3.6$ Hz, 0.4 Hz), 4.43 (d, $J = 8.1$ Hz, 0.6 H), 4.35 (d, $J = 7.6$ Hz, 0.4 H), 4.18 (t, $J = 6.6$ Hz, 1 H), 3.37–4.09 (m, 18 H). ^{13}C NMR (125 MHz, D_2O): δ103.16, 102.48, 101.04, 100.41, 100.33, 98.26, 85.24, 84.36, 80.89, 80.18, 77.63, 77.52, 77.02, 75.54, 74.77, 72.38, 72.23, 70.95, 70.84, 69.27, 69.08, 68.81, 68.72, 66.76, 66.14, 63.98, 63.08, 62.72, 62.52, 60.65, 60.58, 60.37. MS (FAB) 527 ($M + Na^+$); HRMS: calculated for $C_{18}H_{32}O_{16}Na$ ($M + Na^+$) 527.1588, found 527.1601.

*α-D-Galactopyranosyl-(1→4)-β-D-galactopyranosyl-(1→4)-D-gludititol (**13**).* (296 mg, 20%) ^{1}H NMR (400 MHz, D_2O): δ4.91 (d, $J = 3.2$ Hz, 1 H), 4.52 (d, $J = 7.3$ Hz, 1 H), 4.29 (t, $J = 6.5$ Hz, 1 H), 3.50–4.00 (m, 19 H), 1.28 (d, $J = 6.5$ Hz, 1 H). ^{13}C NMR (100 MHz, D_2O): δ103.72, 100.51, 80.15, 77.59, 75.46, 72.46, 72.19, 71.52, 71.28, 71.12, 69.87, 69.33, 69.19, 68.86, 62.76, 62.24, 60.76, 60.51. MS (FAB) 529 ($M + Na^+$).

*Methyl-α-D-galactopyranosyl-(1 → 4)-β-D-galactopyranoside (**14**).* (104 mg, 10%) ^{1}H NMR (500 MHz, D_2O): δ4.81 (d, $J = 3.6$ Hz, 1 H), 4.23 (d, $J = 8.1$ Hz, 1 H), 4.20 (t, $J = 6.1$ Hz, 1 H), 3.88 (m, 2 H), 3.76 (m, 2 H), 3.69 (m, 2 H), 3.54–3.64 (m, 3 H), 3.43 (s, 3 H), 3.38 (m, 2 H). ^{13}C NMR (125 MHz, D_2O): δ 103.98, 100.38, 77.47, 75.17, 72.51, 71.09, 70.94, 69.22, 69.06, 68.79, 60.61, 60.29, 57.30. MS (FAB) 379 ($M + Na^+$); HRMS: calculated for $C_{13}H_{24}O_{11}Na$ ($M + Na^+$) 379.1216, found 379.1218.

Acknowledgments

This work was supported by the NIH Grant AI44040.

[10] Conjugating Low Molecular Mass Carbohydrates to Proteins 1. Monitoring the Progress of Conjugation

By RINA SAKSENA, ANATOLY CHERNYAK, ALEX KARAVANOV, and PAVOL KOVÁČ

Introduction

There are many applications of neoglycoconjugates in the life sciences,[1] of which probably the most promising is their use as immunogenic materials in developing vaccines for infectious diseases and cancer.[2–7] Polysaccharides, such as the O-specific polysaccharide (O-PS) part of lipopolysaccharides (LPS)[8] and capsular polysaccharides,[9] are important natural antigens but they are poor immunogens.[10] They are classified as T-independent (TI) antigens, and the level and the spectrum of antibodies produced after immunization with carbohydrate antigens is insufficient to render protection. Chemical linking of carbohydrates to proteins can transform them into T cell-dependent (TD) antigens. As a result, multiple injections of neoglycoconjugates can sharply boost antibody titers way beyond those observed as a result of priming with TI antigens. This is, essentially, the rationale behind the concept of synthetic vaccines, which pre expected to be free from some of the drawbacks of cellular vaccines.

In this article we describe monitoring the progress of conjugation of synthetic, linker-equipped oligosaccharides to proteins. Practical hints, which may be useful in modifying the protocol described here to better suit a particular situation, are also included. The protocol described here, which

[1] P. Kováč, (Ed.), "Synthetic Oligosaccharides: Indispensable Probes in the Life Sciences," Vol. 560. American Chemical Society, Washington, D.C., 1994.
[2] K. E. Stein, *Int. J. Technol. Assess. Health Care* **10,** 167 (1994).
[3] W. E. Dick, Jr. and M. Beurret, *in* "Conjugate Vaccines" (J. M. Cruse and R. E. Lewis, Jr., eds.), Vol. 10, p. 48. Karger, Basel, 1989.
[4] P. O. Livingston, S. Zhang, and K. O. Lloyd, *Cancer Immunol. Immunother.* **45,** 1 (1997).
[5] P. H. Mäkalä, *FEMS Microbiol. Rev.* **24,** 9 (2000).
[6] S. J. Danishefsky and J. R. Allen, *Angew, Chem., Int. Ed.* **39,** 836 (2000).
[7] J. B. Robbins, R. Schneerson, S. C. Szu, and V. Pozsgay, *Pure Appl. Chem.* **71,** 745 (1999).
[8] O. Lüderitz, M. A. Freudenberg, C. Galanos, V. E. Lehmann, T. Rietschel, and D. H. Shaw, *in* "Current Topics in Membranes and Transport" (F. Bonner and A. Kleinzeller, eds.), Vol. 17, p. 79. Academic Press, New York, 1982.
[9] K. Jann and O. Westphal, *in* "The Antigens" (M. Sela, ed.), p. 1. Academic Press, New York, 1975.
[10] C. T. Bishop and H. J. Jennings, *in* "The Polysaccharides" (G. O. Aspinall, ed.), Vol. 1, p. 291. Academic Press, New York, 1982.

0076-6879/03 $35.00

was developed[11] with synthetic fragments of the O-PS of *Vibrio cholerae* O1, is generally applicable for monitoring conjugation of similar substances, provided their molecular mass is known and they are free of contaminants that could be involved in conjugation, thereby making the interpretation of results unreliable. The conjugations described here were performed according to the single-point attachment model,[3] using linker (spacer)-equipped synthetic oligosaccharides.[12] It allows preparation of well-defined neoglycoconjugates. The conjugation chemistry was performed with squaric acid diesters. (For a description of the chemistry and other aspects of conjugation involving squaric acid diesters, refer to article II dealing with the recovery of excess ligand used in the conjugation reaction.) The conjugations were performed at various hapten–carrier ratios. When conjugates of low hapten–carrier ratios were to be prepared the conjugations were carried out at a 10:1 hapten–BSA ratio. The use of higher hapten–BSA ratios is applicable when a higher carbohydrate–carrier ratio is required, or when working with less reactive carriers and/or haptens. In these situations, recovery of the excess ligand used at the onset of the conjugation may be warranted (see article [11] in this volume[12a]). Neoglycoconjugates similar to those described here were tested and found immunogenic in mice, and also showed protective capacity.[13] Because those materials were prepared with the intention to be used as experimental immunogens, the ability to be able to prepare neoglycoconjugates with narrow, well-defined carbohydrate–protein ratios was important. This required a method allowing routine monitoring of the conjugation reaction, so that the reaction could be terminated when the desired hapten–carrier ratio was achieved. A general protocol, which would allow preparation of neoglycoconjugates with a predicted saccharide–protein ratio, does not exist at this time. Before we addressed the problem of monitoring the progress of conjugation, the carbohydrate–protein ratio in the final conjugate was more or less a trial-and-error situation. The experiment was set up, and the reaction was allowed to proceed for some time based on previous experience or data in the literature describing similar situations. Unfortunately, a direct extrapolation of such data seldom leads to desired results because individual carrier proteins and haptens to be conjugated show different reactivity. To find out the molecular weight of the

[11] A. Chernyak, A. Karavanov, Y. Ogawa, and P. Kováč, *Carbohydr. Res.* **330,** 479 (2001).
[12] Y. Ogawa, P.-S. Lei, and P. Kováč, *Carbohydr. Res.* **293,** 173 (1996).
[12a] R. Saksena, A. Chernyak, E. Poirot, and P. Kováč, *Methods Enzymol.* **362,** 11 (2003) (this volume).
[13] A. Chernyak, S. Kondo, T. K. Wade, M. D. Meeks, P. M. Alzari, J.-M. Fournier, R. K. Taylor, P. Kováč, and W. F. Wade, *J. Infect. Dis.* **185,** 950 (2002).

neoglycoconjugate, the reaction had to be terminated, and the product had to be isolated and purified. Before it could be analyzed, the conjugate had to be obtained in a high degree of purity, to satisfy the requirement of MALDI-TOF mass spectrometry. We found the right tool for more straightforward monitoring of the conjugation reaction to be SELDI-TOF MS in combination with the ProteinChip System (Fig. 1). The technique is closely related to MALDI-TOF MS, but the use of selectively active surfaces allows the sample clean-up to be performed directly on the ProteinChip. In addition to providing information from which the average number of conjugated carbohydrate molecules can be deduced, this technique can also be informative concerning polydispersity of the conjugate.

The ProteinChip technology (Ciphergen Biosystems, Fremont, CH) is based on SELDI-TOF mass spectrometry. It combines solid-phase chromatography (retentate chromatography) with the power of LDI-TOF MS analysis. It uses a variety of chemically active surfaces capable of retaining the proteins or modified derivatives thereof, depending on their intrinsic properties. After proteins have bound to the surface, contaminants may be washed out and the retained sample can be directly analyzed by an LDI-TOF type detector unit. Because of the properties of these surfaces, the ProteinChip system is almost ideally suited for direct monitoring of mass shift changes resulting from chemical modifications of carrier proteins.

The ProteinChip arrays are aluminum strips that contain 8, 16, or 24 active sites. The surface of each site represents a monolayer of chemically active molecules. Each site of the particular type of the array carries an active surface of the same kind, thus permitting the study of one sample under eight different conditions or eight samples under one particular condition. A variety of active surfaces are available (Fig. 2). They can be hydrophilic (based on silicon dioxide), hydrophobic (C_{16} aliphatic hydrocarbon chain), ionic (quaternary amine for positive charge surface or carboxylate for negative charge surface), or metal binding (nitrilotriacetate). The choice of the type of ProteinChip array is determined by the objective of the experiment. A sample withdrawn from the crude conjugation mixture is applied, after dilution, on a site in the strip, which is then air dried. After the binding step, the sample on the chip is washed with water. Proteins are captured or released from the surface according to their affinity for the particular active surface. Therefore, a variety of proteins are retained on the basis of their intrinsic properties, such as hydrophobicity, charge, modification, or primary sequence. Bound proteins are detected in the LDI-TOF-MS type of mass reader (Ciphergen Biosystems) in a matter of just several minutes. Two types of ProteinChip arrays permit covalent binding of proteins via

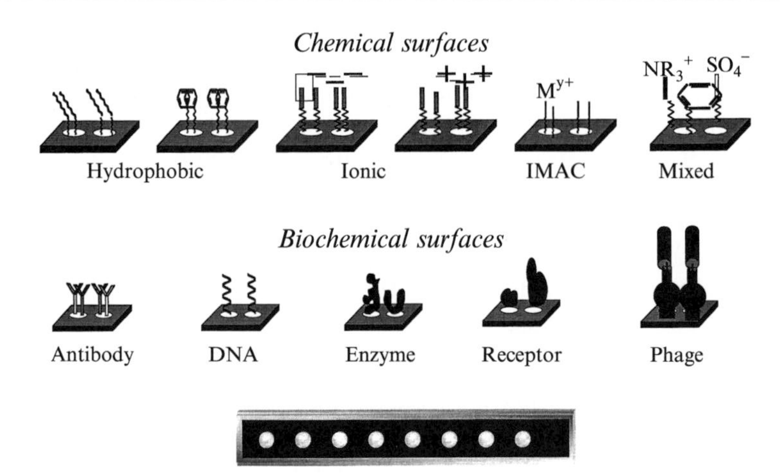

FIG. 1. General scheme of the SELDI technology process. Samples, after dilution, are directly added to the spots with chromatographically active surface; spots are washed, and substances selectively retained on the surface are detected by an LDI-TOF mass reader.

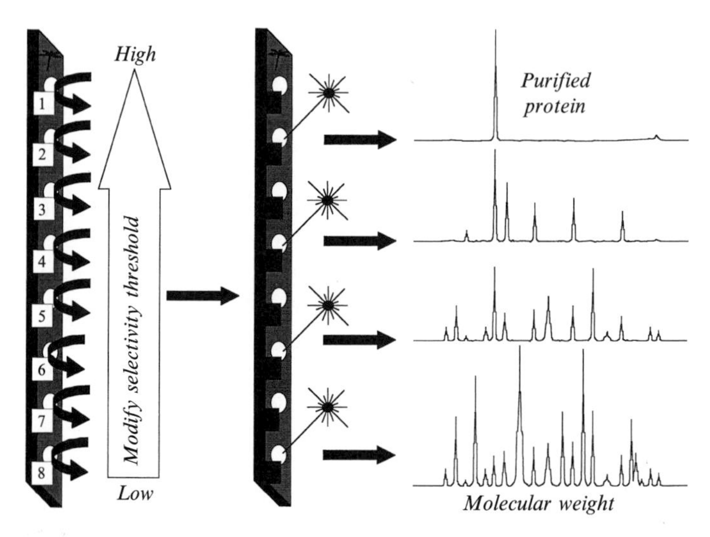

FIG. 2. ProteinChip arrays produced by Ciphergen. Arrays carry a strong anionic-exchange (SAX-2) surface; a weak cationic (WCX2) exchange surface; a hydrophobic surface (H4 equivalent to C_{16} reversed-phase resin); several varieties of IMAC surfaces, which can be loaded with different metal ions; normal-phase (NP-2) silica-coated surfaces; as well as two arrays of covalently binding proteins (PS10 and PS20).

primary amines, thus allowing protein–protein interaction studies. Proteins interacting with the chip surface will be retained on its surface, while other proteins will be removed during the washing step. Apart from the newly discovered[11] utility of this technique for the purpose here described, the ProteinChip system has many other applications.[14–19] Using the above-described technique, we can now withdraw samples from conjugation reaction mixtures, analyze the crude product, and obtain near-real-time information about the increasing molecular weight of the neoglycoconjugate formed.

The O-PS from *Vibrio cholerae* O1, serotypes Inaba and Ogawa, is a homopolymer of α-(1 $\rightarrow$ 2)-linked 4,6-dideoxy-4-amino-D-mannose (perosamine), the amino groups of which are acylated with 3-deoxy-L-*glycero*-tetronic acid. The two serotypes differ in that the upstream terminal perosamine residue in the Ogawa O-PS contains at position O-2 a methyl group, which is not present in the Inaba O-PS (Fig. 3).[20–22] On the basis of extensive solution binding studies, it has been hypothesized[23] that the serological specificity of the Ogawa strain is associated with the methylated, terminal, perosamine residue. The conclusions arrived at in this way have been fully confirmed[24] by X-ray study involving crystalline complexes of synthetic fragments of the O-PS of *Vibrio cholerae* O1, serotype Ogawa

[14] E. F. Petricoin, A. M. Ardekani, B. A. Hitt, P. J. Levine, V. A. Fusaro, S. M. Steinberg, G. B. Mills, C. Simone, D. A. Fishman, E. C. Kohn, and L. A. Liotta, *Lancet* **359,** 572 (2002).

[15] S. R. Weinberger, T. S. Morris, and M. Pawlak, *Pharmacogenomics* **1,** 395 (2000).

[16] E. Tassi, A. Al-Attar, A. Aigner, M. R. Swift, K. McDonnell, A. Karavanov, and A. Wellstein, *J. Biol. Chem.* **276,** 40247 (2001).

[17] G. E. Stoica, A. Kuo, A. Aigner, I. Sunitha, B. Souttou, C. Malerczyk, D. J. Caughey, D. Wen, A. Karavanov, A. T. Riegel, and A. Wellstein, *J. Biol. Chem.* **276,** 16772 (2001).

[18] S. Petruk, Y. Sedkov, S. Smith, S. Tillib, V. Kraevski, T. Nakamura, E. Canaani, C. M. Croce, and A. Mazo, *Science* **294,** 1331 (2001).

[19] C. Rosty, L. Christa, S. Kuzdzal, W. M. Baldwin, M. L. Zahurak, F. Carnot, D. W. Chan, M. Canto, K. D. Lillemoe, J. L. Cameron, C. H. Yeo, R. H. Hruban, and M. Goggins, *Cancer. Res.* **62,** 1868 (2002).

[20] L. Kenne, B. Lindberg, P. Unger, B. Gustafsson, and T. Holme, *Carbohydr. Res.* **100,** 341 (1982).

[21] T. Ito, T. Higuchi, M. Hirobe, K. Hiramatsu, and T. Yokota, *Carbohydr. Res.* **256,** 113 (1994).

[22] K. Hisatsune, S. Kondo, Y. Isshiki, T. Iguchi, and Y. Haishima, *Biochem. Biophys. Res. Commun.* **190,** 302 (1993).

[23] J. Wang, J. Zhang, C. E. Miller, S. Villeneuve, Y. Ogawa, P.-S. Lei, P. Lafaye, F. Nato, A. Karpas, S. Bystrický, S. C. Szu, J. B. Robbins, P. Kováč, J.-M. Fournier, and C. P. J. Glaudemans, *J. Biol. Chem.* **273,** 2777 (1998).

[24] S. Villeneuve, H. Souchon, M. M. Riottot, J. C. Mazie, P. S. Lei, C. P. J. Glaudemans, P. Kováč, J. M. Fournier, and P. M. Alzari, *Proc. Natl. Acad. Sci. USA* **97,** 8433 (2000).

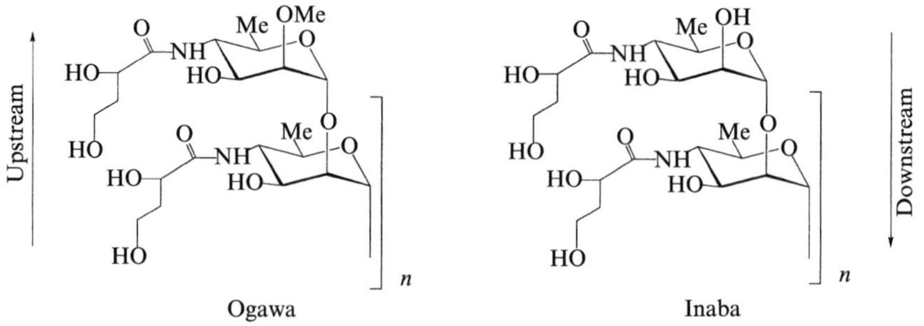

FIG. 3. O-PS of *V. cholerae* O1, serotypes Inaba and Ogawa.

with an Ogawa LPS-specific monoclonal antibody. This laboratory has synthesized fragments of both types of O-PS in the form of methoxycarbonylpentyl glycosides.[12,25] A novel approach to the Inaba disaccharide derivative is described in detail in article [11] in this volume.[12a] Methoxycarbonylalkyl glycosides of oligosaccharides, first introduced by Lemieux *et al.*[26] can be easily converted[27] to derivatives that allow conjugation by squaric acid chemistry.

Practical Considerations

In principle, from a chemical point of view, it should make no difference whether the conjugation involves a larger or a smaller linker-equipped carbohydrate. However, squaric acid derivatives made from monosaccharides are generally more reactive than those made from larger oligosaccharides. This is because for a derivatized hapten to react with the carrier, the former must penetrate into the three-dimensional structure of the latter, where the individual ϵ-amino groups present in the lysine residues are not equally accessible. Therefore, with all other reaction parameters equal, the reaction time required to achieve a certain loading with a monosaccharide derivative is shorter than is the case with a higher oligosaccharide (Tables I–IV). With poorly reactive reactants, when the conjugation reaction is carried out at a relatively low hapten–carrier ratio (e.g., 10:1) the desired hapten–carrier ratio (e.g., 5:1) may not be possible to achieve by simply extending the reaction time, because hydrolysis of the squaric acid monoester takes place during the conjugation at pH 9.

[25] Y. Ogawa, P.-S. Lei, and P. Kováč, *Carbohydr. Res.* **288,** 85 (1996.).
[26] R. U. Lemieux, D. R. Bundle, and D. A. Baker, *J. Am. Chem. Soc.* **97,** 4076 (1975).
[27] V. P. Kamath, P. Diedrich, and O. Hindsgaul, *Glycoconj. J.* **13,** 315 (1996).

TABLE I
MONITORING CONJUGATION OF HAPTEN **3** TO BSA TO OBTAIN CONJUGATE **4**: COMPARISON OF REPRODUCIBILITY OF CONJUGATION[a]

Run	Hapten	Conjugate	Reaction time (min)	Average molecular mass (Da)	Hapten–BSA ratio
1	**3**	**4**	15	67,244	1.5
			30	67,544	2.1
			45	68,011	3.0
			60	68,444	3.8
2	**3**	**4**	15	67,765	2.5
			30	68,107	3.2
			45	69,005	4.9
			60	68,444	5.0

[a]Initial hapten–BSA ratio, 100:1; initial hapten concentration, 20 mmol.

TABLE II
MONITORING CONJUGATION OF HAPTEN **3** TO BSA TO OBTAIN CONJUGATE **4**: COMPARISON OF REPRODUCIBILITY OF CONJUGATION[a]

Run	Hapten	Conjugate	Reaction time (h)	Average molecular mass (Da)	Hapten–BSA ratio
1	**3**	**4**	0.5	66,509	0.14
			3.5	67,072	1.2
			27	68,796	4.5
			96	69,555	6.0
			144	69,988	6.8
			216	70,335	7.4
			288	70,653	8.0
2	**3**	**4**	0.5	66,691	0.5
			3.5	67,352	1.7
			27	68,985	4.8
			96	69,721	6.2
			144	69,911	6.6
			216	70,048	6.9
			288	70,502	7.7

[a]Initial hapten–BSA ratio, 10:1; initial hapten concentration, 15 mmol.

TABLE III
MONITORING CONJUGATION OF INABA DISACCHARIDE[a] TO BSA[b]

Initial hapten–BSA ratio	Reaction time (h)	Average molecular mass (Da)	Hapten–BSA ratio
10:1	1.5	67,006	0.75
	23	69,107	3.5
	40	70,162	5.0
	69[c]	70,520	5.5
100:1	0.25	68,820	2.5
	0.75	69,474	4.4
	1	70,432	5.4
	1.5	71,695	7.1
	2	72,286	7.8
	2.5	73,489	9.5
	3	73,977	10.1
	4	75,140	11.6
	5	75,837	12.6
	6	76,413	13.4
	7	78,283	15.8

[a]For the chemistry involved in the preparation of intermediates, see Ref. 12a.
[b]Initial hapten concentration, 15 mmol.
[c]The hapten–BSA ratio remained, essentially, unchanged.

TABLE IV
MONITORING CONJUGATION OF HAPTEN **6** TO BSA TO OBTAIN CONJUGATE **7**[a]

Hapten	Conjugate	Reaction time (h)	Average molecular mass (Da)	Hapten–BSA ratio
6	**7**	10	71,748	3.1
		20	73,492	4.0
		34	75,147	5.0
		43	75,738	5.3
		59	78,298	6.8
		72	80,219	7.9
		96	81,945	8.9
		144[b]	84,288	10.3

[a]Initial hapten–BSA ratio, 80:1; initial hapten concentration, 20 mmol.
[b]The hapten–BSA ratio remained, essentially, unchanged.

In such situations, a way to achieve a high hapten–carrier ratio is to use a high initial hapten–carrier ratio (see also article [11] in this volume,[12a] on the recovery of excess hapten use in conjugation). The effect of the initial hapten–carrier ratio, hapten concentration, and the size of the hapten on the outcome of the conjugation can be deduced from data in Tables I–IV. On the basis of our experience,[28] different proteins show different reactivity, which adds another fundamental parameter to reckon with. Thus, the use of SELDI-TOF MS for monitoring the progress of conjugation of carbohydrates to protein carriers is generally useful because it minimizes the guesswork concerning the time required to reach the desired loading.

Conjugation of oligosaccharides to proteins depends on many parameters, some of which are difficult to control, and some may not even be known at the present time. Consequently, the reaction rate of the conjugation process is not quite reproducible. A comparison of results obtained in two parallel experiments is shown in Tables I and II. Incomplete reproducibility of the loading of hapten onto carrier becomes an important issue when discrete, low carbohydrate-onto-protein loading is required. A difference in loading of one hapten per carrier can amount to tenths of a percent, and this may be more than the chemist would be prepared to tolerate. This increases the importance of having a near-real-time monitoring technique available, which allows termination of the conjugation reaction when the desired incorporation is reached.

Experiments described here were performed at ligand–protein ratios of 100:1, 80:1, and 10:1. The molecular mass of the neoglycoconjugate versus time was monitored by SELDI-TOF analysis. In addition to yielding higher loading products, experiments involving high ligand–carrier ratios also served to examine the possibility of recovering the ligand used in large excess (see article [11][12a]), with the intention to use it in subsequent conjugation.

Haptens suitable for conjugation to proteins, using squaric acid chemistry, are often obtained by aminolysis or hydrazinolysis of glycosides whose aglycon contains alkoxycarbonylalkyl groups, followed by reaction with squaric acid diesters.[11,13,27,29,30] Conversions leading to neoglycoconjugates **4** and **7** are shown in Schemes 1 and 2.

[28] A. Chernyak, J. Zhang, and P. Kováč, unpublished results (2002).

[29] V. Pozsgay, E. Dubois, and L. Pannell, *J. Org. Chem.* **62,** 2832 (1997).

[30] S. Cohen and S. G. Cohen *J. Am. Chem. Soc.* **88,** 1533 (1966).

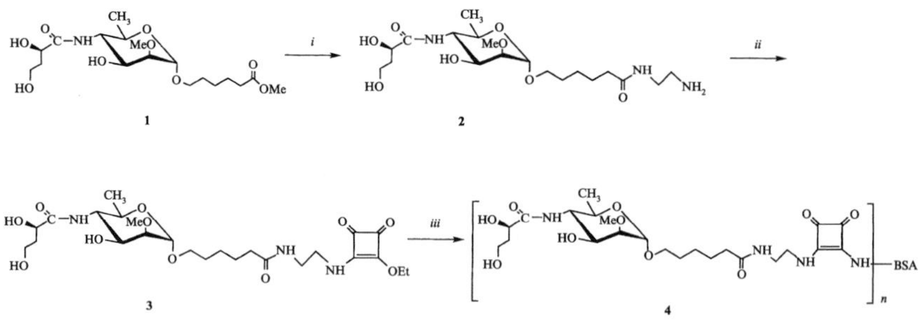

SCHEME 1. Preparation of 1-[(2-aminoethylamido)carbonylpentyl 4-(3-deoxy-L-*glycero*-tetronamido)-4,6-dideoxy-2-*O*-methyl-α-D-mannopyranoside]-2-ethoxycyclobutene-3,4-dione (**3**) and the corresponding BSA-conjugate **4**. (*i*) $H_2NCH_2CH_2NH_2$; (*ii*) 3,4-diethoxy-3-cyclobutene-1,2-dione in ethanol; (*iii*) BSA (pH 9.0).

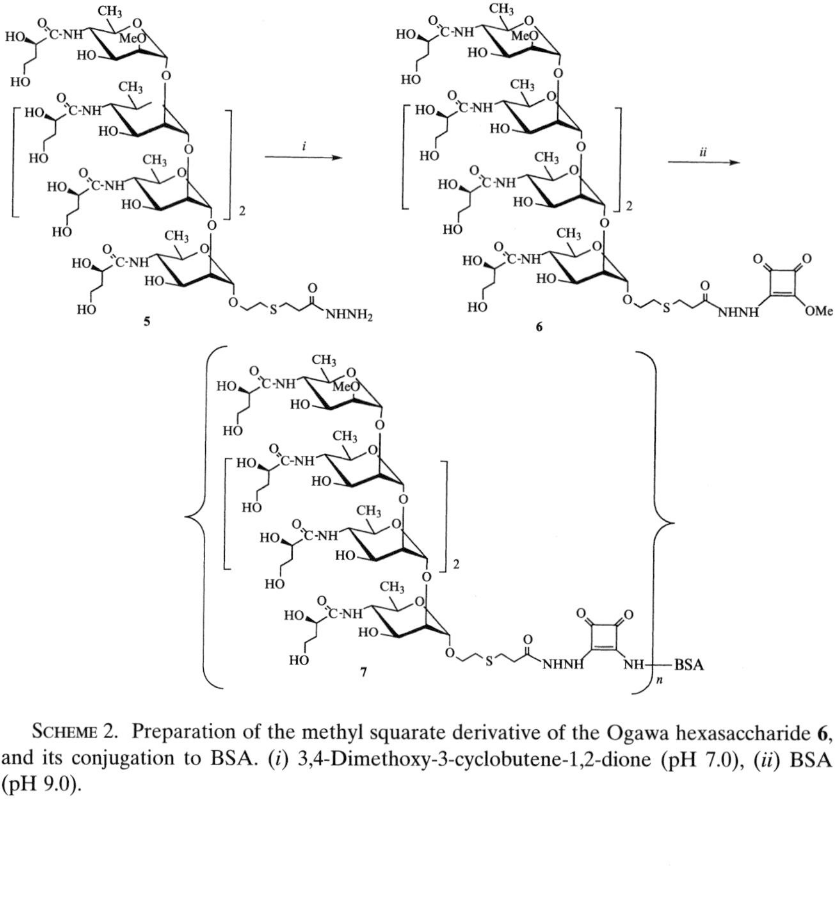

SCHEME 2. Preparation of the methyl squarate derivative of the Ogawa hexasaccharide **6**, and its conjugation to BSA. (*i*) 3,4-Dimethoxy-3-cyclobutene-1,2-dione (pH 7.0), (*ii*) BSA (pH 9.0).

Conversion of Synthetic, Linker-Equipped Low Molecular Mass Carbohydrates to Squaric Acid Derivatives, and Monitoring of Progress of Conjugation to Bovine Serum Albumin

Methoxycarbonylpentyl glycosides of the requisite saccharide are transformed to the corresponding amides of squaric acid monoesters by a two-step process. First, the reaction of the methoxycarbonylpentylglycoside with ethylenediamine or hydrazine gives the corresponding 2-aminoethylamide or a hydrazide, respectively, which is then treated with 3,4-dialkyloxy-3-cyclobutene-1,2-dione (Aldrich, Milwaukee, WI). Subsequent treatment of the squaric acid monoesters formed with bovine serum albumin (BSA) yields the neoglycoconjugates.

(2-Aminoethylamido)carbonylpentyl 4-(3-deoxy-L-glycero-tetronamido)-4,6-dideoxy-2-O-methyl-α-D-mannopyranoside (**2**)

5-Methoxycarbonylpentyl 4-(3-deoxy-L-*glycero*-tetronamido)-4,6-dideoxy-2-*O*-methyl-α-D-mannopyranoside[31] (**1**, 120.8 mg, 0.297 mmol) is dissolved in anhydrous ethylenediamine (Aldrich, 3 ml) and stirred under argon in a closed vial at 70° for 2 days. Thin-layer chromatography [TLC, ethyl acetate–methanol–concentrated NH_4OH, 2:1:0,1 (v/v/v)] then shows the reaction to be complete. Excess reagent is removed by concentration (70°/0.1 torr), the residue is coevaporated with water (five times), and chromatography (silica gel, 0.04–0.06 mm; Merck, Darmstadt, Germany) gives the title compound **2** (116 mg, 90%).

1-[(2-Aminoethylamido)carbonylpentyl 4-(3-deoxy-L-glycero-tetronamido)-4,6-dideoxy-2-O-methyl-α-D-mannopyranoside]-2-ethoxycyclobutene-3,4-dione (**3**)

3,4-Diethoxy-3-cyclobutene-1,2-dione (39 mg, 0.23 mmol) is added with stirring to the above (2-aminoethylamido)carbonylpentyl glycoside (**2**, 100 mg, 0.23 mmol) in ethanol (5 ml). Stirring is continued at room temperature until TLC [ethyl acetate–methanol, 2:1 (v/v)] shows that the starting material is consumed (12–24 h). After concentration, the residue is chromatographed on a column of silica gel and a solution of the product, thus obtained in water (2 ml) is filtered through a Millex GV filter (0.22 μm; Millipore, Bedford, MA) and lyophilized, to give the title compound **3** (97 mg, 75.5%).

[31] J. Zhang, A. Yergey, J. Kowalak, and P. Kováč, *Carbohydr. Res.* **313,** 15 (1998).

*1-{2-[2-(Hydrazinocarbonyl)ethylthio]ethyl 4-(3-deoxy-L-glycero-tetronamido)-4,6-dideoxy-2-O-methyl-α-D-mannopyranosyl-(1 → 2)-tetrakis[4-(3-deoxy-L-glycero-tetronamido)-4,6-dideoxy-2-O-methyl-α-D-mannopyranosyl-(1 → 2)]-4-(3-deoxy-L-glycero-tetronamido)-4,6-dideoxy-2-O-methyl-α-D-mannopyranoside}-2-methoxycyclobutene-3,4-dione (**6**)*

To the 2-[2-(hydrazinocarbonyl)ethylthio]ethylglycoside of the Ogawa hexasaccharide[12] (**5**, 19.85 mg, 11.945 μmol) in 0.1 *M* KH_2PO_4–NaOH buffer (1.85 ml, pH 7.08) is added with stirring 3,4-dimethoxy-3-cyclobutene-1,2-dione (Aldrich; 3.4 mg, 23.9 μmol). More of the reagent (2.5 mg, 17.6 μmol) is added after 2 h, if thin-layer chromatography [$CHCl_3$–methanol–concentrated NH_4OH, 1:3:0.1 (v/v/v)] shows that the reaction is incomplete. The mixture is processed as described above for a similar reaction, and the pure, title compound is obtained after purification on a series of two Sep-Pak C_{18} Plus cartridges (Waters, Milford, MA), which are eluted with 10 through 40% of methanol. After lyophilization, the squarate derivative **6** (17.3 mg, 82%) is obtained as a slightly pink solid, FABMS: *m/z* 1772 $[M+1]^+$.

Conjugation of Squaric Acid Derivatives with BSA

A protocol for a typical experiment can be carried out at a hapten–BSA ratio of 100:1 and at a 20 m*M* concentration of the hapten used. This concentration is within the limits found[31] most useful for achieving high efficiency of conjugation.

A solution of 1-[2-aminoethylamido)carbonylpentyl 4-(3-deoxy-L-*glycero*-tetronamido)-4,6-dideoxy-2-*O*-methyl-α-D-mannopyranoside]-2-ethoxycyclobutene-3,4-dione (**3**, 8.4 mg, 15.05 μmol) in 0.05 *M* boric acid–potassium chloride buffer (pH 9.0, 750 μl; VWR, Willard, OH) is added with stirring (900 rpm) at 22° to BSA (fraction V; Sigma, St. Louis, MO) (additionally purified,[32] 10 mg, 0.15 μmol). The final concentration of sugar hapten is 20 m*M*. Aliquots (1 μl) are withdrawn every 15 min, processed as described below, and analyzed by SELDI-TOF MS (Table I).

After 1 h, when the desired hapten–carrier ratio has been reached, the reaction is terminated by transferring the mixture into a pH 7.0 buffer (3 ml). The reaction vessel is washed (2 × 1 ml and 1 × 0.5 ml of the same buffer), and the combined solution is dialyzed against water (HPLC quality, 60 ml), using the Amicon cell equipped with PM-10 membrane (Millipore). The retained solution is filtered through a Millex GV filter

[32] R. F. Chen, *J. Biol. Chem.* **242,** 173 (1967).

(0.22 μm) and lyophilized, to give conjugate **4** (11 mg, ~100%). For the protocol describing the recovery of the unused hapten, see article [11] in this volume.[12a]

Monitoring of Conjugation Reactions by SELDI-TOF MS

Materials and Reagents

H4 (hydrophobic; C_{16}) ProteinChip arrays
PAP pen (i.e., a hydrophobic pen; Zymed, South San Francisco, CA)
Acetonitrile
Water (HPLC grade)
1% Trifluoroacetic acid (TFA)
Sinapinic acid (SPA)
Buffer, pH 7.0

Overview of Procedure

Aliquots of 1 μl are withdrawn from the reaction mixture of conjugation at the requisite time intervals and diluted with 9 μl of a buffer (pH 7.0; Mallinckrodt Baker, Paris, KY), and 1 μl of the diluted solution is applied to the H4 chip (Ciphergen Biosystems). The chip is cold air dried and 5 μl of HPLC water is applied on the dry spot. Using the same micropipette tip, the water is withdrawn and reapplied several times to wash out the soluble material. This operation is repeated twice. A solution of the matrix [0.5 μl of a freshly prepared mixture of a saturated solution of SPA in acetonitrile and 1% aqueous TFA, 1:1 (v/v); twice with drying between additions] is added to the spot. Chips are read by the SELDI-TOF instrument at the proper sensitivity and laser intensity, as required by the nature of the material. Before measurements, the instrument is calibrated externally, using BSA (molecular mass, 66,430 Da[33]).

Step-by-Step Procedure

1. With a PAP pen, carefully circle the active spots on the H4 strip. Do not touch the surface of the spot.
2. Centrifuge, for 3 min at maximum speed, a freshly prepared saturated solution of SPA in 50% acetonitrile containing 0.5% TFA.

[33] K. Hirayama, S. Akashi, M. Furuya, and K.-I. Fukuhara, *Biochem. Biophys. Res. Commun.* **173,** 639 (1990).

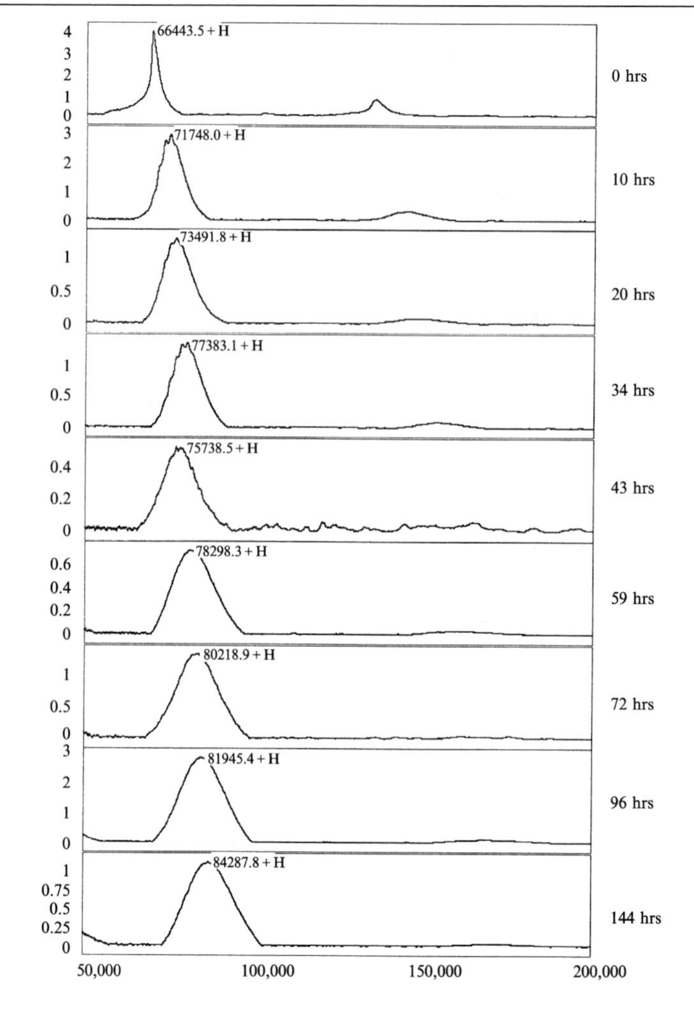

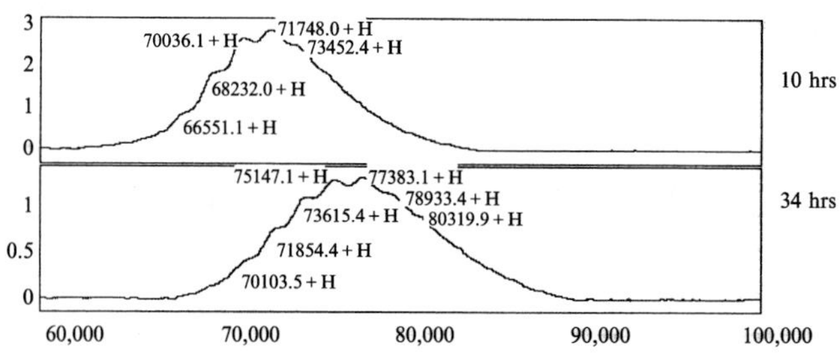

FIG. 4. (*For legend, see opposite page.*)

(For legend, see opposite page.)

3. Transfer a 1-μl aliquot from the conjugation reaction mixture into a polypropylene microtube containing 9 μl of pH 7 buffer. Mix well, and apply 1 μl on the active spot of the chip.

4. Air dry the spot.

5. Wash each spot with 5 μl of water (HPLC grade) two times, using a 10-μl micropipette.

6. Air dry the spot.

7. Apply 0.5 μl of saturated SPA solution per spot and air dry. Repeat the application of 0.5 μl of saturated SPA solution per spot once more and air dry.

8. Read the chip in a PBS II mass reader at laser intensity 245–280 and sensitivity 8–9. These parameters are variable and depend on the molecular mass and concentration of the conjugate under investigation.

The results of the typical analysis of the conjugation reaction at different time points are presented in Fig. 4. The time shift of the average mass of the conjugate formed can be seen and compared with the original mass of BSA. These data permit calculation of the number of carbohydrates conjugated per molecule of BSA. This gives the investigator considerable control over the reaction, which can be terminated when the desired carbohydrate–protein ratio is reached. Valuable haptens, unused during the reaction, if present, can be recovered and used for preparation of similar conjugates (see article [11] in this volume[12a]).

FIG. 4. The progress of conjugation of hexasaccharide **6** (molecular mass, 1772 Da, revealed by SELDI-TOF MS). *Top:* The spectrum taken at $t = 0$ shows the molecular mass of the carrier protein, BSA (molecular mass, 66,443 Da). Spectra taken at the time indicated show the increasing molecular mass of the conjugate formed. *Bottom:* Expansion of spectra taken at reaction times 10 and 34 h, show polydispersity of the material. The presence of the unchanged BSA after the reaction time 10 h, absent after 34 h, can be seen by the shoulder at 66,551 Da (determined with the accuracy of 0.1%).

[11] Conjugating Low Molecular Mass Carbohydrates to Proteins 2. Recovery of Excess Ligand Used in the Conjugation Reaction

By Rina Saksena, Anatoly Chernyak, Emmanuel Poirot, and Pavol Kováč

Introduction

Synthetic glycoconjugates, materials that are often referred to as neoglycoconjugates, were first prepared by Avery and Goebel[1,2] some 70 years ago as products of an intellectual exercise. Avery and Goebel were able to show not only that chemical conjugation of low molecular mass carbohydrates to protein was technically feasible but also that neoglycoconjugates have potential as tools in biology and medicine. Indeed, substances belonging to this new class of compounds have contributed enormously to our better understanding of many fundamental processes in the life sciences.[3,4] The neoglycoconjugates made by Avery and Goebel were not sophisticated from the point of view of carbohydrate chemistry, as they involved only simple mono- and disaccharides. However, with the recognition of the importance of these substances, new conjugation protocols have been developed and, more often than not, the carbohydrates to be conjugated included complex oligosaccharides resulting from intricate chemical syntheses. As such, these materials are generally labor-intensive and, therefore, expensive commodities. The actual expense, and the wasteful nature of the conjugation reaction—as it is often performed—is multiplied by the fact that, in order to maintain a reasonable reaction rate, conjugation of carbohydrates to proteins is normally conducted at a high molar hapten–carrier ratio. When the reaction is terminated, the only material isolated from the conjugation mixture is usually the neoglycoconjugate, even though the mixture may still contain unused hapten. Therefore, when expensive materials are converted into neoglycoconjugates, recovery of unused ligand from the conjugation mixtures ought to be considered.

[1] W. F. Goebel and O. T. Avery, *J. Exp. Med.* **50,** 521 (1929).
[2] O. T. Avery and W. F. Goebel, *J. Exp. Med.* **50,** 533 (1929).
[3] Y. C. Lee and R. T. Lee, "Neoglycoconjugates: Preparation and Application." Academic Press, New York, 1994.
[4] P. Kováč (Ed.), "Synthetic Oligosaccharides: Indispensable Probes in the Life Sciences," Vol. 560. American Chemical Society, Washington, D.C., 1994.

0076-6879/03 $35.00

The purpose of recovery of the excess ligand from the conjugation reaction is, of course, that the material be used in subsequent, similar conjugations. A prerequisite to accomplishing this is that the material not involved in the conjugation remains unchanged. This, unfortunately, is not always the case, because many reagents used for conjugation lack chemospecificity. For example, synthetic oligosaccharides to be linked to proteins by reductive amination are normally prepared in the form of glycosides whose aglycon (linker, or spacer) carries an aldehyde group. When these compounds are allowed to react with the amino groups present in proteins, an intermediate Schiff base (an imine) is formed through a reversible reaction. The unstable Schiff base must be reduced to a secondary amine, to form a stable chemical bond (Scheme 1). This can be achieved with various reducing agents. Among those, sodium cyanoborohydride and pyridine–borane complex have been described as the reagents most suitable for this purpose. However, it appears[5–8] that the chemospecificity[9–11] or high chemoselectivity[3,12,13] of these reagents under the conditions of glycoconjugate preparation to reduce Schiff bases has been largely exaggerated. Clearly, any alcohol formed from the starting aldehyde during conjugation by reductive amination, as a result of chemical nonspecificity of the reduction of the Schiff base, constitutes virtually irreversible loss of the precious material (Scheme 2).

Consequently, notwithstanding the usefulness of reductive amination for conjugating polysaccharides, we do not find reductive amination suitable for economical conjugation of synthetic oligosaccharides to proteins.

A substantially different situation is presented by conjugation using squaric acid chemistry. This method for conjugating carbohydrates was introduced by Tietze *et al.*[14–16] It is based on sequential formation of squaric acid amide ester and squaric acid diamide at pH 7 and 9, respectively,

[5] M. Dubber and T. K. Lindhorst, *Synthesis* (2001).

[6] J. Zhang, A. Yergey, J. Kowalak, and P. Kováč, *Tetrahedron* **54,** 11783 (1998).

[7] R. Roy, E. Katzenellenbogen and H. J. Jennings, *Can. J. Biochem. Cell Biol.* **62,** 270 (1984).

[8] A. Chernyak and P. Kováč, unpublished results.

[9] G. T. Hermanson, "Bioconjugate Techniques." Academic Press, New York, 1996.

[10] G. R. Gray, *Methods Enzymol.* **50,** 155 (1978).

[11] N. Jentoft and D. G. Dearborn, *J. Biol. Chem.* **254,** 4359 (1979).

[12] R. F. Borch, M. D. Berntein, and H. D. Durst, *J. Am. Chem. Soc.* **93,** 2897 (1971).

[13] C. F. Lane, *in* "Selections from the Aldrichimica Chemica Acta," 1968–1982, p. 67. Aldrich Chemical Company, Milwaukee, WI, 1984.

[14] L. F. Tietze, A. Goerlach, and M. Beller, *Justus Liebigs Ann. Chem.* 565 (1988).

[15] L. F. Tietze, M. Arlt, M. Beller, K.-H. Glüsenkamp, E. Jähde, and M. F. Rajewsky, *Chem. Ber.* **124,** 1215 (1991).

[16] L. F. Tietze, C. Schröter, S. Gabius, U. Brinck, A. Goerlach-Graw, and H.-J. Gabius, *Bioconjug. Chem.* **2,** 148 (1991).

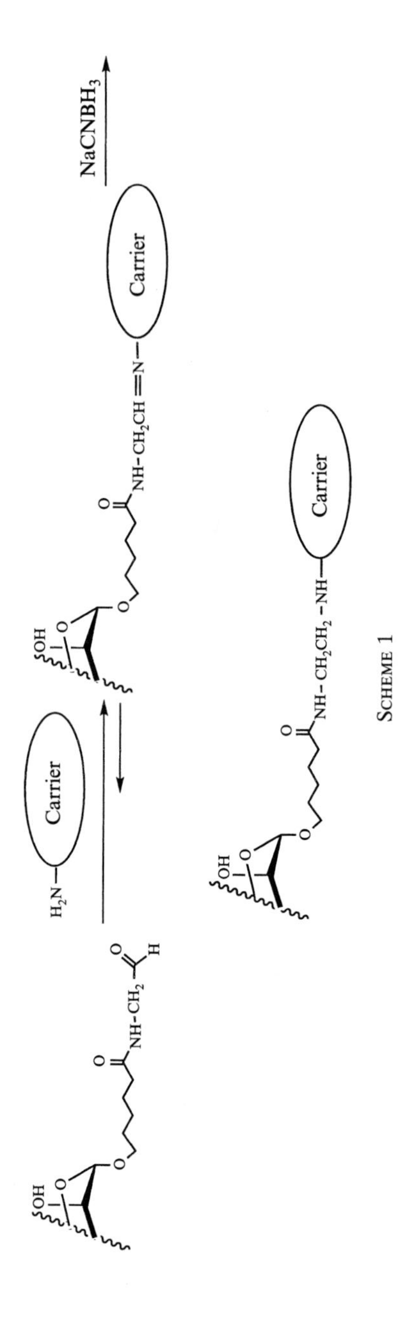

NaCNBH3
Carrier
Carrier
Carrier
H2N
NH-CH2
NH-CH2CH=N
NH-CH2CH2-NH

SCHEME 1

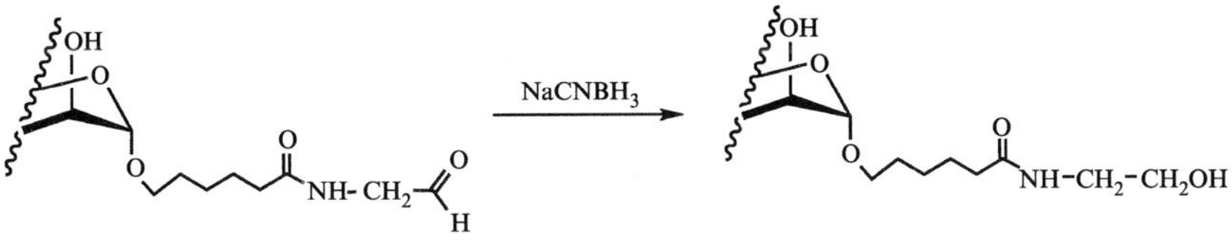

NaCNBH$_3$

NH-CH$_2$-CHO → NH-CH$_2$-CH$_2$OH

SCHEME 2

from a squaric acid diester and an amine. Conjugation of carbohydrates to proteins can be done efficiently by squaric acid chemistry. When the efficiency of conjugation by this method was compared[6,17,18] with that of reductive amination the conjugation involving squaric acid chemistry was found to be superior. Another advantage of conjugation involving squaric acid chemistry is that it can be performed on a small scale.[19] Because synthetic oligosaccharides are expensive materials application of this methodology in the conjugation of such substances is particularly advantageous.

Scheme 3 shows the conjugation of a hypothetical carbohydrate hapten to an amine-containing carrier, such as a protein. The hapten is synthesized in the form of a glycoside whose aglycon contains an amino group. Treatment of such an amine with one of the commercially available[20] squaric acid diesters at pH 7 gives the corresponding amide ester. These materials are remarkably stable under neutral conditions at ambient temperature. Literature data are available[21,22] on the hydrolysis of these esters, showing that the rate of hydrolysis is much slower than that of amide formation. We have been able to confirm[23] the stability of this class of substances, as well as the utility of recovering the unused ligand. We have isolated a substantial amount of the unchanged hapten that had gone through the squaric acid-mediated conjugation process, and used it subsequently for a similar conjugation.[24] During this work we have subjected a squaric acid monoester to hydrolysis under the conditions of conjugation (Scheme 4). The results (Table I) provide information about the rate of

[17] J. Zhang and P. Kováč, *Bioorg. Med. Chem. Lett.* **9,** 487 (1999).

[18] J. Zhang, A. Yergey, J. Kowalak, and P. Kováč, *Carbohydr. Res.* **313,** 15 (1998).

[19] V. P. Kamath, P. Diedrich, and O. Hindsgaul, *Glycoconjug. J.* **13,** 315 (1996).

[20] Squaric acid dimethyl and diethyl esters are available from Aldrich Chemical Co. The analogous dibutyl ester is available from Acros Organics through Fisher Scientific.

[21] K.-H. Glüsenkamp, W. Drosdziok, G. Eberle, E. Jähde, and M. F. Rajewsky, *Z. Naturforsch. C.* **46,** 498 (1991).

[22] S. Cohen and S. G. Cohen, *J. Am. Chem. Soc.* **88,** 1533 (1966).

[23] J. Zhang and P. Kováč, *Carbohydr. Res.* **321,** 157 (1999).

[24] A. Chernyak, A. Karavanov, Y. Ogawa, and P. Kováč, *Carbohydr. Res.* **330,** 479 (2001).

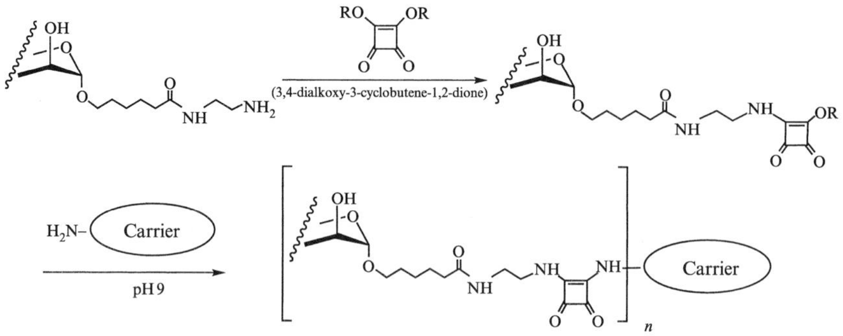

SCHEME 3

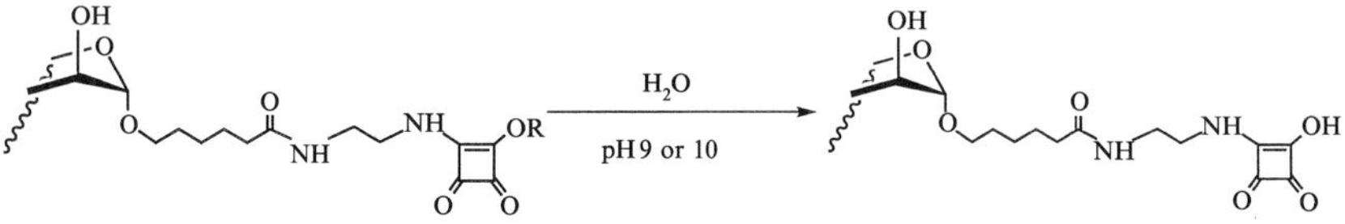

SCHEME 4

hydrolysis of squaric acid amide monoesters during conjugation. Such information is important for designing the most economical conjugation protocol. For example, when a particular conjugation is a slow reaction, the investigator can either start the conjugation at a high hapten–carrier ratio, to maintain a reasonable reaction rate (when a large part of the hapten may hydrolyze during the long reaction time), or add the hapten portionwise.

The protocols for recovery of excess hapten described below have resulted from our experience in conjugating carbohydrates to proteins, using squaric acid diesters. The procedures are simple and can be readily included in overall protocols involving other conjugation chemistry, provided a large part of the hapten used in excess remains unchanged.

Preparation of Linker-Equipped, Upstream, Terminal Disaccharide Determinant of *Vibrio cholerae* O1, Serotype Inaba

General Strategy

In principle, two fundamentally different strategies to assemble oligosaccharides composed of N-acylated perosamine can be envisioned. In the original approach,[25–27] perosamine-containing oligosaccharides that

TABLE I
HYDROLYSIS OF SQUARIC ACID MONOESTER

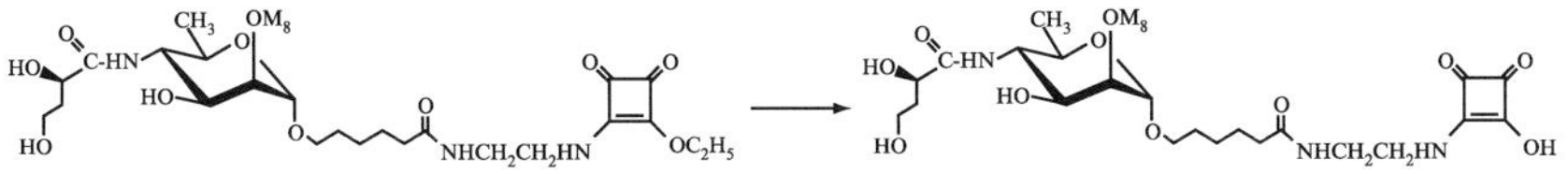

Starting squaric acid monoester (A) → Product acid (B)

Reaction time (h)	A	
	pH 9	pH 10
20	95	50
48	80	30
72	70	10
140	40	5
200	30	
340	10	

[a]At pH 9 and pH 10, to give its product of hydrolysis (B). Expressed in approximate amount (%, determined visually by TLC) of the starting material (A) remaining.

mimic the structure of *Brucella* polysaccharides are assembled with mono- or disaccharide building blocks containing azido groups at position 4, that is, intermediates lacking the *N*-acyl group. We have adopted the same strategy to synthesize some methyl α-glycosides of fragments of O-polysaccharide (O-PS) of *Vibrio cholerae* O1.[28–30]A different assembly strategy uses intermediate building blocks having the *N*-acyl chain already in place. Another aspect that must be taken into consideration when such oligosaccharides are to be used for conjugation to carriers is the stage at which the linker (spacer) should be attached. In our initial synthesis of linker-equipped *N*-acylated perosamine-containing haptens,[31,32] we first prepared the requisite oligosaccharides in the form of 2-(trimethylsilylethyl) (SE) glycosides from building blocks lacking the *N*-acyl group. When the requisite size of the oligosaccharide was assembled, the azido groups were reduced to amino groups and the resulting amines were N-acylated. Subsequently, the SE functionality was exchanged for the aglycon (spacer

[25] D. R. Bundle, M. Gerken, and T. Peters, *Carbohydr. Res.* **174,** 239 (1988).
[26] T. Peters and D. R. Bundle, *Can. J. Chem.* **67,** 497 (1989).
[27] T. Peters and D. R. Bundle, *Can. J. Chem.* **67,** 491 (1989).
[28] P.-s. Lei, Y. Ogawa, and P. Kováč, *Carbohydr. Res.* **279,** 117 (1995).
[29] P.-s. Lei, Y. Ogawa, and P. Kováč, *Carbohydr. Res.* **281,** 47 (1996).
[30] J. Zhang and P. Kováč, *Carbohydr. Res.* **300,** 329 (1997).
[31] Y. Ogawa, P.-s., Lei, and P. Kováč, *Carbohydr. Res.* **288,** 85 (1996).
[32] Y. Ogawa, P.-s. Lei, and P. Kováč, *Carbohydr. Res.* **293,** 173 (1996).

or linker) suitable for linking the hapten to a carrier. The drawback of such an approach is that it requires many chemical manipulations to be performed with the assembled oligosaccharide molecule. We have, in the past, prepared an *N*-acyl-perosamine-containing disaccharide from monosaccharide building blocks that had the requisite *N*-acyl group already in place,[33] and showed that such an approach is feasible. However, because the key glycosyl donor did not bear a selectively removable protecting group at the site of further extension of the oligosaccharide chain, the overall reaction scheme was not applicable to synthesizing higher oligosaccharides. Here we describe a synthesis of the same disaccharide but in the form of glycoside whose aglycon allows conjugation to carriers. The disaccharide is assembled from monosaccharide building blocks that carry the *N*-acyl chain already in place. Also, the glycosyl acceptor used is linker equipped, and the glycosyl donor has a selectively removable protecting group at the site of the further extension of the oligosaccharide chain. Briefly (Schemes 5 and 6), the synthesis starts with the known[27,30] 1,2-di-*O*-acetyl-4-azido-3-*O*-benzyl-4,6-dideoxy-α-D-mannopyranose (**1**) which is converted, conventionally,[34] to the corresponding phenylthio glycoside **2**. Treatment of **2** with hydrogen sulfide[35] selectively reduces the azido function, and the amine **3**, thus obtained, is treated with 3-deoxy-L-*glycero*-tetronolactone to introduce the tetronic acid side chain, giving amide **4**. Subsequent pivaloylation gives the fully protected compound **5**, which is converted to lactol **6** by *N*-iodosuccinimide (NIS)-or *N*-bromosuccinimide (NBS)-mediated hydrolysis.[36,37] The trichloroacetimidate **7**, obtained conventionally[38] from **6**, is then treated with 5-methoxycarbonylpentanol-1[39] to give the spacer-equipped perosamine derivative **8**. Subsequent, selective methanolysis[40,41] gives the 2-OH-free glycosyl acceptor **9**, which is treated with **7** to give the fully protected disaccharide **10**, bearing a selectively removable protecting group at 2-OH, which is potentially useful for making higher oligosaccharides in this series. Compound **10** is sequentially deprotected to give the linker-equipped disaccharides **13**. The nuclear magnetic resenance (NMR) spectra of **13** are similar to those

[33] M. Gotoh and P. Kováč, *J. Carbohydr. Chem.* **13,** 1193 (1994).

[34] R. J. Ferrier and R. H. Furneaux, *Methods Carbohydr. Chem.* **8,** 251 (1980).

[35] T. Adachi, Y. Yamada, I. Inoue, and M. Saneyoshi, *Synthesis* 45 (1977).

[36] J. Ohlsson and G. Magnusson, *Carbohydr. Res.* **329,** 49 (2000).

[37] H. C. Hansen and G. Magnusson, *Carbohydr. Res.* **307,** 233 (1998).

[38] M. Numata, M. Sugimoto, K. Koike, and T. Ogawa, *Carbohydr. Res.* **163,** 209 (1987).

[39] G. Solladie and C. Ziani-Cherif, *J. Org. Chem.* **58,** 2181 (1993).

[40] E. Petráková and J. Schraml, *Collect. Czech. Chem. Commun.* **48,** 877 (1983).

[41] N. E. Byramova, M. V. Ovchinnikov, L. V. Backinowsky, and N. K. Kochetkov, *Carbohydr. Res.* **124,** C8 (1983).

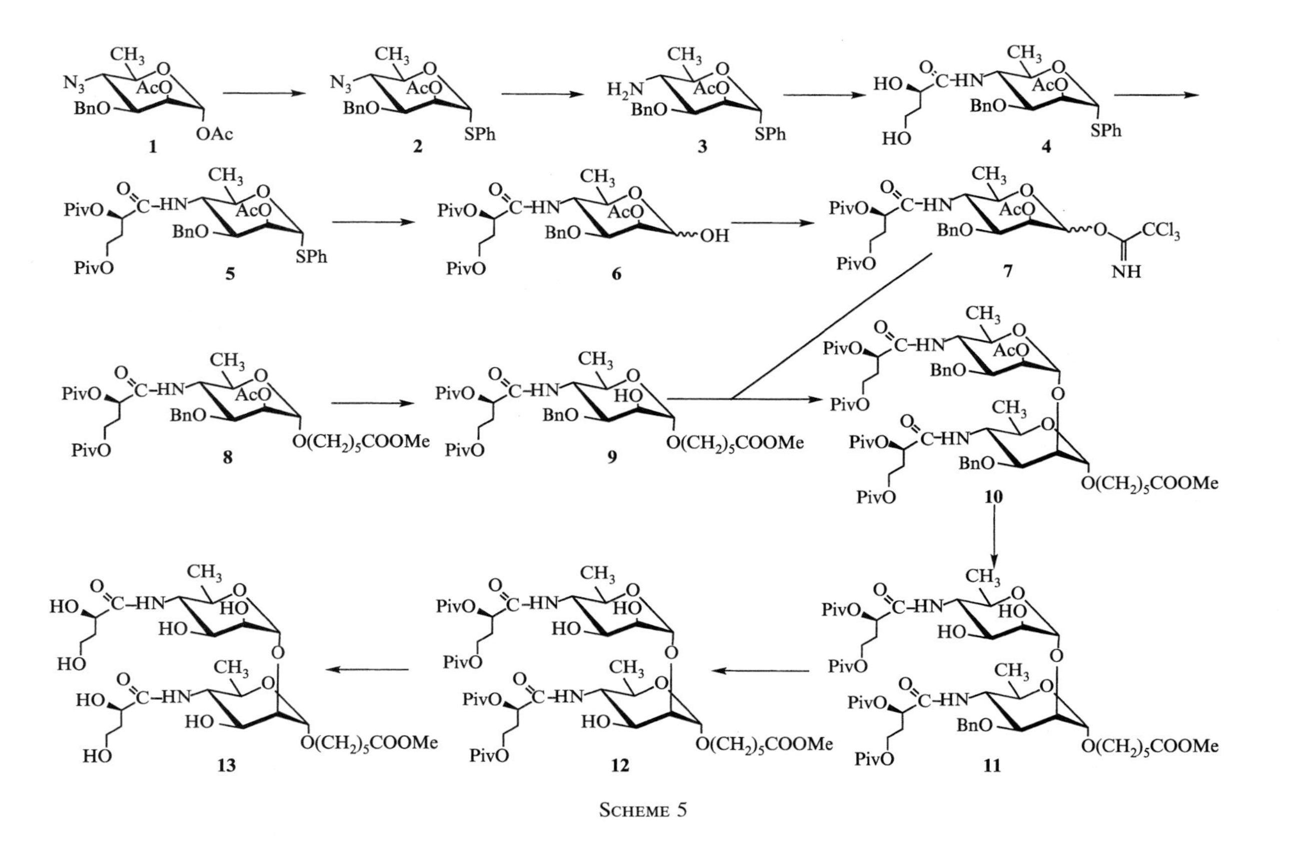
1
2
3
4
5
6
7
8
9
10
11
12
13
OAc
SPh
OH
NH
CCl3
O(CH2)5COOMe
PivO
BnO
AcO
HO
CH3
N3
H2N

SCHEME 5

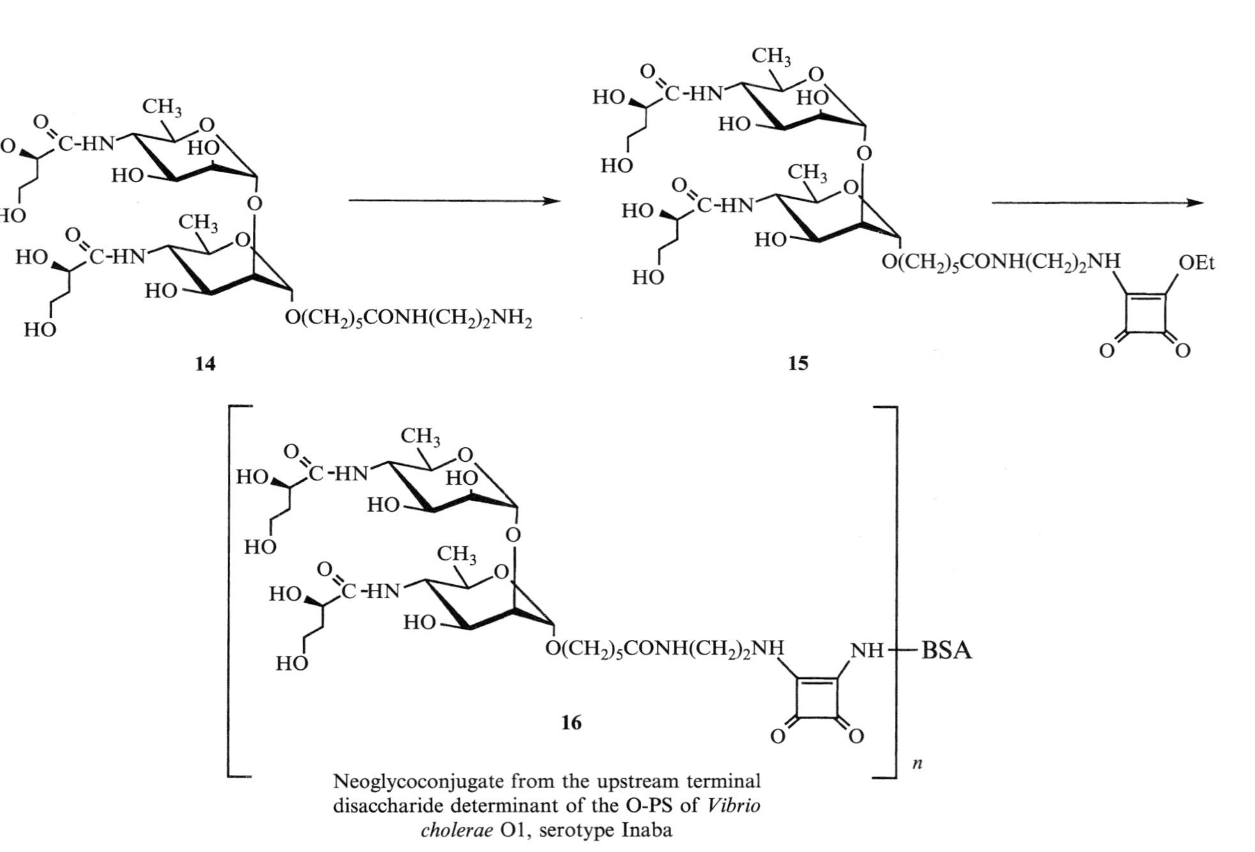

Neoglycoconjugate from the upstream terminal disaccharide determinant of the O-PS of *Vibrio cholerae* O1, serotype Inaba

SCHEME 6

of the corresponding methyl glycoside,[33] the differences reflecting the presence of a different aglycon. Whereas we show here that the selective deacetylation of **10** is feasible, preparation of the disaccharide suitable for conjugation (**13**) does not require such conversion. Simple deacylation of **10** would leave only benzyl groups to be removed before the desired hapten **13** would be obtained. Compound **13** is then treated with ethylenediamine to give amine **14**, which, when treated with a squaric acid diester at pH 7, gives the monoamide ester **15** ready to be conjugated to a suitable carrier.

Conjugation at a high hapten–carrier ratio of the fully functionalized disaccharide **15** to bovine serum albumin (BSA) is described in [10] in this volume.[42] Here we describe recovery of the excess ligand used in the conjugation reaction.

Experimental Methods

General Methods

Instruments and laboratory techniques used but not described here are the same as reported previously.[28] All reactions are monitored by thin-layer chromatography (TLC) on silica gel 60 coated glass slides [Whatman (Clifton, NJ) or Analtech (Newark, DE)]. Column chromatography is performed by gradient elution from columns of silica gel. Solvent mixtures slightly less polar than those used for TLC are used at the onset of development. Assignments of NMR signals (Tables II–IV) are made by first-order analysis of the spectra, and the assignments are supported by homonuclear decoupling experiments or homonuclear and heteronuclear two-dimensional correlation spectroscopy, run with the software supplied with the spectrometers. When reporting assignments of NMR signals of disaccharides, sugar residues are numbered, beginning with the one bearing the aglycone, and are identified by a Roman numeral superscript in listings of signal assignments. Nuclei assignments without a superscript notation indicate that those signals have not been individually assigned. Thus, for example, in a spectrum of a disaccharide, a resonance denoted H-3 could be that of H-3 of either sugar residue. ^{13}C NMR spectra of squaric acid amides show typical splitting of some signals, because of the double bond nature of the vinylogous squaric acid amide group.[15] 3-Deoxy-L-*glycero*-tetronolactone [(*S*)-(−)-α-hydroxy-δ-butyrolactone] and 3,4-diethoxy-3-cyclobutene-1,2-dione are purchased from Aldrich (Milwanker, WI) and used as supplied. Palladium-on-charcoal catalyst (5%, ESCAT 103) is a product of Engelhard Industries (Iselin, NJ). The functionalized

[42] R. Saksena, A. Chernyak, A. Karavanor, and P. Kováč, *Methods Enzymol.* **362** [10] (2003) (this volume).

TABLE II

^{1}H NMR CHEMICAL SHIFTS AND PEAK MULTIPLICITIES[a] IN $CDCl_3$ FOR COMPOUNDS **2–9** and **17**

Compound	Chemical shift (δ)							
	H-1	H-2	H-3	H-4	H-5	H-6	PhCH_2	$COCH_3$
2[b]	5.41d	5.58dd	3.77dd	3.50t	4.03m	1.34d	4.71d; 4.56d	2.12s
3[c]	5.47d	5.60dd	3.56dd	2.96t	5.05m	1.30d	4.72d, 4.42d	2.11s
4[d]	5.46d	5.59dd	3.83–3.69m	4.12q, po	~4.20m, po	1.25d	4.68d, 4.42d	2.12s
5[e]	5.46d	5.66–5.61m[f]	3.77dd	4.03q, po	~4.18m, po	1.24d	4.67d, 4.40d	2.14s
6[g]	~5.18d, po	5.38dd	3.86dd	4.03–3.80m	4.03–3.80m	~1.18d	4.65d, 4.38d	2.14s
7[h]	6.22d	5.53bt	3.86dd, po	4.03bt	~3.83m, po	1,23d	4.68d, 4.42d	2.20
8[i]	4.74d	5.34dd	3.78dd	3.97q	~3.68m, po	~1.21d, po	4.65d, 4.37d	2.14s
9[j]	4.83d	4.00dd, po	3.73dd, po	3.99q, po	~3.70	~1.2d, po	4.66d, 4.53d	
17[k]	4.77d, po	5.77dd	3.50dd	3.43t	3.22m	1.42d	4.77, d, po; 4.49d	2.19s

[a] d, Doublet; dd, doublet of doublets; t, triplet; q, quartet; m, multiplet: b, broad; po, partially overlapped; s, singlet.

[b] $J_{PhCH_2} = 11.3$ Hz.

[c] $J_{PhCH_2} = 11.3$ Hz.

[d] Measured after addition of a drop of D_2O. Before deuteration, the NH signal appears at δ 6.88, $J_{4,NH} = 9.0$ Hz, $\delta_{H\text{-}2'}$ 4.29 (dd, $J = 3.7, 8.2$), $\delta_{H\text{-}3'a,b}$ 2.06m, po, 1.81m, $\delta_{H\text{-}4'a,b}$ 3.83–3.69 (m, po), $J_{PhCH_2} = 11.8$ Hz.

[e] $J_{PhCH_2} = 11.6$ Hz, $\delta_{C(CH_3)_3}$ 1.21s, 1.20s.

[f] Overlapped with the signal of NH.

[g] Data for the α-anomer, largely predominating. δ_{NH} 5.73, $J_{4,NH} = 8.0$ Hz, $\delta_{H\text{-}2'}$ ~5.18 (dd, po, $J = 4.3, 7.3$), $\delta_{H\text{-}3'a,b}$ 2.30–1.91m, po.

[h] δ_{NH} 5.59, $J_{4,NH} = 8.9$ Hz, $\delta_{H\text{-}2'}$ 5.18 (dd, $J = 4.1, 9.3$), $\delta_{H\text{-}3'a,b}$ 2.27–1.98m, po, $\delta_{C(CH_3)_3}$ 1.19s, 1.15s.

[i] δ_{NH} 5.76, $J_{4,NH} = 9.0$ Hz, $\delta_{H\text{-}2'}$ 5.16 (dd, $J = 4.2, 9.4$), $\delta_{H\text{-}3'a,b}$ 2.28–1.95m, po, $\delta_{C(CH_3)_3}$ 1.19s, 1.18s, $\delta_{H\text{-}1''a}$ ~3.64, po, $\delta_{H\text{-}1''b}$ 3.40m, $\delta_{H\text{-}2''}$ ~1.59m, po, $\delta_{H\text{-}3''}$ 1.38m, $\delta_{H\text{-}4''}$ ~1.64m, po, $\delta_{H\text{-}5''}$ 2.32t, $J = 7.3$ Hz, δ_{COCH_3} 3.67s.

[j] δ_{NH} 5.96, $J_{4,NH} = 9.3$ Hz, $\delta_{H\text{-}2'}$ 5.16 (dd, $J = 4.2, 9.4$), $\delta_{H\text{-}3'a,b}$ 2.23m, 2.04m, $\delta_{C(CH_3)_3}$ 1.21s, 1.18s, $\delta_{H\text{-}1''a}$ ~3.65m, po, $\delta_{H\text{-}1''b}$, 3.41 2t, $J = 6.1$ and 9.9 Hz, $\delta_{H\text{-}2'',H\text{-}4''}$, 1.70–1.55m, $\delta_{H\text{-}3''}$ 1.38m, $\delta_{H\text{-}5''}$ 2.32t, $J = 7.3$ Hz.

[k] $J_{PhCH_2} = 11.0$ Hz.

TABLE III
^{1}H NMR Ring-Proton Coupling Constants for Compounds **2–9** and **17**

	Coupling constant (Hz)[a]				
Compound	$J_{1,2}$	$J_{2,3}$	$J_{3,4}$	$J_{4,5}$	$J_{5,6}$
2	1.6	3.2	9.8	9.9	6.3
3	1.4	3.0	9.9	9.9	6.3
4	1.9	3.1	~10	~10	5.9
5	1.5	3.1	10.6	~10	6.2
6	2.0	2.8	10.4	ND[b]	ND[b]
7	1.9	3.1	10.6	~10.2	6.2
8	1.7	3.1	10.6	~10.3	~6.2
9	1.6	~3.0	~10.2	~10.2	ND[b]
17	1.1	3.1	9.6	9.6	6.2

[a] For the conditions of measurements, see Table II.
[b] ND, Not determined because of complex multiplicity or overlapping signals.

terminal, monosaccharide determinant of *Vibrio cholerae* O1, serotype Ogawa, 1-[(2-aminoethylamido)carbonylpentyl 4-(3-deoxy-L-*glycero*-tetronamido)-4,6-dideoxy-2-*O*-methyl-α-D-mannopyranoside]-2-ethoxycyclobutene-3,4-dione, is prepared as described.[18] Conjugation to BSA of this, as well as other haptens described here, is carried out according to the protocol described in [10] in this volume.[42]

Phenylthio 2-O-acetyl-4-azido-3-O-benzyl-4,6-dideoxy-α-D-manno pyranoside (2). To a solution of 1,2-di-*O*-acetyl-4-azido-3-*O*-benzyl-4,6-dideoxyα-α,β-D-mannopyranose[27,30] (**1**, 7.207 g, 19.83 mmol) in dichloromethane (200 ml) is added thiophenol (2.7 ml, 26.4 mmol) followed by boron trifluoride etherate (2.77 ml, 21.82 mmol), and the solution is stirred at room temperature until TLC [hexane–acetone, 4:1 (v/v)] shows that the reaction is complete (3–4 h). The mixture is neutralized with aqueous $NaHCO_3$ and the organic layer is washed with brine, dried over Na_2SO_4, and concentrated. Chromatography gives first the α-anomer (4.4 g), mp 56.5–57.5° (from ethanol–hexane); $[\alpha]_D + 164°$ (*c* 1.2, $CHCl_3$); chemical ionization mass spectrometry (CIMS): *m/z* 431 ($[M+18]^+$).

Analysis: Calculated for $C_{21}H_{23}N_3O_4S$: C, 61.00; H, 5.61; N, 10.16. Found: C, 61.13; H, 5.73; N, 10.09. Eluted next is a mixture of α- and β-anomers (2.4 g), followed by phenylthio 2-*O*-acetyl-4-azido-3-*O*-benzyl-4,6-dideoxy-β-D-mannopyranose (**17**, 0.2 g), mp 87.5–89° (from ethanol); $[\alpha]_D - 10.6°$ (*c* 1.5, $CHCl_3$); fast atom bombardment MS (FABMS): *m/z* 414 ($[M+1]^+$), 436 ($[M+Na]^+$).

TABLE IV
^{13}C NMR CHEMICAL SHIFTS FOR MONOSACCHARIDE DERIVATIVES **2–9** AND **17**[a]

Compound	Chemical shift							
	C-1	C-2	C-3	C-4	C-5	C-6	PhCH_2	CH_3CO
2	86.05[b]	69.01	76.35	64.13	68.23	18.34	71.69	20.93
3	86.43[c]	68.69	77.87	54.11	70.80	17.94	71.31	20.93
4[d]	85.97	68.84	74.43	52.14	69.14	17.77	71.07	20.96
5[e]	86.11	68.73	73.74	52.63	68.96	17.76	70.80	20.98
6[f]	92.16	67.63[g]	73.21	52.35	67.59[g]	17.88	70.69	21.01
7[h]	94.91	65.37	72.23	51.67	70.70	17.90	70.70	20.88
8[i]	97.60	67.23	73.76	52.24	67.66	17.66	70.73	20.82
9[j]	98.84	67.04[g]	76.24	52.03	67.01[g]	17.87	71.02	
17	85.22[k]	68.82	79.24	63.43	75.28	18.77	71.63	20.73

[a] For conditions of measurements, see Table II.
[b] $J_{C\text{-}1,H\text{-}1} = 168.5$ Hz.
[c] $J_{C\text{-}1,H\text{-}1} = 168.1$ Hz.
[d] $\delta_{C\text{-}2'}$ 71.31, $\delta_{C\text{-}3'}$ 35.43, $\delta_{C\text{-}4'}$ 60.22.
[e] $\delta_{C\text{-}2'}$ 70.27, $\delta_{C\text{-}3'}$ 30.93, $\delta_{C\text{-}4'}$ 59.57; $\delta_{C(CH_3)_3}$ 27.09, 26.93; $\delta_{C(CH_3)_3}$ 38.77, 38.68.
[f] $\delta_{C\text{-}2'}$ 70.24, $\delta_{C\text{-}3'}$ 30.90, $\delta_{C\text{-}4'}$ 59.62; $\delta_{C(CH_3)_3}$ 27.04, 26.89; $\delta_{C(CH_3)_3}$ 38.72, 38.66.
[g] The assignment may be reversed.
[h] $\delta_{C\text{-}2'}$ 70.24, $\delta_{C\text{-}3'}$ 30.87, $\delta_{C\text{-}4'}$ 59.57; $\delta_{C(CH_3)_3}$ 27.08, 26.81; $\delta_{C(CH_3)_3}$ 38.71 (2 C).
[i] $\delta_{C\text{-}2'}$ 70.29, $\delta_{C\text{-}3'}$ 30.76, $\delta_{C\text{-}4'}$ 59.54; $\delta_{C(CH_3)_3}$ 26.89, 26.72; $\delta_{C(CH_3)_3}$ 38.57, 38.51. $\delta_{C\text{-}1''}$ 67.31, $\delta_{C\text{-}2''}$ 28.60, $\delta_{C\text{-}3''}$ 25.46; $\delta_{C\text{-}4''}$ 24.28, $\delta_{C\text{-}5''}$ 33.66, δ_{COOCH_3} 51.37.
[j] $J_{C\text{-}1,H\text{-}1} = 167.9$ Hz, $\delta_{C\text{-}2'}$ 70.37, $\delta_{C\text{-}3'}$ 30.97, $\delta_{C\text{-}4'}$ 59.67; $\delta_{C(CH_3)_3}$ 27.13, 26.99; $\delta_{C(CH_3)_3}$ 38.78, 38.71.
[k] $\delta_{C\text{-}1''}$ 67.21, $\delta_{C\text{-}2''}$ 28.86, $\delta_{C\text{-}3''}$ 25.73; $\delta_{C\text{-}4''}$ 24.57, $\delta_{C\text{-}5''}$ 33.94, δ_{COOCH_3} 51.53. $J_{C\text{-}1,H\text{-}1} = 53.5$ Hz.

Analysis: Calculated for $C_{21}H_{23}N_3O_4S$: C, 61.00; H, 5.61; N, 10.16. Found: C, 61.15; H, 5.59; N, 10.18.

Total yield of the phenylthioglycosides, 6.8 g (85%).

Phenylthio 2-O-acetyl-4-amino-3-O-benzyl-4,6-dideoxy-α-D-mannopyranoside (3). Hydrogen sulfide gas is passed at 40° for 30 min through a solution of the azido derivative **2** (2.18 g, 5.29 mmol) in 30 ml of pyridine–water (2:1, v/v). The flask is closed with a rubber septum and the solution is kept at 40° until TLC [hexane–ethyl acetate, 4:1 (v/v)] shows complete disappearance of the starting material (10–16 h). The dark solution is filtered to separate the yellow solid (sulfur) formed, and a weak stream of nitrogen is passed through the solution, which then becomes almost colorless. After concentration, chromatography affords 1.63 g (79%) of **3**, mp 95.5–97.5° (from ethyl acetate $[\alpha]_D$ +41.5° (*c* 1.4, $CHCl_3$); CIMS: *m*/*z* 388 ($[M+1]^+$), 405 ($[M+18]^+$).

Analysis: Calculated for $C_{21}H_{25}NO_4S$: C, 65.09; H, 6.50; N, 3.61. Found: C, 65.29; H, 6.54; N, 3.66.

*Phenylthio 2-O-acetyl-3-O-benzyl-4-(3-deoxy-L-glycero-tetronamido)-4,6-dideoxy-α-D-mannopyranoside (**4**).* (*S*)-(−)-α-Hydroxy-δ-butyrolactone (410 mg, 4 mmol) is added to a solution of the foregoing amine **3** (1.0 g, 2.6 mmol) in dimethylformamide (DMF, 1.3 ml), which is then heated at 100° until TLC (ethyl acetate, undiluted) shows that the reaction is complete (24–48 h). Concentration and chromatography gives **4** (1.3 g, ~100%), which crystallizes as a hemihydrate, mp 98–100° (from ethyl acetate–hexane); $[\alpha]_D$ +51.2° (*c* 1.6, $CHCl_3$); CIMS: *m*/*z* 490 ($[M+1]^+$), 507 ($[M+18]^+$).

Analysis: Calculated for $C_{25}H_{31}NO_7S0.5H_2O$: C, 60.22; H, 6.47; N, 2.81. Found: C, 60.47; H, 6.41; N, 2.82.

*Phenylthio 2-O-acetyl-3-O-benzyl-4-(3-deoxy-L-glycero-2,4-di-O-pivaloyltetronamido)-4,6-dideoxy-α-D-mannopyranoside (**5**).* Trimethylacetyl chloride (0.28 ml, 2.292 mmol) is added to a solution of the foregoing amide **4** (374 mg, 0.764 mmol) in pyridine (5 ml). The solution is stirred at room temperature overnight, when hexane–ethyl acetate (4:1, v/v) shows complete conversion of the starting material. The mixture is neutralized with aqueous $NaHCO_3$ and concentrated, and the residue is partitioned between CH_2Cl_2 and water. The organic layer is dried over Na_2SO_4, concentrated, and chromatographed, to give fully protected compound **5** (493 mg, 96%), mp 125.5–127° (from ethanol); $[\alpha]_D$ +23.0° (*c* 1.1, $CHCl_3$); FABMS: *m*/*z* 658 ($[M+1]^+$), 680 ($[M+Na]^+$).

Analysis: Calculated for $C_{35}H_{47}NO_9S$: C, 63.90; H, 7.20; N, 2.13. Found: C, 64.00; H, 7.17; N, 2.12.

*5-Methoxycarbonylpentyl 2-O-acetyl-3-O-benzyl-4-(3-deoxy-L-glycero-2,4-di-O-pivaloyltetronamido)-4,6-dideoxy-α-D-mannopyranoside (**8**).* A drop of water is added at 0° to a solution of **5** (364 mg, 0.553 mmol) in acetone (10 ml) followed by *N*-bromosuccinimide (108 mg, 0.608 mmol). After 15 min, TLC [hexane–ethyl acetate, 2:1 (v/v)] shows that the reaction is complete. Saturated, aqueous $NaHCO_3$ is added to the reaction mixture, which is then diluted with CH_2Cl_2. The phases are separated, and the aqueous layer is extracted with CH_2Cl_2. The organic layers are combined, dried over Na_2SO_4, and concentrated. Chromatography [hexane–ethyl acetate, 9:1 (v/v)] gives 2-*O*-acetyl-3-*O*-benzyl-4-(3-deoxy-L-*glycero*-2,4-di-*O*-pivaloyltetronamido)-4,6-dideoxy-α,β-D-mannopyranose (**6**) (260 mg, 0.422 mmol, 82%) as a white foam; FABMS: *m*/*z* 566 ($[M+1]^+$), 588 ($[M+Na]^+$).

To a solution of the foregoing lactol **6** (3.73 g, 6.59 mmol) in dichloromethane (66 ml), stirred at room temperature under argon, is added trichloroacetonitrile (6.6 ml, 65.9 mmol) followed by 1,8-diazabicyclo

[5.4.0]undec-7-ene (DBU, 50 μl, 0.33 mmol). After 30 min, the reaction is complete, as shown by TLC [hexane–ethyl acetate, 3:2 (v/v)]. The mixture is concentrated, and chromatography gives the amorphous 2-*O*-acetyl-3-*O*-benzyl-4-(3-deoxy-L-*glycero*-2,4-di-*O*-pivaloyltetronamido)-4,6-dideoxy-α, β-D-mannopyranose trichloroacetimidate (**7**) in virtually theoretical yield; FABMS: *m*/*z* 709 ($[M+1]^+$), 731 ($[M+Na]^+$).

A solution of the trichloroacetimidate (**7**, 755 mg, 1.06 mmol) and 5-methoxycarbonyl-1-pentanol[39] (195 mg, 1.33 mmol) in toluene is concentrated (three times). The residue is dissolved in CH_2Cl_2 (15 ml) and molecular sieves (4 A, 1 g) are added, followed by trimethylsilyl triflate (4.25 μl, 21 μmol). After 3 h TLC [toluene–acetone, 6:1 (v/v)] shows that all glycosyl donor has been consumed. The mixture is neutralized by addition of solid $NaHCO_3$, filtered through Celite, diluted with CH_2Cl_2, washed successively with aqueous $NaHCO_3$, 1 *M* HCl, and $NaHCO_3$, dried over Na_2SO_4, and concentrated. Chromatography [toluene–acetone, 10:1 (v/v)] affords the title, amorphous glycoside **8** (535 mg, 72%); $[\alpha]_D$ -9.1° (*c* 0.9, $CHCl_3$). FABMS: *m*/*z* 694 ($[M+1]^+$), 716 ($[M+Na]^+$).

Analysis: Calculated for $C_{36}H_{55}NO_{12}$: C, 62.32; H, 7.99; N, 2.02. Found: C, 62.19; H, 8.10; N, 1.99.

*5-Methoxycarbonylpentyl 3-O-benzyl-4-(3-deoxy-L-glycero-2,4-di-O-pivaloyltetronamido)-4,6-dideoxy-α-D-mannopyranoside (**9**).* To a solution of **8** (2.98 g, 4.29 mmol) in chloroform (8.5 ml) is added, under argon at 0°, methanolic HCl [19.5 ml, freshly prepared by addition, at 0°, of acetyl chloride (2.0 ml) to dry methanol (50 ml)]. The solution is stirred at ~12° for 24 h with exclusion of moisture, when TLC [hexane–ethyl acetate, 3:2 (v/v)] shows that almost all starting material has been consumed. One major and several minor products are formed. The mixture is treated with solid $NaHCO_3$ and diluted with dichloromethane and then washed with water, dried over Na_2SO_4, and concentrated. Chromatography gives the expected alcohol **9** (1.81 g, 64%) as a colorless foam; $[\alpha]_D$ -14.3° (*c* 0.3, $CHCl_3$); CIMS: *m*/*z* 652 ($[M+1]^+$).

Analysis: Calculated for $C_{34}H_{53}NO_{11}$: C, 62.65; H, 8.20; N 2.15. Found: C, 62.36; H, 8.27; N, 2.15.

*5-Methoxycarbonylpentyl [2-O-acetyl-3-O-benzyl-4-(3-deoxy-2,4-di-O-pivaloyl-L-glycero tetronamido)-4,6-dideoxy-α-D-mannopyranosyl]-(1→2) -3-O-benzyl-4-(3-deoxy-2,4-di-O-pivaloyl-L-glycero tetronamido)-4,6-dideoxy-α-D-mannopyranoside (**10**).* To a solution of the imidate **7** (2.78 g, 3.92 mmol) and alcohol **9** (2.22, g, 3.4 mmol) in dichloromethane (60 ml) is added, under argon and at room temperature, trimethylsilyl triflate (TMSOTf, 23.5 μl, 0.117 mmol). After 15 min, when TLC [hexane–ethyl acetate, 1:1 (v/v)] shows that all glycosyl donor is consumed, the reaction is neutralized by addition of solid $NaHCO_3$, diluted with dichloromethane, washed with a solution of $NaHCO_3$, dried over Na_2SO_4, and concentrated. Chromatography gives the expected disaccharide **10** (3.31 g, 81%) as a colorless foam; $[\alpha]_D$ −20° (*c* 0.9, $CHCl_3$); FABMS: *m*/*z* 1163 ($[M+Cs]^+$). ^1H NMR ($CDCl_3$): δ 5.84, 5.51 (2 d, 2 H, $J_{4,NH} = 9.0$ and 9.3 Hz, respectively, 2 NH), 5.46 (bdd, 1 H, H-2^{II}), ~5.19, ~5.16 (2 dd, partially overlapped, 2 H, $J_{2',3'a} = \sim$ 4.4 Hz, $J_{2',3'b} = 9.3$ Hz, H-$2'^{I,II}$), 4.81 (d, 1 H, $J_{1,2} = 1.8$ Hz, H-1^{II}), 4.71 (d, 1 H, $J_{1,2} = 1.8$ Hz, H-1^{I}), 4.68–4.40 (4 d, 4 H, $J = \sim 12$ Hz, 2 CH_2Ph), 4.14–4.06 (m, partially overlapped, $4'^{I,II}$), ~4.10–3.92 (m partially overlapped, H-$4^{I,II}$), 3.83 (bt, 1 H, H-2^{I}), 3.77–3.71 (2 dd, partially overlapped, H-$3^{I,II}$), 3.71–3.56 (m, partially overlapped, H-$5^{I,II}$), H-1″a, incl s, 3.68 for $COOCH_3$), 3.35 (2 t, 1 H, $J = 6.1$, 9.7 Hz, H-1″b), 2.35–1.93 (m, 7 H, H-$3'^{I,II}$, H-5″a,b, incl 2.12, s, $COCH_3$), 1.70–1.52 (m, 4 H, H-2″ a,b,4″a,b), 1.42–1.32 (m, 2 H, H-3″a,b), 1.24–1.12 [m, 42 H, H-$6^{I,II}$, 4 $C(CH_3)_3$]. ^{13}C NMR ($CDCl_3$): δ 99.67 ($J_{C\text{-}1,H\text{-}1} = 172.5$ Hz, C-1^{II}), 96.43 ($J_{C\text{-}1,H\text{-}1} = 170.0$ Hz, C-1^{I}), 75.29, 72.92 (C-$3^{I,II}$), 75.12 (C-2^{I}), 71.31, 70.53 (2 CH_2Ph), 70.58, 70.32 (C-$2'^{I,II}$), 68.88, 67.77 (C-$5^{I,II}$), 67.33 (C-1″), 66.93 (C-2^{II}), 59.66 (2 C, C-$4'^{I,II}$), 52.36, 51.45 (C-$4^{I,II}$), 51.54 ($COCH_3$), 38.77, 38.72 [$C(CH_3)_3$], 33.89 (C-5″), 30.04, 30.94 (C-$3'^{I,II}$), 28.85 (C-2″), 27.15, 26.98, [4 $C(CH_3)_3$], 25.72 (C-3″), 24.52 (C-4″), 21.06 ($COCH_3$), 18.17, 17.88 (C-$6^{I,II}$).

Analysis: Calculated for $C_{63}H_{94}N_2O_{20}$: C, 63.09; H, 7.90; N, 2.34. Found: C, 62.83; H, 7.69; N, 2.17.

*5-Methoxycarbonylpentyl [3-O-benzyl-4-(3-deoxy-2,4-di-O-pivaloyl-L-glycero-tetronamido)-4,6-dideoxy-α-D-mannopyranosyl]-(1→2)-3-O-benzyl-4-(3-deoxy-2,4-di-O-pivaloyl-L-glycero-tetronamido)-4,6-dideoxy-α-D-mannopyranoside (**11**).* The fully protected disaccharide **10** (1.39 g, 1.16 mmol) is dissolved in chloroform (2.25 ml), and methanolic HCl (5.6 ml, prepared from 2 ml of acetyl chloride and 50 ml of methanol) is added. The solution is kept at 5° for 21 h, when TLC [hexane–ethyl acetate, 1:1 (v/v)] shows that little starting material remains and also that products of further deacylation have formed. The solution is neutralized with solid $NaHCO_3$, diluted with CH_2Cl_2, washed with water, dried over Na_2SO_4, and concentrated, and the residue is chromatographed to give 805 mg (60%) of **11**; $[\alpha]_D$ −15.0° (*c* 0.5, $CHCl_3$); FABMS: *m*/*z* 1163 ($[M+Li]^+$); ^1H NMR ($CDCl_3$): δ 5.82, 5.59 (2 d, 2 H, $J_{4,NH} = 9.6$ Hz, 2 NH), 5.18 (m, 2 H, H-$2'^{I,II}$), 4.96 (d, 1 H, $J_{1,2} = 1.9$ Hz, H-1^{II}), 4.74 (d, 1 H, $J_{1,2} = 1.9$ Hz, H-1^{I}), 4.71–4.47 (m, 4 H, 2 CH_2Ph), 4.17 (bt, 1 H, H-2^{II}), 4.15–4.00 (m, 7 H, H-2^{I},$4^{I,II}$, $4'^{I,II}$a,b), 3.93 (bt, H-2^{I}) 3.73–3.52 (m, 8 H, H-$3^{I,II}$, $5^{I,II}$, 1″a, incl s, 3.67 for $COOCH_3$), 3.42–3.32 (m, 1 H, H-1″b), 2.32 (t, partially overlapped, H-5″a,b), 2.34–1.95 (m, partially overlapped, H-3′a,$b^{I,II}$), 1.70–1.52 (m, 4 H, H-2″a,b,4″a,b), 1.42–1.33 (m, 2 H, H-3″a,b), 1.27–1.15 [m, 42 H, 4 $C(CH_3)_3$, incl 1.16, 1.15, 2 d, $J_{5,6} = 6.2$ Hz,

Hz, H-6^{I,II}]. ^{13}C NMR ($CDCl_3$): δ 100.97 (C-1^{II}), 98.57 (C-1^{I}), 75.55, 75.36 (C-3^{I,II}), 74.19 (C-2^{I}), 71.11, 70.74 (2 CH_2Ph), 70.47, 70.21 (C-2$'^{I,II}$), 68.10, 67.92 (C-5^{I,II}), 67.14 (C-1″), 66.37 (C-2^{II}), 59.60, 59.54 (C-4$'^{I,II}$), 51.90, 51.06 (C-4^{I,II}), 51.44 ($COOCH_3$), 38.66, 38.59 [$C(CH_3)_3$], 33.75 (C-5″), 30.89, 30.85 (C-3$'^{I,II}$), 28.70 (C-2″), 27.01, 26.83 [4 $C(CH_3)_3$], 25.96 (C-3″), 24.36 8(C-4″), 18.00, 17.70 (C-6^{I,II}).

Analysis: Calculated for $C_{61}H_{92}N_2O_{19}$: C, 63.30; H, 8.01; N, 2.42. Found: C, 63.18; H, 8.17; N, 2.38.

*5-Methoxycarbonylpentyl 4-(3-deoxy-L-glycero-tetronamido)-4,6-dideoxy-α-D-mannopyranosyl-(1→2)-4-(3-deoxy-L-glycero-tetronamido)-4,6-dideoxy-α-D-mannopyranoside (**13**).* A mixture of the foregoing alcohol **11** (1 g) and 5% palladium-on-charcoal catalyst (0.35 g) in methanol (50 ml) is stirred under hydrogen overnight, when TLC [CH_2Cl_2–acetone–concentrated NH_4OH, 1:1:0.05 (v/v)] shows that the reaction is complete. After filtration and concentration of the filtrate, the residue is eluted from a small column of silica gel to give **12**; $[\alpha]_D$ −21° (*c* 0.2, $CHCl_3$); FABMS: *m/z* 977 ($[M+1]^+$); ^{1}H NMR ($CDCl_3 + D_2O$): δ 6.87, 6.51 (2 d, 2 H, $J_{4,NH} = 9.0$ Hz, 2 NH), 5.14–5.08 (m, 2 H, H-2$'^{I,II}$), 4.96 (d, 1 H, $J_{1,2} = 1.8$ Hz, H-1^{II}), 4.87 (d, 1 H, $J_{1,2} = 1.5$ Hz, H-1^{1}), 4.17–4.13 (m, 4 H, 4$'^{I,II}$a,b), 4.05 (bt, 1 H, H-2), 3.98 (t, partially overlapped, H-4^{II}), 3.86 (t, partially overlapped, H-4^{I}), 3.94–3.73 (m, partially overlapped, H-3^{II}, 2^{I}, 5^{II}, 3^{I}), 3.68 (s, $COOCH_3$), 3.67–3.50 (m, 2 H, H-5^{I}, 1″a), 3.40–3.33 (m, 1 H, H-1″b), 2.32 (t, 2 H, $J = 7.3$ Hz, H-5″), 2.26–2.10 (m, 2 H, H-3′a,b^{I,II}), 1.70–1.52 (m, 4 H, H-2″a,b,4″a,b), 1.42–1.32 (m, 2 H, H-3″a,b), 1.25, 1.15 [2 s, partially overlapped, 4 $C(CH_3)_3$], 1.14 (2 d, partially overlapped, $J_{5,6} = 6.2$ Hz, H-6^{I,II}). ^{13}C NMR ($CDCl_3$): δ 102.16 (C-1^{II}), 98.38 (C-1^{I}), 79.07 (C-2^{I}), 70.66 (2 C, C-2$'^{I,II}$), 69.38 (C-2^{II}), 68.83 (C-3^{II}), 68.56 (C-3^{I}), 68.19 (C-5^{II}), 67.46 (C-5^{I}), 67.31 (C-1″), 60.04, 59.83 (C-4$'^{I,II}$), 53.71 (C-4^{I}), 53.94 (C-4^{II}), 51.58 ($COOCH_3$), 38.79, 38.69 [$C(CH_3)_3$], 33.86 (C-5″), 30.96, 30.77 (C-3$'^{I,II}$), 28.83 (C-2″), 27.08, 26.98, 26.96 [2 C, C, C, $C(CH_3)_3$], 25.56 (C-3″), 24.43 (C-4″), 17.93, 17.75 (C-6^{I,II}).

Analysis: Calculated for $C_{47}H_{80}N_2O_{19}$: C, 57.77; H, 8.25; N 2.87. Found: C, 57.31; H, 8.18; N, 2.78.

Conventional transesterification of **12**, effected with $NaOCH_3$, in methanol, gives **13** (0.32 g, 57% over two steps). ^{1}H NMR (D_2O): δ 5.02 (d, 1 H, $J_{1,2} = 1.9$ Hz, H-1^{II}), 4.91 (d, 1 H, $J_{1,2} = 1.8$ Hz, H-1^{I}), 4.30, 4.29 (2 dd, 2 H, partially overlapped, $J = 4.0$ and 8.6 Hz, H-2$'^{I,II}$), 4.11 (dd, 1 H, $J_{2,3} = 3.3$ Hz, H-2^{II}), 4.08–4.02 (m, 2 H, H-3^{I,II}), 3.97–3.82 (m, 5 H, H-2^{I}, 4^{I,II}, 5,^{I,II}), 3.76–3.66 (m, 8 H, H-4′a,b^{I,II}, 1″a, incl s, 3.68, $COOCH_3$), 3.57–3.49 (m, 1 H, 1″b), 2.40 (t, 2 H, $J = 7.4$ Hz, H-5″a,b), 2.09–1.78 (2 m, 4 H, H-3′a,b^{I,II}), 1.67–1.56 (m, 4 H, H-4″a,b,2″a,b in that order), 1.43–1.33 (H-3″a,b); ^{13}C NMR (D_2O): δ 102.41 (C-1″), 98.54 (C-1^{I}), 69.24 (C-2^{II}),

69.07 (2 C, C-2′^{I,II}), 68.11 (C-5), 67.97 (C-1″), 67.83 (C-3), 67.71 (2 C, C-3,5), 57.93 (C-4′^{I,II}), 53.16, 52.85 (C-4^{I},4^{II}), 52.20 (COOCH_3), 36.08 (C-3″^{I,II}), 33.69 (C-5″), 28.20 (C-2″), 24.99 (C-3″), 24.10 (C-4″), 17.01, 16.94 (C-6^{I,II}). FABMS: *m*/*z* 641 ($[M+1]^+$), 663 ($[M+Na]^+$).

The foregoing compound can be characterized as the per-*O*-acetyl derivative, 5-methoxycarbonylpentyl 2,3-di-*O*-acetyl-[4-(2,4-di-*O*-acetyl-3-deoxy-L-*glycero*-tetronamido)-4,6-dideoxy-α-D-mannopyranosyl]-(1→2)-3-*O*-acetyl-4-(2,4-di-*O*-acetyl-3-deoxy-L-*glycero*-tetronamido)-4,6-dideoxy-α-D-mannopyranoside, which is obtained by conventional acetylation of **13** with acetic anhydride in pyridine, mp 123.5–124° (from ethanol–hexane); $[\alpha]_D + 8°$ (*c* 0.3, $CHCl_3$). ^{1}H NMR ($CDCl_3$): δ 6.36, 6.35 (2 d, partially overlapped, 2 H, $J = 9.2$ Hz, 2 NH), 5.32 (d, 1 H, $J_{2,3} = 3.5$, $J_{3,4} = 11.2$ Hz, H-3^{II}), 5.22 (dd, partially overlapped, $J_{1,2} = 2.0$ Hz, H-2^{II}), 5.19 (dd, partially overlapped, $J_{2,3} = 3.4$, $J_{3,4} = 11.2$ Hz, H-3^{I}), 5.11, 5.08 (2 d, partially overlapped, 2 H, $J = 4.7$ and 7.5 Hz, H-2′^{I,II}), 4.90 (d, 1 H, H-1^{II}), 4.75 (d, 1 H, $J_{1,2} = 1.7$ Hz, H-1^{I}), 4.32–4.05 (m, 6 H, H-4^{I,II}, 4′^{I,II}a,b), 3.88 (dd, 1 H, H-2^{I}), 3.81 (m, partially overlapped, H-5^{II}), 3.75–3.64 (m, 5 H, H-5^{I}, 1″a, incl s at 3.70, COOCH_3), 3.43–3.36 (m, 1 H, H-1″b), 2.34 (t, 1 H, $J = 7.4$ Hz, Hz, H-5″a,b), 2.23, 2.16, 2.15, 2.12, 2.10, 2.07, 2.05 (7 s, partially overlapped, 7 COCH_3), 2.15–2.07 (m, overlapped, H-3′^{I,II}a,b), 1.73–1.54 (m, 4 H, H-4″a,b,2″a,b), 1.46–1.34 (m, 2 H, H-3″a,b), 1.24 (d, 3 H, $J_{5,6} = 6.4$ Hz, H-6^{I}), 6.20 (d, 3 H, $J_{5,6} = 6.4$ Hz, H-6^{II}); ^{13}C NMR ($CDCl_3$): δ 99.23 (C-1^{II}), 98.47 (C-1^{I}), 76.43 (C-2^{I}), 70.93, 70.84 (C-2′^{I,II}), 69.82 (C-3^{I}), 69.60 (C-2^{II}), 69.13 (C-5^{I}), 68.08 (C-5^{II}), 67.97 (C-3^{II}), 67.13 (C-1″), 59.80 (2 C, C-4′^{I,II}), 51.68 (C-4^{II}), 51.61 (COOCH_3), 51.58 (C-4^{I}), 33.85 (C-5″), 30.60, 30.53 (C-3′^{I,II}), 28.69 (C-2″), 25.43 (C-3″), 24.32 (C-4″), 17.78 (C-6^{I}), 17.62 (C-6^{II}); FABMS: *m*/*z* 935 ($[M+1]^+$), 957 ($[M+Na]^+$).

Analysis: Calculated for $C_{41}H_{62}N_2O_{22}$: C, 52.67; H, 6.68; N, 3.00. Found: C, 52.62; H, 6.74; N, 2.95.

*(2-Aminoethylamido)carbonylpentyl 4-(3-deoxy-L-glycero-tetronamido)-4,6-dideoxy-α-D-mannopyranosyl)-(1→2)-4-(3-deoxy-L-*glycero*-tetronamido)-4,6-dideoxy-α-D-mannopyranoside (**14**).* A solution of **13** (350 mg) in ethylenediamine (5 ml) is stirred under argon at 70° for 48 h, when the mixture becomes dark and TLC [methanol–concentrated NH_4OH, 1C:1 (v/v)] shows complete conversion of the starting material to one more polar product. After concentration at 50°/133 Pa, the residue is chromatographed, to give amorphous **15** (250 mg, 71%). ^{1}H NMR (CD_3OD, 45°): δ 5.10 (d, 1 H, $J_{1,2} = 1.7$ Hz, H-1^{II}), 4.97 (d, 1 H, $J_{1,2} = 1.5$ Hz, H-1^{I}), 4.35–4.31 (2 d, 2 H, partially overlapped, $J = 4.1$ and 7.9 Hz, H-2′^{I,II}), 4.17 (dd, 1 H, $J_{2,3} = 2.8$ Hz, H-2^{II}), 4.14–3.98 (m, 5 H, H-3^{I,II,} 4^{I,II,}5), 3.94 (bdd, 1 H, H-2^{I}), 3.91–3.85 (m, 5 H, H-5,4′^{I,II}a,b), 3.84–3.77 (m, 1 H, H-1″a), 3.59–3.52 (m, 1 H, H-1″b), 3.42 (t, 2 H,

$J = 6.1$ Hz, H-6″a,b), 2.92 (t, 2 H, $J = 6.1$ Hz, H-7″a,b), 2.36 (t, 2 H, $J = 6.1$ Hz, H-5″a,b), 2.21–1.91 (2 m, 4 H, H-3′$^{\mathrm{I,II}}$a,b), 1.82–1.70 (m, 4 H, H-2″a,b,4″a,b), 1.60–1.50 (m, 2 H, H-3″a,b), 1.31, 1.30 (2 d, partially overlapped, $J_{5,6} = \sim 6.0$ Hz, H-6$^{\mathrm{I,II}}$); ^{13}C NMR (CD_3OD): δ 104.09 (C-1^{II}), 100.21 (C-1^{I}), 79.65 (C-2^{I}), 70.92 (C-2^{II}), 69.91 (C-5), 69.53 (C-3), 69.26 (C-3), 68.72 (C-5), 68.74 (C-1″), 59.38 (2 C, C-4′$^{\mathrm{I,II}}$), 54.69, 54.10 (C-4$^{\mathrm{I,II}}$), 41.98 (C-6″), 41.72 (C-7″), 38.19 (2 C, C-3′$^{\mathrm{I,II}}$), 36.93 (C-5″), 30.10 (C-2″), 26.83 C-3″), 26.49 (C-4″), 18.34, 18.24 (C-6$^{\mathrm{I,II}}$); FABMS: m/z 669 ($[M+1]^+$), 691 ($[M+Na]^+$).

*1-[(2-Aminoethylamido)carbonylpentyl 4-(3-deoxy-L-glycero-tetronamido)-4,6-dideoxy-α-D-mannopyranosyl)-(1→2)-4-(3-deoxy-*glycero*-tetronamido)-4,6-dideoxy-α-D-mannopyranoside]-2-ethoxycyclobutene-3,4-dione (**15**).* 3,4-Diethoxy-3-cyclobutene-1,2-dione (18 μl, 0.12 mmol) is added to a solution of the foregoing amine **14** (60 mg, 0.09 mmol) in pH 7.0 buffer (3 ml) and the mixture is stirred at room temperature overnight. TLC [ethyl acetate–methanol, 1:2 (v/v)] shows that the reaction is complete, and that one major UV-positive, less polar product is formed. After concentration, the residue is chromatographed to give **15** (50 mg, 56%). Definite signals in the ^{1}H NMR (CD_3OD) spectrum are at: δ 4.96 (bd, 1 H, $J_{1,2} = \sim 1.6$ Hz, H-1^{II}), 4.84 (bd, 1 H, $J_{1,2} = \sim 1.5$ Hz, H-1^{I}), 4.78–4.68 (m, 2 H, CH_2CH_3), 4.19 (dd, 2 H, $J_{2',3'a} = 4.0$, $J_{2',3'b} = 8.2$ Hz, H-2′$^{\mathrm{I,II}}$), 4.04 (dd, 1 H, $J_{2,3} = 2.9$ Hz, 2^{II}), 3.80 (m, 1 H, H-2^{I}), 2.20, 2.17 (2 t, partially overlapped, 2 H, $J = \sim 7.5$ Hz, H-5″), 2.07–1.77 (2 m, 4 H, H-3′$^{\mathrm{I,II}}$a,b), 1.67–1.36 (m, 9 H, H-2″a,b,3″,4″a,b incl 2 t at 1.47 and 1.45, $J = 7.0$ Hz, CH_2CH_3), 1.17, 1.16 (2 d, partially overlapped, $J_{5,6} = \sim 5.9$ Hz, H-6$^{\mathrm{I,II}}$);^{13}C NMR (CD_3OD): δ 104.12 (C-1^{II}), 100.23 (C-1^{I}), 79.64 (C-2^{I}), 70.96 (C-2^{II}), 70.83 (CH_2CH_3), 70.70 (2 C, C-2′$^{\mathrm{I,II}}$), 69.99 (C-5), 69.63, 69.31 (C-3$^{\mathrm{I,II}}$), 68.73 (C-5), 68.46 (C-1″), 59.43 (2 C, C-4′$^{\mathrm{I,II}}$), 54.73, 54.14 (C-4$^{\mathrm{I,II}}$), 45.23, 45.10 (1 C total, C-6″), 40.93, 40.61 (1 C total, C-7″), 38.24 (2 C, C-3′$^{\mathrm{I,II}}$), 36.93 (C-5″), 30.19 (C-2″), 26.86 C-3″), 26.59, 26.44 (1 C total, C-4″), 18.36, 18.24 (C-6$^{\mathrm{I,II}}$), 16.21, 16.12 (1 C total, CH_2CH_3); FABMS: m/z 793 ($[M+1]^+$).

Recovery of Excess Hapten from Conjugation Reaction

Note: Detailed conjugation protocols can be found in [10] in this volume.[42]

From Conjugation Reaction Using as Hapten the Functionalized Terminal, Monosaccharide Determinant of Vibrio cholerae *Ol, Serotype Ogawa.* The conjugation reaction starts with 10 mg of BSA and 8.4 mg of the hapten, 1-[(2-aminoethylamido)carbonylpentyl 4-(3-deoxy-L-*glycero*-tetronamido)-4,6-dideoxy-2-*O*-methyl-α-D-mannopyranoside]-2-ethoxycyclobutene-3,4-dione,[18] and is aimed at obtaining a conjugate containing five hapten residues per BSA. When the desired hapten–BSA ratio is reached, the neoglycoconjugate is isolated by ultrafiltration. The ultrafiltrate is concentrated, the residue is dissolved in methanol, and examination by TLC [ethyl acetate–methanol–concentrated NH_4OH, 2:1:0.1 (v/v/v)] shows the presence of only one component that chars with 5% H_2SO_4 in ethanol. It cochromatographs with the hapten used for the conjugation. After concentration, the residue is purified with a series of two Sep-Pak Plus C_{18} cartridges (Waters, Milford, MA). The cartridges are first washed with water (2 ml, nine times). The sugar hapten is eluted with a stepwise gradient of aqueous methanol (2-ml fractions, the concentration of methanol changing by 5%, from 0 to 50%). Concentration of fractions containing the charring material (25–45% methanol) followed by lyophilization gives 6.4 mg (79%) of the starting sugar hapten. ^1H NMR confirms the identity of the material.[18]

*From Conjugation Reaction That Leads to Conjugate **16**.* The conjugation reaction starts with 10 mg of BSA and 11.9 mg of hapten **15**. It is aimed at obtaining a conjugate containing 15 hapten residues per BSA. When the desired hapten–BSA ratio of 15.8 is reached, buffer pH 7 is added (3 ml) to terminate the reaction and, by following the protocol described above, hapten **15** (4 mg, 40%) is recovered. ^1H NMR confirms the identity of the material.

From Conjugation Reaction Using as Hapten the Functionalized Terminal, Hexasaccharide Determinant of Vibrio cholerae *OI, Serotype Ogawa.* The conjugation reaction starts with 8.1 mg of BSA and 17 mg of the hexasaccharide hapten. It is aimed at obtaining a neoglycoconjugate containing 10 hapten residues per BSA. When the desired hapten–BSA ratio of 10.1 is reached the neoglycoconjugate and the starting hapten (12.1 mg, 80%) are isolated by following the protocol described above. ^1H NMR confirms the identity of the material.[24]

[12] Neoglycolipid Technology: Deciphering Information Content of Glycome

By WENGANG CHAI, MARK S. STOLL, CHRISTINE GALUSTIAN, ALEXANDER M. LAWSON, and TEN FEIZI

Introduction

Neoglycolipid (NGL) technology[1] was introduced in 1985 to enable direct binding studies to be conveniently performed with oligosaccharides of glycoproteins. At that time there was a lack of microscale methods for examining the carbohydrate–protein interactions, after release of the oligosaccharides from the carrier proteins. Most carbohydrate–protein interactions are of such low affinities that di- or multivalence of both oligosaccharide and of recognition protein is required for detection by precipitation, radioimmunoassay, or enzyme-linked immunosorbent assay (ELISA) experiments. It was clear that a method was required to examine the released oligosaccharides in a multivalent state. The existing methods for examining the recognition of specific oligosaccharides by antibodies and other carbohydrate-recognizing proteins involved purification of the oligosaccharides, often from highly heterogeneous mixtures, and testing them individually as inhibitors of the binding of the recognition proteins to macromolecules or cells. The amounts of oligosaccharides required for inhibition of binding were frequently prohibitive. The approach that we chose to address the problem was chemical conjugation of oligosaccharides to lipid. The artificial glycolipids, neoglycolipids, formed would enable presentation of the oligosaccharides in the clustered state. The conjugation principle selected was reductive amination to a phosphatidylethanolamine-type aminolipid. For *O*-glycans released by reductive alkali treatment, a mild periodate oxidation procedure was included to generate reactive aldehydes for the conjugation. A major advantage of conjugating each oligosaccharide in a mixture to a lipid molecule rather than to a protein, such as bovine serum albumin (BSA), is that each component remains discrete and can be isolated. Thus, the resulting NGLs afforded a means of resolving mixtures by thin-layer chromatography (TLC) before binding experiments, as had already been established for glycosphingolipids. The NGLs could, moreover, be immobilized on other matrices such as the plastic surface of microtiter plate wells. They could also be incorporated into

[1] P. W. Tang, H. C. Gooi, M. Hardy, Y. C. Lee, and T. Feizi, *Biochem. Biophys. Res. Commun.* **132,** 474 (1985).

0076-6879/03 $35.00

liposomes and examined as inhibitors of carbohydrate–protein interactions, such that their inhibitory activities were virtually indistinguishable from those of the ligand-positive glycoproteins, mucins, from which the oligosaccharides had been released.

The lipid initially used for generating NGLs was a natural amino lipid, dipalmitoyl phosphatidylethanolamine (DPPE). The aldehyde group at the reducing end of the oligosaccharide, present in equilibrium form at a low concentration, joins to the amino group of the lipid by reductive amination. A key technical development in NGL technology was the establishment of conditions for a high yield of conjugates from reducing oligosaccharides.[2] Knowing that conjugation by reductive amination involves water elimination,

$$R_1\text{-}NH_2 + CHO\text{-}R_2 \xrightarrow{NaBH_3CN} R_1\text{-}NH\text{-}CH_2\text{-}R_2 + H_2O$$

where R_1-NH_2 is DPPE, and R_2-CHO is reducing sugar, anhydrous conditions were expected to enhance the reaction rate. It was shown that almost complete conjugation of neutral di- to hexasaccharides to DPPE could be achieved in 16 h at 50° by using the solvent mixture chloroform–methanol (1:1, v/v) in the absence of water. With longer oligosaccharides, such as high mannose-type *N*-glycans, that are insoluble in the organic solvent, a limited amount of water (5%, v/v) was included in the reaction mixture, and the incubation temperature was raised to 60°. This resulted in about 80% yield of NGLs. Traces of monoacyl derivatives of DPPE and of the NGLs were formed both under the hydrous and anhydrous conditions. On chromatography, these by-products have R_f values about one-half those of the parent compounds. To avoid these by-products, an alternative synthetic analog, 1,2-dihexadecyl-*sn*-glycero-3-phosphoethanolamine (DHPE),[3,4] is now used.

A landmark development in the application of NGL technology was the sequence determination of NGLs by liquid secondary-ion mass spectrometry (LSIMS), directly from the silica gel surface, following their resolution by TLC and in parallel with binding experiments with carbohydrate-recognizing proteins. This *in situ* TLC/LSIMS method with negative-ion detection takes advantage of the intrinsic negative charge conferred by the phosphate group and the favorable surface activity of the NGL. Intense molecular ions and informative sequence ions of lower abundance are obtained at low picomole levels. From these ions, molecular mass,

[2] M. S. Stoll, T. Mizuochi, R. A. Childs, and T. Feizi, *Biochem. J.* **256,** 661 (1988).

[3] G. Pohlentz, S. Schlem, and H. Egge, *Eur. J. Biochem.* **203,** 387 (1992).

[4] C.-T. Yuen, A. M. Lawson, W. Chai, M. Larkin, M. S. Stoll, A. C. Stuart, F. X. Sullivan, T. J. Ahern, and T. Feizi, *Biochemistry* **14,** 9126 (1992).

monosaccharide composition, and information about sequence, branching pattern, and sulfate and phosphate substitution can be deduced. The method developed permits high-sensitivity detection while retaining the high spatial resolution achieved by high-performance TLC (HPTLC). Electrical continuity between the analyte and the MS sample probe, and optimal matrix solution, are among the important factors for higher sensitivity of detection than previous methods for *in situ* TLC/LSIMS analysis.[5,6] Moreover, it was observed that mild periodate oxidation of the core monosaccharide of reduced *O*-glycans, before NGL formation, yielded derivatives with distinctive mass spectra that permit unambiguous assignments of positions of linkage of oligosaccharide sequences to the core monosaccharide, for example, the identification and sequencing of C3- and C6-linked antennae at the *N*-acetylgalactosaminitol (GalNAcol) core.[7] The high prevalence of yeast-type mannosyl *O*-glycans in the brain[8–10] was discovered by this principle.

In the course of work on NGLs derived from reductively released *O*-glycans, the conditions were refined for mild periodate oxidation such that GalNAcol or mannitol (Manol) core residues are selectively cleaved to generate reactive aldehydes, whereas sialic acid is left largely intact.[9] Thus, the oligosaccharide backbone and peripheral regions of NGLs from GalNAcol- and Manol-terminating neutral and sialylated *O*-glycans are eminently suitable for ligand-binding studies.

A shortfall of the first-generation NGLs is that the amino lipid used, DHPE, does not contain a chromophore. High-performance liquid chromatography (HPLC) of NGLs with ultraviolet (UV) detection has not been convenient because of interference with the UV absorption by the solvents required for separation. A new lipid reagent has now been synthesized that has intense fluorescence under UV light. This is *N*-aminoacetyl-*N*-(9-anthracenylmethyl)-1,2-dihexadecyl-*sn*-glycero-3-phosphoe thanolamine (ADHP),[11] which is derived from DHPE with incorporation of the fluorescent label anthracene (see Preparation of Fluorescent Amino

[5] T. T. Chang, J. O. Lay, and R. J. Francel, *Anal. Chem.* **56,** 109 (1984).

[6] Y. Kushi and S. Handa, *J. Biochem.* **98,** 265 (1985).

[7] M. S. Stoll, E. F. Hounseli, A. M. Lawson, W. Chai, and T. Feizi, *Eur. J. Biochem.* **189,** 499 (1990).

[8] C.-T. Yuen, W. Chai, R. W. Loveless, A. M. Lawson, R. U. Margolis, and T. Feizi, *J. Biol. Chem.* **272,** 8924 (1997).

[9] W. Chai, C.-T. Yuen, T. Feizi, and A. M. Lawson, *Anal. Biochem.* **270,** 314 (1999).

[10] W. Chai, C. T. Yuen, H. Kogelberg, R. A. Carruthers, R. U. Margolis, T. Feizi, and A. M. Lawson, *Eur. J. Biochem.* **263,** 879 (1999).

[11] M. S. Stoll, T. Feizi, R. W. Loveless, W. Chai, A. M. Lawson, and C.-T. Yuen, *Eur. J. Biochem.* **267,** 1795 (2000).

ERRATUM

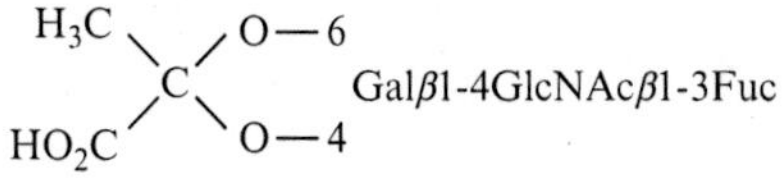

Methods in Enzymology Volume 362, Chapter 12, page 163.

Due to a composition error, an equation was broken incorrectly. The correct equation appears below.

$$\begin{array}{l} HSO_3\text{-Gal}\beta 1\text{-}3/4\text{GlcNAc}\beta 1\text{-}3\text{Gal-} \\ \qquad\qquad\qquad |\,1,4/3 \\ \qquad\qquad\quad \text{Fuc}\alpha \end{array}$$

for the selectins,[4,13] the tetrasaccharide epitope

$$\begin{array}{l} H_3C \diagdown \quad \diagup O-6 \\ \qquad\quad C \qquad\qquad \text{Gal}\beta 1\text{-}4\text{GlcNAc}\beta 1\text{-}3\text{Fuc} \\ HO_2C \diagup \quad \diagdown O-4 \end{array}$$

ISBN: 0-12-182265-6

Lipid ADHP, below). Fluorescent NGLs derived from a variety of neutral and acidic oligosaccharides by conjugation to ADHP, by reductive amination, can be detected and quantified by spectrophotometry and scanning densitometry, and resolved by TLC and HPLC with subpicomole detection. Antigenicities of the fluorescent NGLs are well retained compared with DHPE, and picomole levels can be detected with monoclonal carbohydrate sequence-specific antibodies. Among *O*-glycans from an ovarian cystadenoma mucin, isomeric oligosaccharide sequences, sialyl-Le^a and sialyl-Le^x type, could be resolved by HPLC as fluorescent NGLs, and sequenced by LSIMS.[11] Thus the NGL technology now uniquely combines high sensitivity of immunodetection with a comparable sensitivity of fluorescence detection. Principles are thus established for a streamlined technology whereby an oligosaccharide population can be carried through ligand detection, steps of ligand isolation, sequence determination by mass spectrometry, enzymatic sequencing, and other state-of-the-art carbohydrate analyses.

As reviewed elsewhere,[12] the NGL technology has been powerful in carbohydrate ligand discovery. Among highlights are the discovery, on epithelial glycoproteins, of hitherto unsuspected sulfated carbohydrate ligands of Le^a and Le^x types:

HSO_3-Galβ1-3/4GlcNAcβ1-3Gal-1,4/3
|
Fucα

for the selectins,[4,13] the tetrasaccharide epitope

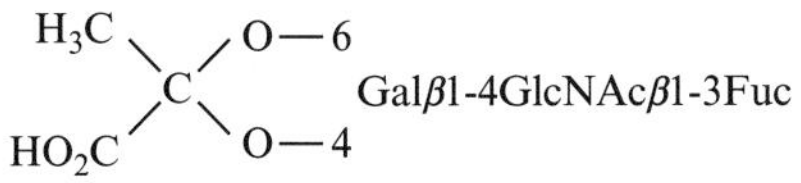

H_3C \ / O—6
C Galβ1-4GlcNAcβ1-3Fuc
HO_2C / \ O—4

for an antibody that blocks aggregation of cells of the sponge, *Microciona prolifera*,[14] and a unique antigenic sequence, ΔUA-GlcN-UA-GlcNAc-, on heparan sulfate[15] (UA, hexuronic acid; ΔUA, 4.5-unsaturated hexuronic acid).

Many of the procedures involved in the preparation and analyses of NGLs have been described.[16,17] Here we present an updated account of

[12] T. Feizi and R. A. Childs, *Methods Enzymol.* **242,** 205 (1994).
[13] W. Chai, T. Feizi, C.-T. Yuen, and A. M. Lawson, *Glycobiology* **7,** 861 (1997).
[14] D. Spillmann, K. Hard, O. J. Thomas, J. F. G. Vliegenthart, G. Misevic, M. M. Burger, and J. Finne, *J. Biol. Chem.* **268,** 13378 (1993).
[15] C. Leteux, W. Chai, K. Nagai, A. M. Lawson, and T. Feizi, *J. Biol. Chem.* **276,** 12539 (2001).

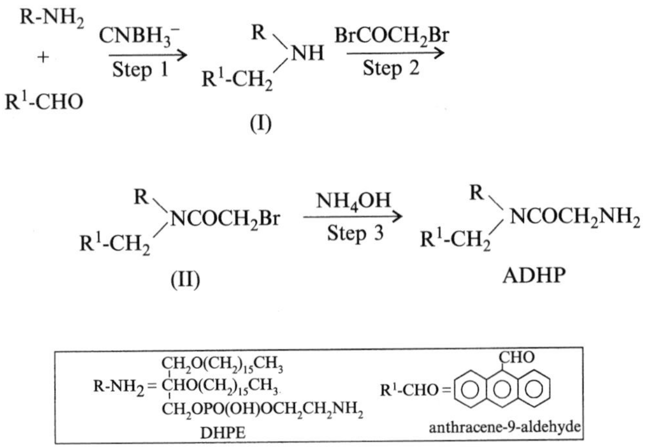

$$R\text{-}NH_2 + R^1\text{-}CHO \xrightarrow[\text{Step 1}]{CNBH_3^-} \underset{(I)}{R(R^1\text{-}CH_2)NH} \xrightarrow[\text{Step 2}]{BrCOCH_2Br}$$

$$\underset{(II)}{R(R^1\text{-}CH_2)NCOCH_2Br} \xrightarrow[\text{Step 3}]{NH_4OH} \underset{ADHP}{R(R^1\text{-}CH_2)NCOCH_2NH_2}$$

R-NH$_2$ = $CH_2O(CH_2)_{15}CH_3$ / $CHO(CH_2)_{15}CH_3$ / $CH_2OPO(OH)OCH_2CH_2NH_2$ DHPE; R^1-CHO = anthracene-9-aldehyde (CHO)

SCHEME 1. Reaction scheme for the synthesis of ADHP.

the procedures, also including assay systems for carbohydrate–protein interactions using NGLs.

Preparation of Fluorescent Amino Lipid ADHP

The fluorescent amino lipid is prepared in three steps (Scheme 1) from DHPE, using anthracene-9-aldehyde as the fluorescent label.[11]

Materials

DHPE (Fluka Chemicals, Dorset, UK)
Anthracene-9-aldehyde
Bromoacetyl bromide
Tetrabutylammonium cyanoborohydride
HPTLC plates: silica gel 60, 5 μm (Merck, Darmstadt, Germany)
Solvent mixtures: chloroform–ethanol (C/E) and chloroform–ethanol–water (C/E/W) of various composition (v/v)

Step 1: Preparation of N-*(9-Anthracenylmethyl)-DHPE*

Dissolve DHPE (20 μmol) and anthracene-9-aldehyde (100 μmol in 6 ml of chloroform), and heat at 60° for 1 h. Add tetrabutylammonium

[16] T. Feizi, M. S. Stoll, C.-T. Yuen, W. Chai, and A. M. Lawson, *Methods Enzymol.* **230,** 484 (1994).

[17] M. S. Stoll and T. Feizi, *in* "A Laboratory Guide to Glycoconjugate Analysis" (P. Jackson, and J. T. Gallagher, eds.), p. 329. Springer-Verlag, New York, 1997.

cyanoborohydride (100 μmol) in chloroform (5 ml), heat the mixture at 60° for 4 h, and concentrate the solution by evaporation under nitrogen to about 1 ml. The main product is a fluorescent band (R_f 0.50, intense blue under 254-nm light) in HPTLC developed with C/E (1:1). The yield is almost quantitative based on disappearance of DHPE. Excess anthracene-9-aldehyde migrates near the solvent front and has a green fluorescence under 254-nm light. Purify the *N*-(9-anthracenylmethyl)-DHPE from the reaction mixture by semipreparative HPTLC developed with C/E (1:1); yield, 40–45%.

The purification can also be carried out with a silica gel column. Prepare the column (1 × 10 cm) from silica gel in C/E/W (50:50:10) as a slurry. When packed, wash the column with 1 bed volume of the same solvent followed by 3 bed volumes of C/E (50:50). After drying under a nitrogen stream at room temperature, redissolve the reaction mixture in C/E (50:50) and apply to the column. Carry out initial elution with the same solvent until a visible yellow band of excess anthracene-9-aldehyde is completely eluted out. Slowly change the mobile phase to C/E/W (50:50:10) to elute ADHP. Monitor progress using long-wavelength UV light illumination; short-wavelength UV is absorbed by glass. Some fluorescent residue remaining on the silica is ignored.

Step 2: Preparation of N-*Bromoacetyl*-N-*(9-anthracenylmethyl)-DHPE*

Add dry pyridine (100 μmol) followed by bromoacetyl bromide (70 μmol) to dry *N*-(9-anthracenylmethyl)-DHPE (10 μmol) in chloroform (2 ml). After 5 min at room temperature, wash the mixture with water (6 × 2 ml). HPTLC of an aliquot of the reaction mixture (containing about 1 nmol of lipid) developed with C/E/W (22:11:1) reveals a major product band, *N*-bromoacetyl-*N*-(9-anthracenylmethyl)-DHPE, R_f 0.53, which exhibits only mild fluorescence, but can be visualized by its strong staining with primulin. There are negligible by-products as assessed by the paucity of additional bands, either fluorescent or primulin stained. Concentrate the reaction mixture to 1 ml by evaporation and use without fractionation.

Step 3: Preparation of N-*Aminoacetyl*-N-*(9-anthracenylmethyl)-DHPE (ADHP)*

Add ethanol (1 ml) and 35% aqueous ammonia (200 μl) to the concentrated reaction mixture containing *N*-bromoacetyl-*N*-(9-anthracenylmethyl)-DHPE (about 10 μmol) in chloroform. After heating at 60° for 4 h, wash the mixture with water (6 × 2 ml) and concentrate by evaporation to about 200 μl. The yield of fluorescent product ADHP (R_f 0.64) is almost quantitative as judged by HPTLC of the washed reaction mixture with C/E/W (22:11:1). This is purified by semipreparative HPTLC developed with the above solvent; yield, 85–90%.

Oligosaccharide Preparations

NGLs can be prepared directly from reducing oligosaccharides by reductive amination (see Preparation of Neoglycolipids of Reducing Oligosaccharides, below) with the amino lipid ADHP or DHPE. The oligosaccharides may be obtained from the following sources.

Free reducing oligosaccharides: Any reducing oligosaccharides, such as those isolated from human or animal milk or chemically synthesized

N-linked glycoprotein oligosaccharides: Released by the enzymes peptide-N^4-(*N*-acetyl-β-glucosaminyl)asparagine amidase (PNGase F) or endo-β-*N*-acetylglucosaminidase F (Endo F)[18] or by hydrazinolysis.[19]

O-linked glycoprotein oligosaccharides: Released by mild alkaline hydrolysis,[13] by hydrazinolysis,[20] and by the sequence-specific enzyme endo-*N*-acetylgalactosaminidase[21]

Natural glycolipids: Released by endoceramidase[22]

Proteoglycans or glycosaminoglycans: Released from proteoglycans or polysaccharides by lyase digestion[23] or nitrous acid degradation[24] and, in the case of hyaluronic acid, also by hydrolase digestion[25]

Bacterial and plant polysaccharides: Released by partial degradation using various chemical methods, including acid or alkaline hydrolysis, acetolysis, and Smith degradation

NGLs can also be prepared from reduced oligosaccharides by mild periodate oxidation of the open-chain vicinal diol followed by conjugation to the amino lipid ADHP or DHPE, through reductive amination

[18] A. L. Tarentino and T. H. Plummer, Jr., *Methods Enzymol.* **230,** 44 (1994).

[19] S. Takasaki, T. Mizuochi, and A. Kobata, *Methods Enzymol.* **83,** 263 (1982).

[20] T. Patel, J. Bruce, A. Merry, C. Bigge, M. Wormald, A. Jaques, and R. Parekh, *Biochemistry* **32,** 679 (1993).

[21] Y. Endo and A. Kobata, *J. Biochem.* **80,** 1 (1976).

[22] T. Osanai, T. Feizi, W. Chai, A. M. Lawson, M. L. Gustavsson, K. Sudo, M. Araki, K. Araki, and C. T. Yuen, *Biochem. Biophys. Res. Commun.* **218,** 610 (1996).

[23] W. Chai, J. Luo, C. K. Lim, and A. M. Lawson, *Anal. Chem.* **70,** 2060 (1998).

[24] J. E. Shively and H. E. Conrad, *Biochemistry* **15,** 3932 (1976).

[25] W. Chai, J. G. Beeson, H. Kogelberg, G. V. Brown, and A. M. Lawson, *Infect. Immun.* **69,** 420 (2001).

(see Preparation of Neoglycolipids of Reduced Oligosaccharides, below). The oligosaccharide alditols may be obtained from the following sources.

O-linked mucin-type glycoprotein oligosaccharides: Released by reductive alkaline hydrolysis[26] and containing the GalNAcol core

O-linked mannosyl glycoprotein oligosaccharides: Released by reductive alkaline hydrolysis[8,10] and containing the Mannol core

Other free oligosaccharide alditols: Reduced for other purposes, for example, for HPLC separation to eliminate the interference caused by α,β-anomeric forms.

Preparation of Neoglycolipids of Reducing Oligosaccharides

Three slightly different procedures are described here, with various solvent compositions and reaction times for conjugation of different oligosaccharides to ADHP or DHPE. Keep the molar ratio of oligosaccharide to ADHP or DHPE at 1:8 and the ratio to the reducing agent, tetrabutylammonium cyanoborohydride, at 1:4 or more.[11] Carry out reactions in glass reaction vials fitted with a polytetrafluoroethylene (PTFE)-lined screw cap. The volume of the reaction mixture should be no less than 25 μl in a 1-ml reaction vial. After conjugation, dilute the mixture with chloroform–methanol–water (25:25:8, v/v/v) to give a concentration of 1 nmol of starting sugar per 5 μl, for analysis and for storage.

Materials

ADHP or DHPE stock solution: 8 nmol/μl in chloroform–methanol (1:1 and 1:3, v/v)

Tetrabutylammonium cyanoborohydride solution: 20 μg/μl in methanol; prepare fresh

Solvents: HPLC grade

Short-Chain Oligosaccharides

For pentasaccharides or shorter oligosaccharides, anhydrous conditions are preferred. Dry oligosaccharides in a reaction vial either by lyophilization or under a stream of nitrogen gas. Add 100 μl of ADHP or DHPE solution [8 nmol/μl chloroform–methanol (1:1, v/v)] and 20 μl of freshly prepared tetrabutylammonium cyanoborohydride (20 μg/μl methanol) to 100 nmol of dried oligosaccharide. Seal the reaction and heat at 60° for 16 h.

[26] R. N. Iyer, and D. M. Carlson *Arch. Biochem. Biophys.* **142,** 101 (1971).

Longer Chain Oligosaccharides

For oligosaccharides larger than hexasaccharides, a small amount of water (~5%) should be included in the reaction mixture to assist the dissolution of the oligosaccharides. Carefully add 5 μl of water to 100 nmol of oligosaccharide, which has been dried either by lyophilization or under a stream of nitrogen gas in a reaction vial, to dissolve the oligosaccharide. Add ADHP solution [100 μl, 8 nmol/μl in chloroform–methanol (1:1, v/v)] and 5 μl of freshly prepared tetrabutylammonium cyanoborohydride (20 μg/μl methanol). Seal the reaction vial and heat at 60° for 16 h.

Highly Acidic Oligosaccharides

For conjugation of oligosaccharides derived from glycosaminoglycans, particularly those with multiple sulfate groups, a modified procedure should be used. Thus, add 5 μl of water to 100 nmol of oligosaccharide, which has been dried by lyophilization in a reaction vial, to dissolve the oligosaccharide. Add ADHP or DHPE solution [100 μl, 8 nmol/μl chloroform–methanol (1:3, v/v)] and 5 μl of freshly prepared tetrabutylammonium cyanoborohydride (20 μg/μl methanol). Seal the reaction vial and heat at 60° for 96 h.

Preparation of Neoglycolipids of Reduced Oligosaccharides

When released from protein by alkaline borohydride treatment, *O*-linked oligosaccharides are in the reduced form and, hence, not suitable for conjugation with amino lipid by reductive amination. A mild periodate oxidation step is required before conjugation[7] to cleave the reduced core residue to create aldehyde groups. This will split the core residue into two aldehyde-containing fragments and these are then converted into two NGLs (Scheme 2). Under our optimized conditions, periodate oxidizes and cleaves only vicinal diols in the open-chain form, not diols on the saccharide ring. A molar ratio of oligosaccharide alditol to periodate of 1:4 is normally employed. However, if more than one diol is present, such as in the case of sialylated oligosaccharide alditols, in which the glycerol side chain contains two additional diols, a molar ratio of 1:2 should be used.[27] Under the latter conditions, although the yield of cleavage products is slightly lower, there is greater selectivity toward the acyclic vicinal diols based on the rate of oxidative cleavage: *threo-diols* > *erythro-diols* > terminal diols. For the sialylated oligosaccharide alditols periodate predominantly cleaves the diols in the *threo* configuration

[27] W. Chai, M. S. Stoll, G. C. Cashmore, and A. M. Lawson, *Carbohydr. Res.* **239,** 107 (1993).

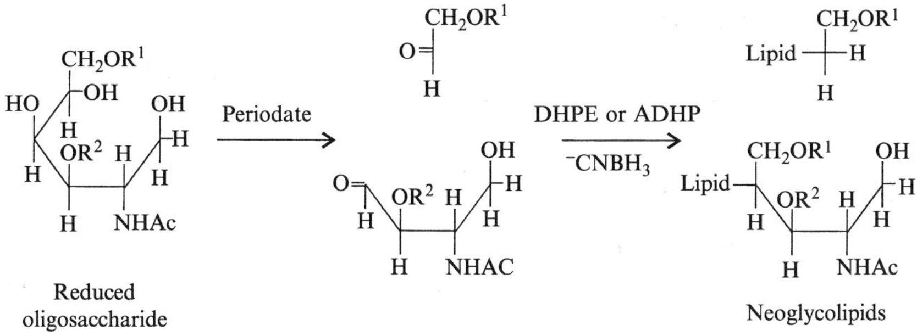

Periodate

DHPE or ADHP

$^{-}CNBH_3$

Reduced oligosaccharide

Neoglycolipids

SCHEME 2. Reaction scheme for mild periodate oxidation of reduced oligosaccharide alditols and subsequent formation of NGL by conjugation to amino lipid. A reduced *O*-glycan containing GalNAcol core residue is depicted here.

(4,5-diol of GalNAcol and 3,4-diol of Manol) in the core sugar, whereas the sialic acid, which contains an *erythro-diol* and a terminal diol, is largely preserved.[9]

Materials

ADHP or DHPE: 1 mg/ml in chloroform–methanol (1:1, v/v)
Imidazole buffer: 40 m*M*; adjust to pH 6.5 with HCl
Sodium periodate solution: 1.25 mg/ml in imidazole buffer; prepare fresh
Butane-2,3-diol reagent: 14.5 mg/ml in water
Tetrabutylammonium cyanoborohydride solution: 20 mg/ml methanol; prepare fresh

Nonsialylated Oligosaccharides

1. Dry the oligosaccharide alditol by lyophilization or by a stream of nitrogen gas and cool to 0° on ice.
2. Add freshly prepared sodium periodate solution (4 molar excess with respect to the oligosaccharide alditol). Vortex the reaction mixture thoroughly and keep on ice in the dark for 5 min.
3. Add butane-2,3-diol (2 molar excess with respect to periodate) and keep on ice in the dark for 40 min.
4. Add ADHP or DHPE solution (1 mg/ml, 5 molar excess with respect to periodate) followed by freshly prepared tetrabutylammonium cyanoborohydride (20 mg/ml, 25 molar excess with respect to periodate).
5. Incubate at 60° for 16 h.

Sialylated Oligosaccharides

Perform as described above, but use a 2 molar excess (with respect to the oligosaccharide alditol) of freshly prepared periodate solution.

Analysis of Neoglycolipid Products by TLC

After conjugation, the NGL products can be analyzed by HPTLC for yield and purity before further analysis by MS and overlay for binding assay.

Materials

HPTLC plates: aluminum backed (5 μm silica; Merck)
Glass tank: For TLC solvent development
Development solvents: chloroform–methanol–water of various composition (v/v):
130:50:9, for NGLs of mono- to trisaccharides
60:35:8, for NGLs of tri- to pentasaccharides
55:45:10, for NGLs of penta- to octasaccharides
105:100:28, for larger NGLs
50:55:18, for large glycosaminoglycan (GAG)-derived NGLs
Primulin reagent: Prepare by diluting primulin (Sigma, Poole, UK) stock solution [100 mg in 100 ml of acetone–water (1:9, v/v)] 1:100 with acetone–water (4:1, v/v)
UV lamp: long-wavelength (e.g., 300 or 366 nm)
UV lamp: short-wavelength (e.g., 254 nm)

Procedure

1. Add a suitable solvent to the TLC tank and allow vapor to equilibrate (>30 min, at room temperature).
2. Cut a TLC plate to the desired size (e.g., 10 cm long and 5–10 cm wide) from the back, using a scalpel blade.
3. Apply NGL solution (1 nmol of starting oligosaccharide) to the TLC plate, 15 mm from the bottom edge as a spot or as a 5-mm band, and allow 15 mm free at both edges.
4. After drying, carefully place the plate into the tank and develop to 5 mm below the top edge.
5. Remove the plate from the tank and dry with warm air.
6. ADHP-derived NGLs can be viewed directly under a UV light. Short-wavelength UV light gives more sensitive detection. DHPE-derived NGLs can be viewed under the UV lamp after staining with the primulin

reagent. The staining can be carried out by spraying with the primulin reagent in a ventilation hood until the silica gel surface appears slightly wet, and then allowing the surface to dry before viewing under long-wavelength UV.

Purification of Neoglycolipid Products

After conjugation, the NGL products may need to be purified to remove the excess aminolipid ADHP or DHPE and salts, such as the reducing agent cyanoborohydride and the buffer solution for periodate oxidation. This permits a better separation by TLC and HPLC, and is also desirable for mass spectrometric analysis and for binding and inhibition assays. Three types of minicolumns can be used: C_{18} cartridge, for removing salts; silica cartridge, for removing excess lipid; phenylboronic acid (PBA) cartridge, for removing both excess lipid and salts, but useful only for oligosaccharides containing a vicinal diol in the reduced linker monosaccharide residue.[28] PBA column separations work best with neutral oligosaccharide NGLs but can work with some acidic sugars if the reaction mixture is made alkaline by addition of ~1% ammonia. Depending on the retention time of particular NGLs, careful elution (collection of smaller volume fractions or use of shallower gradients of polar solvent compositions) may achieve separation of the NGL products from both excess lipid and salt.

Materials

Chromatographic media: Silica cartridge (Sep-Pak, 500 mg silica, 3 ml; Waters, Milford, MA); C_{18} cartridge (Sep-Pak, 500 mg silica, 3 ml; Waters), PBA cartridge (Bond Elut, 1 ml, 100 mg; Varian, Palo Alto, CA), aluminum-backed HPTLC plate (5 μm silica; Merck)

Solvents and solutions (all solvents are HPLC grade): Water, methanol, chloroform, ammonium acetate solution (0.2 *M* in water), chloroform–methanol–25% aqueous ammonia [C/M/A (150:150:1, v/v/v)], chloroform–methanol–0.1 *M* aqueous boric acid [C/M/B (3:7:3, v/v/v)], chloroform–methanol–water (C/M/W) of various composition (v/v)

[28] M. S. Stoll and E. F. Hounsell, *Biomed. Chromatogr.* **2,** 249 (1988).

Removal of Excess Lipid Reagent by Silica Cartridge

1. Wash the column sequentially with 4 ml of methanol, 4 ml of water, 6 ml of 0.2 *M* ammonium acetate, 12 ml of water, 4 ml of methanol, and 6 ml of chloroform.

2. Dissolve the dried NGL conjugation mixture (<100 nmol of starting sugar) in a minimal amount of C/M/W (60:35:8) and apply on the prewashed column (collect the eluent as fraction 1).

3. Elute NGLs of large acidic oligosaccharides with 3×500 μl of C/M/W (60:35:8) (fractions 2–4), 3×500 μl of C/M/W (25:25:8) (fractions 5–7), and 3×500 μl of C/M/W (50:55:18) (fractions 8–10).

4. Before pooling, perform TLC of aliquots of each fraction to identify fractions containing the NGL products.

Desalting by C_{18} Cartridge

1. Wash a C_{18} cartridge with 2 ml of C/M/W (60:35:8) and then 5 ml of C/M/W (15:70:30).

2. Dissolve the dried NGL conjugation mixture (<100 nmol of starting sugar) in a minimal amount of C/M/W (15:70:30). Apply the solution to the prewashed column followed by 1 ml of the same solvent (collect as fraction 1).

3. Further wash with 1 ml of the same solvent (fraction 2).

4. Elute the NGLs and excess lipid with 2×1 ml of C/M/W (60:35:8) (fractions 3 and 4).

Purification of Neutral Neoglycolipids Containing Open-Chain Vicinal Diol by PBA Cartridge

NGLs containing a vicinal diol in the reduced core monosaccharide linked to the aminolipid can be reversibly bonded to PBA, thus providing a means to separate the NGLs from the other components in a reaction mixture. Bonding occurs under mild alkaline conditions whereas release is under mild acidic conditions, especially with boric acid. Highly charged NGLs may not be fully retained by a PBA column. The detailed procedures are as follows.

1. Wash a PBA column with 2 ml of C/M/A (150:150:1).

2. Add 1 μl of ammonia (25% aqueous) to the NGL conjugation solution [to obtain 1 part ammonia solution to 200 parts reaction mixture solution (v/v), containing up to 1 μmol of NGL]. Add water (15%, v/v).

3. Apply to the column and allow to pass through under gravity. Add 1 ml of C/M/A (150:150:1) to wash the column. Collect the fall-through and the wash as fraction 1.

4. Wash further with 1 ml of the same solvent (fraction 2) and 1 ml of chloroform–methanol (1:1, v/v) (fraction 3).

5. Elute with 2 × 2 ml of freshly prepared C/M/B solution (fractions 4 and 5).

6. Evaporate all fractions to dryness under a nitrogen stream below 50°.

7. Remove boric acid in fractions 4 and 5 by repeated coevaporation with methanol (3 × 400 μl). Before the first coevaporation add 1 μl of acetic acid to the methanol solution.

Purification of Neoglycolipids by Semipreparative HPTLC

1. Apply NGL reaction mixture as a long band to an aluminum-backed HPTLC plate.

2. Develop the plate with a suitable solvent.

3. Detect the fluorescent NGL product under a long- or short-wavelength UV light and mark the band of interest. If a fairly high concentration of sample is present, DHPE NGLs are visible as a lightly opaque white band just as the plate dries in air. Alternatively, the band position can be identified and pencil marked after spraying with primulin at the edges of the plate.

4. After localizing the band of interest, cut out the required strips of plate with a scalpel (cut on the reverse side of the plate).

5. Loosen and scrape off the silica gel from the TLC strips and collect the gel in a glass vial.

6. Extract the NGL from the silica gel with C/M/W (25:25:8), 3 × 500 μl; combine the solvent extracts and evaporate under a stream of nitrogen gas.

7. Reextract the residue containing NGL and some silica gel, using less solvent, and transfer to a second glass vial. This leaves most of the silica in the first vial.

8. Alternatively, pack the scraped silica gel as a minicolumn and elute the NGLs with the same solvent.

Fractionation of Neoglycolipid Mixtures

Minicolumn Separations

NGLs may be fractionated into groups (Fig. 1) on the basis of their polarity by methods similar to those described under Purification of Neoglycolipid Products (above). This can be achieved by performing a careful elution, either with a shallower gradient elution or by collecting fractions of smaller volume. As illustrated in Fig. 1, NGLs of sulfated and

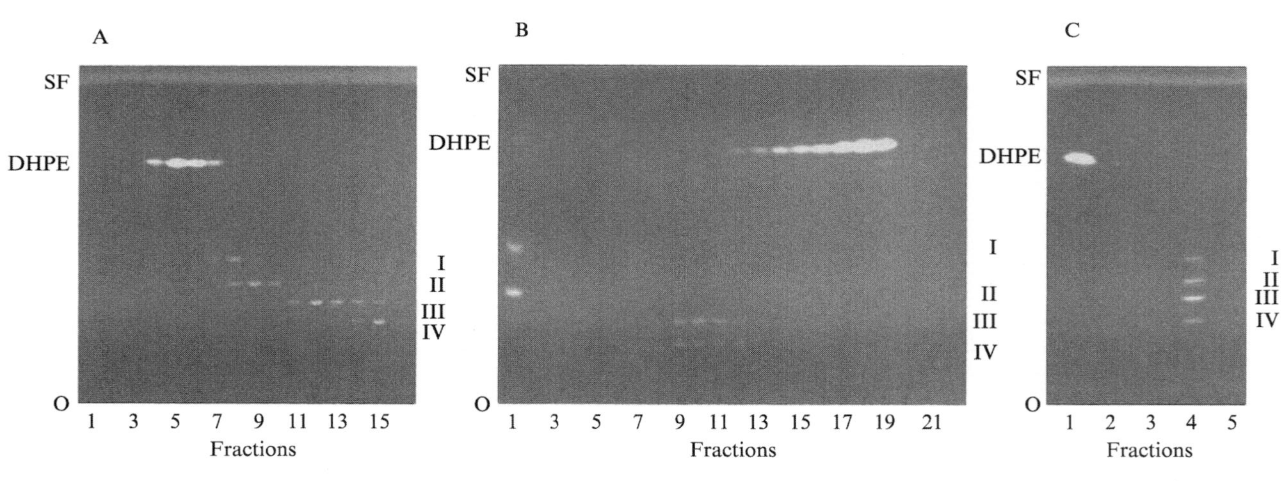

FIG. 1. Fractionation of NGL reaction mixtures by minicolumns. A reaction mixture of DHPE-NGLs derived from HSO_3-6GlcNAcβ1-3Gal (I), 3′-sialyl-Galβ1-4Glc (II), lacto-*N*-fucopentaose II, Galβ1-3(Fucα1-4)GlcNAcβ1-3Galβ1-4Glc (III), and blood group A heptasaccharide, GalNAcβ1-3(Fucα1-2)Galβ1-3(Fucα1-4)GlcNAcβ1-3Galβ1-4Glc (IV), was fractionated with minicolumns [silica in (A); C_{18} in (B); and PBA in (C)]. HPTLC of an aliquot of each eluted fraction was carried out in solvent C/M/W (60:35:8). NGLs and excess DHPE were visualized by staining with primulin and viewing under long-wavelength UV light. Total NGLs on plates A, B, and C were 400, 800, and 100 pmol, respectively. O, Origin; SF, solvent front. Reproduced in part from Ref. 17 with permission.

sialylated disaccharides and of neutral penta- and heptasaccharides can be separated.

Semipreparative HPTLC

For isolating individual neoglycolipids, semipreparative HPTLC can be performed. Application density depends on the required resolution but in general should be less than 10 nmol of NGL per centimeter.

Procedure

1. After development and drying, the positions of fluorescent NGL bands can be detected and pencil marked under a long- or short-wavelength UV light. Nonfluorescent DHPE-derived NGL bands can be found under a long-wavelength UV light after spraying with primulin just at the edges of the plate.
2. Cut out the required strips of plate with a scalpel.
3. Scrape the silica gel from the strips into glass tubes and extract three times with C/M/W (60:35:8), centrifuging between each extraction.
4. Evaporate the pooled extracts to dryness in a glass vial.
5. Reextract the residue containing NGL and some silica with C/M/W (130:50:9) and transfer to a second vial. This provides purified NGL containing only traces of silica; most of the dissolved silica remaining attached to the surface of the first vial.
6. Alternatively, the scraped silica gel may be packed as a minicolumn and NGLs eluted directly with the same solvent.

Overpressured Layer Chromatography

Overpressured layer chromatography (OPLC) employs continuous solvent development and offers advantages over conventional TLC for preparative work, in that components are isolated directly in solution. To minimize background contamination from the silica gel plate in preparative OPLC, washing should be carried out before sample application.

Equipment and Materials

OPLC 50: from Bionisis (Le Plessis Robinson, France), consisting of separation and solvent delivery units

Aluminum-backed silica gel TLC plates with perimeter seals (HTSorb, 5 μm, fluorescent 254, 5×20 cm; Bionisis)

Solvents: Washing, development, and elution are with chloroform–methanol–water (60:35:8, v/v/v)

Automated fraction collector

Procedure

1. Wash the plate overnight (~16 h) with solvent at a flow rate of 125 μl/min.
2. Apply the sample as a line to the origin position on the previously washed and dried TLC plate.
3. Perform elution at 100 or 125 μl/min and collect eluent with an automated fraction collector. The elution of ADHP-derived NGLs may be monitored by UV absorption/fluorescence with an in-line detector or manually by UV fluorescence.

HPLC

ADHP NGLs were designed specifically with an intense UV chromophore/fluorophore within the lipid moiety. These NGLs, therefore, are ideal for HPLC with highly sensitive absorption or fluorescence detection.[11] The detailed procedure for a specific sample type may vary but the following HPLC systems that have so far been used to separate ADHP NGLs can be used as guidelines for other types of samples.

Equipment and Materials

HPLC instrument: With gradient elution
Fluorescence detector: λ_{ex} at 255 nm and λ_{em} at 405 nm
UV detector: Variable wavelength (λ_{max}, 255 nm)
Solvent: HPLC grade, chloroform–methanol–water (C/M/W) of different compositions (v/v), with or without salt (see details below)

Procedure

Neoglycolipids of N-*Glycans.* Amide column (e.g., TSKgel Amide-80, 4.6 × 250 mm; Tosoh Biosep, Montgomeryville, PA) at 35° and a flow rate of 0.5 ml/min with a linear gradient of chloroform–methanol–water, 29:20:4 to 10:20:8 (by volume), over 30 min and isocratically for 10 min with the latter solvent.

Neoglycolipids of Sulfated Oligosaccharides. Amide column (e.g., TSKgel Amide-80, 4.6 × 250 mm) using a linear gradient of C/M/W, 60:40:5 to 25:25:20, both containing 15 m*M* ammonium acetate, pH 5, over 30 min at 0.75 ml/min.

Neoglycolipids of Acidic O-Glycans. Amide column (e.g., TSKgel Amide-80, 4.6 × 250 mm) using a C/M/W gradient from 60:35:5 to

10:20:8, both containing 15 mM ammonium acetate, pH 5, over 35 min at 0.75 ml/min.

Alternatively, an amino column (e.g., 5-μm Hypersil APS-2, 4.6×250 mm; Phenomenex, Macclesfield, UK) can be used. The separation is carried out at 35° and a flow rate of 0.5 ml/min, using a C/M/W gradient of 130:70:9 to 10:20:8, both containing 10 mM sodium chloride and 0.1% formic acid (v/v). In the latter case the NGLs are recovered by desalting on a C_{18} microcartridge as follows: The HPLC fractions are dried in a stream of nitrogen at ambient temperature and the residue is redissolved in methanol–water (70:30, v/v). A microcolumn consisting of a few milligrams of C_{18} (Bond Elut; Varian) in a fine-tipped plastic pipette plugged with a small piece of Teflon wool is washed successively with water, 0.15 M NaCl, water, and methanol–water (70:30, v/v). The redissolved NGLs (up to 5 nmol) are applied to the column and the fall-through plus a wash with the same solvent are discarded. The NGLs are eluted with C/M/W (60:35:8), containing 0.1% ammonia. The process is monitored under UV light at 254 nm. Recovery is 85% from the HPLC column and 75% from the C_{18} desalting process, even at the 100-pmol level.

Visualization of Neoglycolipids

Several reagents are available for the visualization of NGLs using different structural elements, that is, lipid, hexose, or sialic acid. As lipid detection by primulin is nondestructive, sequential staining with primulin spray for DHPE and with orcinol or resorcinol may be performed on the same plate if it is desired to visualize DHPE-derived NGLs and hexose or sialic acid.

Materials

All stock solutions are stored at 4°.

Primulin reagent stock solution: See Analysis of Neoglycolipid Products by TLC (above)

Orcinol reagent: Dissolve 900 mg of orcinol in 25 ml of water. Add 375 ml of ethanol. Cool on ice. Gradually add 50 ml of concentrated sulfuric acid (18 M) with stirring, maintaining the temperature below 10°

Resorcinol reagent: To 5 ml of resorcinol (2% in water) add 45 ml of 5 M HCl. Add 125 μl of 0.1 M $CuSO_4$. Use after 4 h. Stable for 1 week

Visualization of Neoglycolipids and Lipids

Fluorescent ADHP-derived NGLs can be viewed directly under long- or short-wavelength UV light. Nonfluorescent DHPE-derived NGLs can be detected under long-wavelength UV light (300 or 366 nm) after staining with primulin as follows.

1. Prepare working primulin reagent by diluting stock reagent 1:100 with acetone–water (4:1, v/v).
2. Spray with working reagent until the surface appears slightly wet.
3. Dry in a stream of warm air.
4. Lipids fluoresce light blue against a dark blue background under long-wavelength ultraviolet light.

Visualization of Hexose

1. Spray with orcinol reagent until the plates appear slightly wet.
2. Heat in a vented oven at 105° for about 5 min or until the violet color given by hexose is maximal.

Visualization of Sialic Acid

1. Spray with resorcinol reagent until the plate appears slightly wet.
2. Clamp the plate between glass plates to prevent evaporation of reagent.
3. Heat in a vented oven at 140° for about 10 min, or until the purple color given by sialic acid is maximal.

Quantitation of Neoglycolipids

Quantitation of NGLs can be carried out on TLC plates, using a densitometer, after staining essentially as described above (see Visualization of Neoglycolipids, above). However, to obtain better quantitation results, some precautions need to be taken and these are described below. Fluorescent NGLs can be measured directly on TLC plates, using a densitometer, or in solution, using a spectrophotometer.

Equipment and Materials

Spectrophotometer: U-2000 spectrophotometer (Hitachi, Tokyo, Japan)
Densitometer: CS-9000 scanner (Shimadzu, Kyoto, Japan)
Phosphate-buffered saline (PBS): 20 m*M* sodium phosphate buffer containing 150 m*M* NaCl, pH 7.4.
Staining reagents: See Visualization of Neoglycolipids (above)

Staining of Neoglycolipids

DHPE-Derived Neoglycolipids. An immersion procedure that gives a more uniform primulin staining for quantitation of NGLs is the preferred method.

1. Slowly lower the plate into a PBS solution, being careful to avoid trapping air bubbles. Leave to soak for 10 min.
2. Prepare working primulin solution [stock primulin solution–PBS (1:100, v/v)].
3. Transfer the plate without drying to the primulin reagent. Leave to soak for 20 min in the dark.
4. Transfer the plate to a fresh PBS solution without drying, and soak for 10 min in the dark.
5. Drain and air dry the plate.

Hexose-Containing Neoglycolipids. After orcinol staining (see Visualization of Neoglycolipids, above), the plates should be stored in a box containing silica gel, in the dark for 2 h before densitometric measurement.

Sialic Acid-Containing Neoglycolipids. Plates should be left in the dark in a fume cupboard to allow HCl vapor to completely escape before using for densitometry.

Quantitation by Densitometry and Spectrophotometry

Neoglycolipids. Fluorescent NGLs may be quantified by densitometry on TLC plates and by spectrophotometry in solution: medium sensitivity (down to 30 pmol) by flying spot reflectance adsorption at 256 nm and high sensitivity (down to 1 pmol) by linear reflectance fluorescence at λ_{ex} 256 nm and detection through a filter with a cutoff at 400 nm. Spectrophotometry by adsorption at 256 nm in ethanol, using 9-methylanthracene as reference, can be used to prepare standards for TLC densitometry or for larger scale quantitation of purified ADHP NGLs.

For NGLs stained with primulin, the densitometer is used in linear reflectance fluorescence mode with excitation at 370 nm and detection through a filter with a cutoff at 460 nm.

Hexose-Containing Neoglycolipids. For NGLs stained with orcinol, densitometry is performed in flying spot reflectance absorption mode at 550 nm. Test samples are measured against standard NGLs with essentially similar structural features. Because hexose stain is unstable, an average value should be obtained for standards scanned before and after scanning of the test samples.

The orcinol response of the ring-opened core hexose differs with various oligosaccharides. For example, with lactose, and other milk oligosaccharides with a lactose core structure, the ring-opened glucose reacts (surprisingly) as a normal hexose, but with maltose Glcα1-4Glc or isomaltose Glcα1-6Glc and the disaccharide GlcNAcβ1-3Gal, the core hexoses do not react. There is a different response for each hexose; thus if galactose is 1.0, mannose is 1.1, glucose is 0.8, and fucose is 0.6. Thus it is important, when possible, to use reference compounds the core structures of which resemble those of the test compounds.

Sialic Acid-Containing Neoglycolipids. For NGLs stained with resorcinol, densitometry is performed in a similar way as for hexose. Test samples are measured against reference NGLs that contain sialic acid.

Techniques for Studying Carbohydrate–Protein Interactions

Two techniques commonly used to study carbohydrate–protein interactions, using NGLs as oligosaccharide probes, are TLC overlays, in which NGLs are resolved on TLC plates and overlaid with carbohydrate-binding proteins or cells, and microwell assays, in which NGLs are coated on the plastic surface of 96-well microtiter plates. The advantage of TLC overlays is that these can be used with NGL mixtures, and the components are partially or completely resolved before the binding experiments. The bound components can be directly analyzed *in situ* by mass spectrometry. The ligand-positive regions may also be extracted by preparative TLC and further resolved by different TLC solvent systems or by HPLC (fluorescent NGLs). When the starting oligosaccharides are highly heterogeneous, these may be subjected to sequential fractionation procedures, and aliquots converted to NGLs for monitoring of binding activities, as for example in Yuen *et al.*[8] Thus, by this approach, efforts are focused on isolating and characterizing ligand-positive oligosaccharides. Microwell assays are appropriate for purified NGLs, particularly for quantitative comparisons of the intensities of binding of a protein to different oligosaccharide sequences. The microwell assays are ideal for inhibition experiments, in which the potencies of soluble substances, or of lipid-linked oligosaccharides incorporated into liposomes, can be compared as inhibitors of carbohydrate binding. Because of their multivalence, the oligosaccharides on liposomes have 10,000-fold greater inhibitory activities than the corresponding free oligosaccharides.[29] This is therefore the preferred mode of presentation of oligosaccharides, which are available only in limited amounts.

[29] C. Galustian, R. A. Childs, C.-T. Yuen, A. Hasegawa, M. Kiso, A. Lubineau, G. Shaw, and T. Feizi, *Biochemistry* **36,** 5260 (1997).

Preparation of Proteins or Cells for Binding Assays

The carbohydrate-binding assays are most commonly performed with soluble proteins, as we elaborate on here. Whole cells, for example, cultured mammalian cells[30] or bacteria,[31, 32] may also be used after suitable labeling for detection of binding. Alternatively, binding of cells or bacteria may be detected with the aid of antibodies that do not interfere with the binding. It is generally necessary to have the carbohydrate-binding domains of the soluble proteins in an oligomeric or multivalent state. Proteins that are multivalent or have sufficiently high affinities for carbohydrate ligands are used without additional aggregation. Examples include anti-carbohydrate antibodies, collectins, recombinant immunoglobulin G (IgG) Fc fusion proteins that aggregate spontaneously,[29] and IgM Fc fusion proteins. Binding is detected with nonblocking secondary antibodies, for example, anti-Ig Fc, or by prior labeling of the proteins with ^{125}I or with biotin followed by tagged streptavidin. In Fig. 2, results of binding experiments are shown with various NGL mixtures that have been resolved by TLC and overlaid with a radioiodinated protein, human serum mannose-binding protein (Fig. 2A), [^{3}H]thymidine-labeled[30] Chinese hamster ovary (CHO) cells transfected to express E-selectin (Fig. 2B), and [^{14}C]glucose-labeled bacteria, *Pseudomonas aeruginosa*[32] (Fig. 2C). Animal lectins in the form of IgG Fc chimeras are commonly assayed as aggregates after complexing with anti-IgG Fc.[29] It is convenient to use biotinylated forms of the anti-Ig, and to detect binding by ELISA, as described in the procedures below. In Fig. 3, results of binding experiments are shown with NGLs that have been immobilized in plastic microwells and overlaid with a human L-selectin–IgG Fc chimera (90% in dimeric state) (Fig. 3A) and with the precomplexed form of the chimera, using a biotinylated goat anti-IgG (Fig. 3B). The enhancement in binding signals after the aggregation is clearly shown.

To prepare a complex of a recombinant IgG Fc chimera protein with anti-IgG, typically a 1:3 (w/w) ratio of the chimera to the biotinylated goat anti-IgG is used (appropriate volumes are used of concentrated solutions of the chimera and of the anti-Ig, e.g., at 1–2 mg/ml), and the complex is allowed to stand for 1 h before dilution for the binding experiments. For calcium-dependent proteins, Tris-buffered saline (TBS:

[30] M. Larkin, T. J. Ahern, M. S. Stoll, M. Shaffer, D. Sako, J. O'Brien, A. M. Lawson, R. A. Childs, K. M. Barone, P. R. Langer-Safer, A. Hasegawa, M. Kiso, G. R. Larsen, and T. Feizi, *J. Biol. Chem.* **267,** 13661 (1992).

[31] I. J. Rosenstein, M. S. Stoll, T. Mizuochi, R. A. Childs, E. F. Hounsell, and T. Feizi, *Lancet* **ii**, 1327 (1988).

[32] I. J. Rosenstein, M. S. Stoll, and T. Feizi, *Infect. Immun.* **60,** 5078 (1992).

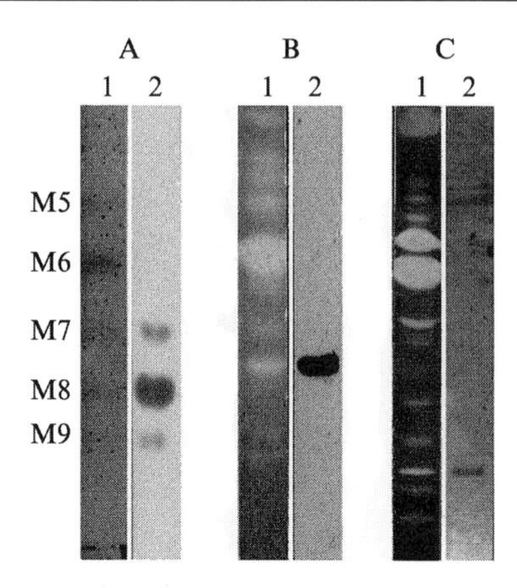

FIG. 2. TLC plate binding experiments showing the binding of a soluble protein (A), cultured cells (B), and bacteria (C) to particular DHPE-NGLs within mixtures. The left lanes are chromatograms of NGL mixtures detected by either orcinol (A) or primulin staining (B and C), and the right lanes are duplicates overlaid to reveal binding. Lanes A contained *N*-linked oligosaccharides released from the recombinant envelope glycoprotein, gp120, of the human immunodeficiency virus; the positions of high mannose-type glycans with five to nine mannose residues (M5 to M9, respectively) are indicated; selective binding to ^{125}I-labeled human serum mannose-binding protein was revealed by autoradiography (reproduced in part from Ref. 40 with permission). Lanes B contained NGLs derived from an acidic oligosaccharide fraction (our unpublished data) of *O*-glycans released by nonreductive alkali hydrolysis from an ovarian cyst mucin glycoprotein,[4] and overlaid with [^{3}H]thymidine-labeled Chinese hamster ovary cells transfected to express E-selectin. The bound component contained a capping sequence of sulfo-Le$^{a/x}$ type, as revealed by *in situ* TLC/mass spectrometry analysis. Lanes C contained NGLs of a fraction of *O*-glycans (desialylated and defucosylated) released by nonreductive alkali treatment from human meconium glycoproteins, and overlaid with [^{14}C]glucose-labeled bacteria, *Pseudomonas aeruginosa*. The bound band contained the sequence Gal-GlcNAc-Gal-GlcNAc- linked to the 6-branch of a GalNAcol core as revealed by *in situ* TLC/mass spectrometry analysis (our unpublished data).

10 m*M* Tris buffer, 150 m*M* NaCl, pH 7.4) is used as buffer. Phosphate-buffered saline (PBS: 50 m*M* phosphate buffer, 150 m*M* NaCl, pH 7.4) may be used as an alternative diluent if calcium is not required for binding.

Blocking of Nonspecific Binding Sites

For blocking of nonspecific binding to TLC plates or microwells, solutions containing 1% (w/v) casein (Pierce, Peterborough, UK) or 3% (w/v) bovine serum albumin (BSA) are frequently used. A variety of commercial blocking solutions may also be used.

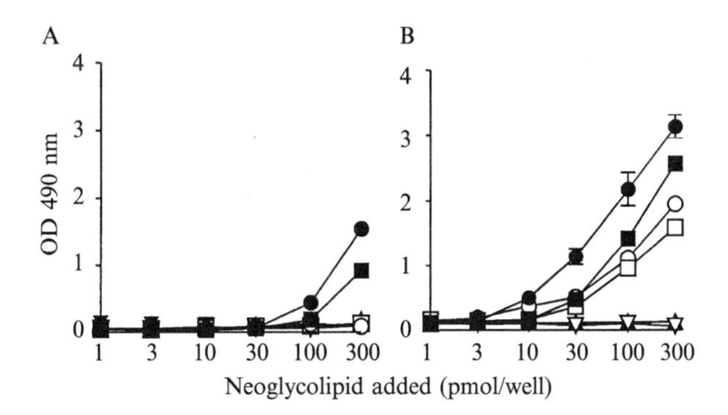

FIG. 3. Microwell binding experiments showing the binding of a human IgG Fc chimera to NGLs immobilized in microwells. Various amounts of the NGL were applied onto plastic microwells, and L-selectin–IgG Fc, 100 ng, was added per well, noncomplexed (A) and precomplexed with biotinylated ant-IgG (B). The binding in (A) was detected with biotinylated anti-IgG followed by a complex of biotinylated horseradish peroxidase made with streptavidin; in (B), a complex of biotinylated horseradish peroxidase and streptavidin was used. Symbols for the oligosaccharide sequences are as follows: (●) 3′-sulfo-Le^{a}; (■) 3′-sulfo-Le^{x}; (○) 3′-sialyl-Le^{a}; (□) 3′-sialyl-Le^{x} (all based on the lacto-*N*- or lacto-*N*-*neo*-tetrasaccharide backbone); (▽) lacto-*N*-tetraose. The NGL of the nonacidic oligosaccharides gave no binding signals with the two L-selectin forms; symbols for these overlap at baseline. From Ref. 29 with permission.

TLC Plate-Binding Assay for Calcium-Dependent Carbohydrate-Binding IgG Fc Chimera

For TLC-binding experiments, DHPE-NGLs are chromatographed in duplicate. One plate is for visualizing the NGL components, and the second for overlays. Binding of some carbohydrate-binding proteins (such as the selectins) is diminished after the nondestructive staining of NGLs with primulin. When fluorescent NGLs are used, or when carbohydrate binding is unaffected by primulin staining of the NGLs, the same plate can be used to advantage for recording the positions of various NGL bands in the chromatogram and the positions of those that are bound.

Materials

Protein–IgG Fc chimera
Aluminum-backed TLC plates
Parafilm
Sandwich box with cover
Solvent for TLC: chloroform–methanol–water (as appropriate)
Staining reagent: e.g., primulin reagent
TBS

TBS containing 50 mM $CaCl_2$ (TBS_{C50})
Blocking and diluent solution: TBS_{C50} containing casein (1%, w/v)
Goat anti-IgG Fc conjugated to biotin from Vector (Peterborough, UK)
Horseradish peroxidase (HRP) conjugated to streptavidin (Sigma)
Fast-DAB solution: Fast-diaminobenzidine (DAB) tablet and hydrogen peroxide tablet (Sigma) dissolved in 5 ml of distilled H_2O
Goat anti-IgG Fc conjugated to HRP; this for use with chimeric proteins that do not require complexing

Procedure for Chimeric Protein Requiring Complexing

1. Cut the TLC plate to the desired size (typically 10 cm long and 5–10 cm wide) and chromatograph NGLs with a suitable solvent, as described in Analysis of Neoglycolipid Products by TLC (above). In the case of DHPE-derived NGLs, duplicate plates are required.
2. Stain one of the plates (e.g., with primulin or orcinol) as described in Visualization of Neoglycolipids (above).
3. Wet the plate for overlay assay by immersion in TBS. This is carried out by slowly immersing 1 cm of plate per 30 s to avoid silica gel flaking.
4. Place the TLC plate onto Parafilm in a sandwich box. Add blocking solution to cover the plate (6 ml will cover a plate of 10 cm^2).
5. Cover the box with a lid to prevent drying of the solution. Incubate for 1 h.
6. Prepare a complex of the IgG Fc chimera with biotinylated goat anti-IgG Fc as described above, and dilute the complex; typically the chimera is diluted to 1–2 μg/ml in diluent solution. For optimization, a range of dilutions of the complex may be tested.
7. Overlay the protein complex onto the plate. Incubate for 2 h (ambient temperature is satisfactory with most carbohydrate-binding proteins).
8. Wash the TLC plate for 5 min by immersion in TBS_{C50} in a sandwich box. Repeat twice.
9. Add streptavidin–HRP, 10 μg/ml in diluent solution; incubate for 30 min.
10. Wash as described above.
11. Add Fast-DAB solution. Develop for 5–10 min.
12. Stop the reaction by rinsing the plate in distilled water (prolonged washing in water will cause the silica to detach from the aluminum backing).

Procedure for Chimeric Protein Not Requiring Complexing

1. Carry out the first five steps as described above.

2. Overlay the IgG Fc chimera (1–2 μg/ml) onto the plate. Incubate for 2 h.

3. Wash the TLC plate for 5 min in TBS_{C50} in a sandwich box as described above. Repeat twice.

4. Add goat anti-IgG Fc conjugated to HRP at the dilution recommended by the manufacturer in diluent solution. Incubate for 1 h.

5. Wash as described above.

6. Add Fast-DAB solution. Develop for 5–10 min.

7. Stop the reaction by rinsing the plate in distilled water.

Microwell-Binding Assay for Calcium-Dependent Carbohydrate-Binding IgG Fc Chimera

For coating onto plastic microwells, NGLs are diluted either in methanol, or in methanol containing the carrier lipids egg lecithin and cholesterol. The binding signals with some proteins but not others are enhanced when carrier lipids are included. It is usual to use duplicate wells for each assay point. For optimization, a checker board titration may be required, that is, titration of both the immobilized NGL and the carbohydrate-binding protein or complex. The procedure described below is for binding experiments with IgG Fc chimeras requiring complexing.

Materials

Protein–IgG Fc chimera
Ninety-six-well Falcon 3912 plates (Marathon Labs, London, UK)
Glass vials, egg lecithin (Nutfield Nurseries, Redhill, UK)
Cholesterol (Sigma)
NGL diluent: Methanol containing egg lecithin (4 μg/ml) and cholesterol (4 μg/ml) TBS
TBS containing 2 m*M* $CaCl_2$ (TBS_{C2})
Blocking solution: TBS_{C2} containing casein (1% w/v)
Diluent solution: TBS_{C2} containing casein (0.1%, w/v)
Goat anti-IgG Fc conjugated to biotin (Vector)
OPD solution: *o*-Phenyldiamine (Sigma) in 100 m*M* phosphate–citrate buffer containing 0.2% (v/v) hydrogen peroxide
H_2SO_4 (3 *M*)

Equipment

Plate reader: ELISA Dynatech MRX microplate reader (Thermo Labsystems, Acton Court, UK)

Procedure

1. Make a range of dilutions of stock NGL solution [stored in C/M/W (25:25:8)] in half-log steps, in NGL diluent; typically, dilutions range from 1 to 6 nmol/ml.
2. Add 50 μl of NGLs per microwell. Leave to dry at 37° for 16 h.
3. Wash each well with 200 μl of TBS per well. Repeat twice.
4. To the NGL-coated microtiter wells add blocking solution (100 μl per well). Incubate for 1 h.
5. Wash each well three times with 200 μl of TBS.
6. Prepare a complex of the IgG Fc chimera with anti-IgG, as described above, and dilute in diluent solution.
7. Add 50 μl of the chimera complex per well, and incubate for 2 h.
8. Wash each well three times with 200 μl of TBS_{C2}.
9. Add horseradish peroxidase–streptavidin (10 μg/ml) diluted in TBS_{C2}–0.1% casein (50 μl/well). Incubate for 30 min.
10. Wash wells as described above.
11. Add OPD solution, 50 μl/well. Leave to develop color for 5–10 min.
12. Stop the color reaction with 3 *M* H_2SO_4, 50 μl/well.
13. Read at 490 nm in the plate reader:

$$OD_{490\ nm}\ \text{of binding} = \text{OD of test NGL} - \text{OD of the negative control NGL}$$

With IgG Fc chimeras not requiring complexing, use goat anti-IgG Fc conjugated to HRP, followed by OPD for color development.

Inhibition Assay Using Neoglycolipids as Reference Ligands Immobilized on Microtiter Wells

The inhibition of the binding of a ligand-positive NGL immobilized on microtiter wells can be assayed using as inhibitors either free oligosaccharides or NGLs displayed in liposomes. The following procedure describes the inhibition of binding of an IgG Fc chimera. For immobilization, 50–100 pmol of NGL ligand is added per well. An initial binding experiment is performed with serial dilutions of the protein, to determine the dilution to be used for the inhibition assay. The protein should be at a limiting rather than a saturating concentration; with an L-selectin–IgG Fc chimera the level selected was 10 ng per well (Fig. 4). With L-selectin, it was found that a better discrimination could be made between the inhibitory activities of oligosaccharides when the noncomplexed, paucivalent form of the protein was used than when the complexed, multivalent form was used (Fig. 4).[29,33]

[33] C. Galustian, A. Lubineau, C. le Narvor, M. Kiso, G. Brown, and T. Feizi, *J. Biol. Chem.* **274,** 18213 (1999).

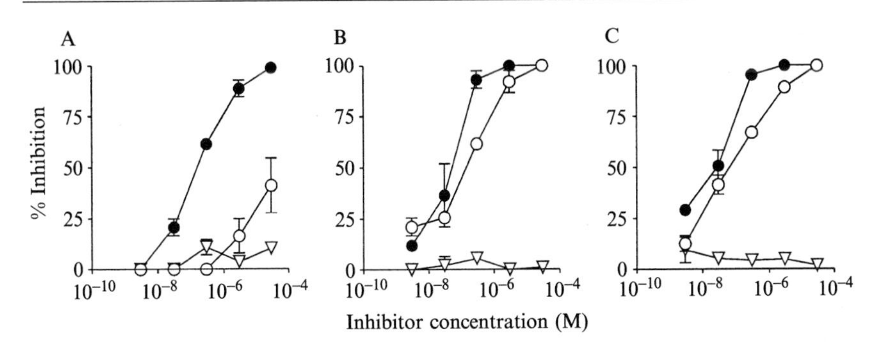

FIG. 4. Inhibition of binding of noncomplexed and precomplexed L-selectin–IgG Fc chimera to NGLs immobilized in microwells. NGLs of sulfo-Le^a (A and B) and sialyl-Le^a (C), 100 pmol/well, were coated onto microwells. The inhibition is shown of the binding of L-selectin, 10 ng, noncomplexed (A) or complexed with biotinylated anti-IgG (B and C), by oligosaccharides displayed on liposomes. Inhibition data are expressed as means of duplicate wells, with the range indicated by error bars. Symbols for oligosaccharides are as follows: (●) 3′-sulfo-Le^a; (○) 3′-sialyl-Le^a; (▽) lacto-*N*-tetraose. From Ref. 29 with permission.

Materials. See the section Microwell-Binding Assay for Calcium-Dependent Carbohydrate-Binding IgG Fc Chimera.

Procedure

1. Immobilize multiples of a ligand-positive NGL, 100 pmol per well, as described above (typically 10 wells are prepared for 5 dilutions of inhibitor in duplicate). Also immobilize 100 pmol of a ligand-negative control NGL per well, in four wells, as negative controls.
2. Block the wells as described above.
3. During 1 h of blocking, prepare inhibitors.

   NGL liposome inhibitors: In glass vials, dispense NGLs, together with egg lecithin and cholesterol (all solutions in methanol), in the ratio of 1 nmol NGL: 0.4 μg cholesterol:0.4 μg egg lecithin. Evaporate the methanol with a gentle stream of nitrogen. Add diluent solution to vials to give an NGL concentration of 10 μM. Sonicate the vials in a sonic water bath (water temperature, 37°) for 10 min to form liposomes. Prepare dilutions (half-log steps) of liposomes in diluent solution

   Oligosaccharide inhibitors: Prepare oligosaccharide solutions in dil uent solution; dilute in half-log steps to give concentrations ranging from 0.03 to 3 mg/ml

4. Wash each well three times with 200 μl of TBS_{C2}.

5. To the wells with the immobilized ligand-positive NGL, add dilutions of the inhibitors at 25 μl per well (also include quadruplicate wells to which diluent solution rather than inhibitor is added at 25 μl per well. These are the uninhibited, positive control wells).

6. To the wells with the immobilized ligand-negative NGL, add diluent solution (25 μl per well).

7. Add to each well the protein–IgG Fc chimera, 25 μl per well at double the concentration selected in the binding experiments, and incubate for 2 h.

8. Wash each well three times with 200 μl of TBS_{C2}.

9. Add horseradish peroxidase-conjugated goat anti-IgG Fc antibody, diluted to 10 μg/ml in diluent solution, 50 μl/well. Incubate for 30 min.

10. Wash the wells as described above.

11. Add OPD solution at 50 μl/well. Leave to develop color for 5–10 min.

12. Stop the color reaction with 3 *M* H_2SO_4, 50 μl/well.

13. Read at 490 nm in a plate reader.

14. Calculate the percentage of inhibition of binding as follows:

$$\% = [(\text{OD without inhibitors} - \text{OD of negative control NGL}) - (\text{OD obtained in the presence of inhibitors} - \text{OD of negative control NGL}) / (\text{OD without inhibitors} - \text{OD of negative control NGL})] \times 100.$$

Structural Characterization of Neoglycolipids by Mass Spectrometry

NGLs can be analyzed by mass spectrometry as described previously[16] to determine their molecular masses, from which the monosaccharide composition together with the presence of substituents can be deduced. In many cases, sequence information can also be derived from the mass spectra. Various ionization techniques have been used successfully and include liquid secondary ion mass spectrometry (LSIMS), electrospray ionization mass spectrometry (ESMS), and matrix-assisted laser desorption ionization mass spectrometry (MALDI-MS). LSIMS has been valuable for obtaining molecular mass and sequence information from NGLs derived from neutral oligosaccharides and acidic oligosaccharides of the type found on glycoproteins. LSIMS is not ideal for NGLs of heavily sulfated glycosaminoglycan-derived oligosaccharides and these are best analyzed by ESMS. Analysis of NGLs by MALDI also provides molecular mass information with high sensitivity. Sequence information about the oligosaccharides can be deduced from collision-induced dissociation (CID) tandem MS (MS/MS) and postsource decay spectra in ESMS and MALDI, respectively. Direct analysis from TLC plates by *in situ* TLC/LSIMS has proved of value for sequence determination of oligosaccharides recognized by carbohydrate-binding proteins and assignments of the specificities of binding (e.g., Refs. 4 and 13–15).

Liquid Secondary Ion Mass Spectrometry

NGLs analyzed by LSIMS give intense molecular ions and a series of fragment ions of lower intensities. Because of the localization of charge, all fragment ions arise by glycosidic bond cleavage and associated fragmentation, and contain the lipid moiety.[34] Hence, the monosaccharide composition, sequence, and branching patterns of the oligosaccharide components can be deduced from the spectra. The presence of other groups such as sulfate or phosphate can also be detected and their locations on individual monosaccharides determined. Spectra are obtained with similar sensitivity in both positive- and negative-ion mode, although the latter are much preferred because of the low chemical background in the mass region that contains the informative sequence ions, and the absence of quasimolecular adductions with alkali salts. In general, sensitivities of detection are in the 100-fmol to 1-pmol range, and for sequence information on the order of 5–10 pmol. Mass spectra of DHPE- and ADHP-derived NGLs show little difference in sensitivity. The molecular and fragment ions are higher by 219 Da for ADHP derivatives. In the case of *O*-glycans, the branching chains at the core GalNAc or Man can be sequenced from the spectra of their NGLs after mild periodate oxidation of the reduced sugars[27] (see Scheme 2). LSIMS has been the analysis method of choice for NGLs of neutral and acidic oligosaccharides containing sialic or hexuronic acids, and sulfated glycoprotein oligosaccharides.

The LSI spectrum in Fig. 5 is that of an ADHP-NGL strongly immunoreactive with a monoclonal antibody, 2D3, directed at the sialyl-Le^a sequence.[11] This NGL was singled out from among HPLC fractions of NGLs derived from the *O*-glycans released by nonreductive alkali treatment of a mucin glycoprotein. The $[M-H]^-$ at *m/z* 1875 is consistent with a dHex.Hex_2.HexNAc.NeuAc-ADHP composition (where dHex is a deoxyhexose; Hex is a hexose; and HexNAc is an *N*-acetylhexosamine). The sequence is assigned as NeuAc-Hex-(dHex)HexNAc-Hex-Hex-ADHP from the fragmentation, that is, *m/z* 1584, M – NeuAc; *m/z* 1422, M – (NeuAc-Hex); *m/z* 1073, M – [NeuAc-Hex-(dHex)HexNAc] corresponding to the core monosaccharide linked to the lipid, Hex-ADHP. The *m/z* 1242 is diagnostic of a branching dHex. These data, coupled with its

[34] A. M. Lawson, W. Chai, G. C. Cashmore, M. S. Stoll, E. F. Hounsell, and T. Feizi, *Carbohydr. Res.* **200,** 47 (1990).

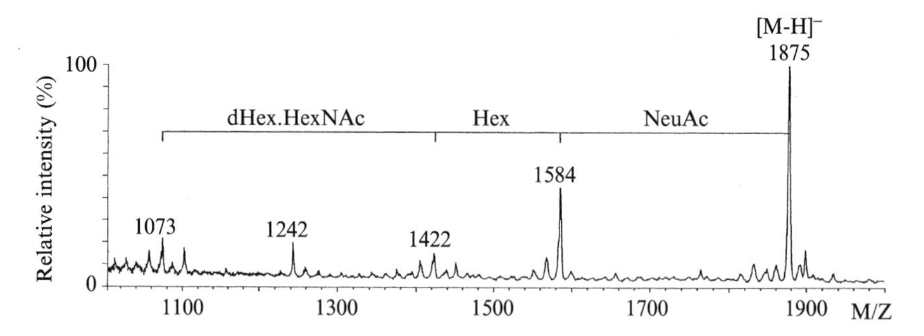

FIG. 5. LSI mass spectrum of the ADHP-NGL of an *O*-glycan released by nonreductive hydrolysis from a mucin glycoprotein. The oligosaccharide was deduced to have the sialyl-Le^a sequence: NeuAcα2-3Galβ1-3(Fucβ1-4)GlcNAc-Hex-ADHP from the fragmentation coupled with binding to anti-sialyl-Le^a but not anti-sialyl-Le^x. Reproduced in part from Ref. 11 with permission.

lack of immunoreactivity with an anti-sialyl-Le^x, established the sequence as NeuAcα2-3Galβ1-3(Fucα1-4) GlcNAc-Hex-ADHP.[11]

Equipment

ZAB2-E mass spectrometer (Micromass-Waters) is used to record spectra in the negative-ion mode (as in Fig. 5). The instrument is fitted with a cesium gun operated at 35 keV and an emission current of 0.5 μA

Procedure

1. Coat the LSIMS stainless steel sample target with 1 μl of a liquid matrix comprising diethanolamine (DEA), tetramethylurea (TMU) and *m*-nitrobenzyl alcohol (NBA) (2:2:1, v/v/v).
2. Apply 1 μl of the sample solution in chloroform–methanol–water (25:25:8, v/v/v).
3. Acquire spectra immediately after insertion of probe. Because of the volatility of the liquid matrix, sample ion signals do not normally last long.

In Situ *TLC/LSIMS*

The development of *in situ* TLC/LSIMS with NGLs[35] has been key to characterizing oligosaccharides of glycoproteins as ligands for

[35] W. Chai, G. C. Cashmore, R. A. Carruthers, M. S. Stoll, and A. M. Lawson, *Biol. Mass Spectrom.* **20,** 169 (1991).

carbohydrate-binding proteins. The approach enables LSIMS analysis of mixtures of NGLs after chromatographic resolution of components on silica gels.[7,11] Thus, NGL bands shown to bind in overlay assays (see Techniques for the Study of Carbohydrate–Protein Interactions, above) can be selected for characterization from among complex mixtures. Regions of interest cut from the TLC plate are prepared for LSIMS by addition of solvent and matrix. Sensitivities are approximately an order of magnitude lower compared with direct ionization from the stainless steel LSIMS probe. The approach is not suited to highly sulfated oligosaccharides derived from glycosaminoglycans, such as heparan sulfate, as these NGLs suffer from adverse adsorption effects and from the loss of sulfate, and would be better analyzed by on-line TLC-ESMS.

TLC-LSIMS has been effective for the structural characterization of *O*-glycan alditols as NGLs.[7,9,10] After the mild periodate oxidation performed before NGL formation, multiple NGL products are formed and TLC separation is required for analysis.

Equipment. As described above.

Procedure.

1. Mark the position of the NGL bands for analysis on the silica surface and aluminum backing of the TLC plate, and place the plate face down on a clean glass surface.
2. Cut out regions of the plate (typically 1.5×5 mm) corresponding to each band, using a scalpel blade and metal ruler.
3. Attach the strips with the aluminum backing to the LSIMS target probe by a mixture of water-soluble glue and glycerol (1:1, v/v).
4. Add 2 μl of chloroform–methanol–water (25:25:8, v/v/v) and 2 μl of liquid matrix DEA/TMU/NBA (2:2:1, v/v/v).
5. Acquire spectra immediately after insertion of probe.

Electrospray Ionization Mass Spectrometry

NGLs derived from acidic oligosaccharides are ionized in negative ion electrospray with a sensitivity similar to that of LSIMS for neutral oligosaccharides (100 fmol–1 pmol). The presence of deprotonation sites, on the phosphate group in DHPE and ADHP and on the acidic groups in the oligosaccharide chain, give rise to multiply charged molecular ions from which the monosaccharide composition can be deduced. Sequence information is deduced from spectra generated by CID MS/MS. The NGLs of neutral oligosaccharides show lower sensitivity of detection in ESMS.

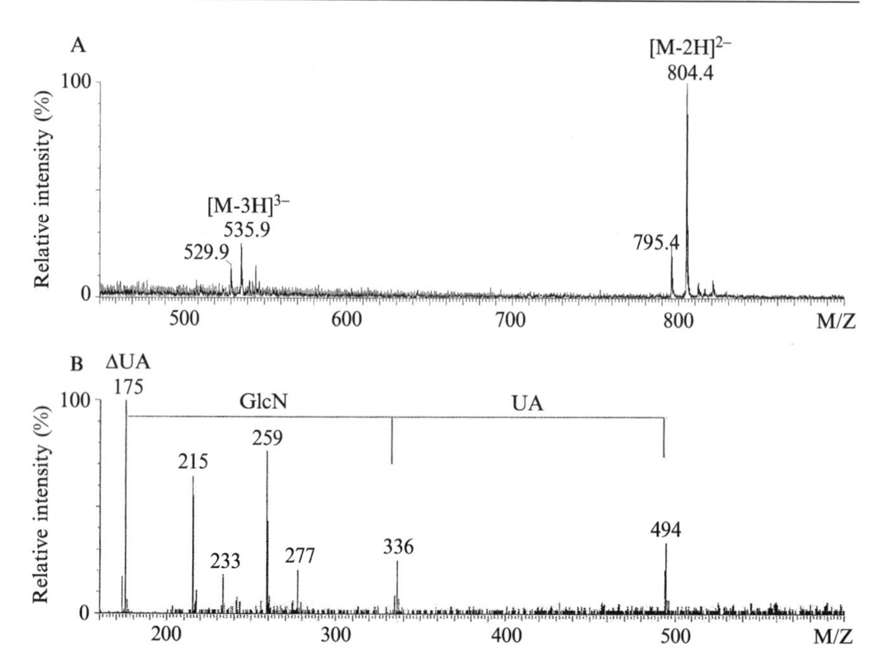

FIG. 6. Negative-ion ES mass spectra of the NGL of a unique tetrasaccharide sequence on heparan sulfate that is recognized by the monoclonal antibody 10E4. (A) Primary mass spectrum of the tetrasaccharide-ADHP, and (B) the low-mass region of the CID MS/MS spectrum of $[M - 2H]^{2-}$ at *m/z* 804.4 of the tetrasaccharide-ADHP from which sequence is derived. Reproduced in part from Ref. 15 with permission.

The primary ES spectrum of a heparan sulfate-derived tetrasaccharide-NGL and the MS/MS spectrum of its doubly charged molecular ion are shown in Fig. 6. This oligosaccharide was identified as a unique sequence on heparan sulfate recognized by the monoclonal antibody 10E4. This antigen is closely associated with prion lesions in the brain.[15] The molecular mass is established as 1610.8 Da from $[M - 2H]^{2-}$ and $[M - 3H]^{3-}$ ions at *m/z* 804.4 and 535.9, respectively, corresponding to a tetrasaccharide containing an *N*-unsubstituted glucosamine (GlcN) residue (Fig. 6A). CID MS/MS of *m/z* 804.4 gives a spectrum showing C- and B-type fragment ions[36] in the low-mass region arising from the nonreducing terminus of the oligosaccharide-NGL (Fig. 6B). The sequence is deduced as follows. The C-type fragment ion at *m/z* 175 identifies a ΔUA residue at the nonreducing end (where ΔUA is the unsaturated glucuronic acid arising from

[36] B. Domon, and C. E. Costello, *Glycoconjug. J.* **5,** 397 (1988).

the enzymatic cleavage of the polysaccharide chain). The ion at *m/z* 336 represents a mass increment of 161, corresponding to a GlcN. The B-type fragment ion at *m/z* 494 represents a mass increment of *m/z* 158, corresponding to a hexuronic acid (UA), and serves to identify unambiguously the sequence ΔUA-GlcN-UA-. The remaining 1115 Da (1609 minus 494 Da) is in accord with the presence of a GlcNAc residue linked to the ADHP. Thus the complete sequence is ΔUA-GlcN-UA-GlcNAc-ADHP.

Equipment. A Q-Tof mass spectrometer (Micromass-Waters) is used to acquire mass spectra (as in Fig. 6). The cone voltage is maintained at 30 V, with a capillary voltage of 3 kV. The CID spectrum is obtained with argon as collision gas at a pressure of 1.7 bar and a collision energy of 35 V.

Procedure.

1. Deliver mobile phase [chloroform–methanol–2 m*M* NH_4HCO_3 (25:25:8, v/v/v)] by syringe pump at a flow rate of 10 μl/min.
2. Loop inject 5 μl of NGL sample solution [1 to 10 pmol/μl chloroform–methanol–water (25:25:8, v/v/v)].
3. Samples may also be introduced into the mass spectrometer by HPLC (or TLC; see below).

On-Line TLC-ESMS

The introduction of on-line TLC-ESMS[37] extends the scope of ESMS in the analysis of NGLs. Sequence information can now be derived from individual NGLs in mixtures derived from acidic oligosaccharides that contain labile sulfate groups (e.g., highly sulfated glycosaminoglycan oligosaccharides). Detection sensitivities are on the order of 5 pmol for TLC-ESMS for composition, and 20 pmol for TLC-ES CID MS/MS to derive saccharide sequence information. A significant feature of on-line TLC-ESMS is that it opens the way to structure characterization of NGLs of acidic oligosaccharide bands that are shown, in parallel experiments, to be recognized by carbohydrate-binding proteins.

Equipment

Overpressured TLC system, OPLC 50 (Bionisis), comprising separation and solvent delivery units.
Mass spectrometer with electrospray ion source.
Dedicated TLC plates with perimeter seals (supplied by Bionsis).

[37] W. Chai, C. Leteux, A. M. Lawson, and M. S. Stoll, *Anal. Chem.* **75,** 118 (2002).

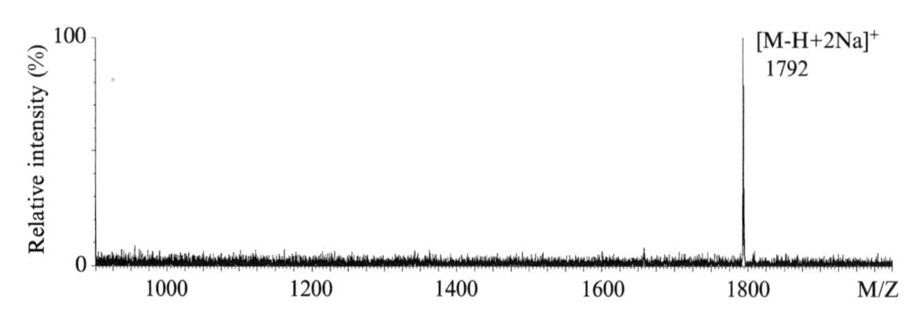

FIG. 7. Positive-ion MALDI spectrum of the ADHP-NGL of the Le[a] sequence Galβ1-3(Fucβ1-4)GlcNAcβ1-3Galβ1-4Glc. The $[M - H + 2Na]^+$ ion at *m/z* 1792 illustrates the simplicity of the spectrum and hence the promise of MALDI for profiling mixtures of NGLs (our unpublished data).

Procedure

1. Wash the TLC plate thoroughly with appropriate solvent and pass eluent to waste.
2. Apply the sample as described in Analysis of Neoglycolipid Products by TLC (above).
3. Place the plate on the cassette and insert into the separation unit.
4. Connect the outlet of the OPLC 50 to the inlet of the electrospray probe by a short length of capillary polyetheretherketone (PEEK) tubing (130-μm i.d. $\times$ 1.6-mm o.d.).
5. Deliver elution solvent by OPLC pump at a flow rate of 60 to 125 μl/min and acquire mass spectra.

Matrix-Assisted Laser Desorption Ionization Mass Spectrometry

We have evaluated MALDI-MS for analysis of NGLs. Quasimolecular ions that establish the monosaccharide composition are obtained in positive mode by formation of adducts with alkali cations from salts present in low amounts in solution. The simplicity of these MALDI spectra should render them suitable for profiling mixtures of oligosaccharides. For example, the quasimolecular ion at *m/z* 1792 given by Galβ1-3(Fucα1-4)GlcNAcβ1-3Galβ1-3Glc-ADHP corresponds to $[M-H+2Na]^+$ (Fig. 7). Spectra can also be obtained of NGLs in negative-ion mode.

Perspectives

The special feature of NGL technology is its versatility in that it is a means of generating oligosaccharide probes with a homogeneous tag, not only from glycoproteins, proteoglycans, and glycolipids, but also from whole cells[22] and organs.[38] Most of the known carbohydrate-recognizing proteins interact with peripheral or backbone sequences of oligosaccharides and in these cases the ring-opened monosaccharide at the site of joining to lipid acts as a flexible linker. However, for those recognition systems that require an intact core residue, it will be desirable to develop a procedure for preserving the closed ring structure. Also, for *O*-glycans and other peptide-linked oligosaccharides, where the sugar–peptide junction and flanking peptides are parts of the recognition motifs, we are considering designs of lipid-linked glycopeptides as probes. There is precedent for successful application of this principle: tyrosine sulfate, which is a part of the recognition motif for P-selectin, was converted to a lipid-linked probe by conjugation to DHPE.[39] This novel probe when used in conjunction with NGLs of the carbohydrate ligands sialyl-Le^x and -Le^a, and sulfo-Le^x and -Le^a, served to demonstrate the synergistic interactions of the two classes of ligands.

We have explored NGL technology as the basis of a carbohydrate microarray system applicable for the generation of large repertoires of immobilized oligosaccharide probes for studying carbohydrate–protein interactions.[38] We have observed that NGLs are robust probes when presented on nitrocellulose membranes, thus enabling sensitive and potentially high-throughput detection of ligands for carbohydrate-binding proteins. We have shown that carbohydrate-recognizing proteins such as monoclonal antibodies, the E- and L-selectins, a chemokine (RANTES), and a cytokine (interferon γ) can single out their ligands, not only in the arrays of homogeneous, structurally defined, oligosaccharides, but also in an array of heterogeneous *O*-glycan fractions derived from brain glycoproteins. The unique feature of the microarray system based on NGL technology is that deconvolution strategies are included, applying the TLC overlay and *in situ* mass spectrometry strategies described in this article, for sequencing ligand-positive components within mixtures. We have proposed[38] that, in conjunction with advanced protein expression systems, mass spectrometry, and bioinformatics, the principle of constructing oligosaccharide arrays from desired sources could form the basis of surveys to identify oligosaccharide-recognizing proteins in the proteome, and to map the repertoire of complementary recognition structures in the glycome.[40]

[38] S. Fukui, T. Feizi, C. Galustian, A. M. Lawson, and W. Chai, *Nat. Biotechnol.* **20,** 1011 (2002).

[39] C. Galustian, R. A. Childs, M. Stoll, H. Ishida, M. Kiso, and T. Feizi, *Immunology* **105,** 350 (2002).

[40] M. Larkin, R. A. Childs, T. J. Matthews, S. Thiel, T. Mizuochi, A. M. Lawson, J. S. Savill, C. Haslett, R. Diaz, and T. Feizi, *AIDS* **3,** 793 (1989).

[13] Preparation and Utility of Neoproteoglycan Probes in Analyses of Interaction with Glycosaminoglycan-Binding Proteins

By HARUKO OGAWA, HARUKO UEDA, AYUMI NATSUME, and RISA SUZUKI

Introduction

Proteoglycans have been attracting increasing biological interest because of their specific signaling roles or modulatory functions in important cellular events, for example, growth factor–receptor interaction and signaling, viral infection, extracellular matrix assembly, and neurite outgrowth and plasticity.[1] Such functions are exhibited mostly by glycosaminoglycan-binding proteins (lectins or receptors) that recognize repeating disaccharide units with microheterogeneities in the sugar residues and the negative charges.

On the other hand, several legume or fungal lectins have been reported to bind acidic polysaccharides.[2,3] Lectins isolated from the fruiting bodies of basidiomycetes or ascomycetes (mushrooms) have been found to exhibit complex binding activities with acidic polysaccharides, glycoproteins, or glycolipids, which cannot be predicted from their specificity to simple sugars. For example, a GlcNAc-specific lectin from *Psathyrella velutina* fruiting bodies (PVL) binds sialoglycoproteins at the same binding site as GlcNAc[4] (as described elsewhere in this volume[3a]), and also exhibits acidic polysaccharide binding activity. This indicates the need to study in detail the specificity of lectins to glycoconjugates, because such multispecificities may call into question the validity of the lectins when used as carbohydrate-specific probes.

In this study, we developed two types of neoproteoglycan probes: one is used to immobilize glycosaminoglycans or polysaccharides effectively on solid substrates, and the other is a biotin- or peroxidase-labeled neoproteoglycan probe that is used to detect binding with high sensitivity. The preparation of the probes is technically feasible. The utility of the neoproteoglycan probe is shown in terms of the interactions of heparin and other

[1] B. Casu and U. Lindahl, *Adv. Carbohyd. Chem. Biochem.* **57,** 159 (2002).
[2] N. Toda, A. Doi, A. Jimbo, I. Matsumoto, and N. Seno, *J. Biol. Chem.* **256,** 5345 (1981).
[3] S. Rosen, J. Bergstrom, K. A. Karlsson, and A. Tunlid, *Eur. J. Biochem.* **238,** 830 (1996).
[3a] H. Ueda and H. Ogawa, *Methods Enzymology* **362**, (2003) (this volume).
[4] H. Ueda, K. Kojima, T. Saitoh, and H. Ogawa, *FEBS Lett.* **448,** 75 (1999).

0076-6879/03 $35.00

glycosaminoglycans with PVL or vitronectin, a multifunctional plasma glycoprotein. A series of neoproteoglycan probes was used to screen lectins from 16 cultivable mushroom extracts, in combination with various hybrid glycoprotein probes having different glycan types.

Preparation Methods

Materials

Heparin (from porcine intestinal mucosa) and *N*-ethoxycarbonyl-2-ethoxy-1,2-dihydroquinoline (EEDQ) are purchased from Wako Pure Chemicals (Osaka, Japan). Colominic acid (from *Escherichia coli*) and pectic acid (from orange) are purchased from Nacalai Tesque (Kyoto, Japan). Horseradish peroxidase (HRP) is from Toyobo (Osaka, Japan), and bovine serum albumin (BSA) is from Iwai Chemicals (Tokyo, Japan). *N*-Hydroxysuccinimidylbiotin from Pierce (Rockford, IL) and biotin hydrazide from ICN Immunobiologicals (Costa Mesa, CA) are used for biotin labeling of the probes. Fucoidin, fetuin (fetal calf serum), human transferrin, and ribonuclease B are purchased from Sigma (St. Louis, MO). Streptavidin-biotinylated HRP complex is from ICN Immunobiologicals. *N*-Acetylchitooligosaccharides (a mixture of pentamers and hexamers, $GlcNAc_{5-6}$), β-galactosidase, and concanavalin A are purchased from Seikagaku Kogyo (Tokyo, Japan). PVL, the mycelium lectin of *P. velutina*, and human plasma vitronectin are prepared in our laboratory as described previously.[4,5]

Preparation of BSA-Conjugated Neoproteoglycan Probes: Glycan–BSA

Glycans are covalently conjugated to BSA before immobilizing them on solid substrate or subsequent biotin labeling of the protein portion. Glycosaminoglycan or polysaccharide (3 mg) is dissolved in 0.4 ml of distilled water, and 3 mg of EEDQ in 0.6 ml of ethanol is added. After preincubation at room temperature for 2 h, 6 mg of BSA is added, and the mixture is incubated at 4° for 2 days with gentle shaking. The formation of BSA-conjugated neoproteoglycan probe is monitored by observing the change in migration position of BSA on sodium dodecyl sulfate–palyacrylamide gel electrophoresis (SDS–PAGE) as detected by Coomassie Brilliant Blue (CBB) staining and by alcian blue staining. Excess reagent is removed by

[5] H. Kitagaki-Ogawa, T. Yatohgo, M. Izumi, M. Hayashi, H. Kashiwagi, I. Matsumoto, and N. Seno, *Biochim. Biophys. Acta* **1033,** 41 (1990).

ultrafiltration (UFP2 LGC, exclusion molecular weight 100,000; Nihon Millipore, Tokyo, Japan) with 10 m*M* Tris-buffered saline (TBS), pH 7.5.[6]

Preparation of Biotin-Labeled Neoproteoglycan Probe: Glycan–BSA–Biotin

Glycan–BSA prepared as described above is biotinylated for detection by the avidin–peroxidase system and for use as a soluble ligand. For biotinylation, 222 μl of 0.1% *N*-hydroxysuccinimidylbiotin is added to glycan–BSA in 3 ml of 50 m*M* sodium bicarbonate buffer (pH 8.5) and the mixture is allowed to stand for 2 h in ice. The biotin-labeled neoproteoglycan probe is dialyzed against TBS.[6]

Preparation of Peroxidase-Conjugated Neoproteoglycan Probe: Glycan–HRP

Glycosaminoglycan is directly coupled to HRP with the aid of EEDQ, using the same procedure as that for the preparation of glycan–BSA.

Preparation of Other Probes: Biotin–PVL, Biotin–GlcNAc$_{5-6}$, Hybrid Glycoprotein

Biotinylation of PVL is performed as described previously[4] by reaction for 4 h in the presence of 10 m*M* GlcNAc in the reaction buffer to protect the binding sites on PVL. Biotinylation of GlcNAc$_{5-6}$ is performed by dissolving 20 mg of biotin hydrazide in 1 ml of dimethyl sulfoxide (DMSO) mixed with 50 mg of GlcNAc$_{5-6}$ in 3 ml of saline and shaking at room temperature for 1 h. For the preparation of hybrid glycoprotein probes, fetuin and transferrin are desialylated with 0.01 *M* HCl at 80° for 1 h, and asialo agalactoglycoproteins are prepared by digestion of asialoglycoproteins (1 mg) with 0.04 unit of β-galactosidase in citrate buffer (pH 3.5) at 37° for 15 h. All glycoproteins are dialyzed and coupled with periodate-oxidized HRP by reductive amination according to the previously described method.[7] The formation of HRP–neoglycoproteins is monitored by observing the change in migration position of glycoproteins on SDS–PAGE and the ability to detect dot-blotted concanavalin A on membranes. The neoglycoproteins are dialyzed against TBS.

[6] H. Matsumoto, A. Natsume, H. Ueda, T. Saitoh, and H. Ogawa, *Biochim. Biophys. Acta* **1526,** 37 (2001).

[7] P. K. Nakane, *Methods Enzymol.* **37,** 137 (1975).

Assay Methods

SDS–PAGE and Binding Assay with Heparin–HRP on Membrane

SDS–PAGE is carried out as described by Laemmli,[8] using an 11% polyacrylamide gel under reducing conditions with 5% mercaptoethanol. The electrophoresed proteins (5 μg) are transferred to a polyvinylidene difluoride (PVDF) membrane (Millipore, Bedford, MA) at room temperature. The membrane is blocked with 3% BSA–TBS, and then incubated with heparin–HRP solution diluted to 1:1000 with TBS for 1 h at room temperature. After washing the membranes three times with TBS, staining is carried out with a substrate mixture solution of 6 mg of 4-chloro-1-naphthol/2 ml of methanol, 10 ml of TBS, and 6 μl of 30% H_2O_2.

Binding Assay with Heparin Neoproteoglycans on Microtiter Plate

Heparin–BSA, $GlcNAc_{5-6}$–BSA, or PVL in 10 m*M* phosphate-buffered saline (PBS), pH 7.0 (100 μl), is added to the wells of an Immulon 1 plate (Dynatech Laboratories, Chantilly, VA) and kept overnight at 4°. All other procedures are performed at room temperature. After washing the wells with PBS and blocking with 3% BSA–PBS for 2 h, 100 μl of HRP- or biotin-conjugated probe is added. After incubation for 1 h, the wells are washed three times with PBS. In the case of HRP-conjugated probes, staining is carried out with 0.04% *o*-phenylenediamine and 0.01% H_2O_2 in 0.1 *M* citrate buffer, pH 5.0, the reaction is stopped by adding 50 μl of 2 *M* H_2SO_4, and absorbance at 490 nm is read, using a microplate reader as described previously.[4] In the case of biotin-labeled probes, 100 μl of streptavidin–biotinylated HRP complex diluted to 1:1000 with PBS is added and incubated for 1 h, followed by washing with PBS. Staining is then carried out by the same procedure as described above. In all cases, the averages of duplicate determinations are plotted. For inhibition assays, inhibitors at various concentrations are preincubated with immobilized PVL, or labeled PVL in some cases, and then incubated with ligands for 1 h. The effect of ionic strength on the interaction of PVL with the ligands is assayed by adding NaCl or Na_2SO_4 to the incubation buffer.

For the interaction of heparin with vitronectin (VN) in plasma, human plasma is diluted 10-fold with PBS–5 m*M* EDTA–0.1 *M* phenylmethylsulfonyl fluoride (PMSF), and treated with 0.8 *M* urea at 37° for 2 h. For control, plasma is treated in the same way but without urea. The treated plasma is diluted and 100-μl aliquots are incubated with the immobilized heparin–BSA (10 μg/ml) in the wells at 25° for 2 h. After washing the

[8] U. K. Laemmli, *Nature* **227,** 680 (1970).

wells with PBS three times, the amount of bound VN is measured with sheep anti-hVN IgG (1 μg/ml) and rabbit HRP-conjugated rabbit anti-sheep IgG (The Binding Site, Birmingham, UK) and an enzyme-linked immunosorbent assay (ELISA), using the color reaction with *o*-phenylenediamine.

Sandwich Affinity Chromatography of Heparin Neoproteoglycan on PVL Preadsorbed by Di-N-acetylchitobiamyl Sepharose

All procedures are carried out with PBS at 4°. Di-*N*-acetylchitobiamyl Sepharose is prepared as described previously.[9] Two milliliters of di-*N*-acetylchitobiamyl Sepharose is incubated with purified PVL (500 μg) overnight, packed into a column (0.75 × 4.5 cm), and washed with 15 ml of PBS. One milliliter of heparin–HRP diluted to 1:1000 with PBS ($\leq$3 ng/μl as heparin) is applied to the column and eluted with 15 ml of PBS, followed by 0.2 *M* GlcNAc. The elution of PVL is monitored by absorbance at 280 nm. To monitor the elution of heparin–HRP with peroxidase activity, a 100-μl aliquot of each eluted fraction (1 ml) is added to the wells of a microtiter plate and protein immobilization is performed overnight. Staining is carried out by the same procedure as that described for the microtiter plate assay. A control experiment is performed without preincubating the di-*N*-acetylchitobiamyl Sepharose gel with PVL.

Binding Assays of Mushroom Extracts with Neoproteoglycan Probes or Hybrid Glycoprotein Probes by Dot Blotting

Fruiting bodies of mushrooms are collected in Gunma Prefecture, Japan and stored at −20° until use. The fruiting bodies (1–4.5 g) are homogenized in 3–15 ml of saline, incubated for 30 min at 4°, and centrifuged at 700 *g* for 20 min and at 10,000 *g* for 5 min at 4°. The supernatants are used as crude extracts and screened for lectins. The following procedures are carried out at room temperature. The protein concentration of each extract is measured with a detergent-compatible (DC) protein assay kit based on the Lowry method (Bio-Rad, Hercules, VA), and an aliquot of each extract (1.0 μg protein/100 μl) is 4-fold serially diluted and dot-blotted onto a PVDF membrane by means of a dot-blotting apparatus (Biodot SF; Bio-Rad). The membrane is then blocked with 3% BSA for 2 h at 4° and allowed to react with a 10-μg/ml concentration of various biotin–neoproteoglycans, or HRP–hybrid glycoproteins, for 1 h. For biotin-labeled neoproteoglycan probes, the membrane is subsequently incubated with streptavidin–biotinylated HRP complex for 1 h. Staining is carried out

[9] I. Matsumoto, H. Kitagaki, Y. Akai, Y. Ito, and N. Seno, *Anal. Biochem.* **116,** 103 (1981).

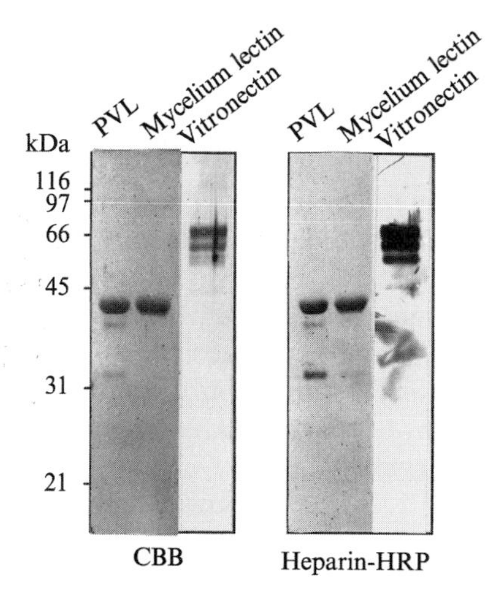

FIG. 1. Reactivities of PVL, mycelium lectin, and vitronectin with heparin–HRP on PVDF membrane. Five micrograms of PVL, mycelium lectin purified from *P. velutina*, or human vitronectin was subjected to SDS–PAGE on 11% polyacrylamide gels and Western blotting was performed on PVDF membranes. Proteins were stained with CBB or allowed to react with heparin–HRP as described in text. Molecular mass markers are shown on the left-hand side.

as described for the membrane assay. The staining intensities are judged visually or quantitatively measured with a refractive densitometer (CS-9300PC; Shimadzu, Kyoto, Japan) at an absorbance at 370 nm for probe staining, or at 550 nm for CBB staining.

Utility of Neoproteoglycan Probes in Interaction Analyses with Glycosaminoglycan-Binding Proteins

Interaction of Heparin–HRP with PVL and Vitronectin on Membrane

As shown in Fig. 1, PVL, the mycelium lectin of *P. velutina*, and vitronectin have been found to bind with heparin–HRP on membrane analysis after SDS–PAGE and electroblotting. The results indicate that the heparin–binding activity is primarily sequence dependent and is recovered by protein refolding into its active conformation after SDS–PAGE under denaturing conditions. Human vitronectin has heparin-binding consensus motifs, XBBXBX and XBBBXXBX, where B is a basic amino acid.[10] PVL contains basic motifs in its peptide sequence, that is, five repeats

[10] A. D. Cardin and J. H. R. Weintraub, *Arteriosclerosis* **9,** 21 (1989).

of RX(D or E)BHXR and four repeats of BB, where B = K or R and X = V, I, L, or P, and these motifs may serve as heparin-binding motifs.

Interaction of Heparin with PVL on Microtiter Plate

PVL is immobilized in the wells of a microtiter plate and bound to heparin–HRP in a concentration-dependent manner at pH ranging from pH 4 to 9; the result obtained at pH 7 is shown as an example in Fig. 2A. PVL exhibits maximum binding to heparin–HRP at pH 7, and to biotinylated $GlcNAc_{5-6}$ or $GlcNAc_{5-6}$–BSA neoglycoprotein at pH 4. Unconjugated HRP does not bind with PVL immobilized in the wells, suggesting that the binding is due to the affinity of PVL for heparin rather than for HRP (data not shown). Alternatively, when heparin–BSA is immobilized in the wells and biotinyl PVL is used as a probe, a similar concentration dependence is observed (Fig. 2B). The binding activity for heparin–BSA is retained in PVL biotinylated in the presence of heparin (open circles, Fig. 2B), whereas it is completely lost in PVL biotinylated in the absence of heparin (open squares, Fig. 2B) and the presence of GlcNAc during biotinylation does not maintain the heparin-binding activity at all. The results suggest that the heparin-binding site is susceptible to biotinylation. In contrast, PVL biotinylated in the absence of GlcNAc retains its GlcNAc-binding activity as detected by a solid-phase assay (data not shown); therefore, biotinylation in the presence of GlcNAc is effective for suppressing the heparin-binding activity, thereby making PVL usable for GlcNAc/NeuAc detection.

Inhibition studies of the heparin-binding activity of PVL at pH 7 reveals the following results: fucoidan [50% inhibitory concentration (IC_{50}, μg/ml) = 40], pectin (IC_{50} = 80), polygalacturonic acid (IC_{50} = 90), dextran sulfate (IC_{50} = 100), and alginic acid (IC_{50} = 700) have higher inhibitory activity than heparin (IC_{50} = 1000), whereas chondroitin, chondroitin sulfates A, B, and C, heparan sulfate, hyaluronic acid, colominic acid, keratan sulfate, dextran, DNA, and GlcNAc do not inhibit heparin binding at the concentration of 6 mg/ml. Pectin and polygalacturonic acid show high inhibitory activity among the polysaccharides tested even though they are unsulfated and esterified. Although the heparin-binding activity of PVL at pH 7 decreases with increasing concentration of salt ions, suggesting that the binding is due mainly to electrostatic interaction, the binding of PVL to polysaccharides is not correlated with the anionic charges. These results suggest that sulfate or carboxyl groups of acidic polysaccharides may contribute to the binding, although the binding does not simply depend on the electrostatic interaction.

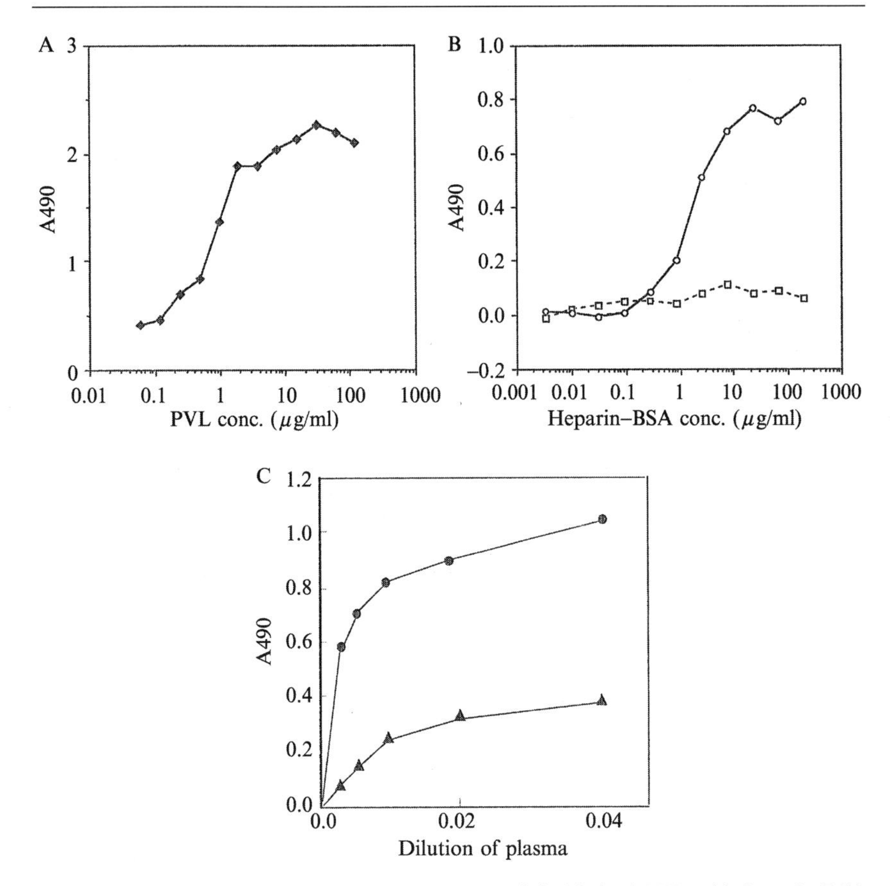

FIG. 2. Reactivities of PVL with heparin–HRP (A), biotinyl PVL with heparin–BSA (B), and plasma vitronectin with heparin–BSA (C) in a microtiter plate assay. (A) PVL was immobilized in wells of microtiter plates, and heparin–HRP diluted to 1:1000 with PBS was added. The subsequent procedures were performed as described in text. (B) Heparin–BSA was immobilized in wells of microtiter plates and a 10-μg/ml concentration of biotinyl PVL in PBS was added. (○) PVL biotinylated in the presence of heparin; (□) GlcNAc. (C) Heparin–BSA (10 μg/ml) was immobilized on microtiter plastes and diluted human plasma with (●) or without (▲) urea treatment was added. The amount of bound VN was measured with sheep anti-hVN IgG and HRP-conjugated rabbit anti-sheep IgG and ELISA, as described in text.

Heparin–BSA immobilized in wells binds to vitronectin in urea-treated or untreated plasma in a concentration-dependent manner, as shown in Fig. 2C. Because plasma vitronectin is converted into its unfolded active form by urea treatment, the amount of bound vitronectin increases on urea treatment of plasma. BSA- and HRP-conjugated neoproteoglycan probes have exhibited their applicability for interaction analyses with a wide range of binding proteins, such as recombinant mutants of vitronectin[11] and various annexins.[12]

Sandwich Affinity Chromatography on Di-N-acetylchitobiamyl Sepharose

Heparin neoproteoglycan is used to analyze the spatial relationship of the binding sites in PVL. Heparin–HRP binds to a di-*N*-acetylchitobiamyl Sepharose column preincubated with PVL and is eluted with 0.2 *M* GlcNAc in association with PVL, as shown in Fig. 3A. Because heparin–HRP does not bind to the column in the absence of PVL (Fig. 3B), it is surmised that the heparin neoproteoglycan binds to the column via PVL that is preadsorbed to the gel. The finding that PVL can bind to heparin- and GlcNAc-containing macromolecules simultaneously at different binding sites suggests that it may serve as an adhesion molecule by forming a multicomponent complex with acidic polysaccharides and other glycoconjugates containing GlcNAc/NeuAc.

Screening of Glycosaminoglycan-Binding Proteins in Mushroom Extract by Dot Blotting

Lectin screening is carried out by reacting dots of 16 cultivable mushroom extracts on a membrane with various hybrid glycoprotein or biotin–neoproteoglycan probes, followed by enzymatic visualization of the bound probe. The results are shown in Fig. 4, and the staining intensities of all binding assays, which are judged visually, are summarized in Table I. The extracts are categorized into three groups according to their reactivities to neoproteoglycan probes: group I extracts react only with certain neoglycoprotein probes but not with any neoproteoglycan probe; group II extracts react with a limited number of neoproteoglycan probes; and group III extracts react almost equally with all probes. The three groups are distributed equally in Orders Agaricoles and Polyporaceae.

Group I. The extracts react only with desialylated neoglycoprotein probes. These extracts are thought to contain lectin(s) specific to oligosaccharides in glycoproteins and exhibit no reactivity to acidic polysaccharide probes. A lectin that would be useful as a specific probe is presumed to be present in group I extracts. On the basis of these presumptions, an asialotransferrin-specific lectin has been purified from *Oudemansiella platyphylla* extract.[6]

[11] A. Yoneda, H. Ogawa, K. Kojima, and I. Matsumoto, *Biochemistry* **37,** 6351 (1998).

[12] R. Ishitsuka, K. Kojima, H. Utsumi, H. Ogawa, and I. Matsumoto, *J. Biol. Chem.* **273,** 9935 (1998).

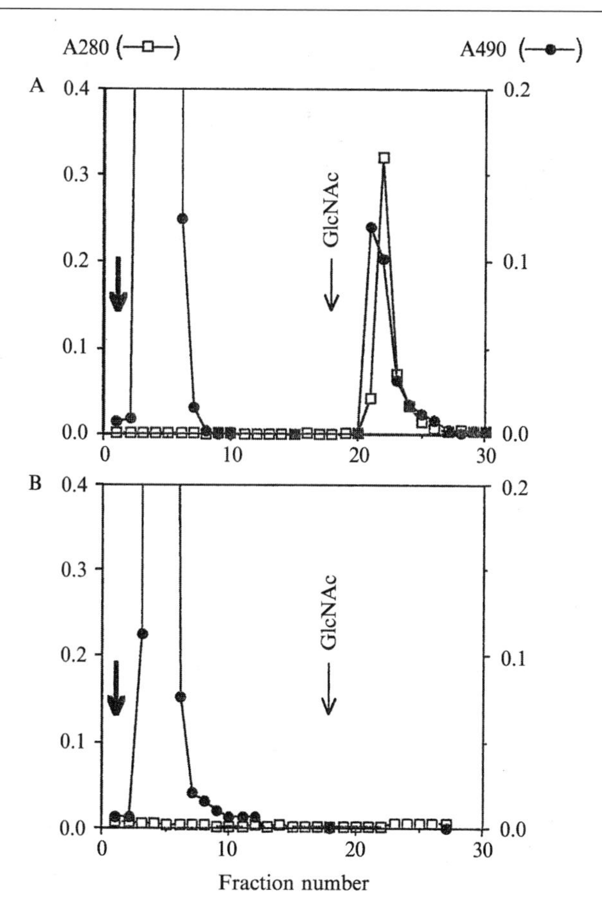

FIG. 3. Sandwich affinity chromatography of heparin–HRP on di-*N*-acetylchitobiamyl Sepharose via PVL. (A) Purified PVL (500 μg) was incubated with 2 ml of di-*N*-acetylchitobiamyl Sepharose in PBS overnight at 4° with shaking and packed into a column (0.75 × 4.5 cm). After washing the column, heparin–HRP diluted to 1:1000 with PBS was applied at the point shown by a boldface arrow and the column was washed again with PBS. The bound material was eluted with 0.2 *M* GlcNAc in PBS at the point indicated by the thin arrow. Elution was monitored at 280 nm for protein (□) and at 490 nm for heparin–HRP (●), as described in text. (B) For the control experiment, heparin–HRP was applied to a di-*N*-acetylchitobiamyl Sepharose column without preincubating the gel with PVL, and elution and monitoring were performed in the same way as in (A).

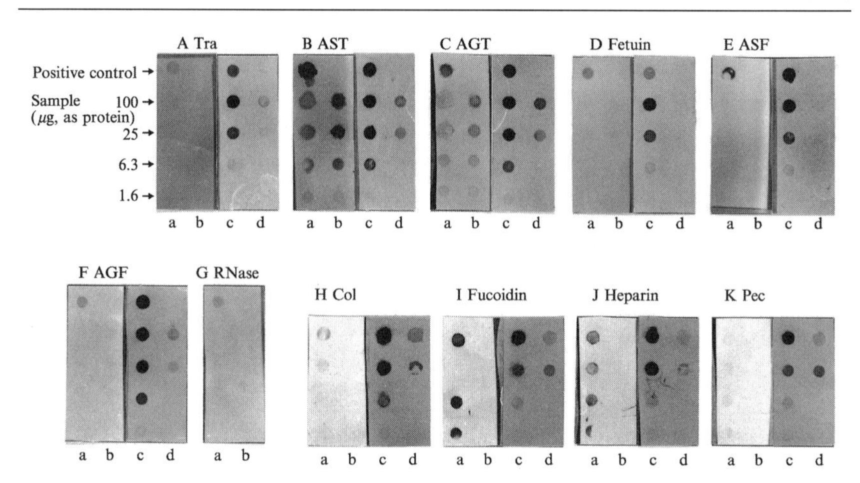

FIG. 4. Reactivities of neoglycoprotein and neoproteoglycan probes with extracts from four cultivable mushrooms. The extract of each mushroom (1.0 mg of protein per 100 μl) was serially diluted four times and dot-blotted onto PVDF membrane with an appropriate positive control. The membrane was then blocked with 3% BSA and allowed to react with HRP–neoglycoprotein or biotin–neoproteoglycan probes (10 μg/ml in TBS) at room temperature for 1 h with gentle shaking. Staining was carried out as described in text. Positive controls used for (A) transferrin (Tra)–HRP probe, *Sambucus sieboldiana* lectin; (B) asialotransferrin (AST)–HRP probe, *Ricinus communis lectin* I; (C) asialo-agalactotransferrin (AGT)–HRP probe, PVL; (D) fetuin–HRP probe, *Sambucus sieboldiana* lectin; (E) asialofetuin (ASF)–HRP probe, *Ricinus communis lectin* I; (F) asialo-agalactofetuin (AGF)–HRP probe, PVL; (G) ribonuclease B (RNase)–HRP probe, concanavalin A. The following neoproteoglycan probes, (H) colominic acid (Col)–BSA–biotin probe, (I) fucoidin–BSA–biotin probe, (J) heparin–BSA–biotin probe, and (K) pectic acid (Pec)–BSA–biotin probe, were used without positive control. Sample extracts: lane a, *Sparassis crispa*; lane b, *Oudemansiella platyphylla*; lane c, *Pluteus atricapillus*; lane d, *Ganoderma lucidum*.

Group II. The dual reactivity of the extracts to hybrid glycoprotein and neoproteoglycan probes suggests the presence of multispecific lectins, like those of *Agaricus bisporus*[3] and PVL,[13] or the presence of multiple lectins reactive to separate probes in one extract. Some preferences have been observed among the neoproteoglycan probes for pectic acid in *Hypsizigus marmoreus*, for heparin and fucoidin in *Lentinus edodes*, and for colominic acid in *Stereum ostrea* extracts.

Group III. The reactivities of extracts to all probes are markedly high. The presence of endogenous peroxidase in the extracts is determined by directly reacting the membrane with the HRP substrates; however, all extracts are shown to have no peroxidase activity. The wide range of

[13] H. Ueda, T. Saitoh, K. Kojima, and H. Ogawa, *J. Biochem.* **126,** 530 (1999).

TABLE I
REACTIVITIES OF MUSHROOM EXTRACTS FOR NEOGLYCOPROTEIN AND NEOPROTEOGLYCAN PROBES[a]

	HRP–glycoprotein probes							Biotin–neoproteoglycan probes				
Mushroom extract	Tra	AST	AGT	Fetuin	ASF	AGF	RNase	Col	Fucoidin	Heparin	Pec	Group
Agaricoles												
Tricholomataceae												
Oudemansiella platyphylla	±	+++	++	−	−	±	−	−	−	−	−	I
Lyophyllum shimeji	±	+	++++	−	−	++	−	−	+	++	+	II
Hypsizigus marmoreus	−	−	++	−	±	−	−	−	−	−	++	II
Collybia dryophila	++	++++	++++	±	+	++++	+	−	+	+	−	II
Pleurotaceae												
Lentinus edodes	++++	++++	++++	+++	+	++++	+	−	++++	++++	−	II
Lentinus lepideus	++++	++++	++++	++++	++++	++++	++++	++++	++++	++++	++++	III
Pluteuceae												
Pluteus atricapillus	+++	++++	++++	+++	+++	++++	NT	+++	+++	++	+++	III
Aphyllophorales												

(*Continued*)

TABLE I (*Continued*)
REACTIVITIES OF MUSHROOM EXTRACTS FOR NEOGLYCOPROTEIN AND NEOPROTEOGLYCAN PROBES[a]

Mushroom extract	HRP–glycoprotein probes							Biotin–neoproteoglycan probes				Group
	Tra	AST	AGT	Fetuin	ASF	AGF	RNase	Col	Fucoidin	Heparin	Pec	
Polyporaceae												
Pycnoporus coccineus	–	++	+++	–	–	++	NT	–	–	–	–	I
Roseofomes subflexibilis	–	++	+++	–	–	++	–	–	–	–	–	I
Fomitopsis pinicola	–	–	±	–	–	±	–	–	–	–	–	I
Fomitella fraxinea	+	++	+++	++	++	+++	NT	+++	++	++	+++	II
Laetiporus sulphureus	++++	++++	++++	++++	++++	++++	++++	++++	++++	++++	++++	III
Ganodermataceae												
Ganoderma neo-japonicum	–	+	++++	–	–	+++	NT	–	–	–	–	I
Ganoderma lucidum	±	++	++	–	–	+	NT	++	++	+	++	II
Sparassidaceae												
Sparassis crispa	±	+++	±	–	–	–	±	++	++++	++++	+	II
Stereaceae												
Stereum ostrea	++	+	+	+	±	+	NT	++	+	+	±	II

Abbreviations: NT, Not tested; Tra, transferrin; RNase, ribonuclease B; Col, colominic acid; Pec, pectic acid; AST, asialotransferrin; AGT, agalactotransferrin; ASF, asialofetuin; AGF, agalactofetuin.

[a] Sixteen cultivable mushroom extracts were dot-blotted onto PVDF membrane and screened for the presence of lectins that are reactive to hybrid glycoprotein or neoproteoglycan probes. Binding and detection procedures were performed as described in text.

reactivities may be caused by the multiple interactions of protein–carbohydrate or protein–HRP. An *N*-acetyllactosamine-specific lectin has been isolated from *Laetiporus sulphureus*, but the results suggest that different lectins are still present in the extract.

Conclusions

Neoproteoglycan and hybrid glycoprotein probes developed in this study were proved to be valuable for elucidating the specificity of lectins and glycosaminoglycan-binding proteins to glycoconjugates. The procedures for lectin screening may be used to probe for proteoglycan receptors in the search for biological ligands. On the other hand, the multispecificity of PVL and other plant lectins may interfere with the interpretation of the results obtained. For PVL, we propose that its binding activity with acidic polysaccharide be selectively diminished by biotinylation for use as a GlcNAc/NeuAc-specific reagent.

[14] Modular Synthesis and Study of Multivalent Carbohydrate Ligands with Long and Flexible Linkers

By ZHONGSHENG ZHANG and ERKANG FAN

Introduction

Protein–carbohydrate interactions on the cell surface are an important class of biological events. Mimicking, modulation, or inhibition of such interactions constitutes an attractive strategy toward the development of novel therapeutics.[1–8] A unique feature of cell surface protein–carbohydrate interactions is that they are usually multivalent—meaning that there are multiple carbohydrates interacting with a single protein at multiple identical binding sites simultaneously.[9] With this multivalent strategy,

[1] M. Mammen, S. K. Choi, and G. M. Whitesides, *Angew. Chem. Int. Ed.* **37,** 2755 (1998).
[2] K. J. Yarema and C. R. Bertozzi, *Curr. Opin. Chem. Biol.* **2,** 49 (1998).
[3] L. L. Kiessling, J. E. Gestwicki, and L. E. Strong, *Curr. Opin. Chem. Biol.* **4,** 696 (2000).
[4] L. L. Kiessling, L. E. Strong, and J. E. Gestwicki, *Annu. Rep. Med. Chem.* **35,** 321 (2000).
[5] M. Okada, *Prog. Polym. Sci.* **26,** 67 (2001).
[6] S. J. Williams and G. J. Davies, *Trends Biotechnol.* **19,** 356 (2001).
[7] D. Wright and L. Usher, *Curr. Org. Chem.* **5,** 1107 (2001).
[8] L. L. Kiessling, *Glycobiology* **11,** 27 (2001).
[9] R. T. Lee and Y. C. Lee, *Glycoconjug, J.* **17,** 543 (2000).

0076-6879/03 $35.00

nature is building strong biological interactions by combining many weaker ones. Not surprisingly, synthetic ligands that mimic cell surface carbohydrates can also be made multivalent, which makes a synthetic ligand more effective in performing its intended task than the corresponding monovalent ligand.[1–8,10] Traditionally, the construction of multivalent ligands relies on generic backbones; and the structural information of the target proteins is usually not used in multivalent ligand design. In contrast, structure-based design of multivalent ligands has attracted increasing attention. By taking into account the spatial arrangement of binding sites on a target protein, high-affinity multivalent ligands that geometrically complement their target can be designed and synthesized.[11–17] The most advanced demonstrations of structure-based multivalent ligand designed to date are the development of pentavalent and decavalent carbohydrate inhibitors that are capable of blocking cell surface binding by AB_5 toxins.[14,15,17] Among these, we have developed a modular synthetic approach that allows rapid synthesis and optimization of large multivalent carbohydrate ligands using long and flexible linkers.

Modular Design and Synthesis of Multivalent Ligands

Cholera toxin (CT) and the closely related *Escherichia coli* heat-labile enterotoxin (LT) are members of the AB_5 family of bacterial toxins. Their B pentamers, which are responsible for host cell surface carbohydrate recognition, are attractive targets and model systems for structure-based design of multivalent ligands that can block the mode of action of the toxin.[18] Figure 1 shows our modular approach toward a multivalent ligand against CT/LT. A complete multivalent ligand is divided into three modules: a *core* that conforms to the symmetry of the binding site

[10] E. Fan and E. A. Merritt, *Curr. Drug Targets Infect. Disord.* **2,** 161 (2002).
[11] R. H. Kramer and J. W. Karpen, *Nature* **395,** 710 (1998).
[12] G. Loidl, M. Groll, H. J. Musiol, R. Huber, and L. Moroder, *Proc. Natl. Acad. Sci. USA* **96,** 5418 (1999).
[13] G. Loidl, H. J. Musiol, M. Groll, R. Huber, and L. Moroder, *J. Peptide Sci.* **6,** 36 (2000).
[14] P. I. Kitov, J. M. Sadowska, G. Mulvey, G. D. Armstrong, H. Ling, N. S. Pannu, R. J. Read, and D. R. Bundle, *Nature* **403,** 669 (2000).
[15] E. Fan, Z. Zhang, W. E. Minke, Z. Hou, C. L. M. J. Verlinde, and W. G. J. Hol, *J. Am. Chem. Soc.* **122,** 2663 (2000).
[16] N. Schaschke, G. Matschiner, F. Zettl, U. Marquardt, A. Bergner, W. Bode, C. P. Sommerhoff, and L. Moroder, *Chem. Biol.* **8,** 313 (2001).
[17] E. A. Merritt, Z. Zhang, J. C. Pickens, M. Ahn, W. G. J. Hol, and E. Fan, *J. Am. Chem. Soc.* **124,** 8818 (2002).
[18] E. K. Fan, E. A. Merritt, C. Verlinde, and W. G. J. Hol, *Curr. Opin. Struct. Biol.* **10,** 680 (2000).

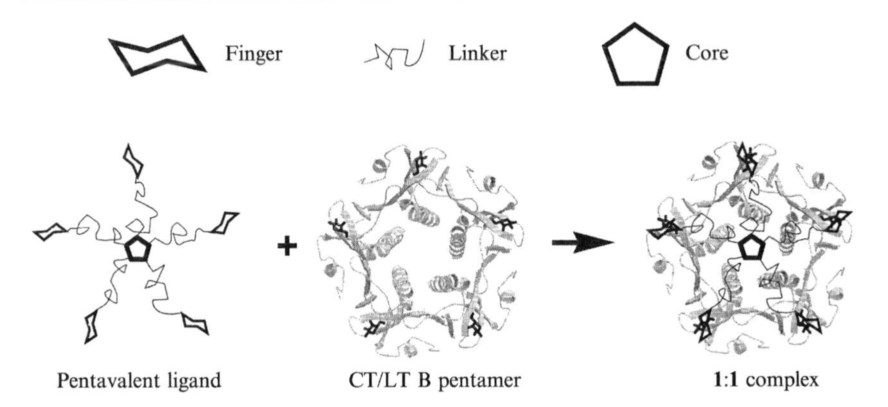

FIG. 1. Structure-based modular design of multivalent carbohydrate ligands against CT and LT.

distribution on the target protein; *linkers* that can optimally position all monovalent ligands for binding to the target; and *fingers* that represent the actual monovalent ligand, which is responsible for binding to the specific target protein. To succeed in using this modular approach, it is necessary to first work out a chemical synthetic route that allows for each module to be efficiently attached to another module. Because we are interested in using long and flexible linkers[11] for our multivalent design, this synthetic method should also address the need for creating a variety of linker lengths.

After testing a variety of chemical reactions, we settled on the squaric acid diester-mediated stepwise coupling of primary amines.[19] The representative activation of individual ligand modules and the assembly of full multivalent ligands are shown in Fig. 2. For connectivity, each module is required to have a terminal amine group. The amine group can be first modified by reaction with a squaric acid diester under neutral conditions to produce the monoester monoamide intermediate (**1** and **2** in Fig. 2). This intermediate is stable, can be isolated in pure form, and is ready to react with another module that contains a free amine group under basic conditions. One notable feature about the reaction of squaric acid diester with an amine is that sugar hydroxyl groups are not interfering with the reaction. Therefore, all finger modules that contain unprotected carbohydrate groups can be used directly for synthesis so long as there is only one free amine group for linkage.

The need for variable linker length is addressed by using a symmetric diamine as the linker unit. As shown in Fig. 2, each linker unit can first

[19] L. F. Tietze, M. Arlt, M. Beller, K.-H. Glusenkamp, E. Jahde, and F. Rajewsky, *Chem. Ber.* **124,** 1215 (1991).

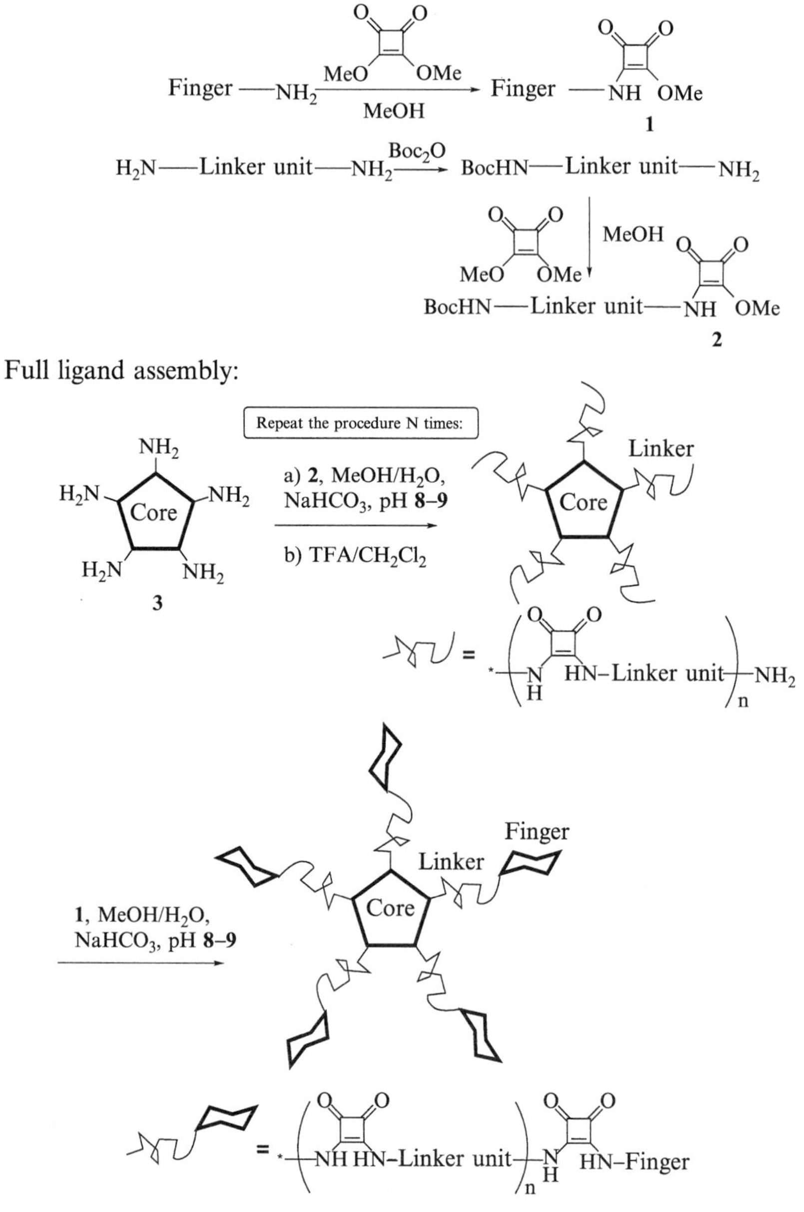

Finger—NH_2 —(MeO, OMe squarate; MeOH)→ Finger—NH OMe (squarate) **1**

H_2N—Linker unit—NH_2 —(Boc_2O)→ BocHN—Linker unit—NH_2

↓ MeO, OMe squarate; MeOH

BocHN—Linker unit—NH OMe (squarate) **2**

Full ligand assembly:

Repeat the procedure N times:

NH_2 H_2N NH_2 Core H_2N NH_2 **3**

a) **2**, $MeOH/H_2O$, $NaHCO_3$, pH **8–9**

b) TFA/CH_2Cl_2

Linker Core

= *—(NH HN–Linker unit)$_n$—NH_2

1, $MeOH/H_2O$, $NaHCO_3$, pH **8–9**

Finger Linker Core

= *—(NH HN–Linker unit)$_n$—NH HN–Finger

FIG. 2. Stepwise activation and coupling of modular multivalent ligand.

be monoprotected with a *t*-butyloxycarbonyl (Boc) group. The other free amine group on the linker unit is then coupled to the squaric acid diester. The resulting compound **2** can be used iteratively for attachment to a growing core–linker module during the full ligand assembly process. The variation of linker length is controlled by the number of times (*n* in Fig. 2) that the linker unit **2** is attached to the core module. Certainly, it is also possible to alter the overall length of the linker module by using many commercially available diamines of variable lengths as the linker unit. Final attachment of the finger module is also straightforward under basic conditions with the activated finger module **1** and the preassembled core–linker module.

One important aspect of our synthetic protocol is the capability of efficiently handling a synthetic scale that is in tens of milligrams in the ligand assembly process. Because each assembly reaction step requires attachment at five growing chains in one molecule, an excess amount of the growing module is needed to push the reaction toward completion. Therefore, at the end of each reaction, product purification could be a bottleneck if it relies solely on reversed-phase high-performance liquid chromatography (HPLC) separation. In our hands, a large pore-sized 22×250 mm preparative C_{18} HPLC column, which is typical in research laboratories, can handle only about 10 mg or less of reaction mixture in each injection. Because of the cleanness of the squaric acid diester-based coupling reaction, we were able to use disposable C_{18} Sep-Pak cartridges[20] to perform one-step reversed-phase purification after each coupling reaction. The resulting product was comparable in purity to HPLC-purified samples in most cases.

A typical experimental protocol for ligand assembly is described below with the following modules: core (**3** in Fig. 2), 1,4,7,10,13-penta(4-aminobutyryl)-1,4,7,10,13-pentaazacyclopentadecane; Boc–linker unit–squaric acid monoamide monoester (**2** in Fig. 2), *N*-Boc-*N*′-(2-methoxy-3,4-dioxo-1-cyclobuten-1-yl)-4,7,10-trioxa-1,13-tridecanediamine; and finger–squaric acid monoamide monoester (**1** in Fig. 2), *N*-(2-methoxy-3,4-dioxo-1-cyclobuten-1-yl)-6-aminocaproyl-β-D-galactosylamine.

Iterative Core–Linker Assembly

A solution of **3** (10 μmol) and **2** (250 μmol, 5-fold excess at each site) in 5 ml of methanol–H_2O (1:1, v/v) is adjusted to pH 8–9 by addition of 0.2 *M* $NaHCO_3$ solution. The mixture is stirred at room temperature overnight

[20] P. D. McDonald (Ed.), "Waters Sep-Pak Cartridge Application Bibliography." Waters, Milford, MA, 1991.

and then neutralized to pH 7 with 0.5% aqueous trifluoroacetic acid (TFA) solution. This sample is then loaded on to Waters (milford, MA) tC_{18} Sep-Pak cartridges. Unwanted starting materials are removed by washing the cartridge with 2–3 bed volumes of 35–50% methanol–H_2O (v/v). The Boc-protected product is eluted out with methanol. After removal of methanol, the residue is treated with 7 ml of TFA–CH_2Cl_2 (1:1, v/v) for 30 min at room temperature. Removal of solvent gives the desired core–linker assembly in 93% overall yield. This sample is ready for further linker–chain growth by repetition of the above-described reaction procedure with **2**. Alternatively, the core–linker assembly is ready for use in full ligand assembly.

Full Ligand Core–Linker–Finger Assembly

A solution of the prepared core–linker (5 μmol) and **1** (125 μmol, 5-fold excess at each reaction site) in 32 ml of aqueous $NaHCO_3$ solution (pH 8–9) is stirred at room temperature overnight. After neutralization with 0.5% aqueous TFA, the sample is loaded onto Waters tC_{18} Sep-Pak cartridges. Unwanted starting materials are removed by elution with a few bed volumes of 10–20% methanol–H_2O (v/v). The desired pentavalent ligand is eluted out with 80% methanol–H_2O (v/v). Rotary evaporation and lyophilization give the full ligand in 88% yield.

Using this modular synthetic approach, we have been able to quickly assemble different combinations of fingers, linkers, and core modules. These ligands have been shown to have high affinity toward CT or LT B pentamers, and are able to inhibit toxin B pentamer binding to ganglioside receptors with up to 100,000-fold better efficiency than the corresponding monovalent ligand.[15,17]

Dynamic Light-Scattering Studies of Modular Pentavalent Carbohydrate Ligands with CT and LT B Pentamer

When a well-designed multivalent ligand interacts with its target protein, it is expected that the multivalent ligand has an increase in intrinsic affinity toward its target as compared with the corresponding monovalent ligand. However, because of the presence of multiple copies of a monovalent ligand, each of which is capable of binding to the intended target independently, there is always the possibility of forming random ligand–protein aggregates[21] in solution rather than discrete complexes.[22] In traditional multivalent ligand design with generic backbones, most ligands are built on polymer systems. The intrinsic sample heterogeneity of a polymer system further complicates biophysical studies that aim to distinguish ligand–protein random aggregation from discrete complex formation. On the other hand, structure-based design of multivalent protein ligands, including our modular ligands with flexible long linkers, often leads to single species multivalent ligands because of the use of stepwise organic synthesis and purification. This offers the opportunity of using dynamic light scattering (DLS) to probe ligand–protein interactions in solution. DLS is an excellent tool with which to detect protein aggregation because of its physical principles. The theoretical and practical aspects of DLS experiments with polymers or biomacromolecules have been reviewed in detail.[23–29] Here, we add a few points of discussion that are directly related to studying multivalent ligand–protein complexes.

Because the goal of DLS in the study of multivalent ligand–protein complexes is to detect or rule out ligand-mediated random protein aggregation, it is best to perform the experiment in a batch mode with mixtures of ligand and protein, without any prior separation of discrete solution species. This will ensure that the measurement reflects the true solution status at equilibrium. From a practical point of view, this means that ligand and protein samples should be filtered separately before mixing. Once filtered individually, ligand and protein solutions can be mixed carefully to the desired ratio, allowed to reach equilibrium, and then subjected to DLS measurements.

Such DLS experiment will always measure signals from a mixture of the unbound multivalent ligand, the unbound target protein, and their complexes in various forms, if present. Although the algorithm for data analysis of multiple solution species can be complicated, for the purpose of detecting protein aggregation there is no need to use DLS to differentiate each and every solution species in this case. Instead, standard equations

[21] S. M. Dimick, S. C. Powell, S. A. McMahon, D. N. Moothoo, J. H. Naismith, and E. J. Toone, *J. Am. Chem. Soc.* **121,** 10286 (1999).
[22] J. E. Gestwicki, L. E. Strong, and L. L. Kiessling, *Angew. Chem. Int. Ed.* **39,** 4567 (2000).
[23] W. Burchard, *Adv. Polym. Sci.* **48,** 1 (1983).
[24] J. M. Schurr and K. S. Schmitz, *Annu. Rev. Phys. Chem.* **37,** 271 (1986).
[25] C. H. Schein, *in* "ACS Symposium Series," Vol. 470, p. 21. ACS Publications, Washington, D.C., 1991.
[26] J. Langowski, W. Kremer, and U. Kapp, *Method Enzymol.* **211,** 430 (1992).
[27] P. Johnson, *Int. Rev. Phys. Chem.* **12,** 61 (1993).
[28] P. Salgi and R. Rajagopalan, *Adv. Colloid Interface Sci.* **43,** 169 (1993).
[29] H. G. Barth and R. B. Flippen, *Anal. Chem.* **67,** R257 (1995).

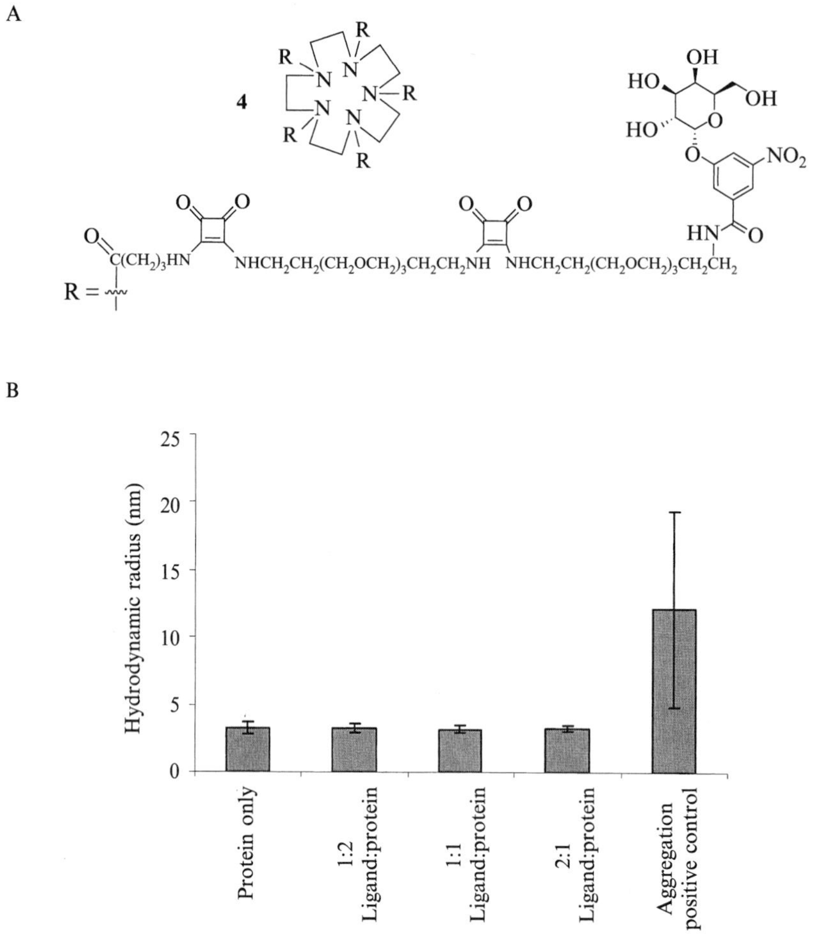

A

B

FIG. 3. (A) Structure of the pentavalent ligand (4) used in a DLS study; (B) DLS results that prove the absence of ligand-mediated random protein aggregation in our modular pentavalent ligand–LT B pentamer system.

(using software provided with commercial instruments) can be used to derive the hydrodynamic radius of solutes, which are now treated as one statistically averaged species. However, in such cases, proper control experiments are critical for drawing conclusions from DLS studies. At the least, it will be necessary to have the target protein sample in solution alone as a control. Preferably, a positive random aggregation control with the target protein is also needed.

An example of a DLS study is given below, using the following samples: Pentavalent ligand (4 in Fig. 3), built from a core of 1,4,7,10,13-penta (4-aminobutyryl)-1,4,7,10,13-pentaazacyclopentadecane with two linker units of 4,7,10-trioxa-1,13-tridecanediamine and 3-nitro-5-(α-D-galactosyloxy)benzoic acid as the finger module[17]; Protein target, LT B pentamer; and LT aggregation reagent, chlorophenyl red-β-D-galactopyranoside (CPRG).

Dynamic Light-Scattering Experiments

Protein or ligand in phosphate-buffered saline (PBS: 150 m*M* NaCl, 10 m*M* phosphate, pH 7.2) is individually filtered through inorganic membrane filters (Anodisc, 0.02 μm; Whatman, Clifton, NJ) into separate vials. Various portions of protein and ligand are then mixed and, without further filtration, transferred into a sample cell for measurement. In this case, the following ligand–protein ratios were reached: **4**–LT (0:11 μM), **4**–LT (11:22 μM), **4**–LT (11:11 μM), and **4**–LT (22:11 μM). Independent experiments for each ligand–protein solution are repeated three or four times. In sample preparation for positive aggregation control, a large amount of precipitate is formed when LT B pentamer ($\sim$10 μM) is mixed with CPRG (560 μM). Therefore, this sample must be filtered after mixing for the DLS measurement. In this case, the exact concentrations of LT B pentamer and CPRG are unknown.

DLS measurement is done on a DynaPro99 instrument (Proterion, Piscataway, NJ) illuminated by a 25-mW, 832.8-nm wavelength, solid-state laser at 25°. Data analysis is performed by the Dynamics software (version 5.25.44) provided with the instrument. Inverse Laplace transform ("regularization fit") analysis is used to find the mean and standard deviation (polydispersity) of the hydrodynamic radius distribution for the molecule/complex species in solution. Figure 3B shows the results for LT B pentamer with various ratios of pentavalent ligand and with aggregating reagent CPRG. It is clear from the graph that our modular pentavalent ligand causes no random aggregation of target protein in solution. Therefore, a specific 1:1 complex is expected for ligand **4** bound to CT/LT B pentamer and this is confirmed by a high-resolution crystal structure.[17]

In summary, structure-based design of multivalent carbohydrate ligands leads to single-species ligands that can conform to the binding site distribution geometry of the target protein. A modular synthetic approach allows for rapid assembly of different combinations of ligand modules, and testing of their functional properties.[15,17] DLS is a useful tool for ruling out ligand-mediated random aggregation in the study of modular single-species multivalent ligands. We expect that structure-based design of highly specific and ultrahigh-affinity multivalent ligands will lead to novel agents for therapeutic and diagnostic applications.

Acknowledgments

We thank Professors W. G. J. Hol, E. A. Merritt, and C. L. M. J. Verlinde for stimulating discussions and collaboration, and thank Dr. Z. Hou and Mr. J. C. Pickens for synthetic assistance. We are grateful for financial support from the National Institutes of Health (AI44954).

[15] Sugar Arrays in Microtiter Plates

By Chi-Huey Wong and Marian C. Bryan

Carbohydrates are the most abundant and diverse family of natural products, universally expressed as monosaccharides, oligosaccharides and glycoconjugates.[1] Several positions on a monosaccharide can be linked to biomolecules such as other saccharides, proteins, or lipids through α or β stereochemistry. This allows for arrays with diversity unlike any other biological molecules, making them ideal as recognition and adhesion factors.[2–4] Therefore, carbohydrates expressed on cellular surfaces are frequently used to differentiate between cell types and states of maturation in individual cell lines.[4] Because individual cell types can have distinct oligosaccharide makeup, carbohydrate targets have also found applications as pharmaceuticals.[1] Structural variability also makes them ideal for mediation of protein stability, transport, and folding as well as cellular storage and stabilization.[3,5] To increase our understanding of this diverse class and their pharmaceutical applications, screening methods to facilitate analysis are critical. One significant issue associated with the screening process is

[1] T. K. Ritter and C.-H. Wong, *Angew. Chem. Int. Ed.* **40,** 3508 (2001).
[2] P. A. Sears and C.-H. Wong, *Angew. Chem. Int. Ed.* **38,** 2300 (1999).
[3] M. Sznaidman, "Bioorganic Chemistry: Carbohydrates." Oxford University Press, Oxford, 1999.
[4] M. Fukuda, "Molecular and Cellular Glycobiology," 1st Ed., Oxford University Press, New York, 2000.
[5] P. Sears and C. H. Wong, *Cell. Mol. Life Sci.* **54,** 223 (1998).

0076-6879/03 $35.00

determination of an effective method for displaying saccharides in biological assays that allows for rapid and facile analysis of the interaction.

Several methods for such manipulations on proteins are commonly performed in microtiter plates. These plates are conducive to a variety of biological applications by binding the analyte of interest to the well and monitoring its response or effect on molecules in solutions. When carbohydrates are displayed in this manner, they are accessible to a variety of chemical and biological treatments including transformations and enzyme-linked immunosorbent assays (ELISAs). With this in mind, techniques both to display and study carbohydrates have been sought. Methods of attachment for saccharides to microtiter plates through noncovalent interaction between a lipid tether on the saccharide and the polystyrene well surface are presented here, as well as a brief summary of work being performed on saccharides displayed in this manner.

Attachment

Ether Linkage

In the following discussion, the lipid moiety is a saturated hydrocarbon between 12 and 14 carbons in length. These hydrocarbons have been found to be ideal because of their retention, uncomplicated attachment, and commercial availability.[6] Methods for attachment of the hydrocarbon to saccharides are summarized in Fig. 1. These reactions are applicable to saccharides displaying acetals, amines, or azides. Formation of ether linkages by displacement of acetates is discussed initially. Transetherification reactions of acetals through S_N1 displacement with a hydroxyl are commonly utilized in carbohydrate chemistry. Acetates make excellent leaving groups, when treated with Lewis acids, because of the stability of the intermediate carbocation.[7] Boron trifluoride diethyl etherate (BF_3 OEt_2) has an added benefit when used as the Lewis acid because it gives only the β adduct when an acetate is ortho to the anomeric position. However, orthoester formation can occur when primary hydroxyls are employed on the hydrophobic linker and alternative Lewis acids can be utilized.[6]

Acetylated saccharides are easily synthesized by addition of acetic anhydride (1.5 equivalents per free hydroxyl) to the sugar in dry pyridine (0.2–0.4 *M*). On completion, the acetylated product can be taken directly into the Lewis acid transetherification. Tetradecanol (2 equivalents) is

[6] M. C. Bryan, O. Plettenburg, P. Sears, D. Rabuka, S. Wacowich-Sgarbi, and C.-H. Wong, *Chem. Biol.* **9,** 713 (2002).

[7] J. March, "Advanced Organic Chemistry: Reactions, Mechanisms, and Structures." John Wiley & Sons, New York, 1992.

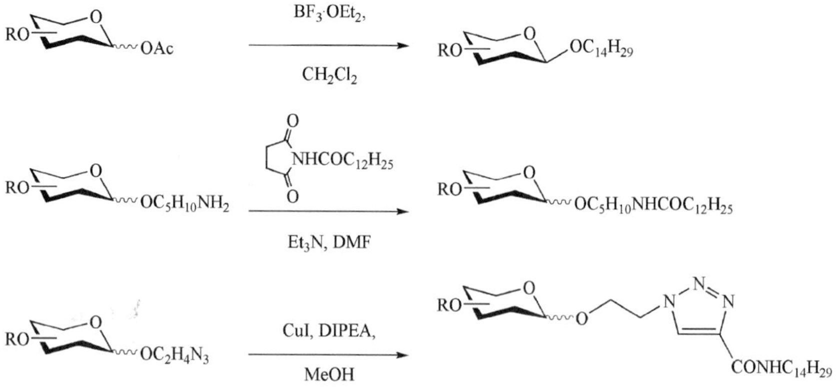

FIG. 1. Coupling of the saccharide to the hydrocarbon tether for noncovalent display on microtiter plates.

added to the saccharide in dry dichloromathane (0.1 *M*). The solution is then brought to 0° and BF_3 OEt_2 (1.5 equivalents) is added slowly. On consumption of the saccharide, the reaction is quenched with saturated aqueous sodium bicarbonate ($NaHCO_3$). The solution is then extracted with ethyl acetate, the organic phase is dried over magnesium sulfate and concentrated, and the glycolipid is purified by column chromatography on silica gel.[6] Sodium methoxide (NaOCH, in water (1 *M*) is then added to the saccharide in methanol (0.1 *M*) until the pH is pH 8–8.5. Deacetylation is quenched by addition of Dowex 50WX4-50 ion-exchange resin until the pH is below pH 6 and the solution can then be filtered and purified, giving the fully deprotected glycolipid. For attachment to the microtiter plate, the glycolipid is dissolved in methanol and the solution is added to the wells. Methanol is then evaporated, leaving the glycolipid-coated plates.[6]

Amine Linkage

Amines present on saccharides can also be used for attachment to hydrophobic tethers containing carboxylic acids. Before reaction with the amino sugar, carboxylic acids can be activated by coupling to *N*-hydroxysuccinimide (NHS). This additional step is not critical but allows for milder conditions for coupling the lipid to the amine on the saccharide. The acid, in this case lauric acid (1 equivalent), and NHS (1.5 equivalents) are dissolved in dichloromethane (0.4 *M*) before the addition of 1,3-diisopropylcarbodiimide (DCC, 1.1 equivalents) at room temperature. Concentration and purification by column chromatography provide quantitative conversion of the acid to the desired ester.[8] Treatment of the amine with the activated ester in the presence of base gives complete conversion to the desired glycolipid. Triethylamine (1.5 equivalents) and the succinimide ester (1.2 equivalents) are added to the amino sugar in dimethylformamide (DMF, 0.1 *M*) at room temperature and the reaction is stirred overnight. Purification of the desired glycolipid is accomplished by concentrating the products by separation between dichloromethane and water. Excess ester dissolves in the organic phase and the deprotected glycolipid remains in the aqueous phase. Lyophilization of water then furnishes the selected glycolipid.[6]

Triazole Linkage

When azide functionalities are present on the carbohydrate, hydrocarbons with terminal alkynes can be attached via triazole formation. As with peptide-based coupling, purification is accomplished by washing and the fully deprotected saccharide can be utilized. This method also has the advantage of being conducive to coupling in the microtiter plate. The alkyne and the azide are coupled by 1,3-dipolar cycloadditions catalyzed by copper iodide (CuI) at room temperature in microtiter plates or by thermal catalysis in glass vials.

When thermal catalysis is used, the azide (1 equivalent) and the alkyne (5 equivalents) are dissolved in methanol (0.1 *M*) and stirred for 24 h at 80°. The reaction is effectively run undiluted, as the methanol evaporates over time. The triazole can then be purified by removing excess alkyne with hot ether washes. Thermal catalysis, however, lacks regiocontrol and the product is a mixture of the 1,4- and 1,5-regioisomers. This issue can be resolved by the addition of trace CuI to the reaction, giving the desired 1,5-regioisomer.[8] CuI catalysis at room temperature is advantageous as the reaction can be performed in the microtiter plate and used directly for screening. In this reaction, the azide (0.1 *M* in methanol) is added to the microtiter plate well followed by the alkyne (0.01 *M* in methanol; 1 equivalent). CuI (5 equivalents) and diisopropylethylamine (DIPEA, 5 equivalents) in methanol are added last. The plate is then covered and shaken at room temperature overnight before the plate is uncovered and the solvent is evaporated. Excess CuI and DIPEA are removed through aqueous washing and the saccharides are ready for screening.[8]

[8] F. Fazio, M. C. Bryan, O. Blixt, J. C. Paulson, and C.-H. Wong, **124,** 14397 (2002).

Biological Assays

Enzyme-Linked Immunosorbent Assay

Since its discovery by Engvall, Perlmann, van Weemen, and Schuurs in 1971,[8a,8b] the ELISA has become one of the most common and important tools for the analysis of recognition processes in protein interactions.[9,10] These assays have been critical in the determination of protein–protein interactions as well as protein interactions with a variety of other substrates.[10] Because carbohydrates are key components in cellular and molecular recognition processes,[1] application of the ELISA to sugars and glycoconjugates is a exceptional tool to further understanding of how such signals are transmitted in nature as well as how they can be altered for drug discovery.

The ELISA has therefore been undertaken for a variety of lectins, illustrated in Fig. 2, with saccharides bound to the surface through the aforementioned lipid. Lectin binding is commonly observed by absorbance or fluorescence of labels attached to the protein. Recognition cascades can also be utilized to observe the bound saccharide. One such example is the recognition of β-galactose by the ricin B chain, which has been used to quantitate retention as well as to differentiate between oligosaccharides

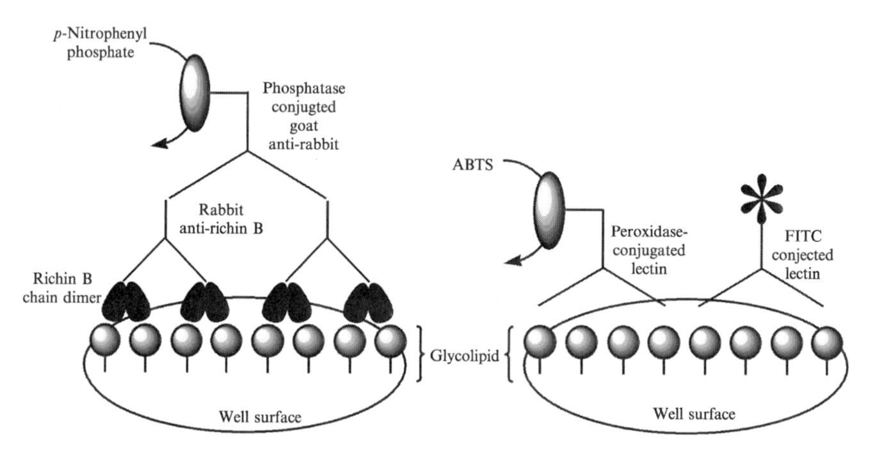

FIG. 2. ELISA Saccharides in microtiter plates. ABTS, 22′-Azinobis(3-ethylbenzo thiazoline-6-sulfonic acid; FITC, fluorescein isothiocyanate.

[8a] E. Engvall and P. Perlmann, *Immunochemistry* **8,** 871 (1971).
[8b] B. K. Van Weemen and A. H. W. M. Schuurs, *FEBS Lett.* **15,** 232 (1971).
[9] M. F. Clark, R. M. Lister, and M. Bar-Joseph, *Methods Enzymol.* **118,** 742 (1986).
[10] J. Gervay and K. D. McReynolds, *Curr. Med. Chem.* **6,** 129 (1999).

through the stereochemistry at the anomeric position. To analyze saccharides with ricin B, the plate is first blocked with TBS buffer [50 mM Tris-HCl (pH 7.5)–150 mM NaCl; 100 μl] containing 1% bovine albumin (BSA) (buffer A). This step is important in the prevention of proteins directly adsorbing to the surface of the plate through hydrophobic interaction and is repeated between every incubation. Ricin B chain from *Ricinus communis* (castor bean) (0.1 mg/ml buffer A) is then incubated in the well for 1 h, followed by incubation with the anti-lectin for *Ricinus communis* from rabbit (5 μg/ml buffer A). Monoclonal anti-rabbit immunoglobulin conjugated with an alkaline phosphatase (5 μg/ml in 0.05 M bicarbonate buffer at pH 9.6; 50 μl) is then incubated last. It is the presence of this last protein that is utilized in photometric observation. When alkaline phosphatase is treated with *p*-nitrophenyl phosphate [1 mg/ml in 10% diethanolamine buffer (pH 9.8)–0.5 mM $MgCl_2$], *p*-nitrophenol is produced in the enzymatic reaction, which is observable at 405 nm.[6]

Lectins can be observed in the plate through conjugation to enzymes that can produce a colorimetric response, therefore bypassing the intermediate cascade as previously mentioned. Two examples of such lectins are concanavalin A (Con A), which binds α-glucosyl residues, and *Tetragonolobus purpureas* (TP), which recognizes L-fucose. Binding of both of these lectins can be observed with the peroxidase-conjugated protein. As before, the plate is blocked to prevent background binding of the protein to the microtiter plate surface. Con A (10 μg/ml buffer) or TP (100 μg/ml buffer) conjugated with a peroxidase is then incubated in the microtiter plate. After washing, 2,2′-azinobis(3-ethylbenzothiazoline-6-sulfonic acid (ABTS) is added to the well. Only wells containing the bound lectin have the peroxidase necessary for observation of the resulting chromophore at 405 nm.[6] Peroxidase-labeled TP has been further utilized to determine binding affinities for saccharides with fucosyl residues as well as to analyze the ability of inhibitors to reduce fucosylation by fucosyltransferase.[8]

Fluorescently labeled lectins are also useful in ELISA analysis. Fluorescein-labeled *Sambucus nigra* (SNA) is one such lectin. SNA is able to bind sialic acid-containing carbohydrates with preferential binding to α-2,6-linked over α-2,3-linked sialic acid and can be used to follow enzymatic transformation *in situ*. Plates are first blocked with BSA and fluorescein-labeled SNA [10 μg/ml 10 mM HEPES buffer (pH 7.5)–150 mM NaCl containing 1% bovine serum albumin] is then incubated in the well. After washing, fluorescence can subsequently be read at 535 nm.[8] This method provides the most direct route for observation of saccharide–lectin binding.

Enzymatic Transformation

Microtiter plates also provide an excellent solid phase for *in situ* biological manipulations such as the enzymatic reactions illustrated in Fig. 3. This can be used to create diversity for libraries of oligosaccharides already present on microtiter plates as well as to study small-molecule inhibition of such transformations. Oligosaccharides with terminal galactose residues have been successfully subjected to oxidation by galactose oxidase (GAO), leading to an aldehyde at the C-6 position.[6] In this enzymatic transformation, horseradish peroxidase (HRP) is utilized in a coupled assay to observe oxidation. The peroxidase (EC 1.11.1.7, type II from horseradish, 1 U in 0.1 *M* Tris buffer, pH 8.0) and *o*-dianisidine (10 μg in methanol) are incubated in the well before addition of the oxidase. This prevents activation or inhibition of GAO by HRP or the indicator.[11] Galactose oxidase (EC 1.1.3.9, from *Dactylium dendroides*, 2 U in buffer) is then added last and the oxidized *o*-dianisidine can be read at 490 nm.[6]

Glycosyl transfer can also be performed *in situ* on saccharides bound to the microtiter plate. Fucosylation by fucosyltransferase is one such reaction. Generation of sialyl Lewis X (SLX) is dependent on addition of a fucose to the precursor NeuAcα2,3Galβ1,4GlcNac at the C-3 position of *N*-acetylglucosamine. The resulting fucosylated product, SLX, is a crucial mediator of inflammation and is often found in the sera of gastrointestinal,

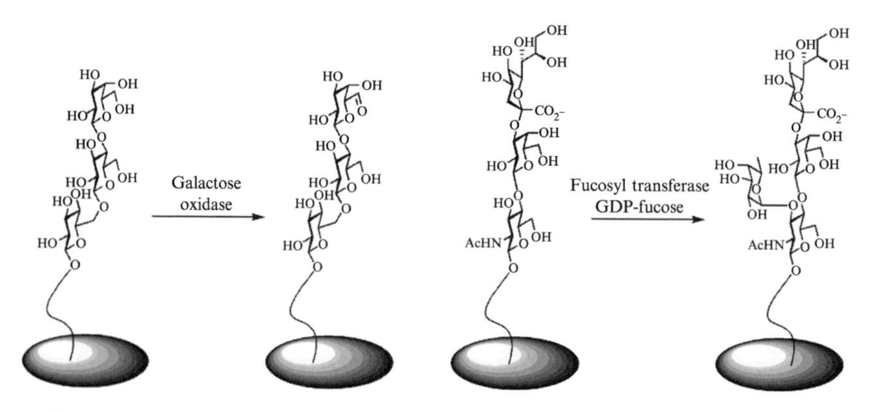

FIG. 3. Enzymatic reactions including oxidation and glycosyl transfer can be performed on saccharides attached to microtiter plates through noncovalent interaction.

[11] P. S. Tressel and D. J. Kosman, *Methods Enzymol.* **89,** 163 (1982).

pancreatic, and breast cancer patients.[12] Therefore, having a method for analysis of inhibition of this pathway has important medicinal applications. NeuAcα2,3Galβ1,4GlcNAc (0.6 μmol) bound to microtiter plate wells is treated with α-1,3-fucosyltransferase [EC 2.4.1.65, 0.1 U in 50 m*M* morpholinepropanesulfonic acid (MOPS), pH 7.4, containing 4 m*M* $MnCl_2$ and 0.5% BSA) and GDP-fucose (5.1 μmol). Production of SLX can be monitored with the fucose-specific lectin TP as previously described.[8] Because the substrate is bound and can be quantitated, inhibition of each enzymatic transformation can also be analyzed. Both these examples of enzymatic transformations as well as the ELISAs illustrate the use of saccharides linked to long-chain hydrocarbons for noncovalent attachment to microtiter plates suitable for high-throughput screening.

Summary

Presented herein are a variety of methods for the attachment of saccharides to the surface of microtiter plates through long-chain hydrocarbons capable of binding to the polystyrene surface through noncovalent interaction. These glycolipids have the benefit of diverse and facile conjugation procedures and the ability to withstanding a range of biological assays. On addition of carbohydrates to the microtiter plate well, a suitable surface for both ELISA and enzymatic transformation is available, making the saccharide arrays useful for high-throughput study of specificity and inhibition in sugar–protein interactions.

[12] Y. Ichikawa, Y.-C. Lin, D. P. Dumas, G.-J. Shen, E. Garcia-Junceda, M. A. Williams, R. Bayer, C. Ketcham, L. Walker, E., J. C. Paulson, and C.-H. Wong, *J. Am. Chem. Soc.* **114,** 9283 (1992).

[16] Comparative Studies of Galactoside-Containing Clusters

By BINGCAN LIU and RENÉ ROY

Introduction

The wide occurrence of β-D-galactopyranoside ligands on mammalian glycoproteins and glycolipids having only subtle differences in topology and substructures is a puzzling phenomenon that must be addressed in the design of potent selective inhibitors.[1] This fact is particularly important if the drug targeting to the C-type lectins, asialoglycoprotein receptors of hepatocytes or macrophages,[2] is not to be confounded with the soluble S-type lectins such as the galectin family.[3] The corollary is also valid: if, for instance, the blockage of the apoptotic behavior of galectin-3 becomes the target receptor,[4] then their rapid clearance by the liver/macrophages would be obstacles to surmount. Moreover, the increasing body of evidence for the distinctive functions of the galectins themselves,[5] all binding to some undefined β-D-galactoside ligands, clearly underscores our need to understand the multiple but selective binding interactions. Given the success of multivalent glycans as powerful inhibitors of carbohydrate–protein interactions,[6] the next strategy ought to address selectivity issues. In fact, few studies point toward a rational understanding of the arrays of multiple receptor–multivalent glycan interactions.[7,8] Trivalent Gal/GalNAc clusters have been shown to markedly distinguish between C-type lectins and galectins.[9]

[1] H. Lis and N. Sharon, *Chem. Rev.* **98,** 637 (1998).
[2] K. Osaki, R. T. Lee, Y. C. Lee, and T. Kawasaki, *Glycoconjugate J.* **12,** 268 (1995).
[3] S. H. Barondes, D. N. W. Cooper, M. A. Gitt, and H. Leffler, *J. Biol. Chem.* **269,** 20807 (1994).
[4] N. L. Perillo, M. E. Marcus, and L. G. Baum, *J. Mol. Med.* **76,** 402 (1998).
[5] H. Leffler, *Trends Glycosci. Glycotechnol.* **9,** 9 (1997).
[6] R. Roy, *Top. Curr. Chem.* **187,** 241 (1997).
[7] S. André, P. J. Cejas Ortega, M. Alamino Perez, R. Roy, and H.-J. Gabius, *Glycobiology* **9,** 1253 (1999).
[8] S, André, R. J. Pieters, I. Vrasidas, H. Kaltner, I. Kuwabara, F. Liu, R. M. J. Liskamp, and H.-J. Gabius, *ChemBioChem* **2,** 822 (2001).
[9] S. André, B. Frisch, H. Kaltner, D. L. Desouza, F. Schuber, and H.-J. Gabius, *Pharm. Res.* **17,** 985 (2000).

0076-6879/03 $35.00

In the present study, we discuss novel synthetic strategies in the design of lactose clusters with "rigidified" core structures.[10] The rationale for conformationally restrained clusters is based on the expectation that more rigid structures would be less entropically penalized on binding to their counterreceptors. In addition, using organometallic cross-coupling chemistry allows for the introduction of more lipophilic linkers, thus potentially limiting solvating water molecules. The use of palladium(0)-catalyzed cross-coupling between prop-2-ynyl lactosides and a wide range of aryl iodide clusters is described.

Discussion

The synthesis of the required prop-2-ynyl lactoside peracetate **2** is described in Scheme 1. The procedure simply involves the treatment of lactose octaacetate **1** with propargyl alcohol and a Lewis acid, preferably boron trifluoride etherate in dichloromethane at room temperature (85%).[11] This protocol is widely applicable to a wide range of other glycosyl acetates.[12] Before attempting the cross-coupling reactions with various aryl iodides, lactoside **2** is treated under oxidative homocoupling conditions to provide dimer **3** in 95% yield.[13,14] Although the homocoupling can be achieved under simpler but harsher conditions (Glaser[15] [Cu(I), O_2] or Eglinton[16] [Cu(II) processes]), the reactions are carried out under milder Sonogashira conditions[17] using dichlorobis(triphenylphosphine)palladium(II) [$(Ph_3P)_2PdCl_2$] and copper(I) iodide in a one-to-one mixture of dimethylformamide and triethylamine at room temperature. However, the choice of catalysts is not critical because other palladium catalysts such as tetrakis(triphenylphosphine)palladium(0) [$(Ph_3P)_4Pd$] or tris(dibenzylideneacetone)dipalladium(0) [$Pd_2(dba)_3$] are equally effective. The formation of homodimers is a typical side reaction of the Sonogashira coupling,

[10] R. Roy, S. Das, R. Dominique, M. C. Trono, F. Hernández-Meteo, and F. Santoyo-González, *Pure Appl. Chem.* **71,** 565 (1999).

[11] H. B. Mereyala and S. R. Gurrala, *Carbohydr. Res.* **307,** 351 (1998).

[12] R. Roy, S. K. Das, F. Santoyo-González, F. Hernández-Meteo, T. K. Dam, and C. F. Brewer, *Chem. Eur. J.* **6,** 1757 (2000).

[13] R. Roy, S. K. Das, F. Hernández-Meteo, F. Santoyo-González, and Z. Gan, *Synthesis* **7,** 1049 (2001).

[14] B. Liu and R. Roy, *J. Chem. Soc. Perkin Trans.* **1,** 773 (2001).

[15] C. Glaser, *Ann. Chem. Pharm.* **154,** 137 (1870); A. S. Hay, *J. Org. Chem.* **27,** 3320 (1962).

[16] G. Eglinton, *J. Chem. Soc.* 889 (1959).

[17] K. Sonogashira, *in* "Metal-Catalyzed Cross-Coupling Reactions" (F. Diederich and P. J. Stang, eds.), p. 203. Wiley-VCH, Weinheim, Germany, 1998.

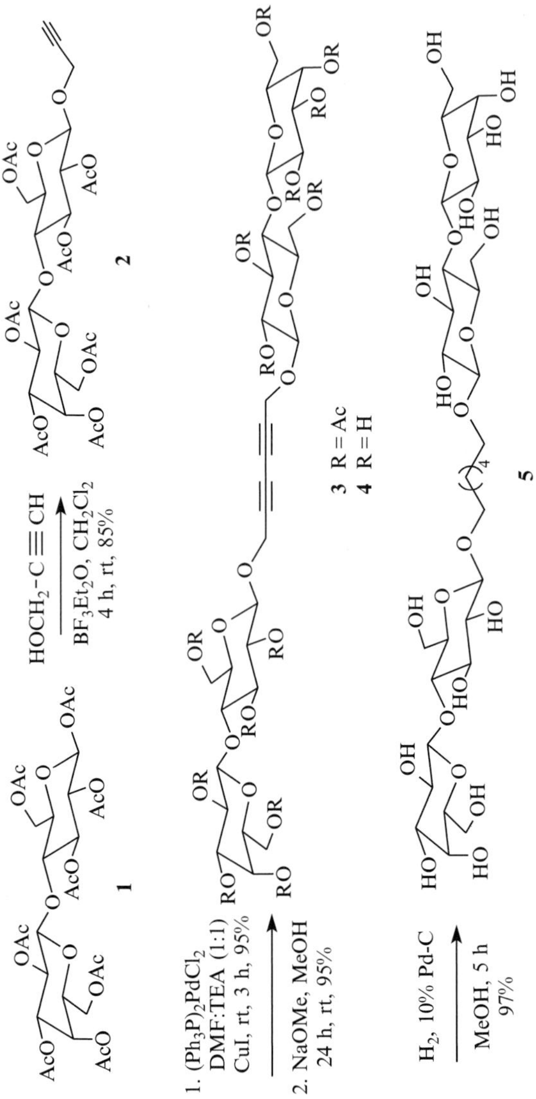

AcO OAc AcO OAc O AcO OAc O AcO OAc

1

$HOCH_2$–C≡CH
BF_3Et_2O, CH_2Cl_2
4 h, rt, 85%

AcO OAc AcO OAc O AcO OAc O AcO O

2

1. $(Ph_3P)_2PdCl_2$
DMF:TEA (1:1)
CuI, rt, 3 h, 95%
2. NaOMe, MeOH
24 h, rt, 95%

RO OR RO OR O O RO OR O RO O O RO OR O RO O RO OR OR RO OR

3 R = Ac
4 R = H

H_2, 10% Pd-C
MeOH, 5 h
97%

HO OH HO OH O O HO OH O HO O O HO OH O HO O HO OH OH HO OH

5

SCHEME 1

especially when copper(I) salts are present. It provides, however, a milder reaction temperature.

Complete de-O-acetylation of dimer **3** under Zemplén conditions ($NaOCH_3$, methanol) affords freely water-soluble dilactoside **4** in 95% yield. Hydrogenation of the conjugated triple bonds with 10% palladium on charcoal provides dimer **5** in essentially quantitative yield. It is worth mentioning that although dimers such as **5** can be directly obtained by bisglycosylation of 1,6-hexanediol, the yields are usually much lower. Because compound **3** is expected as a side product from the next cross-coupling reactions, it serves as a valuable comparative element.

The other cross-coupling reactions are all carried out according to the same protocol described above for **3** [$(Ph_3P)_2PdCl_2$, dimethylformamide–triethylamine (1:1, v/v), CuI, room temperature, 3–5 h]. The simultaneous cross-coupling of prop-2-ynyl lactoside **2** to 1,4-diiodobenzene **6**, to 1,3,5-triiodobenzene **10**, and to both pentaerythritol tetrakis (*meta*- and *para*-iodobenzyl)ethers **13** and **14** provides an efficient entry into this novel family of carbohydrate clusters **7** (dimer, 90%), **11** (trimer, 80%), and tetramers **15** (*meta*) and **17** (*para*) in 78 and 75% yields, respectively (Schemes 2–4).[18,19] As described above, complete de-O-acetylation of the clusters is uneventful and affords freely water-soluble clusters **8**, **12**, **16**, and **18** in more than 90% yields. The solubility of these lactosylated clusters is in striking contrast to those observed for the corresponding β-D-galactoside clusters.[18]

Although dimer **8** and trimer **12** can be further hydrogenated, the benzylic nature of the remaining tetrakis ethers **16** and **18** prevents such treatment because the pentaerythritol moieties would be cleaved. Thus, catalytic hydrogenation of dimer **8** under standard conditions (H2, 10% Pd-C, methanol) affords dimer **9** in 95% yield; dimer **9** possesses a slightly more flexible arm between the lactoside residues in comparison with homodimer **8**.

General Methods and Materials

All chemicals are purchased from Aldrich Chemicals (Milwankee, WI) and used without further purification. Thin-layer chromatography is performed on silica gel F_{254} (Merck, Darmstadt, Germany) precoated aluminum sheets and visualized with molybdenum solution and/or ultraviolet (UV) detection. Column chromatography is run on ultrapure silica gel (Silicycle, Quebec, Canada). Optical rotations are measured with a Perkin-Elmer (Norwalk, CT) 241 polarimeter. All the nuclear magnetic resonance

[18] B. Liu and R. Roy, *Tetrahedron* **57,** 6909 (2001).
[19] B. Liu and R. Roy, *Chem. Commun.* 594 (2002).

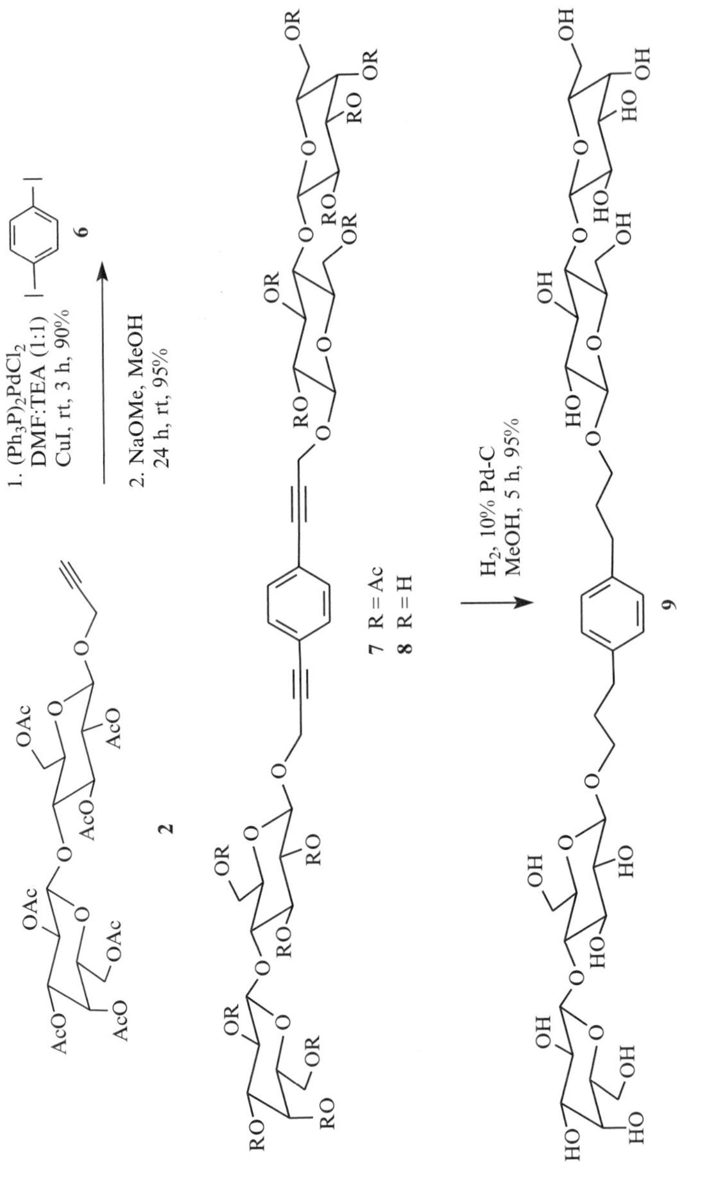

2

1. $(Ph_3P)_2PdCl_2$
DMF:TEA (1:1)
CuI, rt, 3 h, 90%

6

2. NaOMe, MeOH
24 h, rt, 95%

7 R = Ac
8 R = H

H_2, 10% Pd-C
MeOH, 5 h, 95%

9

SCHEME 2

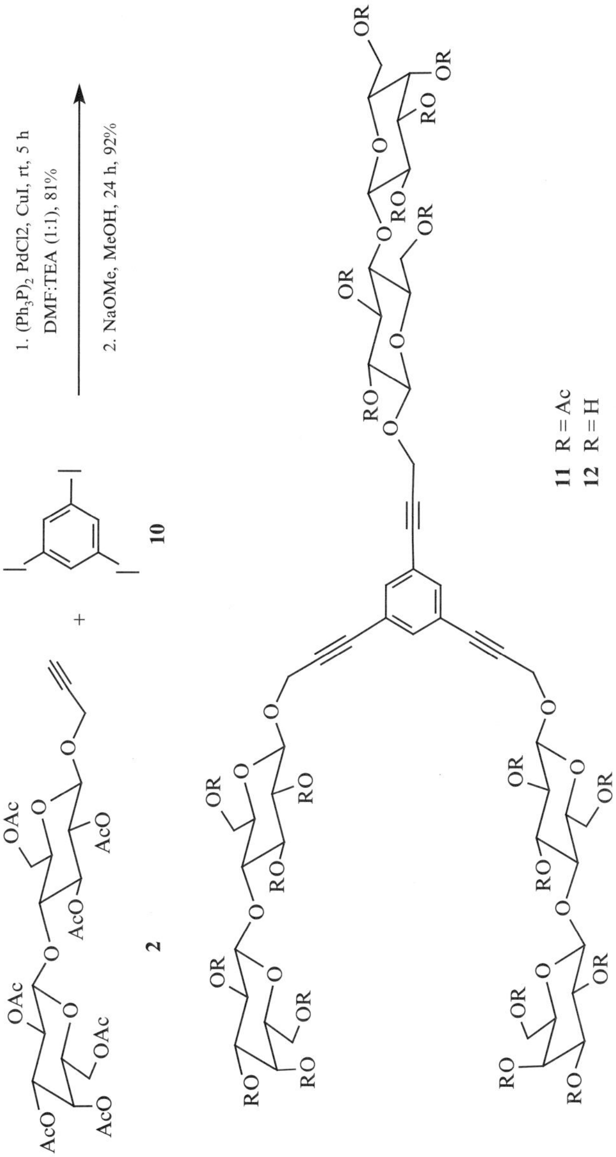

1. $(Ph_3P)_2PdCl2$, CuI, rt, 5 h
DMF:TEA (1:1), 81%

2. NaOMe, MeOH, 24 h, 92%

2 + **10** → **11** R = Ac; **12** R = H

SCHEME 3

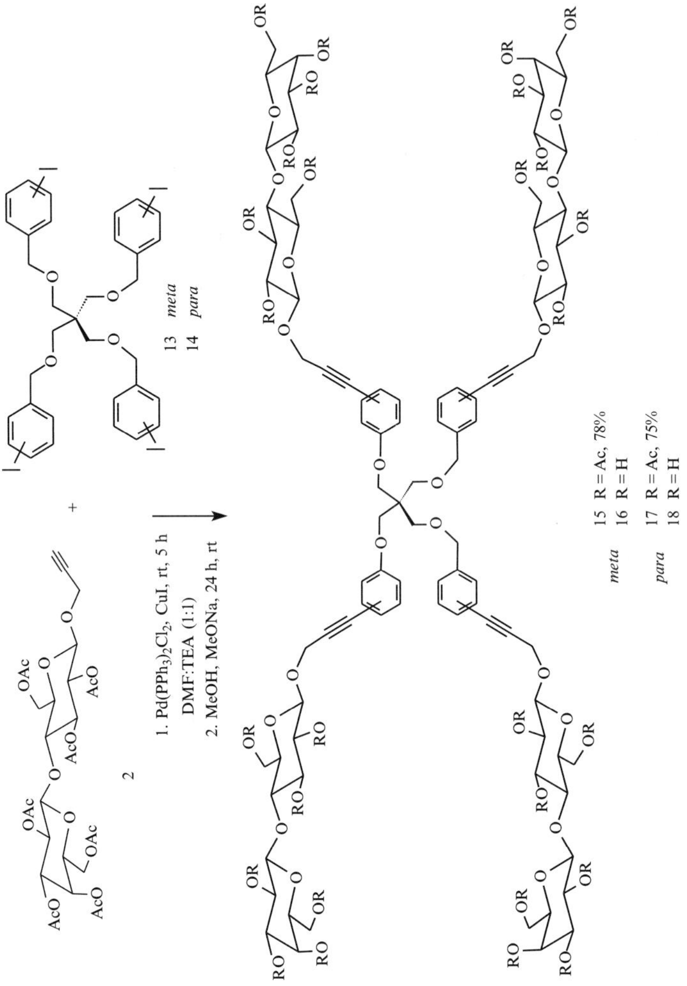

2 + 13 *meta* / 14 *para*

1. $Pd(PPh_3)_2Cl_2$, CuI, rt, 5 h
DMF:TEA (1:1)
2. MeOH, MeONa, 24 h, rt

meta 15 R = Ac, 78%
16 R = H

para 17 R = Ac, 75%
18 R = H

SCHEME 4

(NMR) spectra (500 MHz for ^{1}H and 125.7 MHz for ^{13}C) are recorded on an AMX500 spectrometer (Bruker Biospin, Billerica, MA). The resonances are assigned on the basis of ^{1}H, ^{13}C, and ^{1}H-^{1}H correlation spectroscopy (COSY), distortionless enhancement by polarization transfer (DEPT), and heteronuclear multiple quantum coherence (HMQC) experiments. Chemical shifts are referenced to internal $CDCl_3$ (δ_H 7.27 and δ_C 77.0 ppm). Electrospray ionization-mass spectrometry (ESI-MS) analysis is carried out with a Micromass Quattro LC (Waters, Milford, MA).

*Prop-2-ynyl (2,3,4,6-tetra-O-acetyl-β-D-galactopyranosyl)-(1 → 4)-2,3,6-tri-O-acetyl-β-D-glucopyranoside (**2**)*

Lactose octaacetate (**1**) (1.36 g, 2 mmol) is dissolved in 30 ml of dichloromethane to which are added freshly distilled propargyl alcohol (0.24 ml, 4 mmol) and 1 g of 4-A molecular sieves. The mixture is stirred at room temperature for 30 min. Boron trifluoride diethyl etherate (BF_3-Et_2O, 0.62 ml, 5 mmol) is then added to the flask and the mixture is stirred at room temperature for 4 h. After completion of the reaction, sodium carbonate (0.6 g) is introduced into the flask and the mixture is stirred for an additional 30 min. The mixture is then filtered through a pad of Celite and washed with 50 ml of dichloromethane. The filtrate is washed twice, with 50 ml of water each time, and the organic layer is dried over anhydrous sodium sulfate. After removal of the solvent under reduced pressure, the residue is purified on a silica gel column with hexane and ethylacetate (2:3, v/v) as eluent to provide the title compound **2** in 85% yield. $[\alpha]_D$ $-12.5°$ ($c = 1.5$, $CHCl_3$); ^{1}H NMR (500 MHz, $CDCl_3$) δ 5.31 (1H, bd, $J = 2.4$ Hz, H-4′), 5.19 (1H, t, $J = 9.3$ Hz, H-3), 5.07 (1H, dd, $J = 8.0$ Hz, 10.4, H-2′), 4.91 (1H, dd, $J = 3.4$, 10.4 Hz, H-3′), 4.88 (1H, dd, $J = 8.1$, 9.4 Hz, H-2), 4.71 (1H, d, $J = 8.1$ Hz, H-1), 4.44–4.48 (2H, m, H-6a and H-1′), 4.30 (1H, d, $J = 2.4$ Hz, H-1″), 4.02–4.11 (3H, m, H-6b and H-6′), 3.84 (1H, t, $J = 6.5$ Hz, H-5′), 3.78 (1H, t, $J = 9.6$ Hz, H-4), 3.59–3.61 (1H, m, H-5), 2.43 (1H, t, $J = 2.3$ Hz, H-3″), 2.18, 2.09, 2.03, 2.02, 2.01, 1.93 (21H, 6s, CH_3CO); ^{13}C NMR (125.7 MHz, $CDCl_3$) δ 170.3, 170.1, 170.0, 169.7, 169.6 and 169.0 (CH_3*C*O), 101.0 (C-1′), 97.8 (C-1), 78.0 (alkynyl), 76.1 (C-4), 75.4 (alkynyl), 72.7 (C-5), 72.6 (C-3), 71.3 (C-2), 70.9 (C-3′), 70.7 (C-5′), 69.1 (C-2′), 66.6 (C-4′), 61.8 (C-6), 60.8 (C-6′), 55.8 (C-1″), 20.8, 20.7, 20.6, 20.5, and 20.4 (CH_3CO). ESI-MS calculated for $C_{29}H_{38}O_{18} + (NH_4^{+})$: 692.2; found: 692.2.

*1,6-Bis-[(2,3,4,6-tetra-O-acetyl-β-D-galactopyranosyl)-(1 → 4)-2,3,6-tri-O-acetyl-β-D-glucopyranosyloxy]-2,4-hexadiyne (**3**)*

Prop-2-ynyl (2,3,4,6-tetra-*O*-acetyl-β-D-galactopyranosyl)-(1 → 4)-2,3,6-tri-*O*-acetyl-β-D-glucopyranoside (**2**) (67.4 mg, 0.10 mmol) is dissolved into a solution of 10 ml of dimethyl formamide (DMF) and triethylamine (TEA) (1:1, v/v) to which are added $Pd(PPh_3)_2Cl_2$ (3.6 mg, 5 mol%) and CuI (3.8 mg, 20 mmol%). The solution is stirred at room temperature for 3 h. The solvent and triethylamine are removed under reduced pressure. The residue is purified by silica gel column chromatography, using hexane–ethyl acetate (2:3, v/v), to afford **3** as a white solid in 95% yield. $[\alpha]_D$ −22.1° (c = 1.0, $CHCl_3$); ^1H NMR (500 MHz, $CDCl_3$) δ 5.31 (2H, bd, J = 3.4 Hz, H-4′), 5.19 (2H, t, J = 9.3 Hz, H-3), 5.07 (2H, dd, J = 8.0 Hz, Hz, 10.3, H-2′), 4.91 (2H, dd, J = 3.4 Hz, 10.3, H-3′), 4.88 (2H, dd, J = 8.0, 9.4 Hz, H-2), 4.66 (2H, d, J = 8.0 Hz, H-1), 4.49 (2H, dd, J = 1.6, 12.0 Hz, H-6a), 4.45 (2H, d, J = 8.0 Hz, H-1′), 4.39 (2H, s, H-1″), 4.02–4.11 (6H, m, H-6b and H-6′), 3.84 (2H, t, J = 7.0 Hz, H-5′), 3.79 (2H, t, J = 9.5 Hz, H-4), 3.61–3.64 (2H, m, H-5), 2.12, 2.10, 2.03, 2.02, 2.01, 1.93 (42H, 6s, CH_3CO); ^{13}C NMR (125.7 MHz, $CDCl_3$) δ 170.3, 170.1, 170.0, 169.7, 169.6, and 169.0 (CH_3CO), 101.0 (C-1′), 98.2 (C-1), 76.0 (C-4), 74.2 (alkynyl), 72.8 (C-5), 72.6 (C-3), 71.2 (C-2), 70.9 (C-3′), 70.8 (alkynyl), 70.7 (C-5′), 69.0 (C-2′), 66.6 (C-4′), 61.7 (C-6), 60.8 (C-6′), 56.3 (C-1″), 20.8, 20.7, 20.6, 20.5, and 20.4 (CH_3CO) period. ESI-MS calculated for $C_{58}H_{74}O_{36} + (NH_4^+)$: 1364.4; found: 1364.3.

De-O-acetylation of ***4, 8, 12, 16,*** *and* ***18*** *under Zemplén Conditions*

The fully protected homodimer **3** (68 mg, 0.05 mmol) is suspended in methanol (15 ml), to which is added a catalytic amount of sodium methoxide. The solution is stirred at room temperature for 24 h. After neutralization of sodium methoxide with Amberlite resin (120 H^+), the solution is filtered through a cotton plug. Removal of methanol under reduced pressure affords compound **4** as a white foam in 95% yield. Compounds **8**, **12**, **16**, and **18** are prepared in the same manner in 95, 92, 90, and 91% yields, respectively.

*1,6-Bis-[(β-D-galactopyranosyl)-(1 → 4)-β-D-glucopyranosyloxy]-2,4-hexadiyne (**4**)*

$[\alpha]_D$ −20.8° (c = 1.0, H_2O); ^1H NMR (500 MHz, D_2O) δ 4.70 (2H, J = 8.0 Hz, d, H-1), 4.63 (4H, s, H-1″), 4.49 (2H, d, J = 7.8 Hz, H-1′), 4.03 (2H, dd, J = 1.7, 11.8 Hz, H-6a), 3.98 (2H, d, J = 3.2 Hz, H-4′), 3.96 (2H, dd, J = 6.9, 12.0 Hz, dd), 3.81–3.88 (8H, m, H-5′, H-6b, H-6), 3.67–3.78 (8H, m, H-3, H-3′, H-4, H-5), 3.59 (2H, dd, J = 7.8, 9.9 Hz, H-2′), 3.40 (2H, t, J = 8.3 Hz, H-2); ^{13}C NMR (125.7 MHz, D_2O) δ 102.5 (C-1′), 100.1 (C-1), 77.8, 74.9 (C-5′), 74.4 (C-5), 73.9, 72.4, 72.1, 72.0, 70.5 (C-2′), 69.9 (alkynyl), 68.1 (C-4′), 60.5 (C-6′), 59.5 (C-6), 56.6 (C-1″). ESI-MS calculated for $C_{30}H_{46}O_{22} + (Na^+)$: 781.1; found: 781.2.

1,6-Bis-[(β-D-galactopyranosyl)-(1 → 4)-β-D-glucopyranosyloxy]hexane (**5**)

The fully deprotected homodimer **4** (39 mg, 0.05 mmol) is dissolved into methanol (10 ml) to which is added a catalytic amount of 10% Pd-C. The mixture is stirred at room temperature for 5 h. After filtration through a pad of Celite, the filtrate is concentrated to dryness under reduced pressure to provide compound **5** in 97% yield. $[\alpha]_D$ −21.5° (c = 1.0, H_2O); ^{1}H NMR (500 MHz, D_2O) δ 4.53 (2H, J = 8.0 Hz, d, H-1), 4.50 (2H, J = 7.8 Hz, Hz, d, H-1′), 4.02 (2H, d, J = 12.0 Hz, H-6a), 3.97 (2H, bd, J = 3.2 Hz, H-4′), 3.96 (2H, dd, J = 6.9, 12.0 Hz, H-1a), 3.63–3.90 (18H, m, H-3, H-3′, H-4, H-5, H-5′, H-6b, H-6, H-1b″), 3.59 (2H, dd, J = 7.8, 9.9 Hz, H-2′), 3.45 (2H, t, J = 8.3 Hz, H-2), 1.69 (4H, m, H-2″), 1.44 (4H, bs, H-3″); ^{13}C NMR (125.7 MHz, D_2O) δ 102.5 (C-1′), 101.5 (C-1), 78.0, 74.9, 74.3, 74.0, 72.4, 72.1, 71.3, 70.5, 70.3, 70.1 (C-2), 69.9, 68.1 (C-2′), 67.9 (C-1″), 60.5 (C-6′), 59.7 (C-6), 28.1 (C-2″), 24.3 (C-3″). ESI-MS calculated for $C_{30}H_{54}O_{22}$ + (NH_4^+): 784.3; found: 784.2.

General Procedure for Sonogashira Coupling Reactions

1,4-Diiodobenzene **6** (16.6 mg, 0.05 mmol) is dissolved in a solution of 10 ml of DMF and TEA (1:1, v/v), to which are added $Pd(PPh_3)_2Cl_2$ (3.6 mg, 5 mol%), lactoside **2** (80.9 mg, 0.12 mmol, 2.2 equivalents), and CuI (3.8 mg, 20 mmol%). Under nitrogen, the solution is stirred at room temperature for 5 h. The solvent and triethylamine are removed under reduced pressure. The residue is purified by silica gel column chromatography, using hexane and ethyl acetate (2:3, v/v), to provide **7** as a white solid in 90% yield. The same procedure is applied for the synthesis of compounds **11**, **15**, and **17** in yields of 81, 78, and 75%, respectively.

1,4-Phenylenedi-2-propyne-3,1-diylbis[(2,3,4,6-tetra-O-acetyl-β-D-galactopyranosyl)-(1 → 4)-2,3,6-tri-O-acetyl-β-D-glucopyranoside] (**7**)

$[\alpha]_D$ −31.0° (c = 1.0, $CHCl_3$); ^{1}H NMR (500 MHz, $CDCl_3$) δ 7.35 (4H, s, aromatic), 5.31 (2H, bd, J = 2.9 Hz, H-4′), 5.21 (2H, t, J = 9.2 Hz, H-3), 5.07 (2H, dd, J = 7.9, 10.3 Hz, H-2′), 4.91 (2H, dd, J = 3.5, 10.3 Hz, H-3′), 4.90 (2H, dd, J = 7.9, 9.4 Hz, H-2), 4.76 (2H, d, J = 7.9 Hz, H-1), 4.54 (2H, d, J = 16.1 Hz, H-1a″), 4.50 (2H, d, J = 16.1 Hz, H-1b″), 4.44–4.48 (4H, m, H-6a and H-1′), 4.02–4.11 (6H, m, H-6b and H-6′), 3.84 (2H, t, J = 6.8 Hz, H-5′), 3.80 (2H, t, J = 9.6 Hz, H-4), 3.61–3.64 (2H, m, H-5), 2.11, 2.08, 2.03, 2.02, 2.01, 2.00, 1.99, and 1.93 (42H, 7s, CH_3CO); ^{13}C NMR (125.7 MHz, $CDCl_3$) δ 170.3, 170.2, 170.1, 170.0, 169.7, 169.6, and 169.0 (CH_3CO),

131.6 and 122.5 (aromatic), 101.0 (C-1′), 98.2 (C-1), 86.4 and 85.4 (alkynyl), 76.0 (C-4), 72.8 (C-5), 72.7 (C-3), 71.4 (C-2), 70.9 (C-3′), 70.7 (C-5′), 69.1 (C-2′), 66.6 (C-4′), 61.8 (C-6), 60.8 (C-6′), 56.7 (C-1″), 20.8, 20.7, 20.6, 20.5, and 20.4 (CH_3CO). ESI-MS calculated for $C_{64}H_{78}O_{36} + (Na^+)$: 1445.3; found: 1445.2.

*1,4-Phenylenedi-2-propyne-3,1-diylbis[(β-D-galactopyranosyl)-(1→4)-β-D-glucopyranoside] (**8**)*

Compound **8** is obtained under the Zemplèn conditions described above for **4**: $[\alpha]_D$ −7.0° ($c = 1.0$, DMSO); ^{1}H NMR (500 MHz, DMSO) δ 7.30 (4H, s, aromatic), 4.64 (2H, d, $J = 15.9$ Hz, H-1a″), 4.53 (2H, d, $J = 15.9$ Hz, H-1b″), 4.38 (2H, d, $J = 7.9$ Hz, H-1), 4.17 (2H, d, $J = 7.1$ Hz, H-1′), 3.26–3.79 (22H, m, H-2′, H-3, H-3′, H-4, H-4′, H-5, H-5′, H-6, H-6′), 3.03–3.08 (2H, t, $J = 8.3$ Hz, H-2); ^{13}C NMR (125.7 MHz, DMSO) δ 131.7 and 122.2 (aromatic), 103.8 (C-1′), 100.9 (C-1), 87.7 and 85.0 (alkynyl), 80.6, 75.5, 75.0, 74.9, 73.2, 73.0, 70.5, 68.1, 60.5 (C-6′), 60.3 (C-6), 55.8 (C-1″). ESI-MS calculated for $C_{36}H_{50}O_{22} + (NH_4^+)$: 852.3; found: 852.2.

*1,4-Bis[[[(β-D-galactopyranosyl)-(1→4)-β-D-glucopyranosyl]oxy]propyl]benzene (**9**)*

The fully deprotected homodimer **8** (42 mg, 0.05 mmol) is dissolved into methanol (10 ml), to which is added a catalytic amount of 10% Pd-C. The mixture is stirred at room temperature for 5 h. After filtration through a pad of Celite, the filtrate is concentrated to dryness under reduced pressure to provide compound **9** in 95% yield. $[\alpha]_D$ −6.0° ($c = 1.0$, H_2O); ^{1}H NMR (500 MHz, D_2O) δ 7.30 (4H, s, aromatic), 4.50 (4H, d, $J = 7.8$ Hz, H-1 and H-1′), 3.94–4.06 (6H, m, H-4′, H-1a″, H-6a), 3.58–3.86 (24H, m, H-2′, H-3, H-3′, H-4, H-4′, H-5, H-5′, H-6b, H-6′, H-1″), 3.37 (2H, t, $J = 8.3$ Hz, H-2), 2.74 (4H, t, $J = 7.4$ Hz, H-3″), 1.97 (4H, t, $J = 6.8$ Hz, H-2″); ^{13}C NMR (125.7 MHz, D_2O) δ 139.3 and 128.2 (aromatic), 102.5 (C-1′), 102.4 (C-1), 78.0, 74.9, 74.3, 74.0, 72.4, 72.1, 70.5, 69.2 (C-1″), 68.1, 59.7 (C-6′), 59.4 (C-6), 30.3 and 30.1 (C-2″ and C-3″). ESI-MS calculated for $C_{36}H_{58}O_{22} + (Na^+)$: 865.3; found: 865.2.

*Trimer **11***

$[\alpha]_D$ −22.0° ($c = 1.0$, $CHCl_3$); ^{1}H NMR (500 MHz, $CDCl_3$) δ 7.42 (3H, s, aromatic), 5.31 (3H, dd, $J = 0.9$, 3.4 Hz, H-4′), 5.20 (3H, t, $J = 9.3$ Hz, H-3), 5.07 (3H, dd, $J = 7.9$ Hz, 10.4, H-2′), 4.91 (3H, dd, $J = 3.4$, 10.4 Hz, H-3′), 4.90 (3H, dd, $J = 7.9$, 9.4 Hz, H-2), 4.73 (3H, d, $J = 7.9$ Hz, H-1), 4.55 (3H, d, $J = 16.1$ Hz, H-1a″), 4.50 (3H, d, $J = 16.1$ Hz, H-1b″), 4.44–4.48 (6H, m, H-6a and H-1′), 4.02–4.11 (9H, m, H-6b and H-6′), 3.85 (3H, t, $J = 7.1$ Hz, H-5′), 3.80 (3H, t, $J = 9.2$ Hz, H-4), 3.61–3.66 (3H, m, H-5), 2.11, 2.08, 2.03, 2.02, 2.01, 2.00, 1.99, and 1.93 (63H, 7s, CH_3CO); ^{13}C NMR (125.7 MHz, $CDCl_3$) δ 170.7, 170.5, 170.4, 170.1, 170.0, 169.4 (CH_3*C*O), 135.1 and 123.5 (aromatic), 101.4 (C-1′), 98.6 (C-1), 85.5 and 85.3 (alkynyl), 76.4 (C-4), 73.1 (C-5), 73.0 (C-3), 71.7 (C-2), 71.3 (C-3′), 71.0 (C-5′), 69.4 (C-2′), 66.9 (C-4′), 62.1 (C-6), 61.1 (C-6′), 56.9 (C-1″), 21.2, 21.1, 21.0, and 20.8 (CH_3CO). ESI-MS calculated for $C_{93}H_{114}O_{54}$ + (NH_4^+): 2112.6; found: 2112.4.

*Compound **12***

Deprotected trimer **12** is obtained as described above for **4**: $[\alpha]_D$ −24.8° ($c = 1.0$, H_2O); ^{1}H NMR (500 MHz, D_2O) δ 7.62 (3H, s, aromatic), 4.73–4.78 (9H, m, H-1, H-1″), 4.53 (3H, $J = 7.8$ Hz, d, H-1′), 4.03 (3H, bd, $J = 11.2$ Hz, H-6a), 4.00 (3H, bd, $J = 3.2$ Hz, H-4′), 3.73–3.92 (24H, m, H-3, H-3′, H-4, H-4′, H-5′, H-6b, H-6′), 3.67 (3H, m, H-5), 3.63 (3H, dd, $J = 7.8$ and 9.5 Hz, H-2′), 3.47 (3H, m, H-2); ^{13}C NMR (125.7 MHz, D_2O) δ 134.4 and 122.3 (aromatic), 102.5 (C-1′), 100.3 (C-1), 85.4 and 84.6 (alkynyl), 77.8, 74.9, 74.4, 73.9, 72.2 (C-2), 72.1, 70.5 (C-2′), 68.1 (C-4′), 60.1 (C-6′), 59.6 (C-6), 56.7 (C-1″). ESI-MS calculated for $C_{51}H_{72}O_{33}$ + (NH_4^+): 1230.4; found: 1230.4.

*Compound **15***

Tetramer (*meta*) **15** is obtained by Sonogashira coupling of **2** and tetraiodide **13**: $[\alpha]_D$ − 24.5° ($c = 1.5$, $CHCl_3$); ^{1}H NMR (500 MHz, $CDCl_3$) δ 7.17–7.41 (16H, m, aromatic), 5.29 (4H, bd, $J = 3.1$ Hz, H-4′), 5.18 (4H, t, $J = 9.3$ Hz, H-3), 5.05 (4H, dd, $J = 7.9$, 10.4 Hz, H-2′), 4.91 (4H, dd, $J = 3.2$, 10.3 Hz, H-3′), 4.90 (4H, dd, $J = 7.9$, 9.4 Hz, H-2), 4.73 (4H, d, $J = 7.9$ Hz, H-1), 4.51 (4H, d, $J = 16.0$ Hz H-1a″), 4.50 (4H, d, $J = 16.0$ Hz, H-1b″), 4.44–4.48 (8H, m, H-6a and H-1′), 4.40 (8H, s, benzyl), 4.02–4.11 (12H, m, H-6b and H-6′), 3.83 (4H, t, $J = 6.7$ Hz, H-5′), 3.78 (4H, t, $J = 9.6$ Hz, H-4), 3.61–3.64 (4H, m, H-5), 3.49 [8H, s, C(CH_2OR)$_4$], 2.10, 2.05, 2.00, 1.99, 1.98, 1.96, and 1.91 (84H, 7s, CH_3CO), ^{13}C NMR (125.7 MHz, $CDCl_3$) δ 170.2, 170.0, 169.9, 169.7, 169.6, and 168.9 (CH_3*C*O), 139.0, 130.7, 130.3, 128.3, 127.6, and 121.9 (aromatic), 100.9 (C-1′), 98.1 (C-1), 86.9 and 83.3 (alkynyl), 76.0 (C-4), 72.7 (C-5 and C-3), 72.6 (benzylic), 71.4 and 70.9 (C-3′ and C-2), 70.6 (C-5′), 69.2 [C(CH_2OR)$_4$], 69.0 (C-2′), 66.5 (C-4′), 61.8 (C-6), 60.7 (C-6′), 56.7 (C-1″), 45.6 [*C*(CH_2OR)$_4$], 20.7, 20.6, 20.5, and 20.4 (CH_3CO). ESI-MS calculated for $C_{149}H_{180}O_{76}$ + (2NH_4^+): 1610.5; found: 1610.5.

*Compound **16***

Deprotected *meta*-tetramer **16** is obtained as described above for **4**: $[\alpha]_D$ −7.2° (*c*-1.0, H_2O), ^{1}H NMR (500 MHz, D_2O at 60°) δ 7.47–7.78 (16H, m, aromatic), 4.98–5.10 (12H, m, H-1, H-1″), 4.92 (4H, d, $J = 7.4$ Hz, H-1′), 4.65 (8H, bs, PhCH_2), 4.44 (4H, bs, H-4′), 4.06–4.36 (36H, m, H-2′, H-3, H-3′, H-4, H-5′, H-6, H-6′), 3.93 (4H, bs, H-5), 3.88 (4H, 8, H-2), 3.47 (4H, m, H-2), 3.29 [8H, bs, C(CH_2OR)$_4$]; ^{13}C NMR (125.7 MHz, D_2O) δ 138.7, 131.0, 130.2, 128.5, 127.6, and 121.9 (aromatic), 103.1 (C-1′), 101.1 (C-1), 86.8 and 85.0 (alkynyl), 78.7, 75.3, 74.8 (C-5), 74.5, 72.8, 72.3 (PhCH_2), 71.0, 68.8 [C(CH_2OR)$_4$], 68.6 (C-4′), 61.0 (C-6′), 60.4 (C-6), 57.2 (C-1″), 45.9 [C(CH$_2$OR)$_4$]. ESI-MS calculated for $C_{93}H_{124}O_{48}$ + (Na^+): 2031.7; found: 2031.4.

*Compound **17***

Tetramer (*para*) **17** is obtained by Sonogashira coupling of **2** and tetraiodide **14**: $[\alpha]_D$ − 22.7° ($c = 1.2$, $CHCl_3$); ^{1}H NMR (500 MHz, $CDCl_3$) δ 7.32 (8H, d, $J = 8.1$ Hz, aromatic), 7.19 (8H, d, $J = 8.1$ Hz, aromatic), 5.30 (4H, bd, $J = 2.9$ Hz, H-4′), 5.19 (4H, t, $J = 9.3$ Hz, H-3), 5.06 (4H, dd, $J = 7.8$ Hz, 10.4, H-2′), 4.91 (4H, dd, $J = 3.4$, 10.4 Hz, H-3′), 4.89 (4H, dd, $J = 7.9$, 9.4 Hz, H-2), 4.76 (4H, d, $J = 7.9$ Hz, H-1), 4.53 (4H, d, $J = 16.0$ Hz, H-1a″), 4.50 (4H, d, $J = 16.0$ Hz, H-1b″), 4.48 (4H, m, H-6a), 4.46 (4H, $J = 7.8$ Hz, d, H-1′), 4.40 (8H, s, benzyl), 4.02–4.11 (12H, m, H-6b and H-6′), 3.84 (4H, t, $J = 6.7$ Hz, H-5′), 3.80 (4H, t, $J = 9.6$ Hz, H-4), 3.60–3.65 (4H, m, H-5), 3.52 [8H, s, C(CH_2OR)$_4$], 2.10, 2.07, 2.01, 2.00, 1.99, 1.98, and 1.91 (84H, 7s, CH_3CO); ^{13}C NMR (125.7 MHz, $CDCl_3$) δ 170.2, 170.0, 169.9, 169.7, 169.6, and 168.9 (CH$_3$$C$O), 139.5, 131.6, 127.0, and 121.0 (aromatic), 100.9 (C-1′), 98.1 (C-1), 86.9 and 83.3 (alkynyl), 76.0 (C-4), 72.7 (C-5 and C-3), 72.6 (benzylic), 71.4 and 70.9 (C-3′ and C-2), 70.6 (C-5′), 69.2 [C(CH_2OR)$_4$], 69.1 (C-2′), 66.6 (C-4′), 61.8 (C-6), 60.7 (C-6′), 56.8 (C-1″), 45.6 [C(CH$_2$OR)$_4$], 20.7, 20.6, 20.5, and 20.4 (CH_3CO). ESI-MS calculated for $C_{149}H_{180}O_{76}$ + ($2NH_4^+$): 1610.5; found: 1610.7.

*Compound **18***

Deprotected *para*-tetramer **18** is obtained as described above for **4**: $[\alpha]_D$ 60.0° ($c = 0.5$, H_2O); ^{1}H NMR (500 MHz, D_2O) δ 7.31 (8H, bs, aromatic), 7.00 (8H, bs, aromatic), 4.62–4.78 (12H, m, H-1, H-1″), 4.52 (4H, $J = 7.1$ Hz, d, H-1′), 4.01 (8H, bs, PhCH_2), 3.63–3.91 (44H, m, H-2′, H-3, H-3′, H-4, H-4′, H-5, H-5′, H-6, H-6′), 3.47 (4H, m, H-2), 3.29 [8H, bs, C(CH_2OR)$_4$]; ^{13}C NMR (125.7 MHz, D_2O) δ 138.5, 131.3, 126.8, and 120.6 (aromatic), 102.5 (C-1′), 100.6 (C-1), 86.3 and 84.5 (alkynyl), 77.8, 74.9, 74.2, 74.0, 72.2, 71.7

(Ph*C*H_2), 70.5 (C-2′), 68.1, 65.5, 60.6 (C-6′), 59.7 (C-6), 56.7 (C-1″), 45.9 [*C*(CH_2OR)$_4$]. ESI-MS calculated for $C_{93}H_{124}O_{48} + (Na^+)$: 2031.7; found: 2031.5.

Comments

The compounds described in this article have been evaluated for their relative binding properties with the chimeric recombinant murine galectin-3 by quantitative immunoprecipitation.[20] Although galectin-3 is known to exist as a monomer in solution, the above-described experiments show that it tends to oligomerize in the presence of multivalent clusters. Nonorganized cross-linked lattices are obtained in all cases.[21] Isothermal titration microcalorimetry cannot be achieved because of the insoluble complex formation.

[20] N. Ahmad, H.-J. Gabius, S. Sabesan, R. Roy, F. Macaluso, C. F. Brewer, and B. Liu, 223rd ACS National Meeting, Orlando, FL, April 7–11, 2002 [abstract 28].

[21] C. F. Brewer, *Trends Glycosci. Glycotechnol.* **9,** 155 (1997).

[17] Multivalent Breast Cancer T-Antigen Markers Scaffolded onto PAMAM Dendrimers

By RENÉ ROY and MYUNG-GI BAEK

Introduction

Glycodendrimers are relatively small synthetic biomacromolecules that play key functions in a wide range of biochemical processes involving multivalent carbohydrate–protein interactions.[1–3] They are made of simple core molecules from which several "arms" are extended by repetitive chemical cycles. Commercially available poly(amidoamine) dendrimers (PAMAMs)[4] with surface amine functionalities constitute valuable candidates onto which can be attached various carbohydrates. Some earlier versions of glycodendrimers scaffolded onto these PAMAMs were synthesized using oxime formation,[5] sugar lactones,[6] and isothiocyanate chemistry.[7] Although this conjugation chemistry was efficient, the reactions present some drawbacks. For instance, the first two processes provide structures in which the reducing sugars are sacrificed. Alternatively, the thiourea linkages are often accompanied by *cis/trans* stereoisomers, thus complicating exact structural characterization. We provide herein yet another efficient procedure, which relies on the direct amidation of sugar derivatives containing carboxyl groups onto amine-ending PAMAMs of generations 0 to 3 that contain between 4 (G0) to 32 (G3) amine groups (Scheme 1).

As precursors for the carboxylated sugar haptens, allyl glycoside (**1**) **1** is suitable as it can be readily converted to an acid moiety by a radical reaction involving 3-mercaptopropionic acid (Scheme 2).[8,9] The immunodominant T-antigen [Thomsen-Friedenreich, β-Gal-(1,3)-α-D-GalNAc-OR] has been selected to demonstrate the increased binding affinities of this family of glycodendrimers for both a plant lectin (*Arachis hypogaea*) and a

[1] R. Roy, *Polym. News* **21,** 226 (1996).
[2] N. Röckendorf and T. K. Lindhorst, *Top. Curr. Chem.* **217,** 201 (2001).
[3] J. J. Lundquist and E. J. Toone, *Chem. Rev.* **102,** 555 (2002).
[4] D. A. Tomalia and H. D. Durst, *Top. Curr. Chem.* **165,** 193 (1993).
[5] J. P. Mitchell, K. D. Roberts, J. Langley, F. Koentgen, and J. N. Lambert, *Bioorg. Med. Chem. Lett.* **9,** 2785 (1999).
[6] K. Aoi, K. Itoh, and M. Okada, *Macromolecules* **28,** 5391 (1995).
[7] D. Zanini and R. Roy, *J. Org. Chem.* **63,** 3486 (1998).
[8] P. B. van Seeventer, J. A. L. M. van Dorst, J. F. Siemerink, J. P. Kamerling, and J. F. G. Vliegenthart, *Carbohydr. Res.* **300,** 369 (1997).
[9] R. Roy, M.-G. Baek, and K. Rittenhouse-Olson, *J. Am. Chem. Soc.* **123,** 1809 (2001).

0076-6879/03 $35.00

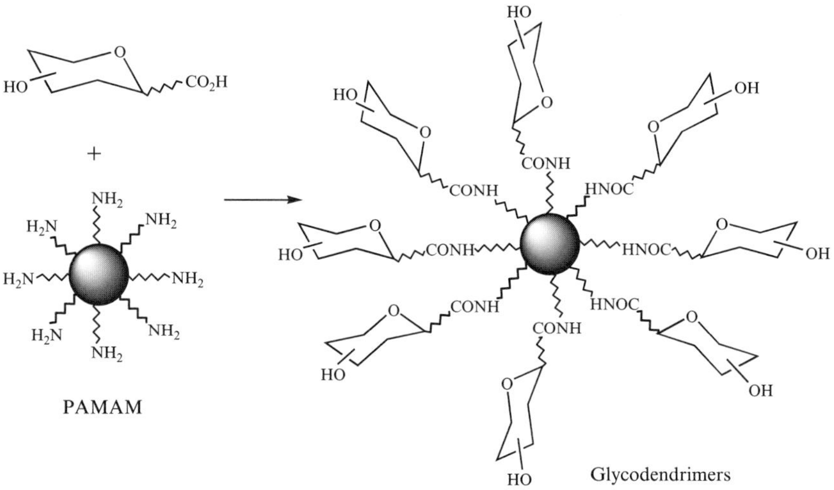

SCHEME 1. General strategy for the amidation of a carboxylated carbohydrate derivative and poly(amido)amine (PAMAM) dendrimers of various generation.

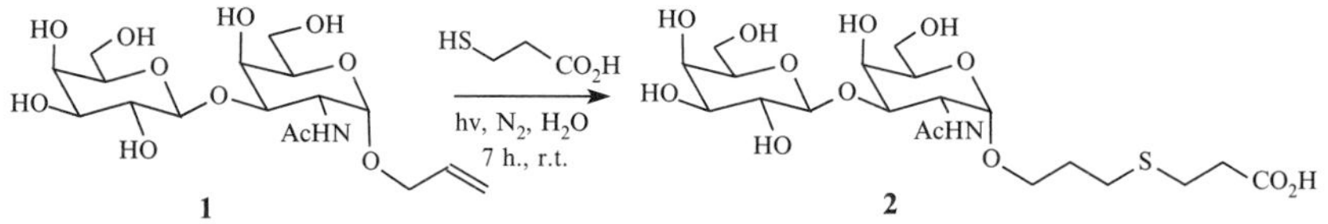

SCHEME 2. Transformation of allyl glycoside (**1**) into an extended acid functionality (**2**) by radical reaction involving 3-mercaptopropionic acid.

mouse monoclonal antibody (IgG) that was generated from a T-antigen neoglycoprotein.[10]

General Procedure

T-antigen (T-Ag)-containing glycoPAMAM dendrimers are prepared by treatment of T-antigen acid derivative **2** with commercially available amine-ending PAMAMs (Sigma-Aldrich St. Lousie, MO) of various generations (**3–6**) containing 4 to 32 amine groups, respectively. As commercial samples of PAMAMs are provided in methanol, the solutions are initially dried under vacuum and the residues are dissolved in dimethyl sulfoxide (DMSO). To a slight excess of **2** (1.1 equivalents per amino group), the

[10] K. Rittenhouse-Diakun, Z. Xia, D. Pickhardt, S. Morey, M.-G. Baek, and R. Roy, *Hybridoma* **17,** 165 (1998).

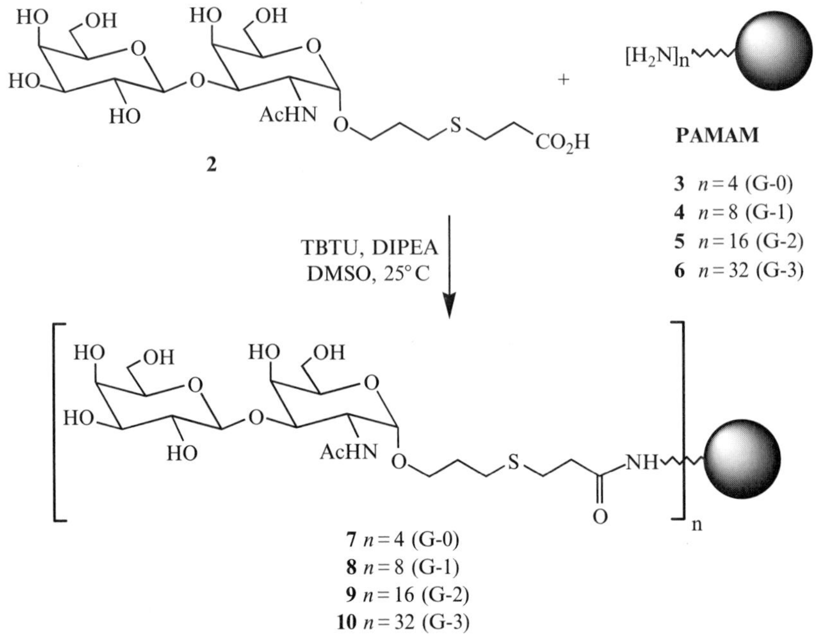

PAMAM

3 $n = 4$ (G-0)
4 $n = 8$ (G-1)
5 $n = 16$ (G-2)
6 $n = 32$ (G-3)

TBTU, DIPEA
DMSO, 25° C

7 $n = 4$ (G-0)
8 $n = 8$ (G-1)
9 $n = 16$ (G-2)
10 $n = 32$ (G-3)

SCHEME 3. Amidation of T-antigen-containing acid functionality **2** and amine-ending PAMAMs, using peptide-coupling reagent TBTU.

popular peptide-coupling reagent 1-[bis(dimethylamino)methylene]-1*H*-benzotriazolium tetrafluoroborate (TBTU) (1.2 equivalents per acid) and diisopropylethylamine (DIPEA) are successively added and the resulting solutions are stirred overnight at room temperature (Scheme 3). The solutions are lyophilized to dryness and the residues are purified by either gel-permeation chromatography (P2 or P4 column, H_2O) or dialysis (H_2O; molecular weight cutoff, 2000) to afford glycoPAMAM conjugates **7–10** having 4, 8, 16, and 32 T-Ag residues in 73, 81, 99, and 79% yield, respectively. Conjugation is judged complete on the basis of negative ninhydrin tests and high-field ^{1}H nuclear magnetic resonance (NMR) spectral data (D_2O).

The efficiency of these glycoPAMAMs for protein-binding interactions is evaluated with peanut lectin, a phytohemagglutinin known for its high selectivity toward the T-antigen,[11] and with a mouse monoclonal antibody generated from an analog of **2**.[10] Initially, the cross-linking ability of glycoPAMAMs **7–10** is established by a kinetic turbidimetric experiment (Fig. 1). The time course for the microprecipitation analysis clearly shows

[11] R. Ravishankar, M. Ravindran, K. Suguna, A. Surolia, and M. Vijayan, *Curr. Sci.* **72,** 855 (1997).

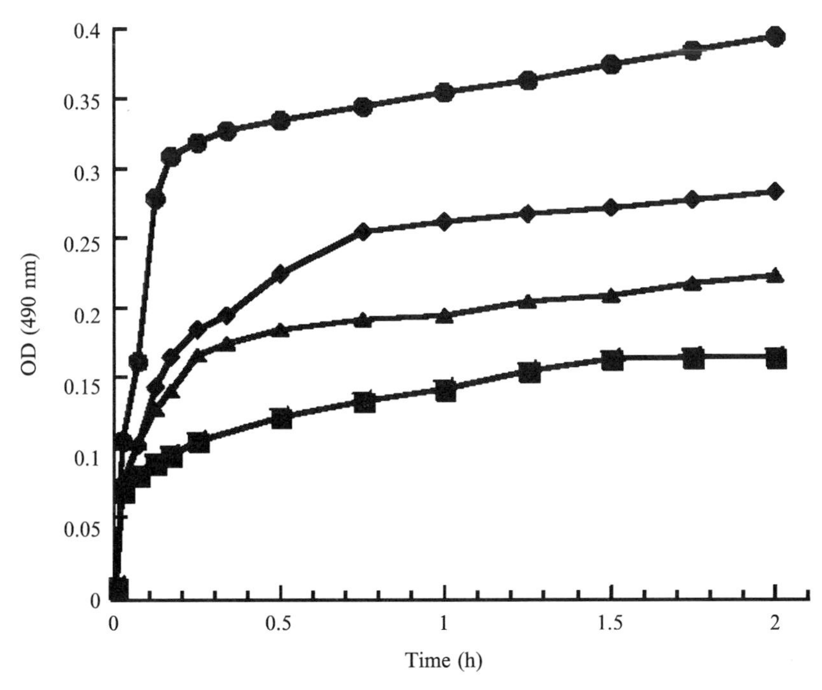

FIG. 1. Turbidimetric analysis of glycoPAMAM dendrimers **7** (■), **8** (▲), **9** (◆), and **10** (●) with peanut lectin from *Arachis hypogaea.*

that insoluble cross-linked lattices are formed rapidly in all four cases and that a time-dependent and almost quantitative precipitation occurs as a function of dendrimer valency. The 32-mer T-antigen **10** has proved to be the most potent antigen in this assay.

The microtiter plate-coating abilities of glycoPAMAMs **7–10** is next evaluated with both protein receptors. This assay demonstrates that the above-described glycodendrimers retain sufficient lipophilic behavior for their adsorption to plastic wells. As seen in Figs. 2 and 3, there is a graded ability for adsorption as a function of dendrimer valency; again, compound **10** has proved to be the most potent.[12,13] Goat anti-mouse IgG labeled with horseradish peroxidase (HRP) is used for monoclonal antibody (mAb) detection, whereas *Arachis hypogaea*–HRP is used for the lectin assay.

The relative inhibitory capacity of the glycodendrimers is next substantiated by competitive double sandwich inhibition in which conjugates **7–10**

[12] M.-G. Baek and R. Roy, *Bioorg. Med. Chem. Lett.* **10,** 11 (2002).

[13] M. G. Baek, K. Rittenhouse-Olson, and R. Roy, *Chem. Commun.* 257 (2001).

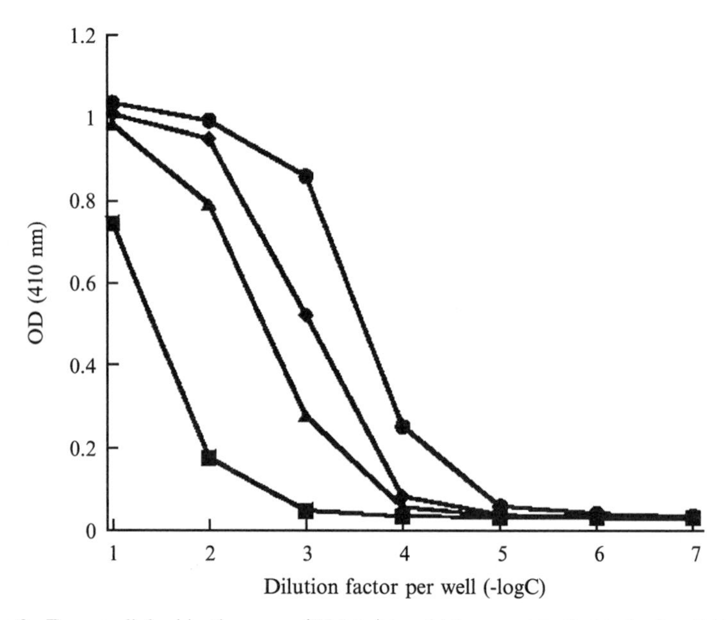

FIG. 2. Enzyme-linked lectin assays (ELLAs) in which peanut lectin binds glycoPAMAM dendrimers **7** (■), **8** (▲), **9** (◆), and **10** (●) coated on microtites plates.

are employed as inhibitors.[14] T-antigen–*co*-polyacrylamide (T-antigen–acrylamide,1:10)[15] is used as a coating antigen (Fig. 4). To this end, the T-antigen polymer [1 μg/well, 0.85 nmol of T-antigen in phosphate-buffered saline (PBS)] is deposited in microtiter plates. In separate plates (Nunclon Delta; Nalge Nunc, Naperville, IL), mouse mAb IgG [50 μl/well, 0.25 μmol in PBS–Tween (PBST)] and glycoPAMAM inhibitors (50 μl/well in PBST with various concentrations from 275 to 1.07 nmol/well) are preincubated. After blocking the microtiter plates with bovine serum albumin (BSA), the wells are filled with antibody–glycoPAMAM inhibitor mixtures (100 μl/well) and the mixtures are then incubated for 1 h at 37°. For the quantitative detection of antibodies, goat anti-mouse mAb–HRP is then added, as described above. On the basis of bulk conjugates, the glycoPAMAMs with highest carbohydrate density **(7)** exhibit the strongest inhibition. Conjugates **7–10** show 50% inhibitory concentration (IC_{50}) values of 5.0, 2.4, 1.4, and 0.6 nmol, respectively, whereas that of monomeric T-antigen

[14] R. Roy and M.-G. Baek, *Rev. Mol. Biotechnol.* **90,** 291 (2002).
[15] M.-G. Baek and R. Roy, *Biomacromolecules* **1,** 768 (2000).

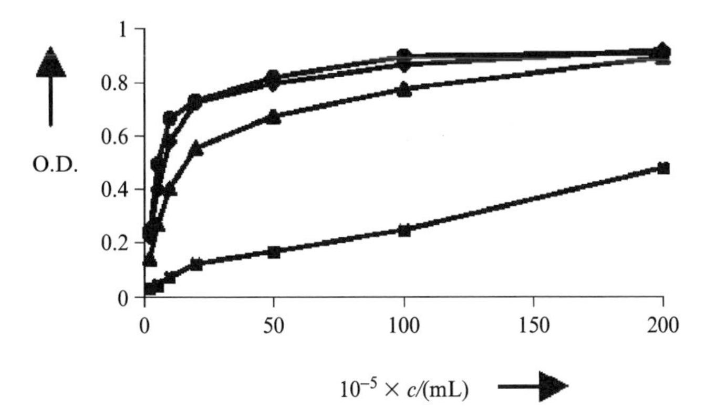

FIG. 3. Enzyme-linked immunosorbent assays (ELISAs) of T-antigen-glycoPAMAMs **7** (■), **8** (▲), **9** (◆), and **10** (●) with mouse monoclonal IgG antibody, goat anti-mouse IgG–horseradish peroxidase conjugate, and ABTS–H_2O_2 as enzyme substrate. OD optical density at 410 nm; ABTS,2,2′-azinobis(3-ethylbenzothiazoline-6-sulfonic acid); *c*; relative concentration.

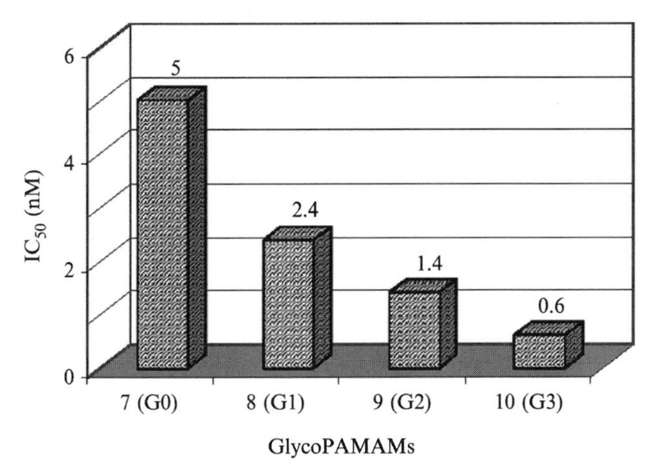

FIG. 4. Relative inhibitory properties of glycoPAMAMs **7–10** toward mouse monoclonal IgG antibody binding to adsorbed T-antigen copolymer, using ELISA. Detection was made with goat anti-mouse IgG–horseradish peroxidase conjugate and ABTS–H_2O_2 as enzyme substrate.

1 is 2.3 μmol. The inhibitory abilities of these conjugates are thus 460, 960, 1700, and 3800 times higher than that of monomer **1**. Interestingly, when expressed on a per-saccharide basis, all dendrimers are equivalent to one another (115-fold better than **1**), thus supporting previous observations that low-density glycoPAMAMs are efficient inhibitors.

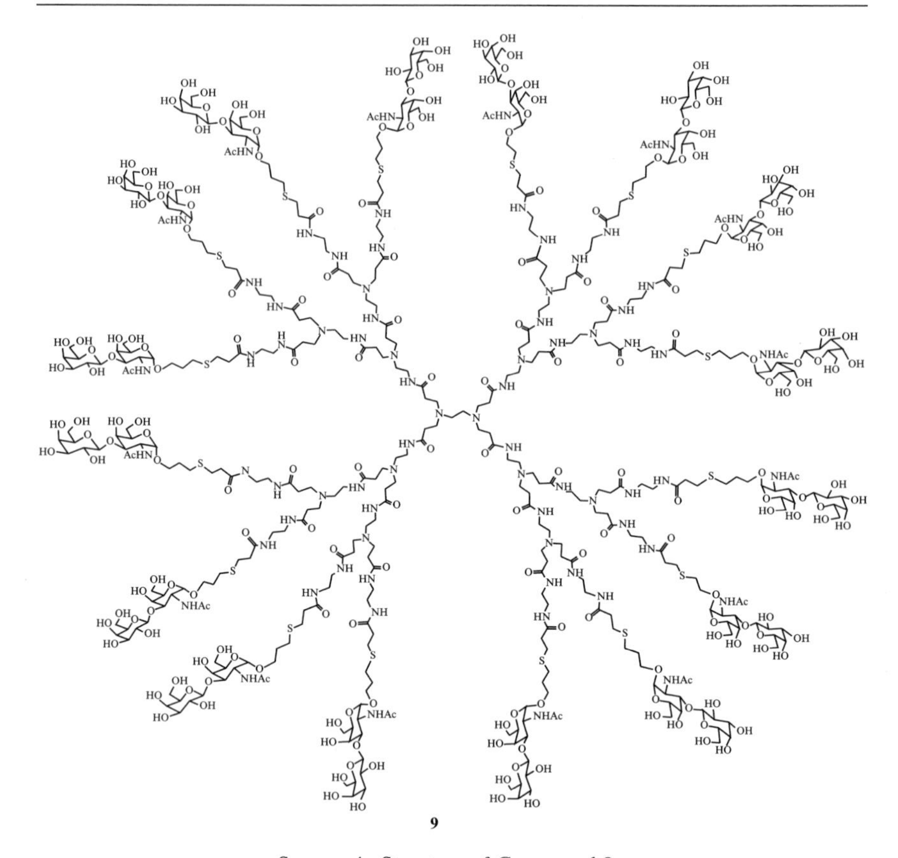

SCHEME 4. Structure of Compound **9**.

Materials and Methods

*3-(2-Carboxyethylthio)propyl β-D-galactopyranosyl-(1→3)-2-acetamido-2-deoxy-α-D-galactopyranoside (**2**)*

A solution of allyl β-D-Gal-(1→3)-α-D-GalNAc (**1**)[9] (100 mg, 0.24 mmol) and 3-mercaptopropionic acid (21 μl, 1 equivalent) in deoxygenated distilled water (2.5 ml) is irradiated (254 nm) for 7 h under an N_2 atmosphere. The reaction mixture is then loaded onto a column (Amberlite IRA 400 OH^-) of anion-exchange resin that has been washed with water. The eluent is then gradually changed to aqueous acetic acid with increasing acetic acid. The fractions containing **2** are collected and evaporated under reduced pressure followed by multiple coevaporation with ethanol. A small amount of water is added and lyophilized to afford a white solid in 83% yield (105.9 mg, 2.0 mmol): mp 90.0–92.5°; $[\alpha]_D$ + 76.0° ($c = 1$, H_2O); R_f 0.33 [$CHCl_3$/methanol/H_2O, 10:9:1 (v/v/v)]; ^{1}H NMR (D_2O) δ 4.96 (d, 1 H, $J_{12} = 3.7$ Hz, H-1), 4.54 (d, 1 H, $J_{1'2'} = 7.7$ Hz, H-1′), 3.89–3.79 (m, 5 H, H-6, H-6′, 1 H of CH_2), 3.65–3.57 (m, 2 H, H-2′, 1 H of CH_2), 2.89 (t, 2 H, $J = 7.5$ Hz, CH_2), 2.80–2.73 (m, 4 H, 2 CH_2), 2.10 (s, 3 H, NAc), 2.01–1.96 (m, 2 H, CH_2); ^{13}C NMR (D_2O) δ 104.2, 96.7 (Cl, Cl′); Calculated for $C_{20}H_{35}O_{13}NS$ (529.4), (+) fast atom bombardinent mass spectrometry FAB-MS (glycerol) m/z: 530.3 (M + 1).

Synthesis of GlycoPAMAM Dendrimers

Tetrameric GlycoPAMAM Dendrimer 7. PAMAM (**3**, $n = 4$) (7.2 mg, 13.9 μmol) and T-antigen derivative **2** (32.4 mg, 61.2 μmol) are dissolved in DMSO (6 ml). TBTU (23.6 mg, 73.5 μmol) and DIPEA are then added. After purification on a P2 column, using water as eluent, glycoPAMAM dendrimer **7** is obtained in 73% yield (26 mg, 10.2 μmol). ^{1}H NMR (D_2O) δ 4.96 (d, 4 H, $J_{12} = 3.8$ Hz, H-1), 4.55 (d, 4 H, $J_{1'2'} = 7.8$ Hz, H-1′), 3.89–3.76 (m, 20 H, H-6, H-6′, CH_2), 3.66–3.58 (m, 8 H, CH_2, H-2′), 3.41 (broad s, 16 H, CH_2, CH_2), 3.29 (broad t, 8 H, CH_2), 2.89 (t, 8 H, $J = 6.9$ Hz, Hz, CH_2), 2.80 (s, 4 H, CH_2), 2.77 (t, 8 H, $J = 7.1$ Hz, CH_2), 2.72 (t, 8 H, $J = 6.7$ Hz, CH_2), 2.62 (t, 8 H, $J = 6.9$ Hz, CH_2), 2.11 (s, 12 H, NAc), 2.01–1.95 (m, 8 H, CH_2); ^{13}C NMR (D_2O) δ 104.2 (4 C), 96.8 (4 C).

Octameric GlycoPAMAM Dendrimer 8. PAMAM (**4**, $n = 8$) (8.37 mg, 5.85 μmol) and T-antigen derivative (**2**, 27.3 mg, 51.6 μmol) are dissolved in DMSO (5 ml) to which TBTU (20 mg, 62.3 μmol) and DIPEA are added. After purification on a P4 column, glycoPAMAM dendrimer **8** is obtained in 81% yield (26.1 mg, 4.7 μmol). ^{1}H NMR (D_2O) δ 4.96 (d, 8 H, $J_{12} = 3.7$ Hz, H-1), 4.55 (d, 8 H, $J_{1'2'} = 7.8$ Hz, H-1′), 3.89–3.77 (m, 40 H, H-6, H-6′, CH_2), 3.74–3.69 (m, 24 H, H-5′, H-3′, CH_2), 3.66–3.52 (m, 24 H, H-2′, CH_2), 3.47 (broad t, 8 H, CH_2), 3.41 (broad s, 32 H, CH_2), 3.28 (broad t, 16 H, CH_2), 2.90–2.83 (m, 24 H, CH_2), 2.80 (s, 4 H, CH_2), 2.79–2.75 (m, 32 H, CH_2), 2.62 (t, 16 H, $J = 6.8$ Hz, CH_2), 2.10 (s, 24 H, NAc), 1.99 (m, 16 H, CH_2); ^{13}C NMR (D_2O) δ 104.2 (8 C), 96.8 (8 C).

16-Mer GlycoPAMAM Dendrimer 9. PAMAM (**5**, $n = 16$) (11.5 mg, 3.52 μmol) and T-antigen derivative (**2**, 32.8 mg, 62.0 μmol) are dissolved in DMSO (10 ml) and TBTU (24.0 mg, 74.7 μmol) and DIPEA are added. After purification by dialysis, glycoPAMAM dendrimer **9** is obtained in 99% yield (40 mg, 3.5 μmol). SF-1 ^{1}H-NMR (D_2O) δ 4.96 (d, 16 H, $J_{12} = 3.7$ Hz, H-1), 4.56 (d, 16 H, $J_{1'2'} = 7.8$ Hz, H-1′), 3.89–3.76 (m, 80 H, H-6, H-6′, CH_2), 3.65–3.58 (m, 64 H, H-2′, CH_2), 3.41–3.30 (m, 120 H, CH_2) 3.28–3.06 (m, 80 H, CH_2), 2.89 (t, 32 H, $J = 7.0$ Hz, CH_2), 2.79–2.75 (m, 92

H, CH_2), 2.62 (t, 32 H, $J = 7.0$ Hz, CH_2), 2.10 (s, 48 H, NAc), 2.09–1.95 (m, 32 H, CH_2); ^{13}C NMR (D_2O) δ 104.2 (16 C), 102.8 (16 C).

*32-Mer GlycoPAMAM Dendrimer **10**.* PAMAM (**6**, $n = 32$) (12.4 mg, 1.80 μmol) and T-antigen derivative (**2**, 33.5 mg, 63.4 μmol) are dissolved in DMSO (8 ml) and TBTU (24.4 mg, 76.0 μmol) and DIPEA are added. After purification by dialysis in water, glycoPAMAM dendrimer **10** is obtained in 79% yield (32.2 mg, 1.38 μmol). ^{1}H NMR (D_2O) δ 4.96 (d, 32 H, $J_{12} = 3.7$ Hz, H-1), 4.56 (d, 32 H, $J_{12'} = 7.7$ Hz, H-1′), 3.95–3.79 (m, 160 H, H-6, H-6′, CH_2), 3.65–3.58 (m, 64 H, H-2′, CH_2), 3.40 (broad s, 192 H, CH_2), 2.92 (broad s, 112 H, CH_2), 2.88 (t, 64 H, $J = 7.0$ Hz, CH_2), 2.77 (m, 128 H, CH_2), 2.62 (t, 64 H, $J = 7.0$ Hz, CH_2), 2.56–2.51 (broad s, 112 H, CH_2), 2.10 (s, 96 H, NAc), 2.00–1.39 (m, 64 H, CH_2); ^{13}C-NMR (D_2O) δ 104.20 (32 C1′), 96.8 (32 C1). Analysis calculated for $C_{942}H_{1664}O_{444}N_{154}S_{32}$ (23,278.7: C, 48.60; H, 7.20; N, 9.27. Found: C, 48.85; H, 7.21; N, 8.79).

*Turbidimetric Analysis between Peanut Lectin and GlycoPAMAMs **7–10**.* Turbidimetric experiments are performed in Linbro (Titertek) microtitration plates (ICN Biomedicals, Irvine, CA) in which stock lectin solutions (50 μl/well) prepared from peanut lectin (2 mg/ml in PBS; Sigma) are mixed with stock solutions (50 μl) of glycodendrimers ***7–10*** (11 nmol/well of T-antigen for all dendrimers) and incubated at room temperature for 3 h. The turbidity of the solutions is monitored by reading the optical density (OD) at 490 nm at regular time intervals until no further noticeable changes are observed. Each test is performed in triplicate.

*Competitive Double-Sandwich Inhibition ELISA Using Mouse mAb IgG_3 and T-Antigen Dendrimers **7–10** as Inhibitors*

Linbro (Titertek) microtiter plates are coated overnight with T-antigen containing *co*-polyacrylamide[15] at 100 μl of a polymer stock solution (10 μg/ml in 0.01 *M* phosphate buffer, pH 7.3) at room temperature. Each well contains polymer at 1 μg/well, which corresponds to 0.36 μg (0.85 nmol) of T-antigen. The wells are then washed three times with 400 μl of PBST [0.01 *M* phosphate buffer, pH 7.3, containing 0.05% (v/v) Tween 20]. BSA solution (1% in PBS, 150 μl/well) is added to each well and incubated for 1 h at 37°. At the same time, 50 μl of mouse monoclonal IgG antibody solution in PBST (10-fold dilution of ascitic fluid, 0.25 μmol/50 μl) and 50 μl of inhibitor solution in PBST (with various concentrations of T-antigen, from 138 to 2.15 nmol/well by 2-fold serial dilution) are mixed in Nunclon (Delta) microtiter plates and preincubated for 1 h at 37°. After excess BSA is washed out with PBST, each well is filled with preincubated mouse IgG MAb–inhibitor solution (100 μl/well) and incubated again for 1 h at 37°. The wells are washed with PBST as described above and then filled with 100 μl of goat-anti-mouse IgG in PBST (1000-fold dilution of ascitic fluid) followed by incubation for 1 h at 37°. The wells are washed with PBST and 50 μl of 2,2′-azinobis(3-ethylbenzothiazoline-6-sulfonic acid) diammonium salt (ABTS) [1 mg/4 ml of citrate–phosphate buffer (0.2 *M*, pH 4.0 with 0.015% H_2O_2)] is added. The reaction is stopped after 20 min with 1 *M* aqueous sulfuric acid solution (50 μl/well). The optical density is measured at 410 nm. Percent inhibitions are calculated as follows:

$$\% \text{ inhibition} = [A_{(\text{no inhibitor})} - A_{(\text{with inhibitor})}/A_{(\text{no inhibitor})}] \times 100$$

IC_{50} values are calculated as the concentration required for 50% inhibition of the coating antigen. All tests are performed in triplicate.

[18] Constructing Azide-Labeled Cell Surfaces Using Polysaccharide Biosynthetic Pathways

By SARAH J. LUCHANSKY, HOWARD C. HANG, ELIANA SAXON, JOCELYN R. GRUNWELL, CHONG YU, DANIELLE H. DUBE, and CAROLYN R. BERTOZZI

Introduction

Cellular perceptions and responses are often initiated by binding events on the cell surface. Carbohydrates have been recognized as central players in these recognition events. For example, B-cell activation and leukocyte homing are processes reliant on carbohydrate epitopes.[1,2] Glycosylation also modulates how efficiently viruses such as human immunodeficiency virus type 1 (HIV-1) and influenza A infect human cells.[3,4]

Glycosylation-mediated recognition events have been studied using targeted gene disruption in mice, mutant cell lines selected for specific glycosylation phenotypes, and small molecule inhibitors.[5–8] Targeted disruption of glycosylation genes has provided significant insight into the function and diversity of carbohydrates. However, in some cases

[1] P. R. Crocker and A. Varki, *Immunology* **103,** 137 (2001).
[2] D. Vestweber and J. E. Blanks, *Physiol. Rev.* **79,** 181 (1999).
[3] A. Varki, *Glycobiology* **3,** 97 (1993).
[4] R. A. Dwek, *Chem. Rev.* **96,** 683 (1996).
[5] T. Hennet and L. G. Ellies, *Biochim. Biophys. Acta* **1473,** 123 (1999).
[6] T. Hennet, D. Chui, J. C. Paulson, and J. D. Marth, *Proc. Natl. Acad. Sci. USA* **95,** 4504 (1998).

0076-6879/03 $35.00

embryonic lethality or upregulation of compensating pathways can interfere with interpretation of the resultant phenotype.[5,9] Cell lines that have been selected on the basis of desired cell surface carbohydrate epitopes are usually not well characterized at the genetic level, and thus the molecular nature of the defect is often unknown. In addition, the generation of a knockout mouse or deficient cell line is time-consuming and not readily reversible, and tissue-specific control can be difficult to exercise in mice. Small molecule inhibitors of glycosylation allow temporal control over the desired phenotype, but there are relatively few that can affect the synthesis of a specific carbohydrate epitope without disrupting the production of an entire class of polysaccharides. The conventional glycosylation inhibitors, such as tunicamycin, produce major perturbations and are some times toxic.

An alternative approach to studying the biology of carbohydrates, pioneered by Reutter and co-workers, utilizes the conversion of unnatural biosynthetic precursors to unnatural cell surface polysaccharides with altered biological function. For example, *N*-acetylmannosamine (ManNAc) derivatives with modified *N*-acyl chains are converted by cells, via the biosynthetic pathway depicted in Fig. 1, to unnatural sialic acids that are displayed on the cell surface.[10,11] These modifications are typically less disruptive to cellular metabolism than small molecule glycosylation inhibitors yet are still reversible and allow temporal control. The ability of unnatural sialic acid residues to modulate receptor-binding events was demonstrated by the inhibition of sialic acid-dependent influenza A virus infection *in vitro*.[12] More recently, sialic acid analogs were shown to be directly taken up by mammalian cells[13,14] and the bacterial

[7] P. Stanley, "*Use of Mammalian Cell Mutants to Study the Functions of N- and O-Linked Glycosylation*" (D. D. Roberts and R. P. Mecham, eds.), p. 181 Academic Press, San Diego, CA, 1993.

[8] M. Alfalah, R. Jacob, U. Preuss, K. P. Zimmer, H. Naim, and H. Y. Naim, *Curr. Biol.* **9,** 593 (1999).

[9] M. Schwarzkopf, K. P. Knobeloch, E. Rohde, S. Hinderlich, N. Wiechens, L. Lucka, I. Horak, W. Reutter, and R. Horstkorte, *Proc. Natl. Acad. Sci. USA* **99,** 5267 (2002).

[10] O. T. Keppler, R. Horstkorte, M. Pawlita, C. Schmidt, and W. Reutter, *Glycobiology* **11,** 11R (2001).

[11] H. Kayser, R. Zeitler, C. Kannicht, D. Grunow, R. Nuck, and W. Reutter, *J. Biol. Chem.* **267,** 16934 (1992).

[12] O. T. Keppler, M. Herrmann, C. W. von der Lieth, P. Stehling, W. Reutter, and M. Pawlita, *Biochem. Biophys. Res. Commun.* **253,** 437 (1998).

[13] C. Oetke, R. Brossmer, L. R. Mantey, S. Hinderlich, R. Isecke, W. Reutter, O. T. Keppler, and M. Pawlita, *J. Biol. Chem.* **277,** 6688 (2002).

[14] C. Oetke, S. Hinderlich, R. Brossmer, W. Reutter, M. Pawlita, and O. T. Keppler, *Eur. J. Biochem.* **268,** 4553 (2001).

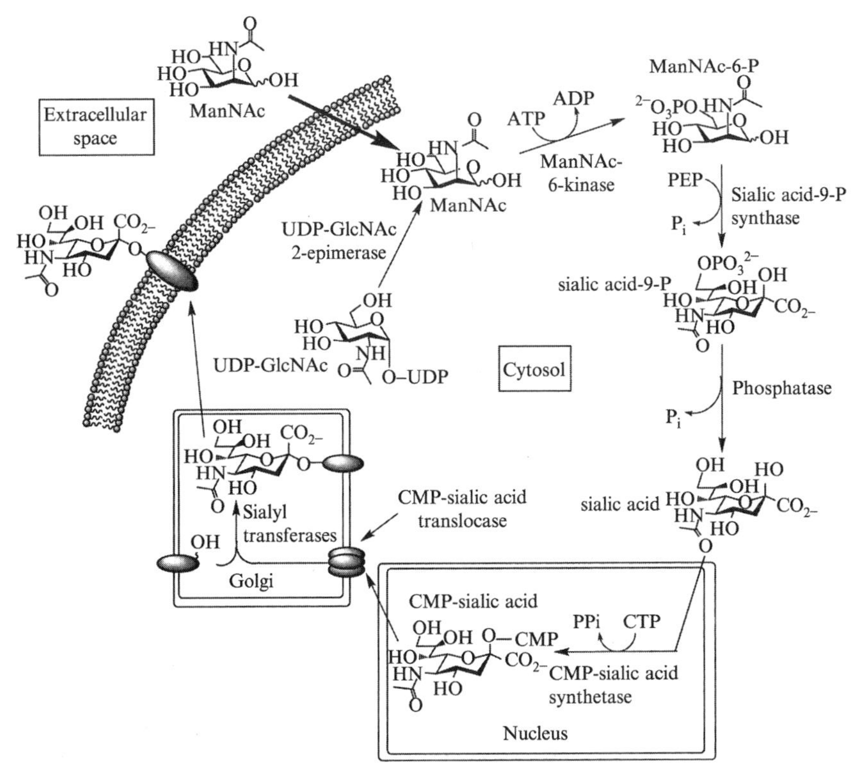

FIG. 1. Sialic acid biosynthesis. *N*-Acetylmannosamine (ManNAc) is taken up by the cell or synthesized from UDP-*N*-acetylglucosamine (UDP-GlcNAc) by UDP-GlcNAc 2-epimerase. ManNAc is phosphorylated by ManNAc-6-kinase to yield ManNAc-6-phosphate (ManNAc-6-P). ManNAc-6-P is subsequently condensed with phosphoenolpyruvate (PEP) to yield sialic acid-9-phosphate (sialic acid-9-P) and inorganic phosphate (P_i) in a reaction catalyzed by sialic acid-9-P synthase. Dephosphorylation of sialic acid-9-P by an unknown phosphatase and transport to the nucleus enable activation to CMP-sialic acid by CMP-sialic acid synthetase. After transport into the Golgi compartment, CMP-sialic acid is utilized by the sialyltransferases that append the residue to glycoconjugates ultimately destined for the cell surface or secretion.

pathogen *Haemophilus ducreyi*,[15] and incorporated into cell surface glycoconjugates.

We have extended this methodology by introducing novel substrates that contain chemical handles and provide reactive functionality to

[15] S. Goon, B. Schilling, M. V. Tullius, B. W. Gibson, and C. R. Bertozzi, *Proc. Natl. Acad. Sci.* **100,** 3089 (2003).

mammalian cell surface glycoproteins. For example, an unnatural analog of ManNAc containing a ketone group, *N*-levulinoylmannosamine (ManLev), is metabolized by cells and can be detected on the cell surface by covalent ligation to a biotin hydrazide probe.[16] This technology enabled glycoform remodeling on an intact cell[17] and engineering of viral infectivity,[18] and has been reviewed previously.[19] More recently, *N*-levulinoyl sialic acid supplied directly to cells has been shown to incorporate into cell surface glycoconjugates.[20]

To expand the repertoire of orthogonal functional groups available for cell surface modification, we developed a novel chemoselective ligation between an azide and a functionalized phosphine, termed the Staudinger ligation. The reaction was first demonstrated on cells using the unnatural ManNAc analog *N*-azidoacetylmannosamine (ManNAz, **1**; Fig. 2).[21] Compound **1** was peracetylated (Ac_4ManNAz, **2a**) to increase cell permeability[22] and incubated with Jurkat cells. *N*-Azidoacetyl sialic acid (SiaNAz, **3**), the biosynthetic product of **1**, was detected on the cell surface with a phosphine probe (Fig. 3).

The Staudinger ligation provides some advantages over the biotin hydrazide reaction with cell surface ketones. Ac_4ManNAz (**2a**) shows lower levels of cellular toxicity than ManLev that is peracetylated (Ac_4ManLev) at high concentration (50 μM). In addition, the conditions for reaction of the azide with the phosphine reagent are milder because the ligation can be performed between pH 7.0 and 7.4. The reaction of cell surface ketones with biotin hydrazide is slow at pH 7.0. Consequently the labeling is performed at pH 6.5, although some cell death is observed at this more acidic pH. Thus, the Staudinger ligation promotes higher cell viability. The azide and phosphine are completely abiotic because no comparable functional groups exist on cells that would compete for these coupling partners. Ketones and aldehydes, on the other hand, are present within cells, and can potentially compete with exogenous hydrazide probes. In addition, on ligation the phosphine is oxidized to a phosphine oxide,

[16] L. K. Mahal, K. J. Yarema, and C. R. Bertozzi, *Science* **276,** 1125 (1997).

[17] K. J. Yarema, L. K. Mahal, R. E. Bruehl, E. C. Rodriguez, and C. R. Bertozzi. *J. Biol. Chem.* **273,** 31168 (1998).

[18] J. H. Lee, T. J. Baker, L. K. Mahal, J. Zabner, C. R. Bertozzi, D. F. Wiemer, and M. J. Welsh, *J. Biol. Chem.* **274,** 21878 (1999).

[19] C. L. Jacobs, K. Y. Yarema, L. K. Mahal, D. A. Nauman, N. W. Charters, and C. R. Bertozzi, *Methods Enzymol.* **327,** 260 (2000).

[20] S. Goon, K. J. Yarema, and C. R. Bertozzi, unpublished results (2002).

[21] E. Saxon and C. R. Bertozzi, *Science* **287,** 2007 (2000).

[22] A. K. Sarkar, T. A. Fritz, W. H. Taylor, and J. D. Esko, *Proc. Natl. Acad. Sci. USA* **92,** 3323 (1995).

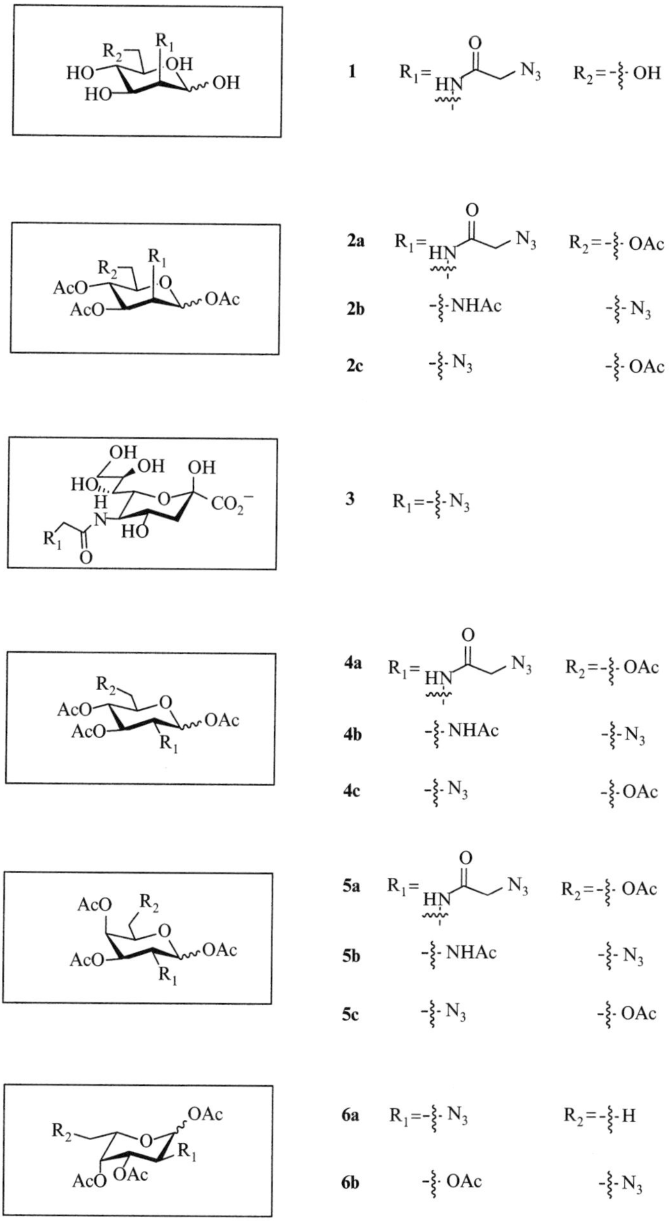

Compound	R_1	R_2
1	R_1 = HN–C(=O)–CH$_2$–N_3	R_2 = OH
2a	R_1 = HN–C(=O)–CH$_2$–N_3	R_2 = OAc
2b	NHAc	N_3
2c	N_3	OAc
3	R_1 = N_3	
4a	R_1 = HN–C(=O)–CH$_2$–N_3	R_2 = OAc
4b	NHAc	N_3
4c	N_3	OAc
5a	R_1 = HN–C(=O)–CH$_2$–N_3	R_2 = OAc
5b	NHAc	N_3
5c	N_3	OAc
6a	R_1 = N_3	R_2 = H
6b	OAc	N_3

FIG. 2. Structures of the azido sugars discussed in this article.

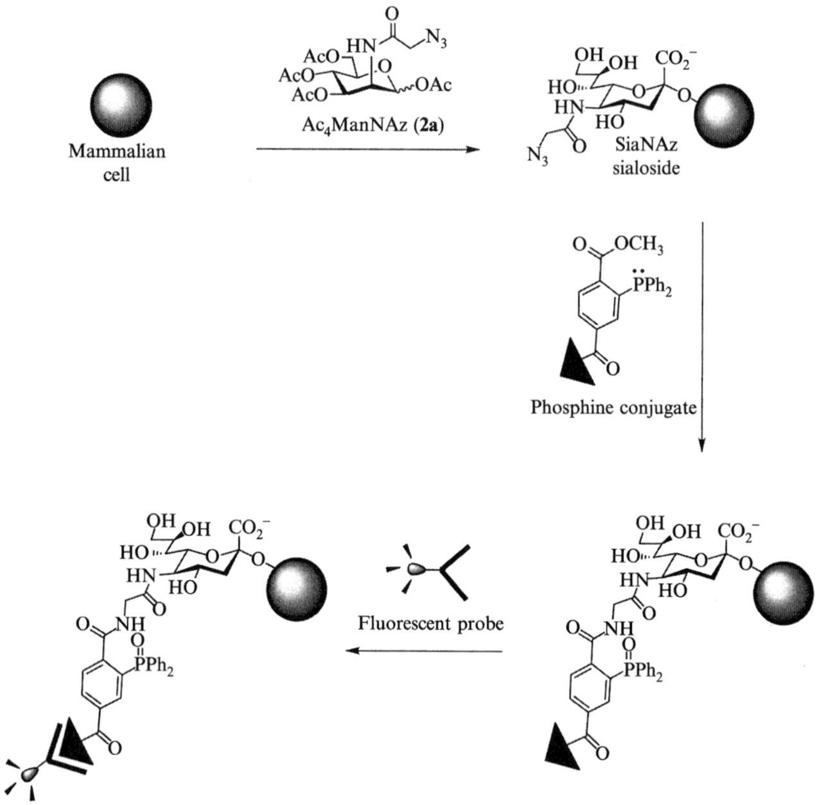

FIG. 3. The Staudinger ligation for detection of cell surface azides. In the first step, a monosaccharide containing an azide, such as $Ac_4ManNAz$ (**2a**), is incubated with mammalian cells. The cells take up the substrate and incorporate SiaNAz (**3**) into cell surface glycoconjugates. In the second step, cell surface azides are reacted with a phosphine conjugate labeled with a detectable probe such as the FLAG peptide (described in text) or biotin.[21] In the final step, the cells are incubated with a fluorescently labeled probe (such as FITC-labeled anti-FLAG or avidin) and analyzed by flow cytometry.

resulting in electronic changes that can be exploited for the design of ligation-sensitive fluorescent probes.[23,24]

This article summarizes the use of azido sugars for metabolic labeling of cell surface glycoconjugates. In addition to ManNAc derivatives, analogs of *N*-acetylglucosamine (GlcNAc), fucose, *N*-acetylgalactosamine

[23] G. A. Lemieux, C. L. de Graffenried and C. R. Bertozzi, *J. Am. Chem. Soc.* **125,** 4708 (2003).

[24] E. Saxon, unpublished results (2002).

(GalNAc), and sialic acid have been tested for metabolic incorporation. Studies of metabolic incorporation and characterization of the cell surface products are discussed.

Introduction of Azido Sugars into Cell Surface Glycoconjugates

Synthetic Methods

All chemical reagents are obtained from commercial suppliers and used without further purification unless otherwise noted. Fluorescein isothiocyanate and anti-FLAG M2 are purchased from Aldrich (Milwaukee, WI) and Sigma (St. Louis, MO), respectively. *N*-Acetylneuraminic acid (NeuAc) aldolase (23 U/mg) is purchased from Toyobo (Osaka, Japan). Fluorenylmethoxycarbonyl (Fmoc)-Lys (±-butoxycarbonyl, Boc-) Wang resin is purchased from Novabiochem (San Diego, CA). All ^{1}H, ^{13}C, and ^{31}P nuclear magnetic resonance (NMR) spectra are measured with a Bruker AMX-300, AMX-400, or DRX-500 MHz spectrometer as noted. Chemical shifts are reported relative to tetramethylsilane for ^{1}H and ^{13}C spectra and relative to H_3PO_4 for ^{31}P spectra. All coupling constants (J) are reported in hertz. Fast atom bombardment (FAB) and electrospray ionization (ESI) mass spectra are obtained at the University of California Berkeley (UC Berkeley) Mass Spectrometry Laboratory. Elemental analysis is obtained at the UC Berkeley Microanalytical Laboratory. High-performance liquid chromatography (HPLC) is performed with a Rainin Dynamax SD-200 HPLC system (Varian, Palo Alto, CA) with 220-nm detection on a Microsorb NH_2 analytical column (4.6 × 250 mm) at a flow rate of 1 ml/min or on a preparative column (25 × 250 mm) at a flow rate of 20 ml/min. Reversed-phase HPLC is performed with detection at 230-nm on a Microsorb C_{18} analytical column (4.6 × 250 mm) at a flow rate of 1 ml/min or on a preparative column (25 × 250 mm) at a flow rate of 20 ml/min. Peptide synthesis is performed on an Applied Biosystems (Foster City, CA) ABI 431 A peptide synthesizer. All cation- and anion-exchange resins are purchased from Bio-Rad Laboratories (Hercules, CA).

Synthesis of Azido Sugars

Two procedures are described for the synthesis of Ac_4ManNAz (**2a**; Fig. 2) from free mannosamine. The same procedures can also be used to synthesize Ac_4GlcNAz (**4a**) and Ac_4GalNAz (**5a**) from the corresponding free amino sugars. The synthesis of **2a** by the second method yields a slight amount of contaminating **4a** and consequently the first route is recommended to obtain **2a**. Compounds **4a** and **5a** can be synthesized in pure from by either route.

The syntheses of unnatural monosaccharides that are not efficiently incorporated into cell surface glycoconjugates are not described here but are detailed in the references cited. The syntheses of $Ac_3$6AzManNAc (**2b**; Fig. 2), $Ac_4$2AzMan (**2c**), $Ac_3$6AzGlcNAc (**4b**), and $Ac_4$2AzGlc (**4c**) have been described.[25] The syntheses of $Ac_3$6AzGalNAc (**5b**) and $Ac_4$2AzGal (**5c**) are discussed elsewhere.[26] Azide-containing fucose analogs, denoted $Ac_3$2AzFuc (**6a**) and $Ac_4$6AzFuc (**6b**), are synthesized as previously described.[27,28]

*First Synthetic Route to ManNAz (*N*-Azidoacetyl-α,β-D-mannosamine, **1**).* Azidoacetic acid is prepared as follows. Sodium azide (5.6 g, 86 mmol) is added to iodoacetic acid (8.0 g, 43 mmol) in H_2O (100 ml) and the reaction mixture is stirred at room temperature. After 4 days, thin-layer chromatography (TLC) analysis using detection by ultraviolet (UV) absorbance and staining with triphenylphosphine/toluene followed by ninhydrin indicates the reaction is complete. The solution is diluted with 1 *M* HCl, the desired product is extracted into ethyl acetate, and the organic phase is washed with saturated $NaHSO_3$ and saturated NaCl. The solution is dried over $MgSO_4$, filtered, and concentrated to afford a yellow oil (2.0 g, 46% crude yield). Azidoacetic acid is volatile and therefore low vacuum should be used when concentrating. The crude oil is used without further purification. Compound **1** is synthesized as follows. D-Mannosamine hydrochloride (1.5 g, 7.0 mmol) is added to azidoacetic acid (0.98 g, 9.7 mmol) in methanol (70 ml). Triethylamine (2.5 ml, 17 mmol) is added and the reaction mixture is stirred for 5 min at room temperature. The solution is cooled to 0 °C and *N*-hydroxybenzotriazole (HOBt, 0.86 g, 7.0 mmol) is added, followed by 1-[3-(dimethylamino)propyl]-3-ethylcarbodiimide hydrochloride (2.68 g, 14.0 mmol). The reaction is allowed to warm to room temperature overnight, at which point TLC analysis with cerric ammonium molybdate (CAM) stain indicates that the reaction is complete. The solution is concentrated and crude **1** is either acetylated directly, or purified when the free sugar is desired for cell experiments. For purification, the crude compound is concentrated and passaged over AG 50W-X8 resin (hydrogen form) and AG 1-X2 resin (acetate form) to remove triethylammonium salts. Compound **1** is eluted with H_2O and concentrated. The resulting light yellow foam is further purified by silica gel chromatography, eluting with $CHCl_3$–methanol (7:1, v/v). Concentration of

[25] E. Saxon, S. J. Luchansky, H. C. Hang, C. Yu, S. C. Lee, and C. R. Bertozzi, *J. Am. Chem. Soc.* **124,** 14893.

[26] H. C. Hang, C. Yu, D. L. Kato, and C. R. Bertozzi, submitted, (2003).

[27] G. Srivastava, K. J. Kaur, O. Hindsgaul, and M. M. Palcic, *J. Biol. Chem.* **267,** 22356 (1992).

[28] M. Ludewig and J. Thiem, *Eur. J. Org. Chem.* **1998,** 1189 (1998).

fractions containing product yields **1** as a white foam (1.4 g, 78% yield). This material is dissolved in phosphate-buffered saline (PBS, pH 7.4, 500 m*M* stock solution) and filtered (0.25-μm pore size sterile filter) before incubation with cells.

*Second Synthetic Route to ManNAz (**1**).* Sodium methoxide (1.25 g, 23.2 mmol) is added to D-mannosamine hydrochloride (5.0 g, 23 mmol) in methanol (110 ml) and the reaction mixture is stirred for 30 min at room temperature. Triethylamine (3.2 ml, 23 mmol) and chloroacetic anhydride (19.8 g, 116 mmol) are added and the solution is stirred overnight at room temperature. Concentration of the solution yields the crude product as a yellow oil that is purified by silica gel chromatography, eluting with $CHCl_3$–methanol (8:1, v/v). Fractions containing product are concentrated to yield *N*-chloroacetyl-D-mannosamine as a yellow foam (3.8 g, 64% yield). Lithium azide (LiN_3) is prepared as follows. Sodium azide (20.0 g, 308 mmol) and lithium sulfate (22.0 g, 200 mmol) are added to H_2O (108 ml) and dissolved by gentle heating. While stirring, 95% ethanol (538 ml) is added slowly and after 10 min the solution is filtered and the precipitate is washed with 95% ethanol. The filtrate is concentrated and dissolved in 95% ethanol (154 ml) by gentle heating at 35 °C. The solution is filtered while hot and the filtrate is concentrated to yield crude LiN_3 as a white solid. LiN_3 (0.48 g, 9.8 mmol) is added to *N*-chloroacetyl-D-mannosamine (0.50 g, 2.0 mmol) in *N,N*-dimethylformamide (DMF, 10 ml) and the reaction mixture is stirred overnight. The solution is concentrated and crude **1** is purified as described above (0.38 g, 75% yield). ^{1}H NMR (300 MHz, D_2O) mixture of anomers: δ 3.56 (2H, app t, $J=9.1$), 3.65–3.72 (1H, m), 3.77–3.83 (1H, m), 3.85–3.87 (2H, m), 3.89–3.91 (2H, m), 3.94 (2H, dd, $J=2.3, 2.7$), 3.99 (1H, app d, $J=3.1$), 4.02 (1H, app d, $J=3.5$), 4.15 (4H, s), 5.23 (1H, d, $J=1.7$), 5.28 (1H, d, $J=3.4$) ppm. ^{13}C NMR (125 MHz, D_2O): δ 51.5, 51.6, 53.2, 54.0, 54.2, 60.2, 66.4, 66.6, 68.7, 69.9, 70.5, 71.5, 71.9, 75.9, 76.3, 92.7, 92.8 ppm. FAB–high-resolution mass spectrometry (HRMS) calculated for $C_8H_{15}N_4O_6$ $(M+H^+)$ 263.0992; found 263.0991.

*Ac$_4$ManNAz (1,3,4,6-tetra-*O*-acetyl-*N*-azidoacetyl-α,β-D-mannosamine, **2a**).* Acetic anhydride (1.0 ml, 11 mmol) is added to a solution of **1** (0.025 g, 0.095 mmol) in pyridine (2 ml) and the reaction mixture is stirred overnight at room temperature. The solution is concentrated, resuspended in CH_2Cl_2, and washed with 1 *M* HCl, saturated $NaHCO_3$, and saturated NaCl. The organic phase is dried over Na_2SO_4, filtered, and concentrated. The crude material is purified by silica gel chromatography, eluting with hexanes–ethyl acetate (2:1, v/v). The fractions containing product are concentrated to yield **2a** (0.039 g, 95% yield). This material is further purified for cell experiments by reversed-phase HPLC, eluting with a gradient of CH_3CN (5–40%) and H_2O. The HPLC-purified compound is dissolved in ethanol (5 mM stock solution) and filtered (0.25-μm pore size sterile filter) before incubation with cells. ^{1}H NMR (500 MHz, $CDCl_3$): δ 2.05 (6H, app s), 2.10 (3H, s), 2.12 (3H, s), 3.82 (1H, ddd, $J=2.4$, 4.5, 9.6), 3.93 (2H, s), 4.13 (1H, dd, $J=2.4$, 12.6), 4.29 (1H, dd, $J=4.5$, 12.6), 5.15 (1H, app t, $J=9.6$), 5.21 (1H, app t, $J=5.7$), 5.78 (1H, app t, $J=4.5$), 6.37 (1H, d, $J=5.7$) ppm. ^{13}C NMR (125 MHz, $CDCl_3$): δ 20.5, 20.5, 20.6, 20.6, 20.6, 20.8, 20.8, 51.2, 52.4, 52.5, 53.2, 61.4, 61.5, 67.4, 67.6, 69.8, 70.3, 72.0, 72.1, 72.9, 90.2, 92.1, 166.9, 168.6, 169.1, 169.2, 170.6, 170.8, 171.5 ppm. FAB-HRMS calculated for $C_{16}H_{22}LiN_4O_{10}$ ($M+Li^+$) 437.1496; found 437.1496.

*Ac_4GlcNAz (1,3,4,6-tetra-*O*-acetyl-*N*-azidoacetyl-α,β-D-glucosamine, **4a**).* ^{1}H NMR (500 MHz, $CDCl_3$): δ 2.04 (2H, s), 2.05 (6H, s), 2.06 (4H, s), 2.09 (3H, s), 3.83 (1H, dddd, $J=2.2$, 2.3, 5.3, 9.8), 3.91 (2H, s), 3.93 (2H, s), 4.00–4.04 (1H, m), 4.07 (1H, dd, $J=2.3$, 12.5), 4.11–4.13 (1H, m), 4.26 (1H, dd, $J=3.9$, 12.5), 4.42–4.47 (1H, m), 5.14 (1H, app t, $J=9.7$), 5.21 (1H, app t, $J=9.6$), 5.29 (1H, app t, $J=10.7$), 5.78 (1H, d, $J=8.7$), 6.20 (1H, d, $J=3.7$), 6.43 (2H, d, $J=8.9$) ppm. ^{13}C NMR (125 MHz, $CDCl_3$): δ 20.5, 20.6, 20.6, 20.8, 51.2, 52.4, 61.4, 67.3, 69.7, 70.3, 90.2, 166.8, 168.6, 169.1, 170.6, 171.5 ppm. FAB-HRMS calculated for $C_{16}H_{22}Li$-N_4O_{10} ($M+Li^+$) 437.1496; found 437.1504.

*Ac_4GalNAz (1,3,4,6-tetra-*O*-acetyl-*N*-azidoacetyl-α,β-D-galactosamine, **5a**).* ^{1}H NMR (500 MHz, $CDCl_3$): δ 2.01 (3H, s), 2.02 (3H, s), 2.02 (3H, s), 2.04 (3H, s), 2.12 (3H, s), 2.16 (3H, s), 2.16 (3H, s), 2.17 (3H, s), 3.92 (2H, s), 3.94 (2H, s), 4.07 (2H, m), 4.08 (1H, m), 4.14 (2H, m), 4.25 (1H, app t, $J=6.7$), 4.36 (1H, app dt, $J=9.1$, 11.2), 4.68 (1H, m), 5.21 (1H, dd, $J=3.3$, 11.3), 5.24 (1H, dd, $J=3.2$, 11.6), 5.39 (1H, app d, $J=2.6$), 5.44 (1H, app d, $J=3.2$), 5.79 (1H, d, $J=8.8$), 6.21 (1H, d, $J=2.0$), 6.28 (1H, d, $J=9.0$), 6.38 (1H, d, $J=9.5$) ppm. ^{13}C NMR (125 MHz, $CDCl_3$): δ 20.6, 20.6, 20.6, 20.7, 20.7, 20.7, 20.9, 20.9, 47.0, 50.1, 52.5, 52.6, 61.2, 61.3, 66.3, 66.6, 67.6, 68.7, 70.0, 71.9, 90.9, 92.6, 167.0, 167.2, 168.9, 169.4, 170.1, 170.2, 170.4, 170.4, 170.5, 171.1 ppm. FAB-HRMS calculated for $C_{16}H_{22}LiN_4O_{10}$ ($M+Li^+$) 437.1496; found 437.1508.

*SiaNAz (*N*-azidoacetyl sialic acid. **3**).* This procedure has been adapted from a protocol previously described.[15] Compound **1** (0.30 g, 1.1 mmol) and sodium pyruvate (1.2 g, 11 mmol) are dissolved in 50 mM K_2HPO_4 (pH 7.2, 11 ml) containing 0.05% NaN_3. NeuAc aldolase (1 mg) is added and the reaction is shaken at 37 °C until ^{1}H NMR analysis indicates the disappearance of **1** (~1–2 days). The reaction is diluted and purified over AG 1-X8 resin (formate form). Compound **3** is eluted with 1 liter of formic acid over a gradient from 1 to 2.5 M. Fractions are assayed for the presence of the sialic acid derivative, using the periodate–resorcinol assay,[29] and then combined and concentrated to yield **3** (0.32 g, 84% yield).

Compound **3** is dissolved in PBS (pH 7.1, 500 m*M* stock solution) and filtered (0.25-μm pore size sterile filter) before use in cell culture. ^{1}H NMR (300 MHz, D_2O): δ 1.77–1.85 (1H, app t, $J = 12.2$), 2.21–2.26 (1H, dd, $J = 4.8$, 12.7), 3.48 (1H, d, $J = 9.2$), 3.51–3.57 (1H, dd, $J = 6.2$, 11.7), 3.65–3.70 (1H, m), 3.74–3.79 (1H, dd, $J = 2.7$, 11.8), 3.90–3.96 (1H, app t, $J = 10.2$), 4.02–4.08 (4H, m) ppm. ^{13}C NMR (125 MHz, D_2O): δ 38.8, 51.9, 52.1, 63.1, 66.4, 68.1, 70.0, 70.1, 95.3, 171.0, 173.3 ppm. FAB-HRMS calculated for $C_{11}H_{19}N_4O_9$ ($M + H^+$) 351.1152; found 351.1152.

Synthesis of Phosphine FLAG

Phosphine FLAG[30] is synthesized by condensation of 1-methyl-2-diphenylphosphinoterephthalate with FLAG peptide. 1-Methyl-2-diphenylphosphinoterephthalate is synthesized in two steps as follows. 1-Methyl-2-aminoterephthalate (0.50 g, 2.6 mmol) is added to a round-bottom flask charged with cold concentrated HCl (5 ml). A solution of $NaNO_2$ (0.18 g, 2.6 mmol) in H_2O (1 ml) is added dropwise, resulting in the evolution of a small amount of orange gas. The mixture is stirred for 30 min at room temperature and then filtered through glass wool into a solution of KI (4.3 g, 25 mmol) in H_2O (7 ml). The dark red solution is stirred for 1 h and then diluted with CH_2Cl_2 (100 ml) and washed with saturated $NaHSO_3$. The organic layer is washed with H_2O and saturated NaCl and then dried over Na_2SO_4, filtered, and concentrated. The crude product is recrystallized from methanol and H_2O to yield 1-methyl-2-iodoterephthalate as a bright yellow solid. Characterization of the intermediate is described elsewhere.[30] In the second step, palladium acetate (2.2 mg, 0.010 mmol) is added to dry methanol (3 ml) containing 1-methyl-2-iodoterephthalate (0.3 g, 1 mmol) and triethylamine (0.3 ml, 2 mmol). The mixture is degassed *in vacuo*. While stirring under argon, diphenylphosphine (0.2 ml, 1 mmol) is added to the flask with a syringe. The resulting solution is heated to reflux overnight, cooled to room temperature, and concentrated. The residue is dissolved in CH_2Cl_2 (125 ml) and washed with H_2O followed by 1 *M* HCl and then concentrated. The crude product is recrystallized from CH_3CN to yield 1-methyl-2-diphenylphosphinoterephthalate as a golden yellow solid (0.37 g, 39% yield over two steps). ^{1}H NMR (400 MHz, $CDCl_3$): δ 3.75 (3H, s), 7.28–7.35 (11H, m), 7.63–7.67 (1H, m), 8.04–8.07 (1H, m) ppm. ^{13}C NMR (125 MHz, $CDCl_3$): δ 52.4, 128.5, 128.6, 128.7, 129.0, 129.7, 130.6, 131.0, 131.8, 131.9, 133.7, 133.9, 135.4, 136.8, 152.7, 169.7, 221.9 ppm. ^{31}P NMR (160 MHz, $CDCl_3$): δ −3.67

[29] G. W. Jourdian, L. Dean, and S. Roseman, *J. Biol. Chem.* **246,** 430 (1971).

[30] K. L. Kiick, E. Saxon, D. A. Tirrell, and C. R. Bertozzi, *Proc. Natl. Acad. Sci. USA*, **99,** 19 (2002).

ppm. FAB-LRMS *m/z* 365.1 ($M+H^+$). Analysis: calculated for $C_{21}H_{17}O_4P$: C, 69.23; H, 4.70; found: C, 69.24; H, 4.78.

FLAG peptide (Asp-Tyr-Lys-Asp-Asp-Asp-Asp-Lys) is synthesized using established automated protocols (user-derived cycles), fluorenylmethoxycarbonyl (Fmoc)-Lys(Boc)-Wang resin (0.83 mmol Lys/g resin), N^{α}-Fmoc-protected amino acids, and 1,3-dicyclohexylcarbodiimide-mediated HOBt ester activation in 1-methyl-2-pyrrolidinone. Five equivalents of *N*-protected amino acid is activated for 30 min with 10 equivalents each of 1,3-diisopropylcarbodiimide and HOBt in DMF. This solution is added to the resin and shaken for 30 min. The N^{α}-Fmoc protecting group of the terminal Asp residue is removed by treatment with piperidine. The resin is washed with DMF followed by CH_2Cl_2 and transferred to a solid-phase reaction vessel for use in the following step.

A solution of 1-methyl-2-diphenylphosphinoterephthalate (0.18 g, 0.50 mmol), HOBt (0.07 g, 0.5 mmol), *O*-benzotriazol-1-yl-*N,N,N,N*-tetramethyluronium hexafluorophosphate (0.19 g, 0.50 mmol), and diisopropylethylamine (0.09 ml, 0.5 mmol) in DMF (2 ml) is added to a solid-phase reaction vessel containing the resin-bound side chain-protected FLAG peptide (0.12 g, 0.10 mmol) in DMF (1 ml). The reaction vessel is agitated for 4 h at room temperature, and then the resin is washed with DMF followed by CH_2Cl_2. The resin is treated with 95% trifluoroacetic acid (3 ml) for 2 h at room temperature. Filtration affords a solution of the crude product that is concentrated and purified by reversed-phase HPLC, eluting with a gradient of CH_3CN (0–50%) and H_2O. ESI-MS *m/z* 680.1 ($M+2H^+$).

Preparation of FITC-Labeled Anti-FLAG Antibody

Fluorescein isothiocyanate (FITC)-labeled anti-FLAG M2 (stock concentration, 1.8 mg/ml) is prepared according to a literature protocol from anti-FLAG M2 and fluorescein isothiocyanate.[31]

Cell Culture Methods

All media for cell culture are purchased from Invitrogen Life Technologies (Carlsbad, CA). Fetal calf serum (FCS) is purchased from HyClone (Logan, UT). Penicillin, streptomycin, monosaccharides used in competition experiments, and glycosylation inhibitors are purchased from Sigma. All flasks and plates for cell culture are purchased from Fisher Scientific (Pittsburgh, PA). Centrifugation of mammalian cells is performed in a

[31] E. Harlow and D. Lane, "Using Antibodies: A Laboratory Manual," Cold Spring Harbor Laboratory Press, Cold Spring Harbor, NY, 1999.

Sorvall centrifuge from Kendro Laboratory Products (Newtown, CT) containing an SH3000 rotor.

All cell lines are maintained in a 5% CO_2, water-saturated atmosphere at 37 °C and media are supplemented with penicillin (100 unit/ml) and streptomycin (0.1 mg/ml). Typically, nonadherent cells (Jurkat, HL-60, and S49) are grown in 25-cm^2 flasks and counted before passaging using a Z2 Coulter counter (Coulter, Miami, FL). Adherent cells (CHO, HeLa, COS7, and NIH 3T3) are grown in 10-cm tissue culture plates. Cells are trypsinized with 0.25% trypsin–EDTA (PBS, pH 7.4), resuspended in medium, and counted before passaging. Although the cell surface labeling experiments to be described are designed for high-throughput analysis using a 96-well plate format, the procedure is easily adaptable to Eppendorf tubes.

Flow cytometry data are acquired with a Coulter EPICS XL-MCL flow cytometer or a FACSCalibur flow cytometer (BD Biosciences Immunocytometry Systems, San Jose, CA) equipped with a 488-nm argon laser. For flow cytometry experiments, 10,000 live cells are analyzed and all data points are collected in triplicate.

Maintenance and Cell Surface Labeling of Nonadherent Cell Lines. Jurkat cells, a human T cell lymphoma line, are maintained in RPMI 1640 medium supplemented with 10% FCS. Jurkat cells stably transfected with human α(1,3/4)-fucosyltransferase VII are cultured under the same conditions.[32] HL-60 cells, a human promyeloid cell line, are maintained in Iscove's minimal essential medium (MEM) containing glutamine and pyruvate supplemented with 20% FCS. S49 cells, a murine T cell lymphoma line, are maintained in RPMI 1640 supplemented with 10% FCS and 1% sodium pyruvate.

Before cell surface labeling experiments, cells are grown to a density of no more than 1.6×10^6 cells/ml. The appropriate volume of the unnatural monosaccharide in ethanol or H_2O is added to each well of a six-well plate. When the compound is added in ethanol, the ethanol is evaporated before addition of the cells. The cells are diluted to a concentration of 200,000 cells/ml and this solution (2 ml) is added to each well. Cells are incubated with the unnatural monosaccharide for 2–3 days, although more recent experiments have shown that cell surface azides can be detected after only 1 day in Jurkat cells.[33]

After incubation, cells from each well of a 6-well plate are concentrated by centrifugation and added to 3 wells of a 96-well V-bottom plate, to be labeled in triplicate. Cells are pelleted at 2500*g* and the medium is

[32] G. S. Kansas, personal communication (2002).

[33] D. H. Dube, unpublished results (2002).

decanted. Cells are resuspended in phosphine FLAG buffer [200 μl, PBS (pH 7.1) containing 1% FCS]. Centrifugation is repeated and cells are washed a second time. After pelleting, the supernatant is decanted, the cells are resuspended in 0.25 mM (or 0.50 mM where indicated) phosphine FLAG (100 μl), and incubated for 1 h at room temperature to label cell surface azides. During each step it is important to fully resuspend the cells, otherwise incomplete labeling is observed.

After the labeling reaction, cells are maintained at 4 °C until analysis in order to slow recycling of the plasma membrane. Cells are pelleted, the supernatant is discarded, and the pellets are resuspended in phosphine FLAG buffer (200 μl) at 4 °C. The washing step is repeated twice. After pelleting the cells and discarding the supernatant, the cells are incubated with FITC-labeled anti-FLAG antibody solution (100 μl, 1:900 dilution in phosphine FLAG buffer). After incubation for 30 min on ice, the cells are washed twice with phosphine FLAG buffer (200 μl) and transferred into the same buffer (400 μl) for analysis by flow cytometry.

Maintenance and Cell Surface Labeling of Adherent Cell Lines. HeLa cells, a human epithelial carcinoma, COS7 cells, a green monkey kidney cell line, and NIH 3T3 cells, a murine fibroblast cell line, are maintained in Dulbecco's modified eagle's medium (DMEM) containing 10% FCS. CHO-K1 cells, a Chinese hamster ovary (CHO) cell line, are grown in HAM F12 nutrient mixture containing glutamine and supplemented with 10% FCS.

For cell surface labeling experiments with adherent cells, the appropriate volume of the unnatural monosaccharide in ethanol or H_2O is placed in the wells of a 12-well plate. When the compound is added in ethanol, the ethanol is evaporated before addition of the cells. To each well is added medium (1 ml) containing trypsinized cells at a density of 125,000 cells/ml. For each concentration of unnatural monosaccharide, three wells are seeded to provide triplicate data points. Adherent cells require greater surface area than nonadherent cells to reach the cell density required for the labeling experiments described here. Cells are incubated for 2–3 days with the unnatural monosaccharide and subsequently analyzed for cell surface azide expression.

Adherent cells are trypsinized for 6 min at room temperature with 0.25% trypsin–EDTA (200 μl) and the reaction is quenched with medium (400 μl). The incubation time required varies with preparations of trypsin–EDTA, but 6 min at room temperature is typically sufficient to free the cells from the plate. Mixing is required to maintain the cells in suspension. It is important that the incubation time remain constant for all adherent cell lines in order to maintain consistent cell surface labeling. For each well of the 12-well plate the solution containing trypsinized cells is concentrated and added to 1 well of a 96-well V-bottom plate. Cells are pelleted as described above and resuspended in phosphine FLAG buffer (200 μl). Labeling with phosphine FLAG and detection with FITC-labeled anti-FLAG antibody are performed as described previously for nonadherent cells.

Detection and Characterization of Unnatural Monosaccharides in Cell Surface Glycoconjugates

The experiments described in this section detail some of the strategies that have been employed to determine the nature of the products derived from azide-containing monosaccharides and their location on the cell surface. In the cell, the peracetylated monosaccharides described above are deacetylated by nonspecific cytosolic esterases.[22] ManNAc analogs are converted into sialic acid analogs through the sialic acid biosynthetic pathway shown in Fig. 1. GalNAc and fucose analogs are designed to be metabolized through the GalNAc and fucose salvage pathways depicted in Figs. 4 and 5, respectively.[34,35] GlcNAc analogs may be converted to sialic acid or GalNAc analogs, or incorporated into GlcNAc-containing glycoconjugates, as depicted in Fig. 6.[34,36]

Ac_4ManNAz (**2a**) is converted into SiaNAz (**3**), which can be detected on the cell surface by staining with phosphine FLAG followed by a FITC-conjugated anti-FLAG antibody as described above. The mean fluorescence intensities (MFI) observed by flow cytometry analysis of various cell lines exposed to **2a** are listed in Table I.[25] $Ac_3$6AzManNAc (**2b**) shows low levels of conversion to the corresponding azido sialic acid, and $Ac_4$2AzMan (**2c**) does not yield cell surface azides that are detectable in our assay. A representative flow cytometry plot comparing the incorporation of **3** and ManNAz (**1**) in Jurkat cells is shown in Fig. 7A.[37] The two substrates lead to nearly identical cell surface fluorescence in a dose-dependent fashion (Fig. 7B and Table I). Because **1** is more accessible synthetically and the compound is metabolized as efficiently as **3**, the ManNAc derivative is preferred for labeling cell surface sialosides with azides. However, by utilizing the sialic acid derivative, three enzymatic steps are bypassed (Fig. 1),

[34] H. H. Freeze, "Monosaccharide Metabolism" (A. C. Varki, J. Esko, H. Freeze, G. Hart, and J. Marth, eds.), p. 69. Cold Spring Harbor Laboratory Press, Cold Spring Harbor, NY, 1999.

[35] M. Tonetti, L. Sturla, A. Bisso, D. Zanardi, U. Benatti, and A. De Flora, *Biochimie* **80,** 923 (1998).

[36] J. M. Ligos, T. L. de Lera, S. Hinderlich, B. Guinea, L. Sanchez, R. Roca, A. Valencia, and A. Bernad, *J. Biol. Chem.* **277,** 6333 (2002).

[37] S. J. Luchansky and C. R. Bertozzi, unpublished results (2002).

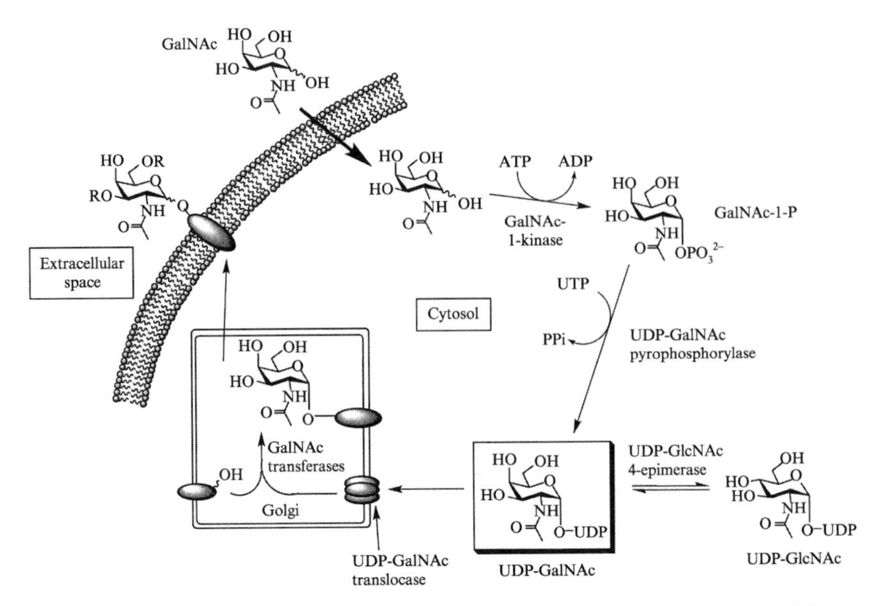

FIG. 4. *N*-Acetylgalactosamine (GalNAc) salvage pathway and UDP-GalNAc biosynthesis. GalNAc is taken up by cells and phosphorylated by GalNAc-1-kinase. GalNAc-1-phosphate (GalNAc-1-P) is then converted by UDP-GalNAc pyrophosphorylase to UDP-GalNAc. UDP-GalNAc is transported into the Golgi compartment and GalNAc is incorporated into glycoconjugates destined for the cell surface or secretion. The UDP-GlcNAc 4-epimerase links the GalNAc and GlcNAc metabolic pathways. "R" indicates that oligosaccharide chains may extend from the positions indicated.

including the phosphorylation of ManNAc, which can be a bottleneck step in the metabolism of unnatural ManNAc analogs.[38] For azide-derivatized ManNAc compounds that are not efficiently incorporated into cell surface glycoproteins, such as **2b**, it is likely that the efficiency of expression on the cell surface can be increased by utilizing the corresponding azido sialic acid derivative directly.

Ac_4GalNAz (**5a**) is utilized by the GalNAc salvage pathway (Fig. 4), although $Ac_3$6AzGalNAc (**5b**) and $Ac_4$2AzGal (**5c**) are not, as shown in Table I.[26] Compound **5a** and Ac_4ManNAz (**2a**) are comparable substrates for introduction of azides into CHO cell glycoconjugates. The data also demonstrate that Ac_4GlcNAz (**4a**) delivers azides to cell surfaces but to a lesser extent than either **2a** or **5a**. $Ac_3$2AzFuc (**6a**) and $Ac_4$6AzFuc (**6b**) are not efficient substrates for the fucose salvage pathway (Fig. 5) as no

[38] C. L. Jacobs and C. R. Bertozzi, *Biochemistry* **40,** 12844 (2001).

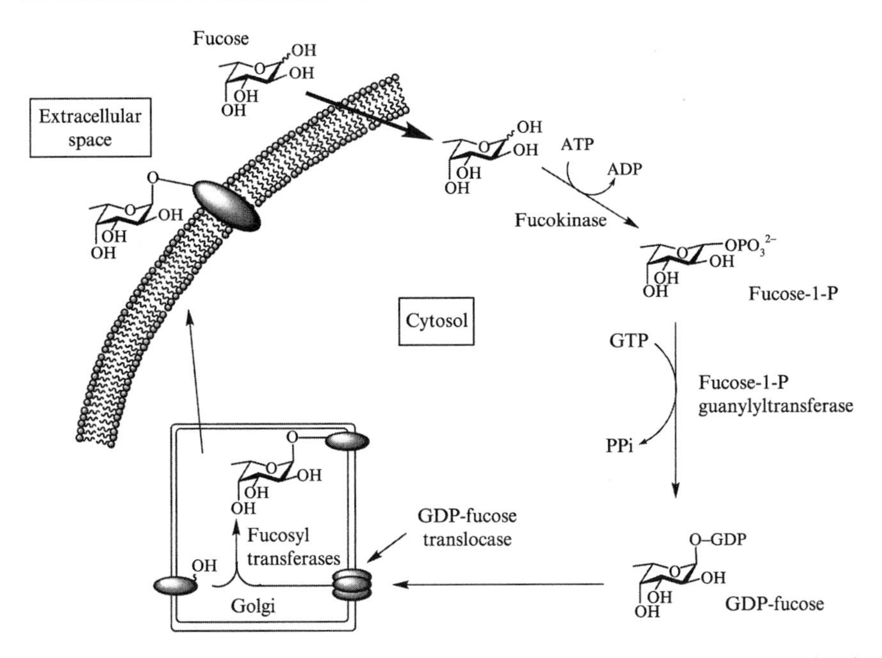

FIG. 5. Fucose salvage pathway. Exogenous fucose is taken up by cells and phosphorylated by fucokinase. Fucose-1-phosphate (fucose-1-P) is then activated by the fucose-1-P guanylyltransferase to yield GDP-fucose along with inorganic pyrophosphate (PP_i). GDP-fucose is transported into the Golgi compartment and fucose is added onto glycoconjugates that are bound for the cell surface or secretion.

significant fluorescence signal is observed when cells are incubated with these substrates.[39]

Choice of Cell Line

The incorporation of unnatural monosaccharides into cell surface glycoconjugates has been analyzed in various cell lines and the data are listed in Table I. The choice of cell line is important when considering the use of analogs that are not incorporated into all types of glycoconjugates. All the cells tested in this study are sialylated and generate mucin-type *O*-linked glycoconjugates with at least the core GalNAc residue in place. These cell lines are therefore relevant to test ManNAc, sialic acid, GlcNAc, and GalNAc analogs (Table I). JF7 cells, Jurkat cells transfected with human α(1,3/4)-fucosyltransferase VII,[32] an enzyme that can generate the fucosylated

[39] S. J. Luchansky, J. R. Grunwell, E. Saxon, and C. R. Bertozzi, unpublished results (2002).

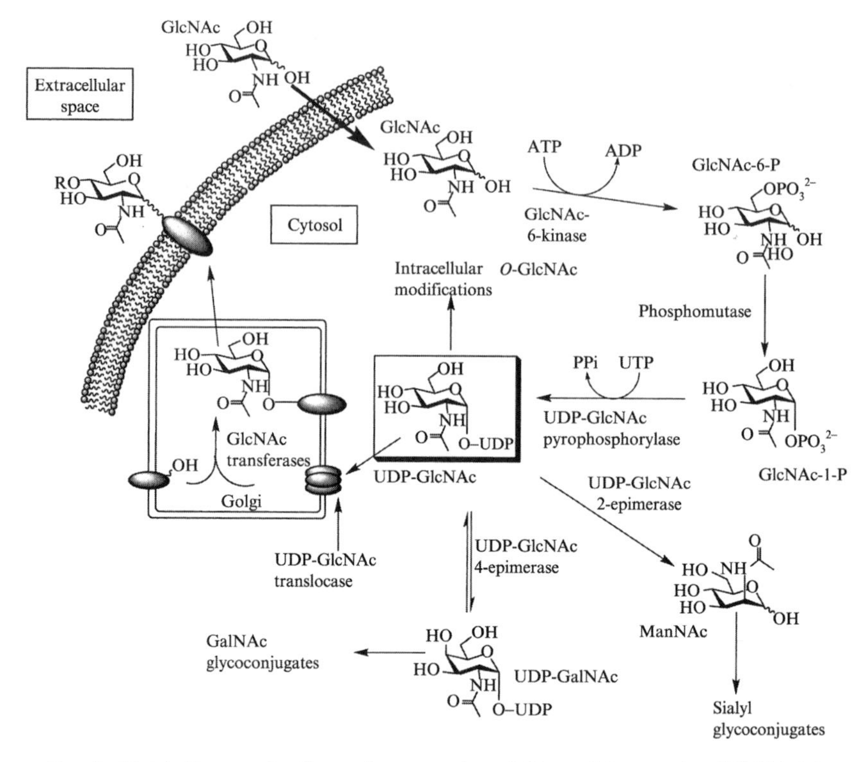

FIG. 6. Metabolic steps leading to incorporation of *N*-acetylglucosamine (GlcNAc) into glycoconjugates. GlcNAc taken up by cells is phosphorylated to GlcNAc-6-phosphate (GlcNAc-6-P) by GlcNAc-6-kinase. The phosphoryl group is transferred by the phosphomutase and subsequent reaction with UDP-GlcNAc pyrophosphorylase yields UDP-GlcNAc and inorganic pyrophosphate (PP_i). UDP-GlcNAc is either transported into the Golgi compartment and appended to glycoconjugates destined for secretion or the cell surface, or acts as an intermediate in one of the pathways depicted above. UDP-GlcNAc is also used by the cytosolic *O*-GlcNAc transferase. "R" indicates that oligosaccharide chains may extend off of the indicated position.

sialyl Lewis X (sLe^X) epitope, and HL-60 cells, which express the enzyme naturally, are good hosts with which to screen fucose analogs. Further information about the glycosylation patterns of the various cell lines employed in these experiments can be obtained from the references cited.[40–48]

[40] P. Stanley, *Glycobiology* **2,** 99 (1992).

[41] N. Jenkins, R. B. Parekh, and D. C. James, *Nat. Biotechnol.* **14,** 975 (1996).

[42] E. Grabenhorst, P. Schlenke, S. Pohl, M. Nimtz, and H. S. Conradt, *Glycoconj. J.* **16,** 81 (1999).

TABLE I
AZIDE-CONTAINING ANALOGS OF MANNAC, SIALIC ACID, GLCNAC, GALNAC, AND FUCOSE TESTED FOR METABOLISM TO CELL SURFACE PRODUCTS IN MAMMALIAN CELLS[a]

Compound	Cell line	MFI (SD)	Ref.
1[b]	Jurkat	38 (3)	37
2a	Jurkat	42 (3)	25
	CHO	40 (1)	25
	HL-60	52 (6)	25
	HeLa	7 (1)	25
	COS7	8 (0)	25
	S49	6 (0)	25
	NIH 3T3	6 (0)	25
	Lec2 CHO	4 (0)	25
2b	Jurkat	2 (0)	25
2c	Jurkat	1 (0)	25
3[b]	Jurkat	38 (2)	37
4a	Jurkat	12 (1)	25
	CHO	4 (0)	25
	HL-60	7 (0)	25
	HeLa	1 (0)	25
	COS7	1 (0)	25
	S49	2 (0)	25
4b	Jurkat	1 (0)	25
4c	Jurkat	1 (0)	25
5a[c]	CHO	34 (1)	26
5b[c]	CHO	1 (0)	26
5c[c]	CHO	1 (0)	26
6a[d]	HL-60	1 (0)	39
	JF7	1 (0)	39
6b[d]	HL-60	1 (0)	39
	JF7	1 (0)	39

[a]Cells were incubated with the unnatural substrate, labeled, and analyzed by flow cytometry as described in text. Cells were labeled with 0.25 m*M* phosphine FLAG. The results are listed as mean fluorescence intensities (MFI), which are reported as a fold change over the background MFI of cells labeled with the phosphine probe but without prior treatment with an azido sugar. The standard deviation from three replicates is given in parentheses. The unnatural substrate concentration was 50 μM.
[b]Unnatural substrate concentration was 10 m*M*.
[c]Cells were labeled with 0.50 m*M* phosphine FLAG.
[d]Unnatural substrate concentration was 250 μM.

[43] M. F. Bierhuizen, and M. Fukuda, *Proc. Natl. Acad. Sci. USA*, **89,** 9326 (1992).
[44] V. Piller, F. Piller, and M. Fukuda, *J. Biol. Chem.* **265,** 9264 (1990).

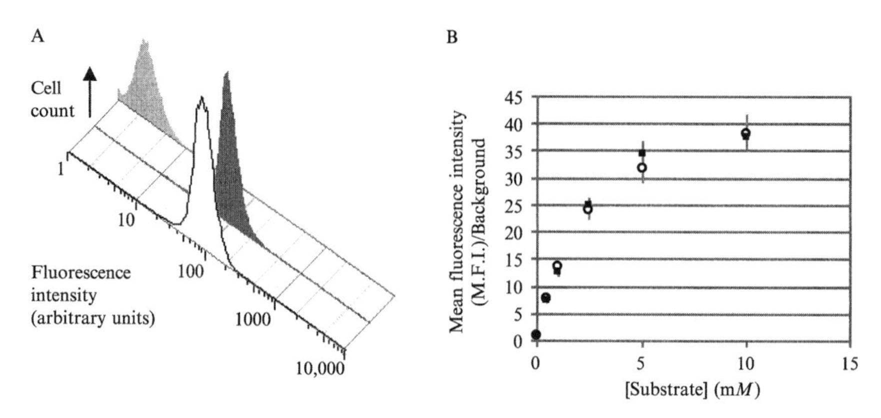

FIG. 7. Representative flow cytometry data. (A) Flow cytometry plot showing cell count versus increasing fluorescence for Jurkat cells incubated in the absence or presence of azido sugars. Cells were incubated with buffer alone (light gray), 10 m*M* ManNAz (**1**, dark gray), or 10 m*M* SiaNAz (**3**, white) and flow cytometry was performed as described in text. (B) Graphic representation of the dependence of fluroescence on the dose of **1** (open circles) or **3** (solid squares). The mean fluorescence intensity (MFI) is reported as a fold change over the background fluorescence of cells that were not exposed to the azido sugars [light gray population in (A)]. The error bars represent the standard deviation of three replicates.

Mutant cell lines provide an opportunity to characterize unnatural substrate metabolism in more detail. Lec2 CHO cells are deficient in the CMP-sialic acid Golgi transporter and consequently display essentially no sialic acid on their glycoconjugates.[7] When incubated with Ac_4ManNAz (**2a**), these cells display few azides on their cell surface compared with wild-type CHO cells (Table I).[49] Ldl-D CHO cells lack the UDP-GlcNAc 4-epimerase,[50] shown in Fig. 4, which interconverts UDP-GlcNAc and UDP-GalNAc. Consequently these cells incorporate GalNAc into cell surface glycoconjugates only when it is available from the medium. This cell line is a good host for GalNAc analogs as there is less endogenous competition with the natural substrate.[26,51]

[45] E. Makatsori, T. Tsegenidis, and N. K. Karamanos, *Biomed. Chromatogr.* **15,** 534 (2001).
[46] E. Makatsori, N. K. Karamanos, N. Papadogiannakis, A. Hjerpe, E. D. Anastassiou, and T. Tsegenidis, *Biomed. Chromatogr.* **15,** 413 (2001).
[47] P. P. Wilkins, R. P. McEver, and R. D. Cummings, *J. Biol. Chem.* **271,** 18732 (1996).
[48] K. E. Norgard, K. L. Moore, S. Diaz, N. L. Stults, S. Ushiyama, R. P. McEver, R. D. Cummings, and A. Varki, *J. Biol. Chem.* **268,** 12764 (1993).
[49] H. C. Hang and C. R. Bertozzi, unpublished results (2002).
[50] D. M. Kingsley, K. F. Kozarsky, L. Hobbie, and M. Krieger, *Cell*, **44,** 749 (1986).
[51] H. C. Hang and C. R. Bertozzi, *J. Am. Chem. Soc.* **123,** 1242 (2001).

Competition Experiments

Competition experiments between the unnatural analog and the natural monosaccharide that it mimicks can be used to determine which biosynthetic pathways are accessed by unnatural substrates. Typically, unnatural acetylated analogs at a concentration between 20 and 50 μM are incubated with Jurkat, HL-60, or CHO cells for 3 days with free monosaccharides such as ManNAc, GalNAc, or sialic acid, varying in concentration from 1 to 20 mM. The cell surface labeling is performed as described above.

ManNAc inhibits the metabolism of Ac_4ManNAz (**2a**) to cell surface SiaNAz as indicated by decreasing cell surface azides, suggesting they compete in the same cellular pathway.[25] Consistent with this observation, exogenous sialic acid also diminishes the metabolism of **2a** in a dose-dependent fashion. By contrast, GalNAc addition has no effect on **2a** metabolism. The data obtained from these experiments are represented in graphic form in Fig. 8A.

Interestingly, ManNAc and sialic acid also compete with Ac_4GlcNAz (**4a**) whereas GalNAc does not, suggesting that **4a** is also a substrate for the sialic acid biosynthetic machinery (Fig. 8B).[25] However, as noted above, the cell surface azide signal observed from **4a** is much lower than that from **2a**. Higher endogenous concentrations of GlcNAc or any of its downstream metabolites (i.e., UDP-GlcNAc, which achieves high micromolar concentrations within the cell[52]) may undermine the metabolic incorporation of exogenous GlcNAc analogs into glycoconjugates. Another possibility is that deacetylated **4a**, like GlcNAc, is subject to competing enzymatic processes within the cell, some of which are depicted in Fig. 6. Compound **4a** may not be fully converted to SiaNAz (**3**) but, rather, incorporated into cell surface glycoconjugates in positions inaccessible to the phosphine probe, or alternatively into *O*-GlcNAc-modified proteins within the cell.

Exogenous GalNAc is able to decrease the metabolism of Ac_4GalNAz (**5a**) to cell surface glycoconjugates. By contrast, neither GlcNAc nor galactose exerts this effect, suggesting that **5a** accesses mainly the GalNAc salvage pathway.[26]

Experiments with Glycosylation Inhibitors

Glycosylation inhibitors are useful tools for identifying the presence of unnatural analogs in various glycoconjugates. Tunicamycin and deoxymannojirimycin can be used to study *N*-linked glycosylation.[40,53,54]

[52] T. Ryll, and R. Wagner, *J. Chromatogr.* **570,** 77 (1991).
[53] J. Bischoff, L. Liscum, and R. Kornfeld, *J. Biol. Chem.* **261,** 4766 (1986).
[54] J. Bischoff and R. Kornfeld, *Biochem. Biophys. Res. Commun.* **125,** 324 (1984).

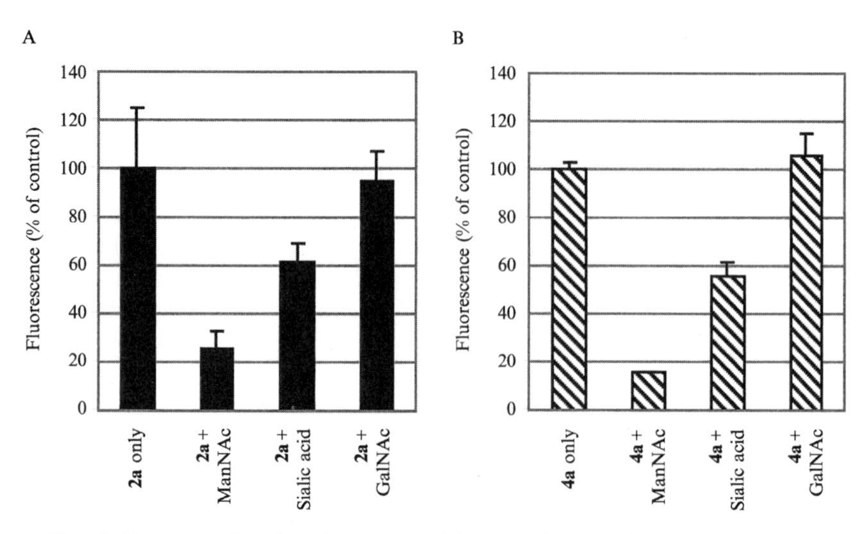

FIG. 8. Representative data from competition experiments with azido and nonazido sugars. (A) Jurkat cells were incubated with $Ac_4ManNAz$ (**2a**, 20 μM) in the absence or presence of various competitive sugars (5 mM). Cell surface labeling and analysis were performed as described in the text. All fluorescence values are reported relative to the fluorescence observed when cells were incubated with **2a** only. (B) Jurkat cells were incubated with $Ac_4GlcNAz$ (**4a**, 40 μM) in the absence or presence of various competitive sugars (5 mM). Cell surface labeling and analysis were performed as described in the text. All fluorescence values are reported relative to the fluorescence observed when cells were incubated with **4a** only. Error bars represent the standard deviation of three replicates.

Tunicamycin inhibits GlcNAc 1-phosphotransferase, the enzyme responsible for the transfer of GlcNAc to dolichol-1-phosphate to initiate the synthesis of the substrate for oligosaccharyltransferase. Deoxymannojirimycin blocks *N*-linked glycan elaboration by inhibition of α-mannosidase I, a trimming enzyme that acts after glycoprotein folding, and appears to have fewer deleterious effects on cells than tunicamycin.

Data for the expression of azides on cells treated with glycosylation inhibitors, as a percentage of that observed on cells without inhibitors, are summarized in Table II. Typically, Jurkat cells are incubated with $Ac_4ManNAz$ (**2a**) and a glycosylation inhibitor for 3 days. Tunicamycin is added at concentrations ranging from 0.25 to 5.0 ng/ml for Jurkat cells. At the lowest concentration of tunicamycin, the number of cell surface azides produced by **2a** is significantly lower than that observed on cells treated with **2a** alone (31% of the control).[25] At higher levels of tunicamycin, cell growth slows significantly. Cells incubated with **2a** and deoxymannojirimycin (0.1–1 mM) show fewer signs of growth inhibition and a decrease in cell surface azide signal that is

TABLE II
INHIBITION OF DELIVERY OF AZIDE-CONTAINING MONOSACCHARIDES TO CELL SURFACE BY VARIOUS GLYCOSYLATION INHIBITORS[a]

Inhibitor	Glycoconjugate target	Quantity	Cell line	MFI (SD) (%)	Ref.
Tunicamycin	*N*-Linked	0.25 ng/ml	Jurkat	31 (12)	25
Deoxymannojirimycin	*N*-Linked	0.25 m*M*	Jurkat	80 (13)	25
	N-Linked	1 m*M*	Jurkat	68 (2)	25
α-Benzyl GalNAc	*O*-Linked	2 m*M*	Jurkat	94 (13)	25
	O-Linked	2 m*M*	HL-60	29 (8)	25
p-Nitrophenyl-β-D-xyloside[b]	GAGs	1 m*M*	CHO	49 (2)	49

Abbreviation: GAG, Glycosaminoglycan.
[a]Cells were incubated with $Ac_4ManNAz$ (**2a**, 20 μ*M*), in the presence of the indicated amounts of glycosylation inhibitor. Analysis by flow cytometry was performed as described in text. Cells were labeled with 0.25 m*M* phosphine FLAG. The results are listed as percent mean fluorescence intensity (MFI), which is reported as a percentage of the MFI observed when cells were incubated with the azido substrate but no inhibitor. The standard deviation from three replicates is given in parentheses.
[b]Cells were incubated with $Ac_4GalNAz$ (**5a**, 50 μ*M*). Cells were labeled with 0.50 m*M* phosphine FLAG.

dependent on the concentration of the inhibitor. Because sialic acid is often a terminal residue on *N*-linked glycoproteins, truncation of these glycoconjugates results in fewer cell surface SiaNAz residues. These results indicate that SiaNAz resides in *N*-linked glycoproteins on Jurkat cells. Because inhibition is not complete (68% of the control, 1 m*M* deoxymannojirimycin), this suggests either that the *N*-linked chains are not fully truncated or that SiaNAz may also be present in other glycoconjugates such as glycolipids.

The small molecules benzyl-2-acetamido-2-deoxy-α-D-galactopyranoside (α-benzyl-GalNAc) and *p*-nitrophenyl-β-D-xyloside act as primers to mucin-type *O*-glycosylation and glycosaminoglycan (GAG) synthesis, respectively.[55,56] By mimicking the core residue attached to a protein, these primers act as substrates for glycosyltransferases and are elaborated in competition with their endogenous substrates. Consequently, elaboration of the core structures on cell surface glycoproteins is diminished.

Jurkat cells incubated with $Ac_4ManNAz$ (**2a**) in the presence of α-benzyl-GalNAc show little change in fluorescence (94% of the control,

[55] S. F. Kuan, J. C. Byrd, C. Basbaum, and Y. S. Kim, *J. Biol. Chem.* **264,** 19271 (1989).
[56] N. B. Schwartz, *J. Biol. Chem.* **252,** 6316 (1977).

2 mM α-benzyl-GalNAc).[25] This is consistent with the observation that Jurkat cells have few elaborated *O*-linked glycoproteins.[44] However, in HL-60 cells, which express higher levels of sialylated *O*-linked glycoconjugates, a more dramatic loss of fluorescence is observed (29% of the control, 2 mM α-benzyl GalNAc). These results suggest that SiaNAz decorates mucin-type *O*-linked glycoconjugates on HL-60 cell surfaces. Experiments with *p*-nitrophenyl-β-D-xyloside indicate that Ac_4GalNAz (**5a**) is incorporated into proteoglycans such as chondroitin sulfate on CHO cells.[49]

Applications of Cell Surface Azide Labeling

We have summarized herein the methods for cell surface glycoconjugate labeling with azido sugars. Using the tools described, such as glycosylation inhibitors and competition experiments, it has been possible to identify the types of glycoproteins that contain azido sugars on cell surfaces. In addition, we have described an alternative chemoselective reaction, the Staudinger ligation,[21] that will afford exciting new applications for these orthogonal cell surface chemical handles. Sialic acid residues derivatized with a ketone have already been exploited both to direct viral infection to ketone-bearing cells and to deliver magnetic resonance imaging (MRI) contrast reagents.[18,57] Further applications using the smaller, completely abiotic azide functionality could include the modulation of binding by sialic acid-sensitive lectins, such as Siglecs, and the study of *O*-GlcNAc-modified proteins within the cell. In addition, Western blotting has been successfully used to detect azide-bearing amino acids within proteins, and this method may also be applicable to the identification of unnatural sugars within glycoproteins.[30] Finally, monosaccharides labeled with an azide may be useful in glycomics studies to identify regulated changes in glycosylation. These applications will advance our understanding of the biological roles of glycosylation.

Acknowledgments

The authors thank Geoffrey S. Kansas (Northwestern University) for JF7 cells and Hector Nolla of the University of California Berkeley flow cytometry facility for assistance. H.C.H. was supported by a predoctoral fellowship from the American Chemical Society (Organic Division), E.S. was supported by a predoctoral fellowship from the Howard Hughes Medical Institute, and D.H.D. was supported by a predoctoral fellowship from the National Science Foundation. This research was supported by grants to C.R.B. from the National Institutes of Health (GM58867) and the Mizutani Foundation.

[57] G. A. Lemieux, K. J. Yarema, C. L. Jacobs, and C. R. Bertozzi, *J. Am. Chem. Soc.* **121,** 4278 (1999).

[19] Cell Wall Engineering of Living Bacteria through Biosynthesis

By REIKO SADAMOTO, KENICHI NIIKURA, KENJI MONDE, and SHIN-ICHIRO NISHIMURA

Introduction

The cell surface plays an important role in intercellular interactions such as recognition, adhesion, and infection. To control these interactions, much work has been done over the past 20 years on techniques for artificial surface display.[1] There have been two main strategies used to study surface display on living cells: the genetic approach and the chemical approach. In the genetic approach, cDNA encoding the target protein is inserted into the host cell, which then expresses the target protein together with an anchoring protein, for example, a glycosylphosphatidylinositol (GPI)-anchored protein, as a chimera protein (Fig. 1A).[2–6] For the chemical approach, Bertozzi *et al.*[7–10] reported that administration of *N*-levulinoylmannosamine to animal cells allowed its metabolic conversion to ketone-bearing sialic acid in mammalian cells (Fig. 1B), which was displayed on the cell surface, and could be modified by chemical reactions. This method is widely applicable to cells that have sialic acid on the surface.

In this article we describe a versatile chemical approach for displaying target compounds on the bacterial cell wall (Fig. 1C). A common structural element of microbial cell walls is murein, which consists of polysaccharides of GlcNAc-β(1$\rightarrow$ 4)muramic acid repeats, in which muramic acid is linked to pentapeptide (Fig. 2). To display target compounds on the cell wall, target compounds may be attached to cell wall precursors. Our group has

[1] S. Stahl and M. Uhlén, *Trends Biotechnol.* **15,** 185 (1997).

[2] T. Murai, T. Yoshino, M. Ueda, I. Haranoya, T. Ashikari, H. Yoshizumi, and A. Tanaka, *J. Ferment. Bioeng.* **86,** 569 (1998).

[3] M. Kainuma, N. Ishida, T. Yoko-o, S. Yoshioka, M. Takeuchi, M. Kawakita, and Y. Jigami, *Glycobiology* **9,** 133 (1999).

[4] E. T. Boder and K. D. Wittrup, *Nat. Biotechnol.* **86,** 569 (1998).

[5] G. Georgiou, C. Stathopoulos, P. S. Daugherty, A. R. Nayak, B. L. Iverson, and R. Curtiss III, *Nat. Biotechnol.* **15,** 29 (1997).

[6] A. Charbit, A. Molla, W. Saurin, and M. Hofnung, *Gene* **70,** 181 (1988).

[7] H. C. Hang and C. R. Bertozzi, *J. Am. Chem. Soc.* **123,** 1242 (2001).

[8] C. L. Jacobs, S. Goon, K. J. Yarema, S. Hinderlich, H. C. Hang, D. H. Chai, and C. R. Bertozzi, *Biochemistry* **40,** 12864 (2001).

[9] E. Saxon and C. R. Bertozzi, *Science* **287,** 2007 (2000).

[10] L. K. Mahal, K. J. Yarema, and C. R. Bertozzi, *Science* **276,** 1125 (1997).

0076-6879/03 $35.00

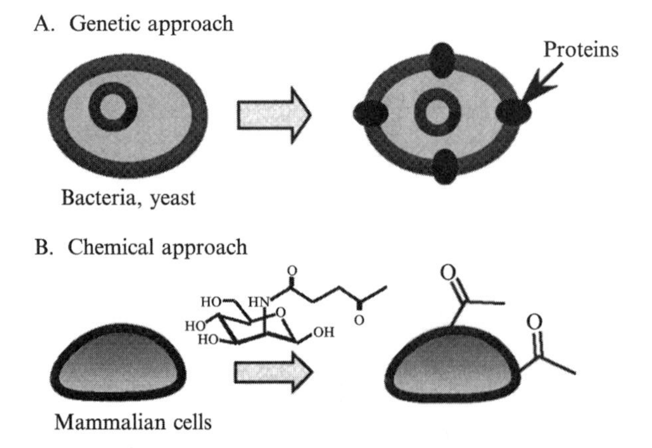

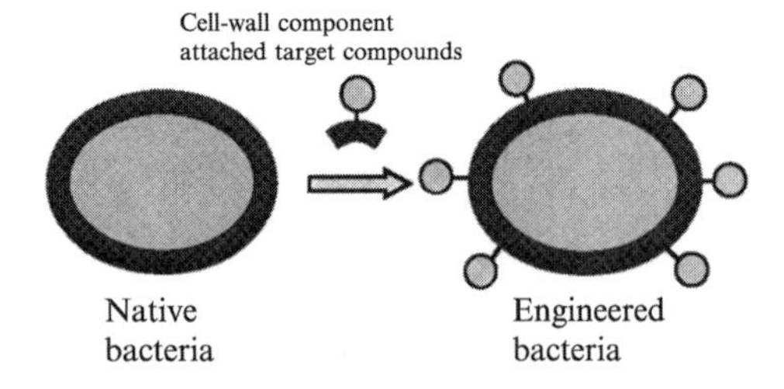

FIG. 1. Genetic and chemical approaches for cell surface display.

reported a chemoenzymatic synthesis of bacterial cell wall precursors,[11] and a new method for displaying target molecules on bacteria.[12] We chemically synthesized target molecules carrying precursors such as UDP-MurNAc, lipid I, and lipid II derivatives **1–4**, and tested whether these compounds could be incorporated into the cell wall with target molecules (Fig. 3). We found that UDP-MurNAc pentapeptide is an efficient metabolic precursor of the cell wall with which to incorporate target molecules. Delivery to the cell wall was confirmed by cell wall extraction and visualization by fluorescence microscopy.

Until more recently, proteins could only be expressed genetically on the surface of bacterial cells as chimeric proteins. Development of a method

[11] H. Liu, R. Sadamoto, P. S. Sears, and C.-H. Wong, *J. Am. Chem. Soc.* **123,** 9916 (2001).

[12] R. Sadamoto, K. Niikura, P. S. Sears, H. Liu, C.-H. Wong, A. Suksomcheep, F. Tomita, K. Monde, and S.-I. Nishimura, *J. Am. Chem. Soc.* **124,** 9018 (2002).

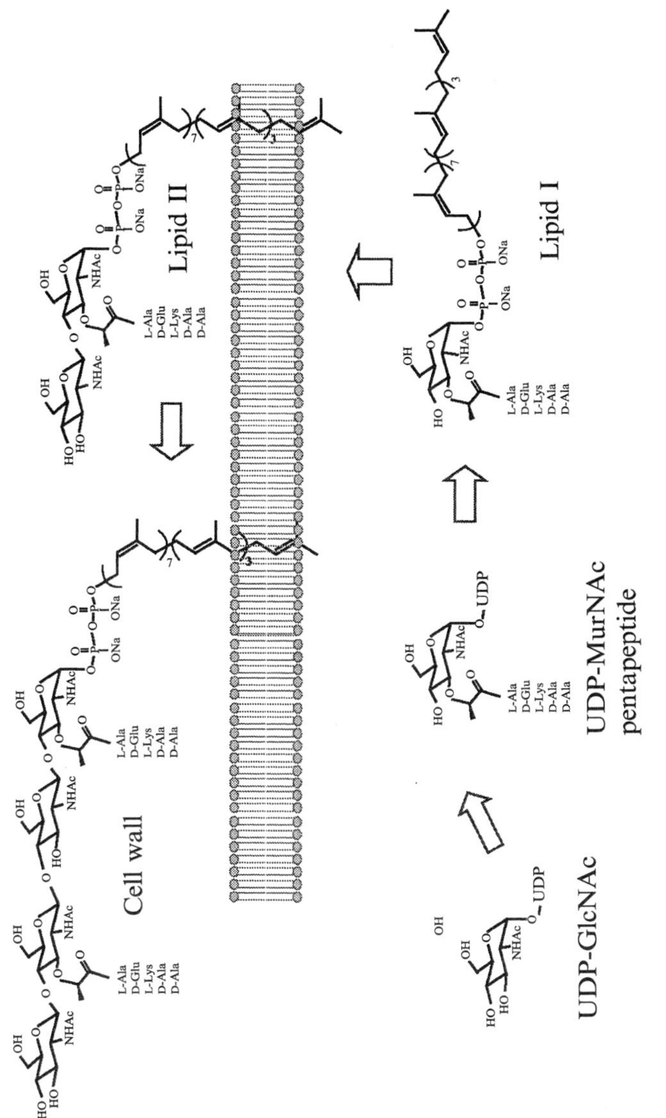

FIG. 2. Bacterial cell wall biosynthesis.

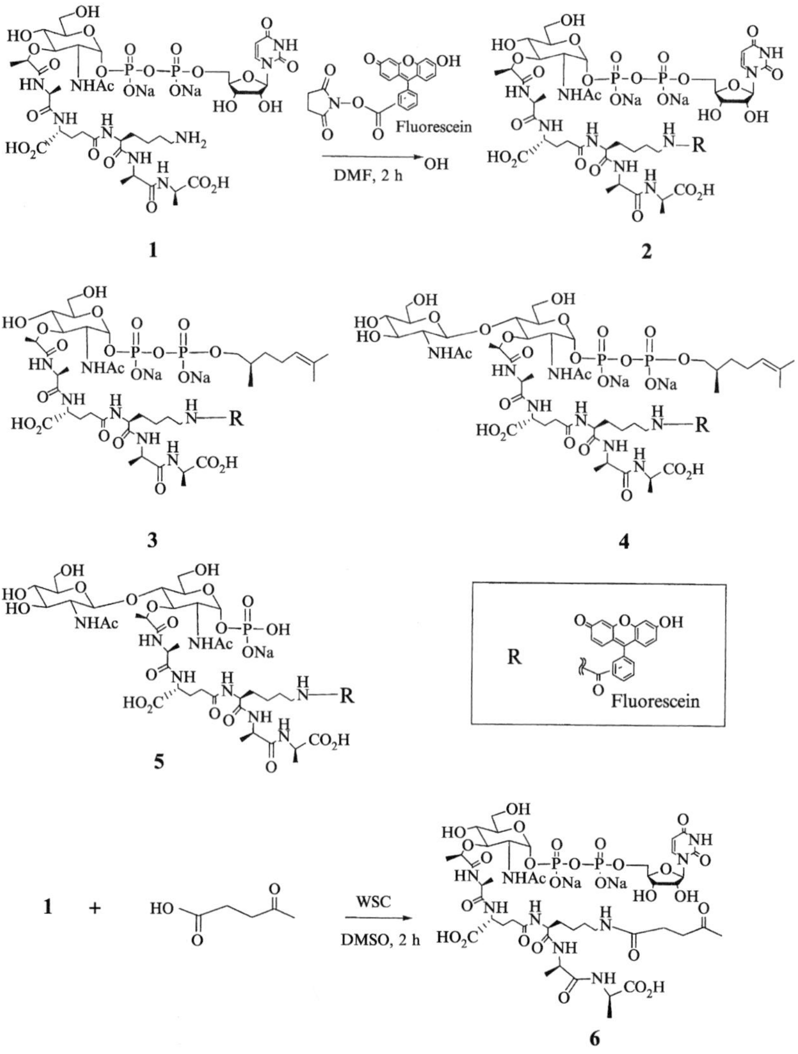

FIG. 3. Chemical structures of synthetic cell wall precursors (**1**–**6**).

for chemical engineering of the bacterial cell surface will now allow a large variety of molecular specimens to be displayed, leading to a wide range of applications, such as in novel drugs or vaccines utilizing lactic acid bacteria. Lactic acid bacteria are gram-positive bacteria that have long been used in food fermentation and drugs. These bacteria have been utilized as antigen-delivering vehicles for mucosal immunization as they are harmless to humans.[13,14] If epitope-bearing bacteria can be created by our strategy and delivered to pathogens, it will allow the development of new, more effective oral vaccines.

Synthesis of Cell Wall Precursors

Cell wall precursor **1** is synthesized according to the procedure reported by Hitchcock *et al.*[15] with slight modifications (Fig. 4). As their reaction between **7** and UMP-morpholidate **8** is slow (2 weeks) and the yield is relatively low (~30%), we add 1*H*-tetrazole as catalyst in the phosphomorpholidate coupling reaction in pyridine,[16] giving rise to a shorter reaction time (2 days) and higher yield (~90%). The coupling reaction between **7** and citronellol morpholidate **11** in the presence of 1*H*-tetrazole also proceeds to give a high yield, generating lipid I derivative **12** after deprotection (two steps, 92% yield). Compound **11** is obtained from citronellol using commercially available tribromoethyl phosphoromorpholinochloridate.

*Compound **10***

Tribromoethyl phosphoromorpholinochloridate (2.9 g, 6.4 mmol) is added to a stirred solution of citronellol **9** (0.9 g, 5.8 mmol) in anhydrous pyridine (10 ml) at -20°. After stirring for 9 h at room temperature, the solution is diluted with CH_2Cl_2 (200 ml) and washed twice with brine. The organic phase is dried (Na_2SO_4) and filtered, and the filtrate is concentrated by rotary evaporator. The crude syrup is purified by silica gel chromatography [elution, CH_2Cl_2–ethyl acetate, 3:1 (v/v)] to give **10** as a colorless syrup in 75% yield. ^1H nuclear magnetic resonance (NMR) (CD_3OD, 500 MHz) 5.10 (1H, m), 4.65 (2H, m), 4.12 (2H, m), 3.68 (4H, m), 3.21 (4H, m), 1.92 (2H, m), 1.75 (1H, m), 1.66 (3H, s), 1.58 (3H, s), 1.5 (1H, m), 1.34 (1H, m), 1.18 (1H, m), 0.9 (3H, dd, $J = 2.1$, 6.3 Hz). Electrospray ionization-mass spectrometry (ESI-MS) (negative) calculated for $C_{15}H_{27}Br_3NO_4P$ $[M\text{-}2H + Na]^-$, 591.09; found, 590.67.

[13] M. Kallioaki, S. Salminen, P. Kero, H. Arvilommi, P. Koskinen, and E. Isolauri, *Lancet* **357,** 1076 (2001).

[14] I. Sakamoto, M. Igarashi, K. Kimura, A. Takagi, T. Miwa, and Y. Koga, *J. Antimicrob. Chemother.* **47,** 709 (2001).

[15] S. A. Hitchcock, C. N. Eid, J. A. Aikins, M. Zia-Ebrahimi, and L. C. Blaszczak, *J. Am. Chem. Soc.* **120,** 1916 (1998).

[16] V. Wittmann and C.-H. Wong, *J. Org. Chem.* **62,** 2144 (1997).

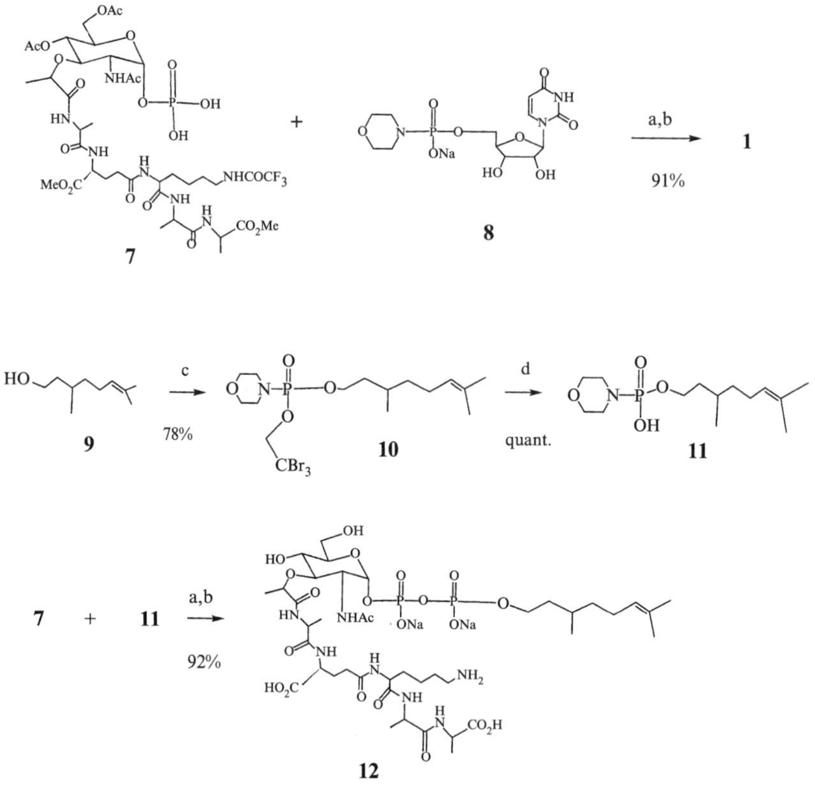

FIG. 4. Synthetic route for cell wall precursors (**7**–**12**). Conditions and reagents: (a) 1*H*-tetrazole, pyridine, 2 days; (b) aqueous NaOH, 30 min; (c) tribromoethyl phosphoromorpholinochloride, pyridine, 9 h; (d) Zn powder, THF.

Citronellol Phosphomorpholidate **11**

Citronellol derivative **10** (50 mg) is dissolved in tetrahydrofuran (THF, 5 ml) at room temperature and Zn powder (100 mg, activated in 0.1 *M* $CuSO_4$ aqueous solution for 5 min) is added in one portion. The suspension is stirred overnight and the solvent is evaporated to generate crude brown solid **11**. The solid is used without purification for the next reaction. ^{1}H NMR (CD_3OD, 500 MHz) δ 5.12 (1H, t, $J = 6.9$Hz), 4.1–3.9 (2H, m), 3.84 (2H, m), 3.64 (2H, t, 4.4 Hz), 3.17 (2H, m), 3.11 (2H, m), 2.0 (2H, m), 1.7 (1H, m), 1.67 (3H, s), 1.6 (3H, s), 1.4 (2H, m), 1.2 (1H, m), 0.9 (3H, m). ESI-MS (neg) calculated for $C_{14}H_{28}NO_4P$ $[M-H]^-$, 304.18; found, 304.3.

Lipid I Derivative **12**

Compound **7** (302 mg, 0.173 mmol) and citronellol phosphomorpholidate **11** (103 mg, 0.33 mmol) are coevaporated with anhydrous pyridine (three 1-ml volumes) and dissolved in pyridine (5 ml) after adding 1*H*-tetrazole (40 mg, 0.57 mmol). The mixture is stirred for 16 h and the reaction is quenched with water. The solvent is evaporated and then an aqueous solution of sodium hydroxide (4.5 ml, 0.7 *N*) is added to the residue and stirred for 30 min at room temperature. The solution is loaded on a BioGel P4 size-exclusion column and eluted with 100 m*M* NH_4HCO_3. The ultraviolet (UV)-active fractions are pooled and lyophilized to generate pure compound **12** (92% yield for two steps) as a white powder. Spectral analysis is in agreement with the data in Ref. 15.

Incorporation of Synthetic Precursors into Bacterial Cell Wall of *Escherichia coli*

Escherichia coli, a gram-negative bacteria, contains a thick lipopolysaccharide layer surrounding the cell wall. To increase the permeability of the lipopolysaccharide layer, *E. coli* is incubated in a 50 m*M* EDTA solution for 30 min before incubation with synthesized cell wall precursors (Fig. 5).

EDTA Treatment

Escherichia coli cells (strain C600) are incubated in LB broth overnight at 37°, collected by centrifugation at 1500 *g* for 3 min 4°, and washed with phosphate-buffered saline (PBS). To improve the permeability of the outer membrane, the cells were then incubated for 30 min in PBS containing 50 m*M* EDTA at 37°, and then washed three times with PBS.

Incubation with Synthesized Compounds

After EDTA treatment, the cells are incubated for 1 h at 37° in PBS containing a 0.5 m*M* concentration of each of the synthesized cell wall precursor components **2–5**. The synthesized cell wall precursors are dissolved in PBS and sterilized by filtration before being added to the bacteria.

Extraction of Cell Wall Fraction of Escherichia coli

Escherichia coli cells are collected by centrifugation at 1500 *g* for 3 min at 4° and washed three times with PBS solution to remove free cell wall components. Cells are disrupted by ultrasonication and the cell wall

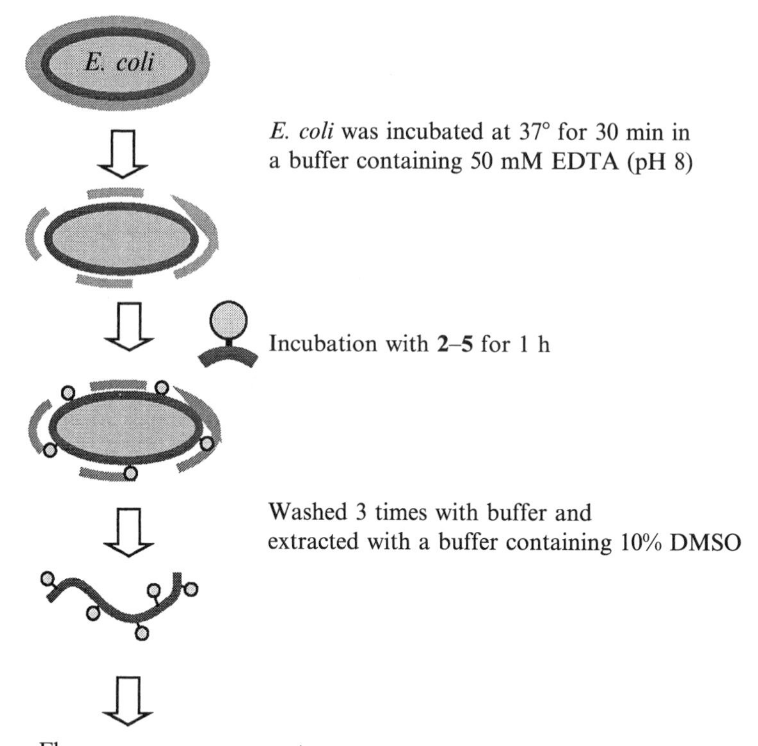

FIG. 5. Experimental procedure for incorporation of fluorescent precursors **2–5** into the cell wall of *E. coli* and their extraction.

fraction is extracted for 2 h with PBS containing 10% dimethyl sulfoxide (DMSO). The insoluble cell debris is then removed by centrifugation at 9000 *g* for 3 min at 4°. The supernatant is diluted with PBS and used for fluorescence intensity measurement.

Fluorescence of Extracted Cell Wall Fraction after Incubation with Precursors

Figure 6 shows the fluorescence intensities of extracted cell walls. The cell wall fraction grown with fluorescein-bearing UDP-MurNAc pentapeptide (**2**) provides a clearly greater intensity than those of the other three components for *E. coli*. Lipid I (**3**) and lipid II (**4**) derivatives give almost the same intensities as the negative control compound (**5**), showing that these two components cannot reach the cell wall. Without EDTA

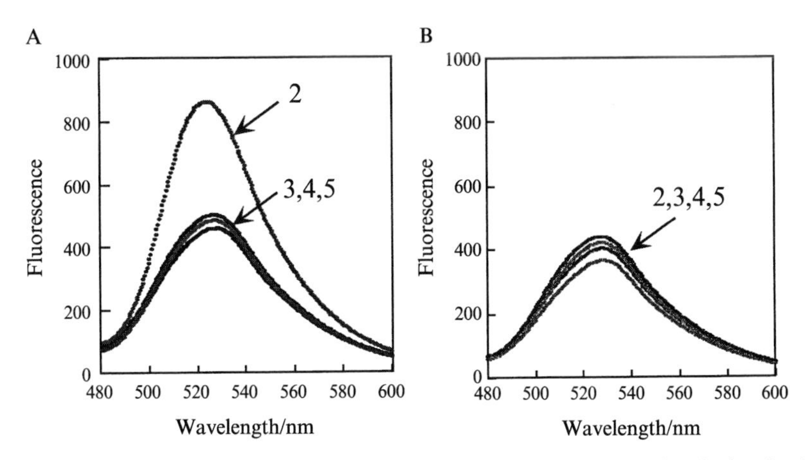

FIG. 6. Fluorescence spectra of extracted cell walls of *E. coli* after incubation in the presence of precursors **2–5**. (A) Precursors were added to *E. coli* after EDTA treatment; (B) precursors were added to *E. coli* without EDTA treatment.

treatment, UDP-MurNAc derivative **2** gives the same intensity as the other three components, indicating that this precursor cannot be incorporated. These results indicate that some UDP-MurNAc derivatives added to the culture medium can be incorporated into the living bacterial cell wall. S. Walker's group reported that citronellol-bearing lipid II is not a proper substrate for the transglycosylation process in cell wall biosynthesis.[17,18] Our results are in agreement with their observation.

The shift in peak wavelength also suggests the successful incorporation of fluorescein into the cell wall, as the microenvironment of fluorescein after incorporation into the cell wall should differ from that of the free form. Indeed, we have found that the maximum wavelength of the fluorescence of extracted cell wall is shifted to a shorter wavelength (blue shift). On the other hand, the addition of **4** to the bacteria does not induce any shift in the spectra, showing that the fluorescence in Fig. 6 is nonspecific contamination remaining after cell wall extraction.

[17] H. Men, P. Park, M. Ge, and S. Walker, *J. Am. Chem. Soc.* **120,** 2484 (1998).

[18] X.-Y. Ye, M.-C. Lo, L. Brunner, D. Walker, D. Kahne, and S. Walker, *J. Am. Chem. Soc.* **123,** 3155 (2001).

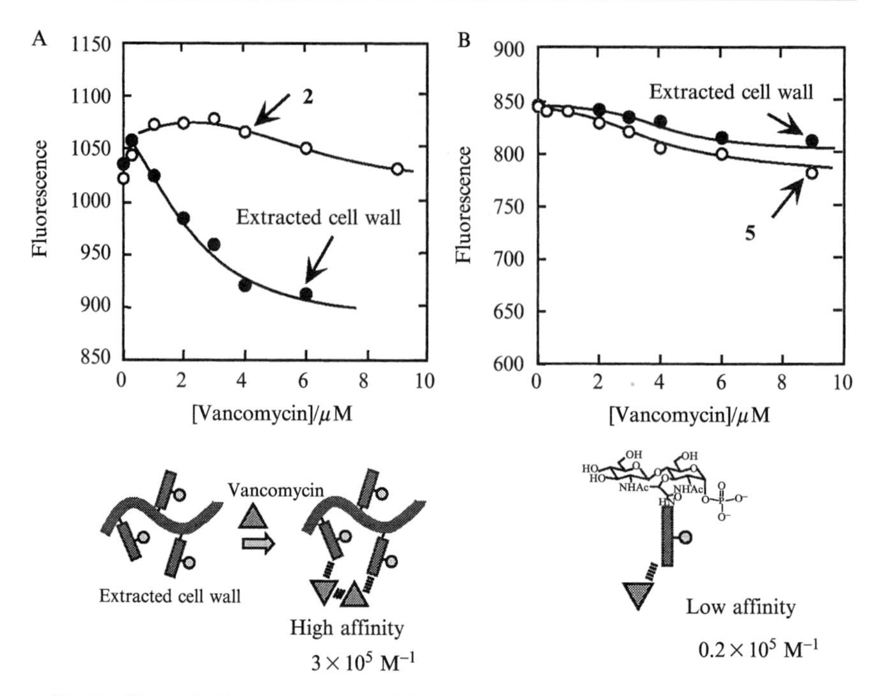

FIG. 7. Change in fluorescence on addition of vancomycin to precursors and extracted cell walls after incubation in the presence of precursors. (A) Addition of vancomycin to precursor **2** and extracted cell wall after incubation with **2**; (B) addition of vancomycin to precursor **5** and extracted cell wall after incubation with **5**.

Binding Behavior of Vancomycin to Extracted Cell Walls

As shown in Fig. 6, there is still some nonspecific contamination of the free components remaining in the cell wall fraction. To clearly distinguish between nonspecific contamination and specific incorporation, we have investigated the binding behavior of vancomycin to extracted cell walls (Fig. 7). Vancomycin is an antibiotic that binds specifically to the D-Ala-D-Ala moiety in pentapeptide, leading to an inhibition of cell wall biosynthesis. Vancomycin shows strong affinity ($\sim 10^6\ M^{-1}$) for the cell wall and cell wall mimics with multivalent binding sites, whereas it shows low affinity for peptides having a single binding site, such as D-Ala-D-Ala dipeptide ($\sim 10^4\ M^{-1}$).[19–22] Utilizing this multivalent effect of vancomycin

[19] D. H. Williams, M. P. Williamson, D. W. Butcher, and S. J. Hammond, *J. Am. Chem. Soc.* **105,** 1332 (1983).

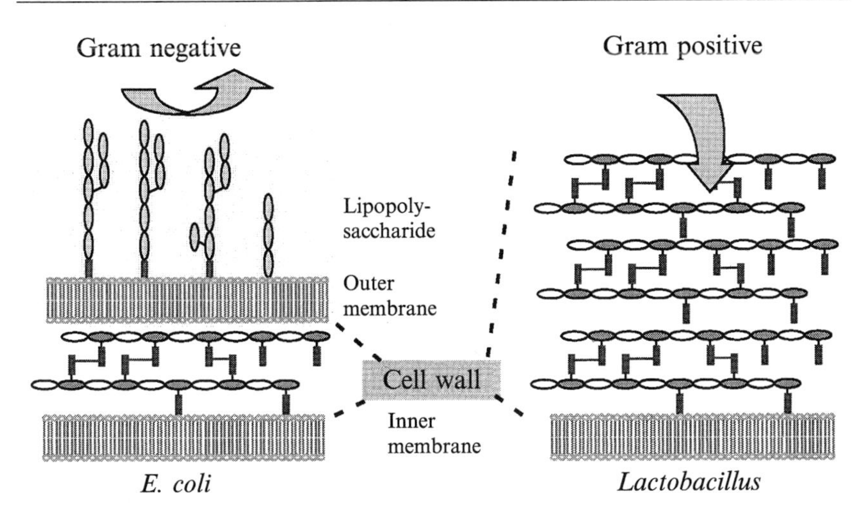

FIG. 8. Structure of gram-negative and positive bacterial surfaces.

binding, we tested whether fluorescein is incorporated into the cell wall or whether it is present in its free form as a contaminant. Addition of vancomycin to extracted cell wall grown with **2** induces a decrease in fluorescence intensity. The binding constant is estimated to be $5 \times 10^5\ M^{-1}$, which is a reasonable value for binding between cell wall and vancomycin. This indicates that fluorescein is incorporated into mature cell wall. On the other hand, extracted cell wall grown with **4** shows weak affinity, with a binding constant of $10^4\ M^{-1}$, showing that the fluorescence of extracted cell wall comes from the contaminated free form of **4** itself.

Incorporation of Synthetic Precursors into Bacterial Cell Wall of Lactic Acid Bacteria

Unlike gram-negative bacteria, gram-positive bacteria have thick cell walls with no outer membrane (Fig. 8). Therefore, more efficient incorporation of target compounds into the cell wall is expected because of the higher permeability.

[20] J. P. Mackay, U. Gerhard, D. A. Beauregard, R. A. Maplestone, and D. H. Williams, *J. Am. Chem. Soc.* **116,** 4573 (1994).
[21] P. J. Loll, R. Miller, C. M. Weeks, and P. H. Axelsen, *Chem. Biol.* **5,** 293 (1998).
[22] Z. Shi and J. H. Griffin, *J. Am. Chem. Soc.* **115,** 6482 (1993).

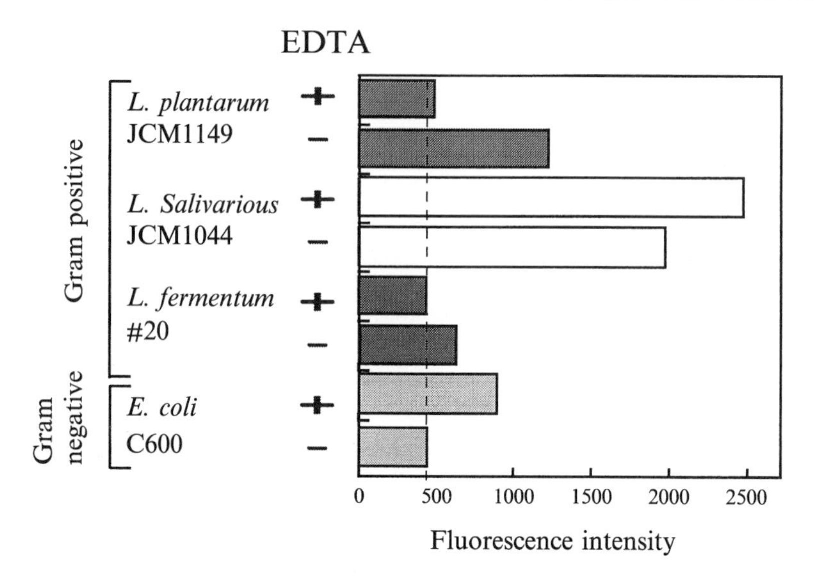

FIG. 9. Effect of EDTA treatment on fluorescence intensity of lactic acid bacteria and *E. coli*.

Comparison of Incorporation between Lactic Acid Bacteria and Escherichia coli

Three strains of lactic acid bacteria are grown in lactobacilli MRS broth (BD Diagnostic Systems, Sparks, MD) under anaerobic conditions at 37° overnight, collected by centrifugation at 1500 *g* for 3 min at 4°, and washed three times with PBS to remove the culture medium. The fluorescence intensity of the cell wall fraction is then measured, using the same procedure as for *E. coli* described above. Fluorescence intensity of extracted cell wall is shown in Fig. 9. All three *Lactobacillus* strains show incorporation without EDTA treatment. The amount of incorporated fluorescein for each of the three strains is larger than that for *E. coli*.

Effective Incorporation of Compounds into Lactic Acid Bacteria

Lactic acid bacteria are incubated in lactobacilli MRS medium containing **1** or **3** (5 m*M*) under anaerobic conditions at 37° overnight. Synthesized cell wall precursors are dissolved in PBS and sterilized by filtration before being added to the bacteria. The bacteria are collected by centrifugation and washed with PBS. After being lysed by ultrasonication, the residue is applied to a centrifuge column (Microcon YM-10; Millipore, Bedfore, MA), and washed three times with PBS. The washed residue is

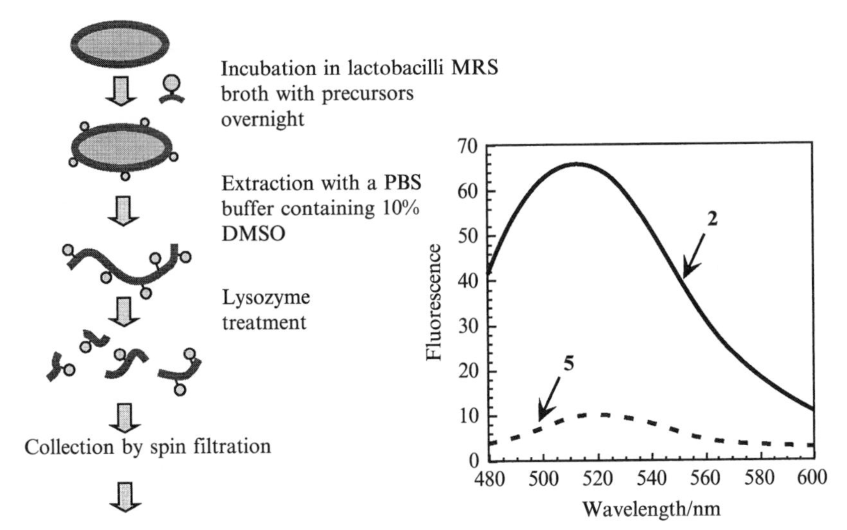

FIG. 10. Experimental procedure for incorporation of fluorescent precursors into lactic acid bacteria in lactobacilli MRS broth and collection of the cell wall fraction.

then treated with 4 ml of lysozyme solution (50 mg/ml in water, lysozyme from chicken egg white; ICN Pharmaceuticals, Costa Muse, CA) for 5 h at 37°. After adding a PBS buffer containing 10% DMSO to the residue, the mixture is filtered through a Microcon YM-10 column. The filtrate, containing digested cell wall, is diluted to the same volume for each sample with buffer and used for fluorescence measurement. Background fluorescence intensity is minimized by lysozyme digestion, which allows incorporated fluorophore to be separated from nonincorporated fluorophore (Fig. 10).

Fluorescence Microscope Observation

For microscopic observations, lactic acid bacteria are incubated in lactobacilli MRS medium containing **6** for 18 h under anaerobic conditions. Compound **6** is dissolved in PBS and sterilized by filtration before being added to the medium. After growth of the bacteria, the cells are collected and washed three times with PBS to remove medium and compounds not incorporated into the cell wall. Alexa 488 hydrazide (1.0 μM in PBS; Molecular Probes, Eugene, OR), which binds specifically to the keto group, is added to the bacteria suspended in the PBS solution. The bacteria collected by centrifugation are washed three times with PBS, suspended in a mixture

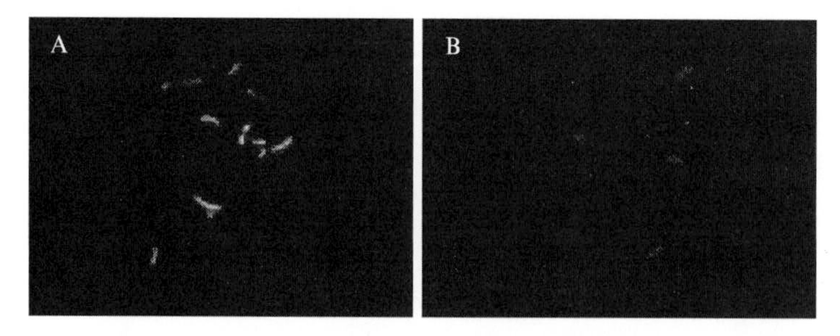

FIG. 11. Fluorescence microscopy images of lactic acid bacteria. (A) Lactic acid bacteria after incubation with **6** and coupling with Alexa 488 hydrazide; (B) lactic acid bacteria after treatment with Alexa 488 hydrazide (control).

of PBS and ethanol (1:1, v/v), and applied to an adhesive-coated glass slide (ADCELL; Erie Scientific, Portsmouth, NH). After dehydration by immersion in ethanol for 5 min, it is mounted, using ProLong antifade (Molecular Probes) mounting medium.

The bacteria, after incubation with **6**, generate strong fluorescence, showing that the keto group is displayed on the cell wall (Fig. 11A). In a control experiment, addition of the hydrazide dye to native bacteria does not provide any surface fluorescence (Fig. 11B). This result shows that ketone-bearing precursor **6** can be metabolically incorporated into the lactic acid bacterial cell wall, providing a platform for the display of a variety of molecules.

Summary

Cell wall precursors that have been modified at their peptide moiety were incorporated into the living bacterial cell wall. Using chemically synthesized bacterial cell wall precursors, a variety of compounds could be attached to the bacterial surface. *Escherichia coli* took the modified precursors into the cell wall after EDTA treatment, whereas lactobacilli took the compounds more effectively without EDTA treatment. Microscopic observation showed that the incorporated ketone moiety retained its reactivity. On the basis of this strategy, any compound can be displayed on the bacterial surface. This strategy for bacterial cell surface engineering will open the door for new technologies and therapies utilizing bacteria.

Acknowledgments

We are grateful to Professor C.-H. Wong (Scripps Research Institute) and Professor F. Tomita (Hokkaido University) for helpful suggestions. This work was supported by a grant for the Glycocluster Project from NEDO.

[20] Molecular Biological Fingerprinting of Human Lectin Expression by RT-PCR

By HARALD LAHM, SABINE ANDRÉ, ANDREAS HOEFLICH, JÜRGEN R. FISCHER, BERNARD SORDAT, HERBERT KALTNER, ECKHARD WOLF, and HANS-JOACHIM GABIUS

Introduction

The diversity of glycan structures in cellular glycoconjugates is reflected in the level of proteins harboring carbohydrate-binding activity. Therefore, distinct oligosaccharide epitopes can be likened to coding units. In regard to lectins linking the sugar code with biological responses,[1] their emerging complexity in human and animal tissues had not been anticipated despite the application of various plant agglutinins as laboratory tools. Research on animal lectins started in 1860 with the observation of S. W. Mitchell that the venom of the rattlesnake (*Crotalus durissus*) contains coagulating activity.[2] Working with washed erythrocytes instead of whole blood, hemagglutination (and not blood coagulation) could be pinpointed as reason for the observed cell aggregation.[3] This assay was crucial during the following decades to detect lectin activity in extracts primarily of plants, bacteria, and viruses. Nearly 30 years ago the presence of Ca^{2+}-independent β-galactoside-specific proteins was found in various mammalian and avian organs.[4,5] At that time, classification of lectins followed the experimental parameters of sugar specificity and dependence of activity on cations.[6,7] By using affinity chromatography, the active principle of such extracts (or the binding partner of the carbohydrate ligand in a neoglycoprotein on cell

[1] H.-J. Gabius, *Naturwissenschaften* **87,** 108 (2000).
[2] S. W. Mitchell, *Smithsonian Contrib. Knowl.* **XII,** 89 (1860).
[3] S. Flexner and H. Noguchi, *J. Exp. Med.* **6,** 277 (1902).
[4] V. I. Teichberg, I. Silman, D. D. Beitsch, and G. Resheff, *Proc. Natl. Acad. Sci. USA* **72,** 1383 (1975).
[5] T. P. Nowak, P. L. Haywood, and S. H. Barondes, *Biochem. Biophys. Res. Commun.* **68,** 650 (1976).
[6] H.-J. Gabius, *Eur. J. Biochem.* **243,** 543 (1997).
[7] D. C. Kilpatrick, "Handbook of Animal Lectins." J. Wiley & Sons, Chichester, 2000.

0076-6879/03 $35.00

surfaces or in tissue sections[8]) was purified to obtain what is now referred to as the first animal galectins.[9–11]

The following work revealed that an organism could express more than one type of lectin sharing this sugar specificity without a Ca^{2+} requirement.[12–16] Because structural analysis of the purified lectins then enabled defining of the general folding pattern and the molecular architecture of the carbohydrate recognition domain, these parameters have superseded sugar specificity for classification of lectin families.[6,7] As experienced with other lectin families, thorough scrutiny for galectins has uncovered a still growing list of members.[17,18] With functions in cell adhesion and growth control, monitoring the presence of galectins has become a topic of interest in cancer research, pathology, and developmental biology. However, the described situation raises serious problems, especially for the interpretation of immunohistochemical data. Because of the often preferential application of antibodies against only certain galectins, for example, galectins 1 and 3 in cancer diagnosis,[19] the correlation of the resulting data to clinical parameters is ambiguous without information on the presence of functionally homologous or antagonistic family members. Accounts on galectins 4, 7, 8, and 9, with an impact on growth control or other characteristics associated with malignancy, epitomize important lectin functionality beyond these two types,[20–25] and the functional divergence noted between

[8] H.-J. Gabius, S. André, A. Danguy, K. Kayser, and S. Gabius, *Methods Enzymol.* **242,** 37 (1994).

[9] A. de Waard, S. Hickman, and S. Kornfeld, *J. Biol. Chem.* **251,** 7581 (1976).

[10] H. Den and D. A. Malinzak, *J. Biol. Chem.* **252,** 5444 (1977).

[11] T. P. Nowak, D. Kobiler, L. E. Roel, and S. H. Barondes, *J. Biol. Chem.* **252,** 6026 (1977).

[12] E. C. Beyer, S. E. Zweig, and S. H. Barondes, *J. Biol. Chem.* **255,** 4236 (1980).

[13] C. F. Roff, P. R. Rosevear, J. L. Wang, and R. Barker, *Biochem. J.* **211,** 625 (1983).

[14] C. F. Roff and J. L. Wang, *J. Biol. Chem.* **258,** 10657 (1983).

[15] H.-J. Gabius, R. Engelhardt, S. Rehm, and F. Cramer, *J. Natl. Cancer Inst.* **73,** 1349 (1984).

[16] R. F. Cerra, M. A. Gitt, and S. H. Barondes, *J. Biol. Chem.* **260,** 10474 (1985).

[17] D. N. W. Cooper and S. H. Barondes, *Glycobiology* **9,** 979 (1999).

[18] H.-J. Gabius, *Anat. Histol. Embryol.* **30,** 3 (2001).

[19] H.-J. Gabius, *Cancer Invest.* **15,** 454 (1997).

[20] K. Polyak, Y. Xia, J. L. Zweier, K. M. Kinzler, and B. Vogelstein, *Nature* **389,** 300 (1997).

[21] J. Wada, K. Ota, A. Kumar, E. I. Wallner, and Y. S. Kanwar, *J. Clin. Invest.* **99,** 2452 (1997).

[22] F. Bernerd, A. Sarasin, and T. Magnaldo, *Proc. Natl. Acad. Sci. USA* **96,** 11329 (1999).

[23] Y. R. Hadari, R. Arbel-Goren, Y. Levy, A. Amsterdam, R. Alon, R. Zakut, and Y. Zick, *J. Cell Sci.* **113,** 2385 (2000).

[24] Y. Hippo, M. Yashiro, M. Ishii, H. Taniguchi, S. Tsutsumi, K. Hirakawa, T. Kodama, and H. Aburatani, *Cancer Res.* **61,** 889 (2001).

[25] N. Nagy, Y. Bronckart, I. Camby, H. Legendre, H. Lahm, H. Kaltner, Y. Hadari, P. Van Ham, P. Yeaton, J.-C. Pector, Y. Zick, I. Salmon, A. Danguy, R. Kiss, and H.-J. Gabius, *Gut* **50,** 392 (2002).

galectins 1 and 3 in a tumor model with competition for the same ligand emphasizes the requirement for comprehensive monitoring.[26] To address this issue a convenient and reliable method for profiling the expression of all members of a defined protein family is required. Thus, we have developed a routine assay to determine, as an example, the profile of galectin expression using human tumor cell lines as a model.[27]

Rationale for PCR Approach and Primer Selection

This assay is based on the availability of cDNA sequence information for the presently defined family members; it is thus not necessary to have information on the 5′ or 3′ untranslated regions. Because the N- and C-terminal stretches are not involved in ligand contacts, sequence variations in these regions are expected to be sufficient for devising discriminatory primer sets (see below). The positions for the primers are also selected to be at the beginning and the end of published complete cDNA sequences to take technical benefit from the presence of introns in the homologous galectin genes. This primer placement maximizes their distance at the level of the genomic DNA, thereby minimizing the risk of amplification of residual genomic DNA in RNA preparations. The applied strategy is depicted in Fig. 1 for galectin 3, illustrating also the exon–intron structure of the human gene.[28] Sense and antisense primers anneal to complementary sequences in exons 2 and 6, respectively, yielding a product of 719 bp from cDNA. Because of the presence of intervening sequences the primer sites are separated by more than 7.5 kb of genomic DNA, rendering it impossible for the applied *Taq* polymerase to amplify residual genomic DNA. To take advantage of this possibility of preventing amplification of genomic DNA we consistently select primer sequences for the human galectins at the beginning and the end of the corresponding coding sequence. Experimentally, the primer pairs are tested directly on genomic DNA as a control to prove the validity of our approach. Negative results infer the presence of introns for those galectin genes not analyzed in detail up to now. Indeed, no amplification has occurred in any instance except for one case. The exception is galectin 7: in that case the primers amplify a product of approximately 1200 bp from genomic DNA. Sequence analysis of the product reveals that 5′ and 3′ coding regions are separated by

[26] J. Kopitz, C. von Reitzenstein, S. André, H. Kaltner, J. Uhl, V. Ehemann, M. Cantz, and H.-J. Gabius, *J. Biol. Chem.* **276,** 35917 (2001).

[27] H. Lahm, S. André, A. Hoeflich, J. R. Fischer, B. Sordat, H. Kaltner, E. Wolf, and H.-J. Gabius, *J. Cancer Res. Clin. Oncol.* **127,** 375 (2001).

[28] M. M. Kadrofske, K. P. Openo, and J. L. Wang, *Arch. Biochem. Biophys.* **349,** 7 (1998).

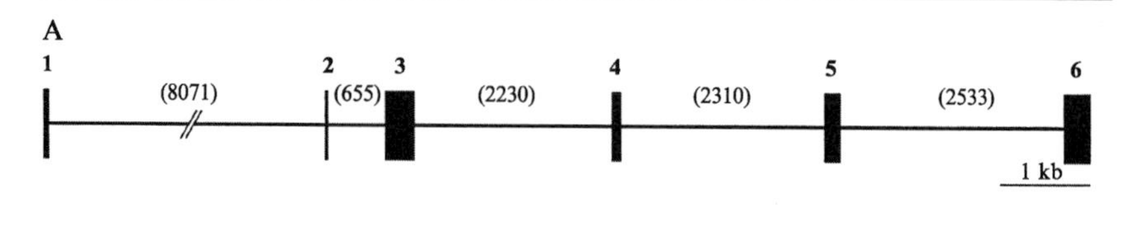

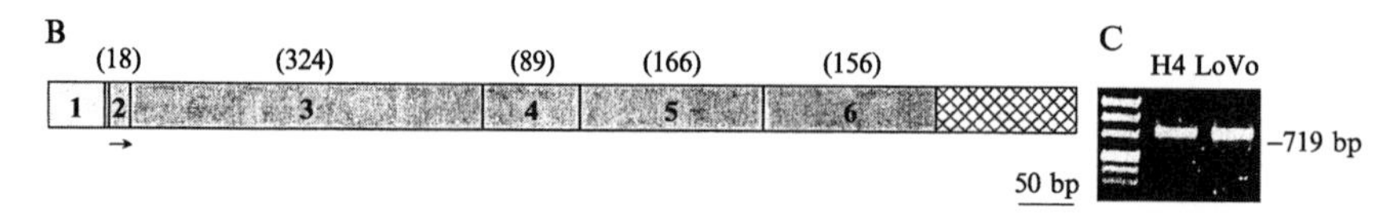

FIG. 1. (A) Genomic organization of the human galectin-3 gene. Numbers indicate exons, numbers in parentheses refer to the length of introns. Data were derived from sequence AL139316, which comprises the complete genomic sequence of the human galectin-3 gene. (B) Full-length mRNA of human galectin 3. Arrows indicate the position of primers used in PCR analysis. Numbers in parentheses indicate the number of coding base pairs in each exon. Open bar, 5′ UTR; screened bar, coding region; cross-hatched bar, 3′ UTR. (C) RT-PCR analysis of galectin-3 expression in H4 (neuroglioma cell line) and LoVo (cell line derived from a supraclavicular metastasis of a colorectal carcinoma), using primers Galectin-3#1 and Galectin-3#2.

intronic sequences that exhibit the typical splice site consensus sequences at the 5′ end (GT) and the 3′ end (AG). However, even in this case the expected product of 282 bp from cDNA amplification can be spotted easily and reliably by its characteristic length in gel electrophoretic analysis. Having addressed the problem of how to avoid mispriming by contaminating genomic DNA, it is essential in the next step to ensure primer specificity for the target. For this purpose, we have performed an extensive BLAST analysis. We have also deliberately taken into account the fact that the T_m values of the primer-dependent duplexes should be similar. This consideration allows entering the same annealing temperature into the cycling program in all cases. Table I lists the sequences of the primers used and the predicted length of each amplification product.

To validate that the positive signals obtained by reverse transcription-polymerase chain reaction (RT-PCR) are indicative of the actual presence of the protein, we first perform analyses in cell lines in which the expression of galectin-1 and -3 protein has previously been investigated. The results of PCR analysis are almost completely in accordance with data obtained by Western blot or fluorescence-activated cell sorting (FACS) analyses. In only two instances do RT-PCR and Western blots give a positive result for galectin 3 whereas FACS analysis detects no protein (Table II). These examples show that cell surface presentation will not be invariably detectable for a galectin that also has intracellular functions. In principle, these

TABLE I
SEQUENCES OF PRIMERS USED IN RT-PCR ANALYSES[a]

Primer	Sequence	Location	Accession no.	Size (bp)
Galectin-1#1	AAC CTG GAG AGT GCC TTC GA	45–64	X14829	321
Galectin-1#2	GTA GTT GAT GGC CTC CAG GT	367–348		
Galectin-2#1	ATG ACG GGG GAA CTT GAG GTT	27–47	M87842	358
Galectin-2#2	TTA CGC TCA GGT AGC TCA GGT	384–364		
Galectin-3#1	ATG GCA GAC AAT TTT TCG CTC C	1–22	S59012	719
Galectin-3#2	ATG TCA CCA GAA ATT CCC AGT T	719–698		
Galectin-4#1	GCT CAA CGT GGG AAT GTC TGT	101–121	U82953	609
Galectin-4#2	GAG CCC ACC TTG AAG TTG ATA	709–689		
Galectin-7#1	ATG TCC AAC GTC CCC CAC AAG	19–39	L07769	282
Galectin-7#2	TGA CGC GAT GAT GAG CAC CTC	300–280		
Galectin-8#1	GTT GTC CTT AAA CAA CCT ACA G	45–66	X91790	608
Galectin-8#2	TAA CGA CGA CAG TTC GTC CAG	652–632		
Galectin-8#3	CGA ATG GCA GGC TAA GCT GG	605–585		561
Galectin-9#1	ACT ATT CAA GGA GGT CTC CAG	145–165	Z49107	571
Galectin-9#2	GGA TGG ACT TGG ATG GGT ACA	715–695		

[a]Reproduced from Ref. 27 with permission.

TABLE II
DETECTION OF GALECTIN EXPRESSION BY VARIOUS METHODS[a]

Cell line	Western blot		FACS analysis		RT-PCR	
	Gal-1	Gal-3	Gal-1	Gal-3	Gal-1	Gal-3
Caco-2	–	+	ND	ND	–	+
Colo205	ND	ND	+	+	+	+
HT29	–	+	ND	ND	–	+
SW620	+	+	+	+	+	+
DU145	+	+	+	–	+	+
LNCaP	–	–	–	–	–	–
PC-3	+	+	+	–	+	+
NIH-OVCAR3	ND	ND	(+)	(+)	+	+

[a]Reproduced in part from Ref. 27 with permission. ND, Not determined; +, strong; (+), weak; –, undetectable expression.

results confirm the expectation that the established approach is reliable to detect galectin expression in a given cell line. However, two cautionary notes should be kept in mind at this point. First, the presence of detectable mRNA may not necessarily lead to significant production of the protein. Second, the occurrence of point mutations in the primer region might result in false-negative results. Immunodetection can additionally suffer from such events in the coding sequence, which may alter or truncate the protein, yielding the synthesis of variants. In fact, we have detected point mutations that will lead to premature termination of the protein in the tandem repeat-type galectin-9 sequence in several colorectal carcinoma cell lines.[29] If an eye is kept on these possibilities, the PCR approach is a reliable method to quickly screen a panel of cell lines or any other type of specimen for galectin gene expression.

Technical Protocol and Results of Profiling for Human Tumor Cells

Total RNA is extracted from cells grown to subconfluence by the addition of TriPure reagent (Roche Diagnostics, Mannheim, Germany) according to the manufacturer's recommendation. Before cDNA production, 10 μg of total RNA is routinely incubated for 30 min at 37° in 1 m*M* Tris-HCl buffer (pH 7.4) containing 1 m*M* $MgCl_2$, with 10 U of RNase-free DNase 1 (Roche Diagnostics) in a final volume of 20 μl to degrade genomic DNA. The reaction is stopped by incubation for 10 min at 75° to inactivate the enzyme. As a template for the first-strand synthesis, 2.5 μg of total RNA is used. Reverse transcription is performed for 1 h at 37° in RT buffer [50 m*M* Tris-HCl (pH 8.3), 75 m*M* KCl, 3 m*M* $MgCl_2$, provided by the supplier], 10 m*M* dithiothreitol (DTT), dNTPs (1 m*M* each), random hexamer primers (0.6 μg), and 20 U of Moloney murine leukemia virus (Mo-MuLV) reverse transcriptase (GIBCO, Karlsruhe, Germany) in a final volume of 20 μl. The reaction is terminated by incubation at 95° for 10 min. The amplification of galectin-7 sequences has proved to be somewhat tricky, because the amplified sequence is relatively GC rich (65.6%). This difficulty can be overcome by the use of Q solution [containing dimethyl sulfoxide (DMSO), provided by Qiagen (Hilden, Germany)] together with a *Taq* polymerase from Qiagen. This combination works well for amplification of the other galectin-specific transcripts and, therefore, this enzyme is used in all of the PCR analyses. The 20-μl reactions contain 2 μl of cDNA, 0.5 U of *Taq* polymerase, Qiagen PCR buffer, 3 m*M* $MgCl_2$, dNTPs (50 μM each), sense and antisense primers (0.1 μM each), and 4 μl of Q solution. By following the rules noted above for primer selection, the amplification of galectin-specific transcripts can be performed under identical cycling conditions: samples are heated to 94° for 4 min followed by 36 cycles of 94° for 1 min (denaturation), 60° for 1 min (annealing), and 72° for 2 min (synthesis). After a final extension period of 10 min at 72° amplified

[29] H. Lahm, A. Hoeflich, S. André, B. Sordat, H. Kaltner, E. Wolf, and H.-J. Gabius, *Int. J. Oncol.* **17,** 519 (2000).

TABLE III
GALECTIN EXPRESSION BY HUMAN TUMOR CELL LINES OF VARIOUS HISTOGENETIC ORIGINS[a]

Tissue	Total	Gal-1	Gal-2	Gal-3	Gal-4	Gal-7	Gal-8	Gal-9
Breast	9	9	0	9	0	5	9	0
Colon	22	8	7	22	14	16	22	16
Lung	10	4	0	5	0	2	8	0
Brain	8	8	2	8	3	0	8	3
Skin	9	8	0	9	0	0	8	0
Kidney	2	1	0	1	0	0	2	0
Prostate	3	2	0	2	0	0	3	0
Ovary	2	2	0	2	0	0	2	1
Blood	17	11	3	5	3	0	15	14
Connective tissue	1	1	0	1	0	1	1	0
Σ:	83	54	12	64	20	24	78	34

[a]A detailed galectin expression profile of the majority of cell lines can be found in Ref. 27.

products are separated on 2% Tris–acetic acid–EDTA (TAE) gels and visualized by ethidium bromide staining under ultraviolet (UV) light.

We have subjected a large number of human tumor cell lines of different histogenetic origin to PCR analysis to provide initial insight into the complexity of galectin expression of various tumor cell types and to compare their individual galectin gene expression patterns. Figure 2 depicts typical results of a screening for galectins 1, 2, and 4 in breast and colorectal cancer cell lines. At present, more than 80 cell lines have been investigated, and a summary of the results is shown in Table III. Several conclusions can be derived from inter- and intratumor type comparison. Galectins 1, 3, and 8 are widely expressed in different tumor entities. In particular, galectin-8 mRNA is detected in all but three tested cell lines. In contrast, galectins 2, 4, 7, and 9 show more restricted expression. In cell lines derived from skin, kidney, and the urogenital system none of these galectins is found. The screening also reveals that galectin 9 shows preferential expression in colonic and hematopoietic cell lines. The majority of cell lines express at least three or four different genes for members of the galectin family. This result is consistent with the concept that galectins may exert their biological activity as a consequence of the sum of agonistic and antagonistic actions of the family members. Implications for further immunohistochemical investigations are obvious. In addition to taking stock of intergalectin comparison, monitoring of the amplification products also provides insights into

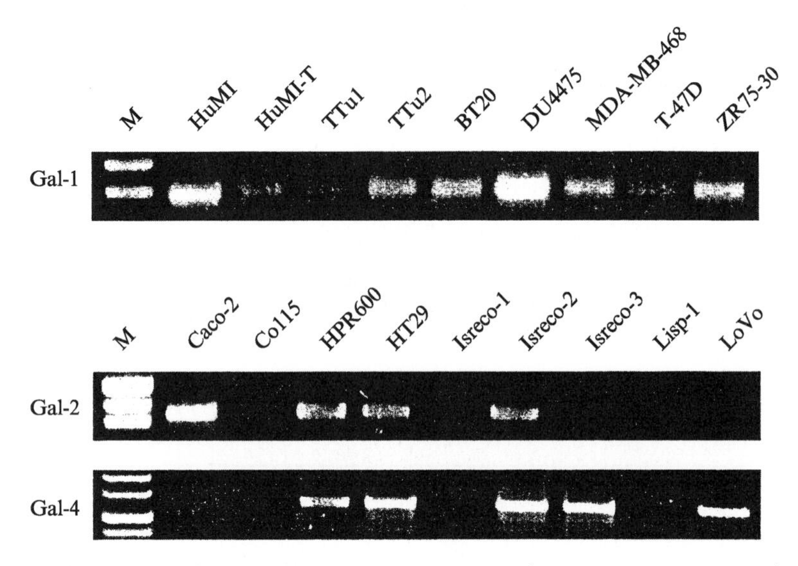

FIG. 2. Representative RT-PCR analyses for galectin 1 (breast carcinoma cell lines) and galectins 2 and 4 (colorectal carcinoma cell lines). M, Marker.

diversity of expression for an individual galectin, for example, detection of isoforms.

Detection and Analysis of Galectin Isoforms

In two cases in the course of this profiling the characteristics of the RT-PCR products did not completely match the expected lengths. Colorectal carcinoma cell lines differentially expressed either the galectin 9-specific transcript of the expected size, a longer product, or both.[29] Sequencing of the shorter product revealed that these cell lines expressed ecalectin, which is closely related to galectin 9 with amino acid differences at positions 5, 88, 135, 238, and 281, rendering it likely to be an allelic variant.[30,31] The longer transcript is constituted by a galectin-9 isoform characterized by an insert of 96 bp in the linker region. For mouse (rat) galectin 9, a 31 (32)-amino acid insertion at this place characterizes the intestinal isoform.[32] When we investigated additional cell lines we also found transcripts of the longer isoform in tumor cell lines derived from the urogenital or hematopoietic

[30] Ö. Türeci, H. Schmitt, N. Fadle, M. Pfreundschuh, and U. Sahin, *J. Biol. Chem.* **272,** 6416 (1997).

[31] R. Matsumoto, H. Matsumoto, M. Seki, M. Hata, Y. Asano, S. Kanegasaki, R. L. Stevens, and M. Hirashima, *J. Biol. Chem.* **273,** 16976 (1998).

[32] J. Wada and Y. S. Kanwar, *J. Biol. Chem.* **272,** 6078 (1997).

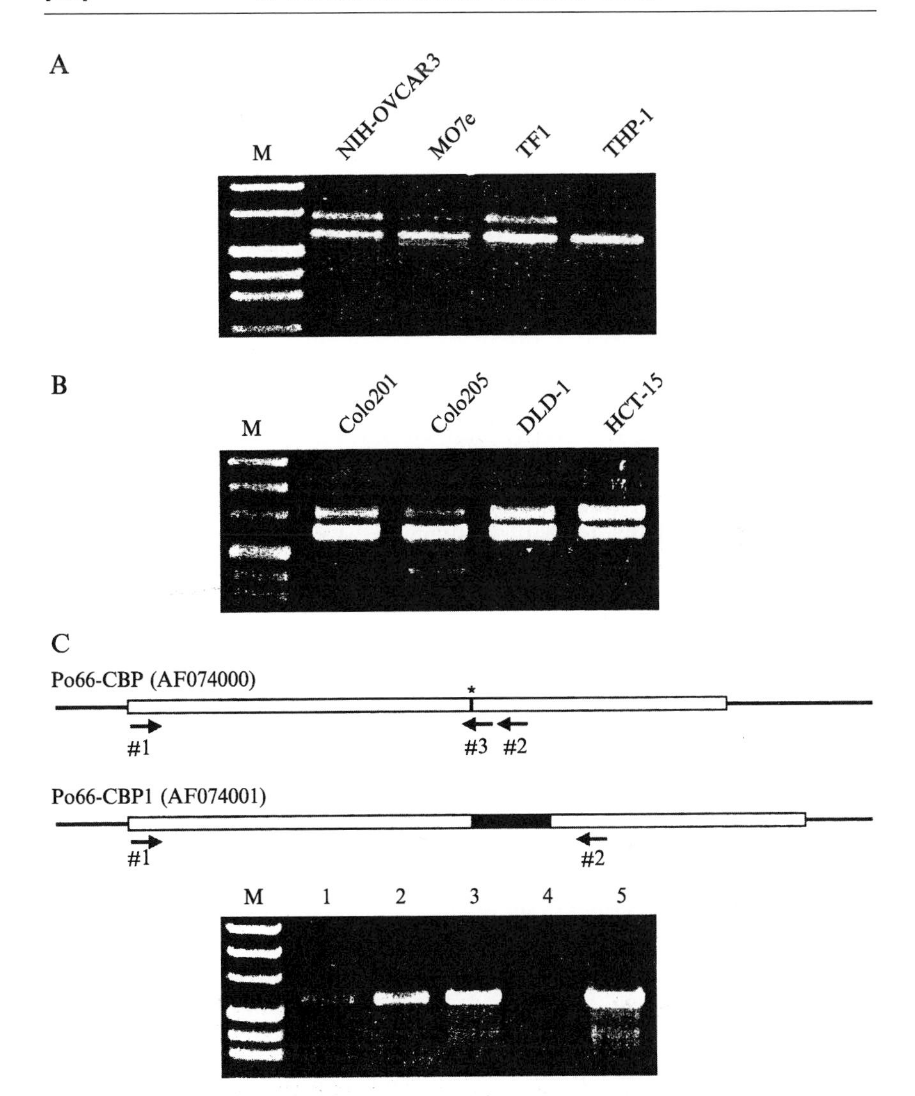

FIG. 3. Expression of galectin-8 and -9 isoforms. (A) RT-PCR analysis for galectin-9 expression, using primers Galectin-9#1 and Galectin-9#2. (B) RT-PCR analysis for galectin-8 expression, using primers Galectin-8#1 and Galectin-8#2. (C) Selective detection of Po66-CBP transcripts in colorectal carcinoma cell lines using primers Galectin-8#1 and Galectin-8#3. Lane 1, LS411N; lane 2, LS1034; lane 3, Isreco-1; lanes 4 and 5, plasmids containing the complete sequences of galectin 8-related cDNAs (accession numbers AF342816 and AF342815). The asterisk indicates the position of the insert (solid bar) of the longer isoform. M, Marker.

system (Fig. 3A). Interestingly, the occurrence of inserts is not confined to this tandem repeat-type galectin. We likewise obtained an additional band in galectin 8-specific RT-PCR analyses (Fig. 3B). However, in all positive cell lines both transcripts were concomitantly expressed and differential expression of only one transcript was seen in no instance. Sequence analysis revealed that these products are identical to carbohydrate-binding protein Po66-CBP or an isoform with an additional 42-amino acid insert in the linker region (Po66-CBP1) closely related to prostate carcinoma tumor antigen 1 (PCTA-1).[33,34] To selectively amplify the shorter Po66-CBP product we used an antisense primer that is partly complementary to sequences at the 3′ end of the insert. Using this primer, only the short transcript was amplified (Fig. 3C). Analysis suggests that Po66-CBP and its isoforms are derived from a single gene located on chromosome 1q42 as a result of alternative splicing events.[34] These results lead to the conclusion that the processing of transcripts from genes for tandem repeat-type galectins apparently adds a further level of complexity to galectin expression. RT-PCR analysis provides pertinent information about the occurrence of these isoforms (Fig. 3).

Conclusions

The described procedure is used to determine the profile of galectin expression in cell lines and tumor tissue with standard equipment and reasonable investments in reagents. By using discriminatory primer sets it also allows the detection of variant mRNAs with different coding lengths for the linker peptide of tandem repeat-type galectins. By all means, the given results emphasize that interpretation of immunohistochemical data on the basis of only one or a few family members may well lead to questionable conclusions. This situation encourages studies with an adequately extended panel of antibodies.[35,36] Their application in Western blot analysis is a standard means to monitor the expression levels of individual proteins. Using quantitative RT-PCR in combination with that method, analysis of regulation of lectin production will comprise transcriptional and

[33] R. V. Gopalkrishnan, T. Roberts, S. Tuli, D.-c. Kang, K. A. Christiansen, and P. B. Fisher, *Oncogene* **19,** 4405 (2000).

[34] N. Bidon, F. Brichory, S. Hanash, P. Bourguet, L. Dazord, and J. P. Le Pennec, *Gene* **274,** 253 (2001).

[35] I. Camby, N. Belot, S. Rorive, F. Lefranc, C.-A. Maurage, H. Lahm, H. Kaltner, Y. R. Hadari, M.-M. Ruchoux, J. Brotchi, Y. Zick, I. Salmon, H.-J. Gabius, and R. Kiss, *Brain Pathol.* **11,** 12 (2001).

[36] U. Wollina, T. Graefe, S. Feldrappe, S. André, K. Wasano, H. Kaltner, Y. Zick, and H.-J. Gabius, *J. Cancer Res. Clin. Oncol.* **128,** 103 (2002).

translational processes. Technically, this described method can be taken to the histochemical level by performing *in situ* PCR analysis. When isoforms or truncated transcripts occur, as, for example, is known to happen with the type II transmembrane C-type lectins dectin-2 and plasmacytoid dendritic cell-specific antigen 2 (BDCA-2)[37–39] the selection of appropriate primers can compensate for the lack of availability of discriminatory monoclonal antibodies to define the expression profile at the level of transcription. Finally, in this era of database mining with commensurate growth in appreciation of the diversification of protein families, it is clear that the given protocol can readily be adapted to the detection of message for newly found open reading frames. Members of any other well-characterized protein family in or beyond glycosciences fall into its scope.

Acknowledgments

We are grateful to R. Ohl for excellent processing of the manuscript, to A. Helfrich for expert technical assistance, and to the Wilhelm Sander-Stiftung (Munich) for generous financial support.

[37] K. Ariizumi, G. L. Shen, S. Shikano, R. Ritter III, P. Zukas, D. Edelbaum, A. Morita, and A. Takashima, *J. Biol. Chem.* **275,** 11957 (2000).

[38] A. Dzionek, Y. Sohma, J. Nagafune, M. Cella, M. Colonna, F. Facchetti, G. Gunther, I. Johnston, A. Lanzavecchia, T. Nagasaka, T. Okada, W. Vermi, G. Winkels, T. Yamamoto, M. Zysk, Y. Yamaguchi, and J. Schmitz, *J. Exp. Med.* **194,** 1823 (2001).

[39] C. G. Figdor, Y. van Kooyk, and G. J. Adema, *Nat. Rev. Immunol.* **2,** 77 (2002).

Section II

General Techniques

[21] Visualization and Characterization of Receptor Clusters by Transmission Electron Microscopy

By JASON E. GESTWICKI, CHRISTOPHER W. CAIRO, M. JACK BORROK, and LAURA L. KIESSLING

Introduction

Multivalency is a feature common to many natural glycoconjugates, including glucosaminoglycans, mucins, and lipopolysaccharides.[1–8] Multivalent ligands often bind avidly to their target receptors. In addition, they can simultaneously bind multiple receptors and cause them to become clustered (Fig. 1).[2,9–13] Receptor clustering by multivalent ligands modulates the cell adhesion and signaling functions of many carbohydrate-binding proteins, including galectins and selectins.[14,15] Not all clusters, however, are equivalent in signaling potency or adhesion properties. Clusters that contain multiple copies of a receptor often elicit greater responses; therefore, the stoichiometry of receptor–ligand clusters can be an important determinant of activity. The relationship between stoichiometry and activity is evident in the systems that govern cellular proliferation,[16–18]

[1] C. R. Bertozzi and L. L. Kiessling, *Science* **291,** 2357 (2001).
[2] L. L. Kiessling, J. E. Gestwicki, and L. E. Strong, *Curr. Opin. Chem. Biol.* **4,** 696 (2000).
[3] M. Mammen, S.-K. Choi, and G. M. Whitesides, *Angew. Chem. Int. Ed. Engl.* **37,** 2755 (1998).
[4] Y. C. Lee and R. T. Lee, *Acc. Chem. Res.* **28,** 321 (1995).
[5] R. Roy, *Curr. Opin. Struct. Biol.* **6,** 692 (1996).
[6] T. Feizi, *Immunol. Rev.* **173,** 79 (2000).
[7] E. J. Toone, *Curr. Opin. Struct. Biol.* **4,** 719 (1994).
[8] A. Imberty and K. Drickamer, *Curr. Opin. Struct. Biol.* **9,** 547 (1999).
[9] C.-H. Heldin, *Cell* **80,** 213 (1995).
[10] J. D. Klemm, S. L. Schreiber, and G. R. Crabtree, *Annu. Rev. Immunol.* **16,** 569 (1998).
[11] R. N. Germain, *Curr. Biol,* **7,** R640 (1997).
[12] B. Sulzer and A. S. Perelson, *Mol. Immunol.* **34,** 63 (1997).
[13] S. Damjanovich, R. Gaspar, Jr., and C. Pieri, *Q. Rev. Biophys.* **30,** 67 (1997).
[14] H. Lis and N. Sharon, *Chem. Rev.* **98,** 637 (1998).
[15] J. C. Sacchettini, L. G. Baum, and C. F. Brewer, *Biochemistry* **40,** 3009 (2001).
[16] A. C. Rapraeger, A. Krufka, and B. B. Olwin, *Science* **252,** 1705 (1991).
[17] L. Pellegrini, D. F. Burke, F. von Delft, B. Mulloy, and T. L. Blundell, *Nature* **407,** 1029 (2000).
[18] A. D. DiGabriele, I. Lax, D. I. Chen, C. M. Svahn, M. Jaye, J. Schlessinger, and W. A. Hendrickson, *Nature* **393,** 812 (1998).

METHODS IN ENZYMOLOGY, VOL. 362
0076-6879/03 $35.00

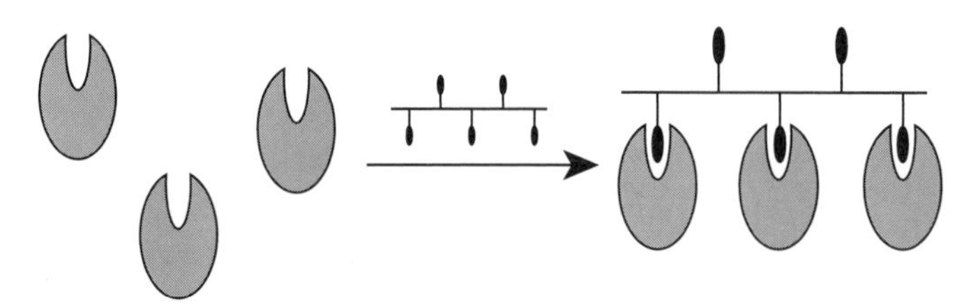

FIG. 1. Multivalent ligand-induced clustering of a receptor.

immune responses,[12,19] apoptosis,[20–23] and cell aggregation[24]: ligands that cluster more receptors are more potent. Thus, there is a need for assays that can be used to characterize multivalent ligand–receptor clusters.

Despite the importance of stoichiometry in multivalent carbohydrate–receptor interactions, methods for measuring functional valency, or the number of receptors bound to a single multivalent ligand, are not general. These methods, including immunoprecipitation,[25] spin labeling,[26] fluorescence and scanning near-field optical microscopy,[27–30] analytical ultracentrifugation, circular dichroism,[31] electrospray ionization mass spectrometry,[32,33] capillary and gel electrophoresis,[34–36] and light-scattering

[19] H. M. Dintzis, R. Z. Dintzis, and B. Vogelstein, *Proc. Natl. Acad. Sci. USA* **73,** 3671 (1976).

[20] A. S. Varadhachary, M. Edidin, A. M. Hanlon, M. E. Peter, P. H. Krammer, and P. Salgame, *J. Immunol.* **166,** 6564 (2001).

[21] J.-L. Bodmer, P. Schneider, and J. Tschopp, *Trends Biochem. Sci.* **27,** 19 (2002).

[22] Y. Liu, L. Xu, N. Opalka, J. Kappler, H.-B. Shu, and G. Zhang, *Cell* **108,** 383 (2002).

[23] C. H. Weber and C. Vincenz, *FEBS Lett.* **492,** 171 (2001).

[24] J. E. Gestwicki, L. E. Strong, C. W. Cairo, F. J. Boehm, and L. L. Kiessling, *Chem. Biol.* **9,** 163 (2002).

[25] F. A. Stephenson, *Curr. Drug Targets* **2,** 233 (2001).

[26] D. Marsh and L. I. Horváth, *Biochim. Biophys. Acta* **1376,** 267 (1998).

[27] P. Nagy, A. Jenei, A. K. Kirsch, J. Szöllösi, S. Damjanovich, and T. M. Jovin, *J. Cell Sci.* **112,** 1733 (1999).

[28] S. Oleskevich, F. J. Alvarez, and B. Walmsley, *J. Neurophysiol.* **82,** 312 (1999).

[29] P. Hinterdorfer, H. J. Gruber, J. Striessnig, H. Glossman, and H. Schindler, *Biochemistry* **36,** 4497 (1997).

[30] J. Matko and M. Edidin, *Methods Enzymol.* **278,** 444 (1997).

[31] T. A. Isenbarger and M. P. Kreb, *Biochemistry* **40,** 11923 (2001).

[32] T. D. Veenstra, *Biophys. Chem.* **79,** 63 (1999).

[33] Y. Ho, A. Gruhler, A. Heibut, G. D. Bader, L. Moore, S.-L. Adams, A. Millar, P. Taylor, K. Bennett, K. Boutilier, L. Yang, C. Wolting, I. Donaldson, S. Schandorff, J. Shewnarane, M. Vo, J. Taggart, M. Goudreault, B. Muskat, C. Alfarano, D. Dewer, Z. Lin, K. Michalickova, A. R. Willems, H. Sassi, P. A. Nielsen, K. J. Rasmussen, J. R. Andersen, A. Podtelejnikov, E. Nielsen, J. Crawford, V. Poulsen, B. D. Sørensen, J. Matthiesen, R. C. Hendrickson, F. Gleeson, T. Pawson, M. F. Moran, D. Durocher, M. Mann, C. W. V. Hogue, D. Figeys, and M. Tyers, *Nature* **415,** 180 (2002).

experiments,[37] are generally well-suited to the examination of high-affinity interactions between homogeneous protein populations, such as complexes formed with monoclonal antibodies. In contrast, carbohydrate receptor clustering, although a key component of signaling, is not necessarily a high-affinity event.[38] Most methods that are generally applicable to the study of carbohydrate binding tend to ignore the impact of lectin clustering and, instead, focus on the role of multivalency in enhancing functional affinity (apparent binding affinity). Affinity-based assays often fail to identify ligands with potent clustering abilities. New methods are needed for the routine analysis of glycoconjugate–receptor clusters.

Several features of carbohydrate ligand–lectin clustering must be taken into account when developing new assays. When multivalent carbohydrate binding occurs, the resultant complexes can have a wide range of functional affinities. Consequently, methods used to investigate the stoichiometries of these complexes must allow analysis of either strong or weak binding. Moreover, when the binding of a multivalent ligand results in receptor clusters that vary in stoichiometry and abundance, it is useful to have a sensitive method that can be used to visualize the entire range of clusters, including low-abundance species. Further, complexation of receptors and multivalent ligands can give rise to assemblies with high molecular weights. These large complexes often precipitate,[39,40] leading to significant constraints on solution-based approaches.

Advances in methods for single particle detection are providing new opportunities for investigating proteins.[41–45] We envisioned that applying single particle techniques to the analysis of multivalent carbohydrate–

[34] K. Shimura and K. Kasai, *Electrophoresis* **19,** 397 (1998).

[35] J. R. Cann, *Electrophoresis* **19,** 127 (1998).

[36] J. C. Byrd, J. H. Y. Park, B. S. Schaffer, F. Garmroudi, and R. G. MacDonald, *J. Biol. Chem.* **275,** 18647 (2000).

[37] A. V. Timoshenko, I. V. Gorudko, S. N. Cherenkevich, and H.-J. Gabius, *FEBS Lett.* **449,** 75 (1999).

[38] J. E. Gestwicki, C. W. Cairo, L. E. Strong, K. A. Oetjen, and L. L. Kiessling, *J. Am. Chem. Soc.* **124,** 14922 (2002).

[39] J. B. Corbell, J. J. Lundquist, and E. J. Toone, *Tetrahedron Asymmetry* **11,** 95 (2000).

[40] C. W. Cairo, J. E. Gestwicki, M. Kanai, and L. L. Kiessling, *J. Am. Chem. Soc.* **124,** 1615 (2002).

[41] D. P. Felsenfeld, D. Choquet, and M. P. Sheetz, *Nature* **383,** 438 (1996).

[42] M. Edidin, S. C. Kuo, and M. P. Sheetz, *Science* **254,** 1379 (1991).

[43] R. Iino, I. Koyama, and A. Kusumi, *Biophys. J.* **80,** 2667 (2001).

[44] J. C. Loftus and R. M. Albrecht, *J. Cell Biol.* **99,** 822 (1984).

[45] S. Damjanovich, J. Matko, L. Matyus, G. Szabo, Jr., J. Szollosi, J. C. Pieri, T. Farkas, and R. Gaspar, Jr., *Cytometry* **33,** 225 (1998).

protein interactions would be synergistic with existing methods.[46] Specifically, we have applied transmission electron microscopy (TEM) to directly visualize individual carbohydrate ligand–lectin clusters. Like other highly sensitive methods, TEM has the advantages of nanometer resolution and a requirement for only small quantities of reagents. In addition, because individual clusters can be visualized, we anticipated that even minor species could be detected in our analyses.[46]

To explore the generality of our TEM method, we sought to investigate a series of carbohydrate receptors with diverse characteristics. Toward this goal, we chose to study the clustering of the plant lectin concanavalin A (Con A), the bacterial periplasm resident, glucose/galactose-binding protein (GGBP), and the human lectin, mannose-binding protein (MBP). In addition to their varied origins and functions, these proteins possess a range of affinities: Con A and MBP bind mannose with a K_d of approximately 1 mM,[47,48] whereas GGBP has a K_d of 0.5 μM for galactose.[49] Further, the quaternary structures of these proteins are diverse: Con A is a homotetramer, GGBP is a monomer, and MBP is a homotrimer.[50] Finally, these proteins may have important therapeutic or biotechnological potential: Con A has potent apoptotic and mitogenic activity,[51] GGBP is being explored as a potential biosensor,[52] and MBP is involved in the function of the innate immune system.[53]

We hypothesized that our TEM method would be particularly useful in exploring the effects of ligand valency on the stoichiometry of the resulting clusters (Fig. 2). We chose to focus on multivalent ligands derived from ring-opening metathesis polymerization (ROMP) because of the strong clustering activity of ligands derived from this scaffold.[38,46] To explore the influence of ligand valency on the stoichiometry of receptor clusters by TEM, we generated multivalent carbohydrate-bearing ligands with distinct valencies by ROMP (Fig. 4).[54–56] We generated ROMP-derived multivalent ligands displaying a defined number of mannose or galactose

[46] J. E. Gestwicki, L. E. Strong, and L. L. Kiessling, *Angew. Chem. Int. Ed. Engl.* **39,** 4567 (2000).
[47] R. V. Weatherman and L. L. Kiessling, *J. Org. Chem.* **61,** 534 (1996).
[48] M. S. Quesenberry, R. T. Lee, and Y. C. Lee, *Biochemistry* **36,** 2724 (1997).
[49] W. Boos, A. S. Gordon, R. E. Hall, and H. D. Price, *J. Biol. Chem.* **247,** 917 (1972).
[50] W. I. Weis and K. Drickamer, *Annu. Rev. Biochem.* **65,** 441 (1996).
[51] D. H. Cribbs, V. M. Kreng, A. J. Anderson, and C. W. Cotman, *Neuroscience* **75,** 173 (1996).
[52] J. S. Marvin and H. W. Hellinga, *J. Am. Chem. Soc.* **120,** 7 (1998).
[53] W. I. Weis, M. E. Taylor, and K. Drickamer, *Immunol. Rev.* **163,** 19 (1998).
[54] T. M. Trnka and R. H. Grubbs, *Acc. Chem. Res.* **34,** 18 (2001).
[55] L. E. Strong and L. L. Kiessling, *J. Am. Chem. Soc.* **121,** 6193 (1999).
[56] M. Kanai, K. H. Mortell, and L. L. Kiessling, *J. Am. Chem. Soc.* **119,** 9931 (1997).

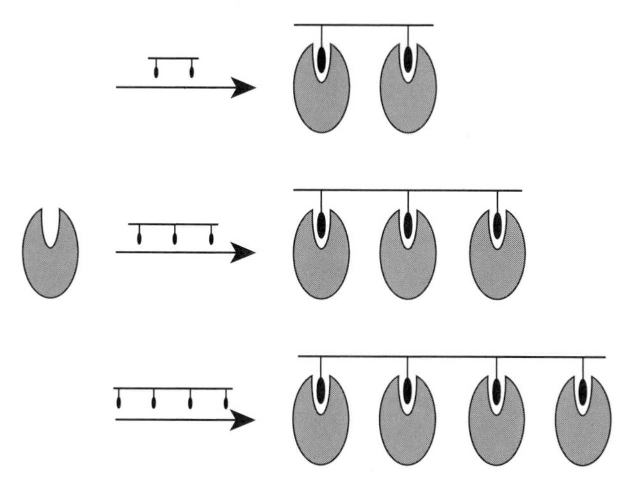

FIG. 2. Illustration of the ligand valency dependence of receptor cluster sizes. Increases in the valency of the multivalent ligand increase the maximum stoichiometry of the receptor and ligand in the resulting complex.

derivatives. Mannose is a ligand for Con A and MBP and galactose binds to GGBP. We envisioned that, if the ligands promote lectin clustering, increasing the valency of the synthetic ligands would increase the number of copies of lectin incorporated into the resulting clusters.[46]

Transmission Electron Microscopy Protocol

General Considerations

The resolution of TEM is greatly increased by using particles with high electron density, such as heavy metals. Therefore, we label biotinylated receptors with an electron-dense streptavidin–gold nanoparticle by taking advantage of the tight interaction between biotin and streptavidin (Fig. 3). Here, we describe the biotinylation of Con A, GGBP, and MBP and present a protocol for the subsequent TEM-based investigation of multivalent ligand interactions with these biotinylated receptors. The sources of the receptors are as follows: Con A is obtained from Vector Laboratories (Burlingame, CA), GGBP is produced in *Escherichis coli* by the osmotic shockate method as previously described,[57] and recombinant MBP is produced in *E. coli* and purified according to previously described procedures.[58]

[57] Y. Anraku, *J. Biol. Chem.* **243,** 3116 (1968).

[58] W. I. Weis, G. V. Crichlow, H. M. K. Murthy, W. A. Hendrickson, and K. Drickamer, *J. Biol. Chem.* **266,** 20678 (1991).

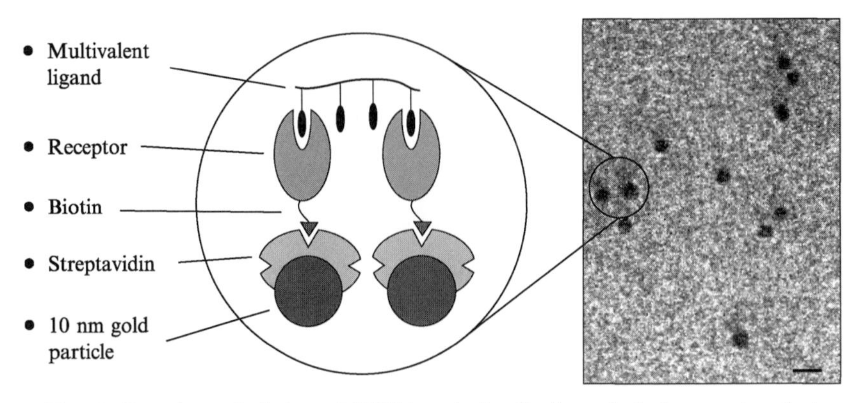

FIG. 3. Experimental design of TEM-based visualization of single receptor clusters. Minimally biotinylated receptor is bound to a multivalent ligand via its carbohydrate-binding site and a gold nanoparticle via its biotin group. At the right is a sample field from an experiment in which biotinylated Con A was treated with steptavidin–gold conjugates and a multivalent ligand. A cluster of two is shown in the circle. Bar: 20 nm.

Biotinylation Reactions

To minimize nonspecific effects, the receptor of interest should be modified with a low level of biotin (ideally one copy of biotin per receptor). Commercial reagents are available for biotinylating receptors and the subsequent determination of the extent of biotinylation. Here we provide a brief synopsis of these protocols for the biotinylation of the target receptors (Con A, GGBP, and MBP).

A solution of receptor (1 mg ml^{-1}) in 0.1 *M* sodium borate at pH 8.8 is biotinylated for 12 h at room temperature, using final biotinylating reagent concentrations ranging from 0 to 500 μg ml^{-1} in a 1-ml final volume. The biotinylating reagent is sulfosuccinimidyl-6-(biotinamido)hexanoate (EZ-Link sulfo-NHS-LC-biotin; Pierce, Rockford, IL). The biotinylation reactions are quenched with a 1 *M* aqueous solution of NH_4Cl. Excess biotinylating reagent is removed by extensive dialysis against 10 m*M* HEPES, pH 7.0, at 4°. The concentration of receptor after dialysis is determined by the Bradford assay, using bovine serum albumin (BSA) as a standard. Molar ratios of biotin to receptor are determined with 2-(4′-hydroxyazobenzene)benzoic acid (HABA; Pierce) according to the manufacturer's specifications.

A Con A–biotin conjugate with a biotin-to-Con A tetramer ratio of 2:1 is derived from a reaction with a starting biotinylating reagent concentration of 5 μg ml^{-1}. Modified GGBP and MBP are labeled with a protein-to-biotin ratio of 1:1. These results are obtained with a biotinylating

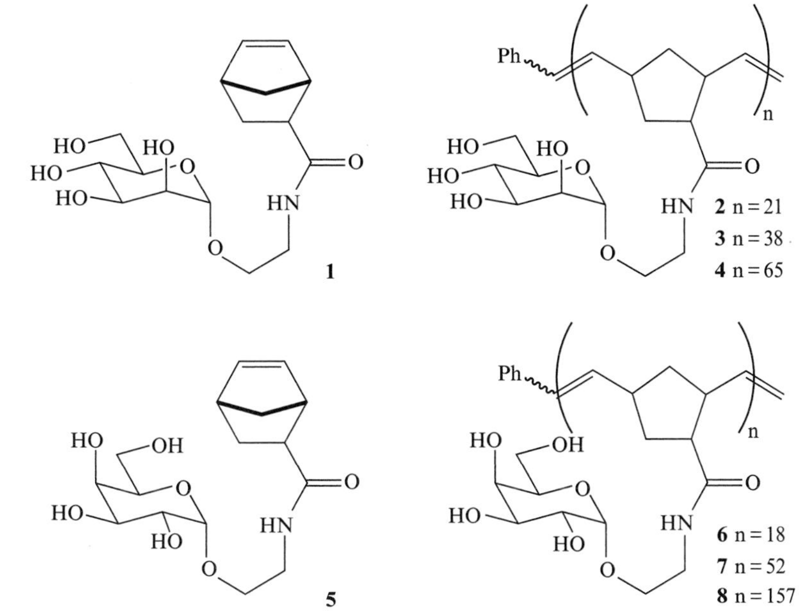

FIG. 4. Chemical structures of mannose- and galactose-bearing synthetic glycoconjugates. Monomers **1** and **5** were used to generate **2–4** and **6–8**, as previously described.[56] The average valency (degree of polymerization) of the polymers (n) is shown.

reagent concentration of 25 and 100 μg ml^{-1}. The conjugates are stored at -20° and are used for TEM experiments.

Electron Microscopy

Complexes between ligand and biotinylated receptor are assembled in solution and then visualized by placing them on grids for electron microscopy. Ligand (0.75 μM saccharide) is added to biotinylated receptor (Con A, 2.3 μM; GGBP, 1.8 μM; MBP, 1.9 μM) in 5 μl of phosphate-buffered saline (PBS), pH 7.2. Ligand concentrations are based on the total saccharide residue concentration, not on polymer concentration. Complexes between ligand and receptor are allowed to form for 15 min at room temperature before streptavidin–10 nm gold (3.0 μM) is added. This mixture is transferred to Formvar-treated copper grids. After an additional incubation of 10 min, excess liquid is removed. TEM is performed on a LEO (Oberkochen, Germany) Omega 912 energy-filtering electron microscope (EFTEM) outfitted with a ProScan slow scan charge-coupled device (CCD) camera. A convenient number of particles per field is obtained at a magnification of $\times$12,500.

Stoichiometry Optimization

Monomers **1** and **5** (Fig. 4) serve as important controls for optimizing the experimental conditions. These monovalent ligands are incapable of nucleating clusters; therefore, these compounds can be used to determine conditions for optimal particle distribution. High concentrations of gold particles should be avoided, as it can be difficult to discern coincidental adjacent placement from multivalent ligand-induced clustering. In addition, low concentrations should be avoided, as this prevents collection of statistically meaningful data. Subsequently, a short series of dilution experiments are performed in the presence of monomeric ligands **1** and **5** to optimize the reagent concentrations. Under optimal conditions, a molar ratio of 1:1.3 (receptor:gold particles) is used. A molar ratio of 1:3 (ligand:receptor) is used for Con A and a molar ratio of 1:2.5 is used for GGBP and MBP. These conditions yield approximately 10–100 particles per field.

Image Analysis

Complex formation is determined manually, using the measuring tools in Adobe Photoshop 5.0. The stoichiometry of individual clusters is scored by determining the number of gold particles that can be simultaneously contacted by a 25-nm line. Thus, clusters of two are defined as two gold particles within 25 nm and clusters of three, four, and five proteins are defined as three, four, or five gold particles within this distance, respectively. This distance (25 nm) is determined on the basis of the maximum distance between the terminal saccharides on a fully extended polymer generated by ROMP, as previously described.[46,56] Although the measurement distance used is specific for a polymer of intermediate size, the results of the manual counting experiments are not significantly altered when the distance is doubled (data not shown). This independence is likely due to the gold particles being well-distributed in the observation fields.

Three experiments are performed on separate days, using freshly prepared grids. On each day, 80–150 gold particles are counted from approximately 10–20 images of random fields for each treatment. The percentage of each complex type (unclustered, cluster of two, cluster of three, or cluster of four) is determined on each day. The final reported percentages are averages of the three experiments (Fig. 5). The 10-nm gold particles are used as a convenient internal size standard. On occasion, large aggregates of gold are observed in the preparations; these particles are discounted from the analysis and can be removed by centrifugation before sample preparation.

Discussion

Analysis of the TEM images revealed that increasing the length of the multivalent ligand increases the number of copies of each of the bound lectins (Fig. 5). For example, three copies of Con A could assemble on multivalent ligand **4**, whereas the shorter oligomer **2** typically complexed only two receptors. Previous experiments have indicated that high-valency ligands have a greater avidity for Con A.[56] These TEM results suggest that enhanced avidity could be derived from the ability of these materials to assemble larger clusters of Con A. In addition, these results provide a means to quantitate the stoichiometry of these clusters and provide visual images of the complexes proposed to mediate high-avidity binding.

As expected, each receptor–ligand pair produced a distribution of possible stoichiometries. For example, addition of the 157-mer galactose ligand **8** to GGBP yielded an increase in the percentage of clusters of two (12% of the total particles), but also a significant number of clusters of three (2.4%). Long polymers were also required to alter the proportions of clustered MBP; the longest mannose polymer used (**4**) resulted in an increase in the proportion of clusters of two and three. In addition, compounds **2** and **4** were able to cluster three and four copies of MBP; no clusters of these sizes were observed in the absence of the multivalent ligands. The presence of these low-abundance clusters (less than 5–10%) would be difficult to detect in other assays, but individual clusters are discernible by TEM. This is a significant advantage, because subsets of the population could make disproportionately high contributions to the biological activity of the sample. Therefore, resolving populations may be important for understanding the consequences of receptor clustering.

We had previously attempted to explore the clustering of Con A by light-scattering experiments; however, these experiments were unsuccessful because of the insolubility of the resulting clusters. Our TEM method, in contrast, provides a simple means to investigate and visualize ligand-mediated Con A clustering. The low protein concentrations required for these experiments may favor soluble clusters. In addition, because receptors are viewed on surfaces and not in solution, we do not anticipate that solubility is required.

In the application of the TEM method described here, information about the approximate size of the ligand was used to analyze the data. In less structurally characterized systems, access to this information may be limited. However, for chemically defined glycoconjugates, which are used in many applications, structural approximations are often available. Moreover, if some aspect of the ligand is systematically varied, useful information about its ability to promote clustering can be obtained in the absence of structural approximations.

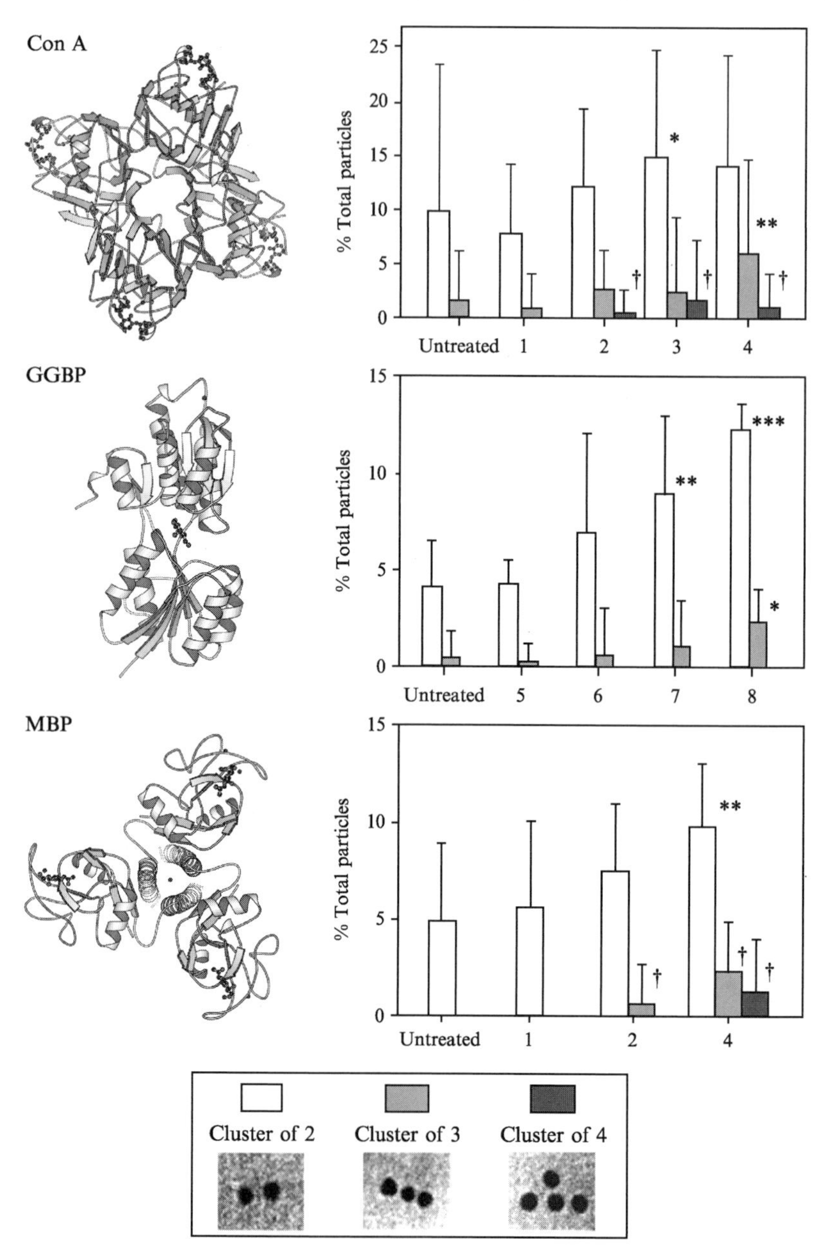

FIG. 5. Ligand valency determines the number of receptors bound to a ligand. The protein structures of receptors Con A, GGBP, and MBP are shown next to the data relevant to their clustering by multivalent ligands. The graphic legend displays possible cluster sizes from clusters of two to four and the images are representative clusters of that type. The data (±SD) are presented as a percentage of the total number of particles counted. For populations with

Conclusions

The prevalence of multivalency in protein–carbohydrate interactions mandates that new methods to characterize multivalent binding be developed. Although many assays focus on assessing the apparent affinity of a multivalent carbohydrate–lectin interaction, many multivalent ligands promote lectin clustering. Because the clustering of a lectin can interfere with or augment its function,[15,24] methods to characterize lectin–ligand assemblies are needed. The TEM method described here facilitates the visualization and characterization of such clusters. In addition, using this technique, the effects of ligand valency on cluster size can be investigated. We envision that future applications may focus on the effects of other aspects of ligand architecture on receptor clustering, such as binding epitope density[40] or shape of the multivalent display.[38] We anticipate that our approach is general and that diverse synthetic glycoconjugates and natural multivalent ligands can be subjected to TEM analysis.

The results of TEM experiments should be interpreted as providing a lower bound to the stoichiometry of individual clusters. As performed here, we have optimized conditions to visualize individual clusters; to be useful this must be performed at dilute conditions so that second-order clustering is not observed and individual clusters are resolved. Therefore, these results may not represent total populations of bound versus free ligand. Instead, they represent the number of receptors that can bind to individual multivalent ligands. This is distinct from methods, such as quantitative precipitation, that yield only the final stoichiometry of precipitated proteins that may be incorporated into an extended lattice.[59]

Another potential use of this TEM method is in exploring heterogeneous multireceptor clusters. Gold nanoparticles of various sizes are commercially available; therefore, each component of a proposed multiprotein assembly could be individually labeled. Subsequent reconstitution of the labeled proteins could allow visualization of the stoichiometry of the heterogeneous complex. Results from such experiments can provide insight into the function of multireceptor clusters.

overlapping error, p values were calculated and are noted (Con A: $^{*}p < 0.3$; $^{**}p < 0.2$; $^{***}p < 0.1$. GGBP and MBP: $^{*}p < 0.05$; $^{**}p < 0.01$; $^{***}p < 0.005$). Species that are not found in untreated experiments are noted (†). The PDB codes for the protein structures shown are as follows: Con A is 1CVN,[60] GGBP is 1GLG,[61] and MBP is 1BCH.[62]

[59] L. R. Olsen, A. Dessen, D. Gupta, S. Sabesan, J. C. Sacchettini, and C. F. Brewer, *Biochemistry* **36,** 15073 (1997).

[60] J. H. Naismith and R. A. Field, *J. Biol. Chem.* **271,** 972 (1996).

[61] M. N. Vyas, N. K. Vyas, and F. A. Quiocho, *Biochemistry* **33,** 4762 (1994).

[62] A. R. Kolatkar, A. K. Leung, R. Isecke, R. Brossmer, K. Drickamer, and W. I. Weis, *J. Biol. Chem.* **273,** 19502 (1998).

Acknowledgments

The authors thank K. Dickson and C. Lavin for experimental assistance and Dr. L. E. Strong, F. J. Boehm, and R. M. Owen for preparation of polymers. We thank Dr. K. Drickamer for generously supplying the plasmid encoding MBP-A. This work was supported in part by the NIH (GM 55984 and GM 49975). J.E.G. thanks the NIH-sponsored Biotechnology Training Program for a predoctoral fellowship (GM 08349), and M.J.B. acknowledges the Molecular Biosciences Training Program (GM 07215) for a predoctoral fellowship.

[22] Exploring Kinetics and Mechanism of Protein–Sugar Recognition by Surface Plasmon Resonance

By MILI KAPOOR, CELESTINE J. THOMAS, KIRAN BACHHAWAT-SIKDER, SHILPI SHARMA, and AVADHESHA SUROLIA

Introduction

The study of biomolecular recognition is of basic importance in understanding processes of molecular recognition and biological function. Lectins are a class of nonenzymatic carbohydrate-binding proteins found ubiquitously in nature. These lectin–ligand attachments are critical in several biological processes such as cell signaling, life cycling of pathogens, fertilization, and inflammatory responses and are put to use in biomedical research as carbohydrate probes based on the binding to surface sugars.[1,2] The kinds of forces involved in lectin–sugar interactions are generally weak and include noncovalent forces, yet the specificity required for a given cellular adhesive event is great.

Determination of kinetic/thermodynamic parameters involved in protein–sugar interactions at the molecular level provides a basic framework for understanding the mechanism of recognition.[3–6] Conventional analysis often relies on techniques requiring large amounts of proteins and/or ligands or both and often entails modification to incorporate a

[1] I. E. Liener, N. Sharon, and I. J. Goldstein, "The Lectins: Properties, Functions and Applications in Biology and Medicine." Academic Press, London, 1986.
[2] N. Sharon and H. Lis, "Lectins." Chapman & Hall, London, 1989.
[3] M. I. Khan, M. V. K. Sastry, and A. Surolia, *J. Biol. Chem.* **261,** 3013 (1986).
[4] F. P. Schwarz, S. Misquith, and A. Surolia, *Biochem. J.* **316,** 123 (1996).
[5] K. Bachhawat, C. J. Thomas, B. Amutha, M. V. Krishnasastry, M. I. Khan, and A. Surolia, *J. Biol. Chem.* **276,** 5541 (2001).
[6] P. Adhikari, K Bachhawat-Sikder, C. J. Thomas, R. Ravishankar, A. A. Jeyaprakash, V. Sharma, M. Vijayan, and A. Surolia, *J. Biol. Chem.* **276,** (2001).

0076-6879/03 $35.00

chromophore or fluorophore, which may alter the mechanism of interaction. Biosensor-based techniques, such as surface plasmon resonance (SPR), provide an alternative method of studying lectin–carbohydrate interactions that overcomes these shortcomings. Detection in this case depends on changes in optical properties of a surface layer with increasing mass of macromolecules at the biospecific interface and does not require any labeling of interactants. In addition, SPR yields real-time information about the course of binding and can be applied to interactions with equilibrium dissociation constants (K) in the range of 10^{-4} to 10^{-10} M. It also uses small sample volumes. SPR essentially needs immobilization of one of the interacting molecules, over which the complementary molecule is allowed to flow. Information about both the association (k_1) and dissociation (k_{-1}) rate constants can be obtained in a single experiment. SPR biosensors are made by Biacore (Uppsala, Sweden) (Biacore), Intersens Instruments (Amersfoort, The Netherlands) (IBIS) and affinity-based biosensor instruments are made by Labsystems Affinity Sensors (Cambridge, UK) (IAsys) and Artificial Sensing Instruments (Zurich, Switzerland) (BIOS-1). These biosensors are not constructed in the same way as the Biacore, and use different detector technologies.[7]

There are many wide-ranging studies in the literature in which SPR has been used successfully to deduce interactions between biomolecules. These include cell adhesion molecules,[8] receptor–ligand interactions,[9,10] T cell antigen receptor and MHC-encoded molecules,[11–13] signal transduction,[14–16] antibody–antigen interactions,[17,18] virus research,[19] protein–DNA and

[7] J. Hodson, *Biotechnology* **12,** 31 (1994).

[8] P. A. van der Merwe and A. N. Barclay, *Curr. Opin. Immunol.* **8,** 257 (1996).

[9] A. C. Cunningham and J. A. Wells, *J. Mol. Biol.* **234,** 554 (1993).

[10] K. Johanson, E. Appelbaum, M. Doyle, P. Hensley, B. Zhao, S. S. Abdel-Meguid, P. Young, R. Cook, S. Carr, R. Matico, D. Cusimano, E. Dul, M. Angelichio, I. Brooks, E. Winborne, P. McDonnell, T. Morton D. Bennett, T. Sokoloski, D. McNulty, M. Rosenberg, and I. Chaiken, *J. Biol. Chem.* **270,** 9459 (1995).

[11] S. M. Alam, P. J. Yravers, J. L. Wung, W. Nasholds, S. Redpath, S. C. Jameson, and N. R. Gascoigne, *Nature* **381,** 616 (1996).

[12] J. J. Boniface and M. M. Davis, *Methods Enzymol.* **6,** 168 (1994).

[13] M. Corr, A. E. Slanetz, L. F. Boyd, M. T. Jelonek, S. Khilko, B. K. al-Ramadi, Y. S. Kim, S. E. Maher, A. L. Bothwell, and D. H. Margulies, *Science* **265,** 946 (1994).

[14] S. Felder, M. Zhou, P. Hu, J. Urena, and A. Ullrich, M. Chaudhuri, M. White, S. E. Shoelson, and J. Schlessinger, *Mol. Cell. Biol.* **13,** 1449 (1993).

[15] J. E. Ladbury, M. A. Lemmon, M. Zhou, J. Green, M. C. Botfield, and J. Schlessinger, *Proc. Natl. Acad. Sci. USA.* **92,** 3199 (1995).

[16] M. M. Morelock, R. H. Ingraham, R. Betagiri, and S. Jakes. *J. Med. Chem.* **38,** 1309 (1995).

[17] P. Holliger, T. Prospero, and G. Winter, *Proc. Natl. Acad. Sci. USA* **90,** 6444 (1993).

[18] R. F. Kelley and M. P. O'Connell, *Biochemistry* **32,** 6828 (1993).

[19] R. W. Glaser and G. Hausdorf, *J. Immunol. Methods* **189,** 1 (1996).

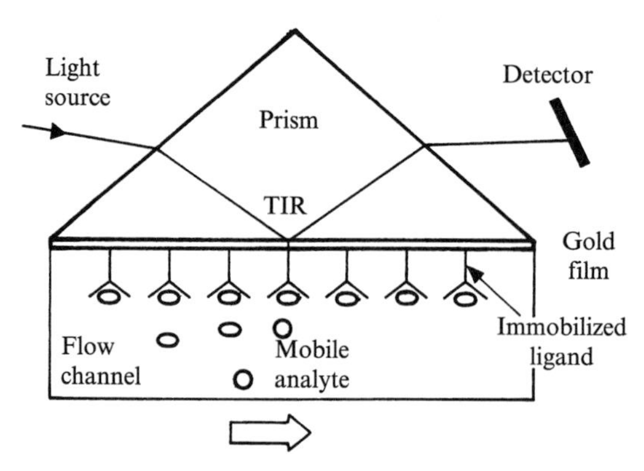

FIG. 1. Schematic representation of the principle of Biacore's SPR technology with a ligand-coated sensor chip. As the analyte binds to the coated sensor chip, the resulting changes in refractive index of the solution in contact with the chip can be detected by monitoring the change in SPR angle. TIR, Total internal reflection.

DNA–DNA interactions,[20–23] and protein–carbohydrate interactions.[5,6] We provide here a critical evaluation of this method in studies of protein–carbohydrate recognition by describing a few representative examples.

SPR Technology

Surface plasmon resonance (SPR), originally proposed by Kretschmann and Raether,[24] is an optical phenomenon arising in thin metal films under conditions of total internal reflection (Fig. 1). Light traveling through a glass prism is totally internally reflected on reaching the interface of the glass (higher refractive index) with another medium such as a gas or liquid (lower refractive index). However, an electromagnetic field component called the evanescent wave penetrates a short distance (on the order of one wavelength) into the medium of lower refractive index. If the glass surface is coated with a thin layer of metal such as silver or gold, the

[20] P. J. Bates, H. S. Dosanjh, S. Kumar, T. C. Jenkins, C. A. Laughton, and S. Neidle, *Nucleic Acids Res.* **23,** 3627 (1995).

[21] K. Bondeson, Å. Frostell-Karlsson, L. Fägerstam, and G. Magnusson, *Anal. Biochem.* **214,** 245 (1993).

[22] M. Buckle, R. M. Williams, M. Negroni, and H. Buc, *Proc. Natl. Acad. Sci. USA* **93,** 889 (1996).

[23] R. J. Fisher and M. Fivash, *Curr. Opin. Biotechnol.* **5,** 389 (1994).

[24] B. E. Kretschmann and H. Raether, *Z. Naturforch. A* **23,** 2135 (1968).

evanescent wave sets up an oscillation (resonance) of the electrons (plasmons) in the gold. The energy lost to the evanescent wave is a function of the refractive index or concentration of molecules in a layer immediately adjacent to the gold film. This phenomenon produces a sharp dip in the intensity of reflected light at a specific angle, the resonance angle. Adsorption or desorption of macromolecules at the sensor surface changes the local refractive index and produce a shift in the refractive index. By keeping other factors constant, SPR can be used to measure changes in the concentration of molecules in a surface layer of solution in contact with the sensor surface.

The biomolecular interaction analysis (BIA) system has three essential components: the sensor chip, the optical system, and the liquid-handling system with integrated microfluidic cartridge (IFC). The sensor chip is a glass slide coated on one side with a thin gold film, to which a surface matrix, usually a layer of carboxymethylated dextran, is covalently attached. The matrix is the means by which biomolecules can be immobilized on the sensor chip surface. The optical system uses fixed array diode detectors to monitor the position of the resonance angle that changes once the interaction between the ligand and analyte takes place. The resonance angle is expressed in resonance units (RU). A response of 1000 RU corresponds to a change in surface concentration on the sensor chip of about 1 ng/mm^2 for proteins. The IFC is made up of flow channels, sample loops, and pneumatic valves and is responsible for controlling the delivery of the sample and eluent to the sensor surface. A microfluidic system is employed to generate a constant flow of mobile phase across two (Biacore X) or four (Biacore 2000) flow paths in series, or it can address each flow path individually. Microfluidics ensures controlled contact time of the sample with the surface, accurately and reproducibly.

Experimental Methods

Ligand

The ligand to be immobilized should be at least 90% pure and homogeneous. The concentration of the stock solution should be between 0.5 and 5 mg/ml. For immobilization on a CM5 chip, using the amine coupling kit, 10–50 μg of ligand is normally sufficient, although variations in this amount will depend on the biochemical properties of the molecule to be immobilized, the immobilization chemistry, and the applications to be performed. The composition of the stock solution should have a salt content below 200 m*M* and should not contain reactive compounds such as Tris, glycine, or bovine serum albumin (BSA).

Buffer

The choice of buffer used during the experiment depends on the stability of the ligand and the analyte in that buffer. The standard buffer used in Biacore contains 10 m*M* HEPES (pH 7.4), 150 m*M* NaCl, 3.4 m*M* EDTA, and 0.005% polyoxyethylene (20) sorbitan monooleate (P20). The salt prevents the nonspecific binding of the analyte, the detergent P20 avoids adsorption of the analyte to the flow channels, and EDTA chelates the contaminating ions. All the buffers to be used should be filtered and degassed. The same buffer should be used in sample preparation and dialysis also. Phosphate-buffered saline (PBS) is another commonly used buffer.

How Much to Immobilize

Depending on the application a known amount of ligand must be immobilized on the sensor chip surface. Affinity ranking can be done with low- to moderate-density sensor chips. The important factor is that the analyte should saturate the ligand within a proper time frame. Kinetics are done with the lowest ligand density that still gives a good response without being disturbed by secondary factors such as mass transfer or steric hindrance. Low molecular weight binding is done with high-density sensor chips to bind as much as possible of the analyte to obtain a proper signal.

Preconcentration

Preconcentration is done with the dextran-carboxyl group-based sensor chips and gives an idea about the optimal immobilization conditions for the ligand. In a preconcentration experiment the ligand is diluted in several immobilization buffers, each differing by 1 pH unit. The buffer of choice for immobilization will be the one that gives the fastest preconcentration, but has the highest pH. However, there are occasions when the preconcentration is efficient but the covalent linking is not, because of the low pH. In these cases choose buffer that is 1 pH unit higher. However, high preconcentration values do not guarantee that the ligand will bind in large amounts as other factors, such as activation and availability of binding sites on the ligand, also have an effect.

Sensor Surfaces and Immobilization

The sensor surface and the immobilization technique used are important in determining the correct evaluation of the various binding constants, as both the interacting molecules should be present in their native conformation and should be optimally exposed for the interaction. The macromolecule should be attached with uniform orientation and unrestrained accessibility for the mobile reactant. In SPR, one can immobilize either of the reacting moieties depending on the feasibility of the chemistries involved in the immobilization and the molecular weight of the ligand.

Several groups have used SPR for the analysis of lectin–carbohydrate binding.[25–31] Only a few have immobilized the lectin to the surface,[30,31] rather than the oligosaccharide or glycoprotein. However, immobilization of lectin, rather than the glycoprotein or oligosaccharide, would enable rapid screening of a large number of glycan-containing compounds in a relatively short time, by passing them across the same flow cell by repeated regeneration of the surface. This would appear more practical when a lectin of unknown specificity requires characterization. This also ensures that the determining moieties are optimally available for the interaction by obviating the possibility of steric hindrance. This may become a serious limitation when sugars are immobilized on the sensor surface. On the other hand, immobilization of sugars ensures easy monitoring of the binding event on passage of protein because of the greater molecular mass of the protein and hence a greater change in the resonance units. However, immobilization of sugars to the dextran surface requires derivatization of sugars. Mann *et al.*[32] used *R-C*-(ethylamino)mannose for coupling to the sensor surface and observed nonspecific binding of concanavalin A (Con A) to the dextran surface. Con A has a 3 to 4-fold higher affinity for mannose over glucose, but background binding to the dextran surface did not allow determination of the Con A–mannose interaction.

[25] Y. Shinohara, F. Kim, M. Shimizu, M. Goto, M. Tosu, and Y. Hasegawa, *Eur. J. Biochem.* **233,** 189 (1994).
[26] A. M. Hutchison, *Anal. Biochem.* **220,** 303 (1994).
[27] Y. Hasegawa, Y. Shinohara, and S. Hiroyuki, *Trends Glycosci. Glycotechnol.* **9,** S15 (1997).
[28] K. Yamamoto, C. Ishida, Y. Shinohara, Y. Hasegawa, Y. Konami, T. Osawa, and T. Irimura, *Biochemistry* **33,** 8159 (1994).
[29] P. Adler, S. J. Wood, Y. C. Lee, R. T. Lee, W. A. Petri, Jr., and R. L. Schnaar, *J. Biol. Chem.* **270,** 5164 (1995).
[30] I. Blikstad, L. G. Fägerstam, R. Bhikabhai, and H. Lindblom, *Anal. Biochem.* **233,** 42 (1996).
[31] S. M. Haseley, P. Talaga, J. P. Kammerling, and J. F. G. Vliegenhart, *Anal. Biochem.* **274,** 203 (1999).
[32] D. A. Mann, M. Kanai, D. J. Maly, and L. L. Kiessling, *J. Am. Chem. Soc.* **120,** 10375 (1998).

Covalent Coupling of Lectins to Sensor Surfaces

SPR chips have a gold surface, which is usually covered by a monolayer of alkyl thiols to minimize nonspecific binding and to create a hydrophilic surface.[33] Carboxymethylated (CM5) chips have a matrix of carboxymethylated dextran covalently attached (1–3 ng/mm^2) to form a flexible hydrogel (estimated thickness, 100–200 nm). This matrix provides an environment suitable for studies of biomolecular interactions and enhances the ligand immobilization capacity of the surface, from typically 1–5 ng/mm^2 on untreated gold to up to 50 ng/mm^2 in the matrix. The sensor chip can be used repeatedly, depending on the properties of the immobilized ligand and the regeneration conditions used. One of the widely used immobilization procedures utilizes the activation of carboxyl groups of the dextran matrix with *N*-hydroxysuccinimide (NHS) and *N*-ethyl-*N′*-(dimethylaminopropyl)carbodiimide (EDC) to form NHS esters, which enables coupling to the amino groups on proteins.[34] However, the technique requires the proteins to be at pH values much below their p*I* to ensure effective electrostatic interaction between the protein molecules and the negatively charged dextran surface. Another disadvantage of this procedure is the mixture of populations that emerge with differing accessibilities and reactivities to binding sites, caused by the binding event being a random process with any of the protein lysines participating in it.[35]

Binding Process

After the immobilization of ligand, buffer is allowed to flow until a stable baseline is attained. Next, a mobile second reactant (referred to as the analyte) is injected into the buffer flow above the sensor surface, and the progress of complex formation is monitored. This is followed by the dissociation phase, in which the free mobile reactant is absent from the buffer and the time course of the complex dissociation is recorded.

Regeneration

Regeneration is the procedure in which the analyte is washed away from the ligand but leaves the ligand unharmed. The regeneration conditions must be evaluated empirically because the combination of physical forces responsible for binding is often unknown, and the regeneration conditions must not cause irreversible damage to the ligand. Choose the

[33] S. Löfas and B. Johnsson, *J. Chem. Soc. Chem. Commun.* **21,** 1526 (1990).
[34] A. Johnsson, S. Löfas, and G. Lindquist, *Anal. Biochem.* **198,** 268 (1991).
[35] D. J. O'Shannessy, M. Brigham-Burke, and K. Peck, *Anal. Biochem.* **205,** 132 (1992).

mildest regeneration conditions that completely dissociate the complex. In addition, other molecules, such as ligand and ligand analogs that bind to the analyte to diminish rebinding of the analyte during dissociation, can be added. In the case of lectin–sugar interactions, the surfaces can be easily regenerated by passing over the surface a higher concentration of a weakly binding analog of the sugar such that it out competes the bound sugar from the surface.

Data Analysis

The data obtained from a typical experiment can be evaluated by two main approaches for the determination of binding constants: kinetic data analysis and equilibrium analysis.

Kinetic Data Analysis. Kinetic data analysis is the analysis of binding kinetics from the time course of binding, using commercial software. In this analysis, association (k_1) and dissociation (k_{-1}) rate constants are obtained by nonlinear fitting of the primary sensorgram data, using BIA evaluation software version 3.0. The dissociation rate constant is derived using Eq. (1):

$$R_t = R_{ta} e^{-k_{-1}(t - t_0)} \tag{1}$$

where R_t is the response at time t, R_{t0} is the amplitude of the initial response, and k_{-1} is the dissociation rate constant. The association rate constant k_1 can be derived by Eq. (2), from the measured k_{-1} values:

$$R_t = R_{\max}\left[1 - e^{-(k_1 C + k_{-1})(t - t_0)}\right] \tag{2}$$

where $R_{\max}$ is the maximum response, C is the concentration of the analyte in the solution, and k_1 and k_{-1} are the association and dissociation rate constants, respectively. The ratio of k_1 and k_{-1} yields the value of association constant K_a (k_1/k_{-1}).

Equilibrium Analysis. Equilibrium analysis is the analysis of equilibrium surface binding isotherms from plateau signals, which is analogous to a standard Scatchard plot.[36] Equilibrium constants are determined by measuring the concentration of free interactants and complex at equilibrium. In BIA, the concentration of free analyte is equal to that of bulk analyte and the concentration of free ligand is derived from the concentration of the complex if the total surface binding capacity is known. Also, the concentration of complex can be measured directly as the steady state response. Thus,

$$K_{eq} = R_{eq} / C(R_{\max} - R_{eq}) \tag{3}$$

[36] G. Scatchard, *Ann. N. Y. Acad. Sci.* **51,** 660 (1949).

where C is the analyte concentration, R_{max} is the total surface binding capacity in response units (RU), and R_{eq} is the steady state binding level in response units or

$$R_{eq}/C = K_{eq}R_{max} - K_{eq}R_{eq} \tag{4}$$

Thus, a plot of R_{eq}/C against R_{eq} at different analyte concentrations gives a straight line from which R_{max} and K_{eq} can be calculated. This plot is analogous to a Scatchard plot.[36] In this method, no information about binding progress is used and it is independent of mass transport limitations or other transient artifacts of surface binding. One limitation of this approach is the relatively long contact time that may be required to reach equilibrium.

To completely evaluate a sensor curve, many other additional factors should be included in the data evaluation, such as bulk contribution of the sample refractive index, baseline drift, and residuals, that is, nondissociable analyte that remains bound to the surface. The bulk contribution can be minimized by careful preparation of the analyte, for example, by dialyzing the analyte in the same buffer that is to be used during continuous flow. The contribution from this term can be ascertained by injection of analyte on the reference surface. Also, one of the participating moieties may be immobilized on the surface; this can lead to differences in the binding properties of the two moieties. The immobilized ligand may also have the potential for oligomerization. If the binding properties depend on oligomerization status, the result of the biosensor experiment would depend on the surface density of the immobilized species. Thus, it is possible to immobilize two or three different concentrations of the ligands on different flow cells during the initial phases of standardization, in order to determine the right concentration of ligand required for the study.

Determination of Thermodynamic Parameters by SPR

SPR has been used to determine not only activation parameters but also thermodynamic quantities in many interesting studies. Patil *et al.*[37] studied the temperature dependence of the rate constants by employing Arrhenius plots. These plots of $\ln(k/T)$ versus $(1/T)$ yielded straight lines for both the association and dissociation rate constants. The value of the slope of an Arrhenius plot yields the energy of activation E_A, which can be used to determine the activation parameters ($\Delta H^{\ddagger}$, $\Delta S^{\ddagger}$) using Eqs. (5)–(7)[38]:

[37] A. R. Patil, C. J. Thomas, and A. Surolia, *J. Biol. Chem.* **275,** 24348 (2000).

[38] K. J. Laidler, "Theories of Chemical Reaction Rates." R. E. Krieger Publishing, Huntington, NY, 1979.

$$\Delta H^{\ddagger} = E_A - RT \tag{5}$$

$$\ln(k/T) = -\Delta H^{\ddagger}/RT + -\Delta S^{\ddagger}/R + \ln(k'/h) \tag{6}$$

$$\Delta G^{\ddagger} = \Delta H^{\ddagger} - T\Delta S^{\ddagger} \tag{7}$$

where k is the appropriate rate constant, k' is Boltzmann's constant, and h is Planck's constant. Thomas *et al.*[39–41] observed close correspondence between the value of parameters determined by SPR and those determined by more direct and accurate methods such as fast reaction kinetics by stopped flow fluorescence spectrometry and equilibrium thermodynamics measurements by isothermal titra calorimetry (ITC).

Some Representative Examples

Investigating Garlic Lectin Mannooligosaccharide Recognition by SPR

Bulbs of garlic (*Allium sativum*) accumulate two types of mannose-binding lectins: the heterodimeric *Allium sativum* agglutinin (ASA) I and the homodimeric ASA II. Because of the strict specificity for mannose, these lectins find use as affinity ligands for the purification of glycoproteins such as IgM, α_2-macroglobulin, and β-lipoprotein. Bachhawat *et al.*[5] have studied the kinetics of binding of mannooligosaccharides to ASA I by surface plasmon resonance.

Materials. Garlic lectin is isolated from its natural source by a standard protocol.[42] Sugars such as mannose, methyl-α-mannose, and Man_3-$GlcNAc_2$ are obtained from Sigma (St. Louis, MO). Man_3 and Man_5 are procured from Dextra Laboratories (Reading, UK). $Man_9GlcNAc_2$ and $Man_9GlcNAc_2Asn$ are prepared in the laboratory by hydrazinolysis and Pronase digestion of soybean agglutinin, respectively.[43,44] A CM5 certified grade chip and coupling reagents are from Biacore. All biosensor data are collected on a Biacore 2000 optical biosensor.

Method

1. ASA I is purified by affinity chromatography on mannose–Sepharose matrix and gel filtration on a BioGel P200 column.

[39] C. J. Thomas, B. P. Gangadhar, N. Surolia, and A. Surolia, *J. Am. Chem. Soc.* **120,** 12428 (1998).

[40] C. J. Thomas and A. Surolia, *J. Biol. Chem.* **274,** 29624 (1999).

[41] C. J. Thomas and A. Surolia, *FEBS Lett.* **445,** 420 (1999).

[42] T. K. Dam, K. Bachhawat, P. G. Rani, and A. Surolia, *J. Biol. Chem.* **273,** 5528 (1998).

[43] H. Lis and N. Sharon, *J. Biol. Chem.* **253,** 3468 (1978).

[44] A. K. Mandal and C. F. Brewer, *Biochemistry* **31,** 2602 (1992).

2. The sensor chip surface is activated by a 7-min pulse of an equal mixture of 0.1 *M* NHS and 0.4 *M* EDC at a flow rate of 5 μl/min, leading to the creation of primary amine linkable groups. The standard HEPES buffer is used during immobilization.

3. Purified ASA I (0.1 mg/ml) in 5 m*M*, sodium acetate buffer (pH 4.5) is allowed to flow over the surface at a rate of 1 μl/min for 50 min. Nearly 1500 RU is coupled under these conditions. During this phase the ligand forms covalent attachment with the sensor surface.

4. The unreacted NHS esters of the surface are blocked by flowing 1 *M* ethanolamine dissolved in water at a rate of 5 μl/min for 7 min.

5. For the determination of association rate constants, the solution of ligands (1–80 m*M*) in PBS is passed over the chip at a flow rate of 5 μl/min.

6. The dissociation rate constant is determined by subsequently passing PBS over the chip at a flow rate of 5 μl/min.

7. The surface is regenerated by a 10-s pulse of 500 m*M* methyl-α-mannopyranoside.

Before injection the ASA I samples are dialyzed extensively against the same buffer to avoid buffer mismatch. Another flow cell is kept as the reference surface and is treated in the same way as the ligand surface, except that ASA I is not passed over this surface. This is important to normalize the chemistries between the two flow cells. A representative sensorgram depicting interaction of increasing amounts of trimannoside to immobilized ASA I at 25° is shown in Fig. 2. The sensorgram demonstrates how the dissociation rate constant follows zero-order kinetics, that is, the rate does not depend on the concentration of the sugars passed over the surface. On the other hand, the association rate constant follows pseudo-first-order rate kinetics, in which the rate depends on the concentration of the flowing analyte. The curve is fitted by mass transport analysis at 25° to yield values of k_1 and k_{-1} of 77.3 $M^{-1}s^{-1}$ and 0.61 s^{-1}, respectively, thus estimating a K_a of 127 M^{-1}. These parameters are in agreement with an earlier isothermal titration calorimetry study (Table 1[44a]). Scatchard analysis of the binding data also determines a similar association constant for the interaction. As can be observed from the data, the binding affinity for the mannose-containing oligosaccharides increases with increasing chain length, with a significant jump observed when the oligosaccharides are extended with α-1,2-linked mannose residues at their nonreducing end. The use of a certified grade sensor chip in this study allows satisfactory recording of response unit changes corresponding to oligosaccharide

[44a] A. R. Patil, S. Misguith, T. K. Dam, V. Sharma, M. Kapoor, and A. Surolia, *Methods in Enzymology*. **362** [38] (2003) (this volume).

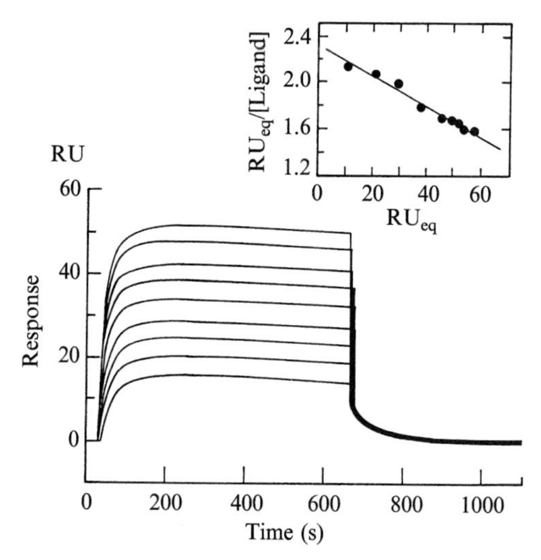

FIG. 2. A representative sensorgram depicting interaction of increasing amounts of trimannoside to the immobilized ASA I at 25°. The sugar (concentrations of 5, 10, 20, 30, 40, 50, 60, 70, and 80 m*M*, reading from bottom to top) was injected for 600 s at a flow rate of 5 μl/min. The dissociation reaction was recorded by flowing buffer at 5 μl/min. The sensor ship surface was regenerated by a 10-s pulse of methyl-α-mannopyranoside. *Inset*: Scatchard analysis of the sensorgram. Reproduced by permission from K. Bachhawat, C. J. Thomas, B. Amutha, M. V. Krishnasastry, M. I. Khan, and A. Surolia, *J. Biol. Chem.* **276,** 5541 (2001) and from the publisher, the American Society for Biochemistry and Molecular Biology (ASBMB).

TABLE 1

BINDING KINETICS OF ASA I–MANNOOLIGOSACCHARIDE INTERACTION[a]

Oligosaccharide	k_1 (M^{-1} s^{-1})	k_{-1} (s^{-1})	K_a (k_1/k_{-1}) (M^{-1})	K_a (Scatchard) (M^{-1})	K_a (ITC) (M^{-1})
Man_3	77.3	0.61	126.7	120.2	144
Man_3 $GlcNAc_2$	74.1	0.59	125.6	119.8	—
Man_5	98.3	0.58	169.5	163.4	162
$Man_7GlcNAc_2$	5051	0.060	8.4×10^4	8.5×10^4	—
$Man_9GlcNAc_2Asn$	6.1×10^4	0.049	1.2×10^6	1.1×10^6	9.1×10^5

[a] At 298 ± 1 K. The association constants obtained from the k_1/k_{-1} ratio, Scatchard analysis, and ITC measurements are in complete agreement. ITC could not be conducted for $Man_3GlcNAc_2$ and $Man_7GlcNAc_2$ because of sample limitation. Structures of sugars mentioned are shown elsewhere in this volume.[44a] Reproduced by permission from K. Bachhawat, C. J. Thomas, B. Amutha, M. V. Krishnasastry, M. I. Khan, and A. Surolia, *J. Biol. Chem.* **276,** 5541 (2001) and from the publishers, ASBMB.

having a molecular mass greater than 400 Da. The binding constant is determined by the ratio of k_1 and k_{-1}. For most of the binding reactions studied, it is the decrease in the dissociation rate constant that has been shown to be responsible for increased binding affinities. However, in studies of the ASA I–mannooligosaccharide interaction, it is the unprecedented increase in the association rate constant that has been implicated in the observed enhancement in affinities.

Tethering of Liposomes to Mimic the Cell Surface

Immobilization of lectin on the sensor surface and monitoring sugar binding or vice versa does not mimic biological recognition per se, as these interactions do not occur between free sugar or lectin but on cell surfaces. Thomas and Surolia[45] have described a novel method of studying lectin–sugar interactions in which the liposomes bearing sugar residues are tethered to the biosensor surface, using another noncompeting lectin (of a different specificity), mimicking the *in vivo* cell scenario to a large extent. *Ulex europeas* agglutinin 1 (UEA1) is a 68-kDa protein having specificity for L-fucose. The unique specificity of UEA1 for fucose residue-containing oligosaccharides helps identify fucosylated sites on cell surfaces. It has also been identified as a key marker for cancerous cells and thus determination of kinetic/thermodynamic parameters involved in the UEA1–fucose-containing oligosaccharide interaction would be useful in understanding the mechanism of recognition.

Materials. Ricin[46] and UEA1[47] are isolated in the laboratory by established protocols.

Method

1. L-α-Dimyristoylphosphatidylcholine (DMPC) or egg phosphatidylcholine (PC), 1 m*M*, and ganglioside M1 (G_{M1}, 5 mol% of total lipid) containing 1–20% H-fucolipid in (2:1, v/v) chloroform–methanol is dried in glass vials under nitrogen and further dried under vacuum for 2 h. PBS is added and vials are vortexed and sonicated, and the suspension of liposomes is passed through 50-nm polycarbonate material. The unincorporated material is separated from the liposomes by Sepharose CL-4B column chromatography.

[45] C. J. Thomas and A. Surolia, *Arch. Biochem. Biophys.* **374,** 8 (2000).

[46] P. S. Appukuttan, A. Surolia, and B. K. Bachhawat, *Indian J. Biochem. Biophys.* **14,** 382 (1979).

[47] M. E. Pereira, E. C. Kisailus, F. Gruezo, and E. A. Kabat, *Arch. Biochem. Biophys.* **185,** 108 (1978).

2. Ricin is covalently immobilized on the CM5 sensor chip at concentrations of 40 μg/ml in 10 mM sodium acetate buffer, pH 4.8, using the amine coupling kit. Nearly 1500 RU is immobilized under these conditions.

3. Liposomes containing H-fucolipid and G_{M1} are passed over the ricin surface at a flow rate of 2 μl/min for 30 min and washed with PBS. The observed binding of liposomes to the surface is due to the interaction between ricin and G_{M1}. Flow of liposomes followed by PBS is repeated several times until no further increase in response units is observed. The chip is washed with a 5-s pulse (10 μl/min) of 2 mM NaOH to remove the multilamellar structures. The surface is further washed with PBS at a flow rate of 100 μl/min.

4. UEA1 is passed over these "tethered" liposomes at 5 μl/min and association rates are determined for the binding of UEA1 (5–50 μM) to H-fucolipid present in the liposomes. Dissociation rates are evaluated by passing a solution of 1 mM L-fucose in PBS at a flow rate of 5 μl/min. The surface is regenerated by a 10-s pulse of 10 mM L-fucose flowing at 50 μl/min.

UEA1 does not bind to tethered liposomes, which lack the H-fucolipid, or to ricin per se, demonstrating that the binding of lectin to liposome is strictly sugar mediated. UEA1 incubated with an excess of L-fucose also does not bind to the liposomes. If the glycolipid concentration of the liposomes is also varied, it is found that a 5% glycoprotein concentration gives the most satisfactory results. Rate constants are determined for the interaction of H-fucolipid and UEA1 at different temperatures and the linearity of the Arrhenius plots demonstrates that the interaction is only a single-step bimolecular association reaction, atleast in the temperature range studied. Thus tethering of liposomes using lectins provides a new platform for monitoring lectin–sugar interactions.

Incorporation of Glycolipids to Generate Synthetic Surfaces: Competition Experiments

Evaluation of the association constants of protein–carbohydrate interactions is usually difficult because of the weak affinities ($K_a = 10^3$ to 10^4 M^{-1}) involved. Complications could also arise from the contribution of multivalency in these interactions. In certain instances, the dissociation rate constants obtained by SPR do not correlate with the dissociation rate constants in solutions.[48] Several studies have explored the advantages of competition assays in this regard. In these assays, the ability of a substrate to inhibit the interactions of a soluble receptor with an immobilized ligand

[48] L. Nieba, A. Krebber, and A. Plückthun, *Anal. Biochem.* **234,** 155 (1996).

is explored.[48–51] Because in these studies a ligand competes for a receptor in solution, it minimizes the differences associated with surface composition. Mann *et al.*[32] studied the binding of monovalent and multivalent ligand binding to concanavalin A (Con A). They described the incorporation of glycolipids into synthetic surfaces to control the available ligand density on the surface. In this approach the gold surface is modified with a long-chain alkane attached through a thiol group, on which a new self-assembled monolayer (SAM) of defined composition can be generated. The synthetic glycolipid used was *R*-*O*-ethylaminomannosylphosphatidylethanolamine, which displays a terminal mannose residue with an R-anomeric linkage, the preferred configuration for Con A binding. By combining the synthetic glycolipid with phosphatidylcholine in various molar ratios, it was possible to control the density of saccharide ligands presented.

Adhikari *et al.*[6] have demonstrated how a modification of SPR can be used to elucidate the inhibition of binding of recombinant peanut agglutinin (rPNA) to immobilized T-antigen by various sugars and hence a determination of their binding affinities (Fig. 3). This competitive binding assay is important in determining binding affinities of low molecular weight sugars that do not lead to a detectable change in response units on binding to protein immobilized on the sensor surface.

Method

1. In this particular assay, biotinylated T-antigen is immobilized on a streptavidin chip followed by injection of protein. Sensorgrams of the concentration-dependent binding of rPNA and its mutants to immobilized T-antigen provide an estimate of the kinetic parameters involved in the interaction.

2. Subsequently, the sugars whose binding affinities to rPNA must be estimated are preincubated with rPNA and passed over the immobilized T-antigen. The sugar-binding affinity determines the extent and rate of binding of rPNA to the immobilized T-antigen. The concentration of a sugar required for 50% inhibition of the on-rate (k_1) for binding of the sugar to rPNA defines the potency with which that sugar binds to the lectin. The IC_{50} values obtained from SPR match well with earlier hemagglutination inhibition studies.

[49] R. Karlsson, *Anal. Biochem.* **221,** 142 (1994).

[50] L. D. Ward, G. J. Howlett, A. Hammacher, J. Weinstock, K. Yasukawa, R. J. Simpson, and D. J. Winzor, *Biochemistry* **34,** 2901 (1995).

[51] M. M. Morelock, R. H. Ingraham, R. Betagiri, and S. J. Jakes, *J. Med. Chem.* **38,** 1309 (1996).

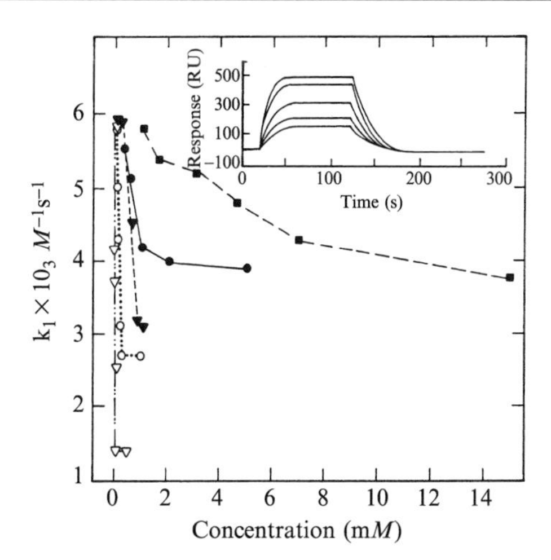

FIG. 3. Inhibition of the binding of rPNA to immobilized T-antigen by various sugars. A plot of the decrease in on-rate in the presence of increasing concentrations of various sugars is shown: ●, Lactose; ○, methyl-α-galactose; ▼, methyl-β-galactose; ▽, T-antigen; ■, N-acetyllactosamine). *Inset*: Sensorgram depicting the decrease in the rate of binding of rPNA preincubated with various concentrations of lactose. The concentrations of lactose used were 0, 0.25, 0.5, 1, and 2 mM from top to bottom, respectively. RU, Response units. Reproduced by permission from P. Adhikari, K. Bachhawat-Sikder, C. J. Thomas, R. Ravishankar, A. A. Jeyaprakash, V. Sharma, M. Vijayan, and A. Surolia, *J. Biol. Chem.* **276,** 40734 (2001) and from the publisher, the ASBMB.

Problems Related to Immobilization

High local concentrations of the immobilized reactant at the sensor surface can lead to steric hindrance in binding to neighboring binding sites,[52] as well as pronounced rebinding events.[35,53,54] When the association rate constant is greater than $1 \times 10^6\ M^{-1}\ s^{-1}$ the measured binding rate in some cases may reflect the transport of analyte into the matrix rather than the reaction rate itself.[54] The limitation of the binding kinetics by the rate of transport of the mobile reactant to the sensor surface is a potential problem in the correct determination of binding parameters. Inconsistencies in the derived equilibrium and rate constants compared with those derived by other methods have been observed.[55] In certain instances the deviations were attributed to heterogeneity of the binding sites, which could

[52] P. R. Edwards, A. Gill, D. V. Pollard-Knight, M. Hoare, P. E. Buckle, P. A. Lowe, and R. J. Leatherbarrow, *Anal. Biochem.* **231,** 210 (1995).

[53] P. Schuck and A. P. Minton, *Anal. Biochem.* **240,** 262 (1996).

[54] I. Chaiken, S. Rose, and R. Karlsson, *Anal. Biochem.* **201,** 197 (1992).

[55] W. Ito and Y. Kurusawa, *J. Biol. Chem.* **268,** 20668 (1993).

be intrinsically present or could be produced by the immobilization procedure.[56] Mass transport-related ambiguities can be identified by varying the flow rate, as mass transport is influenced by flow whereas the intrinsic reaction rate is flow independent. The analysis of binding data by conventional methods might lead to determination of parameters that lie between the mass transport rate constant and the true intrinsic chemical binding rate constants. Schuck and Minton[53] investigated this problem and developed a new kinetic analysis method; they tested it on experimental data and on simulated data generated with a computer model for combined mass transport and reversible binding to a single class of immobilized sites.

Another issue of general concern is the cost of sensor chips. This is of particular importance in the phase of method development of an assay for the optimization of immobilization conditions for a novel peptide or protein. Chatelier *et al.*[57] addressed this by developing methods for reuse of Biacore sensor chips on which peptide or protein had been covalently immobilized. They suggest a method that uses a combination of both enzymatic (Pronase E) and chemical (bromoacetic acid) treatments of used sensor chips. Regeneration requires an overnight incubation of the sensor chip *ex situ*. The results demonstrate that this method not only substantially removes immobilized proteins such as IgG, protein G, and gp120, but also a high density of proteins can be immobilized onto the reconditioned surfaces, because reactive carboxylic acids are regenerated by this procedure.

Comparison of SPR Data with Data Produced by Other Bioanalytical Methods

Many studies in the literature compare the values of equilibrium and kinetic rate constants obtained from biosensor experiments with the constants obtained for the same macromolecule in other laboratories[17] and also with constants determined by employing other techniques such as calorimetry[5,41,58] sedimentation equlibrium,[59] fluorescence quenching,[55] enzyme linked immunosorbent assay (ELISA),[48] and filter binding assays.[60] Shinohara *et al.*[25] investigated the interaction of a glycopeptide derived from asialofetuin with six lectins. The binding activity of each lectin was almost the same as reported previously on the basis of affinity

[56] D. J. O'Shannessy, M. Brigham-Burke, K. K. Soneson, P. Hensley, and I. Brooks, *Anal. Biochem.* **212,** 457 (1993).

[57] R. C. Chatelier, T. R. Gengenbach, H. J. Griesser, M. Brigham-Burke, and D. J. O'Shannessy, *Anal. Biochem.* **229,** 112 (1995).

[58] A. Bernad and H. R. Bosshard, *Eur. J. Biochem.* **230,** 416 (1995).

[59] L. C. Gruen, A. A. Kortt, and E. Nice, *Eur. J. Biochem.* **217,** 319 (1993).

[60] A. Takano, M. Hatanaka, and M. Maki, *FEBS Lett.* **352,** 247 (1994).

chromatography. Patil *et al.*[37] determined the thermodynamic parameters for the interaction of calreticulin, a lectin-like molecular chaperone of the endoplasmic reticulum, with monoglucosyl IgG by SPR. The interaction was completely inhibited by free oligosaccharide, $Glc_1Man_9GlcNAc_2$, whereas $Man_9GlcNAc_2$ did not bind to the calreticulin–substrate complex, attesting to the exquisite specificity of this interaction. The values obtained by SPR were in agreement with those obtained by immunoassay. On the basis of these data the authors provided a mechanism of the calreticulin chaperone activity.

However, there are reports in which differences in the values were apparent,[54,61] thus attesting to the fact that, while conducting SPR experiments, it is necessary to perform proper controls and also to analyze the data rigorously.

Conclusions

SPR offers a wide range of advantages to its users for analyzing biomolecular interactions. It is a label-free, real-time, optical detection method that has been employed to characterize many macromolecular interactions. In addition to giving qualitative data, it is also possible to obtain quantitative data by careful design and analysis of experiments. For example, not only the binding constants of protein–carbohydrate interactions can be determined with ease and at high sensitivity but also the activation and thermodynamic parameters of the reaction involved therein, the precision of which in most cases is comparable to that produced by more direct biophysical techniques, such as fast reaction kinetics and isothermal titration calorimetry.[39–41] However, care needs to be taken in interpreting the results. As one of the reactants is immobilized on the sensor surface, it can lead to several immobilization-related artifacts, such as steric hindrance, immobilization in multiple conformations, and mass transport-related difficulties. However, careful selection of immobilization chemistries and experimental controls for potential artifacts allow for rigorous measurement of kinetic and equilibrium constants.

Acknowledgments

This work was supported by grants from the Department of Biotechnology and the Department of Science and Technology, Government of India (to A.S.). The Biacore facility was funded by the Department of Biotechnology, Government of India for program support to the Indian Institute of Science, Bangalore in the area of Drug and Molecular Design.

[61] D. R. Hall, J. R. Cann, and D. J. Winzor, *Anal. Biochem.* **235,** 175 (1996).

[23] Quantitative Lectin–Carbohydrate Interaction Analysis on Solid-Phase Surfaces Using Biosensor Based on Surface Plasmon Resonance

By HIROYUKI SOTA, REIKO T. LEE, YUAN C. LEE, and YASURO SHINOHARA

Introduction

The importance of lectin–carbohydrate interactions in typical multicellular organism events is becoming widely recognized. Considering that oligosaccharides usually exist as glycoconjugates and most interactions with sugar-recognizing molecules occur on a solid-phase surface, it is important to measure the interaction on solid-phase surfaces to probe possible enhanced avidity of lectins for cell surface carbohydrates. With Biacore (Biacore, Uppsala, Sweden), a biosensor based on surface plasmon resonance (SPR), one interactant to be studied is immobilized onto the sensor surface, while the other interactant is passed over the surface in solution. The sensor device is composed of a sensor chip, which is a thin film of gold covering a thin glass plate and coated with carboxymethylated dextran, and a prism that is placed on the glass surface of the chip. When a binding event occurs on the sensor surface, the resonance angle of the reflected light changes because the SPR response correlates with changes in the refractive index through the prism. Through the use of Biacore data acquisition software, the resonance angle is expressed in terms of resonance units (RU).[1]

In this article, we report a method for oligosaccharide immobilization onto the sensor surface and the results of quantitative analysis of the interactions between lectins and carbohydrates, using such sensor surfaces.

Materials

Oligosaccharides M5, M6, M7D1, M7D3, M8D1D3, M9, NGA2, NA2, NA2B, NA3, and NA4 (see Fig. 1 for structures) are obtained from Oxford GlycoSystems (Abingdon, UK). A biotinylating reagent, 4-(biotinamido)-phenylacetylhydrazide (BPH), is synthesized from *p*-aminophenylacetic acid and biotin via three steps as described.[2] Legume lectins E_4-phytohemagglutinin (PHA) and L_4-PHA from *Phaseolus vulgaris* are purchased

[1] R. Granzow and R. Reed, *Biotechnology* **10,** 390 (1992).

[2] Y. Shinohara, H. Sota, M. Gotoh, M. Hasebe, M. Tosu, J. Nakao, Y. Hasegawa, and M. Shiga, *Anal. Chem.* **68,** 2573 (1996).

Manα1-6
Manα1-3Manα1-6
Manβ1-4GlcNAcβ1-4GlcNAc
Manα1-3
M5

Manα1-6
Manα1-3Manα1-6
Manβ1-4GlcNAcβ1-4GlcNAc
Manα1-2Manα1-3
M6

Manα1-2Manα1-6
Manα1-3Manα1-6
Manβ1-4GlcNAcβ1-4GlcNAc
Manα1-2Manα1-3
M7D1

Manα1-6
Manα1-3Manα1-6
Manβ1-4GlcNAcβ1-4GlcNAc
Manα1-2Manα1-2Manα1-3
M7D3

Manα1-2Manα1-6
Manα1-3Manα1-6
Manβ1-4GlcNAcβ1-4GlcNAc
Manα1-2Manα1-2Manα1-3
M8D1D3

Manα1-2Manα1-6
Manα1-2Manα1-3Manα1-6
Manβ1-4GlcNAcβ1-4GlcNAc
Manα1-2Manα1-2Manα1-3
M9

GlcNAcβ1-2Manα1-6
Manβ1-4GlcNAcβ1-4GlcNAc
GlcNAcβ1-2Manα1-3
NGA2

Galβ1-4GlcNAcβ1-2Manα1-6
Manβ1-4GlcNAcβ1-4GlcNAc
Galβ1-4GlcNAcβ1-2Manα1-3
NA2

Galβ1-4GlcNAcβ1-2Manα1-6
GlcNAcβ1-4Manβ1-4GlcNAcβ1-4GlcNAc
Galβ1-4GlcNAcβ1-2Manα1-3
NA2B

Galβ1-4GlcNAcβ1-2Manα1-6
Manβ1-4GlcNAcβ1-4GlcNAc
Galβ1-4GlcNAcβ1-2Manα1-3
Galβ1-4GlcNAcβ1-4
NA3

Galβ1-4GlcNAcβ1-6
Galβ1-4GlcNAcβ1-2Manα1-6
Manβ1-4GlcNAcβ1-4GlcNAc
Galβ1-4GlcNAcβ1-2Manα1-3
Galβ1-4GlcNAcβ1-4
NA4

FIG. 1. Structures and abbreviations for oligosaccharides used in this study.

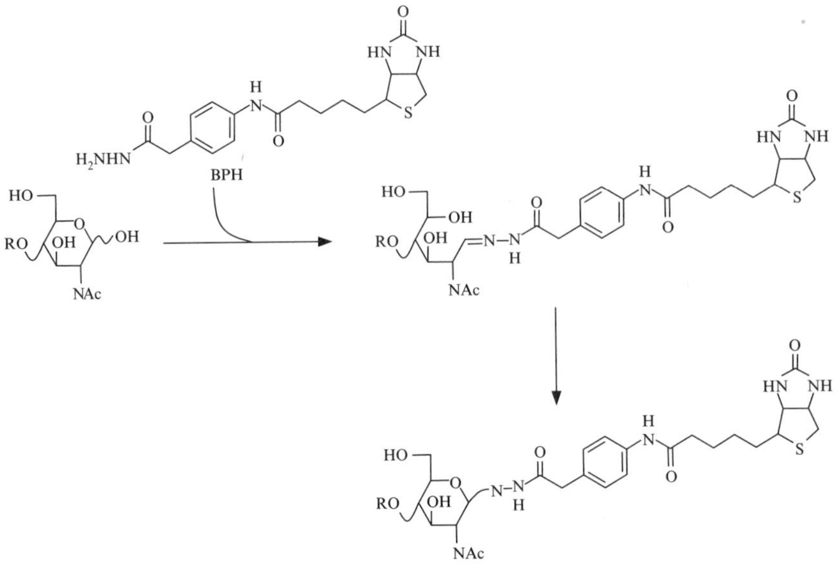

FIG. 2. Structure of BPH and the chemistry of adduct formation with carbohydrate.

from Honen (Tokyo, Japan). Preparation of chimeric lectins between E_4-PHA and L_4-PHA has been described.[3] A fragment of rat mannose-binding protein C (MBP-C) that comprises all the carbohydrate recognition domain (CRD) and a neck portion (MBP-C CRD) is prepared as described.[4]

Biotinylation of Oligosaccharides

Procedure

The oligosaccharide (1 pmol–10 nmol) in 10 μl of water is incubated with a 4-fold molar excess of BPH in 30% acetonitrile (10 μl) at 90° for 1 h. After the reaction, 20 μl of formate buffer (50 m*M*, pH 3.5) is added and the mixture is stored at 4° for 12 h to promote tautomerization from acyclic Schiff base-type hydrazone to stable β-glycoside (Fig. 2).

[3] Y. Kaneda, R. F. Whittier, H. Yamanaka, E. Carredano, M. Gotoh, H. Sota, Y. Hasegawa, and Y. Shinohara, *J. Biol. Chem.* **277,** 16928 (2002).
[4] K. Drickamer, *Biochem. Soc. Trans.* **17,** 13 (1989).

Notes

The key to efficient adduct formation is to carry out the reaction in a small volume (~20 μl) in a tightly sealed flat-bottomed reaction tube, in which almost all solvent is vaporized on heating at 90°.[5] Formate buffer is added after cooling the reaction tube on ice to let the vaporized gas liquefy. This method can be applied to sialylated oligosaccharides without any loss of sialyl residues. One of the advantages of BPH labeling of oligosaccharide is that it produces β-*N*-glycoside, which mimicks the naturally occurring *N*-glycan structure.

Purification of Oligosaccharide–BPH Adducts

Procedure

BPH-labeled oligosaccharides are purified by reversed-phase chromatography on a TSK gel ODS_{80} column (4.6 × 250 mm; Tosoh, Tokyo, Japan) with ultraviolet (UV) detection at 252 nm. Separation is performed at 40° at a flow rate of 0.5 ml/min. Elution is performed isocratically with 70 m*M* phosphate buffer (pH 6.8) containing 14% acetonitrile.

Notes

The BPH-labeled oligosaccharides possess superior properties for chromatographic separation. The BPH-labeled oligosaccharides can be purified by other optimized chromatographic modes[2] including anion-exchange, gel-filtration, and reversed-phase chromatography under gradient conditions, when the labeling is performed for a complex mixture of oligosaccharides (e.g., glycan pool derived from natural sources).

Immobilization of BPH-Labeled Oligosaccharide onto Sensor Surface

Procedure

The purified BPH-labeled oligosaccharide (~10 pmol) is injected onto a streptavidin-preimmobilized Biacore sensor surface (sensor chip SA; Biacore). Typically, 10 μl of a BPH-labeled oligosaccharide solution (~1 μM) is passed over the surface at a flow rate of 2 μl/min.

[5] Y. Shinohara, H. Sota, F. Kim, M. Shimizu, M. Gotoh, M. Tosu, and Y. Hasegawa, *J. Biochem.* **117,** 1076 (1995).

Notes

The injection of ~10 pmol of biotinyl glycan is sufficient to saturate the binding sites of immobilized streptavidin on sensor chip SA. Because the amount of streptavidin is constant and an excess of purified BPH–oligosaccharide is introduced, the molar amount of each immobilized oligosaccharide is presumed to be nearly constant. The surface-immobilized BPH–oligosaccharides are fairly stable: almost no decrease in response is detected during 20 repeated injections of lectins. The immobilized BPH-labeled oligosaccharide has also been confirmed to be feasible for *in situ* enzymatic digestion on the sensor surface.[2]

Quantitative Interaction Analysis of Rat Liver Mannose-Binding Protein with Immobilized High Mannose-Type Oligosaccharides

MBP-C[6] belongs to the collection group of C-type (Ca^{2+}-dependent) lectins, which also includes the serum mannose-binding protein (MBP-A) and pulmonary surfactant apoproteins A and D. Both MBP-A and MBP-C have similar specificity of recognizing terminal mannose, glucose, *N*-acetylglucosamine and L-fucose, but they show different affinities toward high mannose-type glycans.[7] The biological role of MBP-C is unclear at present, whereas it is known that MBP-A functions in immune surveillance by recognizing mannose and other sugar structures present on pathogenic organisms.[8] The cloned fragment of MBP-C, MBP-C CRD, that exists as a stable trimer is used to elucidate quantitatively the fine specificity of MBP-C for different oligosaccharides.

Procedure

High mannose-type oligosaccharides M5, M6, M7D1, M7D3, M8D1D3, and M9 (Fig. 1) are labeled by BPH as described above. The BPH oligosaccharide (~10 pmol), after reversed-phase high-performance liquid chromatography (RP-HPLC) purification, is introduced onto sensor chip SA. For SPR analyses, HBS buffer [10 m*M* HEPES (pH 7.4), 0.15 *M* NaCl, 1 m*M* $CaCl_2$, and 0.05% Biacore surfactant P20] is used throughout. MBP-C CRD is purified by Superose 12 (Amersham Biosciences, Piscataway, NJ) before introduction onto the oligosaccharide surfaces at a flow rate of 20 μl/min. After monitoring for 3 min, dissociation is initiated by introduction of the buffer. The chip surface is regenerated by treatment with

[6] K. Drickamer, *J. Biol. Chem.* **261,** 6878 (1986).
[7] R. T. Lee, Y. Shinohara, Y. Hasegawa, and Y. C. Lee, *Biosci. Rep.* **19,** 283 (1999).
[8] W. I. Weis, M. E. Taylor, and K. Drickamer, *Immunol. Rev.* **169,** 19 (1998).

50 mM H_3PO_4. The apparent association and dissociation of MBP-C CRD to surfaces immobilized with high-mannose glycans are found to be fairly rapid (Fig. 3A). For this reason, the equilibrium binding level (R_{eq}) is used to obtain the affinity constant (K_a). R_{eq} is obtained from the apparent equilibrium bound level (in resonance units) by correcting for the background resonance units, which is obtained by injecting the same concentrations of MBP-C CRD onto an unmodified sensor surface. The K_a and R_{max}, and the

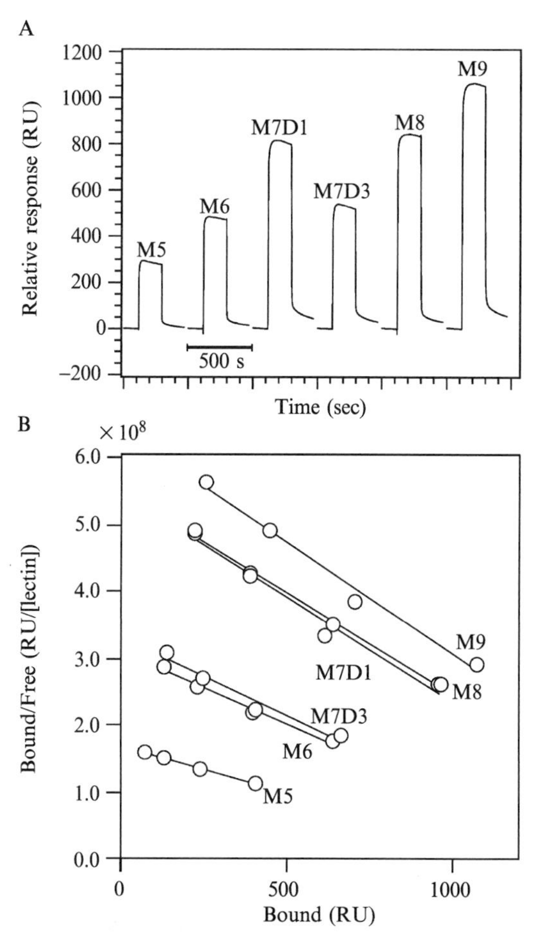

FIG. 3. Interaction between MBP-C CRD and immobilized high mannose-type oligosaccharides. (A) Sensorgrams showing the interactions. Each lectin was introduced onto the surface at a concentration of 3.7 μM. RU, Response units. (B) Scatchard plot of the interactions. x axis, R_{eq} in response units; y axis, R_{eq}/C_0, where C_0 is the injected MBP-C concentration, which varied from 0.46 to 3.7 μM.

TABLE I
PARAMETERS OBTAINED FROM SCATCHARD PLOT ANALYSIS FOR INTERACTION OF MBP-C CRD WITH IMMOBILIZED OLIGOSACCHARIDES[a]

	MBP-C	
Oligosaccharide	K_a (M^{-1}) $\times 10^4$	R_{max} (RU)
M5	13.5	1222
M6	21.0	1457
M7D1	30.8	1769
M7D3	23.6	1403
M8D1D3	30.4	1806
M9	33.7	1902

[a] From Ref. 4.

maximum lectin–oligosaccharide complex concentration (in resonance units), are determined by linear least-squares curve fitting of the Scatchard equation:

$$R_{eq}/C_0 = K_a R_{max} - K_a R_{eq} \tag{1}$$

where C_0 is the constant concentration of injected lectin.[9] To do this, the R_{eq} values collected at several MBP concentrations (0.46–3.7 μM) are plotted against R_{eq}/C_0, and K_a and R_{max} are calculated from the slope and intercept, respectively.

Notes

In SPR analysis, the difference in oligomeric form of the lectin often produces a completely different sensorgram. Even nominal contamination of the analyte solution with different oligomeric forms can make the sensorgram complicated. Gel chromatographic purification of MBP-C CRD before SPR analysis is therefore performed to remove oligomerically heterogeneous components.

From the Scatchard plot analysis, the K_a and R_{max} values are calculated (Fig. 3B and Table I). This analysis reveals that both K_a and R_{max} tend to increase as the number of terminal Manα1,2 residues increases. The solution-phase assay and the microplate assay also show that MBP-C CRD exhibits a slightly higher K_a for Manα1,2Man than for Manα1,3 and Manα1,6Man.[7,10] The result observed here may be explained by this

[9] R. Karlsson, A. Michaelsson, and L. Mattsson, *J. Immunol. Methods.* **145,** 229 (1991).
[10] R. T. Lee and Y. C. Lee, *Glycoconi. J.* **14,** 357 (1997).

relatively small but possibly augmentable affinity under nonhomogeneous conditions.

Quantitative Interaction Analysis of Chimeric Lectins of *Phaseolus vulgaris* Erythro- and Leukoagglutinating Lectins

E_4-PHA and L_4-PHA[3] are homotetrameric legume lectins from *Phaseolus vulgaris*. They both bind complex-type glycan structures, but with somewhat different specificity. For instance, E_4-PHA and L_4-PHA exhibit greatest affinity for complex *N*-glycans containing either bisecting GlcNAc or β1,6-linked LacNAc, respectively. To decipher the mechanism involved in this fine specificity difference, chimeric lectins are prepared by loop-swapping mutagenesis. Among four loops located near the sugar-binding site of these lectins,[11] loop B and C differ considerably between the two isolectins. Therefore, six chimeric proteins, E-b_L, E-c_L, E-b_L/c_L, L-b_E, L-c_E, and L-b_E/c_E, are prepared to investigate the roles of loops B and C in defining fine sugar-binding specificities. *Note*: Letters after a hyphen denote the mutated loop. For instance, E-b_L means that loop B of the E subunit is substituted with loop B of the L subunit.

Procedure

Oligosaccharides NGA2, NA2, NA2B, NA3, and NA4 (Fig. 1) are labeled with BPH and immobilized onto a sensor chip SA as described above. Lectins (10 μg/ml in HBS) are introduced onto the surface at a flow rate of 20 μl/min. The interaction between lectin and oligosaccharide is monitored as the change in SPR response at 25°. After 3 min, flow is switched from sample to HBS buffer in order to initiate dissociation. Sensor surfaces are regenerated with 50 m*M* H_3PO_4. The area under the curve for the dissociation phase ($AUC_d^{0 \rightarrow \infty}$)[3] is calculated using the trapezoidal rule, by extrapolating over an infinite time interval using nonlinear regression (Fig. 4).

Notes

We use $AUC_d^{0 \rightarrow \infty}$ as a quantitative index for evaluating sensorgrams obtained from interactions between lectins and immobilized oligosaccharides. Because this method does not require model fitting except for extrapolation using nonlinear regression, the obtained values are free from miscalculations due to inadequate models. The calculated $AUC_d^{0 \rightarrow \infty}$ values for native E_4-PHA, L_4-PHA, and six chimeric lectins are summarized

[11] V. Sharma and A. Surolia, *J. Mol. Biol.* **267,** 433 (1997).

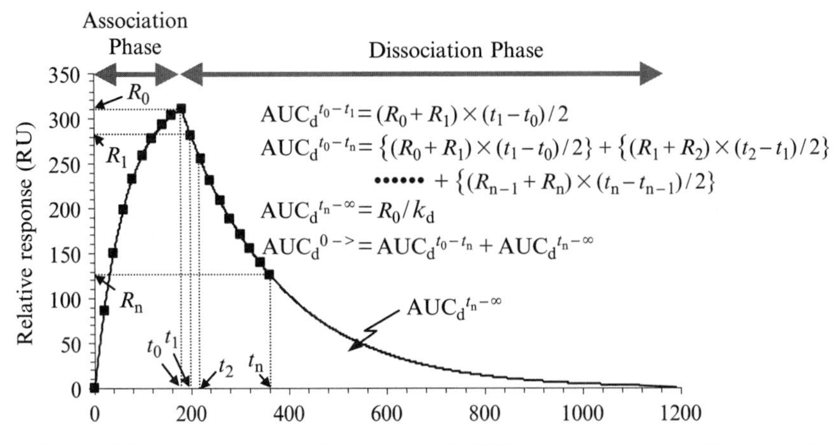

FIG. 4. Calculation of area under the curve (AUC), using the trapezoidal rule. In the analysis presented in Fig. 5, the time interval used was 1 s. The dissociation rate constant (k_d) was calculated by using the dissociation phase of, typically, the last 30 s by directly fitting the dissociation curve to the equation: $R_t = R_0 \exp(-k_d t)$.

in Fig. 5. We observed good correlation between $AUC_d^{0 \to \infty}$ and elution order derived by immobilized lectin affinity chromatography. The $AUC_d^{0 \to \infty}$ is a meaningful parameter, which can be described as Eq. (2) if the interaction can be approximated by a one-to-one interaction model.

$$AUC_d^{0 \to \infty} = \int_{t=0}^{t=\infty} R_0 \exp(-k_d t) dt = R_0/k_d \qquad (2)$$

where R_0 represents the amplitude of the dissociation process and k_d is the dissociation rate constant. The results shown in Fig. 5 indicate that the binding preference of E_4-PHA and L_4-PHA for bisecting GlcNAc and β1,6-linked branch structures, respectively, is almost solely attributable to loop B. The contribution of the central portion of loop C to the recognition of those structural motifs is found to be negligible. Instead, it modulates affinity toward *N*-acetyllactosamine residues present at the nonreducing terminus.

General Comments

The use of SPR allows easy access to the kinetic data of binding in addition to the determination of affinity. The interaction analysis between oligosaccharide and lectin on the solid support, however, sometimes produces complexity in the analysis because the avidity effect may complicate the interpretation of data. This is especially pronounced when the

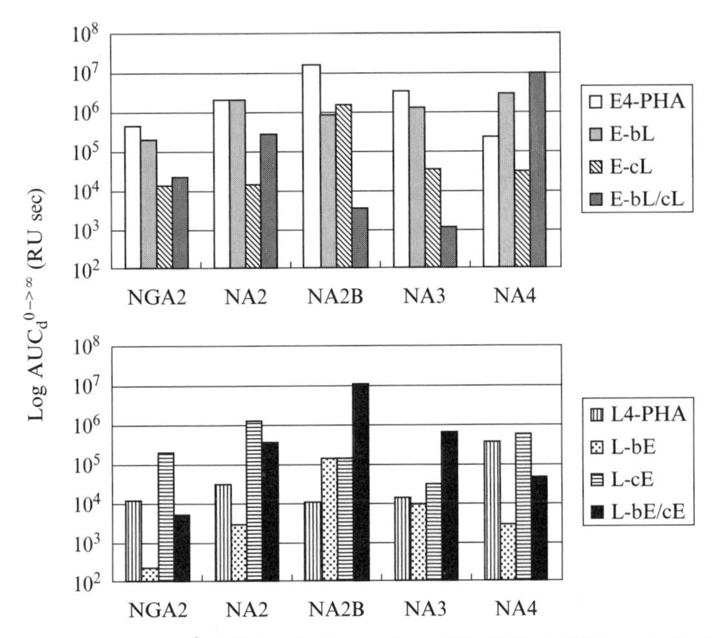

FIG. 5. Calculated $AUC_d^{0 \to \infty}$ for the interaction of E_4-PHA, L-PHA, and six chimera lectins with immobilized oligosaccharides. E-b_L, chimeric lectin of E_4-PHA, in which loop B is replaced with that of L_4-PHA; E-c_L, chimeric lectin of E_4-PHA, in which the central portion of loop C is replaced with that of L_4-PHA; E-b_Lc_L, chimeric lectin of E_4-PHA, in which loop B and the central portion of loop C are replaced with those of L_4-PHA; L- b_E, chimeric lectin of L_4-PHA, in which loop B is replaced with that of E_4-PHA; L-c_E, chimeric lectin of L_4-PHA, in which the central portion of loop C is replaced with that of E_4-PHA; L-b_Ec_E, chimeric lectin of L_4-PHA, in which loop B and the central portion of loop C are replaced with those of E_4-PHA.

oligosaccharide is immobilized and a multivalent lectin is introduced as an analyte.[12] This effect could be explained by the nonhomogeneous reaction conditions, wherein a molecule possessing a higher diffusion coefficient (e.g., oligosaccharide) is immobilized at high density and a multivalent molecule having a lower diffusion coefficient (e.g., lectin) is introduced.[13] Notably, this effect should also be considered in several biological states in which the ligand is immobilized. By carefully designing the experimental conditions, for example, site-specific oligosaccharide immobilization, controlled carbohydrate density, and use of purified lectin, analysis of the interaction on the solid-phase surface provides precious quantitative information as described here.

[12] Y. Shinohara, Y. Hasegawa, H. Kaku, and N. Shibuya, *Glycobiology* **7,** 1201 (1997).
[13] R. Karlsson, H. Roos, L. Fagerstam, and B. Persson, *Methods* **6,** 99 (1994).

Acknowledgments

This original work was partly supported by the New Energy and Industrial Technology Development Organization in the Ministry of Economy, Trade, and Industry.

[24] Surface Plasmon Resonance Study of RNA–Aminoglycoside Interactions

By Chi-Huey Wong and Fu-Sen Liang

There is growing interest in developing small molecules to specifically target RNA because of potential therapeutic applications.[1–7] RNA usually folds into unique structures, like proteins, that provide the possibility of specific recognition by small molecules.[8–10] Several antibiotics are known to target bacterial ribosomal RNA. Because RNA functional domains are more highly conserved than are protein active sites, RNA may experience a slower development of drug resistance. Usually there are several potential targeting sites in a given RNA sequence, whereas proteins often contain only one targeting site. In addition, it is possible to achieve tissue- or disease-specific targeting by selecting specific spliced transcripts.[11] Generally, translation associated with protein synthesis, mRNA maturation, and ribonucleoprotein–RNA interactions are considered the most promising points of intervention.

An increasing number of small molecules with different structures have been discovered that bind to RNA.[12,13] With the advanced techniques available to elucidate RNA structures and to probe RNA–small molecule interactions, the underlying recognition principles regarding molecular details are just beginning to be revealed. For rational design of selective

[1] S. J. Sucheck and C.-H. Wong, *Curr. Opin. Chem. Biol.* **4,** 678 (2000).
[2] W. D. Wilson and K. Li, *Curr. Med. Chem.* **7,** 73 (2000).
[3] Y. Tor, *Angew. Chem. Int. Ed.* **38,** 1579 (1999).
[4] T. Hermann and E. Westhof, *Curr. Opin. Biotechnol.* **9,** 66 (1998).
[5] N. D. Pearson and C. D. Prescott, *Chem. Biol.* **4,** 409 (1997).
[6] S. R. Eddy, *Nat. Rev. Genet.* **2,** 919 (2001).
[7] G. Storz, *Science* **296,** 1260 (2002).
[8] E. Westhof and V. Fritsch, *Structure* **8,** R55 (2000).
[9] T. Hermann and D. J. Patel, *J. Mol. Biol.* **294,** 829 (1999).
[10] P. Moore, *Annu. Rev. Biochem.* **68,** 287 (1999).
[11] D. Ecker and R. H. Griffey, *Drug Discov. Today* **4,** 420 (1999).
[12] T. Hermann, *Angew. Chem. Int. Ed.* **39,** 1890 (2000).
[13] J. Gallego and G. Varani, *Acc. Chem. Res.* **34,** 836 (2001).

0076-6879/03 $35.00

RNA binders, detailed knowledge of structure–affinity relationships between RNA and small molecules is required, as well as analysis of the binding specificity of various RNA sequences.

Several methods have been used to investigate the binding between RNA and small molecules.[14] The gel mobility shift assay,[15,16] nitrocellulose filter binding assay,[17,18] and scintillation proximity assay[18,19] are usually performed by competition between small molecules and RNA-binding proteins with RNA and have been used for small molecule library screening. Circular dichroism can provide thermodynamic information as well as the binding mode of small molecules interacting with RNA.[20] Chemical and nuclease footprinting can be used to determine the small molecule binding sites.[21,22] Fluorescence-based methods have been used for the study of binding and for high-throughput screening.[23,24] Probing the recognition by chemically synthesized molecules can reveal essential structural features required for the binding activity.[25] Nuclear magnetic resonance (NMR)[22,26–28] and X-ray crystallography[29–33] are powerful tools to access the most detailed structural information about RNA or RNA–small

[14] K. A. Xavier, P. S. Eder, and T. Giordano, *Trends Biotechnol.* **18,** 349 (2000).

[15] P. L. Molloy, *Methods Mol. Biol.* **130,** 235 (2000).

[16] N. Gelus, C. Bailly, F. Hamy, T. Klimkait, W. David Wilson, and D. W. Boykin, *Bioorg. Med. Chem.* **7,** 1089 (1999).

[17] K. B. Hall and J. K. Kranz, *Methods Mol. Biol.* **118,** 105 (1999).

[18] H.-Y. Mei, D. P. Mack, A. A. Galan, N. S. Halim, A. Heldsinger, J. A. Loo, D. W. Moreland, K. A. Sannes-Lowery, L. Sharmeen, H. N. Truong, and A. W. Czarnik, *Bioorg. Med. Chem.* **5,** 1173 (1997).

[19] S. Udentfriend, L. Gerber, and H. Nelson, *Anal. Biochem.* **161,** 494 (1987).

[20] S. Sehlstedt, P. Aich, J. Bergman, H. Vallberg, B. Norden, and A. Graslund, *J. Mol. Biol.* **278,** 31 (1998).

[21] D. Moazed and H. Noller, *Nature* **327,** 389 (1987).

[22] M. I. Recht, D. Fourmy, S. C. Blanchard, K. D. Dahlquist, and J. D. Puglisi, *J. Mol. Biol.* **262,** 421 (1996).

[23] W. C. Lam, J. M. Seifert, F. Amberger, C. Graf, M. Auer, and D. P. Millar, *Biochemistry* **37,** 1800 (1998).

[24] K. Hamasaki and R. R. Rando, *Anal. Biochem.* **261,** 183 (1998).

[25] M. Hendrix, P. B. Alper, E. S. Priestly, and C.-H. Wong, *Angew. Chem. Int. Ed.* **36,** 95 (1997).

[26] S. Yoshizawa, D. Fourmy, and J. D. Puglisi, *EMBO J.* **17,** 6437 (1998).

[27] D. Fourmy, S. Yoshizawa, and J. D. Puglisi, *J. Mol. Biol.* **277,** 333 (1998).

[28] W. H. Gmeiner, *Curr. Med. Chem.* **5,** 115 (1998).

[29] Q. Vicens and E. Westhof, *Chem. Biol.* **9,** 747 (2002).

[30] Q. Vicens and E. Westhof, *Structure* **9,** 647 (2001).

[31] V. Ramakrishnan, *Cell* **108,** 557 (2002).

[32] A. P. Carter, W. M. J. Clemons, D. E. Brodersen, R. J. Morgan-Warren, B. T. Wimberly, and V. Ramakrishnan, *Nature* **407,** 340 (2000).

[33] N. Ban, P. Nissen, J. Hansen, P. B. Moore, and T. A. Steitz, *Science* **289,** 905 (2000).

molecule complexes. Mass spectrometry is another new tool, which has been used to determine binding constants, binding sites, and binding specificity.[34–37]

Surface plasmon resonance (SPR) biosensor has been used to measure RNA–small molecule interactions and provides real-time monitoring without a labeling requirement.[38,39] SPR is an optical phenomenon that occurs when a near-infrared light is reflected under certain conditions from a conducting film at the interface between two media having different refractive indices. SPR causes a reduction in the intensity of reflected light at a specific angle of reflection. This angle varies with the refractive index close to the surface on the side opposite the reflected light.[40] In Biacore (Uppsala, Sweden) real-time biosensor systems, the ligands are immobilized onto the glass of the sensor chip, and the analytes flow over the surface with immobilized ligands. When analytes bind to ligands on the sensor surface, the concentration and therefore the refractive index at the surface changes and an SPR response is detected. The observed SPR signal is correlated to the mass concentration change on the chip surface, which allows the binding events to be monitored. Both kinetic and thermodynamic information can be derived on the basis of the experimental design, and the required amounts of materials are usually small (picomoles to nanomoles).

Although the SPR-based method has been applied to monitor macromolecular interactions, only more recently, with improved instrument detection accuracy, can bindings between small molecules and macromolecules be monitored successfully by SPR. Some examples of RNA–small molecule binding analysis by SPR include binding between aminoglycosides or derivatives and the wild-type or mutant RNA of the bacterial ribosome 16S A-site,[41–44] the Rev responsive element (RRE) of human

[34] S. A. Hofstadler and R. H. Griffey, *Curr. Opin. Drug Discov. Dev.* **3,** 423 (2000).
[35] S. A. Hofstadler and R. H. Griffey, *Chem. Rev.* **101,** 377 (2001).
[36] K. A. Sannes-Lowery, R. H. Griffey, and S. A. Hofstadler, *Anal. Biochem.* **280,** 264 (2000).
[37] J. A. Loo, V. Thanabal, and H.-Y. Mei, *in* "Mass Spectrometry in Biology and Medicine" (A. L. Burlingame, S. A. Carr, and M. A. Baldwin, eds.), p. 73. Humana Press, Totowa, NJ, 2000.
[38] P. Schuck, *Annu. Rev. Biophys. Biomol. Struct.* **26,** 541 (1997).
[39] Biacore, "BIAcore BIAapplications Handbook." Biacore, Uppsala, Sweden, 1998.
[40] F. Markey, *BIA J.* **6,** 14 (1999).
[41] C.-H. Wong, M. Hendrix, E. S. Priestly, and W. A. Greenberg, *Chem. Biol.* **5,** 397 (1998).
[42] S. J. Sucheck, A. L. Wong, K. M. Koeller, D. D. Boehr, K.-A. Draker, P. S. Sears, G. D. Wright, and C.-H. Wong, *J. Am. Chem. Soc.* **122,** 5230 (2000).
[43] C.-H. Wong, M. Hendrix, D. D. Manning, C. Rosenbohm, and W. A. Greenberg, *J. Am. Chem. Soc.* **120,** 8319 (1998).
[44] P. B. Alper, M. Hendrix, P. S. Sears, and C.-H. Wong, *J. Am. Chem. Soc.* **120,** 1965 (1998).

immunodeficiency virus type 1 (HIV-1),[45] and the Bcr-Abl translocation breakpoint.[46]

Aminoglycosides are a group of structural diverse aminocyclitol conjugates with potent antibiotic activities, and are able to bind selectively to specific RNA sequences. They are found to bind the bacterial ribosome 16S A-site,[47] the regulatory domains of HIV-1 mRNA,[48,49] the mRNA sequence of the oncogenic Bcr-Abl fusion protein,[46] hammerhead ribozyme,[50] group I intron,[51] hepatitis delta virus ribozyme,[52] and thymidylate synthase mRNA.[53] The high binding affinity and specificity have made the study of aminoglycoside–RNA binding the paradigm for RNA–small molecule interactions.

The bindings between aminoglycosides and bacterial ribosomal RNA are of particular interest. For five decades aminoglycosides have been used clinically as antibiotics, and they function as protein translation inhibitors. They interfere with proofreading during protein synthesis by binding to the highly conserved A-site sequence of the small 30S ribosomal subunit.[54–56] With the emerging problems of resistance against existing aminoglycoside antibiotics, synthetic aminoglycoside derivatives with potent antibiotic activities and resistant to modification enzymes[57] are desired. Several aminoglycoside derivatives have been synthesized,[58] and SPR-based methods are efficient and powerful tools for the binding studies. The discovery of neamine binding to the 16S A-site in 2:1 stoichiometry by the SPR method has resulted in the development of bivalent aminoglycoside-derived antibiotics that target the A-site with high affinity and also inhibit the resistance-causing enzyme (Fig. 1).[42]

[45] M. Hendrix, E. S. Priestly, G. F. Joyce, and C.-H. Wong, *J. Am. Chem. Soc.* **119,** 3641 (1997).

[46] S. J. Sucheck, W. A. Greenberg, T. J. Tolbert, and C.-H. Wong, *Angew. Chem. Int. Ed.* **39,** 1080 (2000).

[47] D. Fourmy, M. I. Recht, S. C. Blanchard, and J. D. Puglisi, *Science* **274,** 1367 (1996).

[48] M. L. Zapp, S. Stern, and M. R. Green, *Cell* **74,** 969 (1993).

[49] S. Wang, P. W. Huber, M. Cui, A. W. Czarnik, and H.-Y. Mei, *Biochemistry* **37,** 5549 (1998).

[50] T. Hermann and E. Westhof, *J. Mol. Biol.* **276,** 903 (1998).

[51] U. von Ahsen, J. Davies, and R. Schroeder, *Nature* **353,** 368 (1991).

[52] J. Rogers, A. H. Chang, U. von Ahsen, and R. Schroeder, *J. Mol. Biol.* **259,** 916 (1996).

[53] A. R. Ferre-D'Amare, K. Zhou, and J. A. Doudna, *Nature* **395,** 567 (1998).

[54] S. Campuzano, D. Vazquez, and J. Modolell, *Biochem. Biophys. Res. Commun.* **87,** 960 (1979).

[55] J. Woodcock, D. Moazed, D. Cannon, J. Davies, and H. F. Noller, *EMBO J.* **10,** 3099 (1991).

[56] M. J. Cabanas, D. Vazquez, and J. Modolell, *Eur. J. Biochem.* **87,** 21 (1978).

[57] G. D. Wright, *Curr. Opin. Microbiol.* **2,** 499 (1999).

[58] L. P. Kotra and S. Mobashery, *Curr. Org. Chem.* **5,** 193 (2001).

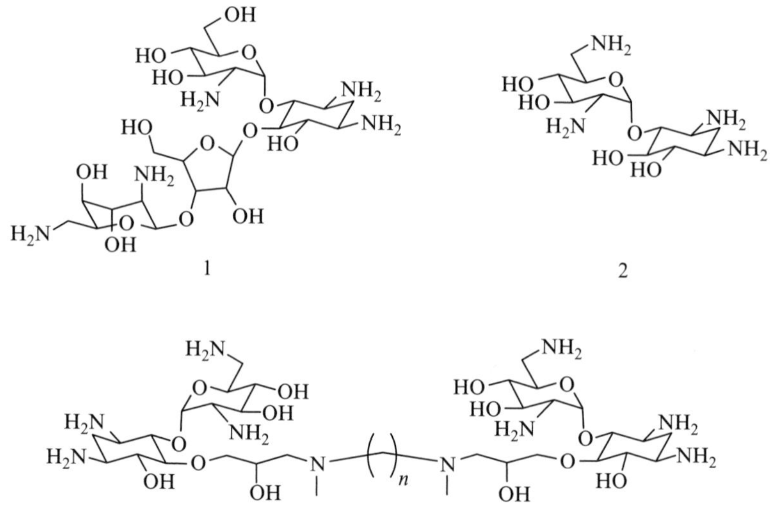

FIG. 1. Aminoglycosides used in this study: paromomycin (1), neamine (2), and neamine dimer (3). From S. J. Sucheck, A. L. Wong, K. M. Koeller, D. D. Boehr, K.-A. Draker, P. S. Sears, G. D. Wright, and C.-H. Wong, *J. Am. Chem. Soc.* **122,** 5230 (2000).

This article focuses on describing procedures performed in the SPR binding assay, using aminoglycoside–bacterial ribosome 16S A-site binding as an example. Methods of data analysis to derive the equilibrium constant and binding stoichiometry are also discussed.

Preparations of Biotinylated RNA

Chemoenzymatic Synthesis

5′-Biotinylated bacterial 16S A-site RNA is prepared by *in vitro* transcription[59,60] in the presence of guanosine 5′-monophosphorothioate (5′-GMPS)[61] followed by biotinylation with biotin iodoacetamide derivative (Fig. 2).[45] Detailed procedures are described below.

The DNA templates (5′-GGC GAC TTC ACC CGA AGG TGT GAC GCC CTA TAG TGA GTC GTA TTA-3′) are annealed to a 2-fold excess of 18-mer T7 promoter (TAA TAC GAC TCA CTA TAG) in H_2O by heating to 65° and slowly cooling to below 37°. The 5′-phosphorothioate

[59] J. R. Wyatt, M. Chastain, and J. D. Puglisi, *Biotechniques* **11,** 764 (1991).
[60] J. F. Mulligan and O. C. Uhlenbeck, *Methods Enzymol.* **180,** 51 (1989).
[61] A. B. Burgin and N. R. Pace, *EMBO J.* **9,** 4111 (1990).

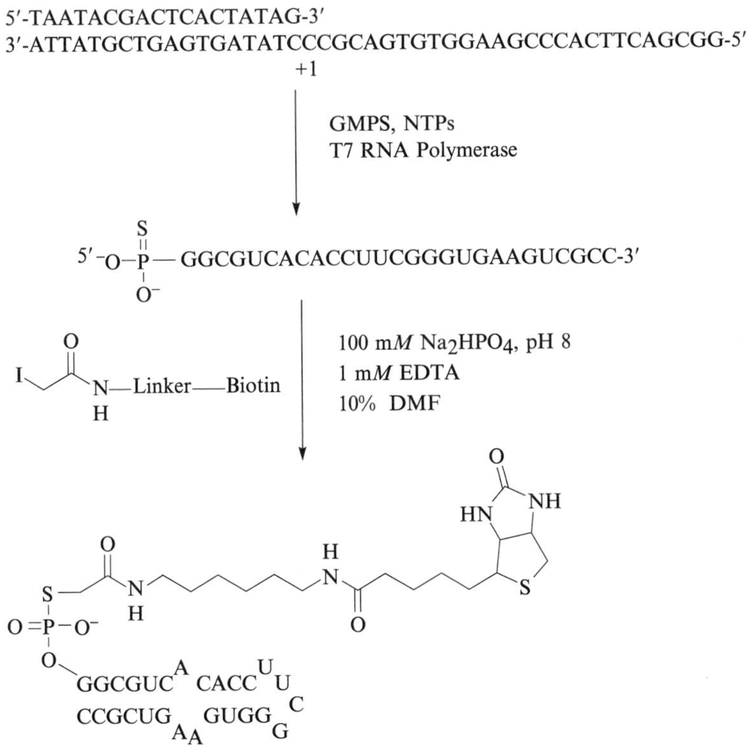

5′-TAATACGACTCACTATAG-3′
3′-ATTATGCTGAGTGATATCCCGCAGTGTGGAAGCCCACTTCAGCGG-5′
+1

GMPS, NTPs
T7 RNA Polymerase

5′-O–P(=S)(O⁻)—GGCGUCACACCUUCGGGUGAAGUCGCC-3′

I–CH2–C(=O)–NH—Linker—Biotin

100 mM Na_2HPO_4, pH 8
1 mM EDTA
10% DMF

GGCGUC A CACC U U C G GUGG AA CCGCUG

FIG. 2. Chemoenzymatic synthesis of 5′-biotinylated bacterial 16S A-site RNA. 5′-Phosphorothioates were synthesized by *in vitro* transcription with T7 RNA polymerase in the presence of GMPS, followed by biotinylation with biotin iodoacetamide derivative to produce 5′-biotinylated bacterial 16S A-site RNA.

transcripts are then prepared by incubating the annealed templates (0.2 μM) in 50 mM tris(hydroxymethyl)aminomethane (Tris) (pH 7.5), 15 mM $MgCl_2$, 2 mM spermidine, 5 mM dithiothreitol (DTT), 2 mM adenosine 5′-triphosphate (ATP), 2 mM cytidine 5′-triphosphate (CTP), 2 mM uridine 5′-phosphate (UTP), 0.2 mM guanosine 5′-phosphate (GTP), and 4 mM 5′-GMPS with T7 RNA polymerase (5 U/μl) and inorganic pyrophosphatase (0.001 U/μl) for 2 h at 37° in a total volume of 100 μl. Additional 1-μl aliquots of 20 mM GTP are added at 20-min intervals, and 500 U of T7 RNA polymerase is added after 1 h. Reactions are quenched by addition of ethylenediaminetetraacetic acid (EDTA), extracted with phenol, and precipitated with ethanol. Electrophoresis is carried out on 20% denaturing polyacrylamide gels. Full-length transcripts are excised from the gel, eluted into 200 mM NaCl, 10 mM Tris (pH 7.5), and 0.5 m*M* EDTA, and desalted on a Nensorb column (PerkinElmer Life Sciences, Boston, MA). Typically, a 100-μl transcription reaction produces 150–500 pmol of RNA-5′-phosphorothioate. The purified RNA transcripts are then resuspended in 18 μl of 100 m*M* NaH_2PO_4 (pH 8.0) containing 1 m*M* EDTA and are treated with 2 μl of 20 m*M* biotin iodoacetamide in dimethylformamide (USB, Cleveland, OH). After 2 h at room temperature, 2 μl of the biotinylation reagent is added and incubation is continued for 1 h. RNA is then precipitated with ethanol in the presence of glycogen, purified by gel, and desalted as described above.

Directly Purchased RNA

5′-Biotinylated RNA of bacterial ribosome 16S A-site (5′-biotin-GGC GUC ACA CCU UCG GGU GAA GUC GCC) can be obtained directly from Dharmacon Research (Lafayette, CO).[62] RNA is available as the 2′-*O*-bis(2-acetoxyethoxy)methyl (ACE) protected form. Performing deprotection steps per the instructions from the vendor (described below) is required before use.

RNA pellets are spun down in tubes and 400 μl of deprotection buffer [100 m*M* acetic acid, adjusted to pH 3.8 with *N*, *N*, *N*′, *N*′-tetramethylethylenediamine (TEMED)] is added to each tube of RNA (~90 nmol). After pellets have been completely dissolved by pipetting up and down, vortex the tubes for 10 s and spin down for 10 s. Incubate at 60° for 30 min, and then lyophilize or centrifuge (SpeedVac; Thermo Savant, Holbrook, NY) to dryness. After deprotection, RNA is further purified by ethanol precipitation and gel purification.

SPR Binding Experiments

The SPR binding experiments are performed with a Biacore 2000 system from Biacore (Uppsala, Sweden). All solutions and sensor chips are equilibrated to room temperature before use. Buffer solutions are prepared in water produced by a Milli-Q system (Millipore, BedFord, MA), and the working area is pretreated with RNase*Zap* (Ambion, Austin, TX) to inactivate any ribonuclease.

Preparation of RNA-Immobilized Sensor Chip

A new streptavidin-functionalized Biacore sensor chip (Sensor chip SA; Biacore) is docked into the Biacore 2000.[39] The system is primed with running buffer, and normalized with BIAnormalizing solution [40% (w/w)

[62] S. A. Scaringe, *Methods Enzymol.* **317,** 3 (2000).

glycerol in H_2O; Biacore]. Then, start a new sensorgram with HBS running buffer flowing at a rate of 2 μl/min and allowing the buffer to run for about 10 min, or wait until the baseline is stable, before injecting RNA.

Before immobilizing RNA on the sensor chip, 80 μl of 125 n*M* 5′-biotinylated RNA solution in running buffer, the HEPES-buffered saline (HBS) buffer [10 m*M* *N*-(2-hydroxyethyl) piperazine-*N*′-(2- ethanesulfonic acid) (HEPES), 3.4 m*M* EDTA, 150 m*M* NaCl (pH 7.4), filtered through a 0.22-μm pore size Millipore Stericup (Millipore)] is renatured by heating to 80° for 2 min and slowly cooling to room temperature. RNA is then immobilized onto the chosen flow cell by injecting 60 μl of RNA solution by QUICKINJECT command at a flow rate of 2 μl/min. After finishing RNA injection, keep running the buffer through the cell until the response units on the sensorgram become stable. A typical sensorgram is shown as Fig. 3. Record the response units before injecting RNA and at the end of the sensorgram. The response unit difference between these two points (RU_{RNA}) corresponds to the amount of RNA immobilization.

The above-described procedures are repeated for different RNAs loaded on other flow cells. Usually one of the flow cells is left unmodified as a control. The functionalized RNA sensor chip is stored at 4–8°, and is used for assays within 2 weeks.

Collection of RNA–Aminoglycoside Binding Data

Paromomycin sulfate is obtained from Fluka (Sigma-Aldrich) (Buchs, Switzerland). Neamine is obtained by acid-catalyzed cleavage of neomycin B and is purified by ion-exchange chromatography on Amberlite CG-50. All the aminoglycoside stock solutions are made in Milli-Q water or diethyl

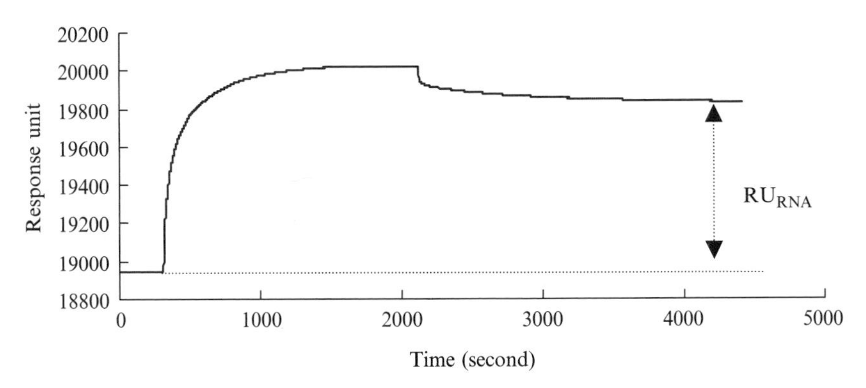

FIG. 3. Sensorgram collected when immobilizing bacterial ribosome 16S A-site RNA to sensor chip SA on the Biacore 2000 system. RU_{RNA} is shown.

pyrocarbonate (DEPC)-treated water (Ambion). The following typical procedures are from a 16S A-site–paromomycin binding experiment.

Paromomycin sulfate samples are prepared by serial dilutions from 20 m*M* stock solutions to 80 μ*M*, 50 μ*M*, 20 μ*M*, 10 μ*M*, 8 μ*M*, 5 μ*M*, 2 μ*M*, 1 μ*M*, 800 n*M*, 500 n*M*, 200 n*M*, 100 n*M*, 80 n*M*, 50 n*M*, 20 n*M*, and 0 n*M* (blank control) with running buffer in RNase-free microcentrifuge tubes (Ambion). The series of concentrations used in the experiment are designed to cover the expected dissociation constant (K_d) value. If the K_d is unknown, a wider concentration range with fewer data sets is used to obtain the appropriate K_d value, and then more detailed studies are performed for more accurate K_d values. Usually at least 10 data points are required to obtain well-fitting results. All the samples are centrifuged at 14,000 rpm (room temperature, 2 min.) for degassing and precipitating of any undissolved materials. Aminoglycoside samples (100 μl) of different concentrations are transferred to autoclaved 7-mm plastic vials, and are capped with pierceable plastic crimp caps. All aminoglycoside samples and buffers are placed in the positions as indicated in the method below before performing the assay.

The procedures for binding studies are automated as methods using repetitive cycles. The method used in this experiment is shown below.

```
MAIN
        RACK 1 thermo_b
        RACK 2 thermo_a
        FLOW 5
        LOOP paromomycin ORDER
    APROG RNA_binding %concentration %position %regeneration
%sample
        ENDLOOP
        APPEND continue
END

DEFINE LOOP paromomycin
LPARAM %concentration %position %regeneration %sample
       0.00u          r2f1      r2e1          buffer
       0.02u          r2a1      r2e1          paromomycin
       0.05u          r2a2      r2e1          paromomycin
                      .
                      .
                      .
                      .
       100.00u        r2b6      r2e4          paromomycin
END
```

```
DEFINE APROG RNA_binding
        PARAM %concentration %position %regeneration %sample
        CAPTION %concentration M %sample
        KEYWORD Concentration_(M) %concentration
        KEYWORD Sample %sample
                          FLOW        10 – f
                          *KINJECT    %position 45 270
                 –0 : 15  RPOINT      baseline-b-w10
                  2 : 00  RPOINT      2_min-w10
                  4 : 00  RPOINT      4_min-w10
                  5 : 00  RPOINT      diss_30_sec
                  5 : 30  RPOINT      diss_1_min
                  8 : 30  RPOINT      diss_4_min
                          *INJECT     %regeneration 20
                  2 : 30  RPOINT      regeneration

END
```

Before starting, the flow cells for the experiment are chosen. Samples are injected by KINJECT command at a flow rate of 5–10 μl/min. To minimize carryover, samples are injected in the order of increasing concentration. In the first one or two cycles, only running buffer is injected as a control and to establish a stable baseline value. In each cycle, running buffer should flow for 2 min before injecting sample. The following association and dissociation take 4.5 min each, and a 2-min regeneration step with regeneration buffer (150 m*M* Na_2SO_4) is performed to remove any analyte that is still bound to the surface. Data points are collected 15 s before sample injection; 2, 4, 5, 5.5, and 8.5 min after sample injection; and 2.5 min after regeneration buffer injection. Data from two independent experiments are averaged to obtain more reliable information. A typical sensorgram is shown in Fig. 4.

Data Processing and Analysis

To obtain good quality data for analysis, a few processing steps are performed.[63,64] First, relative response units are obtained by zeroing the response unit just before the start of the association phase (15 s before sample injection). All the data points are adjusted relative to this point. Second, data collected from functionalized flow cells is subtracted by the

[63] R. L. Rich and D. G. Myszka, *Curr. Opin. Biotechnol.* **11,** 54 (2000).
[64] D. G. Myszka, *J. Mol. Recognit.* **12,** 279 (1999).

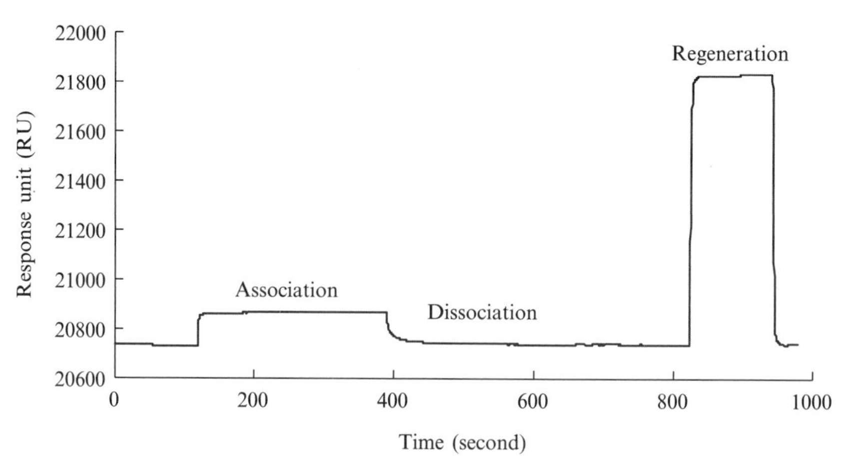

FIG. 4. Sensorgram of SPR binding experiment. Data collected from the binding of paromomycin (50 μM) to immobilized 16S A-site. Association, dissociation, and regeneration phases are shown.

corresponding data from the control flow cell. Third, data from the buffer cycle is subtracted from the control cell-corrected sample data. These steps remove most system artifacts. RU_{max}, the maximum binding capacity of the surface ligand (RNA) for analyte (aminoglycoside) in response units, or the expected value for the maximum equilibrium response of analyte, is calculated according to Eq. (1).[39,65] RU_{max} is required for the following data analysis.

$$RU_{max} = RU_{RNA}\left(\frac{MW_{analyte}}{MW_{RNA}}\right)\left[\frac{\left(\frac{\partial n}{\partial C}\right)_{analyte}}{\left(\frac{\partial n}{\partial C}\right)_{RNA}}\right]N \quad (1)$$

where RU_{max} is the expected values for the maximum equilibrium response of analyte, RU_{RNA} is the amount of RNA immobilized on the sensor chip (in response units), $MW_{analyte}$ is the molecular weight of aminoglycoside, MW_{RNA} is the molecular weight of RNA, n is the refractive index of aminoglycoside or RNA, C is the concentration of aminoglycoside or RNA bound at the surface (in mass/volume), $(\partial n/\partial C)_{analyte}$ is the refractive index increment of aminoglycoside, $(\partial n/\partial C)_{RNA}$ is the refractive index increment of RNA, and N is the number of analyte-binding sites on RNA (binding stoichiometry).

[65] T. M. Davis and W. D. Wilson, *Anal. Biochem.* **284,** 348 (2000).

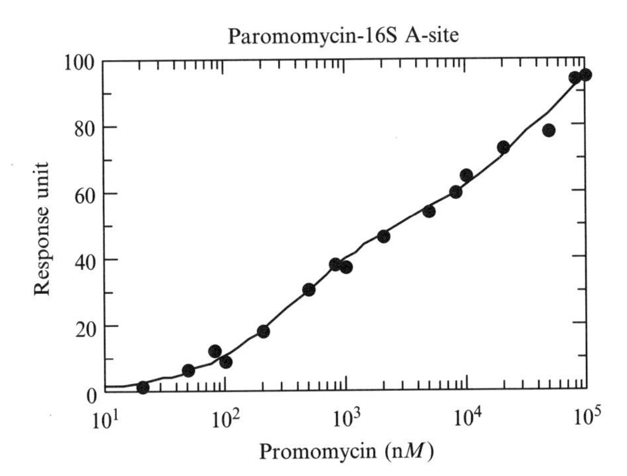

FIG. 5. Collected and processed response units of paromomycin–16S A-site are plotted as a function of paromomycin concentration. The resulted isotherm is fitted to Eq. (2). The calculated K_d is 395 n*M*.

RU_{RNA} is derived by subtracting the response units before loading RNA from the response units after loading RNA (Fig. 3). The refractive index increments ratio of analyte and ligand (RNA), $[(\partial n/\partial C)_{analyte}/(\partial n/\partial C)_{ligand}]$, which arises from the different molar refractive index increments of the analyte and RNA, is required to correct the SPR data. The determination of this correction factor is discussed in detail by Davis and Wilson.[65] The correction factor for aminoglycosides and bacterial 16S A-site RNA binding is determined as reported.[41,43]

Dissociation constants (K_d) are calculated by fitting the recorded binding isotherm to Eq. (2), shown below, using KaleidaGraph (Synergy Software, Reading, PA) or pro Fit (QuantumSoft, Zurich, Switzerland). Isotherms of aminoglycoside response unit versus aminoglycoside concentration are fitted to Eq. (2) (Fig. 5):

$$RU_{obs} = RU_{fit1}[AG]/(K_{d1} + [AG]) + RU_{fit2}[AG]/(K_{d2} + [AG]) + \cdots \quad (2)$$

where [AG] is the concentration of aminoglycoside in the flow solution and RU_{obs} is the corrected observed response units of bound aminoglycoside. Response units from the association phase are used to calculate K_d. RU_{fit} is the fitted value for maximum binding of the surface (in response units). K_{d1} is the dissociation constant of the specific binding site. K_{d2} or more (K_{dn}) represent other weaker or nonspecific binding. The number of K_{dn} used in the fitting is adjusted depending on the fitting pattern and on the observed range of binding equivalents.

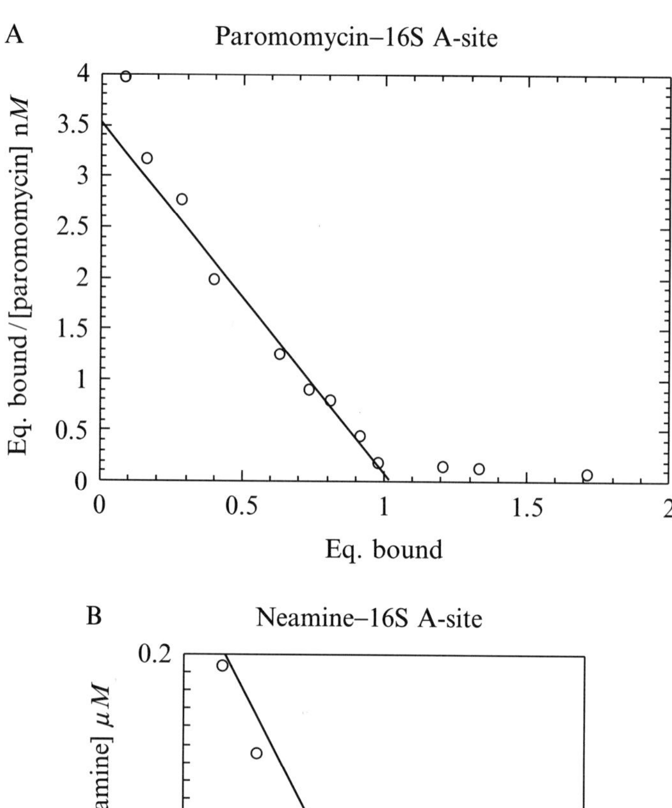

FIG. 6. Scatchard plots of aminoglycoside–RNA binding. (A) Scatchard plot of paromomycin–16S A-site binding. The binding stoichiometry is 1:1. Calculated K_d = 255 nM (slope = $-1/K_d$). (B) Scatchard plot of neamine–16S A-site binding. The binding stoichiometry of neamine:RNA is 2:1. Calculated K_d = 10 μM. From S. J. Sucheck, A. L. Wong, K. M. Koeller, D. D. Boehr, K.-A. Draker, P. S. Sears, G. D. Wright, and C.-H. Wong, *J. Am. Chem. Soc.* **122,** 5230 (2000).

Isotherms of equivalents of aminoglycoside bound versus aminoglycoside concentration are fitted to Eq. (3).[44]

$$\text{Equivalents bound} = (K_a[\text{AG}] + 2K_aK_{ns}[\text{AG}]^2 + 3K_aK_{ns}^2[\text{AG}]^3)/(1 + K_a[\text{AG}] + K_aK_{ns}[\text{AG}]^2 + K_aK_{ns}^2[\text{AG}]^3) \quad (3)$$

Equivalents of aminoglycoside bound are calculated by dividing the observed response units of bound aminoglycoside by RU_{max} of one equivalent aminoglycoside bound. K_a is the equilibrium constant for specific binding. $K_d = 1/K_a$. K_{ns} is the equilibrium constant for multiple site binding.[44]

K_d can also be obtained by fitting the linear part of the Scatchard plot (equivalents of aminoglycoside bound/free aminoglycoside concentration versus equivalents of aminoglycoside bound) to a straight line (slope $= -1/K_d$) (Fig. 6).

The binding stoichiometry is obtained by fitting the linear part of the Scatchard plot to a straight line and calculating the *x*-axis intercepts as an indication of binding ratio (Fig. 6). Additional nonspecific bindings usually occur at higher concentration because cationic aminoglycosides are likely to bind anionic phosphate backbone of RNA nonspecifically.

Further information regarding kinetic and thermodynamic analysis of SPR binding data can be found in the reviews by Myszka *et al.*[66] and Davis and Wilson.[67]

Acknowledgments

We thank Dr. Lac Lee and Dr. Thomas J. Tolbert for helpful discussions.

[66] D. G. Myszka, *Methods Enzymol.* **323,** 325 (2000).
[67] T. M. Davis and W. D. Wilson, *Methods Enzymol.* **340,** 22 (2001).

[25] Frontal Affinity Chromatography as a Tool for Elucidation of Sugar Recognition Properties of Lectins

By Jun Hirabayashi, Yoichiro Arata, and Ken-ichi Kasai

The principle and practice of frontal affinity chromatography (FAC) as an analytical tool for biospecific interactions were established in the 1970s by Kasai *et al.*[1–4] It was first applied to enzyme–substrate analog systems,

0076-6879/03 $35.00

and later to lectin–sugar systems.[5–7] Because most researchers had regarded affinity chromatography only as one of many new effective purification methods, the idea of using it as a tool for the analysis of biospecific interactions was not widespread. Affinity chromatography is effective not only for preparative but also for analytical purposes, because it is a combination of a powerful separation principle (chromatography) and a significant biological phenomenon, that is, molecular recognition carried out by proteins (bioaffinity). It has the potential to provide us with significant information about functional proteins, that is, their ability to specifically recognize their counterparts. Only its formulation needed to be achieved to give birth to a new research tool. Adoption of not zonal chromatography but frontal chromatography was a critical requirement for successful application.

The remarkable characteristics of FAC are as follows: (1) the theoretical basis is simple and straightforward because this method requires only an equilibrium state (more precisely, a dynamic equilibrium state) established between an immobilized ligand and a soluble analyte; (2) FAC utilizes a relatively weak interactions in contrast to almost all currently available techniques; (3) neither special equipment nor sophisticated skill is needed for operation because the simplest elution mode, isocratic elution, is applied throughout the chromatographic runs. This also makes FAC robust; (4) an accurate elution volume can be determined easily by processing integrated multiple data points, thus minimizing the influence of noise in the measurement of signals and resulting in acquisition of reliable equilibrium constants (K_d values); (5) once a minimum set of basic parameters for a given column is established, K_d values for other analytes can be determined without knowing their exact concentrations. This is a remarkable advantage of FAC; (6) there is no need to use a high concentration of analytes for the purpose of facilitating complex formation and acquisition of larger signals even for weak interactions. The minimum concentration detectable by an appropriate procedure is sufficient regardless of its affinity; and (7) it is economically advantageous because only ordinary laboratory equipment is needed.

[1] K. Kasai and S. Ishii, *J. Biochem.* **77,** 261 (1975).
[2] M. Nishikata, K. Kasai, and S. Ishii, *J. Biochem.* **82,** 1475 (1977).
[3] K. Kasai and S. Ishii, *J. Biochem.* **84,** 1051 (1978).
[4] K. Kasai and S. Ishii, *J. Biochem.* **84,** 1061 (1978).
[5] Y. Oda, K. Kasai, S. Ishii, *J. Biochem.* **89,** 285 (1981).
[6] Y. Ohyama, K. Kasai, H. Nomoto, Y. Inoue, *J. Biol. Chem.* **260,** 6882 (1985).
[7] K. Kasai, M. Nishikata, Y. Oda, and S. Ishii, *J. Chromatogr.* **376,** 323 (1986).

At its primitive stage, when high-performance liquid chromatography (HPLC) was not yet popular, FAC was rather time-consuming and required a relatively large amount of analyte. However, significant improvements in both instruments and materials have now made FAC a highly sensitive, accurate, rapid, stable, and convenient procedure for the determination of equilibrium constants between an immobilized ligand and a soluble analyte.[8,9] In its application to the field of glycobiology, the use of a small column of immobilized lectin and a variety of fluorescent oligosaccharides [*Pyridylaminated* lectin (PA)-sugars] has made this reinforced FAC an extremely effective and high-throughput tool for profiling lectins in terms of their binding properties (both binding specificity and binding strength[8–13]). It is now possible to determine systematically K_d values of a large set of PA-oligosaccharides for many kinds of lectins, and also for other sugar-binding proteins.

Information obtained by FAC is only that at equilibrium (static, not dynamic). Although FAC does not give rate constants for interaction (k_{ON} and k_{OFF}), which some modern apparatuses such as surface plasmon resonance (SPR)-based biosensors can provide, this can be adequately compensated for because of the extremely high sensitivity of FAC in determining K_d values. This article focuses on the procedure for systematic determination of K_d values of a variety of PA-oligosaccharides for lectins, that is, for the profiling of lectins. Another remarkable application of FAC that makes full use of mass spectrometry (MS) detection has been reported,[14] and can be found elsewhere in this volume.[14a]

Principle

Theory for Determination of Affinity Constants

In frontal affinity chromatography, a relatively large volume of a dilute analyte solution is continuously applied to a small column packed with an

[8] J. Hirabayashi, Y. Arata, and K. Kasai, *J. Chromatogr.* **890,** 261 (2000).

[9] Y. Arata, J. Hirabayashi, and K. Kasai, *J. Chromatogr.* **905,** 337 (2001).

[10] Y. Arata, J. Hirabayashi, and K. Kasai, *J. Biol. Chem.* **276,** 3068 (2001).

[11] J. Hirabayashi, T. Hashidate, Y. Arata, N. Nishi, T. Nakamura, M. Hirashima, T. Urashima, T. Oka, M. Futai, W. E. G. Mueller, F. Yagi, and K. Kasai, *Biochim. Biophys. Acta* **1572,** 232 (2002).

[12] J. Hirabayashi and K. Kasai, *J. Chromatogr.* **771,** 67 (2002).

[13] M. Sato, N. Nishi, H. Shoji, M. Seki, T. Hashidate, J. Hirabayashi, K. Kasai, Y. Hata, S. Suzuki, M. Hirashima, and T. Nakamura, *Glycobiology* **12,** 191 (2002).

[14] D. C. Schriemer, D. R. Bundle, L. Li, and O. Hindsgaul, *Angew. Chem. Int. Ed.* **37,** 3383 (1998).

[14a] M. M. Palcic, B. Zhang, K. Qian, B. Rempel, and O. Hindogonl, *methods for Enzymology.* **362** [26] (2003) (this volume).

affinity adsorbent (e.g., PA-oligosaccharide solutions are applied to an immobilized lectin column). It is important that a relatively weak affinity adsorbent that does not tightly bind the target analyte should be used. During passage through the column, the analyte molecules interact with the immobilized ligands on the adsorbent and, therefore, they exit the column late. When the amount of the applied analyte molecules exceeds the ability of the column to retain them, leakage occurs and the concentration of the analyte in the eluate finally reaches a plateau where the concentration is equal to that of the initial solution. After this a dynamic equilibrium state is maintained in the column, and the amount of eluted analyte from the outlet of the column becomes equal to that of newly applied analyte at the top of the column.

Unlike ordinary chromatography, the elution curve is composed of the front and the plateau as shown in Fig. 1A. The amount of analyte molecules retarded during the passage through the column is equal to the area within the two elution curves: the right curve (I) being that of the analyte and the left curve (II) being that of a reference substance having no affinity for the adsorbent. This area corresponds to the amount of the analyte molecules that are interacting with the immobilized ligand in the column, and is equal to the rectangle, $[\mathrm{A}]_0(V - V_0)$, where $[\mathrm{A}]_0$ is the initial concentration of the analyte, V is the elution volume of the analyte, and V_0 is that of a reference substance. Therefore, $[\mathrm{A}]_0(V - V_0)$ indicates the degree of saturation of the immobilized ligand, and is a function of the dissociation constant (K_d), the amount of immobilized ligand in the column (B_t), and $[\mathrm{A}]_0$.

$$[\mathrm{A}]_0(V - V_0) = \frac{B_\mathrm{t}[\mathrm{A}]_0}{[\mathrm{A}]_0 + K_\mathrm{d}} \tag{1}$$

Equation (1) is equivalent to the Michaelis–Menten equation of enzyme kinetics and in principle to Langmuir's adsorption isotherm. It gives a hyperbolic curve, indicating that the column becomes saturated at an infinite concentration of analyte, and the maximum binding ability of the column, B_t. Equation (1) can be derivatized to various linear-type equations, but the following Woolf–Hofstee-type equation is convenient for the present purpose:

$$[\mathrm{A}]_0(V - V_0) = B_\mathrm{t} - K_\mathrm{d}(V - V_0) \tag{2}$$

If we apply various concentrations of the analyte ($[\mathrm{A}]_0$), measure V, and make an $[\mathrm{A}]_0(V - V_0)$ versus $(V - V_0)$ plot, the slope gives $-K_\mathrm{d}$, and the intercept on the ordinate is B_t. B_t is the amount of immobilized ligand molecules (e.g., a lectin) actually retaining binding ability.

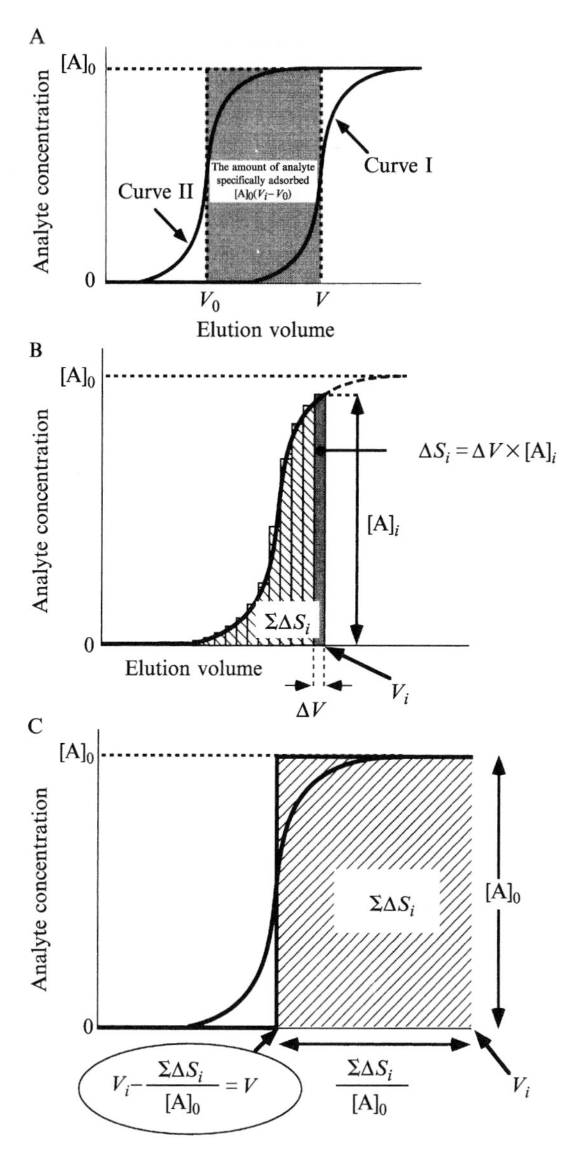

FIG. 1. Principle of frontal affinity chromatography and explanation of data-processing procedure. (A) Curve I indicates the elution profile of an analyte that specifically interacts with the immobilized ligand (elution volume, V). Curve II indicates the elution profile of an analyte that does not interact with the immobilized ligand (elution volume, V_0). (B) Calculation of the area under the elution curve by integration of small rectangles corresponding to ΔV times $[A]_i$ up to V_i. (C) When elution of the analyte reaches the plateau ($[A]_0$), the value $V_i - \Sigma\Delta S_i / [A]_i$ will converge to the true elution volume (V).

If $[A]_0$ is set negligibly low in comparison with K_d, Eq. (1) can be simplified as follows:

$$(V - V_0) = \frac{B_t}{K_d} \tag{3}$$

Therefore, once the B_t value of a given affinity column is determined by concentration dependence analysis, using an appropriate ligand, K_d values for other ligands can be determined by only one measurement of the V value provided that the ligand concentration is adequately low (e.g., less than 1% of the K_d). This feature is analogous to enzyme kinetics (that is, at low concentrations of a substrate, the ratio between reaction rate and substrate concentration becomes constant), and provides a great advantage from an experimental point of view; that is, it is not necessary to pay too much attention to the concentration of the analyte. Even use of an analyte solution of unknown concentration is possible. Moreover, even for a weakly interacting analyte (having a large K_d), there is no need to increase its concentration.

Calculation of V *Value from Elution Profile*

Frontal chromatography allows accurate determination of the elution volume (V). V can be considered the volume at which the hypothetical boundary of the analyte solution would appear if the boundary were not disturbed at all. If an ideal elution curve is obtained, the V value can be simply estimated from the elution volume corresponding to $[A]_0/2$, although it is not often the case. However, even if an ideal elution curve cannot be obtained, it is possible to calculate an accurate V value as follows: the area under the elution curve, $\Sigma\Delta S_i$, is calculated by summing small rectangles ΔS_i, which is equal to ΔV for base and $[A]_i$ for height up to V_i (Fig. 1B). When V_i reaches the plateau region, $\Sigma\Delta S_i$ becomes equal to the shaded rectangle shown in Fig. 1C, that is, $[A]_0$ times $(V_i - V)$. The V value can then be estimated as the left side of the rectangle. Any elution volume can be adopted as V_i provided that it is taken from the plateau region. This relation can be summarized as Eq. (4):

$$V = V_i - \frac{\Sigma\Delta S_i}{[A]_0} \tag{4}$$

Actually, the calculation was made on a personal computer using commercially available table calculation software such as Excel. It is also possible to estimate the V value automatically by using macro commands. For such a purpose, it is necessary to judge whether V_i is really taken from the plateau region. For this, the following principle applies[9]: when the elution curve reaches the plateau, $\Sigma\Delta S_i/(V_i - V)$ should become constant ($[A]_0$) regardless of V_i and, consequently, the left-hand side of Eq. (4) converges to a constant value that is the true V value. Therefore, calculation according to Eq. (4) is done for every datum point; and when this value converges to a constant value, it can be judged that V_i is in the plateau region and the true V value is given. The V value thus obtained is reliable because the influence of noise is minimized by integration of multiple data.

Construction of Frontal Affinity Chromatography System

The analytical system for determination of K_d values of PA-sugars for an immobilized lectin can be easily constructed with commercially available parts for liquid chromatography. Figure 2 shows a diagrammatic example of the system. It is composed of a pump, an injector connected to a sample loop of relatively large volume (e.g., 2 ml, composed of a poly ether ether ketone (PEEK) tube of 0.75-mm inner diameter), a small column (e.g., 4.0×10 mm, 126 μl), and a fluorescence detector. Because equilibrium constants are sensitive to temperature change, the column and principal part of the sample loop are immersed in a water bath. Output from the fluorescence detector is transformed into digital signals and sent to a computer for spread-sheet calculations.

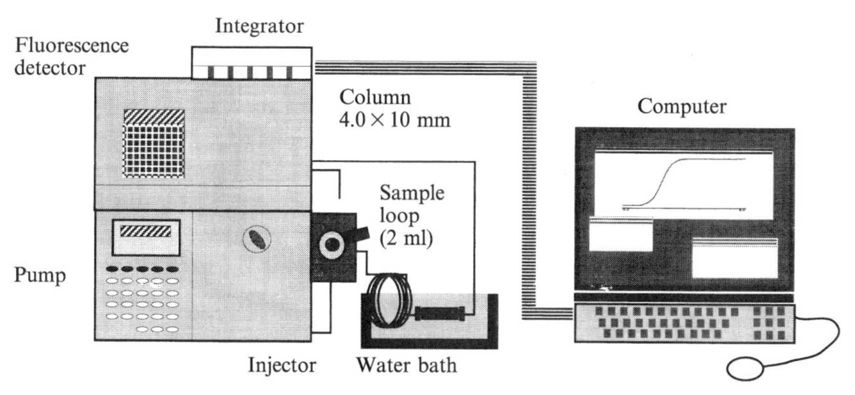

FIG. 2. Diagram of FAC system. A conventional HPLC system is used with a relatively large (2-ml) sample loop and relatively small column (4.0×10 mm) so that an excess volume of analyte is continuously applied to the column at a constant concentration (10 nM in the case of PA-oligosaccharides) and at a constant flow rate (0.25 ml/min). The sample loop and column are immersed in a water bath to maintain the temperature (standard temperature, 20°). Elution of PA-oligosaccharides is monitored by a fluorescence detector (excitation at 320 nm and emission at 400 nm), and signals are sent to a computer every 2 s. Elution volumes, V, are calculated by commercially available software such as Excel.

Experiments

Preparation of Affinity Adsorbent. Preparation of a suitable affinity adsorbent is essential for successful FAC. It must be noted that a relatively weak adsorbent is needed. Affinity adsorbents used for preparative purposes are inappropriate because they bind sugars too strongly, which makes it impossible to observe their elution. A weak affinity adsorbent, which probably leaks target sugars, is desirable. Therefore, a relatively smaller amount of lectin molecules should be immobilized on the supporting matrix. Because B_t is equal to v[B], where v is the bed volume of the column and [B] is the concentration of the immobilized ligand, Eq. (3) can be rewritten as follows:

$$\frac{(V - V_0)}{v} = \frac{[B]}{K_d} \quad (5)$$

This indicates that the extent of retardation relative to the column volume is equal to the ratio of the concentration of the immobilized ligand and K_d. Therefore, if [B] is equal to K_d, the analyte molecules will be retarded by 1 column volume, and if [B] is 10 times the K_d, they will be retarded by 10 column volumes. In the present system, an analyte solution that is 2.0 ml in total volume is usually applied to a column bed volume of 0.126 ml. Therefore, it is desirable to keep [B] less than 10 times the K_d of a typical analyte. Otherwise, it becomes impossible to reach the plateau and obtain an accurate V value. It is recommended to assess in advance how much lectin should be immobilized if some information about the binding strength of the target lectin is available. For example, if a target lectin with a molecular weight of 20,000 has a K_d value of about $10^{-4}M$ for a typical PA-oligosaccharide, [B] of $10^{-4}M$ can be attained by immobilizing 2 mg of the lectin on 1 ml of agarose gel provided that all immobilized lectin molecules remain active. By using this adsorbent, retardation of the above analyte by 1 column volume would be expected.

For the supporting matrix, a commercially available preactivated material such as HiTrap *N*-hydroxy-succinimide (NHS)-activated Sepharose (Amersham Biosciences, Piscataway, NJ) packed in a cartridge column is convenient. Nonspecific interaction between agarose gel and PA-oligosaccharides has been found to be negligible. A typical procedure for preparation of a lectin column is as follows: Lectin proteins (0.1–5 mg, depending on the binding strength) are dissolved in 1 ml of 0.2 *M* $NaHCO_3$, pH 8.3, containing 0.5 *M* NaCl, and coupled to 1 ml of the activated agarose according to the manufacturer's manual. To protect the binding site, a competitive sugar (e.g., 0.1 *M* lactose in the case of galectins) is usually added to the reaction mixture. In this example, because too many activated groups have been introduced into HiTrap agarose, almost all of the lectin molecules are likely to be immobilized only on the narrow top layer when the lectin solution is applied to the cartridge column, resulting in an extremely heterogeneous product. For more homogeneous coupling, the cartridge is chilled on ice, and the lectin solution, also chilled, is injected into the cartridge. The coupling reaction is allowed to proceed at the temperature of ice for the first 15 min, and then for another 15 min at room temperature. After the reaction the cartridge is washed, and unreacted activated groups are deactivated by treatment with 0.5 *M* ethanolamine in 0.5 *M* NaCl, pH 8.3, for 1 h at room temperature. After thorough washing, the resultant adsorbent is suspended in an appropriate buffer for analysis and packed in a column (a guard column for usual HPLC columns, 4.0×10 mm).

PA-Oligosaccharides. A variety of PA-oligosaccharides are commercially available [Takara Shuzo (Kyoto, Japan); Seikagaku Kogyo (Tokyo, Japan)]. It is also possible to prepare PA derivatives of oligosaccharides having the desired structure.[15]

Operation of Chromatography. An appropriate buffer (20 m*M* phosphate buffer, pH 7.2, containing 2 m*M* EDTA and 150 m*M* NaCl in the case of an immobilized galectin column) is continuously applied to the column at a flow rate of 0.25 ml/min. PA-oligosaccharide is dissolved in 2 ml of the same buffer at a concentration of $\sim$10 n*M*. The sample injector is turned to the load position, and the sample loop is completely evacuated by injection of about 20 ml of air, using a hypodermic syringe. This process is important in order to make a sharp boundary between the elution buffer and analyte solution. The sample loop is then filled with 2 ml of PA-oligosaccharide solution (the volume of the sample loop is set slightly smaller than 2 ml in order to be filled completely). The bulb of the injector is reversed to the injection position so that the PA-oligosaccharide solution in the sample loop is introduced to the top of the column. Elution of PA-oligosaccharide is monitored with a fluorescence detector (excitation, 320 n*m*; emission, 400 n*m*), and signals are collected every 2 s, sent to a computer, and processed.

First, it is necessary to measure V_0. For this purpose, a PA-sugar that has no affinity for the immobilized lectin (e.g., PA-rhamnose for immobilized galectins) is usually applied.

Because the column is automatically washed after the application of 2 ml of a PA-oligosaccharide solution after the plateau has been reached, it is necessary to wait only until the signals return to the base level before applying the next sample. However, time can be saved by temporally

[15] S. Hase, T. Ikenaka, and Y. Matsushima, *Biochim. Biophys. Res. Commun.* **85,** 257 (1978).

increasing the flow rate (e.g., to twice that of an ordinary run) and accelerating the washing process. When the signal returns to base level, the flow rate is reduced to its initial value, and the next PA-oligosaccharide is applied. Usually, different PA-oligosaccharide solutions can be applied every 10 min.

Determination of B_t*: Analysis of Concentration Dependency.* Experimental determination of the B_t value for a given column is essential for the determination of K_d values. Otherwise, only relative values of affinity would be obtained. The B_t value is not equal to the amount of immobilized protein in the column, however, because not all lectin molecules immobilized on the supporting matrix necessarily retain binding ability. Some lectin molecules may have been inactivated, and some others may have been immobilized unfavorably, for example, in a manner by which access of ligand molecules to the binding site of the lectin is sterically prevented. Such availability of the immobilized ligand varies from lectin to lectin, because it depends on the position of reactive groups such as the amino group of lysine residues on the protein molecule. If the lysine content of the target lectin is relatively high, relatively low availability might result, because the probability of immobilization at a site near the binding site is rather high. Determination of the B_t value is made by analysis of concentration dependence. For this purpose, different concentrations of an appropriate PA-oligosaccharide are applied to the column and corresponding V values are collected. From a plot created according to Eq. (3), both B_t and K_d values can be calculated.

It is desirable to collect datum points corresponding to concentrations of the PA-oligosaccharide around its K_d. However, it is sometimes unrealistic to perform such an experiment when the K_d value of a target PA-oligosaccharide is relatively large (weak binding ability), because PA-oligosaccharides are usually expensive. In such a case, the target oligosaccharides mixed with differing concentrations of an appropriate oligosaccharide having the same saccharide structure but lacking the PA moiety (unlabeled oligosaccharide, less expensive), and applied to the affinity column. It is assumed that the PA moiety has no influence on the interaction of the saccharide moiety with the lectin; so far, this procedure has been found effective. Alternatively, oligosaccharides labeled with *p*-nitrophenol or *p*-nitroaniline, which are detectable by ultraviolet (UV) absorption and less expensive, can be used. An example of determination of K_d and B_t performed for an immobilized galectin-7 is shown in Fig. 3.

Systematic Determination of K_d *Values of Lectin.* Once the B_t value of a given lectin column is obtained, K_d values of other PA-oligosaccharides can be determined one by one with only a single run for each by using an

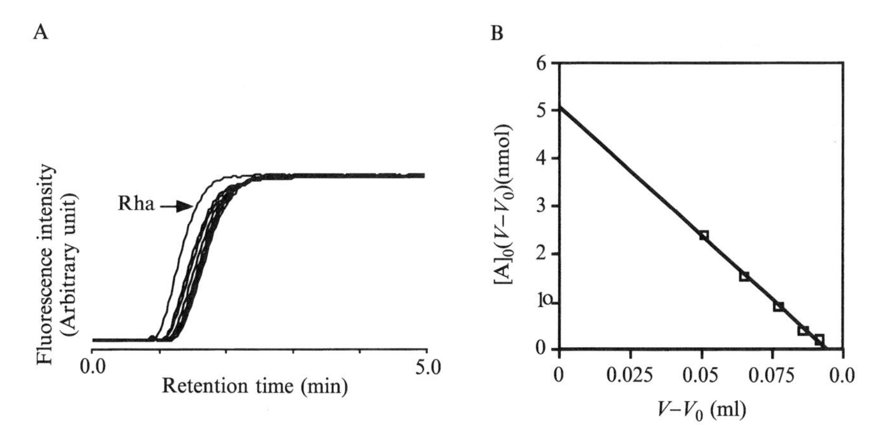

FIG. 3. An example of determination of B_t (amount of functional immobilized ligand in the column) and K_d by concentration dependence analysis. Lactofucopentaose-1 solution (2, 5, 11, 23, and 47 μM) dissolved in 20 mM phosphate buffer, pH 7.2, containing 2 mM EDTA, 150 mM NaCl, and 10 nM PA-lactofucopentaose-1 was applied to a column packed with immobilized human galectin-7 (1.5 mg of protein per milliliter of agarose), and elution curves were made by measuring fluorescence. The temperature was 20°. (A) Elution curves of PA-lactofucopentaose-1 in the presence of excess concentrations of nonfluorescent oligosaccharide. Elution curve of PA-rhamnose (noninteracting reference, indicated as Rha) is also shown. (B) Woolf–Hofstee-type plot of the obtained data. Calculated B_t and K_d values were 5.1 nmol and 54 μM, respectively. The B_t value obtained indicated that almost all immobilized galectin-7 molecules retained binding ability.

~10 nM solution. Because it takes only about 10 min for 1 run, determination of several tens of K_d values per day is not difficult, and the profiling of 1 lectin in terms of its binding property is possible in a few days. An example of profiling of a lectin is shown in Figs. 4 and 5. Figure 4 shows elution patterns of various PA-oligosaccharides, structures of which are shown schematically in Fig. 5, from a column of immobilized human galectin-7. Elution patterns of each PA-oligosaccharide are overlaid with that of PA-rhamnose, which has no affinity for galectin-7. Table I summarizes the results obtained.

Key Points for Success in FAC

One of the weak points of FAC is its rather narrow window in terms of the range of measurable K_d values. In the present system, two orders of magnitude of K_d are covered because measurable $(V - V_0)$ values range from approximately 5 to 500 μl. If the affinity of a particular PA-oligosaccharide for a given lectin column is too strong, even its elution front is not observable. For example, in the case of a K_d value less than

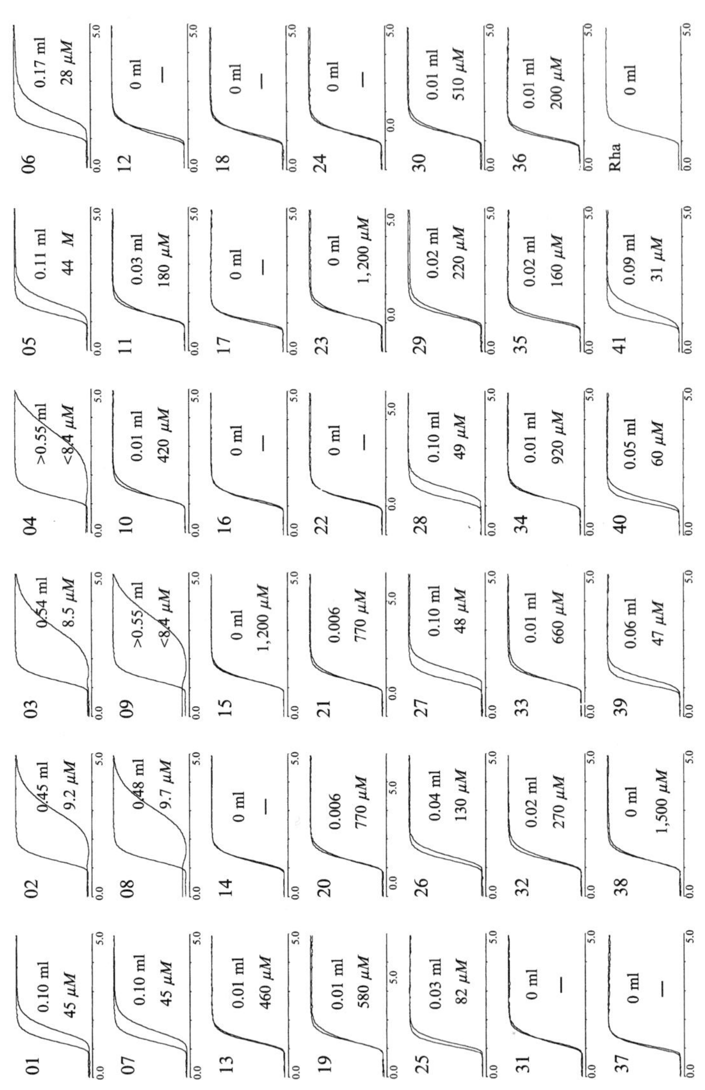

FIG. 4. Elution profiles of various PA-oligosaccharides from an immobilized galectin-7 column. Each elution profile is overlaid with that of PA-rhamnose. The number in the upper left-hand corner of each profile indicates the identification number of PA-oligosaccharides shown in Fig. 5. The numbers under the curves give the extent of retardation (upper number, $V - V_0$, ml) and K_d value (lower number, μM).

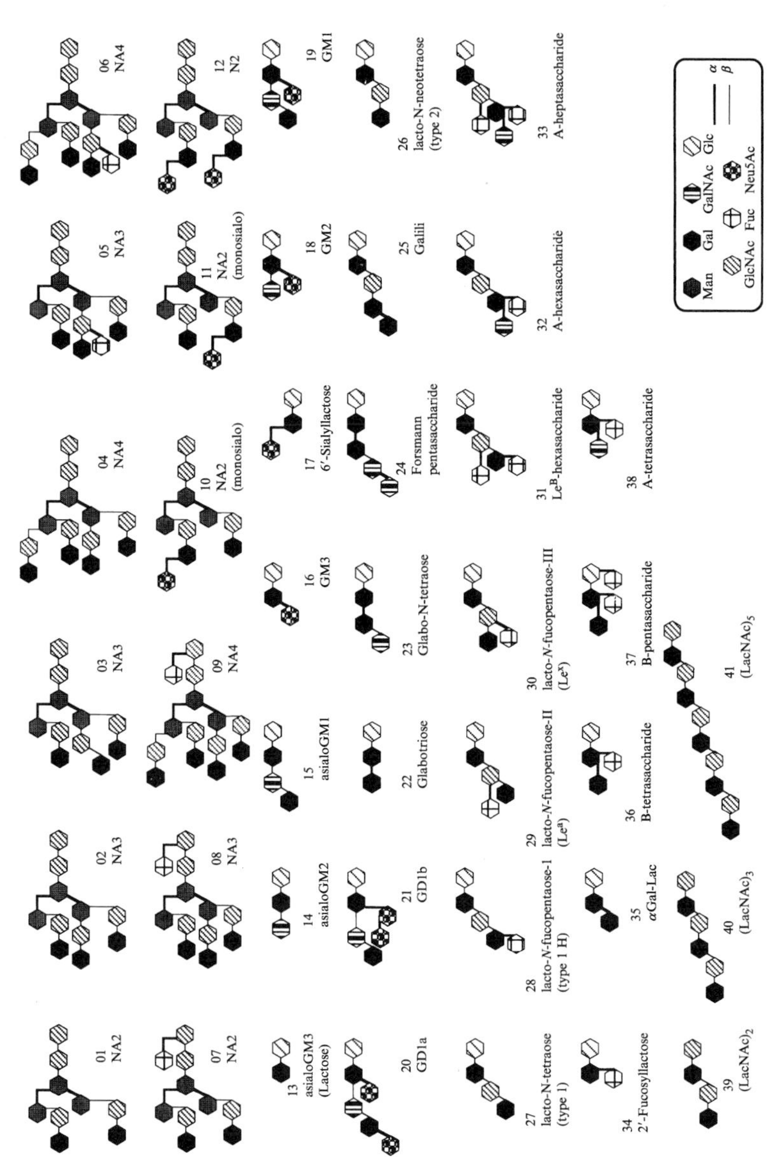

FIG. 5. Structures of PA-oligosaccharides used for profiling of galectin-7.

TABLE I
K_d VALUES OF PA OLIGOSACCHARIDES FOR HUMAN GALECTIN–7[a]

No.	Trivial name	K_d (μM)
1	NA2	45
2	NA3	9.2
3	NA3 (type I)	8.5
4	NA4	7.4
5	NA3 (LeX)	44
6	NA4 (LeX)	28
7	NA2 (α1-6Fuc)	45
8	NA3 (α1-6Fuc)	9.7
9	NA4 (α1-6Fuc)	<8.4
10	NA2 (monosialo)	420
11	NA2 (monosialo)	180
12	NA2 (disialo)	—
13	GA3	460
14	GA2	—
15	GA1	1200
16	GM3	—
17	6′-SiaLac	—
18	GM2	—
19	GM1	580
20	GD1a	770
21	GD1b	770
22	Gb3	—
23	Gb4	1200
24	Forsmann	—
25	Galili	82
26	LNnT	130
27	LNT	48
28	LNFP-I	49
29	LNFP-II	220
30	LNFP-III	510
31	LNDFH	—
32	A-hexa	270
33	A-hepta	660
34	2′-FucLac	920
35	α-GalLac	160
36	B-tetra	200
37	B-penta	—
38	A-tetra	1500
39	LN2	47
40	LN3	60
41	LN5	31

[a] At pH 7.2 and 20°.

1/100 of [B], $(V - V_0)$ will exceed 100 times the column volume, and application of only 2 ml of analyte to a 0.126-ml column is totally inadequate. In such a case, we recommend using affinity adsorbent mixed with underivatized agarose beads in order to reduce the apparent [B]. If the binding strength is still too high, it is necessary to prepare another lot of adsorbent containing a smaller amount of the lectin.

The reverse is also the case. For analysis of weakly interacting PA-oligosaccharides, preparation of a stronger affinity adsorbent may be needed, that is, either the amount of immobilized lectin should be increased or the availability of the immobilized lectin should be improved. If an affinity adsorbent having an effective [B] value of 1 m*M* (in the case of a lectin of molecular weight 20,000, 20 mg/ml) can be prepared, determination even of K_d values in the millimolar range seems possible.

Even if a prepared affinity adsorbent is not adequately strong and, consequently, difficulty in determination of K_d and B_t arises, such a situation can sometimes be improved by carrying out experiments at lower temperatures, because the binding strength of a lectin has often been found to increase at lower temperatures. From our experience, a decrease in temperature from 20 to 10° approximately double the affinity.

Some Considerations on Choice of Procedures

The unique characteristic of FAC is its preference for weak interactions. Such a characteristic is not expected for almost any other presently available method. Strongly interacting pairs are not favorable targets of FAC from the viewpoints of both principle and practice, as discussed below. Because FAC presumes that a dynamic equilibrium state is maintained in the columns, it is necessary for every analyte molecule to repeat the association with the immobilized ligands and the dissociation from them during passage through the column. This, however, is difficult for a strongly interacting system. If an analyte having strong affinity (i.e., having a small K_d) for the immobilized ligand is applied to the column, it will be adsorbed too tightly to the top of the column and will not dissociate within the relatively short time required for one run of frontal analysis. In such a case, not only the K_d but also the k_{OFF}, the rate constant for dissociation should be taken into consideration.

As an example, assume the k_{OFF} of a complex formed between an immobilized lectin and a particular PA-oligosaccharide is 1 s^{-1}. Because $t_{1/2} = \ln 2/k_{OFF} = 0.693/k_{OFF}$ for first-order reactions, the half-life of a complex ($t_{1/2}$), in other words, the average time required for dissociation will be 0.7 s. Therefore, it will take 70 s to repeat association–dissociation 100 times. Because the solution passes through the column in about 30 s under the conditions applied to the FAC system described in this article, such a k_{OFF} value would seem to be the lower limit tolerable for the measurement. In other words, a k_{OFF} greater than 1 s^{-1} is desirable for the present FAC system.

So, what is the magnitude of K_d measurable in the present FAC system? The association rate constant (k_{ON}) of a small ligand such as an oligosaccharide for a protein seems to range from 10^5 to $10^7 M^{-1}s^{-1}$. Because $K_d = k_{OFF}/k_{ON}$, if the k_{ON} and k_{OFF} of an interacting system of interest are about $10^7 M^{-1}s^{-1}$ and 1 s^{-1}, respectively, the measurement of a K_d as low as $10^{-7} M$ will be possible. On the other hand, if k_{ON} is about $10^5 M^{-1}s^{-1}$, it will be difficult to measure a K_d less than 10^{-5} M. However, an interacting system in which k_{OFF} is much larger (e.g., 10–100) allows measurement of smaller K_d values. Unlike strongly interacting system such as antigen–antibody and avidin–biotin systems, in which only binding is important, interactions between a sugar-binding protein and oligosaccharides seem to be weak and rapid, having relatively large k_{ON} and large k_{OFF} values, probably because dissociation is also important in addition to specific association (e.g., rolling of leukocytes). Therefore, FAC seems to be an especially useful analytical tool for glycobiology.

Although slowly dissociating analytes having a k_{OFF} of less than 1 s^{-1} are not desirable objects for the FAC system described here, there may be some possible solutions using appropriate modifications, for example, by a decrease in flow rate or elevation of temperature. If the flow rate is decreased to one-tenth, interactions having a 10-fold slower dissociation rate can be analyzed, although one run would take more than 2 h including the washing process.

Concluding Remarks

Although the application of FAC only to lectin–PA-oligosaccharide interactions is highlighted in this article, the potential of FAC is not limited to such interactions. FAC will become much more versatile as a tool for investigations in a wide area of biorecognition phenomena. We strongly encourage reader to explore and develop FAC for use in his/her own field.

[26] Evaluating Carbohydrate–Protein Binding Interactions Using Frontal Affinity Chromatography Coupled to Mass Spectrometry

By MONICA M. PALCIC, BOYAN ZHANG, XIANGPING QIAN, BRIAN REMPEL, and OLE HINDSGAUL

Introduction

Frontal affinity chromatography coupled online to electrospray mass spectrometry (FAC/MS) was originally developed for high-throughput screening of mixtures of compounds.[1–4] In FAC/MS the protein of interest, an antibody, lectin, or enzyme, is immobilized on a microcolumn. This is readily accomplished by biotinylation of the target protein and then binding to streptavidin-coated beads although, in principle, any immobilization technique can be used. A sample with potential ligands is continuously infused through the column containing immobilized protein. The order of elution of the ligands correlates with their affinities for the protein, with the weakest binding ligands eluting first. The point of breakthrough of compounds from the column can be determined by monitoring ultraviolet–visible (UV–Vis) absorbance or fluorescence, or by electrospray mass spectrometry. The former detection methods are covered in another article in this volume.[5] Online detection by electrospray ionization–mass spectrometry (ESI-MS) is described in this article. ESI-MS is a two-dimensional analysis and even trace components in mixtures can be monitored as long as their m/z values are distinct from those of other compounds in the solution. The dissociation constants (K_d values) of active ligands in mixtures can be evaluated from their breakthrough times, using the FAC theory as developed by Kasai *et al.*[6] In this article we cover the methodology for lectin immobilization on a microcolumn, the determination of K_d values for lectin ligands, and the screening of synthetic oligosaccharide mixtures by FAC/MS.

[1] D. C. Schriemer, D. R. Bundle, L. Li, and O. Hindsgaul, *Angew. Chem. Int. Ed.* **37,** 3383 (1998).
[2] D. C. Schriemer and O. Hindsgaul, *Comb. Chem. High Throughput Screen.* **1,** 155 (1998).
[3] B. Zhang, M. M. Palcic, D. C. Schriemer, G. Alvarez-Manilla, M. Pierce, and O. Hindsgaul, *Anal. Biochem.* **299,** 173 (2001).
[4] B. Zhang, M. M. Palcic, H. Mo, I. J. Goldstein, and O. Hindsgaul, *Glycobiology* **11,** 141 (2001).
[5] J. Hirabayashi, Y. Grata, and K. Kasai, *Methods Enzymol.* **362,** [25] (2003) (this volume).
[6] K. Kasai, Y. Oda, M. Nishikawa, and S. Ishii, *J. Chromatogr.* **376,** 33 (1986).

0076-6879/03 $35.00

Experimental

Lectin Immobilization on a Microcolumn

Microcolumns are prepared by packing controlled porous glass beads (37–74 μm) bearing covalently coupled streptavidin (185 μmol/g, CPG-SA; CPG, Lincoln Park, NJ) into polyether ether ketone (PEEK) tubing (i.d., 0.5 mm; length, 5 cm). The internal column volumes are 9.8 μl. Commercially available biotinylated *Arachis hypogaea* (peanut agglutinin, PNA; Sigma, St. Louis, MO) or *Griffonia simplicifolia* I B_4-isolectin (GS-I-B_4; GY Laboratories, San Mateo, CA) biotinylated by reaction with sulfosuccinimidyl-6-(biotinamido)hexanoate (sulfo-NHS-LC-biotin; Pierce, Rockford, IL)[3] is loaded onto the packed streptavidin columns by infusion of a 0.1-mg/ml solution in phosphate-baffered saline (PBS) at 8 μl/min until the column is saturated. After lectin adsorption is complete, the columns are blocked by infusion of 1 ml of d-biotin (0.2 mg/ml in PBS) at 8 μl/min. Unreacted biotin is removed by washing the columns with PBS buffer and the columns are stored at 4° until use. In parallel, identical blank columns are prepared by blocking packed CPG-SA beads with d-biotin and washing as described above. These are used to detect nonspecific binding of compounds to the columns.

FAC/MS

The FAC/MS system has three 1-ml syringes on a multisyringe pump (PDH 2000; Harvard Apparatus, Holliston, MA) with a switching valve (model 9725; Rheodyne, Rohnert Park, CA) connected to the inlet of the microcolumn, as shown in Fig. 1. The column outlet is connected to a

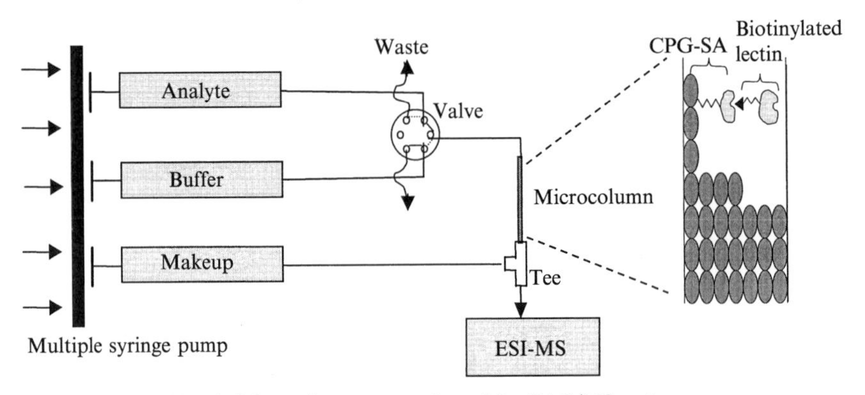

FIG. 1. Schematic representation of the FAC/MS system.

mixing tee for the addition of makeup solution [high-performance liquid chromotography (HPLC)-grade acetonitrile] with combined flow directly into an electrospray mass spectrometer (single-quadrupole 1100 MSD; Hewlett-Packard, Palo Alto, CA). For frontal chromatography analysis the column is initially flushed with 10 m*M* ammonium acetate buffer, pH 6.8, containing 1 m*M* NaCl for PNA or 2 m*M* ammonium acetate buffer, pH 7.2, containing 0.1 m*M* calcium acetate for GS-I-B_4 lectin. Flow is then switched to the second solution that contains carbohydrate ligands in the same buffers. For initial characterization of the column eluent, the spectrometer is scanned from m/z 100 to 1500 in 1.5 s in positive ion mode. When ligand ions are found, they are monitored in selected ion monitoring mode (SIM) locked on the m/z values for the individual ligands. A chamber voltage of −4000 V with a grounded electrospray needle, an N_2 drying gas flow rate of 4 liters/min, and an N_2 nebulizer pressure of 480 mbar were used. Breakthrough volumes are calculated from the midpoints in the extracted ion chromatograms.

Column Capacity (B_t) *and* K_d *Determination by FAC/MS*

The dynamic column binding capacity B_t is the amount of active immobilized lectin with ligand-binding capacity. From frontal affinity chromatography theory,[5,6] the relationship between B_t, the dissociation constant for a ligand (K_d), the concentration of the ligand $[X]_0$, and the retention volume ($V_x - V_0$) is given by

$$V_x - V_0 = B_t/(K_d + [X]_0) \tag{1}$$

V_0, the void volume of the column, is obtained with compounds that have no affinity for the protein whereas V_x depends on the concentration of ligand in the mixture. Figure 2 shows the extracted ion chromatograms of the eluent from a PNA column that has specificity for terminal β-Gal residues. The column was infused at 8 μl/min with a 1 μM solution of a void marker Manα1→3[Manα1→6]Manβ-O$(CH_2)_7CH_3$ monitored at $m/z = 639.3$ $(M + Na)^+$ and various concentrations of T-disaccharide ligand, Galβ1→3GalNAcα-O$(CH_2)_8CO_2CH_3$ monitored at $m/z = 576.3$ $(M + Na)^+$. The breakthrough times increase from 2.33 to 2.53 min (V_x, 18.6 to 20.2 μl) as the concentration of ligand decreases from 10 to 2.5 μM (Fig. 2). B_t and K_d values of 89 pmol and 18.1 μM, respectively, were obtained from the y intercept and slope of plots of $1/(V_x - V_0)$ versus $1/[X_0]$ (Fig. 3). These values can also be determined by non linear regression analysis of the data fit to Eq. (1).[3]

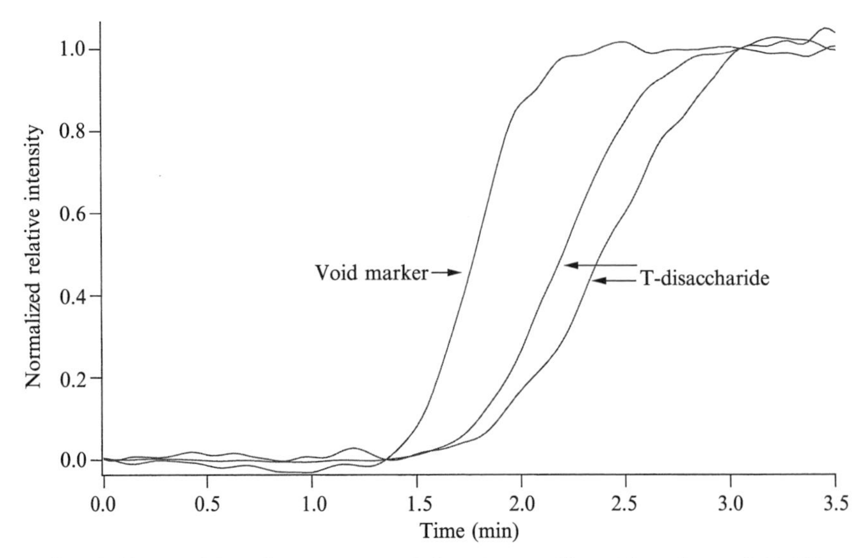

FIG. 2. Extracted ion chromatogram of the trimannoside void marker and Galβ1 $\rightarrow$ 3GalNAcα-OR (10 and 2.5 μM) in 10 mM ammonium acetate buffer, pH 6.8, with 1 mM NaCl run at 8 μl/min through an immobilized peanut agglutinin column.

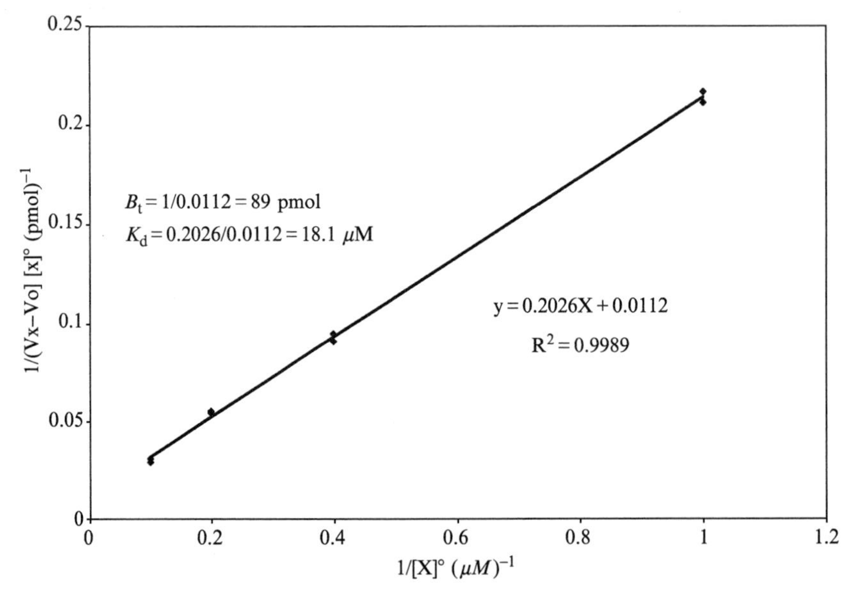

FIG. 3. Determination of the column binding capacity (B_t) and K_d of Galβ1$\rightarrow$3GalNAcα-OR for a peanut agglutinin column from duplicate runs at each concentration of ligand.

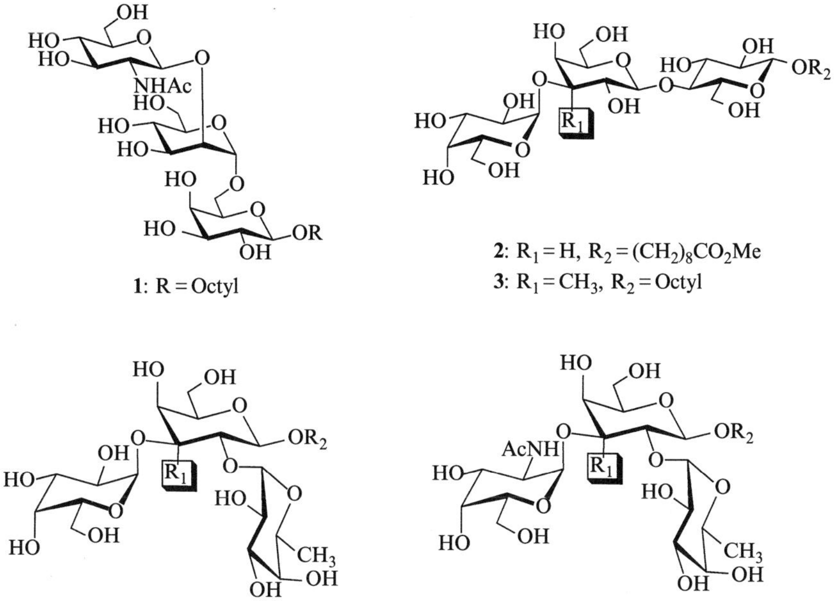

1: R = Octyl

2: $R_1 = H$, $R_2 = (CH_2)_8CO_2Me$
3: $R_1 = CH_3$, R_2 = Octyl

4: $R_1 = H$, R_2 = Octyl
5: $R_1 = CH_3$, R_2 = Octyl
6: R_1 = Pr, R_2 = Octyl

7: $R_1 = H$, $R_2 = (CH_2)_8CO_2Et$
8: $R_1 = CH_3$, R_2 = Octyl
9: R_1 = Pr, R_2 = Octyl

FIG. 4. Structures of synthetic trisaccharides **1–9** present in the screening mixture.

Screening Mixtures of Compounds

Once the B_t of a microcolumn and the concentration of a ligand are known, the dissociation constant for that ligand can be determined from a single frontal analysis chromatogram by measurement of $V_x - V_0$. This is possible even for mixtures of compounds as long as their m/z values are unique.

Isolectin B_4 isolated from *Griffonia simplicifolia* has an affinity for terminal α-Gal residues.[7,8] A GS-I-B_4 microcolumn with a B_t value of 189 pmol was used to screen a synthetic trisaccharide-based compound mixture for a stronger ligand. The structures of the nine compounds in the mixture are given in Fig. 4. Among them, the inactive ligand, compound **1**, was used as a void volume marker. The compounds were continuously

[7] C. Wood, E. A. Kabat, L. A. Murphy, and I. J. Goldstein, *Arch. Biochem. Biophys.* **198,** 1 (1979).

[8] I. J. Goldstein, D. A. Blake, S. Ebisu, T. J. Williams, and L. A. Murphy, *J. Biol. Chem.* **256,** 3890 (1981).

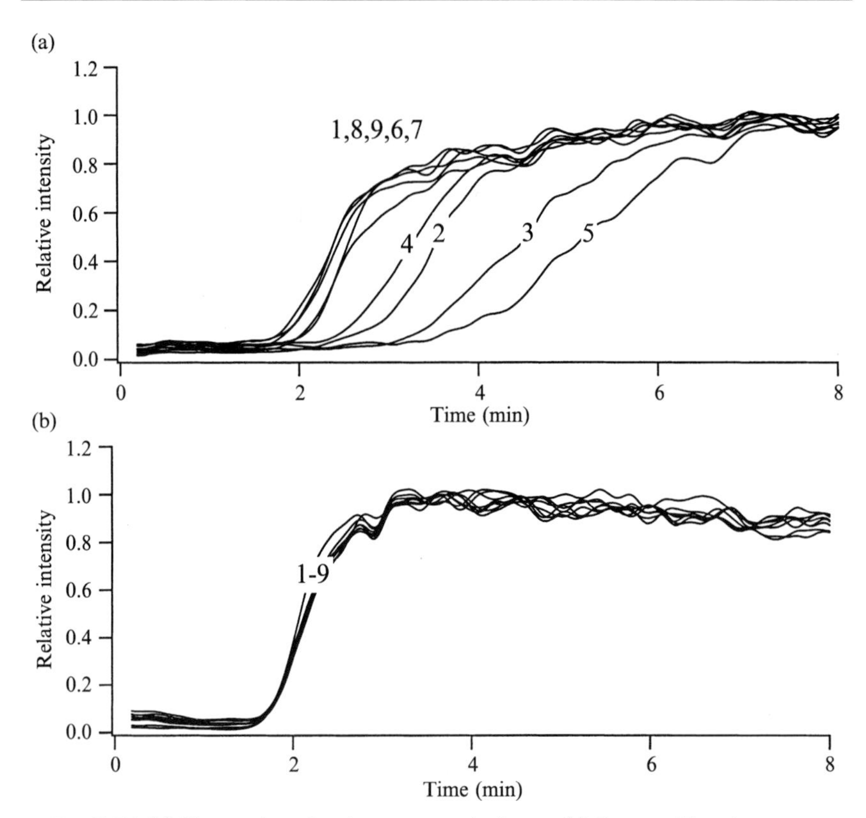

FIG. 5. FAC/MS screening of a nine-compound mixture. (a) Extracted ion chromatogram of trisaccharides **1–9** flowing through the *Griffonia simplicifolia* I B_4-isolectin microcolumn. (b) Control experiment showing the extracted ion chromatogram of the same mixture flowing through a blank column. All the signals are normalized to the intensity of the signal for void volume marker **1**.

infused at 2 μM each in 2 mM ammonium acetate buffer, pH 7.2, containing 0.1 mM calcium acetate, through the GS-I-B_4 microcolumn while the elution profile at each m/z value was monitored by ESI-MS detection in the positive ion mode. Figure 5a shows the extracted ion chromatogram of the mixture.

The elution front of the void volume marker **1** appeared first, followed by compounds **8**, **9**, **6**, and **7** in that order, with the fronts of **4** and **2** coming later. Compounds **3** and **5**, methyl-branched derivatives of **2** and **4** (blood group B antigen),[9] were strongly retarded and came out last, indicating that the introduction of a methyl group at the critical Galα1$\rightarrow$3Gal linkage

[9] X. Qian, K. Sujino, A. Otter, M. M. Palcic, and O. Hindsgaul, *J. Am. Chem. Soc.* **121,** 2063 (1999).

TABLE I
STRUCTURES, m/z VALUES, AND DISSOCIATION CONSTANTS OF TRISACCHARIDES **1–9**[a]

Compound	Saccharide	m/z $(M+Na)^{+b}$	K_d (μM) Mixture[c]	K_d (μM) Individual[d]
1	GlcNAcβ1→2Manα1→6Glcβ-OR_1[e]	680.3	—	—
2	Galα1→3Galβ1→4Glcβ-OR_2	697.3	19	15.5 ± 0.4
3	Galα1→3-[3-C-Me]Galβ1→4Glcβ-OR_1	653.3	10	10.0 ± 0.8
4	Galα1→3-[L-Fucα1→2]-Galβ-OR_1	623.3	25	20.4 ± 0.9
5	Galα1→3-[L-Fucα1→2]-[3-C-Me]Galβ-OR_1	637.3	7	5.5 ± 0.2
6	Galα1→3-[L-Fucα1→2]-[3-C-Pr] Galβ-OR_1	665.3	~150	105 ± 1.0
7	GalNAcα1→3-[L-Fucα1→2]-Galβ-OR_3	736.5	~140	118 ± 6.0
8	GalNAcα1→3-[L-Fucα1→ 2]-[3-C-Me] Galβ-OR_1	678.3	>300	288 ± 12
9	GalNAcα1→3-[L-Fucα1→2]-[3-C-Pr] Galβ-OR_1	706.4	>300	264 ± 11

[a] Screened against *Griffonia simplicifolia* I B_4 isolectin.
[b] Monoisotopic molecular weight of the singly charged sodium adduct.
[c] Dissociation constants estimated from a single infusion of the nine-compound mixture.
[d] Dissociation constants with standard deviations determined from infusion of each ligand at a single concentration in triplicate experiments.
[e] R_1 = Octyl, $R_2 = (CH_2)_8CO_2Me$, $R_3 = (CH_2)_8CO_2Et$.

enhanced their binding affinity for GS-I-B_4 lectin. The control experiment with the blank column is shown in Fig. 5b. The eight trisaccharide derivatives **2–9** do not exhibit nonspecific absorption to the column because they break through at the same time as the void volume marker **1**. This indicates that compounds **2–9** bind to the GS-I-B_4 affinity column specifically. The dissociation constants of the ligands in the mixture were each estimated from a single run (Fig. 5a), based on Eq. (1), and are presented in Table I. Because ligands **2–9** in the mixture are competing for the same binding site on GS-I-B_4, their K_d^{mix} values determined in a mixture normally underestimate their potency. Therefore, the K_d values for all the ligands were also determined by triplicate runs of a single ligand concentration fitted to Eq. (1). These values are also listed in Table I for comparison. In our experience the GS-I-B_4 microcolumn was stable for at least 6 months, as long as the column was stored at 4° in the presence of PBS buffer containing 0.1 m*M* calcium acetate and 0.05% sodium azide.

FAC/MS Overview

FAC/MS is an affinity selection method that is useful for screening mixtures of compounds. The affinities of compounds and their relative ranking are readily obtained after column characterization. In most cases, the K_d values that can be determined range from 0.1 to 300 μM for 5-cm columns; however, weaker binders can be evaluated with longer columns with a larger column capacity. Only small quantities of ligand samples are required, typically a few milliliters of 1 μM solutions. However, nonspecific binding of compounds to the column can be problematic; this is estimated by running compound mixtures through blank biotin-blocked columns. FAC/MS also requires a volatile buffer such as ammonium acetate, causing limitations in buffer additives that might be essential for ligand binding or protein stabilization.

Acknowledgments

M.M.P. and O.H. acknowledge funding from the Natural Sciences and Engineering Research Council of Canada (NSERC). B.Z. was supported by a postdoctoral fellowship from the Alberta Heritage Foundation for Medical Research (AHFMR), X.Q. by an Alberta Research Council carbohydrate graduate scholarship, and B.R. by summer studentships from NSERC and AHFMR.

[27] Determination of Protein–Oligosaccharide Binding by Nanoelectrospray Fourier-Transform Ion Cyclotron Resonance Mass Spectrometry

By WEIJIE WANG, ELENA N. KITOVA, and JOHN S. KLASSEN

Introduction

Diverse biological functions, including cellular growth and adhesion, bacterial and viral infections, inflammation, and the immune response, depend on the recognition of specific carbohydrates by proteins. A detailed understanding of carbohydrate recognition requires analytical methods that can provide information about the specificity and affinity of these interactions. There are a number of established analytical techniques suitable for the study of protein–carbohydrate binding. High-field nuclear magnetic resonance (NMR) and X-ray analysis have been used to characterize the three-dimensional structure of a number of protein–carbohydrate complexes. The binding affinity can be evaluated by semi-quantitative methods such as inhibition of hemagglutination, precipitation assays, and enzyme-linked immunosorbent assays (ELISAs), or by quantitative methods such as isothermal titration microcalorimetry (ITC) and

0076-6879/03 $35.00

surface plasmon resonance or Biacore assays. Each of these techniques has advantages and disadvantages.[1] ITC is the best method for determining the association thermochemistry for protein–carbohydrate complexes and has found wide application; however, it fails to distinguish binding to different protein quaternary structures and, in certain cases, to provide direct information about binding stoichiometry. It also requires milligram quantities of protein and ligand for each analysis.

Mass spectrometry (MS), with its speed, sensitivity, specificity, and ability to directly determine binding stoichiometry, is a powerful tool for studying noncovalent biomolecular complexes. Although a number of ionization techniques, such as matrix-assisted laser desorption/ionization and laser-induced liquid beam ionization/desorption, have been shown to be amenable to the production of specific noncovalent complexes in the gas phase,[2] electrospray (ES) and its low-flow variant, nanoflow electrospray (nanoES), are the predominant ionization techniques. An important feature of the ES (and nanoES) technique is the ability to maintain the solution close to physiological conditions, at neutral pH and ambient temperature, which is important for preserving the native protein structure and equilibrium between bound and unbound species in solution, up to the formation of the gas-phase ions. NanoES, which generally operates at solution flow rates of 1–10 nl/min, is particularly well suited for the study of weakly bound complexes[3] and has the added advantage of consuming only picomoles of analyte.

Beyond the detection of noncovalent complexes, ES-MS is increasingly being used to evaluate binding affinity and stoichiometry. A number of quantitative binding studies have appeared, dealing with protein–protein and protein–small molecule complexes,[4] protein–oligonucleotide complexes,[5] and peptide and RNA-binding antibiotics[6] and small molecule–RNA complexes.[7] Daniel and co-workers have described these quantitative

[1] J. J. Lundquist and E. J. Toone, *Chem. Rev.* **102,** 555 (2002); D. R. Bundle, *Methods Enzymol.* **247,** 288–305 (1994); D. R. Bundle, *in* "Carbohydrates" (S. Hecht, ed.), p. 370. Oxford University Press, Oxford, 1998.

[2] J. M. Daniel, S. D. Friess, S. Rajagopalan, S. Wendt, and R. Zenobi, *Int. J. Mass Spectrom.* **216,** 1 (2002).

[3] M. S. Wilm and M. Mann, *Anal. Chem.* **68,** 1 (1996).

[4] A. Ayed, A. N. Krutchinsky, W. Ens, K. G. Standing, and H. W. Duckworth, *Rapid Commun. Mass Spectrom.* **12,** 339 (1998).

[5] M. J. Greig, H. Gaus, L. L. Cummins, H. Sasmor, and H. R. Griffey, *J. Am. Chem. Soc.* **117,** 10765 (1995).

[6] T. J. D. Jorgensen, P. Roepstorff, and A. J. R. Heck, *Anal. Chem.* **70,** 4427 (1998); K. A. Sannes-Lowery, R. H. Griffey, and S. A. Hofstadler, *Anal Biochem.* **280,** 264 (2000).

[7] R. H. Griffey, K. A. Sannes-Lowery, J. J. Drader, V. Mohan, E. E. Swayze, and S. A. Hofstadler, *J. Am. Chem. Soc.* **122,** 9933 (2000).

studies in a review.[2] Although far from being an established method, the aforementioned studies indicate that ES-MS can, in certain cases, provide quantitative binding information.

The first direct observation of a protein–oligosaccharide complex, hen egg white lysozyme and a hexasaccharide of N-acetylglucosamine, by ES-MS was reported in 1991.[8] More recently, van Dongent and Heck described a study of carbohydrate ligands with the lectin apo-concanavalin (Con A),[9] which exists in both dimer–tetramer forms in solution. From the ES-MS data, the authors determined the binding stoichiometry of the different quaternary Con A complexes and demonstrated that dimer and tetramer forms exhibit similar affinities for the carbohydrates investigated. This binding information could not be obtained by any other analytical technique. Despite the success of these earlier studies, ES-MS has not found wide application in the area of protein–carbohydrate binding because the detection of protein–carbohydrate complexes is generally more challenging than other protein–-ligand complexes. A major cause of this is the low binding affinity, generally in the range of 10^3 to 10^5 M^{-1}, characteristic of protein–carbohydrate complexes.[10] With few exceptions, ES-MS studies of protein–ligand complexes have been restricted to moderately or strongly bound complexes, with association constants $>10^5$ M^{-1}. The detection of low-affinity complexes requires the use of high ligand concentrations, which tend to suppress the formation of gas-phase protein and protein–ligand ions and lead to the formation of nonspecific complexes. Further, many carbohydrate-binding proteins are heterogeneous in structure and composition, resulting in a distribution of ions with similar mass-to-charge ratios (m/z). This, combined with the high m/z typical of protein and complex ions produced from solutions at neutral pH ($m/z > 3000$), and the low molecular weight of many model carbohydrate ligands (mono- to tetrasaccharides), requires the use of mass analyzers with high m/z and high-resolution capabilities.

In this article, we describe the application of nanoES and Fourier-transformion cyclotron resonance (FT-ICR) MS to evaluate the binding affinity and stoichiometry for two carbohydrate-binding proteins. The first part of this work focuses on results obtained in our laboratory for the binding of a single-chain variable fragment (scFv) of a monoclonal antibody, Se155–4,[11] with its native trisaccharide ligand. The influence of several

[8] B. Ganem, Y.-T. Li, and J. D. Henion. *J. Am. Chem. Soc.* **113,** 7818 (1991).

[9] W. D. van Dongent and A. J. R. Heck, *Analyst* **125,** 583 (2000).

[10] E. J. Toone, *Curr. Opin. Struct. Biol.* **4,** 719 (1994); T. K. Dam, R. Roy, S. K. Das, S. Oscarson, and C. F. Brewer, *J. Biol. Chem.* **275,** 14223 (2000).

[11] A. Zdanov, Y. Li, D. R. Bundle, S.-J. Deng, C. R. MacKenzie, S. A. Narang, N. M. Young, and M. Cygler, *Proc. Natl. Acad. Sci. USA* **91,** 6423 (1994).

experimental parameters on the mass spectrometry-derived binding constants is discussed. Second, we describe a binding study of the multivalent B_5 homopentamer of the Shiga-like toxin 1 [SLT-1(B_5)] with the P^k trisaccharide.[12] The excellent agreement between the mass spectrometry and ITC-derived association constants obtained in these two studies demonstrates that, under appropriate experimental conditions, nanoES-MS can provide quantitative information about solution binding for protein–oligosaccharide complexes.

Experimental Methods

Materials

The carbohydrate-binding antibody single-chain fragment[11] (molecular mass, 26,539 Da) and the Shiga-like toxin type 1 homopentamer[13] (molecular mass, 38,440 Da) are produced by recombinant technology. Sodium dodecyl sulfate–polyacrylamide slab gel electrophoresis (SDS–PAGE) in a 12% gel with reducing agent (dithiothreitol, DTT) is performed to confirm the purity of the proteins after they are isolated by affinity chromatography.[13,14] Both proteins are concentrated and dialyzed against deionized water, using MicroSep microconcentrators (molecular mass cutoff, 10,000 Da; Sin-Can, Calgary, Alberta), and lyophilized before MS analysis. The trisaccharide ligands are synthesized at the University of Alberta (Edmonton, Canada) in the laboratory of D. R. Bundle.

Sample Preparation

Proteins are stored at -20° as lyophilized samples or as concentrated aqueous solutions (>0.1 m*M*) in 50–100 m*M* ammonium acetate buffer.

The protein sample is weighed immediately after removing it from the lyophilizer and dissolved in a known volume of aqueous buffer solution. When lyophilizing the protein is undesirable, the concentration of dissolved protein can be determined from the absorbance measured at 280 nm and an extinction coefficient calculated from the amino acid sequence of the protein. Because oligosaccharides are highly hygroscopic, they may contain adsorbed water if stored as dry samples. This water can be removed in a "drying pistol", wherein the sample is gently dried in a vacuum chamber maintained at $\sim$5 torr and 56°.

[12] E. N. Kitova, P. I. Kitov, D. R. Bundle, and J. S. Klassen, *Glycobiology* **11,** 605 (2001).
[13] G. Mulvey, R. Vanmaele, M. Mrazek, M. Cahill, and G. D. Armstrong, *J. Microbiol. Methods* **32,** 247 (1998).
[14] D. R. Bundle, E. Eichler, M. A. J. Gidney, M. Meldal, A. Ragauskas, B. W. Sigurskjold, B. Sinnot, D. C. Watson, M. Yaguchi, and N. M. Young, *Biochemistry* **33,** 5172 (1994).

The pH of the nanoES solutions is maintained near pH 7 with 50 m*M* ammonium bicarbonate (pH 7.2) or ammonium acetate (pH 6.8) buffer. The actual pH is confirmed by measurements performed with a pH meter with microelectrodes (710Aplus pH/ISE meter and microcombination pH electrodes; Thermo Orion, Beverly, MA) which is suitable for solution volumes of a few microliters.

Mass Spectrometry

All experiments are performed with an Apex 47e Fourier-transform ion cyclotron resonance (FT-ICR) mass spectrometer (Bruker-Daltonics, Billerica, MA) equipped with an external nanoelectrospray source (Analytica of Branford, Branford, CT). A simplified illustration of the instrument is shown in Fig. 1. NanoES tips (3-μm o.d. and $\leq$1.0-μm i.d.) are pulled from aluminosilicate tubes (1-mm o.d., 0.68-mm i.d.), using a P-97 or P-2000 laser puller (Sutter Instrument, Novato, CA). A platinum wire, inserted into the nanoES tip, is used to establish electrical contact with the analyte solution. The tip is positioned <1.0 mm from a stainless steel sampling capillary, using a microelectrode holder. A potential of 800 to 1100 V is applied to the Pt wire in the nanotip in order to spray the solution. Typically, a stable electrospray ion current of $\sim$0.1 μA is obtained. The solution flow rate ranges from 1 to 10 nl/min depending on the diameter of the nanotip, the electrospray voltage, and the composition of the solution.

The droplets and gaseous ions produced by nanoES are introduced into the vacuum chamber of the mass spectrometer through a heated stainless steel sampling capillary (0.43-mm i.d.), maintained at a temperature of 150°. The gaseous ions sampled by the capillary (52 V) are transmitted through a skimmer (4 V) and accumulated in a trapping hexapole. Unless otherwise noted, the ions are accumulated in the hexapole for 1.5 s. After

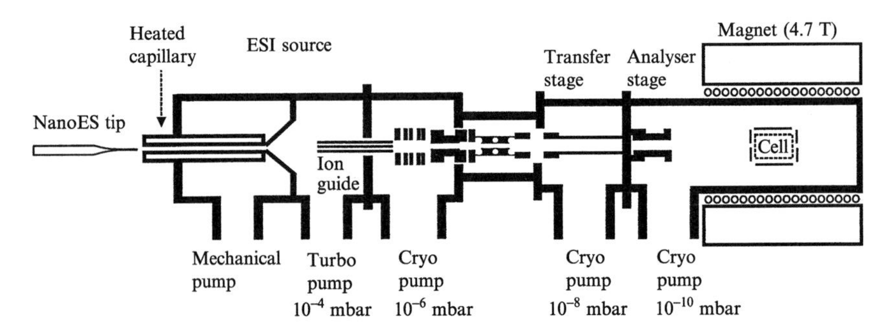

FIG. 1. Schematic illustration of the Fourier transform ion cyclotron resonance (FT-ICR) mass spectrometer with nanoelectrospray ion source.

accumulation, ions are ejected and accelerated (about −2700 V) through the fringing field of a 4.7-T superconducting magnet, decelerated, and introduced into the ion cell. The typical pressure for the instrument is $\sim 5 \times 10^{-10}$ mbar. Data acquisition is performed with Bruker Daltonics (Billerica, MA) XMASS software (version 5.0). The time-domain spectra consist of the sum of 30 transients containing 128,000 data points, per transient.

Calculating K_{assoc}

The equilibrium constant, K_{assoc}, for the association reaction involving a protein (P) and a ligand (L) [Eq. (1)] is given by the following expression:

$$\mathrm{P} + \mathrm{L} \rightleftharpoons \mathrm{P} \cdot \mathrm{L} \tag{1}$$

$$K_{\text{assoc}} = [\mathrm{PL}]_{\text{equil}} / [\mathrm{P}]_{\text{equil}} [\mathrm{L}]_{\text{equil}} \tag{2}$$

The equilibrium concentrations, $[\mathrm{PL}]_{\text{equil}}$, $[\mathrm{P}]_{\text{equil}}$, and $[\mathrm{L}]_{\text{equil}}$, can be deduced from the initial concentration of protein and ligand in solution, $[\mathrm{P}]_0$ and $[\mathrm{L}]_0$, and the relative abundance of the bound and unbound protein ions, $\mathrm{P} \cdot \mathrm{L}^{n+}$ and P^{n+}, measured in the mass spectrum. Assuming that the ionization and detection efficiencies for the $\mathrm{P} \cdot \mathrm{L}^{n+}$ and P^{n+} ions are similar, which is reasonable when the molecular weight of ligand is small compared with the protein,[2] the ratio of the ion intensities (I) of the bound and unbound protein ions, determined from the mass spectrum, should be equivalent to the equilibrium concentrations in solution, that is, $[I(\mathrm{P} \cdot \mathrm{L}^{n+})/I(\mathrm{P}^{n+})] = [\mathrm{PL}]_{\text{equil}}/[\mathrm{P}]_{\text{equil}}$. Ideally, this ratio ($R$) should be independent of charge state; however, this is sometimes not the case. The reason for this is not understood, but may reflect the statistical nature of the charging mechanism in the ES process. Consequently, it is advisable to average the ratios measured for all of the observed charge states (n). Because the ion signal in FT-ICR MS is proportional to the abundance of the ion as well as the charge state of the ion, the average value of R should be calculated using the charge-normalized ion intensities:

$$R = \frac{[\mathrm{PL}]_{\text{equil}}}{[\mathrm{P}]_{\text{equil}}} = \frac{\sum_n [I_{(\mathrm{P \cdot L})^{n+}}/n]}{\sum_n [I_{(\mathrm{P})^{n+}}/n]} \tag{3}$$

The equilibrium concentration, $[\mathrm{PL}]_{\text{equil}}$, can be determined from the measured R value and $[\mathrm{P}]_0$.

$$[\mathrm{PL}]_{\text{equil}} = \frac{R[\mathrm{P}]_0}{1 + R} \tag{4}$$

Once $[PL]_{equil}$ has been calculated, equilibrium concentration $[L]_{equil}$ can be found from

$$[L]_{equil} = [L]_0 - [PL]_{equil} \tag{5}$$

The equilibrium concentration of free ligand in solution cannot be determined directly from the mass spectrum because the ionization and detection efficiencies for the protein and ligand ions are expected to be different because of the large difference in m/z and, potentially, other factors.

K_{assoc} can then be solved using Eqs. (6a) and (6b):

$$K_{assoc} = \frac{[PL]_{equil}}{[P]_{equil}([L]_0 - [PL]_{equil})} \tag{6a}$$

$$K_{assoc} = \frac{R}{[L]_0 - \frac{R[P]_0}{1+R}} \tag{6b}$$

When the protein (or protein assembly) can bind to N ligands (where $N > 1$), there are N reactions to be considered:

$$P + L \rightleftharpoons P \cdot L \tag{7a}$$

$$P \cdot L + L \rightleftharpoons P \cdot L_2 \tag{7b}$$

$$P \cdot L_2 + L \rightleftharpoons P \cdot L_3 \tag{7c}$$

$$\vdots \quad \vdots \quad \vdots \qquad \vdots$$

$$P \cdot L_{N-1} + L \rightleftharpoons P \cdot L_N \tag{7N}$$

Here, we describe only the simplest case, in which all N binding sites are equivalent, with identical binding constants. The treatment of more complicated cases has been discussed elsewhere.[15] The equilibrium concentrations, $[PL]$, $[PL_2]$, $\cdots$, $[PL_N]$, can be determined from relative abundance of the corresponding ions observed in the mass spectrum and Eq. (8a). Then, using these values, the equilibrium concentration of L can be found from Eq. (8b):

$$[P] + [PL] + [PL_2] + \cdots + [PL_{N-1}] + [PL_N] = [P]_0 \tag{8a}$$

[15] C. Tanford, "Physical Chemistry of Macromolecules." John Wiley & Sons, New York, 1961.

$$[L] + [PL] + 2[PL_2] + \cdots + (N-1)[PL_{N-1}] + (N)[PL_N] = [L]_0 \quad (8b)$$

K_{assoc} can be determined from any of the following equations, which are based on the general expression $K_i = K_{assoc}\ (N - i + 1)/i$, where i is the number of occupied binding sites[15]:

$$\frac{[PL]}{[P][L]} = (N)K_{assoc} \quad (9a)$$

$$\frac{[PL_2]}{[PL][L]} = \frac{(N-1)K_{assoc}}{2} \quad (9b)$$

$$\vdots \qquad \vdots \qquad \vdots$$

$$\frac{[PL_N]}{[PL_{N-1}][L]} = \frac{K_{assoc}}{N} \quad (9N)$$

An average K_{assoc} can be determined from the binding constant determined for each of the binding reactions, Eqs. (9a)–(9N).

Results and Discussion

I. *Binding of scFv and Gal[Abe]Man*

Using nanoES-FT-ICR/MS, we have investigated the binding affinity of the single-chain variable domain fragment (scFv), based on the carbohydrate-binding antibody Se155-4,[11] for the trisaccharide ligand α-D-Gal(1→2)[α-D-Abe(1→3)] α-D-Man$p$→OMe, where Gal is galactose, Abe is abequose, and Man is mannose. We refer to this ligand as Gal[Abe]Man. A K_{assoc} of $(1.6 \pm 0.2) \times 10^5\ M^{-1}$ for this interaction in 50 mM Tris and 150 mM NaCl at pH 8.0 and 298 K has been previously determined by ITC.[14]

A typical nanoES mass spectrum obtained from an aqueous buffered solution of scFv and Gal[Abe]Man is shown in Fig. 2. The dominant ions observed in the mass spectrum correspond to the protonated, unbound scFv ion, $(\text{scFv} + n\text{H})^{n+}$, and the protonated complex, $(\text{scFv} \cdot \text{L} + n\text{H})^{n+}$, with charge states of $n = 9$–12. As described in Experimental Methods, K_{assoc} can be deduced from the ion intensity ratio, $R = I(\text{scFv}\cdot\text{L})^{n+}/I(\text{scFv})^{n+}$, determined from the mass spectrum and the initial analyte concentrations. However, the value R, and, consequently, K_{assoc}, is sensitive to experimental parameters/conditions, such as analyte concentration, number and duration of MS measurements, and source conditions. We have investigated the influence of these experimental parameters on the MS-derived binding affinity and have identified optimal conditions for

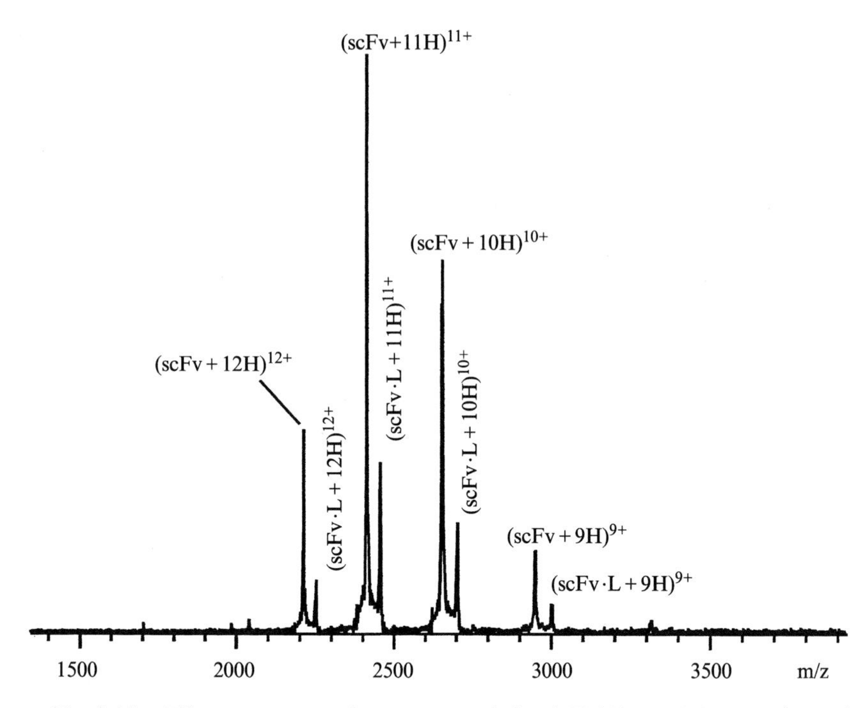

FIG. 2. NanoES mass spectrum of an aqueous solution (pH 6.8) containing scFv (12 μM) and its native ligand, Gal[Abe]Man ≡ L (6 μM).

the determination of K_{assoc}. The results from this study are described below.

Influence of experimental conditions on MS-derived K_{assoc}

Requirement for Multiple Measurements. For a given nanoES tip, the R values obtained from sequential measurements performed under identical conditions were found to fluctuate, significantly in some cases. Shown in Fig. 3a are the R values obtained from five nanoES tips pulled under identical conditions. For each tip, R was determined from six sequential measurements. The duration of each measurement was approximately 1 min. It can be seen in Fig. 3a that the value of R obtained from a given tip fluctuates by as much as 15%, compared with the average value (see Fig. 3b). The reason for this fluctuation is not understood but likely reflects changes in the spray characteristics of the nanoES tip with time. The difference in behavior observed for the different tips may reflect small differences in tip geometry. Shown in Fig. 3b are the average values of R, determined from the six measurements, for each tip. Despite fluctuations in the single

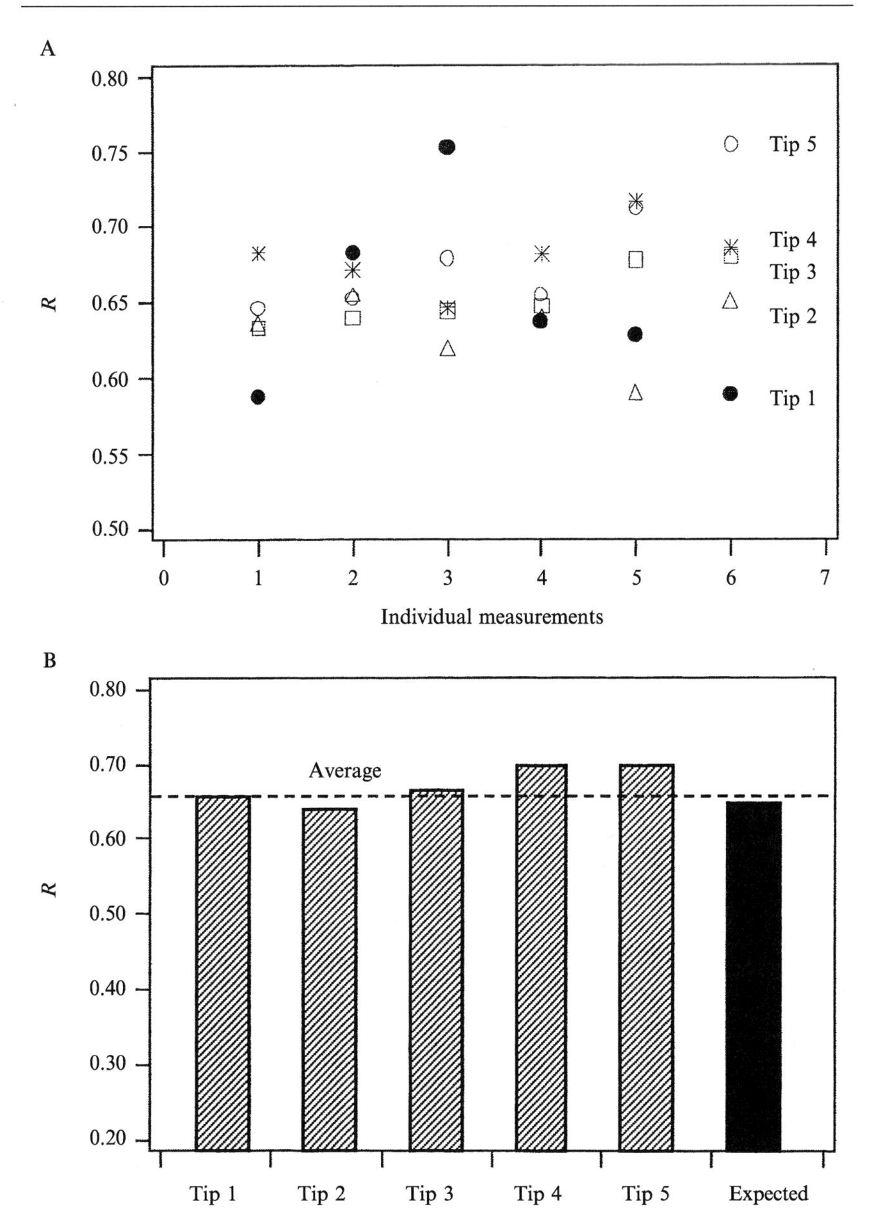

FIG. 3. (A) Distribution of R obtained from five different nanoES tips and six individual measurements. (B) Comparison of the averaged value of R for each nanoES tip (dashed line indicating the average value of R calculated for all five tips), and the expected R value derived from the reported K_{assoc} (solid column).

measurements, the average R values were found to be similar, ranging from 0.63 to 0.68. An R value of 0.66 was obtained as the average for all five tips; this is close to the expected value of 0.64, which can be calculated from the ITC-derived K_{assoc} and the initial protein and ligand concentrations. The average R, value, for all five tips, corresponds to a K_{assoc} of $1.8 \times 10^5\ M^{-1}$, which is in excellent agreement with ITC-derived value of $1.6 \times 10^5\ M^{-1}$. These results indicate that binding constants determined from single MS measurements might be unreliable; the use of multiple measurements carried out with different tips is recommended. However, on the basis of results discussed in the proceeding section, the use of sequential measurements, lasting more than several minutes with a single tip, is not advisable when the stability of the protein–carbohydrate complex is sensitive to the solution pH.

pH Changes in NanoES Solution. It is well established that electrochemical reactions, which occur at the electrode in the nanoES tip, can alter the pH of the nanoES solution. For aqueous solutions in positive ion mode, the dominant electrochemical reaction at the platinum electrode used in the present work is the oxidation of H_2O, leading to the production of H_3O^+ [Eq. (10)] and a decrease in pH:[16]

$$2H_2O \rightleftharpoons O_2 + 4H^+ + 4e^- \tag{10}$$

In the present experiments, spraying for approximately 30 min resulted in a decrease in the bulk pH of the nanoES solution from 6.8 to ~6.0. Van Berkel and co-workers have shown that the end of the nanoES tip, where the droplets are formed, experiences a more significant drop in pH compared with the bulk solution.[17] Therefore, the actual pH of the nanoES droplets produced after 30 min of spray is expected to be substantially less than 6.0. In Fig. 4a, the R values determined from a single tip are plotted versus spray time. The magnitude of R decreases with increasing spray duration, from a value of 0.62 to 0.45, corresponding to a decrease of 45% in K_{assoc}. This behavior is consistent with a reduction in the stability of the scFv-Gal[Abe]Man complex, which decreases significantly below pH 5.5, resulting from a decrease in solution pH.[14]

The influence of spray duration on the binding of scFv to the structural analog, Tal[Abe]Man, for which a binding constant of $1.17 \times 10^5\ M^{-1}$ has been determined by ITC,[18] was also investigated. The bulk pH of the solution decreased from pH 6.8 to 5.6 after 35 min of spraying and R decreased

[16] G. J. Van Berkel, F. Zhou, and J. T. Aronson, *Int. J. Mass Spectrom.* **162,** 55 (1997).
[17] G. J. Van Berkel, K. G. Asano, and P. D. Schnier, *J. Am. Soc. Mass Spectrom.* **12,** 853 (2001).
[18] D. R. Bundle, unpublished results (1996).

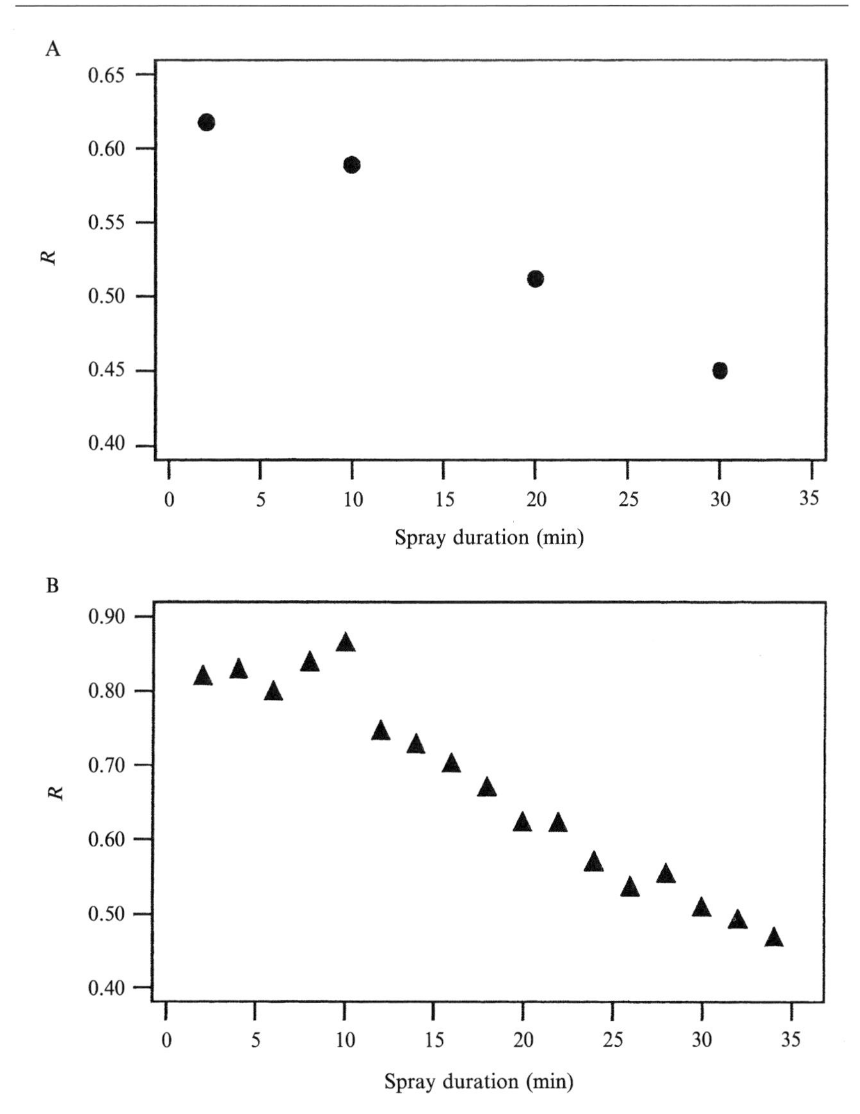

FIG. 4. Value of R plotted versus the spray duration for scFv and (A) its native ligand, Gal[Abe]Man (●); (B) Talα[Abe]Man (▲).

from 0.82 to 0.45 (see Fig. 4b), corresponding to a decrease of 50% in the K_{assoc}.

Influence of Ligand Concentration. In general, MS-derived binding constants are determined, not from measurements at a single set of analyte concentrations, but from titration experiments in which the concentration of one analyte is fixed and the concentration of the other is varied.[5,19,20] In the present study, binding constants were determined over a range of ligand concentrations, 6×10^{-6} to 1.9×10^{-5} M, with the protein concentration fixed at 1.81×10^{-5} M. In practice, an R of 0.1 is the smallest value that can be reliably determined from the MS data; this places a lower limit on the ligand concentration. For the given protein concentration, this lower limit corresponds to a ligand concentration of 3×10^{-6} M. The observation of scFv·(Gal[Abe]Man)$_2^{n+}$ ions at ligand concentrations exceeding that of the protein by more than a factor of 2 establishes the upper limit for the ligand concentration; see Fig. 5. Because the scFv has only one carbohydrate-binding site, these ions must be artifacts of the nanoES process. The observation of nonspecific complexes that result from the formation of random intermolecular interactions between analyte molecules as the nanoES droplets shrink due to evaporation of solvent is not uncommon, particularly when working at high analyte concentrations.[9,21] The appearance of nonspecific complexes obscures the true solution composition, making the reliable determination of K_{assoc} all but impossible.

As expected, increasing the ligand concentration results in an increase in the relative abundance of scFv·(Gal[Abe]Man)$^{n+}$ ions (Fig. 5). Shown in Fig. 6 are the values of R measured at different ligand concentrations. The predicted values, based on the ITC-derived K_{assoc}, are also shown in Fig. 6 and are found to be in good agreement with the MS values. The MS-derived K_{assoc} values determined at the different ligand concentrations are summarized in Table I.

In-Source Dissociation. The influence of several source parameters, such as inlet capillary temperature, source voltages, and hexapole accumulation times, on the relative abundance of scFv·L^{n+} and scFv^{n+} ions was also examined. Of these, only the accumulation time in the hexapole was found to have a significant effect on R. Shown in Fig. 7 are mass spectra, obtained from a single nanoES tip, recorded with accumulation times of 1.0, 3.0, and 6.0 s. The fraction of scFv·L^{n+} ions, relative to scFv^{n+} ions, decreases significantly with increasing accumulation time. The R values, determined over a range of accumulation times, are shown in Fig. 8. Under our experimental conditions, R decreased by 59% on increasing the

[19] K. Hirose, *J. Inclusion Phenomena Macrocyclic Chem.* **39,** 193 (2001).

[20] J. A. Loo, P. Hu, P. McConnel, W. T. Mueller, T. K. Sawyer, and V. Thanabal, *J. Am. Soc. Mass Spectrom.* **8,** 234 (1997); H.-K. Lim, Y. L. Hsieh, B. Ganem, and J. Henion, *J. Mass Spectrom.* **30,** 708 (1995); R. H. Griffey, S. A. Hofstadler, K. A. Sannes-Lowery, D. J. Ecker, and S. T. Crooke, *Proc. Natl. Acad. Sci. USA* **96,** 10129 (1999).

[21] C. V. Robinson, E. W. Chung, B. B. Kragelund, J. Knudsen, R. T. Aplin, F. M. Poulsen, and C. M. Dobson, *J. Am. Chem. Soc.* **118,** 8646 (1996); V. Gabelica, E. De Pauw, and F. Rosu, *J. Mass Spectrom.* **34,** 1328 (1999).

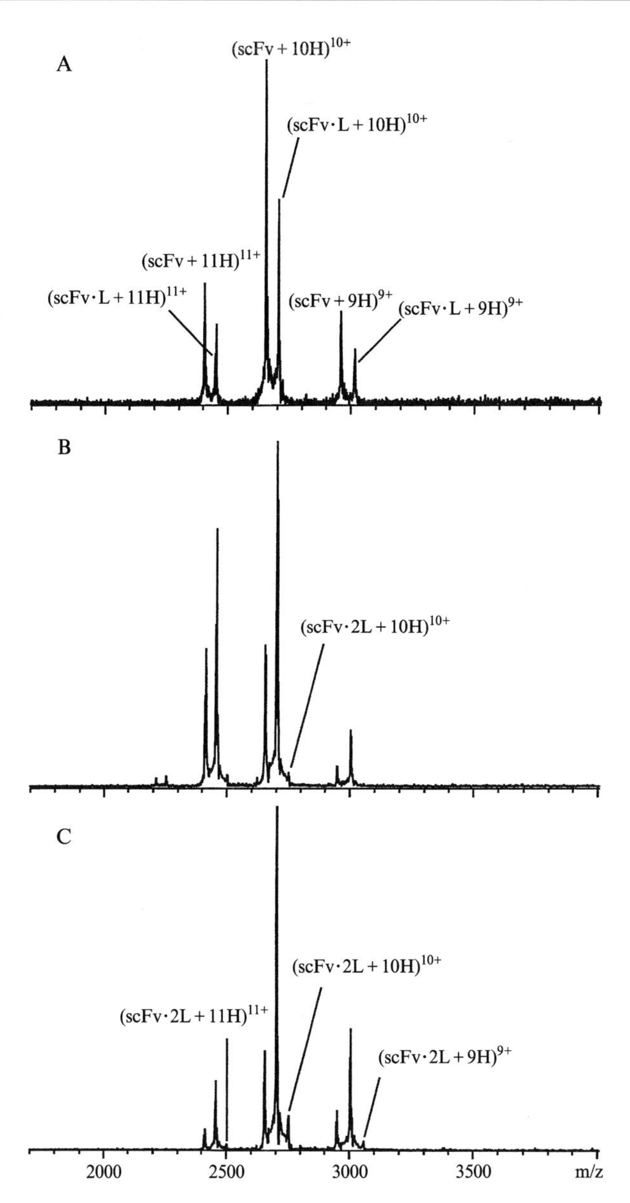

FIG. 5. NanoES mass spectra obtained for aqueous solutions containing scFv (18.1 μM) and Gal[Abe]Man (L) at increasing concentrations: (A) 10.8 μM; (B) 35.6 μM; and (C) 45.1 μM.

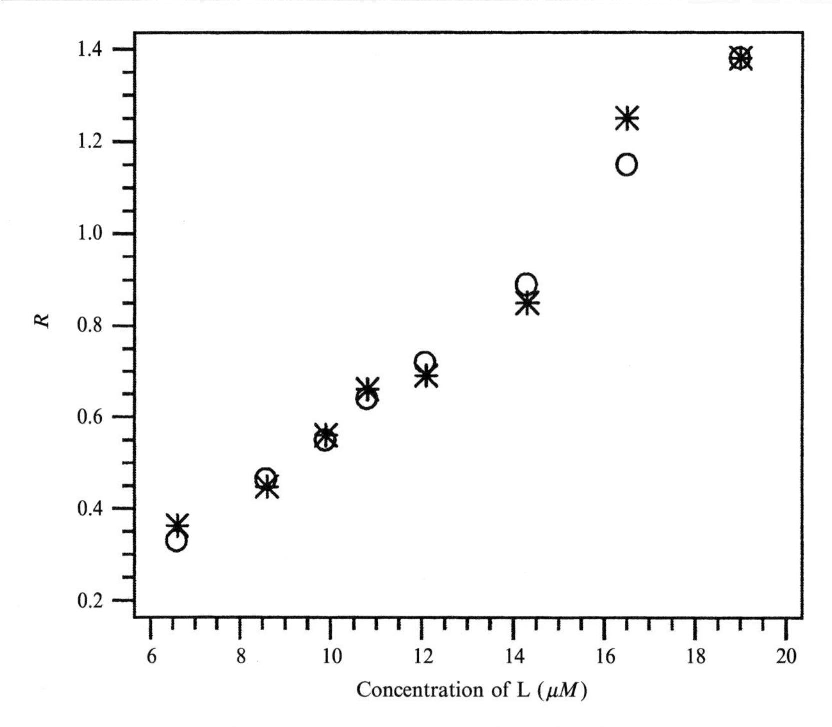

FIG. 6. Dependence of R on the concentration of ligand (L). (∗) MS-derived value, (○) value predicted from the ITC-derived K_{assoc}.

TABLE I
R VALUES AND ASSOCIATION CONSTANTS FOR scFv AND Gal[Abe]Man[a]

Concentration of Gal[Abe]Man (μM)	*R*		$K_{assoc} \times 10^{-5}$ (M^{-1})(MS)	Difference[c] (%)
	MS	ITC[b]		
6.6	0.36	0.33	1.98	+23.8
8.6	0.45	0.47	1.50	−6.3
9.9	0.56	0.55	1.62	+1.3
10.8	0.66	0.64	1.84	+5.7
12.1	0.69	0.72	1.47	−8.1
14.3	0.85	0.91	1.42	−11.3
16.5	1.25	1.12	1.95	+21.9
19.0	1.38	1.36	1.62	+1.3

[a] Determined by nanoES-FT-ICR/MS (MS) at different ligand concentrations.
[b] R values calculated from the ITC-derived K_{assoc} of $(1.6 \pm 0.2) \times 10^5 M^{-1}$ from Ref. 14.
[c] Values correspond to $\{[K_{assoc}\text{ (MS)} - K_{assoc}\text{ (ITC)}]/K_{assoc}\text{ (ITC)}\} \times 100\%$

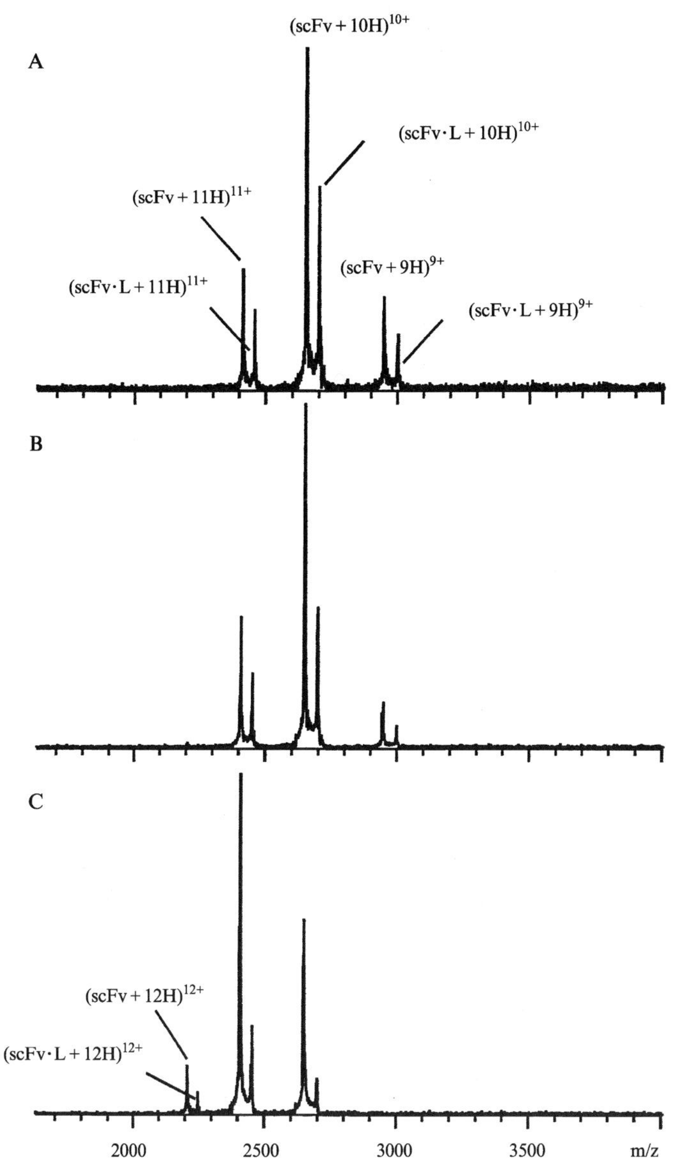

FIG. 7. NanoES mass spectra of an aqueous solution of scFv (18.1 μM) and Gal[Abe]Man (L) (10.8 μM) obtained at different hexapole accumulation times: (A) 1 s; (B) 3 s; (C) 6 s.

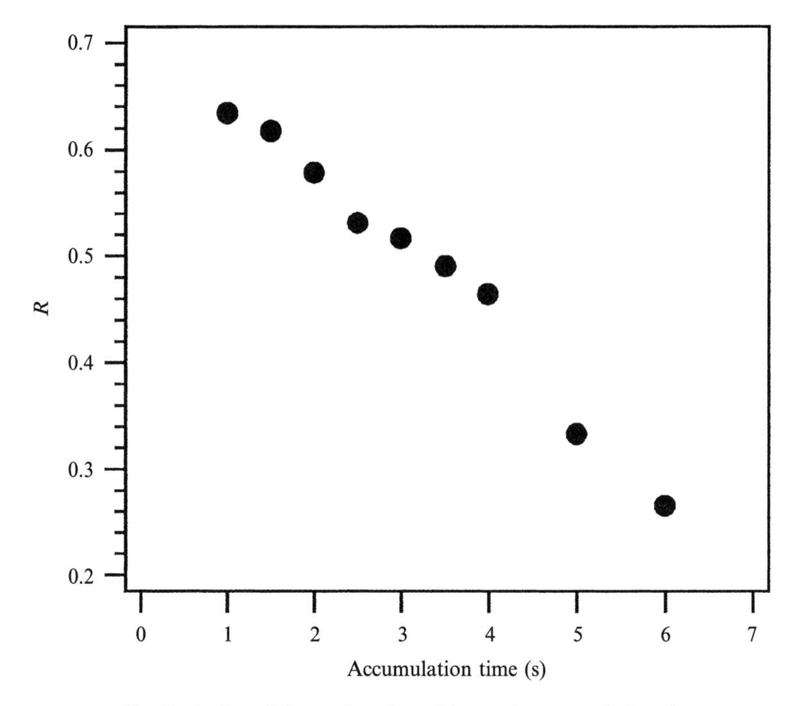

FIG. 8. A plot of R as a function of hexapole accumulation time.

accumulation time from 1 s ($R = 0.63$) to 6 s ($R = 0.26$), leading to a decrease of 77% in K_{assoc}. This effect is attributed to the dissociation of the scFv·L^{n+} ions while stored in the hexapole. The pressure in the hexapole is not uniform but ranges from 10^{-3} to 10^{-5} torr. Acceleration of the trapped ions by the high radio-frequency (rf) field (1 kV) applied to the hexapole rods results in some of the complex ions being collisionally heated and subsequently dissociating. These results clearly indicate that shorter accumulation times, which minimize the extent of collision-induced dissociation, will lead to more reliable binding constants. However, spectra acquired with short accumulation times, less than 1 s, suffer from poor signal-to-noise(S/N) ratios. Therefore, an accumulation time of 1.5 s was chosen as a compromise between enhancing the S/N ratio of the spectra and minimizing in-source dissociation.

II. *Binding of SLT-1(B_5) and P^k Trisaccharide*

Our laboratory has also applied nanoES-FT-ICR MS to investigate the binding of the P^k trisaccharide to the multivalent protein complex

SLT-1 (B_5).[12] SLT-1(B_5) is a homopentamer consisting of five subunits, each with a molecular mass of 7688 Da. From solution NMR studies,[22] it is known that each subunit has one dominant P^k-binding site (referred to as "site 2"), with two additional binding sites (sites 1 and 3), per subunit, suggested from the crystal structure.[23] The crystal structure of the SLT-1(B_5) complex and the location of binding site 2 are shown in Fig. 9. From ITC measurements, it has been shown that binding between SLT-1(B_5) and P^k at site 2 is noncooperative with a K_{assoc} of $(1.5 \pm 0.5) \times 10^3\ M^{-1}$.[24] The K_{assoc} values for sites 1 and 3 have not been determined but are believed to be significantly smaller than for site 2.

A nanoES spectrum obtained for a solution of SLT-1(B_5) at pH 7.2 is shown in Fig. 10. The dominant ions correspond to the intact pentamer (i.e., B_5^{n+}) with charge states of 11 to 13. The absence of monomer or smaller oligomer ions indicates that the pentamer is stable in the nanoES solution and that its quaternary structure is preserved throughout the nanoES process. The long accumulation times (in the hexapole), up to 15 s, which were necessary to obtain adequate S/N ratios for the B_5^{n+} ions, were not found to promote dissociation of the complex.

To establish the binding constant for P^k, nanoES-MS measurements were performed on aqueous solutions (pH 7.2) containing 5.0 μM SLT-1(B_5) and three different concentrations of P^k (45, 154, and 310 μM). Despite the low binding constant, complexes of SLT-1(B_5) and P^k, $(B_5{\cdot}P_i^k)^{n+}$, were readily observed (Fig. 11a–c) and, as expected, the degree of complexation increased with the concentration of the ligand. The number of bound ligands and even nature of the charging agents are easily identified from the FT-ICR mass spectra. At the highest concentration investigated (310 μM) complexes with up to five bound P^k ligands were observed (Fig. 11c). In contrast to the behavior observed for the scFv-Gal[Abe]Man system, nonspecific complexes of SLT 1(B_5) and P^k were not observed in the mass spectra. The reason for this difference in behavior is not known. It may be that the nonspecific complexes, if present, dissociate before detection.

On the basis of the relative abundance of the protonated $(B_5{\cdot}P_i^k)^{n+}$ ions and the initial concentration of protein and ligand, the equilibrium concentration of B_5 and the $B_5{\cdot}P_i^k$ complexes [Eqs. (11a–11e)] could be

[22] H. Shimizu, R. A. Field, S. W. Homans, and A. Donohue-Rolfe, *Biochemistry* **37,** 11078 (1998).

[23] H. Ling, A. Boodhoo, B. Hazes, M. D. Cummings, G. D. Armstrong, J. L. Brunton, and R. J. Read, *Biochemistry* **37,** 1777 (1998); D. J. Bast, L. Banerjee, C. Clark, R. J. Read, and J. L. Brunton, *Mol. Microbiol.* **32,** 953 (1999).

[24] P. M. St. Hilaire, M. K. Boyd, and E. J. Toone, *Biochemistry* **33,** 14452 (1994).

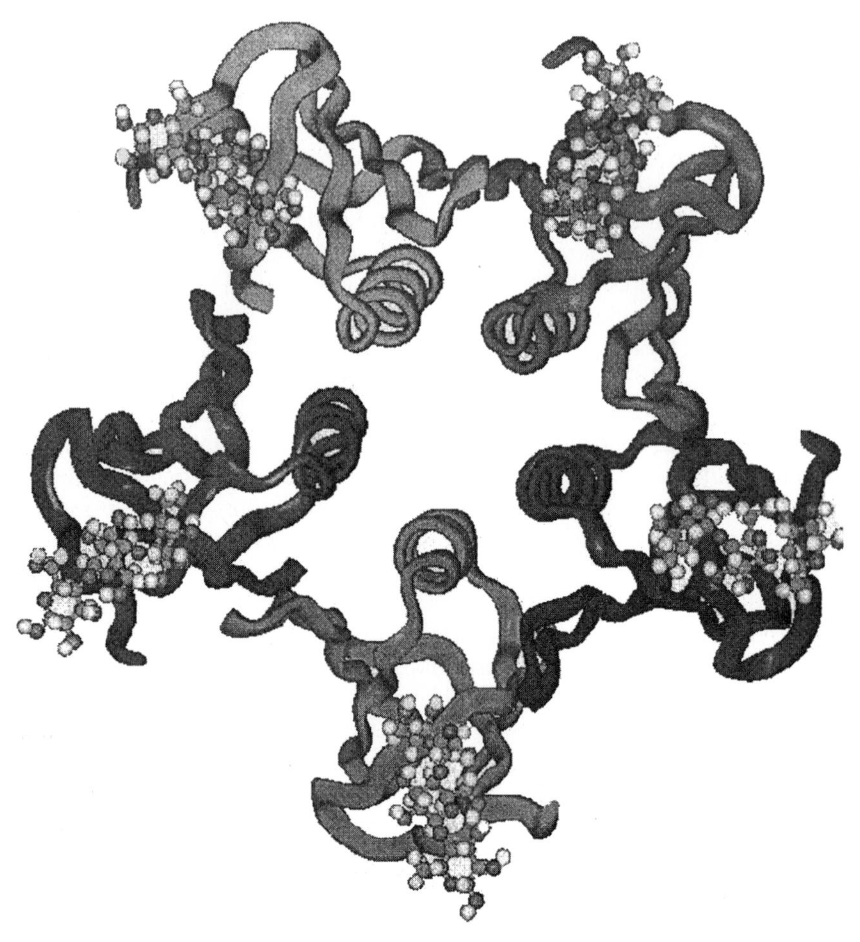

FIG. 9. Crystal structure of the complex SLT-1(B_5)·(P^k)$_5$.

calculated. This in turn allowed for the determination of K_{assoc} for the five noncooperative binding sites of SLT-1(B_5):

$$B_5 + P^k \rightleftharpoons B_5 \cdot P^k \tag{11a}$$

$$B_5 \cdot P^k + P^k \rightleftharpoons B_5 \cdot P^k_2 \tag{11b}$$

$$B_5 \cdot P^k_2 + P^k \rightleftharpoons B_5 \cdot P^k_3 \tag{11c}$$

$$B_5 \cdot P^k_3 + P^k \rightleftharpoons B_5 \cdot P^k_4 \tag{11d}$$

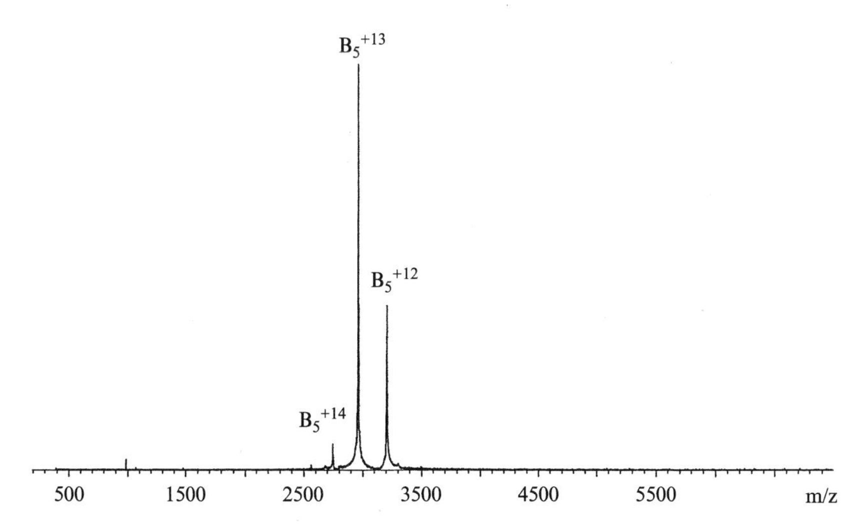

FIG. 10. NanoES mass spectrum of an aqueous solution (pH 7.2) of 5.0 μM SLT-1(B_5).

$$B_5 \cdot P_4^k + P^k \rightleftharpoons B_5 \cdot P_5^k \tag{11e}$$

Reliable K_{assoc} values could be determined only when the relative ion abundance was determined from MS data with high S/N ($S/N > 3$). For example, at the highest ligand concentration (310 μM), complexes with up to five P^k ligands were observed in the mass spectrum, allowing for the determination of K_{assoc} for $i = 1$–5. However, because of the poor S/N for the $B_5{\cdot}P_5{}^k$ ion, its relative abundance and, consequently, K_{assoc} for $i = 5$, could not be accurately determined. The MS-derived K_{assoc} values determined for the +12 and +13 charge states are listed in Table II. The values in parentheses were calculated from MS data with low S/N and were not used to calculate the average association constant. The individual binding constants reported in Table II are similar, ranging from 1×10^3 to 4×10^3 M^{-1}, and independent of the ligand concentration and charge state. The overall K_{assoc}, corresponding to the average of the binding constants obtained for the +12 and +13 charge states and the three ligand concentrations, $(2.5 \pm 0.8) \times 10^3$ M^{-1}, is in excellent agreement with the ITC-derived value of $(1.5 \pm 0.5) \times 10^3$ M^{-1}.[24]

In the present study, the number of binding sites and the value of K_{assoc} were known in advance. To confirm by MS that there were only five dominant binding sites, nanospectra acquired at significantly higher ligand concentrations, so that the five binding sites become saturated, are necessary. However, a dramatic decrease in the S/N for the $(B_5{\cdot}P_i{}^k)^{n+}$ ions was

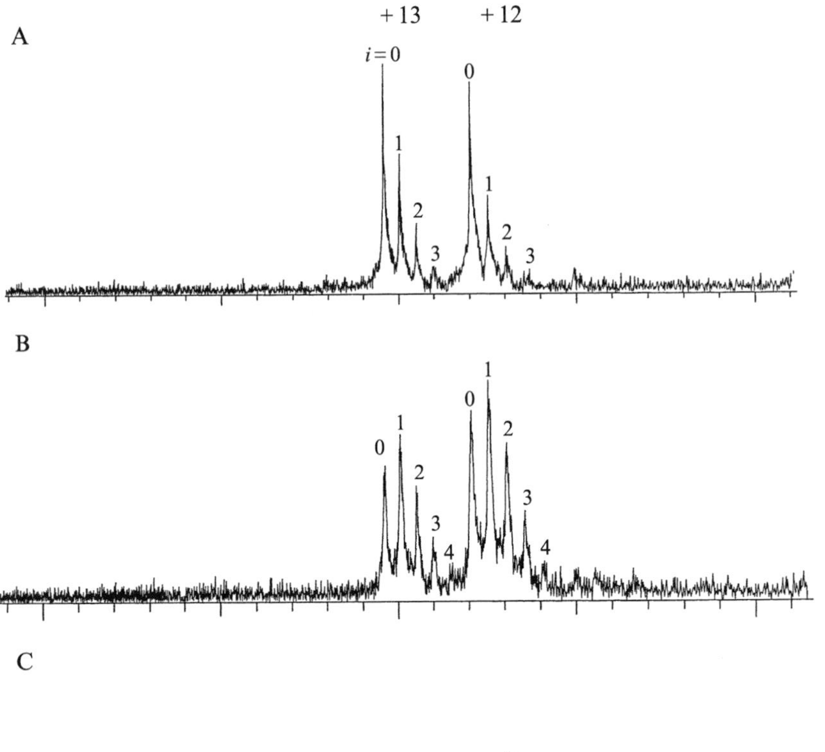

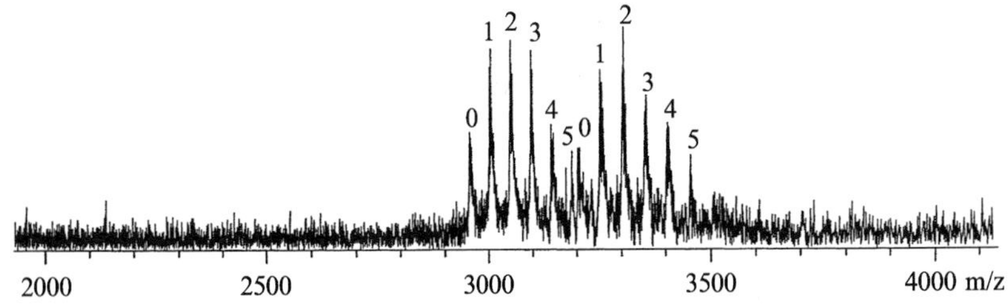

FIG. 11. NanoES mass spectra of aqueous solutions (pH 7.2) containing 5.0 μM SLT-1(B_5) and three different concentrations of P^k: (A) 45 μM, (B) 154 μM, and (C) 310 μM.

observed at ligand concentrations of >500 μM and the relative ion abundance could not be accurately determined.

This study represents the first direct determination by nanoES-MS of the binding stoichiometry and affinity for a protein–ligand complex with an association constant in the 10^3 M^{-1} range. On the basis of the spectrometric data obtained for solutions of SLT-1(B_5) and P^k, it was confirmed

TABLE II
ASSOCIATION CONSTANTS FOR BINDING OF SLT1(B_5) AND P^k[a]

Concentration of P^k (μM)	Number of bound P^k	$K_{assoc} \times 10^{-3}$ (M^{-1})	
		Charge state +12	Charge state +13
45	1	2.0	2.8
	2	3.5	3.8
	3	(4.4)	(5.0)
154	1	1.6	1.8
	2	2.2	2.0
	3	3.1	2.7
	4	1.8	2.5
310	1	1.7	1.6
	2	2.3	1.7
	3	1.9	3.2
	4	4.6	3.2
	5	(21.4)	(9.9)
			Average = 2.5 ± 0.8

[a] Determined by nanoES-FT-ICR/MS.

that the five dominant P^k binding sites (i.e., site 2) of SLT-1(B_5) operate in a noncooperative manner. Further, from the relative abundance of the $(B_5{\cdot}P_i{}^k)^{n+}$ ions observed in the mass spectra, an association constant, in excellent agreement with the ITC-derived value, was determined.

Summary

The two studies presented here demonstrate that nanoES-FT-ICR MS is a powerful method for studying the association of oligosaccharide ligands with monomeric and multimeric proteins. It permits the facile identification of the occupancy of binding sites, information that is not readily available by other techniques. Its high-resolution capability is ideally suited to the observation of interactions between a large protein receptor and a relatively small oligosaccharide ligand. The sensitive and rapid determination of association constants for protein carbohydrate complexes is expected to find wide application.

[28] Analysis of Lectin–Carbohydrate Interactions by Capillary Affinophoresis

By KIYOHITO SHIMURA and KEN-ICHI KASAI

Introduction

Electrophoresis has long been applied to the analysis of interactions between proteins and ligands. Among these, lectin–carbohydrate interactions represent a typical area where affinity constants have been determined by observing the migration of lectins in gel matrices to which carbohydrates are immobilized.[1–3] The migration of lectins is diminished by interaction with the immobilized carbohydrates in these applications. We describe a different approach, in which carbohydrate ligands are attached to soluble ionic polymers. In this case, the electrophoretic migration of lectins is enhanced by the interactions. We refer to the ligand–ionic polymer conjugate as an "affinophore" and electrophoresis using the affinophore as "affinophoresis."[4,5] Typically, polyliganded affinophores are prepared by coupling *p*-aminophenyl glycosides to an anionic polymer, succinylpolylysine, at a glycoside:succinyllysine ratio of about 10%.[6]

Capillary electrophoresis has many characteristics that recommend it in the analysis of molecular interactions, that is, the ability to analyze interactions in free solutions, a short analysis time, precise temperature control, and a small sample size.[7–10] Although polyliganded affinophores are also effective in detecting lectin–carbohydrate interactions in a capillary, they are not suitable for the determination of affinity constants because of the multivalency of most lectins.[11] Monoliganded affinophores were developed to solve this problem, and affinity constants between divalent lectins and

[1] K. Takeo, *in* "Advances in Electrophoresis" (A. Chrambach and M. J. Dunn, eds.), Vol. 1, p. 229. VCH, New York, 1987.
[2] T. C. Bøg-Hansen, *in* "Affinity Chromatography and Molecular Interactions: INSERM Symposia Series" (J. M. Egly, ed.), Vol. 86, p. 399. INSERM, Paris, 1979.
[3] V. Hořejši and M. Tichá, *J. Chromatogr.* **376,** 49 (1986).
[4] K. Shimura, *J. Chromatogr.* **510,** 251 (1990).
[5] K. Shimura and K. Kasai, *Methods Enzymol.* **271,** 203 (1996).
[6] K. Shimura and K. Kasai, *J. Chromatogr.* **400,** 353 (1987).
[7] S. Honda, A. Taga, K. Suzuki, S. Suzuki, and K. Kakehi, *J. Chromatogr.* **597,** 377 (1992).
[8] Y.-H. Chu, L. Z. Avila, J. Gao, and G. M. Whitesides, *Acc. Chem. Res.* **28,** 461 (1995).
[9] K. Shimura and K. Kasai, *Anal. Biochem.* **251,** 1 (1997).
[10] N. H. H. Heegaard and R. T. Kennedy, *Electrophoresis* **20,** 3122 (1999).
[11] K. Shimura and K. Kasai, *Electrophoresis* **19,** 397 (1998).

0076-6879/03 $35.00

carbohydrates were determined by capillary affinophoresis performed in a competitive manner.[12,13]

Outline of Experimental Procedures

Interactions with an anionic affinophore increase the mobility of a lectin toward the positive electrode and result in a change in detection time for the lectin. The mobility change can be calculated from the change in the detection time. Because mobility change is proportional to the degree of saturation of the binding site of the lectin with the affinophore, the dissociation constants (K_d) between the lectin and the affinophore, and the maximum mobility change ($\Delta\mu_{max}$) of the lectin, can be determined by affinophoresis, using different concentrations of affinophore.

Once the K_d and $\Delta\mu_{max}$ are determined, the affinity constants for other sugars can be determined by competition experiments.[12,13] A neutral sugar competes with the affinophore for binding to the lectin and diminishes the effect of the affinophore. As in enzyme kinetics, the competition apparently increases the K_d value between the lectin and the affinophore. The affinity constant of the lectin and the neutral sugar can be calculated as a K_i value in manner identical to that used for the competitive inhibition of enzyme kinetics.

Preparation of Monoliganded Affinophore

Glutathione is used as an affinophore matrix for monoliganded affinophores. A carbohydrate ligand is attached by utilizing the unique reactivity of the thiol group of glutathione with iodoacetylated *p*-aminophenylglycosides.[12,13] Succinylation of the amino group of the glutathione moiety confers three negative charges to the affinophore that can produce a sufficient change in the mobility of the lectins, on binding to it (Fig. 1).

p-Aminophenylglycoside (AP-glycoside, 58 μmol; Sigma, St. Louis, MO) is dissolved in 1 ml of 2-(*N*-morpholino)ethanesulfonic acid (MES)–NaOH buffer (pH 6.0), and *N*-iodoacetoxysuccinimide[14] (25 mg, 88 μmol) in 80 μl of *N,N*-dimethylformamide is then added to the solution. After a 1-h reaction at room temperature in the dark, a mixed-bed ion-exchange resin (45 mg, AG-X8 resin; Bio-Rad, Hercules, CA) is added to the mixture and it is vortexed mainly to remove free iodoacetic

[12] K. Shimura and K. Kasai, *J. Biochem.* **120,** 1146 (1996).

[13] K. Shimura, Y. Arata, N. Uchiyama, J. Hirabayashi, and K. Kasai, *J. Chromatogr. B* **768,** 199 (2002).

[14] A. Hampton, L. A. Slotin, and R. R. Chawla, *J. Med. Chem.* **19,** 1279 (1976).

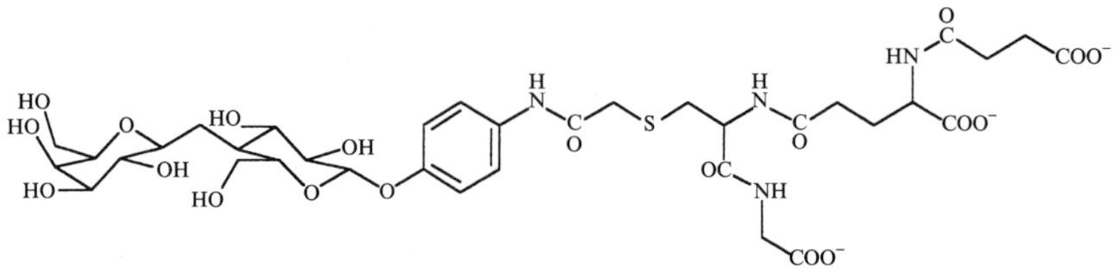

FIG. 1. Structure of a monoliganded affinophore bearing a lactoside moiety (sGS-AP-Lac). Reproduced from Ref. 13.

acid formed by the hydrolysis of *N*-iodoacetoxysuccinimide. The supernatant is recovered and 300 μl of 0.5 *M* sodium phosphate buffer (pH 7.5) containing 25 m*M* EDTA is added. Glutathione (GSH, reduced form, 23 mg, 75 μmol) is added to the mixture, followed by a 2-h reaction at room temperature in the dark. The mixture is acidified to pH 2–3 by the addition of 1 *M* hydrochloric acid. The coupling product (GS-AP-glycoside) is purified by high-performance liquid chromatography (HPLC) on a reversed-phase chromatographic column (TSK-Gel ODS-80TS, 4.6 mm i.d. $\times$ 25 cm) with a cartridge guard column (TSK guard gel ODS-80TS, 3.2 mm i.d. $\times$ 1.5 cm) (Tosoh, Tokyo, Japan). About one-fifth of each of the reaction mixtures is applied to the column, which had been equilibrated with 0.1% trifluoroacetic acid, and eluted with a gradient of 0 to 25% acetonitrile in 0.1% trifluoroacetic acid over a period of 25 min at a flow rate of 1 ml/min at room temperature, with detection by measurement of the absorption at 280 nm. GS-AP-glycoside, a new peak, is collected, combined, and dried by evaporation. The GS-AP-glycoside preparation is dissolved in 700 μl of 0.1 *M* sodium phosphate buffer (pH 7.5) containing 25 m*M* EDTA and succinic anhydride, 20 mg, is added to the solution. The mixture is allowed to react for 10 min, while maintaining the pH at 7–8 by the addition of 5 *M* NaOH. Aliquots, 100 μl, of the reaction mixture are acidified to pH 2–3 by the addition of hydrochloric acid, and applied to the chromatographic column, followed by elution under the same conditions as described above. The affinophore, succinylated GS-AP-glycoside (sGS-AP-glycoside), eluting as a new peak, is collected, combined, and dried by evaporation. It is then dissolved in 1 ml of water and stored in a freezer. The concentration of the affinophore is determined by means of the phenol–sulfuric acid method with corresponding sugar solution as a standard. The molecular extinction coefficient of the affinophores in water at 248 nm has been determined to be 10,400 M^{-1} cm^{-1}. The overall yield of the affinophore from the *p*-aminophenylglycoside is 40–50%.

Electroendosmosis and Capillary Affinophoresis

A distinctive feature of free solution capillary electrophoresis, in comparison with the electrophoresis in gels, is the apparent contribution of electroendosmosis to the results of the electrophoresis. The inner surface of the capillaries used in the affinophoresis is negatively charged, and the application of an electric field generates an asymmetric migration of the counter ions of the immobilized negative charges on the surface. The migrating counter ions push water molecules and produce a plug flow, in which the flow speed is identical irrespective of the distance from the wall. The speed of the flow can be sufficiently high to carry all the molecular species participating in the interactions toward the negative electrode. This enables the detection of the entire sample components injected at the positive end of the capillary in a relatively short time with a fixed detector on the capillary.

Electrically neutral species are transported only by electroendosmosis to the detection point (Fig. 2). On the other hand, negatively charged species migrate toward the positive electrode against the overwhelming electroendosmosis, and are detected after the neutral species. The faster the migration toward the positive electrode, the later they are detected. When an anionic affinophore is involved in the electrophoresis, the interaction should increase the detection time for the lectin. Anionic affinophores migrate toward the positive electrode much faster than lectins but not fast enough to move backward against the electroendosmotic flow. The affinophore solution needs to be loaded only in the capillary, and this allows the lectin, injected at the positive end, to interact with the affinophore from the beginning of the electrophoresis to its detection. In competition experiments, a competing sugar needs to be added only in the positive electrode buffer. The neutral sugar is transported along the capillary only by the electroendosmosis. The lectin and the affinophore electrophoretically migrate into the zone of the neutral sugar, and competition can then be observed (Fig. 2).

Calculation of Mobility Change

To determine the affinity constant, the mobility changes of a lectin must be calculated from the detection time obtained under different conditions of affinophoresis. In free solution capillary electrophoresis, the detection time is a function of both the electrophoretic mobility of a sample and the electroendosmotic mobility of a capillary. Coelectrophoresis with a reference molecule, that is, an electrophoresis marker that does not interact

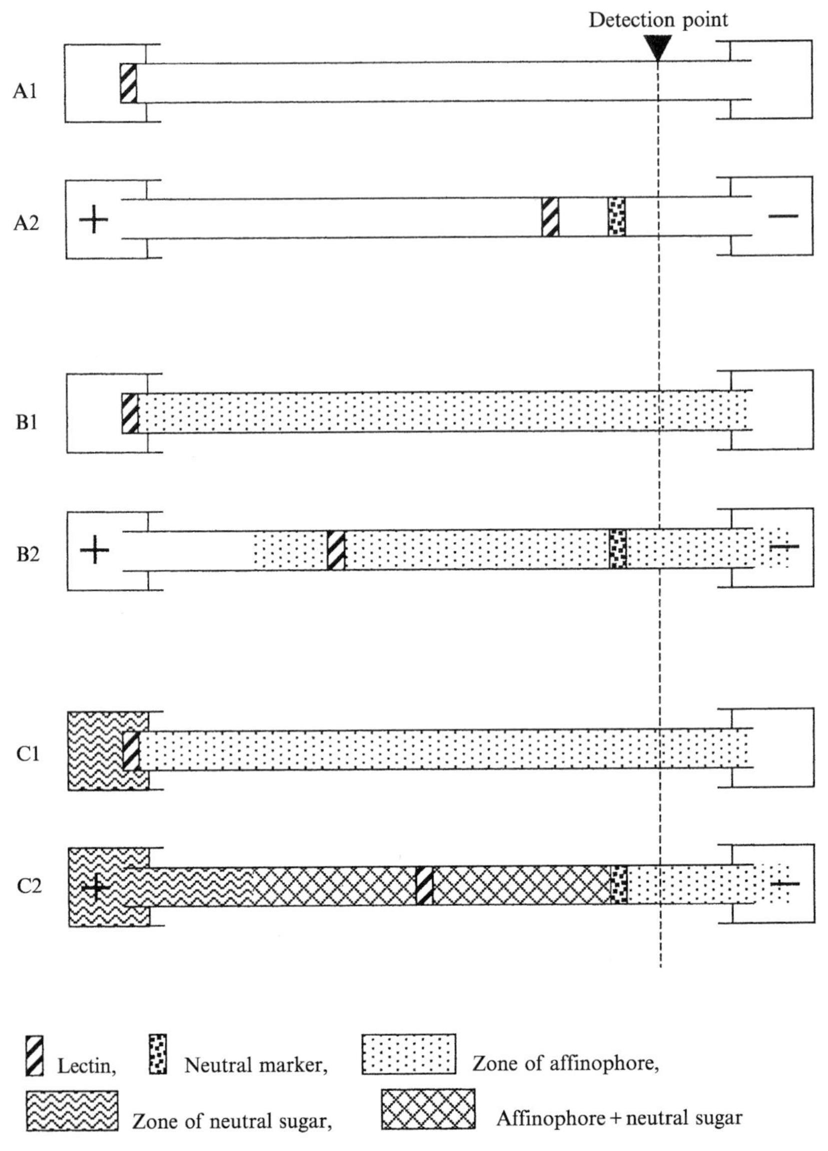

FIG. 2. Capillary affinophoresis and electroendosmosis. (A1) A sample solution is injected at the positive end of the capillary. (A2) Application of an electric field generates electroendosmosis due to the negatively charged surface of the capillary. If the lectin is negatively charged and migrates toward the positive side, a neutral marker is detected before the lectin. (B1) In the case of affinophoresis, an affinophore is included only in the capillary. (B2) The affinophore migrates toward the positive end much faster than the lectin, and the lectin migrates in the affinophore solution until it is detected. Interaction between the lectin

with the affinophore, allows the calculation of the mobility change from a change in the detection time as shown below.[15,16]

The detection time (t) of a lectin is a function of its electrophoretic mobility (μ) as well as of the electroendosmotic mobility (μ_{eo}) of a capillary:

$$t = \frac{L}{E} \cdot \frac{1}{\mu + \mu_{eo}} \tag{1}$$

where L is the distance (cm) from the injection end to a detection point, and E is the field strength (V/cm). The same relation should be applied for the detection time of an electrophoresis marker (t_r):

$$t_r = \frac{L}{E} \cdot \frac{1}{\mu_r + \mu_{eo}} \tag{2}$$

where μ_r represents the electrophoretic mobility of the marker. When the lectin and the marker are subjected to simultaneous electrophoresis, μ_{eo}, E, and L are the same in Eqs. (1) and (2). As a result, the difference between the reciprocals of each detection time is proportional to the difference between the mobility of the lectin and the marker, that is, $\mu - \mu_r$.

$$\frac{1}{t} - \frac{1}{t_r} = \frac{E}{L}[(\mu + \mu_{eo}) - (\mu_r + \mu_{eo})] = \frac{E}{L}(\mu - \mu_r) \tag{3}$$

Thus the mobility difference between the lectin and the marker in the same run can be expressed as follows:

$$\mu - \mu_r = \frac{L}{E}\left(\frac{1}{t} - \frac{1}{t_r}\right) \tag{4}$$

The mobility difference in the presence of the affinophore can also be calculated using Eq. (4). When the mobility of the lectin changes from μ_0 to μ by the affinophoresis, the mobility change $\Delta\mu$ ($\Delta\mu = \mu - \mu_0$) induced by the affinophoresis can be extracted from the two mobility differences from the marker in the presence and the absence of the affinophore, that is,

---

and the affinophore increases the mobility of the lectin and retards its detection. (C1) To observe competition by a neutral sugar, it is included only in the positive electrode solution. (C2) The neutral sugar is rapidly transported toward the negative side by electroendosmosis, and the lectin migrates in a mixed zone of the affinophore and the neutral sugar. Effective competition suppresses the acceleration of the electrophoresis of the lectin by the affinophore.

[15] K. Shimura and K. Kasai, *Anal. Biochem.* **227,** 186 (1995).

[16] F. A. Gomez, L. Z. Avila, Y.-H. Chu, and G. M. Whitesides, *Anal. Chem.* **66,** 1785 (1994).

$$\begin{aligned}\Delta\mu &= \mu - \mu_0 \\ &= (\mu - \mu_r) - (\mu_0 - \mu_r) \\ &= \frac{L}{E}\left[\left(\frac{1}{t} - \frac{1}{t_r}\right) - \left(\frac{1}{t_0} - \frac{1}{t'_r}\right)\right]\end{aligned} \quad (5)$$

where t_0 is the detection time of lectin in the absence of the affinophore. The values of t_r and t'_r might be different because of variations in the electroendosmosis between the two runs. The calculation of $\Delta\mu$ by Eq. (5) permits the variation in electroendosmosis between the runs to be determined.

Instrument and Capillary

A capillary electrophoresis instrument with an ultraviolet absorption detector is required. The instrument should have a temperature-controlling system for the capillary, because the interactions occur during the electrophoresis run.

Fused silica capillaries (25–50 μm i.d., 375 μm o.d.) with a polyimide covering (Polymicro Technologies, Phoenix, AZ) should be suitable for most applications. A fused silica capillary has a tendency to adsorb proteins mainly because of the anionic characteristics of its surface caused by the dissociation of silanol groups. The adsorption largely depends on the nature of the particular protein under consideration. If the protein does not pose any problems related to adsorption, that is, the electropherograms are reproducible without obvious tailing and broadening of peaks, the use of a fused silica capillary is the most straightforward. Coating of the inner surface of the capillary may alleviate the problem, should it occur.

Capillary Coating

We typically use capillaries that are coated with an anionic polymer, succinylpolylysine. An epoxy group is introduced onto the inner surface of the capillary by reaction with an epoxysilane and polylysine is then reacted to cover the surface. Finally, the amino function of polylysine is succinylated.[15]

An aqueous solution of 3-glycidoxypropyltrimethoxysilane [5% (v/v)] is adjusted to pH 5.5–5.8 with 5 m*M* KOH and clarified by centrifugation. The fused silica capillary (60 cm long) is filled with the solution and the ends are closed by inserting them into rubber septa used for gas chromatography. The capillaries are heated in a boiling water bath for 30 min and rinsed by passing 0.1 ml each of water and acetone through them, followed by drying by passing air by suction. The capillaries are filled with a poly(L-lysine) solution (Sigma; average degree of polymerization, 270; 10 mg/ml in 1 *M* NaCl; adjusted to pH 10 with NaOH) and reacted overnight at room temperature. They are then washed with 0.1 ml of water and filled with a mixture [1:1 (v/v)] of succinic anhydride in dimethylformamide (25 mg/ml) and *N*-ethylmorpholine. They are allowed to react for 30 min at room temperature and, finally, rinsed with 0.1 ml of water.

The coating procedure can be facilitated by the use of a syringe (50–100 μl; a pressure-lock or gas-tight syringe with a 26-gauge needle with a tip cut at 90°), with a segment of Teflon tubing (2 cm long, 0.3 mm i.d. × 1.5 mm o.d.; Alltech Associates, Deerfield, IL) attached to the tip of the needle. The tube is fitted to both the needle of the syringe and the capillary (375 μm o.d.) by thrusting the tips of them into each end of the tube after softening by mild heating with an electric heater. After cooling, the capillary can be detached and reconnected without heating, and the connection permits the rapid injection of solutions into capillaries with the syringe.

Equations for Affinophoresis

Most lectins are divalent and must be treated accordingly in the analysis. We start with the equations for affinophoresis of monovalent proteins and then return to divalent systems. For a monovalent lectin, L, and an affinophore, A, we assume the following binding equilibrium:

$$\mathrm{L} + \mathrm{A} \rightleftharpoons \mathrm{LA} \tag{6}$$

The original electrophoretic mobility of the lectin is μ_0 and that of the complex with the affinophore is μ_c. Because the lectin and the affinophore are in dynamic binding equilibrium, the macroscopic mobility of the lectin, μ, in equilibrium should be the weighted average of μ_0 and μ_c:

$$\mu = (1 - \alpha)\mu_0 + \alpha\mu_c \tag{7}$$

The weight, α, is the degree of saturation of the lectin with the affinophore and is a function of the dissociation constant K_d ($K_d = [\mathrm{L}][\mathrm{A}]/[\mathrm{LA}]$) of the complex and the concentration of the affinophore, [A]:

$$\alpha = \frac{[\mathrm{A}]}{K_d + [\mathrm{A}]} \tag{8}$$

From Eq. (7) and (8), the mobility change $\Delta\mu$ ($\Delta\mu = \mu - \mu_0$) of the lectin by affinophoresis can be expressed as follows:

$$\Delta\mu = \Delta\mu_{max} \frac{[A]}{K_d + [A]} \tag{9}$$

where $\Delta\mu_{max}$ ($\Delta\mu_{max} = \mu_c - \mu_0$) represents the maximum mobility change of the lectin. It should be noted that the equation has the same form as the Michaelis–Menten equation for enzyme kinetics and, thus, the K_d and $\Delta\mu_{max}$ can be determined in a manner similar to that used for determining K_m and V_{max} in enzyme kinetics. A plot of $\Delta\mu$ versus $\Delta\mu/[A]$ according to the following linear equation yields a straight line that intercepts the ordinate at $\Delta\mu_{max}$ with a slope of $-K_d$:

$$\Delta\mu = \Delta\mu_{max} - K_d \frac{\Delta\mu}{[A]} \tag{10}$$

We call this a "Woolf–Hofstee plot" after the investigators who contributed to the development of the corresponding plot in enzyme kinetics.[17] The values of $\Delta\mu$ and $\Delta\mu/[A]$ for each run can be easily calculated with spreadsheet software by providing t, t_r, and [A] as input data.

Divalency

The lectin is assumed to be monovalent in the above-described treatments. In the case of divalent lectins, two monoliganded affinophores can simultaneously bind to the lectin. An identical analysis is applicable to such dimeric lectins under the following conditions: (1) the two binding sites are equivalent; (2) the two binding sites are independent; and (3) the mobility change induced by the binding of the second affinophore is identical to that induced by the first one.[12] In the case of a homodimeric lectin, the two binding sites are apparently equivalent. The independency of the binding sites cannot be known in advance. It will, however, be revealed by a linear relation of the plot according to Eq. (10). The mobility change caused by the successive binding of each affinophore can be considered to be approximately identical, because the electrophoretic mobility is proportional to the number of charges on a particle with an identical electrophoretic drag, that is, a negligible change in the molecular size of the lectin on binding with the affinophores. The relatively small molecular size of the monoliganded affinophore in comparison with a protein should warrant this approximation. In the analysis of divalent lectins, it should be noted that the K_d that appears in the equations is that for a single binding site and not for the entire protein.

[17] M. Dixon and E. C. Webb, *in* "Enzymes," 3rd ed., p. 62. Longman, London, 1979.

Affinophoresis

The buffer for the electrophoresis should not have a large electric conductivity. When 0.1 *M* Tris–acetic acid buffer (pH 7.9) is used as in the example described below, the current is about 25 μA at 300 V/cm with a 50 μm i.d. capillary. As a criterion, electric power at less than 1 W/m of capillary length can be considered safe. At 350 V/cm, 30 μA corresponds to about 1 W/m.

An affinophore solution is included only in the capillary (Fig. 2). The volume of the capillary is less than 1 μl and this is the volume of affinophore solution actually consumed in a single run of affinophoresis. Experimentally, however, it would be better to prepare affinophore solutions in a greater volume (about 50 μl) in order to ensure precision in the concentration of the affinophore against the effect of evaporation, as well as possible errors in the use of micropipettes. Cooling of the vial rack is effective in minimizing the evaporation of water. Evaporation depends on the humidity of the surrounding air and overcooling might result in the condensation of water under conditions of high humidity. An overlay of a small quantity of mineral oil has been shown not to interfere with the injection processes and to be effective in suppressing evaporation.[18,19]

A lectin zone formed by injecting a sample does not contain affinophore and there should be a presteady state period before achieving a binding equilibrium between a lectin sample and the affinophore. During this period, the macroscopic mobility of the lectin gradually changes from μ_0 to μ as expressed by Eq. (7). The length of the period is a function of the relative concentration of the lectin to that of affinophore. When this period is not negligibly small, the mobility change of the lectin becomes smaller than would be predicted from Eq. (9), especially at low concentration ranges of affinophore. This results in a downward curvature of the plot according to Eq. (10) in the vicinity of the abscissa. The concentration of the lectin sample should be nearly identical to or lower than that of the affinophore on a molar basis. In the following experiments, lectin solutions of 0.2–1.0 μg/μl are used with absorption detection at 214 nm. This phenomenon might be problematic, when the affinity for an affinophore is high, where considerable binding occurs, even at low affinophore concentration.

As an example, the affinophoresis of pea lectin (*Pisum sativum*) using a mannoside affinophore is shown in Fig. 3. Pea lectin is a homodimeric lectin

[18] N. H. H. Heegaard, J. W. Sen, and M. H. Nissen, *J. Chromatogr. A* **894,** 319 (2000).
[19] K. Shimura, N. Uchiyama, and K. Kasai, *Electrophoresis* **22,** 3471 (2001).

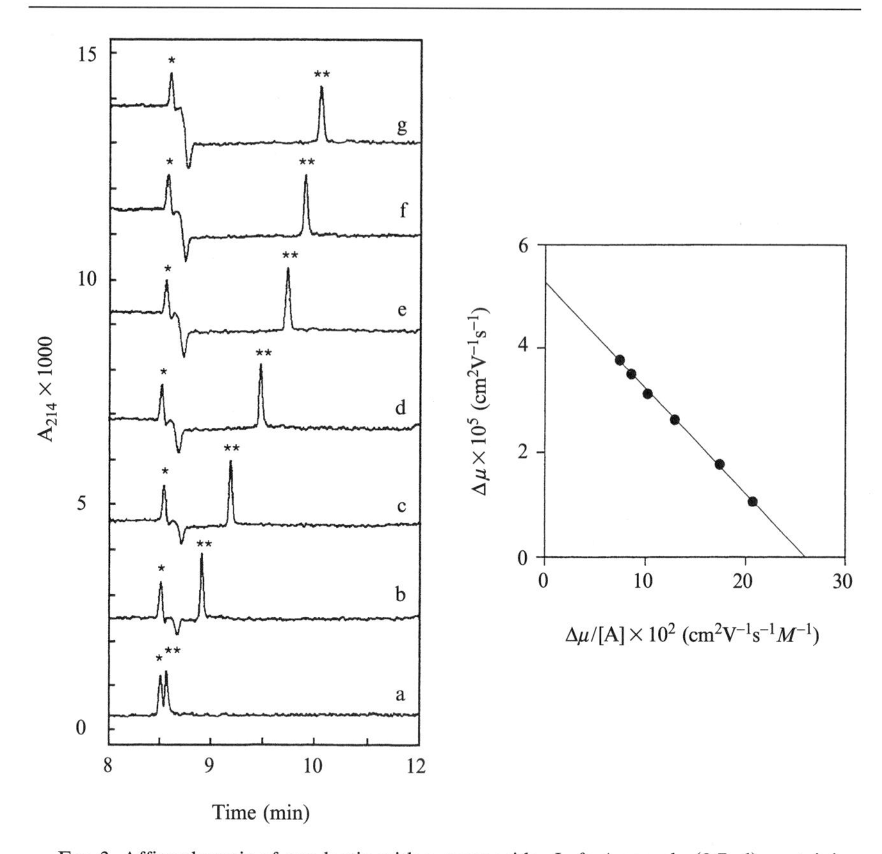

FIG. 3. Affinophoresis of pea lectin with a mannoside. *Left*: A sample (0.7 nl) containing pea lectin (**, 0.2 μg/μl) and cytidine (*, 0.2 m*M*) was injected by means of pressure at the positive end of a capillary (25 μm i.d., 375 μm o.d., 57 cm total length, 50 cm separation distance, and inner-coated with succinylpolylysine) filled with a solution of the affinophore (sGS-AP-Man). Electrophoresis was carried out at a field strength of 350 V/cm (7 μA) at 25°, with detection at A_{214}. Tris–acetate buffer (0.1 *M*, pH 7.9) was used throughout the system. The concentration of the affinophore was (a) 0 m*M*; (b) 0.05 m*M*; (c) 0.1 m*M*; (d) 0.2 m*M*; (e) 0.3 m*M*; (f) 0.4 m*M*; (g) 0.5 m*M*. *Right*: A plot according to Eq. (10) was prepared, using the data shown on the left, in order to determine the K_d and $\Delta\mu_{max}$ values. Reproduced from Ref. 12.

with a molecular mass of 49,000 Da and has two carbohydrate-binding sites.[20] Its structure is closely related to concanavalin A and the affinity constants for mannose and some other carbohydrates have been reported.[20,21]

[20] I. S. Trowbridge, *J. Biol. Chem.* **249,** 6004 (1974).
[21] F. P. Schwarz, K. D. Puri, R. G. Bhat, and A. Sorolia, *J. Biol. Chem.* **268,** 7668 (1993).

The main component of pea lectin has been purified by anion-exchange chromatography and this preparation is used in the experiments. Electrophoresis is carried out with an automated capillary electrophoresis instrument (P/ACE 2210 with a UV detector; Beckman, Fullerton, CA). A fused silica capillary (25 μm i.d., 375 μm o.d.) coated with succinylpolylysine is installed in a capillary cartridge with a separation distance of 50 cm, from the positive end to the detection point. The capillary is filled with an electrophoresis buffer (0.1 *M* Tris–acetic acid buffer, pH 7.9, containing 0.02% NaN_3 as a preservative) with or without the mannoside affinophore (sGS-PA-Man), containing *p*-aminophenyl-α-D-mannoside as an affinity ligand, using a pressure of 20 lb/in^2 for 3 min. The sample, which contains pea lectin (0.2 μg/μl) and 0.2 m*M* cytidine as a neutral marker in the electrophoresis buffer, is injected at the positive end for 10 s under a pressure of 0.5 lb/in^2 (injection volume of 0.67 nl, which corresponds to 0.13 ng of pea lectin). Electrophoresis is carried out at a field strength of 350 V/cm with an electric current of 6.7 μA. The cartridge temperature is set at 25° and pea lectin and cytidine are detected by absorption measurement at 214 nm. The electrophoresis buffer is also used as an electrode solution in each electrode vessel. The solution of pea lectin (20 μl) and that of the affinophore (50 μl) at different concentrations are placed in polypropylene microvials and the vials are set in a cooled (typically 10°) rotating tray to minimize evaporation. After electrophoresis, the capillary is rinsed with 0.1 *M* sodium carbonate buffer (pH 10) containing 1 *M* NaCl and 10 m*M* EDTA in the high-pressure rinse mode for 3 min and then with water for 3 min.

In the absence of affinophore, the mobility of the lectin is low as indicated by the close separation of the peaks from that of the neutral marker, cytidine. The detection time of the lectin gradually increases as the concentration of the affinophore becomes higher, indicating binding of the affinophore. The mobility change, $\Delta\mu$, at each concentration of the affinophore is calculated using Eq. (5), and is plotted against $\Delta\mu/[A]$ according to Eq. (10) (Fig. 3, right). A linear regression provides $-K_d$ as the slope of the line and $\Delta\mu_{max}$ at the intercept with the ordinate, that is, $K_d = 0.202$ m*M* and $\Delta\mu_{max} = 5.25 \times 10^{-5}$ cm^2 V^{-1} s^{-1}. Five identical runs yield an average value for K_d of 0.199 m*M* (coefficient of variation, 6.0%) and $\Delta\mu_{max}$ of 5.35×10^{-5} cm^2 V^{-1} s^{-1} (coefficient of variation, 1.6%).

Competitive Capillary Affinophoresis

The addition of a neutral sugar to the affinophoresis system results in suppression of the mobility change induced by the affinophore. As described in the previous section, a neutral sugar can be added only in the buffer in the positive electrode reservoir. The electroendosmosis in the capillary transports the sugar solution toward the negative end, and a lectin and an affinophore migrate in the neutral sugar zone (Fig. 2). The volume of the positive electrode buffer containing the neutral sugar can be as small as 100 μl. The concentration of an affinophore should be set so as to be between the K_d and $2 \times K_d$. Other conditions should be the same as those used for the determination of K_d and $\Delta\mu_{max}$ of the affinophore.

The detailed experimental conditions for a competitive capillary affinophoresis experiment using a lactoside affinophore are described for the determination of affinity constants of recombinant human galectins for neutral sugars. Recombinant human galectin-1 (rhGal-1) is a dimeric protein composed of two identical 14-kDa polypeptide chains with 134 amino acid residues.[22] The apparent difference between rhGal-1 and the native galectin-1 found in human tissues is the absence of an acetyl group at the N terminus of rhGal-1. The C2S mutant of rhGal-1 has one amino acid substitution at the second residue from the N terminus, from cysteine to serine. This mutation at one of the six cysteine residues found in galectin-1 has been reported to substantially increase the stability of sugar-binding activity of the lectin in the absence of reducing agents.[22]

A fused silica capillary (50 μm i.d., 375 μm o.d.) coated with succinylpolylysine is used, with a separation distance of 20 cm. The electrophoresis instrument and the detection system are the same as those used in the experiments described for pea lectin, and the cartridge temperature is set at 20°. Solutions, 20 μl each of a lactoside affinophore (sGS-AP-Lac) bearing *p*-aminophenyl-β-lactoside as an affinity ligand, a lectin at 1 μg/μl, and references of 1 m*M* acrylamide and 0.3 m*M* acetyltryptophan in the electrophoresis buffer, are placed in polypropylene minivials (200 μl) and overlaid with 20 μl each of mineral oil (light white oil, density = 0.84 g/ml; g/ml; Sigma) to prevent the evaporation of water.[18,19] The capillary is filled with the affinophore solution at a pressure of 20 lb/in^2 for 30 s. A solution of electrophoresis markers and that of the lectin are consecutively injected at the positive end for 1 s each under a pressure of 0.5 lb/in^2 (injection volume of 2.2 nl of each solution). Electrophoresis is carried out at a field strength of 300 V/cm with an electric current of about 25 μA, using a neutral sugar solution of 200 μl in the electrophoresis buffer as the positive electrode solution. After electrophoresis, the capillary is rinsed with water at a pressure of 20 lb/in^2 for 3 min.

The electropherograms shown in Fig. 4 were obtained by the affinophoresis of the C2S variant of rhGal-1 with the lactoside affinophore. The K_d value of the lectin for the affinophore was determined to be

[22] J. Hirabayashi and K. Kasai, *J. Biol. Chem.* **266,** 23648 (1991).

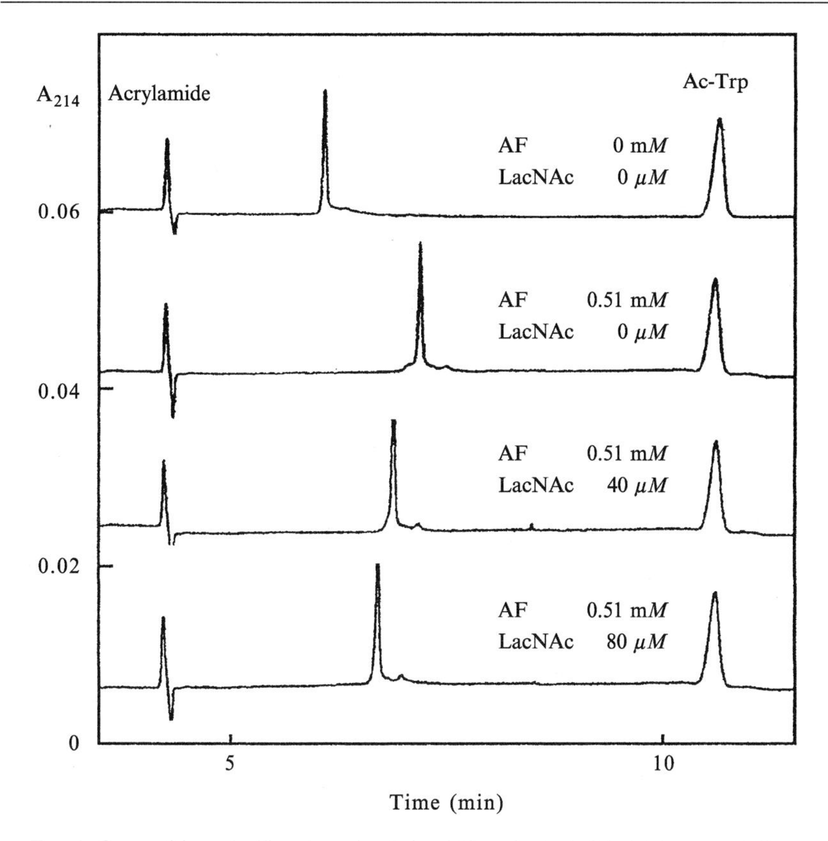

FIG. 4. Competition of affinophoresis of the C2S variant of rhGal-1 by *N*-acetyllactosamine. Affinophoresis of C2S (1 μg/μl) was carried out with the lactoside affinophore (sGS-AP-Lac) at a concentration of 0.51 m*M* in a succinylpolylysine-coated capillary (50 μm i.d., 375 μm o.d., 23 cm total length, 20 cm separation distance) with a field strength of 300 V/cm at 20°. *N*-Acetyllactosamine (LacNAc) was added to the electrophoresis buffer at the anode at the concentrations indicated above each electropherogram. Acrylamide and acetyltryptophan (Ac-Trp) were injected with the lectin sample as electrophoresis markers. Detection was carried out by absorption measurement at 214 nm. Reproduced from Ref. 13 .

0.40 m*M*, with a $\Delta\mu_{\max}$ value of 4.69×10^{-5} cm^2 V^{-1} s^{-1}. The affinophore at a concentration of 0.51 m*M* lengthens the detection time for the lectin by about 1 min (Fig. 4, second electropherogram from top). The addition of *N*-acetyllactosamine to the affinophoresis system cancels the effect of the affinophore, and the detection time is reduced as a function of the concentration of *N*-acetyllactosamine (Fig. 4, third and fourth electropherograms from top).

Calculation of Affinity Constants for Neutral Sugars

The mobility change of the lectin in the presence of the affinophore and the neutral sugar, $\Delta\mu_i$, can be written as follows:

$$\Delta\mu_i = \Delta\mu_{max} \frac{[A]}{K_{d\,app} + [A]} \tag{11}$$

where the $K_{d\ app}$ is the apparent dissociation constant for the affinophore in the presence of the neutral sugar, and has a relation with the dissociation constant, K_i ($K_i = [L][I]/[LI]$), for a neutral sugar, I:

$$K_{d\,app} = K_d\left(1 + \frac{[I]}{K_i}\right) \tag{12}$$

Because $\Delta\mu_{max}$ has already been determined, $K_{d\ app}$ can be calculated by measurement of $\Delta\mu_i$ in the presence of a neutral sugar according to Eq. (11). K_d is also already known by the plot of Eq. (10) and K_i can be calculated from Eq. (12). Alternatively, Eq. (12) can be rearranged as

$$\frac{K_{d\,app}}{K_d} = \frac{[I]}{K_i} + 1 \tag{13}$$

A plot of $K_{d\ app}/K_d$ against different [I] should result in a straight line with a slope of $1/K_i$. It is preferable to use the competing sugar at concentrations around its K_i value.

The results of competition experiments with some neutral sugars in the affinophoresis of pea lectin with the mannoside affinophore are plotted according to Eq. (13) (Fig. 5). The reciprocal of the slope of the lines corresponds to the dissociation constant K_i for the lectin–neutral sugar complexes. The values are compared with those determined by calorimetry[21] (Table I). The affinophoresis is carried out at 25° but some of the calorimetry values are determined at different temperatures. A close agreement is found for methyl-α-D-mannoside and D-mannose, which are determined at identical temperatures. Other calorimetry values are determined at 16 or 14° and are considerably smaller than those obtained by affinophoresis. This discrepancy can be largely attributed to the difference in temperature used in the different determinations and the positive ΔH° values for the dissociation of the pea lectin–sugar complexes, that is, 10 and 13.6 kJ/mol kJ/mol for D-glucose and methyl-α-D-glucoside, respectively.[21]

The dissociation constants of recombinant human galectin-1 (rhGal-1), its C2S variant,[22] and recombinant human galectin-3 (rhGal-3)[23,24] for

[23] Y. Oda, H. Leffler, Y. Sakakura, K. Kasai, and S. H. Barondes, *Gene* **99,** 279 (1991).

[24] J. Hirabayashi, Y. Sakakura, and K. Kasai, *in* "Lectins and Glycobiology" (H.-J. Gabius and S. Gabius, eds.), p. 474. Springer-Verlag, New York, 1993.

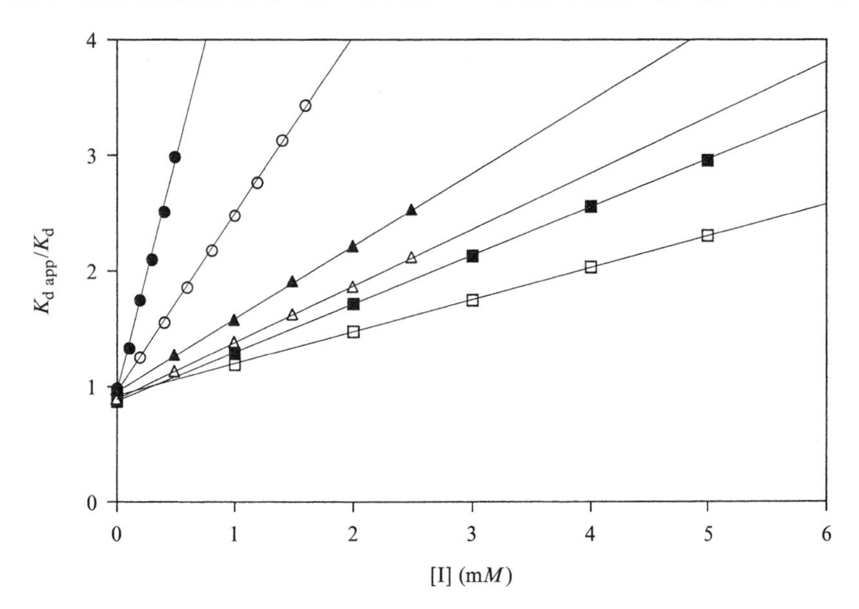

FIG. 5. Determination of the dissociation constants of pea lectin for neutral sugars (I) through competition experiments of affinophoresis using the mannoside affinophore. The results are plotted according to Eq. (13), (●) *p*-Aminophenyl-α-D-mannoside; (○) methyl-α-D-mannoside; (▲) D-mannose; (△) maltose; (■) methyl-α-D-glucoside; (□) D-glucose. Reproduced from Ref. 12 .

various neutral sugars, as determined by capillary affinophoresis, are summarized in Fig. 6. The striking preference of rhGal-1 and its C2S variant for *N*-acetyllactosamine [Gal(β1–4)GlcNAc] and lactose [Gal(β1–4)Glc] is clear, as has been repeatedly reported.[25,26] The affinity of the two lectins for *N*-acetyllactosamine, lactose, and galactose is hardly distinguishable, indicating that the effect of the substitution of the second cysteine residue of rhGal-1 to a serine residue on affinity is negligible, as has been previously noted.[22] The most relevant work that can be compared with the present results for rhGal-1 is that reported by Lee *et al.*[26] They have reported the concentrations, I_{50}, for various sugars that cause a 50% reduction in the binding of ^{125}I-labeled galectin-1 to lactose-immobilized agarose beads. The values of I_{50}, with relative values of $1/I_{50}$ to lactose in parentheses, are as follows: *N*-acetyllactosamine, 0.13 mM (350%); lactose, 0.46 mM (100%); methyl-α-galactoside, 26 mM (1.8%); melibiose [Gal(α1–6)Glc],

[25] C. P. Sparrow, H. Leffler, and S. H. Barondes, *J. Biol. Chem.* **262,** 7383 (1987).
[26] R. T. Lee, Y. Ichikawa, H. J. Allen, and Y. C. Lee, *J. Biol. Chem.* **265,** 7864 (1990).

TABLE 1
AFFINITY CONSTANTS OF PEA LECTIN FOR NEUTRAL SUGARS

Sugar	Affinophoresis[a] (mM)	Calorimetry[b]
p-Aminophenyl-α-D-mannoside	0.25	Not determined
Methyl-α-D-mannoside	0.65	0.61 mM[c]
D-Mannose	1.6	1.3 mM[d]
Maltose	2.1	Not determined
Methyl-α-D-glucoside	2.4	1.3 mM[e]
D-Glucose	3.6	2.8 mM[f]

[a] 0.1 *M* Tris–acetic acid buffer (pH 7.9) containing 0.02% NaN_3, 25° (Ref. 12).
[b] 0.02 *M* sodium phosphate buffer (pH 7.4) containing 0.15 *M* NaCl (Ref. 21).
[c] 25°.
[d] 24.3°.
[e] 16°.
[f] 14.2°.

30 m*M* (1.5%); galactose, 63 m*M* (0.7%) and methyl-β-galactoside, 85 m*M* (0.5%). The K_i values of C2S and its relative affinities determined by affinophoresis are as follows, in the same order as above: *N*-acetyllactosamine, 0.067 m*M* (352%); lactose, 0.23 m*M* (100%); methyl-α-galactoside, 16 m*M* (1.5%); melibiose, 21 m*M* (1.1%); galactose, 30 m*M* (0.78%); methyl-β-galactoside, 39 m*M* (0.6%). Although the I_{50} values are about twice as large as the K_i values determined by capillary affinophoresis, the consistency of the relative magnitude of the affinity is remarkable. When lactose-immobilized beads are used at a concentration equivalent to the K_d value for the lectin and the modified beads, as was the case in their experiments,[26] the I_{50} value for each sugar should be twice as large as the corresponding K_i value. This theory may simply explain the difference, although there are some additional factors to be considered, that is, the divalency of the lectin in the interaction with the lactose-immobilized beads and the difference in temperature between their experiments, 25°, and ours, 20°.

rhGal-3 also has an exclusively high affinity for *N*-acetyllactosamine and lactose. rhGal-3 was shown to have a 2 to 10 times higher affinity than rhGal-1 for sugars containing galactose at their nonreducing ends, whereas the affinities for sucrose [Glc(α1-β2)Fru] are almost the same. This result indicates a higher preference of galectin-3 for galactosides than galectin-1. The preference for methyl-α-galactoside over methyl-β-galactoside is not observed for rhGal-3. Sparrow *et al.* report I_{50} values for a variety of sugars

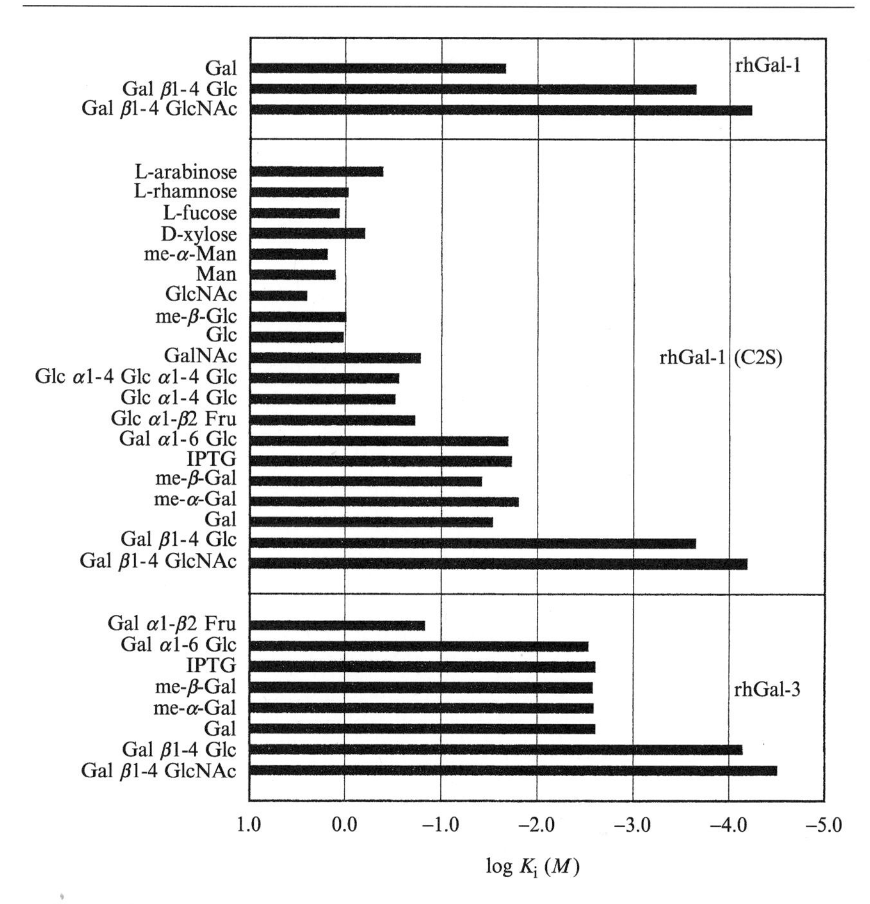

FIG. 6. Dissociation constants of recombinant human galectins for neutral sugars determined by capillary affinophoresis. The K_i values (M) are the dissociation constants for a single binding site of lectins. rhGal-1, Recombinant human galectin-1; C2S, a C2S variant of rhGal-1; rhGal-3, recombinant human galectin-3; Gal, D-galactose; Man, D-mannose; Glc, D-glucose; Fru, D-fructose; GalNAc, *N*-acetyl-D-galactosamine; GlcNAc, *N*-acetyl-D-glucosamine; IPTG, isopropyl-β-D-thiogalactoside; me-, methyl. Original data were taken from Ref. 13 .

and oligosaccharides in the binding of human galectin-3 (called HL-29 at that time) to asialofetuin–agarose beads.[25] A portion of the data that are relevant to our results is as follows (with a relative value of $1/I_{50}$ to lactose in parentheses): lactose (100%), *N*-acetyllactosamine (1130%), galactose (1.8%), methyl-α-galactoside (2.6%), and methyl-β-galactoside (1.6%). Overall agreement in the relative affinity data based on the I_{50} values and K_i values of our determinations by capillary affinophoresis is apparent, although the affinity for *N*-acetyllactosamine has been emphasized in their data.

Divalency Revisited

The monoliganded affinophore was developed to circumvent difficulty related to the divalent nature of lectins. After the determination of affinity constants for neutral sugars had been established by capillary affinophoresis, we attempted a new approach for estimating the degree of contribution of divalent interactions between lectins and polyliganded matrices. The interactions between a divalent pea lectin and polyliganded affinophores, succinylpolylysine bearing a *p*-aminophenyl-α-D-mannoside at different ligand densities, were investigated.[11] The overall dissociation constant between the lectin and the affinophore was determined by capillary affinophoresis to show a higher affinity for the affinophore with higher ligand densities. The overall interactions should comprise both monovalent and divalent interactions. When a neutral sugar, methyl α-D-mannoside, interferes with these interactions, the contribution of divalent interactions should decrease more rapidly than monovalent interactions. The reason for this is that one binding site is already occupied with the neutral sugar, and the lectin should behave as a monovalent molecule. The ratio of the temporary monovalent lectins to the divalent lectins with two free binding sites increases with increasing degree of saturation by the neutral sugar.

When affinophoresis using a polyliganded affinophore was interfered with by a neutral sugar, the plot according to Eq. (13) became nonlinear with a gradual increase in the slope with increasing concentration of neutral sugar, indicating the decreasing contribution of divalency[11]: the more curvature, the higher the contribution of divalent interactions. Because the K_i value of the neutral sugar had already been determined, nonlinear regression allows the determination of two elemental equilibrium constants, one for the monovalent binding ($K_1 = [L][A]/[LA]$, where LA denotes the monovalent complexes) and the other for the divalent interactions ($K_2 = [LA]/[L{=}A]$, where L = A denotes the divalent complexes). The divalency value, representing the mole fraction of the divalent species in the complex, can be calculated from K_2 values. Polyliganded affinophores bearing a mannoside ligand at different ligand densities from 4.2 to 17.5% of mannoside over a succinyllysine residue were found to interact with pea lectin with divalency values of 24 to 62%.[10] This principle, based on competition with a monovalent ligand, should be applicable to the analysis of interactions with polyliganded cell surface or extracellular matrices in conjunction with a wide variety of techniques for analyzing molecular interactions.

Conclusion

Capillary affinophoresis, performed with an automated electrophoresis instrument, is a useful technique for determining the affinity constants of divalent lectins for neutral sugars. The affinity constants obtained are strictly for a single binding site and are free from the components related to multivalent interactions. These values should be useful in comparing affinities of various lectins and sugars. Once a suitable affinophore is established for a target lectin, the method can be carried out with a relatively small amount of lectin sample and time for a variety of sugars with good precision and reproducibility.

[29] Describing Topology of Bound Ligand by Transferred Nuclear Overhauser Effect Spectroscopy and Molecular Modeling

By HANS-CHRISTIAN SIEBERT, JESÚS JIMÉNEZ-BARBERO, SABINE ANDRÉ, HERBERT KALTNER, and HANS-JOACHIM GABIUS

Introduction

The perspective needed to translate progress in unraveling the role of protein–carbohydrate interactions in clinically relevant processes (e.g., initiation of infections by pathogen attachment or lectin-dependent endocytosis) into therapeutic concepts is one reason for the structural analysis of lectin–ligand complexes.[1–4] Despite the exquisite sensitivity of powerful X-ray diffraction techniques, concerns about whether the crystal structures fully reflect structural details in solution cannot be ignored.[5,6] To address this issue, nuclear magnetic resonance (NMR) spectroscopy can provide salient input on ligand topology even without meeting the requirement for a complete signal assignment. This is especially helpful when the size

[1] Y. C. Lee and R. T. Lee, *J. Biomed. Sci.* **3,** 221 (1996).
[2] K.-A. Karlsson, *Mol. Microbiol.* **29,** 1 (1998).
[3] H. Rüdiger, H.-C. Siebert, D. Solís, J. Jiménez-Barbero, A. Romero, C.-W. von der Lieth, T. Díaz-Mauriño, and H.-J. Gabius, *Curr. Med. Chem.* **7,** 389 (2000).
[4] N. Yamazaki, S. Kojima, N. V. Bovin, S. André, S. Gabius, and H.-J. Gabius, *Adv. Drug Deliv. Rev.* **43,** 225 (2000).
[5] G. Wagner, S. G. Hyberts, and T. F. Havel, *Annu. Rev. Biophys. Biomol. Struct.* **21,** 167 (1992).
[6] M. W. MacArthur, P. C. Driscoll, and J. M. Thornton, *Trends Biotechnol.* **12,** 149 (1994).

0076-6879/03 $35.00

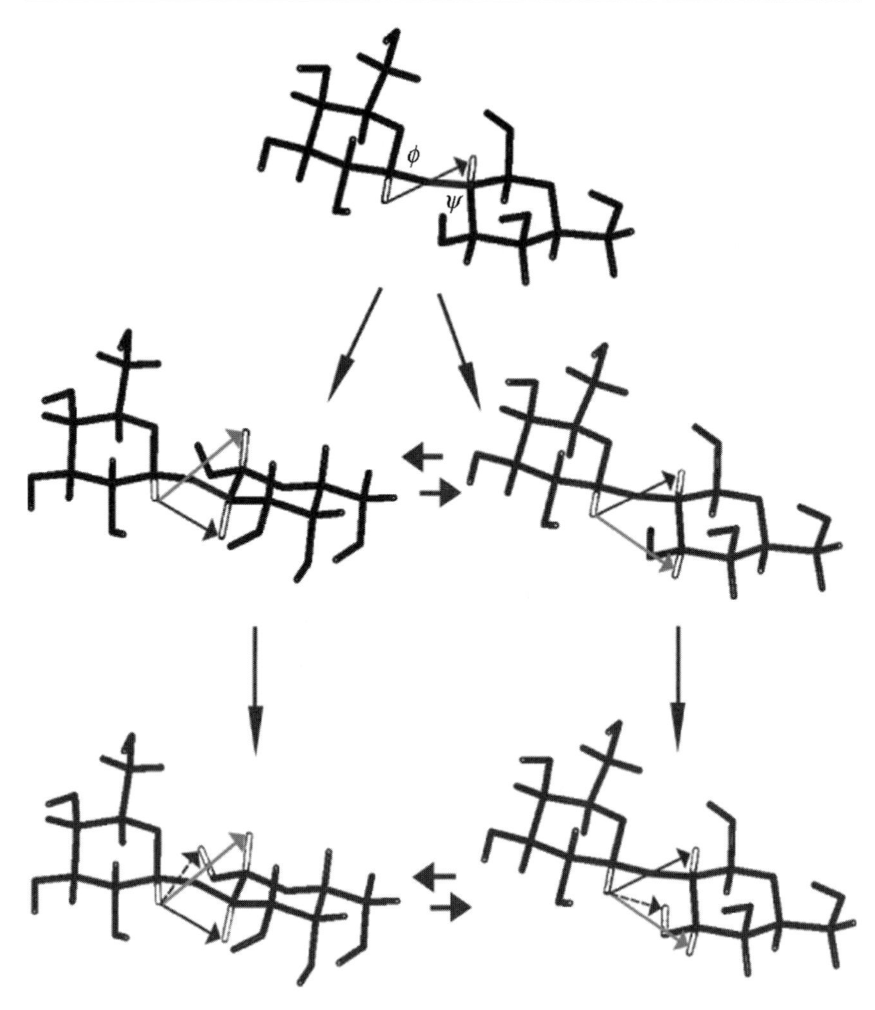

FIG. 1. Illustration of the impact of experimentally accessible distance constraints of a lectin ligand (Galβ1,2Gal) by monitoring through-space magnetization transfer on defining the conformation of the ligand. The number of ϕ,ψ-angle combinations of the glycosidic bond of the given disaccharide that can theoretically satisfy one measured interproton distance will be reduced by additional interresidual nuclear Overhauser effects (NOEs) in the free state and, if the ligand is bound to a lectin, transferred nuclear Overhauser effects (trNOEs). It is evident that a single interresidual contact is not sufficient for reliable definition of a distinct set of ϕ,ψ angles of the glycosidic bond (see also Fig. 3c). However, combined with molecular modeling such a result can be a sensitive indicator for occurrence of selection of cognate ligand topology from distinct conformational families (see also Fig. 4 for prediction of exclusive NOEs). Simultaneously measuring more than one contact will stepwisely limit the conformational space of the disaccharide (see also Fig. 3c for illustration of the consequences of being able to work with two contacts in the ϕ,ψ,E plot). Addition of water-sensitive

of the protein precludes attaining this aim with currently available technology. In combination with two-dimensional (2D) NMR spectroscopy of the free ligand, transferred nuclear Overhauser effect (trNOE) experiments are a rich source from which to collect data on the bound-state conformation of a ligand.[7,8]

As outlined in detail in this series previously,[9] this type of analysis depends on rapid exchange of the ligand between bound and free states within the relaxation time scale. In principle, this approach then extends the range of information beyond that made available by measuring the nuclear Overhauser effect (NOE) of free ligands. Its analytical character can be compared to a molecular ruler defining proton pairs in contact and the distance between the two protons (Fig. 1). The detection of a certain (exclusive) signal, as commented on below in the sections on combined application of molecular modeling, NMR spectroscopy, and ligand docking, can be squared with occurrence of a distinct low-energy conformation, like a fingerprint. In fact, the systematic monitoring of this through-space magnetization transfer was crucial to reveal an often limited flexibility of saccharide structures in solution.[10–16] In other words, a glycan epitope might be rather rigid or oscillate between few, energetically preferred conformations. The precision with which the relative positions of the two pyranose rings connected by a glycosidic bond (defined in space by its dihedral angles ϕ and ψ) can be described definitely hinges on the number of detected interresidual contacts (Fig. 1). Instances in which two or even more interresidual contacts can be picked up are a boon for

[7] A. A. Bothner-By and R. Gassend, *Ann. N. Y. Acad. Sci.* **222,** 668 (1973).

[8] F. Ni, *Prog. NMR Spectrosc.* **26,** 517 (1994).

[9] L. Y. Lian, I. L. Barsukov, M. J. Sutcliffe, K. H. Sze, and G. C. K. Roberts, *Methods Enzymol.* **239,** 657 (1994).

[10] O. Jardetzky, *Biochim. Biophys. Acta* **621,** 227 (1980).

[11] J. P. Carver, *Pure Appl. Chem.* **65,** 763 (1993).

[12] H.-J. Gabius, *Pharmaceut. Res.* **15,** 23 (1998).

[13] A. Poveda and J. Jiménez-Barbero, *Chem. Soc. Rev.* **27,** 133 (1998).

[14] J. Jiménez-Barbero, J. L. Asensio, F. J. Cañada, and A. Poveda, *Curr. Opin. Struct. Biol.* **9,** 549 (1999).

[15] A. Imberty and S. Pérez, *Chem. Rev.* **100,** 4567 (2000).

[16] T. Peters, *in* "Carbohydrates in Chemistry and Biology," Part I: "Chemistry of Saccharides" (B. Ernst, G. W. Hart, and P. Sinaÿ, eds.), Vol. 2, p. 1003. Wiley-VCH, Weinheim, Germany, 2000.

hydroxyl protons to the common C–H sensors (*top* and *middle*) increases the chances to track down new contact(s) (*bottom*). Molecular modeling will reveal whether the contacts can be accommodated into a distinct "real" conformation or should be assigned to independent conformational states, because no single conformation satisfies the measured constraints ("virtual" conformation).[10,11]

limiting the graphic representation of the conformational space of the ligand to actually populated area segments (Fig. 1). In a broader context, the preferential occupancy of certain sections of the conformational space of a di- or oligosaccharide engenders notable consequences for the thermodynamics of ligand binding.[3,12] The selection of preformed bioactive conformations helps in avoiding a severe entropic penalty in the energetic balance sheet.

Principle of trNOE Measurements

The trNOE spectroscopy (trNOESY) experiment is a regular NOE spectroscopy (NOESY) experiment for a receptor–ligand system in the presence of a molar excess of ligand. As schematically shown in Fig. 2, on binding of the ligand some of its own protons are positioned in spatial vicinity for NOE contacts. The cross-relaxation rates of the bound molecule are negative, thus being subject to a change in sign. In addition to alterations in positions of chemical shifts or in line broadening of ligand resonances by the binding to the lectin the appearance of negative cross-peaks from the carbohydrate in trNOESY spectra is definitive evidence for the interaction process. The intensity of the respective NOE signals reflects the distance (≤ 3.5 Å) to be covered by the transfer process in a proton pair with a $1/r^6$ relationship. The build-up curves of the trNOEs can further be processed quantitatively in a full matrix relaxation approach with a relaxation matrix comprising the properties of the free and bound states of the ligand. The full relaxation matrix is a function of the correlation times and of the interproton distances of the corresponding state. In addition, a kinetic matrix defines the exchange process, which includes the molar fractions of free and bound ligand and the exchange rates between the two states. The diagonal and off-diagonal elements of the relaxation matrix are affected by these exchange contributions. In this manner, a quantitative analysis of the observable cross-relaxation rates and their translation into accurate interproton distances for the bound state are feasible.[8]

Before starting to draw conclusions from the data of this analysis, it is noteworthy that the trNOE experiment is not devoid of inherent problems, one of them being the presence of resonances from protein protons in the spectrum. Depending on the size of the protein, these proton signals can be selectively removed by using T_2 or $T_{1}\rho$ filters.[17,18] Because magnetization transfer can proceed ("diffuse") to a third proton, establishing a three-spin

[17] C. P. Glaudemans, L. Lerner, G. D. Daves, Jr., P. Kovac, R. Venable, and A. Bax, *Biochemistry* **29,** 10906 (1990).

[18] T. Scherf and J. Anglister, *Biophys. J.* **64,** 754 (1993).

system, this spin diffusion process that is typical for large molecules adds signals to the trNOESY spectrum produced by a transfer pathway.[19] In this case, apart from direct enhancements in a proton pair close in space, the typical and negative cross-peaks originate from an indirect connection via "silent" mediators. These peaks have the same sign as those coming from direct contacts and cannot be distinguished from them without adequate precautions. Thus, a trNOESY signal from such an indirect resonance process may lead to the calculation of a conformation incorrectly referred to as bioactive. As indicated above, it is possible to take spin diffusion into account by using a quantitative analysis based on the extended full relaxation matrix approach. Other effective techniques are the combination of trNOESY data with trROESY (transferred rotating-frame Overhauser effect spectroscopy), MINSY (mixing irradiation during NOESY), and/or QUIET (quenching undesirable, indirect external trouble)-NOESY experiments.[20–22] In trROESY, as in the regular ROESY set-up,[23–25] direct cross-peaks are always positive, that is, they show the opposite sign relative to the diagonal signals (Fig. 2). Spin diffusion via one relay proton leads to a negative cross-peak that can now easily be distinguished from direct interactions. In trROESY experiments usually more than one relay proton from the binding site of the protein is involved and the respective cross-peak is no longer detectable. Explicitly, if a cross-peak is observed in the trNOESY spectrum and no cross-peak appears in the trROESY spectrum, the respective cross-peak in the trNOESY spectrum will most likely be caused by spin diffusion (Fig. 2). Therefore, it has become common practice for an analysis of bound-state ligand topology to be performed as an integrated set of trNOESY/trROESY. In addition, stepwise variation of mixing times helps to pinpoint the emergence of cross-peaks by relatively time-consuming three-spin processes.

By taking these considerations into account, information about bioactive conformations of carbohydrate ligands can be obtained in solution. The combination with molecular modeling, as briefly indicated above, has

[19] A. A. Bothner-By and J. H. Noggle, *J. Am. Chem. Soc.* **101,** 5152 (1979).

[20] S. R. Arepalli, C. P. J. Glaudemans, G. D. Daves, P. Kovac, and A. Bax, *J. Magn. Reson. B* **106,** 195 (1995).

[21] W. Massefsky and A. G. Redfield, *J. Magn. Reson.* **78,** 150 (1988).

[22] S. J. Vincent, C. Zwahlen, C. B. Post, J. W. Burgner, and G. Bodenhausen, *Proc. Natl. Acad. Sci. USA* **94,** 4383 (1997).

[23] A. A. Bothner-By, R. L. Stephens, J.-M. Lee, C. D. Warren, and R. W. Jeanloz, *J. Am. Chem. Soc.* **106,** 811 (1984).

[24] A. Bax and D. G. Davis, *J. Magn. Reson.* **63,** 207 (1985).

[25] H. Kessler, C. Griesinger, R. Kerssebaum, K. Wagner, and R. R. Ernst, *J. Am. Chem. Soc.* **109,** 607 (1987).

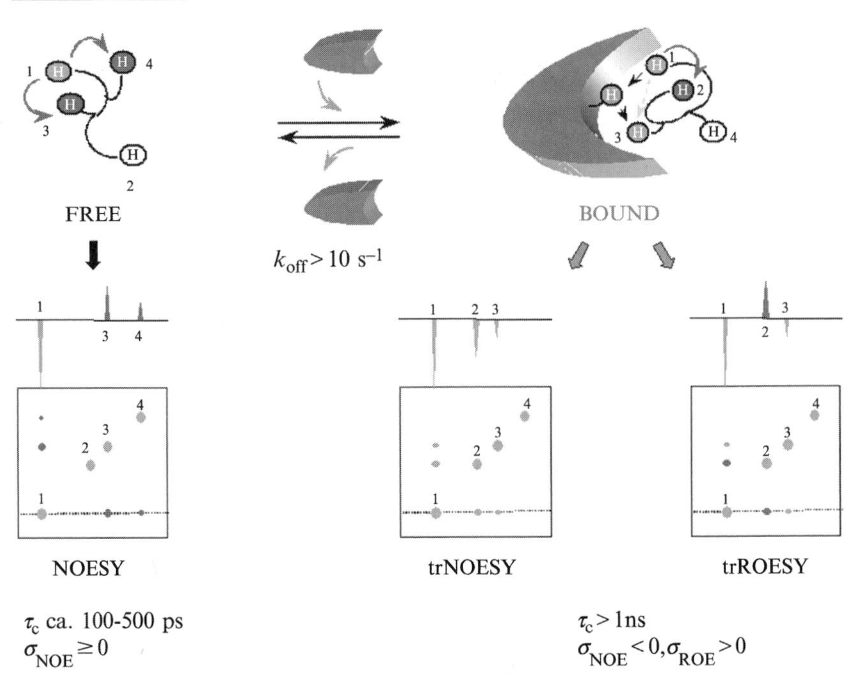

FIG. 2. Schematic explanation of principles of trNOESY/trROESY experiments for a ligand with a proton H1 (gray) in spatial vicinity to protons H3/H4 (black). In the free state (*left*), NOE cross-peaks for a small ligand molecule with fast tumbling rate are positive or close to zero and, therefore, the cross-peaks have opposite sign to the diagonal peaks. The intensity of the cross-peaks can be related to the distance (r) between the corresponding cross-relaxing protons (H1 is closer in space to H3 than to H4 and, therefore, the corresponding cross-peak is stronger; H2 is far from H1 and does not produce any cross-peak) with a $1/r^6$ relationship. For a ligand–receptor system in chemical exchange, in which the ligand is present in excess, there will be a fraction of bound ligand molecules. Its size will depend on the association constant. In the bound state, the ligand is accommodated into the binding site of the macromolecule, and its tumbling rate and dipolar relaxation properties are markedly affected. Provided that the off-rate is fast in the relaxation time scale, trNOESY cross-peaks might be picked up at the resonance frequencies of the free ligand. These cross-peaks will be negative (gray, same sign as the diagonal peaks), because the effective correlation time is now in the nanosecond region. Unfortunately, spin diffusion processes, mediated either by protons of the protein (H1→HP→H3) or by other protons of the ligand itself, can be operative under these conditions and may lead to spurious cross-peaks (H1/H3). Therefore, the use of trROESY experiments is recommended. Under spin-locking conditions and independent of the correlation time of the molecule or molecular complex, direct cross-peaks (H1/H2) are always positive and indirect signals mediated by a third spin are negative (H1/H3). The peaks mediated by a fourth or a fifth spin are almost negligible.

proved valuable to correlate experimental conformational constraints with the topology of the ligand at computationally calculated low-energy levels, to dock the ligand in the derived topology into binding sites of lectins, and to evaluate the extent of molecular fit including estimations of ensuing changes in enthalpy and entropy. Details first on data acquisition by NMR spectroscopy and then on the interplay between experimental and computational methods are outlined below.

Practice of NOESY/ROESY Experiments

In the first step, complete assignment of chemical shifts for the protons of the ligand is carried out by standard NMR techniques including correlated, total correlated, and relayed coherence transfer spectroscopy. By repeatedly dissolving the saccharide compound in D_2O followed by lyophilization in each cycle, H_2O contamination will be avoided. For NOESY/ROESY monitoring, 500-MHz ^{1}H NMR spectra are recorded, and the assignment facilitates defining the proton pairs of the interresidual contacts shown in Fig. 3a. The tested disaccharide interacts specifically with an intestinal galectin (CG-14), a member of a large family of animal lectins involved in regulation of cell proliferation/apoptosis, cell–cell (matrix) interactions, and mediator release (for further information about galectins, see article elsewhere in this volume[25a]). As initially shown in the case of a galectin with Galβ1,2Galβ1,R as ligand (see Fig. 1 for illustration of the two bioactive conformations),[26] interresidual contacts (in this case two respective cross-peaks) are recorded (Fig. 3b). When starting work on less well-studied lectins, it is generally helpful to perform systematic studies at 360 MHz to determine optimal conditions for the acquisition of experimental information with the lectin–ligand mixture. The way in which changes in the molar sugar (trimannoside)–protein (C-type lectin domain of rat mannan-binding lectin) ratio from 12:1 to 6:1, and in the temperature from 35° and 15° to 5°, pave the way toward intense and negative trNOESY peaks has been graphically described.[27] Stability and solubility of the protein are key factors for completing such a series.

After setting these parameters (temperature, molar ligand excess relative to the receptor) recording spectra at increasing mixing times (e.g., 25,

[25a] H. Lahm, S. André, A. Hoeflich, J. R. Fischer, B. Sordat, H. Kaltner, E. Wolf, and H.-J. Gabius, *Methods Enzymol.* **362,** [20] (2003) (this volume).

[26] H.-C. Siebert, M. Gilleron, H. Kaltner, C.-W. von der Lieth, T. Kozár, N. V. Bovin, E. Y. Korchagina, J. F. G. Vliegenthart, and H.-J. Gabius, *Biochem. Biophys. Res. Commun.* **219,** 205 (1996).

[27] E. W. Sayers and J. H. Prestegard, *Biophys. J.* **82,** 2683 (2002).

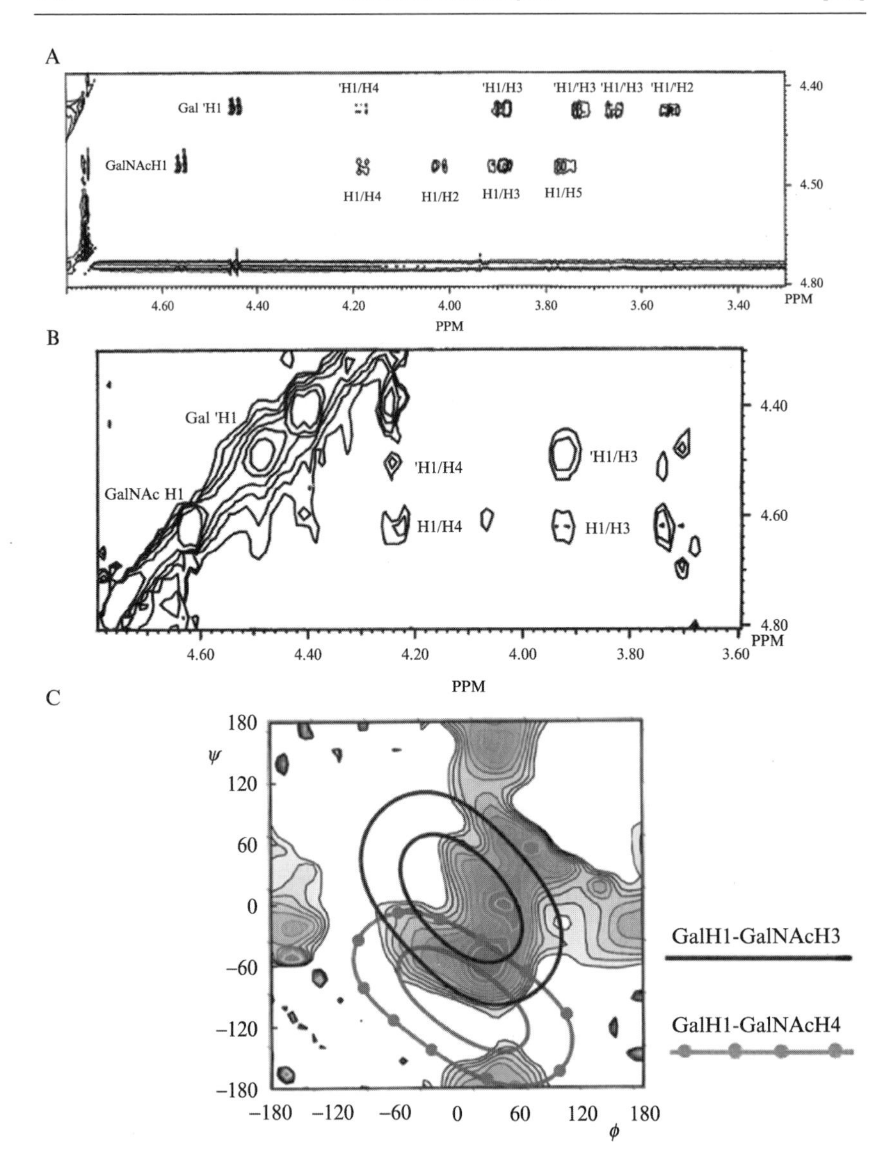

FIG. 3. Relevant parts of a two-dimensional (2D) ROESY spectrum of Galβ1,3GalNAc in D_2O at 500 MHz and 303 K with interresidual cross-peaks (A), a 2D trNOESY spectrum of this disaccharide in the presence of an avian galectin (CG-14; molar ratio of ligand to galectin monomer, 10:1) under identical conditions and mixing time of 200 ms (B), and the distance map introducing the measured constraints from the two interresidual contacts into the ϕ,ψ, E plot, based on previously presented data from molecular mechanics calculations with the dielectric constant ε set to 4 (C; with help from C.-W. von der Lieth, Heidelberg).[28] An area of overlap for the two pairs of contour lines lies within the central low-energy valley (see also Fig. 1).

50, 75, and 100 ms) helps identify spin diffusion events. Two examples of how spin diffusion effects show up in such experiments and what build-up curves of NOESY cross-peaks with/without differences in transverse relaxation of the bound ligand can tell us about interproton distances and the levels of immobilization of individual pyranose units by entering the binding site of a lectin have been provided for a mannose- and a galactose-binding lectin, respectively.[27,28] The route from measuring the distance constraint by position and intensity of the trNOE signal (Fig. 3b) to proposing a ligand conformation will now be sketched for the galactoside: the random walk molecular mechanics protocol (for program details, see Kozár and von der Lieth[29]) for correlating ϕ,ψ-angle combinations of the disaccharide with energy levels has been applied to draw a "topological" map (Fig. 3c). In combination with distance mapping, which introduces the inevitably averaged experimental NMR information into the ϕ,ψ,E plot, the conformational space of the ligand is weighed against two parameters: (1) experimentally determined distance range of the two distinct protons in contact per cross-peak and (2) relative position of the respective ϕ,ψ-angle combinations in the energy contour profile.[30–32] The way these two parameter sets give shape to the conformational space of the bound ligand is illustrated in Fig. 3c. In the next section, the potential of this approach to combine experimental and computational protocols is further underscored. Modeling is extended to ligand docking into the combining site of a galectin. In detail, we focus on galectin-1, a prototype galectin as CG-14,[33,34] and present information about its conformer selection when binding lactose. A comparison to the properties of *C*-lactose is added because of its intriguing conformational properties and the pharmacological advantage to be nonhydrolyzable.[35]

[28] M. Gilleron, H.-C. Siebert, H. Kaltner, C.-W. von der Lieth, T. Kozár, K. M. Halkes, E. Y. Korchagína, N. V. Bovin, H.-J. Gabius, and J. F. G. Vliegenthart, *Eur. J. Biochem.* **252,** 416 (1998).

[29] T. Kozár and C.-W. von der Lieth, *Glycoconj. J.* **14,** 925 (1997).

[30] L. Poppe, C.-W. von der Lieth, and J. Dabrowski, *J. Am. Chem. Soc.* **112,** 7762 (1990).

[31] H.-C. Siebert, G. Reuter, R. Schauer, C.-W. von der Lieth, and J. Dabrowski, *Biochemistry* **31,** 6962 (1992).

[32] C.-W. von der Lieth, H.-C. Siebert, T. Kozár, M. Burchert, M. Frank, M. Gilleron, H. Kaltner, G. Kayser, E. Tajkhorshid, N. V. Bovin, J. F. G. Vliegenthart, and H.-J. Gabius, *Acta Anat.* **161,** 91 (1998).

[33] H. Kaltner and B. Stierstorfer, *Acta Anat.* **161,** 162 (1998).

[34] H.-J. Gabius, *Biochimie* **83,** 659 (2001).

[35] J. Jiménez-Barbero, J.-F. Espinosa, J. L. Asensio, F. J. Cañada, and A. Poveda, *Adv. Carbohydr. Chem. Biochem.* **56,** 235 (2001).

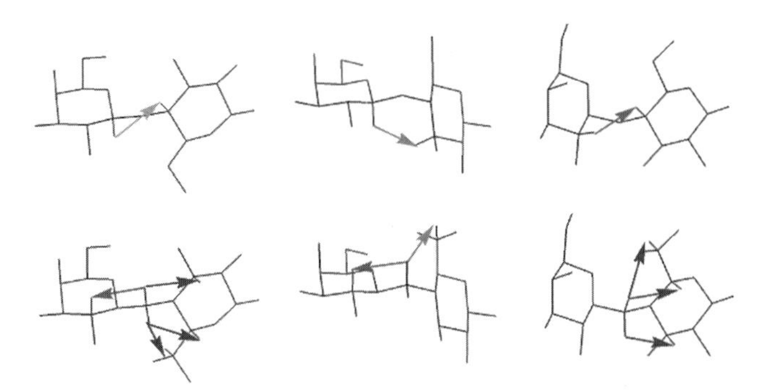

FIG. 4. Illustration of the exclusive NOE contacts in the three conformational families for *O*/*C*-lactose. *Top*: The *syn*-ϕ (H1 Gal/H4 Glc), *anti*-ψ (H1 Gal/H3 Glc), and *anti*-ϕ (H2 Gal/H4 Glc) conformers are shown from left to right with their crucial short-distance contacts depicted by arrows. *Bottom*: The additional contacts in C-lactose between protons of the methylene bridge and the pyranose rings.

Combined Application of Molecular Modeling, NMR Spectroscopy, and Ligand Docking

The populated conformational space of the disaccharide lactose (Galβ1,4Glc) is limited to about 12% of the total ϕ,ψ-defined energy surface based on calculations with MM3*($\varepsilon = 80$) as integrated in MACROMODEL as described.[36] The three conformational families comprise the *syn* configuration (ϕ, ψ: 54°, 18°) which is preferentially populated (about 90%), the *anti*-ψ configuration (ϕ, ψ: 36°, 180°) and the only slightly populated *anti*-ϕ configuration (ϕ, ψ: 180°, 0°), allowing rapid fluctuations of about ±20° around the given angle combinations.[37] Each topological arrangement is characterized by an exclusive NOE contact, initially referred to as such by Dabrowski *et al.*[38] These contacts are compiled in the top portion of Fig. 4. The occurrence of a respective signal predicted from modeling is thus a fingerprint-like indication for a population of the respective area in the ϕ,ψ,E plot as alluded to in the Introduction and in Fig. 1. The presence of an H1 Gal/H4 Glc trNOE signal would thus strongly argue in favor of selection of the *syn* conformer by binding (Fig. 4).

As shown by systematic studies with lactose, its β1,3-isomer, and β-galactosylxylosides and illustrated in Fig. 5, the β-galactosides indeed

[36] J. F. Espinosa, F. J. Cañada, J. L. Asensio, M. Martin-Pastor, H. Dietrich, M. Martin-Lomas, R. R. Schmidt, and J. Jiménez-Barbero, *J. Am. Chem. Soc.* **118,** 10862 (1996).

[37] J. L. Asensio and J. Jiménez-Barbero, *Biopolymers* **35,** 55 (1995).

[38] J. Dabrowski, T. Kozár, H. Grosskurth, and N. E. Nifant'ev, *J. Am. Chem. Soc.* **117,** 5534 (1995).

adopt this conformation when interacting with galectin-1.[39,40] Molecular docking analysis can gauge whether and why this selection takes place. Using the complete set of coordinates of the bovine galectin-1 crystal structure with precise definition of the binding site, including the central Trp residue, a sphere of nearly 1000 water molecules to solvate interface and galactose unit and the AMBER force field (see also Fig. 5), the docking of the *syn*-conformer appears to be favored over that of the *anti*-ψ form by an energy difference of 10.2 kcal/mol in the case of a β-galactosylxyloside (Fig. 5; and see below).[39,40] The *anti*-ϕ conformer comes into steric conflict with the backbone of galectin, rendering contact formation impossible.[40]

Relative to *O*-lactose, its C-derivative harbors a larger extent of flexibility around the C-aglyconic bond than is detectable for the O-glycosidic bond.[35] This factor translates into an increase in the conformationally accessible space to about 23% and a shift of population density to the *anti*-ψ conformer (to about 55%).[36] Moreover, the presence of C–H bonds between the two rings has important consequences for the number of NOE contacts. The methylene bridge adds five NOE contacts to the group of interresidual signals of the *O*-glycoside. The respective list is useful to attribute occurrence of key (exclusive) signals to a fitting ϕ,ψ-combination unless binding shifts the conformation of the ligand away from any free-state topology (Fig. 4, bottom). With these sensors at hand for the interpretation of spectra, the analysis of the bound-state conformation of *C*-lactoside provides a striking example of differential conformer selection by different types of carbohydrate-binding proteins: galectin-1 binds the *syn*-conformer. Hereby, a total of three hydrogen bonds are possible in the binding site involving Glu-71 and Arg-48 and the O-3 of the penultimate Glc moiety supplementing the primary contacts with the Gal unit.[39] In the *anti*-ψ conformation these interactions are no longer possible because of the altered spatial presentation of the Glc moiety relative to the Gal unit. Now only one hydrogen bond between Arg-48 and Glc O-6 can be formed, helping explain the enthalpic difference in docking analysis.[39] Offering a topologically matching set of contact points also for this part of the disaccharide, that is, the Glc (or Xyl; see Fig. 5) moiety, the binding site of galectin-1 is thus designed to accommodate the energetically favored *syn*-conformer of *O*-lactose and also that of *C*-lactose, as indicated in the preceding section by quoting the calculated energy difference.[39,40] To

[39] J. L. Asensio, J. F. Espinosa, H. Dietrich, F. J. Cañada, R. R. Schmidt, M. Martín-Lomas, S. André, H.-J. Gabius, and J. Jiménez-Barbero, *J. Am. Chem. Soc.* **121,** 8995 (1999).

[40] J. M. Alonso-Plaza, M. A. Canales, M. Jimenez, J. L. Roldan, A. Garcia-Herrero, L. Iturrino, J. L. Asensio, F. J. Cañada, A. Romero, H.-C. Siebert, S. André, D. Solís, H.-J. Gabius, and J. Jiménez-Barbero, *Biochim. Biophys. Acta* **1568,** 225 (2001).

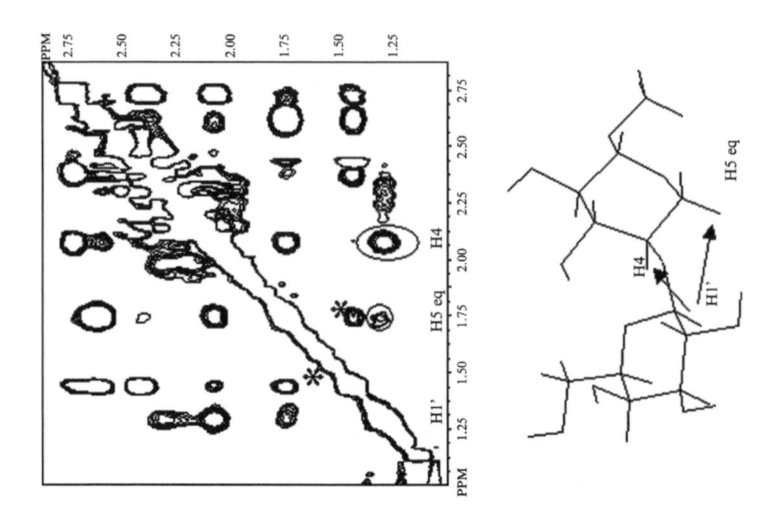

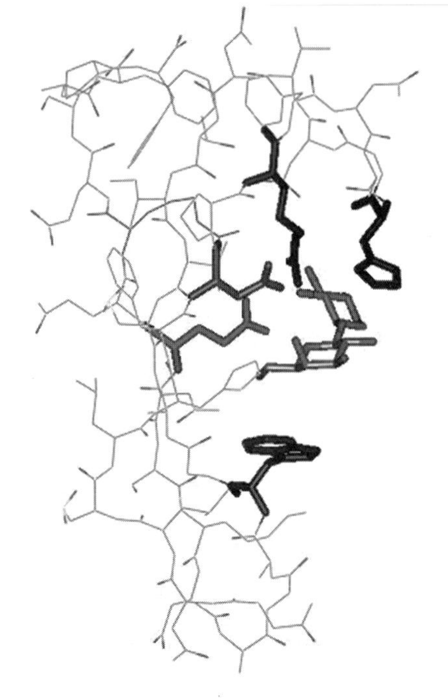

FIG. 5.

answer the question concerning whether the binding sites of nonhomologous galactoside-specific lectins share this selectivity, the behavior of a plant lectin is analyzed under the same conditions.

The lectin subunit of the toxin ricin maintains preference for the *syn*-conformer of *O*-lactose but binds the *anti*-ψ conformation of *C*-lactose.[36,41] In this case, docking analysis provides no obvious reasons why the *anti*-ψ conformation of *C*-lactoside is favored.[36,40,41] Even more intriguing than this selection process by ricin, the H2 Gal/H4 Glc NOE signal (see Fig. 4, top), which is only weakly visible in the spectrum of the free ligand characterizing the high-energy *anti*-ϕ conformation, becomes the strongest interresidual signal of *C*-lactoside bound to *Escherichia coli* β-galactosidase.[42] Fittingly, the three predicted NOE contacts with the protons

FIG. 5. Combination of molecular modeling for the ligand, its docking into the binding site of the lectin (left) and trNOE experiments on lectin-ligand mixtures (right) to show that galectin-1 binds the *syn* conformer of β-galactosides (here Galβ1, 4Xyl; 303 K; molar ratio 28:1, mixing time 250 ms; 500 MHz). Exclusively the interresidual H1 Gal/H4 Xyl and H1 Gal/H5eq Xyl cross-peaks are observable (see Fig. 4 for listing of exclusive NOE contacts for lactose). Cross-peaks originating from spin diffusion (marked with asterisks) are also observed herein, following the pathway H1 Xyl$\rightarrow$H5ax Xyl$\rightarrow$H5eq Xyl. They can be unequivocally identified by trROESY experiments (see Fig. 2 for schematic explanation of the principle). The structure of the complex (left) has been generated by docking experiments using the X-ray structure of bovine galectin-1 complexed with O-lactose (1SLA) and superimposing the Gal moiety of Galβ1, 4Xyl in both the *syn* and *anti*-Ψ conformers on that of O-lactose (*syn*-type conformation). 40 Energy calculations were done using the AMBER force field, and atomic charges were AMBER charges. The region close to the carbohydrate recognition site of the known protein was considered to comprise the amino acid residues within a radius of 10 Å. A sphere of more than 800 water moleules was added. The sphere was centered on the Gal moiety to solvate the interface and the binding site. A template force potential was introduced to avoid major movements of the polypeptide backbone during the calculations. The disaccharide and the lateral chains of amino acids were left mobile during the minimization processes. No cutoffs for nonbonding interactions were used. Energy minimizations were then conducted using 2000 conjugate gradient iterations. The best complex in terms of energy level is shown, which corresponds to the experimental NMR results. The typical CH-π stacking of the galactose moiety with the Trp residue is evident. Key interactions between the Xyl moiety and serveral polar amino acid residues are possible only when the sugar topology adopts the *syn* conformer. Therefore, the establishment of three hydrogen bonds betwen Glc O3 and these lateral chains contributes to the recognition of this conformer (for further details, see Alonso-Plaza *et al.* 40).

[41] J.-F. Espinosa, F. J. Cañada, J. L. Asensio, H. Dietrich, M. Martín-Lomas, R. R. Schmidt, and J. Jiménez-Barbero, *Angew. Chem.* **108,** 323 (1996).

[42] J. F. Espinosa, E. Montero, A. Viãn, J. L. García, H. Dietrich, R. R. Schmidt, M. Martín-Lomas, A. Imberty, F. J. Cañada, and J. Jiménez-Barbero, *J. Am. Chem. Soc.* **120,** 1309 (1998).

of the methylene bridge are also recorded on the basis of the scheme in Fig. 4, bottom (left). Their detection and the additional analysis of the *S*-lactoside add convincing support to the assumption that this glycosidase accommodates the high-energy conformation of the nonhydrolyzable substrate mimetics in contrast to the studied plant and animal lectins.[42,43] Moving from the primary structure of the ligand to its topological features, this result is not without implications for drug design.[3,12] It is along this line toward rational drug development that an improved definition of the topology of the ligand in the binding site is desirable. So far in our description of NOESY, solely protons of C–H bonds have been exploited as reporter units. For the given long-range purpose, access to information about ligand conformation provided by protons of water-sensitive hydroxyl groups would be welcome to broaden the experimental basis (see also Fig. 1).

Hydroxyl Groups of Ligand as Sensors

As a means to eliminate the problem of proton exchange between hydroxyl groups of the ligand and the bulk solvent the binding reaction could be studied in an aprotic solvent. To achieve the intended refinement it is indispensable that the protein retains the architecture of its binding site and its ligand affinity. Quantitative activity assessments, for example, by solid-phase assays, have been reported for galectin-1, a ricin-related lectin from mistletoe, and a galactoside-specific polyclonal immunoglobulin G fraction from human serum that prove the stability of these sugar receptors in dimethyl sulfoxide (DMSO).[44] Besides the routine precautions described above, the quality of the recorded spectra benefits from the addition of Al_2O_3 that has been heated at about 600° to let all water evaporate.[45] Treated as such, this compound acts as water and/or proton acceptor, thereby suppressing exchange effects between protons of hydroxyl groups and H_2O/H^+ in the aprotic solvent. The hygroscopic nature of heat-treated Al_2O_3 diminishes the exchange constant k so that the condition $k << 1/T_1$ (where T_1 in the longitudinal relaxation time) is fulfilled.[45]

Up to 10 mg of this substance added per 0.4 ml solvent reduces signal detection to genuine OH...HO contacts in a study with an immunoglobulin G fraction as sugar receptor.[44] The addition of heat-treated Al_2O_3 is thus as

[43] A. Garcia-Herrero, E. Montero, J. L. Muñoz, J. F. Espinosa, A. Vián, J. L. García, J. L. Asensio, F. J. Cañada, and J. Jiménez-Barbero, *J. Am. Chem. Soc.* **124,** 4804 (2002).

[44] H.-C. Siebert, S. André, J. L. Asensio, F. J. Cañada, X. Dong, J.-F. Espinosa, M. Frank, M. Gilleron, H. Kaltner, T. Kozár, N. V. Bovin, C.-W. von der Lieth, J. F. G. Vliegenthart, J. Jiménez-Barbero, and H.-J. Gabius, *ChemBioChem* **1,** 181 (2000).

[45] J. Dabrowski, H. Grosskurth, C. Baust, and N. E. Nifant'ev, *J. Biomol. NMR* **12,** 161 (1998).

effective to reduce illegitimate signals from contacts other than the new two-spin processes as systematic variation of the mixing time is to distinguish direct contacts from spin diffusion-dependent cross-peaks (Fig. 6).[44] Looking at the basic scheme in Fig. 1 and the way one more contact improves our view of ligand topology (Fig. 3c), the ability to include hydroxyl group protons into mapping the topology of a ligand is sure to make its mark on the overall precision of the process. When a hydroxyl group proton of the ligand in the complex dissolved in the aprotic solvent becomes engaged in a trNOE contact, this new water-sensitive conformational constraint will make the accessible ϕ,ψ-space for the ligand notably smaller, as documented in the initial report on this technique.[44] Combined studies using molecular modeling and NMR spectroscopy protocols together with

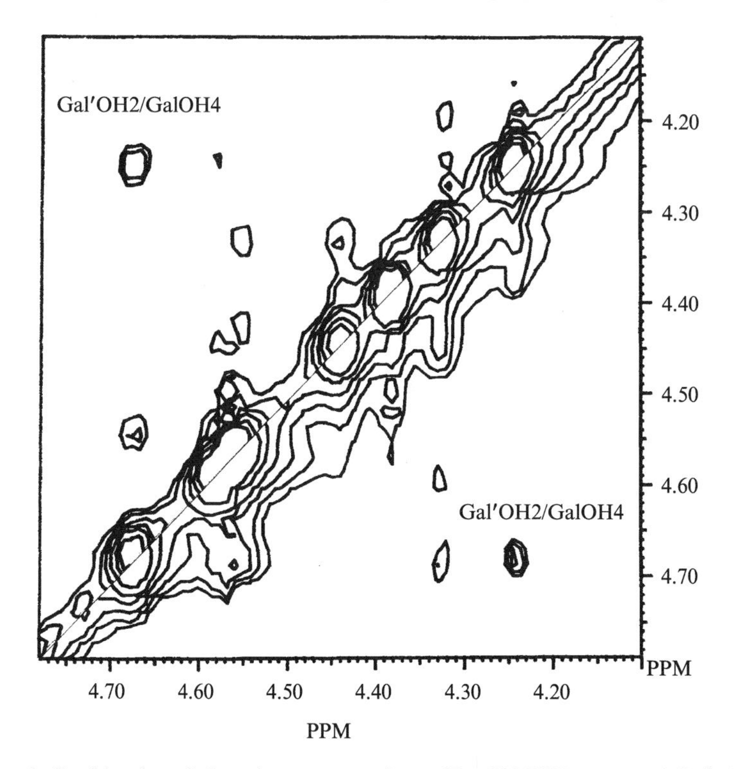

FIG. 6. Combination of the relevant parts of two 2D trNOESY spectra of Gal'α1,3Gal bound to α-galactoside-specific immunoglobulin G fraction from human serum in [D_6]DMSO at 303 K (molar ratio, 20:1) to illustrate the relative effects of short mixing time and presence of heat-treated Al_2O_3 as suppressor of proton exchange to block magnetization transfer between hydroxyl groups. The diagonal line separates results obtained in the absence of Al_2O_3 (mixing time, 20 ms) (*bottom right*) from those in the presence of Al_2O_3 (mixing time, 80 ms) (*top left*).

biochemical activity assays in water and aprotic solvent thus offer the potential to significantly refine definition of bound-state ligand topology.

Conclusions

A key insight from structural analysis of oligosaccharides by NMR spectroscopy is that—in contrast to peptides—they can often adopt only a limited set of distinct conformations in solution.[10–16] When these ϕ,ψ-combinations of dihedral angles are correlated to the energy levels of the respective glycoside conformation, central or side valleys in the energetic landscape are generally attributed to them.[15,32,46] Because dynamic fluctuations are possible between conformers not separated by high barriers of the energy potential, a new paradigm emerges that the same sugar determinant as coding unit for biological information might be presentable for a distinct lectin type in bioactive and bioinert versions.[47]

Alluding to E. Fischer's famous lock-and-key analogy,[48] this situation for the free ligand has been likened to a bunch of keys.[49] Explicitly, one saccharide structure can form more than one key. As concluded by S. Roseman, "it is this interplay between protein and different conformers (*of a single oligosaccharide*) that likely allows a single carbohydrate structure, such as hyaluronan, to be used in many different ways."[50] The predicted differential conformer selection by lectins has up to now experimentally been proved for galectins and selectins.[26,32,39,40,51,52] By synthesis of isotope-enriched carbohydrate ligands, that is, the sialylated Le^x tetrasaccharide, a means to improve signal dispersion and assignment for ligands longer than disaccharides has been exploited in the course of the studies. The herein given account on trNOESY/trROESY to detect conformer selection by endogenous lectins involved in, for example, cell adhesion processes during inflammation or malignant progression epitomizes the principle that one of the major conformers in solution can fit into the binding site without marked distortion, a result relevant for the design of potent inhibitors.

[46] S. W. Homans, *Prog. NMR Spectrosc.* **22,** 55 (1990).
[47] H.-J. Gabius, *Naturwissenschaften* **87,** 108 (2000).
[48] E. Fischer, *Ber. Dt. Chem. Ges.* **27,** 2985 (1894).
[49] B. J. Hardy, *J. Mol. Struct.* **395–396,** 187 (1997).
[50] S. Roseman, *J. Biol. Chem.* **276,** 41527 (2001).
[51] L. Poppe, G. S. Brown, J. S. Philo, P. V. Nikrad, and B. H. Shah, *J. Am. Chem. Soc.* **119,** 1727 (1997).
[52] R. Harris, G. R. Kiddle, R. A. Field, M. J. Milton, B. Ernst, J. L. Magnani, and S. W. Homans, *J. Am. Chem. Soc.* **121,** 2546 (1999).

A deviation from this rule appears to be possible in the following case: when the distribution of charge density in a carbohydrate ligand becomes a significant factor in the binding process, mutual adaptation of surface contacts to reach optimal charge complementarity might be a factor to cause conformational changes. In the case of plasma antithrombin and a highly charged heparin-derived tetrasaccharide, especially the dihedral angles between the GlcN-6-SO_4 and GlcA moieties are altered, presumably to allow establishment of a charge distribution closely matching that of the protein.[53] Corresponding changes take place in the architecture of its binding site to reach an optimal fit.[54] In conclusion, together with tailoring parameters of the spatial presentation of the ligand by branching, maxicluster formation, or dendrimeric supports,[1,55,56] the ligand topology inferred by trNOESY/ trROESY experiments is valuable information for the design of high-affinity binding partners to therapeutically block unwanted interactions of endogenous lectins or to target drugs by cell type-selective endocytosis.

Following the route to further increase knowledge about carbohydrate ligand topology, measurements on receptor–ligand complexes in an aprotic solvent, as described above, or the combination of selective water inversion with ^{13}C filtering (tested for [$^{13}C_6$]-α-methyl-D-mannopyranoside and the C-type lectin domain of rat mannan-binding lectin)[57] turn water-sensitive hydroxyl groups from silent bystanders into active sensors. Eventually, the aim in NMR application is the complete structural analysis of the complex, as briefly mentioned in the Introduction. Thereby, the solution architecture of the binding site, too, is thoroughly mapped. This procedure has been completed for hevein domain-containing plant lectins, the mannoside-specific cyanobacterial protein cyanovirin-N, and the adhesin domain of P-pili from uropathogenic *E. coli* (PapGII).[58–63] With both the

[53] M. Hricovini, M. Guerrini, and A. Bisio, *Eur. J. Biochem.* **261,** 789 (1999).

[54] R. Skinner, J.-P. Abrahams, J. C. Whisstock, A. M. Lesk, R. W. Carrel, and M. R. Wardell, *J. Mol. Biol.* **266,** 601 (1997).

[55] R. Roy, *Trends Glycosci. Glycotechnol.* **8,** 79 (1996).

[56] S. André, R. J. Pieters, I. Vrasidas, H. Kaltner, I. Kuwabara, F.-T. Liu, R. M. J. Liskamp, and H.-J. Gabius, *ChemBioChem* **2,** 822 (2001).

[57] E. W. Sayers, J. L. Weaver, and J. H. Prestegard, *J. Biomol. NMR* **12,** 209 (1998).

[58] J. L. Asensio, F. J. Cañada, M. Bruix, A. Rodriguez-Romero, and J. Jiménez-Barbero, *Eur. J. Biochem.* **230,** 621 (1995).

[59] J. L. Asensio, F. J. Cañada, M. Bruix, C. Gonzalez, N. Khiar, A. Rodriguez-Romero, and J. Jiménez-Barbero, *Glycobiology* **8,** 569 (1998).

[60] J. L. Asensio, H.-C. Siebert, C.-W. von der Lieth, J. Laynes, M. Bruix, U. M. Soedjanaatmadja, J. J. Beintema, F. J. Cañada, H.-J. Gabius, and J. Jiménez-Barbero, *Proteins* **40,** 218 (2000).

[61] J. F. Espinosa, J. L. Asensio, J. L. García, J. Laynez, M. Bruix, C. Wright, H.-C. Siebert, H.-J. Gabius, F. J. Cañada, and J. Jiménez-Barbero, *Eur. J. Biochem.* **267,** 3965 (2000).

ligand topology and the positioning of amino acid side chains in the binding site becoming identifiable in solution, help in answering the question concerning which topological factors are involved in enthalpy/entropy changes in the thermodynamics of the binding process is likely not to be long in coming.[12,64] These issues properly addressed, the resulting data will guide synthetic strategies to produce high-affinity ligands, turning such basic insights into the how and why of protein–carbohydrate interactions into a hot spot of drug design, as programmatically stated at the beginning of this article.

[62] C. A. Bewley, *Structure* **9,** 931 (2001).
[63] M.-a. Sung, K. Fleming, H. A. Chen, and S. Matthews, *EMBO Rep.* **2,** 621 (2001).
[64] T. K. Dam and C. F. Brewer, *Chem. Rev.* **102,** 387 (2002).

[30] Studies of Carbohydrate–Protein Interaction by Capillary Electrophoresis

By SUSUMU HONDA and ATSUSHI TAGA

Introduction

Among various methods for molecular binding studies, capillary electrophoresis (CE) is a unique method that can be used to directly observe the interaction between molecules under conditions similar to physiological state without prior immobilization of the reactants to a solid phase. It is suitable in searching for carbohydrate-binding proteins and protein-binding carbohydrates. It also allows reliable micro/ultramicro-scale estimation of association/dissociation constants. Furthermore, CE enables direct observation of enzymatic reactions occurring in a capillary and assay of enzyme activity. This article describes the characteristic features of CE and shows how CE has been applied for various kinds of studies related to carbohydrate–protein binding.

Principle of Carbohydrate–Protein Binding Studies by Capillary Electrophoresis

Capillary electrophoresis is a general name for electrophoresis performed in a narrow tube filled with an electrolyte solution and is an ultimate form of electrophoresis. Early types of electrophoresis were performed with matrices such as filter paper, cellulose acetate membrane, and

0076-6879/03 $35.00

polyacrylamide and agarose gels, which serve to prevent convective flow induced by Joule heating. These matrices can also be used to immobilize reactants, and electrophoresis using matrices resembles affinity chromatography. CE is different from these classic types of electrophoresis in that separation can be performed in free solution without any matrices. It has a high capability in separation owing to efficient dissipation of Joule heat and the plug flow nature of electroosmotic flow (EOF). Detection is reproducible and sensitive, requiring only a small amount of sample. Thus, CE is a most powerful tool in modern separation science.

There are several types of capillary electrophoresis differing in potential distribution: moving-boundary electrophoresis, zone electrophoresis, isotachophoresis, and isoelectric focusing.[1] Zone electrophoresis utilizes linear distribution of potential across a capillary. This kind of CE is currently the most popular, and a number of separation modes have been developed for a wide range of analytes. All these modes can be done simply by changing the additive to the electrolyte solution. The mode used for molecular binding studies is affinity capillary electrophoresis (ACE). This mode is based on the bioaffinity between molecules and has been employed for analysis and binding studies of various kinds of biomedical and pharmaceutical substances. Its usefulness has also been demonstrated for carbohydrate separation[2] and observation of carbohydrate–protein interaction.[3–6]

In ACE the mobility of a solute changes when an interacting substance is present as an additive in the electrophoretic solution. Because the change can be easily observed as the migration time shift of the solute, the presence of such interacting substances can readily be observed. Therefore, ACE is a convenient tool for searching for interacting substances. On the other hand, the magnitude of this change depends on the concentration of the interacting substance, and mathematical treatment of the relationship between mobility change and additive concentration can lead to a simplified equation, from which the association/dissociation constant can easily be calculated. Thus, ACE is also an excellent tool for estimating association/dissociation constants at a micro/ultramicro scale. If the protein, one of the counterparts, is an enzyme, it specifically associates with the substrate to give an adduct, which then yields a product. Because this series of

[1] S. W. Compton and R. G. Brownlee, *Biotechniques* **6,** 432 (1988).

[2] A. Taga, Y. Yabusako, A. Kitano, and S. Honda, *Electrophoresis* **19,** 2645 (1998).

[3] S. Honda, A. Taga, K. Suzuki, S. Suzuki, and K. Kakehi, *J. Chromatogr.* **597,** 377 (1992).

[4] A. Taga, S. Yasueda, M. Mochizuki, H. Itoh, and S. Honda, *Analysis* **26,** M36 (1998).

[5] A. Taga, K. Uegaki, Y. Yabusako, A. Kitano, and S. Honda, *J. Chromatogr. A* **837,** 221 (1999).

[6] K. Uegaki, A. Taga, Y. Akada, S. Suzuki, and S. Honda, *Anal. Biochem.* **309,** 269 (2002).

reactions can be directly observed in a capillary, CE can be used not only for the assay of enzyme activity but also for observation of the mechanism of the enzymatic reaction.

Search for Carbohydrate-Binding Proteins by CE

When a carbohydrate sample is introduced into an electrophoretic solution containing a protein having affinity to the carbohydrate, the mobility, accordingly, the migration time, of the carbohydrate changes. On the contrary, if a protein sample is introduced into a buffer containing an interacting carbohydrate, the migration time of the protein is altered. Such changes are an indication of affinity between the sample and the additive. A typical example[7] is shown below.

Disaccharide-Binding Proteins in Serum

Figure 1a shows an electropherogram of a mixture of simple disaccharides (maltose, cellobiose, gentiobiose, lactose, and melibiose) derivatized with 8-aminonaphthalene 1,3,6-trisulfonate (ANTS), analyzed in a neutral phosphate buffer in an uncoated fused silica capillary. Separation is poor, with only a small peak and a large peak being observed for the disaccharides (Fig. 1a). The small peak is gentiobiose (GB) and the large peak is due to the overlap of all the other disaccharides. This occurs because zone electrophoresis, the exerting mode, does not allow separation of isomers having the same charge-to-mass ratio, as do these disaccharide derivatives. Addition of human serum to the running buffer (Fig. 1b–d), however, causes changes of separation pattern and peak range, and these changes become more pronounced as serum concentration increases. The disaccharide giving the greatest change is gentiobiose (GB). This means that GB most strongly interacts with constituent(s) of serum, an unexpected finding.

To examine which constituent(s) interact with GB, serum is analyzed in the same buffer (Fig. 2, bottom electropherogram). The capillary inner wall is coated with linear polyacrylamide in this case, to avoid adsorption of proteins. A huge peak of albumin (HAS) is observed at ~17 min, together with weak, faster moving peaks of globulins in a range of ~ 11–15 min, as shown in Fig. 2.

Addition of GB modified with mercaptoethane sulfonate (MerES)[8] to the electrophoretic solution reveals an interesting phenomenon: the β-globulin peak is retarded and split into a few peaks (Fig. 2, middle

[7] A. Taga, Y. Morioka, R. Maruyama, and S. Honda, paper presented at HPLC Kyoto (September 2001, Kyoto, Japan).

[8] A. Taga, M. Mochizuki, H. Itoh, S. Suzuki, and S. Honda, *J. Chromatogr. A* **839,** 157 (1999).

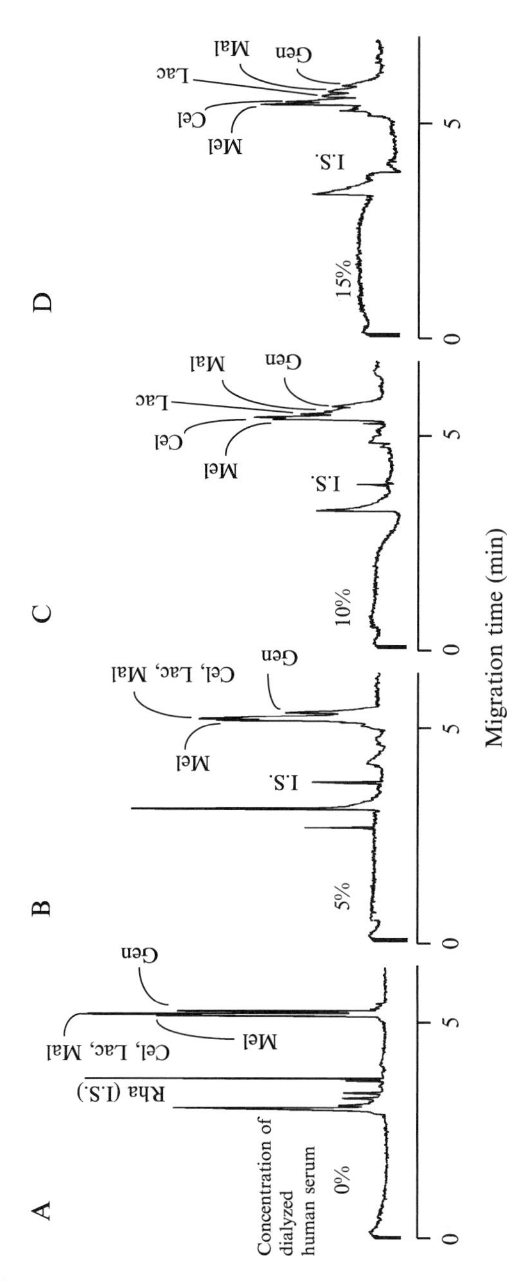

FIG. 1. Separation of simple disaccharides [maltose (Mal), lactose (Lac), cellobiose (Cel), melibiose (Mel), and gentobiose (Gen)] derivatized with 8-aminonaphthalene 1,3,6-trisulfonate in the presence of human serum.[7] Capillary, polyacrylamide-coated fused silica (50 μm i.d., 50 cm); electrophoretic solution, 50 m*M* phosphate buffer (pH 6.8) not containing (A) or containing human dialyzed pool serum [(B) 5%; (C) 10%; (D) 15%]; capillary oven temperature, 30°; applied voltage, 15 kV; detection, UV absorption at 280 nm. Rha (1.5.), rhamnose (internal standard, I. S.).

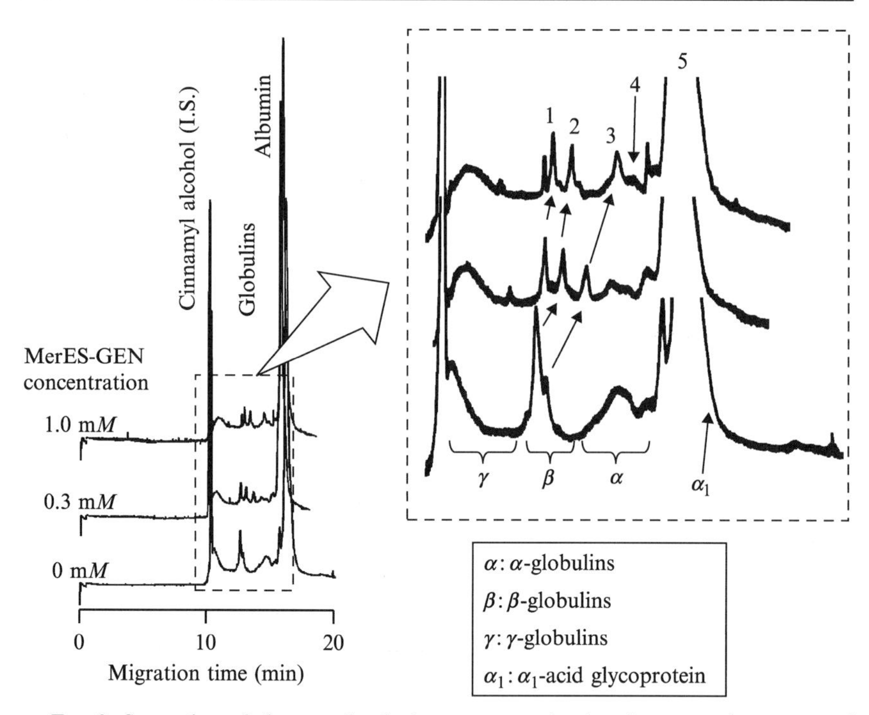

FIG. 2. Separation of the proteins in human serum in the absence and presence of gentiobiose derivatized with MerES.[7] Capillary, Polybrene/polyacrylic acid double-coated fused silica (50 μm i.d., 72 cm); electrophoretic solution, 50 mM phosphate buffer (pH 6.8) not containing or containing MerES-GEN (0.3 or 1.0 mM); capillary oven temperature, 30°; applied voltage, 15 kV; detection, UV absorption at 280 nm. *Inset*: Magnification of the globulin peaks.

electropherogram), and retardation and splitting become more marked as the concentration of the added serum increases (Fig. 2, top electropherogram). Modification with MerES confers charge to GB; otherwise the serum proteins having only a small charge would not move the adduct. Two MerES molecules react with GB to give a bis(MerES) derivative, which does not interfere with the detection of proteins, although existing at a rather high concentration, because it does not absorb UV light at the wavelength for protein detection (280 nm). The retardation and splitting of the β-globulin peak is strong evidence that there are proteins in the β-globulin fraction that interact with GB, and thus the specificity between GB and a few proteins in the β-globulin fraction is confirmed.

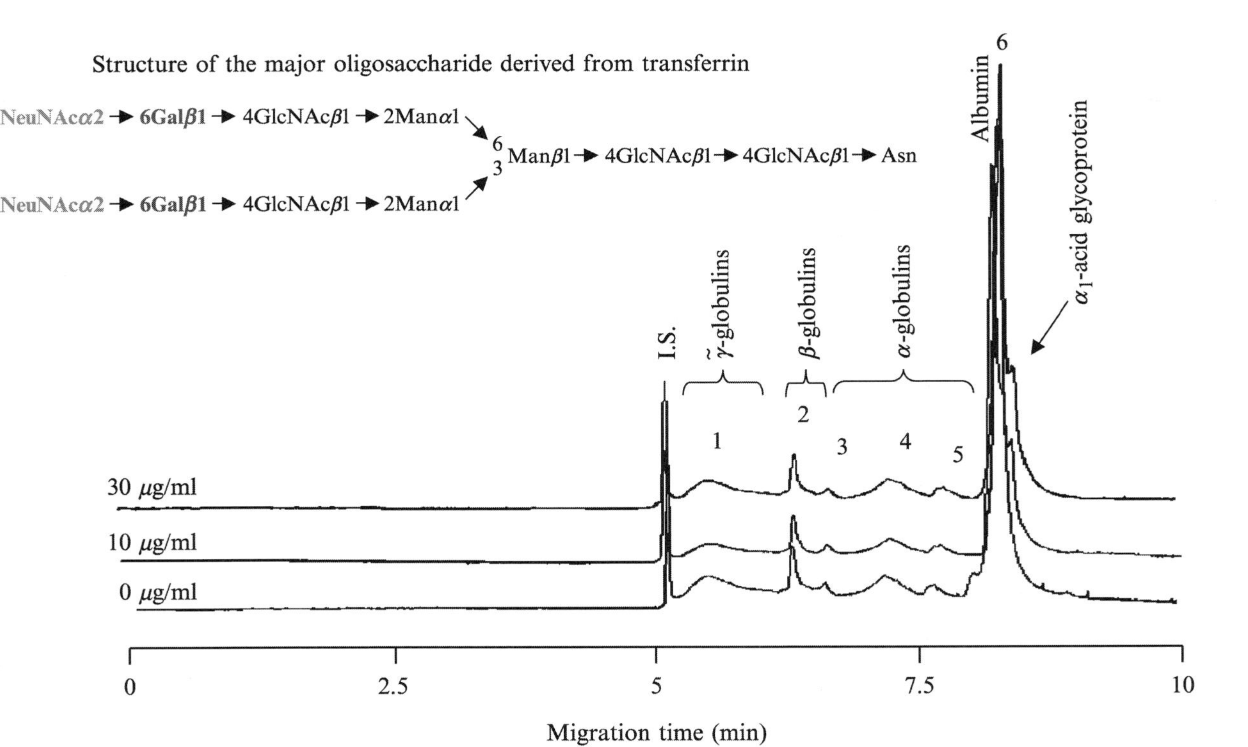

FIG. 3. Separation of the proteins in human serum in the absence and presence of TF.[7] Capillary, Polybrene/polyacrylamide double-coated fused silica (50-μm i.d., 72 cm); electrophoretic solution, 50 m*M* phosphate buffer (pH 7.5) not containing or containing (10 or 30 μg) TF; applied voltage, 15 kV; capillary oven temperature, 25°; detection, UV absorption at 200 nm. Internal standard (I.S.), cinnamyl alcohol.

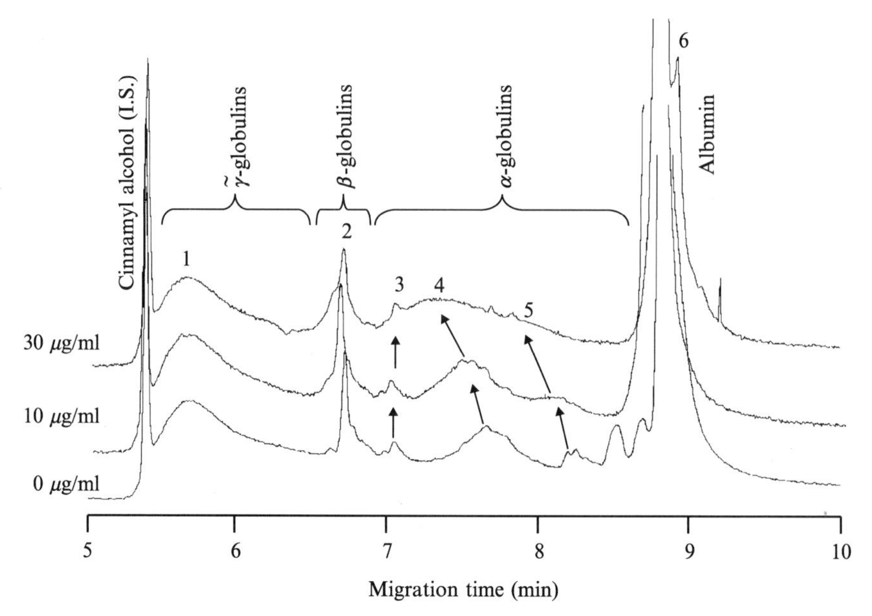

FIG. 4. Separation of the proteins in human serum in the absence and presence of ATF.[7] Capillary, Polybrene/polyacrylamide double-coated fused silica (50 μm i.d., 72 cm); electrophoretic solution, 50 m*M* phosphate buffer (pH 7.5) not containing or containing (10 or 30 μg) ATF; applied voltage, 15 kV; capillary oven temperature, 25°; detection, UV absorption at 200 nm. Internal standard (I.S.), cinnamyl alcohol.

Asialotransferrin-Binding Protein in Serum

The following is another example to demonstrate the presence of a carbohydrate-binding protein in human serum.[7] Transferrin (TF), an iron-transporting protein, is known to have a biantennary *N*-glycan as the main glycan chain, and both termini of the antennae are known to be modified by the *N*-acetylneuraminic acid (NANA) residue. Analysis of human serum in a neutral phosphate buffer containing TF does not cause any change in the protein peaks without regard to TF concentration, as shown in Fig. 3.

The NANA residue in TF can easily be removed by digestion with sialidase to give asialotransferrin (ATF). Addition of ATF to the running buffer, however, results in retardation of the α-globulin peak (Fig. 4). This series of analyses indicate that α-globulin specifically recognizes ATF. Because it does not recognize TF, the NANA residue is considered to play an important role in the recognition of α-globulin. Thus, ACE gives direct evidence demonstrating the specific binding of ATF to α-globulin, and thereby suggests the control of binding through an on–off mechanism by the NANA residue.

Estimation of Association–Dissociation Constants by CE

Although carbohydrate–protein interaction is relatively weak compared with other types of interactions such as protein–protein and protein–DNA interactions, its importance in various cell events as well as carbohydrate-mediated transport and metabolism of biological substances has been well recognized and has been studied from various aspects. In such studies the estimation of the magnitude of interaction is of primary importance. Because such interaction obeys the law of mass action, the magnitude of interaction is properly expressed as the association constant (K_a), which is the ratio of the adduct concentration to the product of the carbohydrate and protein concentrations. For the dissociation constant, the ratio is the reverse. Various methods have been used for K_a estimation, including equilibrium dialysis,[9] centrifugation,[10] UV wavelength shift,[11] UV absorption/fluorescence intensity change,[12,13] depolarization angle change,[14] affinity chromatography,[15] and surface plasmon resonance.[16] The CE method that is the subject of this article is also a method of choice for K_a estimation.

There are two systems for the study of carbohydrate–protein interaction by CE, depending on whether the carbohydrate or the protein is the sample or the additive in the electrophoretic polution.

Normal System

In this system a protein sample is introduced into a carbohydrate-containing electrophoretic solution. The carbohydrate as additive must have an electric charge strong enough to move the protein sample, so that the change in migration time can be measured accurately. Otherwise, an electric charge must be introduced on the carbohydrate additive by suitable chemical means.

The principle of K_a estimation has been described previously.[4] The outline is shown below.

[9] P. N. Pincard, "Handbook of Experimental Immunology," Vol. 1, Chapter 17. Blackwell, Oxford, 1978.

[10] W.-H. Wu and J. H. Rockey, *Biochemistry* **8,** 2719 (1969).

[11] R. J. Doyle, E. P. Pittz, and E. E. Woodside, *Carbohydr. Res.* **9,** 89 (1968).

[12] A. Jimbo, N. Seno, and I. Matsumoto, *J. Biochem.* (*Tokyo*) **95,** 267 (1984).

[13] K. A. Kronis and J. P. Carver, *Biochemistry* **21,** 3050 (1982).

[14] S.-F. Chao, W. J. Rocque, S. Daniel, L. E. Czyzyk, W. C. Phelps, and K. A. Alexander, *Biochemistry* **38,** 4586 (1999).

[15] D. S. Hage, T. A. Noctor, and I. W. Wainer, *J. Chromatogr. A* **693,** 23 (1995).

[16] Y. Shinohara, F. Kim, M. Shimizu, M. Goto, M. Tosu, and Y. Hasegawa, *Eur. J. Biochem.* **223,** 189 (1994).

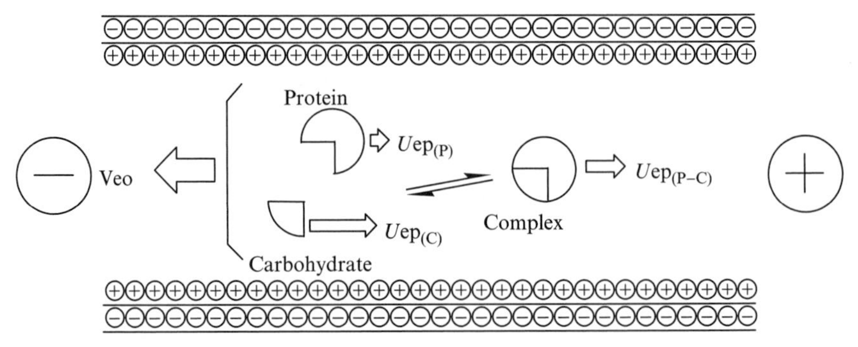

FIG. 5. Schematic illustration of carbohydrate–protein interaction in a capillary. $Uep_{(P)}$, $Uep_{(C)}$, and $Uep_{(P-C)}$ designate the velocities of a protein, a carbohydrate, and the adduct thereof, respectively.

As illustrated in Fig. 5, the interaction between a protein and a carbohydrate in an electrophoretic solution is an equilibrium reaction, and the K_a value for this reaction can be expressed by Eq. (1):

$$K_a = [PC][P]^{-1}[C]^{-1} \tag{1}$$

where [P], [C], and [PC] are the concentrations of the protein, carbohydrate, and adduct formed, respectively. On the one hand, the molar fraction of the adduct to the total protein (α) is

$$\alpha = [PC]([P] + [PC])^{-1} \tag{2}$$

Combining Eqs. (1) and (2) gives

$$K_a^{-1}[C]^{-1} + 1 = \alpha^{-1} \tag{3}$$

On the other hand, the apparent migration velocities of the protein in the absence (V_0) and the presence (V_P) of the carbohydrate are given by Eqs. (4) and (5), respectively,

$$V_0 = V_{e0} - U_P \tag{4}$$

$$V_P = V_{e0} - [\alpha U_{PC} + (1 - \alpha)U_P] \tag{5}$$

where V_{e0} is the velocity of electroosmotic flow (EOF), and U_P and U_{PC} are the velocities of electrophoretic migration of the protein and the adduct, respectively. Subtraction of Eq. (5) from Eq. (4) gives Eq. (6):

$$V_0 - V_P = \alpha(U_{PC} - U_P) \tag{6}$$

and substitution of Eq. (6) by α in Eq. (3) gives

$$K_a^{-1}[C]^{-1} + 1 = (U_{PC} - U_P)(V_0 - V_P)^{-1} \quad (7)$$

Because

$$V_0 = \iota\, t_1^{-1} \quad \text{and} \quad V_P = \iota\, t^{-1}$$

Eq. (6) can be rewritten as

$$U_{PC} - U_P = \iota[(t_0^{-1} - t_2^{-1}) - (t_0^{-1} - t_1^{-1})] = \iota(t_2 - t_1)t_1^{-1}t_2^{-1} \quad (8)$$

where t_0, t_1, and t_2 are the migration times of the neutral marker, the protein, and the adduct, respectively, and ι is the effective length of a capillary. The term t_1 is the migration time of the protein in the absence of the carbohydrate and t_2 can be approximated as the time when further addition of the carbohydrate does not cause change in protein migration time any more. From Eqs. (6)–(8), a key equation[9] [Eq. (9)] is derived:

$$(t - t_1)^{-1} = t_1^{-1}t_2(t_2 - t_1)^{-1}K_a^{-1}[C]^{-1} + (t_2 - t_1)^{-1} \quad (9)$$

Therefore, the reciprocal of migration time difference, $(t - t_1)^{-1}$, is the first-order function of the reciprocal of carbohydrate concentration, $[C]^{-1}$, and gives a straight line having a slope $t_1^{-1}t_2\,(t_2 - t_1)^{-1}K_a^{-1}$ (A) and a y intercept $(t_2 - t_1)^{-1}$ (B). Consequently K_a can be obtained as follows:

$$K_a = (Bt_1 + 1)A^{-1}t_1^{-1} \quad (10)$$

Because the migration time difference is small (usually within 1 min) and the migration time itself may vary due to the fluctuation of the velocity of EOF, migration time should be measured as accurately as possible.

Correction for the variation of EOF is also important to obtain a reliable value of association constant. Migration time should be corrected using an equation, $t_{abs}^{-1} = t_{pre}^{-1} + [(t_{ref})_{abs}^{-1} - (t_{ref})_{pre}^{-1}]$, where t_{abs} is the corrected migration time of a protein sample in the absence of a carbohydrate. The term t_{pre} is the migration time of a protein in the presence of a carbohydrate, that is, the observed t value, and $(t_{ref})_{abs}$ and $(t_{ref})_{pre}$ are the migration times of an internal reference in the absence and presence, respectively, of the carbohydrate.

Figure 6 shows the change in migration time of *Ricinus communis* agglutinin, 60 kDa, (RCA_{60}) introduced into a phosphate buffer containing lactobionic acid, a carboxylated derivative of lactose, having a specificity to galactose residue.[3]

The faster moving peak in each electropherogram is that of mesityl oxide as a neutral marker. It is indicated that the RCA_{60} peak is retarded more and more markedly as the lactobionic acid concentration increases.

Figure 7a and b shows the normal and double-reciprocal plot, respectively, of $(t - t_1)$ versus [C]. Good linearity (correlation coefficient, 0.98) is

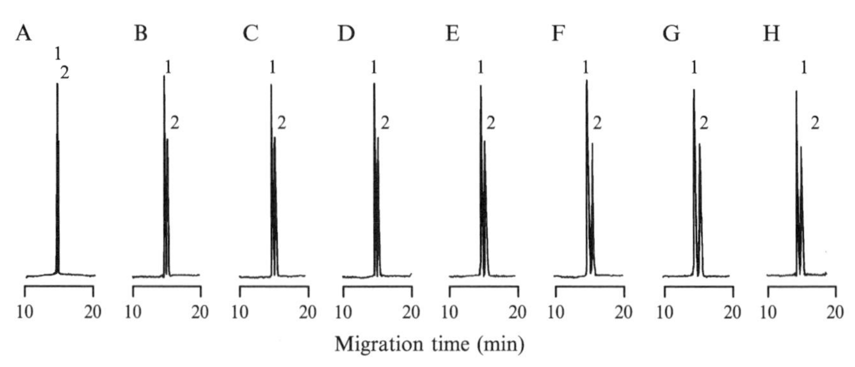

FIG. 6. Concentration-dependent retardation of the RCA_{60} peak in the presence of lactobionic acid.[3] Capillary, fused silica (50 μm i.d., 85 cm); electrophoretic solution, 50 mM phosphate buffer (pH 6.8) not containing lactobionic acid (A) or containing lactobionic acid at (B) 0.2 mM, (C) 0.3 mM, (D) 0.4 mM, (E) 0.5 mM, (F) 1.0 mM, (G) 5.0 mM, or (H) 10.0 mM. Capillary oven temperature, 30°; applied voltage, 20 kV; detection, UV absorption at 220 nm. The first and the second peaks of each electropherogram are of mesityl oxide as a neutral marker and RCA_{60}, respectively. Reproduced from Ref.3 with permission of the publisher.

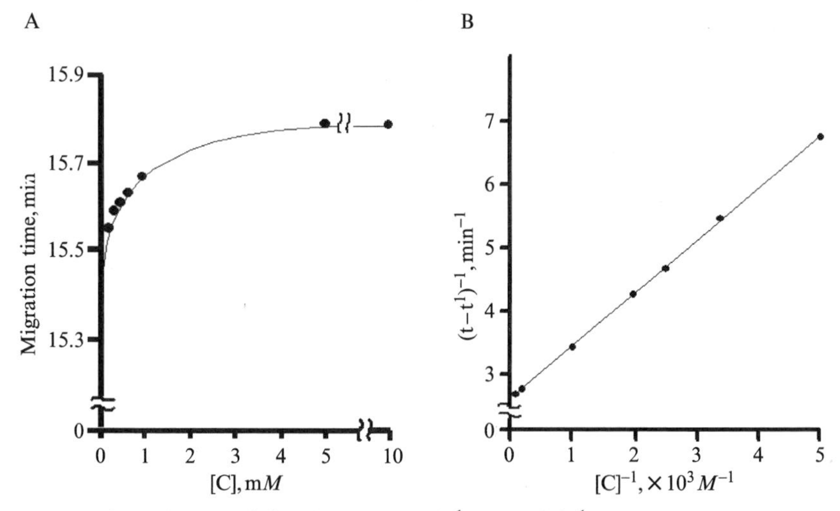

FIG. 7. $(t - t_1)$ versus [C] (A) and $(t - t_1)^{-1}$ versus $[C]^{-1}$ (B) plots for the RCA_{60}–lactobionic acid interaction.[3] Analytical conditions are as in Fig. 6. Reproduced from Ref. 3 with permission of the publisher.

observed and the K_a constant can be obtained as $3.2 \times 10^3 M^{-1}$ at 30° from the slope (8.1×10^{-1} mM min^{-1}) and the y intercept (2.7 min^{-1}) using Eq. (10).

This method allows estimation of K_a with small amounts of protein and carbohydrate. Although the concentration of RCA_{60} is rather high

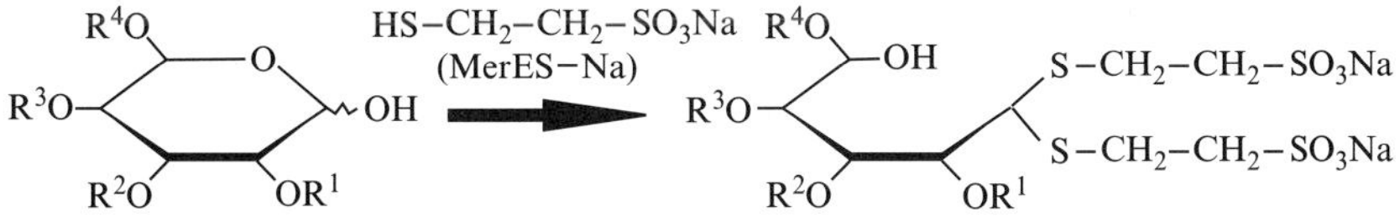

R^4O $HS-CH_2-CH_2-SO_3Na$ R^4O
$(MerES-Na)$ OH $S-CH_2-CH_2-SO_3Na$
R^3O OH R^3O
$S-CH_2-CH_2-SO_3Na$
R^2O OR^1 R^2O OR^1

FIG. 8. Introduction of MerES groups to a neutral carbohydrate to confer electric charge without introducing chromophore.

(1 mg/ml), the total injected amounts for seven injections are as small as 70 ng (~ 1.5 pmol). The protein concentration can be reduced to the μ mol level by introducing a fluorescence tag such as a substituted benzolazane[17] and monitoring by laser-induced fluorescence. The total amount of lactobionic acid is about 2 μmol. In binding studies by CE the anodic and cathodic reservoirs as well as the capillary are filled with the additive-containing electrophoretic solution, but our experiment indicates that filling the reservoirs can be omitted. By this omission the amount of additive can be reduced by 100-fold. The proposed method allows K_a estimation with high reproducibility, the relative standard deviation ($n = 5$) being 3.6%. Note that Eq. (9) does not contain the protein concentration term, implying that K_a can be obtained without knowing the exact concentration of the protein sample.

In the above example of the RCA_{60}–lactobionic acid combination the carbohydrate additive had a negative charge, which facilitates migration of RCA_{60}. When the carbohydrate additive does not possess an electric charge, it must be converted to an ionic species by introducing an ionic tag. One example is dithioacetylation with MerES.[8] Two MerES groups are rapidly introduced to each reducing carbohydrate by dissolving a neutral carbohydrate in an MerES solution in trifluoroacetic acid to give a bis(MerES) derivative in quantitative yield at room temperature (Fig. 8).

Trifluoroacetic acid serves as an acid catalyst, and the product can easily be isolated after evaporation of the reaction mixture. A similar experiment is done by replacing lactobionic acid with the bis(MerES) derivative of lactose and by replacing mesityl oxide with cinnamyl alcohol. The result indicates that the RCA_{60} peak is retarded as the concentration of the MerES derivative of lactose increases, and the plot of $(t - t_1)^{-1}$ versus $[C]^{-1}$ gives a straight line. The K_a value estimated from the slope and the y intercept is $2.5 \times 10^3 M^{-1}$ at 30°, slightly lower than the K_a value for the RCA_{60}–lactobionic acid interaction ($3.2 \times 10^3 M^{-1}$). Therefore, the

[17] S. Honda, J. Okeda, H. Iwanaga, S. Kawakami, A. Taga, S. Suzuki, and K. Imai, *Anal. Biochem.* **284,** 99(2000).

influence of the tag is not so great, and this method can also be used for K_a estimation involving neutral carbohydrates.

The K_a values of maltooligosaccharides to *Lens culinaris* agglutinin (LCA) are similarly estimated as bis(MerES) derivatives.[8] The results (maltose, $6.6.7 \times 10^2 M^{-1}$; maltotriose, $9.64 \times 10^2 M^{-1}$; maltopentaose, $1.37 \times 10^3 M^{-1}$) indicate that the K_a value increases gradually as the degree of polymerization increases, the same tendency observed in our previous articles obtained by high-performance affinity chromatography[18] and also by CE in the reversed system described below. This phenomenon is presumably due to increasing binding sites with degree of polymerization.

It has been shown that the normal system allows reliable estimation of K_a, utilizing the linear relationship between $(t - t_1)^{-1}$ and $[\mathrm{C}]^{-1}$. It does not require immobilization of one of the reactants, unlike affinity chromatography and surface plasmon resonance methods. The estimation can be done on a small scale and with high reproducibility. When the carbohydrate additive does not carry an electric charge, MerES groups can be introduced easily and quantitatively, and the resultant derivative can be subjected to K_a estimation. However, the mathematical treatment mentioned above is performed on an assumption that the protein sample and the carbohydrate additive interact in a monovalent mode, that is, each of these reactants has only one binding site. When there are multiple binding sites differing in the magnitude of interaction, mathematical treatment will become much more complex, and the apparent K_a values estimated by the above calculation represent only the overall K_a values. In some cases of carbohydrate–lectin interaction, the lectin side is composed of multiple subunits each having almost the same structure. In such cases binding sites are multiple and the overall K_a value should be divided by the number of binding sites to obtain the K_a value for individual bindings.

Reverse System

In the reverse system a carbohydrate sample is introduced to an electrophoretic solution containing a protein. Because of the high concentration of the protein additive, which is liable to be adsorbed onto fused silica to alter the ζ potential between the capillary inner wall and the electrophoretic solution, and thus the velocity of EOF, the capillary inner wall must be protected from adsorption by coating it with relevant materials. Among various materials linear polyacrylamide has proved suitable, completely covering the silanol groups on the surface of the inner wall. EOF is completely diminished, and ionic solutes move only on the basis of their mobilities. Coating

[18] K. Kakehi, Y. Kojima, S. Suzuki, and S. Honda, *J. Chromatogr.* **502,** 297 (1990).

with linear polyacrylamide has a drawback because it is not durable in strongly acidic and alkaline media. Dynamic coating with Polybrene followed by polyacrylic acid can also be used for studies of carbohydrate–protein interaction. It is much more stable over a wider pH range.

Because the equilibrium between the carbohydrate sample and the protein additive is similar to that in the normal system, Eq. (11), which is similar to Eq. (9), can be derived by a mathematical treatment that replaces all the migration time terms for the protein sample by those for the carbohydrate sample, and the carbohydrate concentration by the protein concentration. Namely,

$$(t' - t'_1)^{-1} = t'^{-1}_1 t'_2 (t'_2 - t'_1)^{-1} K_a^{-1} [\mathrm{P}]^{-1} + (t'_2 - t'_1)^{-1} \tag{11}$$

where t' designates the general migration time term of the carbohydrate sample in this system, t'_1 is the migration time of the carbohydrate sample in the absence of the protein, and t'_2 is the migration time of the adduct. [P] is the concentration of the protein additive. A double reciprocal plot of $(t' - t'_1)^{-1}$ versus $[\mathrm{P}]^{-1}$ gives a straight line and K_a can be estimated by Eq. (12), using the slope A' and the y-intercept B':

$$K_a = (B' t'_1 + 1) A'^{-1} t'^{-1}_1 \tag{12}$$

In a reverse system tagging of carbohydrate samples is often required because they usually do not have chromophores that can be directly detected. Derivatization with 1-phenyl-3-methyl-5-pyrazolone (PMP) is recommended because the procedure is the simplest and gives quantitative yields of derivatives under mild conditions, but the electric charge of the derivatives is not large. Derivatization with ANTS has a stronger negative charge, but it has a tendency to give slightly higher K_a values, because of the influence of the ANTS group introduced into the molecule.

In binding studies by CE a combination of a single ligand and a single protein is often used. However, we can recall the excellent capability of CE in separation. If a mixture of ligands is introduced to a suitable electrophoretic solution, the ligands tend to be separated from each other. If the electrophoretic solution contains a protein and the ligands have different affinities for this protein, the separation of the ligands is affected by the affinities of individual ligands. We have used this principle for simultaneous estimation of K_a values for multiple carbohydrates.

Figure 9a shows the separation of various disaccharides derivatized with PMP in a neutral phosphate buffer.[2]

Because the exerting mode is zone electrophoresis and the disaccharide derivatives are positional isomers to each other, having the same charge-to-mass ratio, only incomplete group separation occurs, giving glucobiose and galactosylglucose peaks. The reason for group separation is presumably

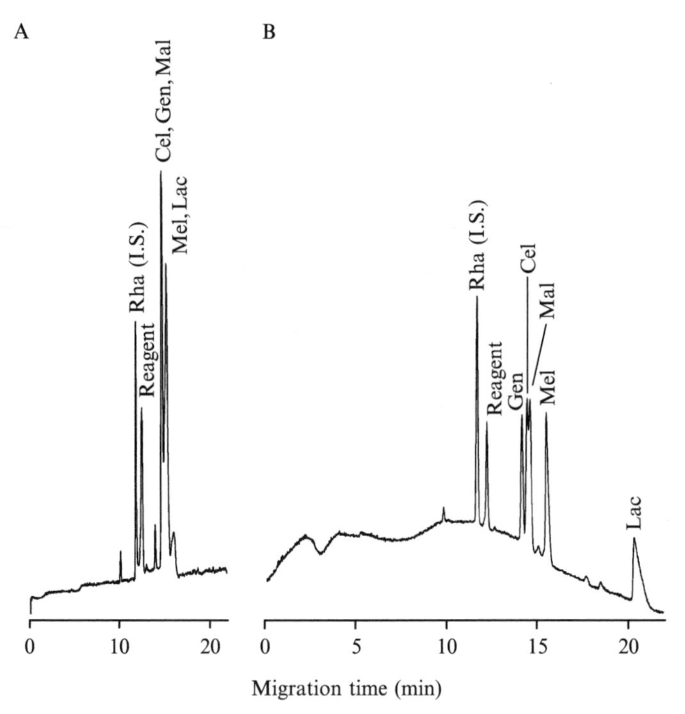

FIG. 9. Separation of disaccharide derivatives with PMP in the absence (A) and presence (B) of RCA_{60}.[2] Capillary, uncoated (A) or linear polyacrylamide-coated (B) fused silica (50 μm i.d., 50 cm for both); electrophoretic solution, 50 mM phosphate buffer (pH 6.8) not containing (A) or containing RCA_{60} (5 mg/ml) (B); capillary oven temperature, 30°; applied voltage, 18 kV; detection, UV absorption at 245 nm. Mal, Cel, Gen, Lac, and Mel denote maltose, cellobiose, gentiobiose, lactose, and melibiose, respectively. Reproduced from Ref. 2 with permission of the publisher.

due to the different magnitude of intramolecular hydrogen bonding, as observed also for the PMP derivatives of monosaccharides.[19] Addition of RCA_{60} to the running buffer results in retardation of the galactosylglucose peaks, especially the lactose peak. This is plausible because RCA_{60} is a galactose-recognizing lectin. The K_a values for these galactosylglucoses can simultaneously be estimated using Eq. (12).[5] The obtained values of A', B', and K_a for lactose are $9.2 \times 10^{-6} M$ min^{-1}, 3.6×10^{-2} min^{-1}, and $1.1 \times 10^{4} M^{-1}$, respectively. Those for melibiose are $3.0 \times 10^{-5} M$ min^{-1}, 2.0×10^{-1} min^{-1}, and $8.9 \times 10^{3} M^{-1}$, respectively.

Figure 10 shows another example of the simultaneous estimation of K_a values for the interaction of isomaltooligosaccharides having various

[19] S. Honda, K. Togashi, and A. Taya, *J. Chromatogr. A* **791,** 307 (1997).

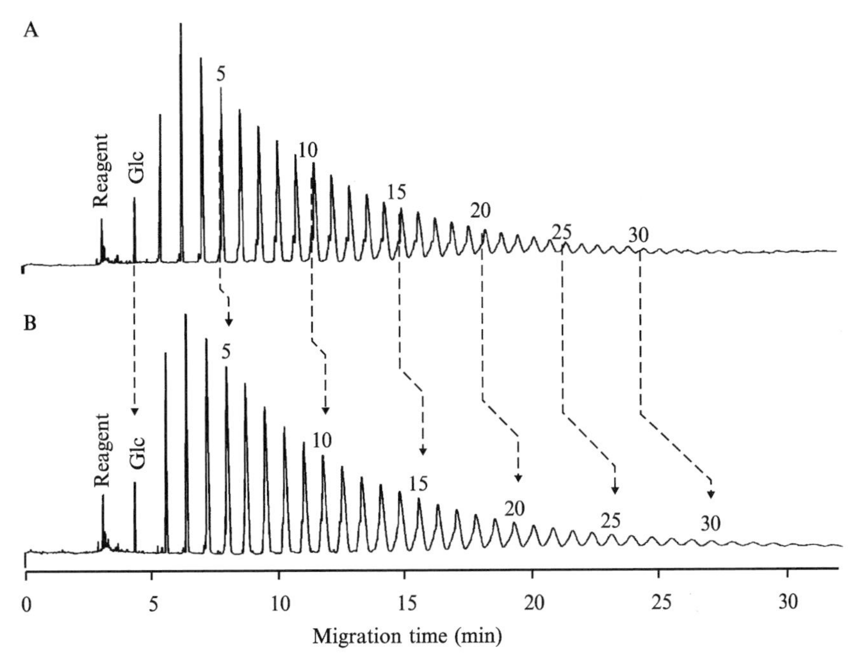

FIG. 10. Retardation of the peaks of maltooligosaccharides derivatized with ANTS in the presence of LCA.[5] Capillary, linear polyacrylamide-coated fused silica (50 μm i.d., 50 cm); electrophoretic solution, 50 mM phosphate buffer (pH 6.8) not containing (A) or containing LCA (0.6 mg/ml) (B); capillary oven temperature, 30°; applied voltage, 15 kV; detection, UV absorption at 280 nm. Reagent, ANTS. The peak numbers correspond to degree of polymerization. Reproduced from Ref. 5 with permission of the publisher.

degrees of polymerization with LCA. The ANTS derivatives of these oligosaccharides are well separated from each other and the migration delay increases with degree of polymerization. Figure 11 shows the relationship between the obtained K_a values and degree of polymerization.

Interestingly, K_a increases almost linearly with degree of polymerization, consistent with our previous findings for the interaction of maltooligosaccharides with LCA studied by high-performance affinity chromatography[18] and CE.[8]

The capability of CE in the reverse system for simultaneous K_a estimation has been further extended to the study of glycoprotein glycoforms. Every glycoprotein has a polypeptide core that is glycosylated to various extents during processing in cells to give glycoforms. Because various cell events are considered to be the result of the association of glycoprotein glycans with a specific protein, K_a values for individual glycoforms should be measured by a reliable method. Here, the CE method, which allows

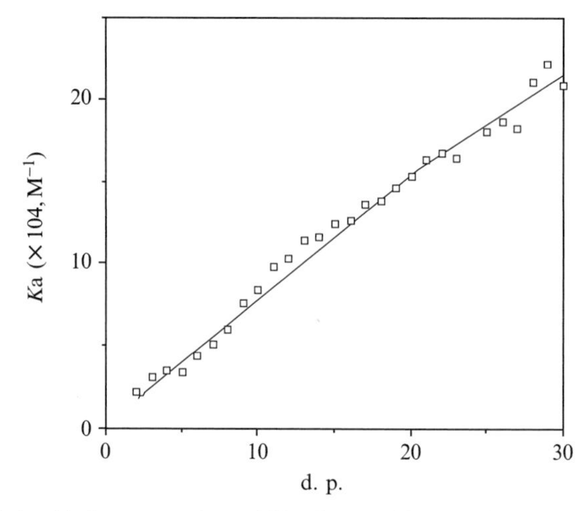

FIG. 11. Relationship between estimated K_a values and degree of polymerization, based on the result in Fig. 10. Reproduced from Ref. 5 with permission of the publisher.

simultaneous K_a estimation, is most suited for this purpose. The following is a typical example of such studies. Hen ovalbumin is known to have various asialoglycans belonging to the high-mannose and hybrid types. Previous work on high-performance affinity chromatography using an ovalbumin glycan-immobilized silica column indicated that individual glycans behave in different ways in the recognition of proteins and allowed efficient separation of several lectins by stepwise elution with competing sugars.[20] Therefore, it is expected that glycoforms may have various K_a values for lectins.

Figure 12 shows the electropherograms of a purified ovalbumin preparation in the presence of various concentrations of LCA. Electropherograms a, b, c, d, and e (Fig. 12) are obtained in the presence of LCA at 0, 0.4, 0.8, 1.4, and 2.0 mg/ml, respectively. Separation of glycoforms is not good but addition of LCA causes peak splitting and change in migration profile. The change in profile implies retardation of specific peaks by interaction with this lectin. The ovalbumin glycan chains are known to stem from the Asn-293 residue of the polypeptide core and to have structures as shown in Fig. 13.

Because LCA has weak specificity for the mannose residue, glycoforms having high mannose-type glycans, such as A, B, and D (Fig. 13) will be more retarded. In addition, larger glycans will be more retarded, because affinity is generally considered to increase with molecular mass, that is,

[20] S. Honda, K. Suzuki, S. Suzuki, and K. Kakehi, *Anal. Biochem.* **169,** 239 (1988).

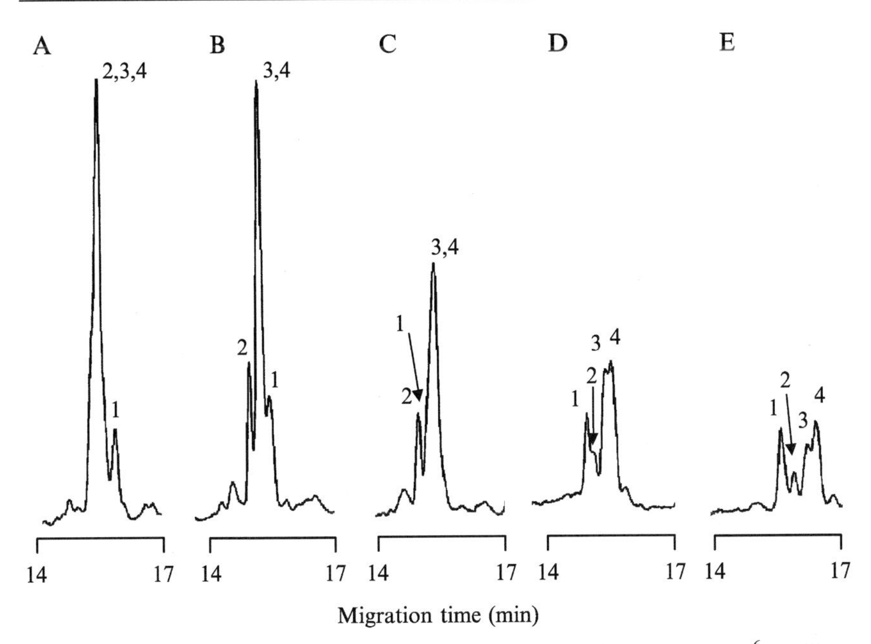

FIG. 12. Change in the migration profile of ovalbumin glycoforms with LCA.[6] Capillary, linear polyacrylamide-coated fused silica (50 μm i.d., 58 cm); electrophoretic solution, 100 mM phosphate buffer (pH 6.8) not containing LCA (A) or containing LCA at 0.4 mg/ml mg/ml (B), 0.8 mg/ml (C), 1.4 mg/ml (D), or 2.0 mg/ml (E). Capillary oven temperature, 30°; applied voltage, 20 kV; detection, UV absorption at 214 nm. The numbering of peaks is tentative. Reproduced from Ref. 6 with permission of the publisher.

degree of polymerization, as evidenced by the result of isomaltooligosaccharides–LCA interaction mentioned above. On the basis of these assumptions and known relative abundance, peaks 1–4 (Fig. 12) are assigned as follows: peak 1, C + E + H + 1; peak 2, E + G; peak 3, A; peak 4, B + D. This assignment is tentative and will be improved by further experiments using more lectins having other types of specificity. Approximate K_a values estimated as described above are as follows: peak 1, $5.4 \times 10^4 M^{-1}$; peak 2, $6.4 \times 10^4 M^{-1}$; peak 3, $6.9 \times 10^4 M^{-1}$; peak 4, $8.3 \times 10^4 M^{-1}$. The values for peaks other than peak 3 will be only the overall values for the plural glycoforms.

The above-mentioned results highlight the advantage of the reverse system for simultaneous estimation of K_a values of multiple carbohydrates to a common protein. The simultaneous K_a estimation for glycoprotein glycoforms should especially be evaluated, because such estimation will give useful information for understanding the mechanism of carbohydrate–protein interaction in living bodies.

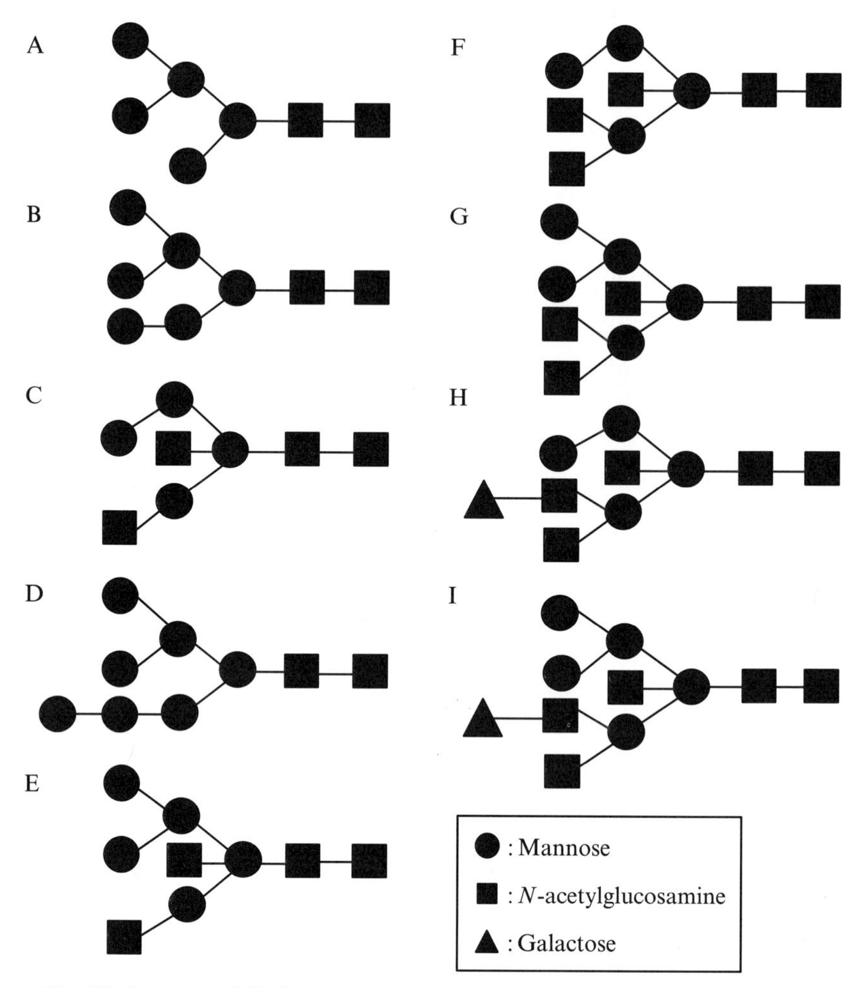

FIG. 13. Structures of *N*-glycans in the ovalbumin glycoforms. Reproduced from Ref. 6 with permission of the publisher.

Observation of Process of Enzymatic Reaction

An enzymatic reaction usually proceeds slowly between an enzyme and a substrate to give a product. If a substrate is introduced into a capillary filled with an electrophoretic solution containing a specific enzyme and an electric field is applied, the substrate will first associate with the enzyme to form an adduct, which will subsequently change to the product while moving through the electrophoretic solution. If the electrophoretic solution has an appropriate salt composition, the product will move to the detection

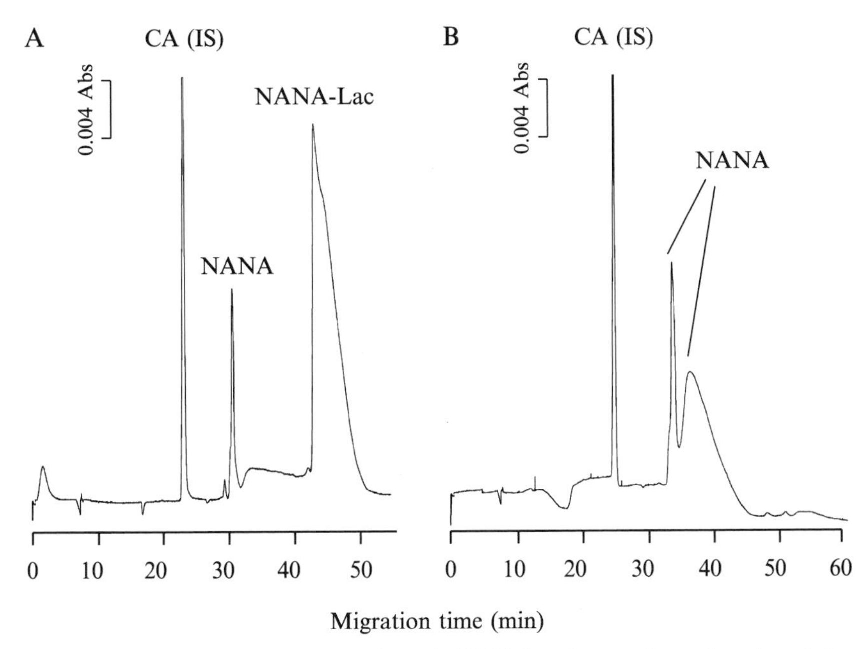

FIG. 14. *In situ* enzymatic digestion of NANA-Lac in an electrophoretic solution containing sialidase. Capillary, linear polyacrylamide-coated fused silica (50 μm i.d., 50 cm); electrophoretic solution, 50 m*M* acetate buffer (pH 5.0) not containing sialidase (A) or containing sialidase (250 mU/ml) (B); capillary oven temperature, 37°; applied voltage, 5 kV; detection, UV absorption at 200 nm. NANA-Lac was introduced to the electrophoretic solutions from the anodic end, and voltage was applied between the inlets of the capillary. The large peaks are NANA-Lac (A) or NANA formed by digestion (B). The sharp small peaks at ~ 30 min in both electropherograms are due to NANA present as an impurity in NANA-Lac. Reproduced from Ref. 21 with permission of the publisher.

window and will be detected there. Figure 14a and b shows the electropherograms of *N*-acetylneuraminyllactose (NANA-Lac), obtained when introduced to an acetate buffer (pH 5.0) and the same buffer containing sialidase from *Clostridium perfringens*, respectively, filled in a linear polyacrylamide-coated capillary.[21] In the absence of sialidase the electropherogram shows the presence of NANA-Lac as a large peak at ~45 min, accompanied by a minor sharp peak of NANA as an impurity in NANA-Lac at ~30 min. In the presence of sialidase, however, the NANA-Lac peak disappears, and instead a broad peak appears at a shorter migration time centered at ~37 min. A sharp peak of NANA as an impurity is also observed at ~30 min. On application of voltage, the NANA-Lac begins

[21] A. Taga, M. Sugimura, S. Suzuki, and S. Honda, *J. Chromatogr. A* **954,** 259 (2002).

to move through the sialidase-containing buffer and undergoes an enzymatic reaction with the sialidase to gradually give a broad peak of the product. It appears faster than the NANA-Lac peak in Fig. 14a and the shape is rather fronting. This broad peak is due to the formation of NANA from NANA-Lac by hydrolytic cleavage at the zero-order function, and the rate constant can be estimated by the [C]–t plot based on this electropherogram. In this CE method higher applied voltage causes faster movement of the substrate (NANA-Lac), allowing shorter contact time. Based on the optimization of applied voltage, a two-step application (5 kV followed by 20 kV) has proved to be the optimum, and under these optimized conditions the conjugated NANA in NANA-Lac can accurately be determined. These conditions also allow quantitative determination of the conjugated NANA in fetuin, a typical sialoglycoprotein in serum. Because this CE method gives information about the amount of NANA released in a specified time, sialidase activity can also be estimated, the obtained activity being 65 U/mg for a nominal 85 U/mg. This is a reasonable value considering some extent of deactivation during long storage.

Concluding Remarks

In this article we have focused on the efficient application of CE for studies related to carbohydrate–protein interaction. It includes a search for carbohydrate-binding proteins in a biological fluid, K_a estimation in both a normal and a reverse system, and observation of an *in situ* enzymatic reaction. All these applications demonstrate the usefulness of the CE-based method, and the results obtained are considered to provide important information about various kinds of biological phenomena induced by carbohydrate–protein interaction.

The area of separation science is rapidly growing and the age of microchip technology has dawned. Based on this trend, microchip electrophoresis (ME) is expected to become a tool for rapid analysis. A few articles have already been published for carbohydrate analysis, that report successful separation of several carbohydrates within 1 min by ME.[22] Application of ME to studies of carbohydrate–protein interaction will appear soon and might promise a future for ME in glycobiology. Although the speed-up of the analytical procedure will bring forth inherent problems to be solved, we must keep our eyes on this trend of separation science.

[22] S. Suzuki, N. Shimotsu, S. Honda, A. Arai, and H. Nakanishi, *Electrophoresis* **22,** 4023 (2001).

[31] Carbohydrate–Lectin Cross-Linking Interactions: Structural, Thermodynamic, and Biological Studies

By TARUN K. DAM and C. FRED BREWER

Introduction

The carbohydrate moieties of glycoproteins and glycolipids have been shown to be involved in a variety of biological recognition processes including cell–cell and cell–substratum interactions, fertilization, immunity, apoptosis, and metastasis of tumor cells.[1–3] The composition and structures of the oligosaccharides correlate with cell differentiation and transformation. For example, the expression of oligosaccharides possessing specific Lewis blood group antigenic determinants is developmentally regulated and altered as a result of differentiation and oncogenic transformation.[4]

The molecular recognition properties of the oligosaccharide chains of glycoproteins and glycolipids are often characterized in terms of their interactions with lectins. Binding of multivalent lectins to cells often leads to cross-linking and aggregation of specific glycoprotein and glycolipid receptors, with concomitant biological responses. For example, cross-linking of glycoconjugates on the surface of cells has been implicated in the mitogenic activities of lectins including concanavalin A (Con A),[5] in the arrest of bulk transport in ganglion cell axons,[6] in the molecular sorting of glycoproteins in the secretory pathways of cells,[7] and in the apoptosis of activated human T cells.[2] Furthermore, lectin-induced cross-linking of transmembrane glycoproteins leads to changes in their interactions with cytoskeletal proteins and concomitant alterations in the mobility and aggregation of other surface receptors.[8] A number of mammalian lectins are involved in receptor-mediated endocytosis of glycoproteins,[1] whereas others have been implicated in cellular recognition processes including apoptosis[2] and metastasis.[9]

[1] K. Drickamer and M. E. Taylor, *Annu. Rev. Cell Biol.* **9,** 237 (1993).
[2] N. L. Perillo, K. E. Pace, J. J. Seilhamer, and L. G. Baum, *Nature* **378,** 736 (1995).
[3] I. Brockhausen and W. Kuhns, "Glycoproteins and Human Disease." R. G. Landes, Austin, TX, 1997.
[4] S. Hakomori, *Annu. Rev. Immunol.* **2,** 103 (1984).
[5] G. L. Nicolson, *Biochim, Biophys. Acta* **457,** 57 (1976).
[6] B. T. Edmonds and E. Koenig, *Cell Motil. Cytoskel.* **17,** 106 (1990).
[7] K.-N. Chung, P. Walter, G. W. Aponte, and H.-P. Moore, *Science* **243,** 192 (1989).
[8] K. L. Carraway and C. A. C. Carraway, *Biochim. Biophys. Acta* **988,** 147 (1989).
[9] K. N. Konstantinov, B. A. Robbins, and F.-T. Liu, *Am. J. Pathol.* **148,** 25 (1996).

0076-6879/03 $35.00

Lectins undergo two general types of cross-linking interactions with multivalent carbohydrates, designated as type 1 and type 2 complexes.[10] In a type 1 complex, binding between a divalent lectin and a divalent carbohydrate results in one-dimensional cross-linking (e.g., helical). In a type 2 complex, binding between a multivalent lectin and multivalent carbohydrate, where the valency of either the lectin or carbohydrate is greater than two, results in two-dimensional (planar or tubular) or three-dimensional cross-linking (crystalline, in some cases). Importantly, type 2 interactions can lead to the formation of homogeneous cross-linked complexes, even in the presence of mixtures of the molecules.[10] Hence, type 2 interactions can be an important source of binding specificity between lectins and glycoconjugate receptors. There are several reports of X-ray crystal structures of lectin–carbohydrate cross-linked complexes.[11–15]

In general, multivalent protein–carbohydrate interactions increase the affinity and specificity of monovalent interactions by two means. First, multivalent carbohydrate binding to a lectin possessing multiple subsites per subunit results in large increases in the affinity of interaction.[16] Second, binding of a multivalent carbohydrate to a lectin possessing a single binding site per subunit can result in smaller increases in affinity, and, under the appropriate stoichiometric conditions, cross-linking of lectin molecules by the carbohydrate leading to type 1 or 2 complexes.[10] Because of the importance in understanding the structural and thermodynamic basis of lectin–carbohydrate cross-linking interactions and their biological significance, a combination of methods including quantitative precipitation analysis, precipitation kinetics, electron microscopy (EM), X-ray powder diffraction, model building, X-ray crystallography, isothermal titration calorimetry (ITC), and confocal microscopy has been employed to study these interactions. These methods, their applications, and findings are discussed below.

[10] C. F. Brewer, *Chemtracts Biochem. Mol. Biol.* **6,** 165 (1996).

[11] W. I. Weis, K. Drickamer, and W. Hendrickson, *Nature* **360,** 127 (1992).

[12] Y. Bourne, B. Bolgiano, D.-I. Liao, G. Strecker, P. Cantau, O. Herzberg, T. Feizi, and C. Cambillau, *Nat. Struct. Biol.* **1,** 863 (1994).

[13] A. Dessen, D. Gupta, S. Sabesan, C. F. Brewer, and J. C. Sacchettini, *Biochemistry* **34,** 4933 (1995).

[14] L. R. Olsen, A. Dessen, D. Gupta, S. Sabesan, J. C. Sacchettini, and C. F. Brewer, *Biochemistry* **36,** 15073 (1997).

[15] C. S. Wright and G. Hester, *Structure* **4,** 1339 (1996).

[16] R. Lee and Y. C. Lee, *Glycoconj. J.* **4,** 317 (1987).

Quantitative Precipitation Analysis

Methods

Multivalent Lectin–Oligosaccharide and Lectin–Glycopeptide Cross-Linked Complexes. Increasing amounts of an individual oligosaccharide or glycopeptide are added to a series of tubes containing fixed amounts of lectin. The final volume of each tube is made identical with buffer and the solution is allowed to stand until equilibrium is attained. The precipitate is collected after centrifugation, washed with cold buffer, and dissolved with competing monosaccharide. The amount of total lectin precipitated is determined from the absorbance at 280 nm or from radioactivity (if the lectin is labeled). In addition, the concentration of lectin in the supernatants is measured to check the percentage of lectin precipitated (Fig. 3A and B).[17,18] The concentration of oligosaccharide or glycopeptide in the precipitates is determined by phenol sulfuric acid assay or radioactivity (if the ligand is labeled).

If a binary mixture of oligosaccharides or glycopeptides is used to precipitate a lectin, single or parallel experiments can be performed, depending on the labeling status of the ligands and lectin. One or both oligosaccharides or glycopeptides are labeled with ^{3}H or ^{14}C in each experiment (Figs. 3D and 4B).[19,20] If a single oligosaccharide or glycopeptide is used for a binary mixture of two lectins, one or both lectins may be labeled with ^{3}H or ^{14}C (Fig. 3C).[20]

Multivalent Lectin–Glycoprotein Cross-Linked Complexes. Increasing amounts of ^{3}H or ^{14}C-labeled glycoprotein are placed into a series of tubes and buffer is added to make the volumes equal. A fixed amount of labeled lectin (^{3}H or ^{14}C, depending on the label of glycoprotein) is then added to each tube. Precipitation is allowed until equilibrium is reached. The supernatant is removed after centrifugation and the precipitate is washed three times with cold buffer. After the final washing, the precipitate is dissolved in buffer containing competitive monosaccharide. The solutions are then analyzed for total protein content (absorbance) as well as radioactivity (Fig. 4A).[21] When a binary mixture of two glycoproteins is assayed with a lectin, parallel experiments are performed with the glycoproteins alternately labeled and tested against labeled lectin.[22]

[17] L. Bhattacharyya, M. Haraldsson, and C. F. Brewer, *Biochemistry* **27,** 1034 (1988).
[18] L. Bhattacharyya and C. F. Brewer, *Eur. J. Biochem.* **178,** 721 (1989).
[19] L. Bhattacharyya, M. I. Khan, and C. F. Brewer, *Biochemistry* **27,** 8762 (1988).
[20] L. Bhattacharyya and C. F. Brewer, *Eur. J. Biochem.* **208,** 179 (1992).
[21] M. I. Khan, D. K. Mandal, and C. F. Brewer, *Carbohydr. Res.* **213,** 69 (1991).
[22] D. K. Mandal and C. F. Brewer, *Biochemistry* **31,** 12602 (1992).

TABLE I
RADIOACTIVE LABELING CONDITIONS IN DIFFERENT QUANTITATIVE PRECIPITATION ASSAYS

Components of assay	Radioactive label (^{14}C or ^{3}H)
1 oligosaccharide–1 lectin	No label
2 equimolar oligosaccharides–1 lectin	No label
2 equimolar oligosaccharides–1 lectin	One oligosaccharide labeled at a time
1 oligosaccharide–2 equimolar lectins	One lectin labeled
1 oligosaccharide–2 equimolar lectins	Two lectins labeled
2 glycopeptides–1 lectin	Both glycopeptides labeled
1 glycoprotein–1 lectin	Both labeled
2 glycoproteins–1 lectin	One glycoprotein labeled at a time and the lectin labeled

Quantitative precipitation experiments can be performed under different conditions including different buffers, temperature, and salt concentrations. The reaction time can vary from 1 to 96 h. The final reaction volumes range from 100 to 1000 μl. Oligosaccharides, glycopeptides, glycoproteins, and lectins are labeled with ^{3}H or ^{14}C by reductive methylation.[23] Table I presents some of the labeling requirements in different types of quantitative precipitation assays.

Results

Various aspects of cross-linking interactions of multivalent oligosaccharides and glycoproteins with lectins can be studied by quantitative precipitation analysis.

Cross-Linking of Lectins with Oligosaccharides and Glycopeptides. Quantitative precipitation assays show that D-galactose-specific lectins from *Ricinus communis* (agglutinin I) (RCA-I), *Erythrina indica* (EIL), *Erythrina arborescens* (EAL), and *Glycine max* (soybean) (SBA) are cross-linked and precipitated by tri- and tetraantennary complex-type oligosaccharides containing nonreducing terminal galactose residues (Fig. 1). The presence or absence of a bisecting GlcNAc in biantennary complex-type oligosaccharides has little effect on the binding activities and valencies of the carbohydrates, in contrast to the results found with Con A as described below. Asialofetuin glycopeptide **7** also precipitates RCA-I and EIL, and is trivalent in both cases, as is corresponding oligosaccharide **4**.

The ratio of concentration of oligosaccharide or glycopeptide and the concentration of lectin monomer at the equivalence zone (region of maximum precipitation) gives the valency of the carbohydrate to the lectin

[23] N. Jentoft and D. G. Dearborn, *Methods Enzymol.* **91,** 570 (1983).

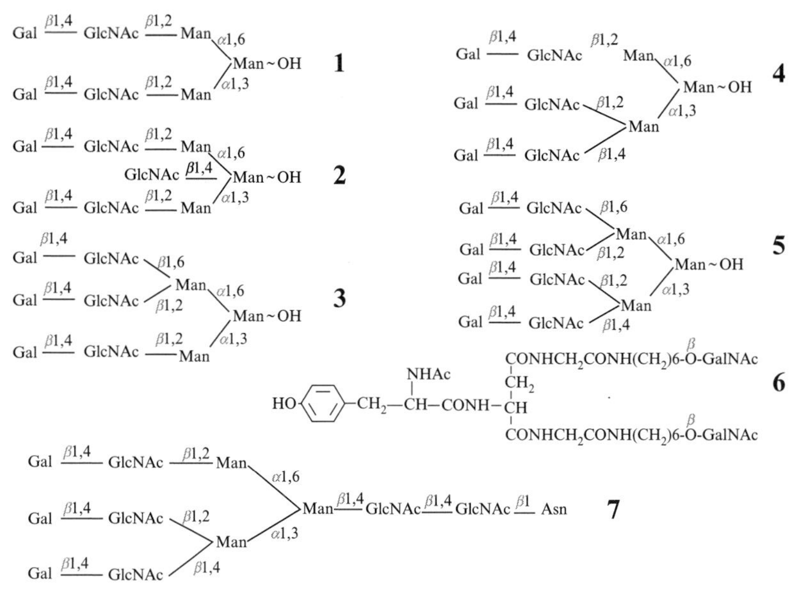

FIG. 1. Structures of oligosaccharides **1–6** and of glycopeptide **7**.

(Fig. 3).[24] For RCA-I, EIL, EAL, and SBA, the ratios of the concentrations of the triantennary oligosaccharides **3** and **4** to protein monomers are approximately 1:3 (trivalent). Compound **5** is tetravalent for EIL, EAL, and SBA but trivalent for RCA-I.[17] The valency of a carbohydrate is often equal to the number of its different chains containing the binding epitope, although in some instances structural valency may be different from functional valency. Branching patterns, chain lengths, affinity, and valency of the carbohydrate and the structures of lectins influence the extent of their precipitation.

High mannose-type glycopeptides **9**, **10**, and **11** and the bisected hybrid-type glycopeptide **12** (Fig. 2) are capable of specifically binding and precipitating Con A and are divalent for the lectin (Fig. 3B). Nuclear magnetic relaxation dispersion (NMRD) measurements show that these glycopeptides as well as **13** bind primarily by the trimannosyl moiety on their α(1-6) arms (primary site). However, unlike **9–12**, **13** fails to precipitate the protein under the same conditions because **13** has the primary (high-affinity)

[24] E. A. Kabat, "Structural Concepts in Immunology and Immunochemistry," 2nd ed. Holt, Rinehart and Winston, New York, 1976.

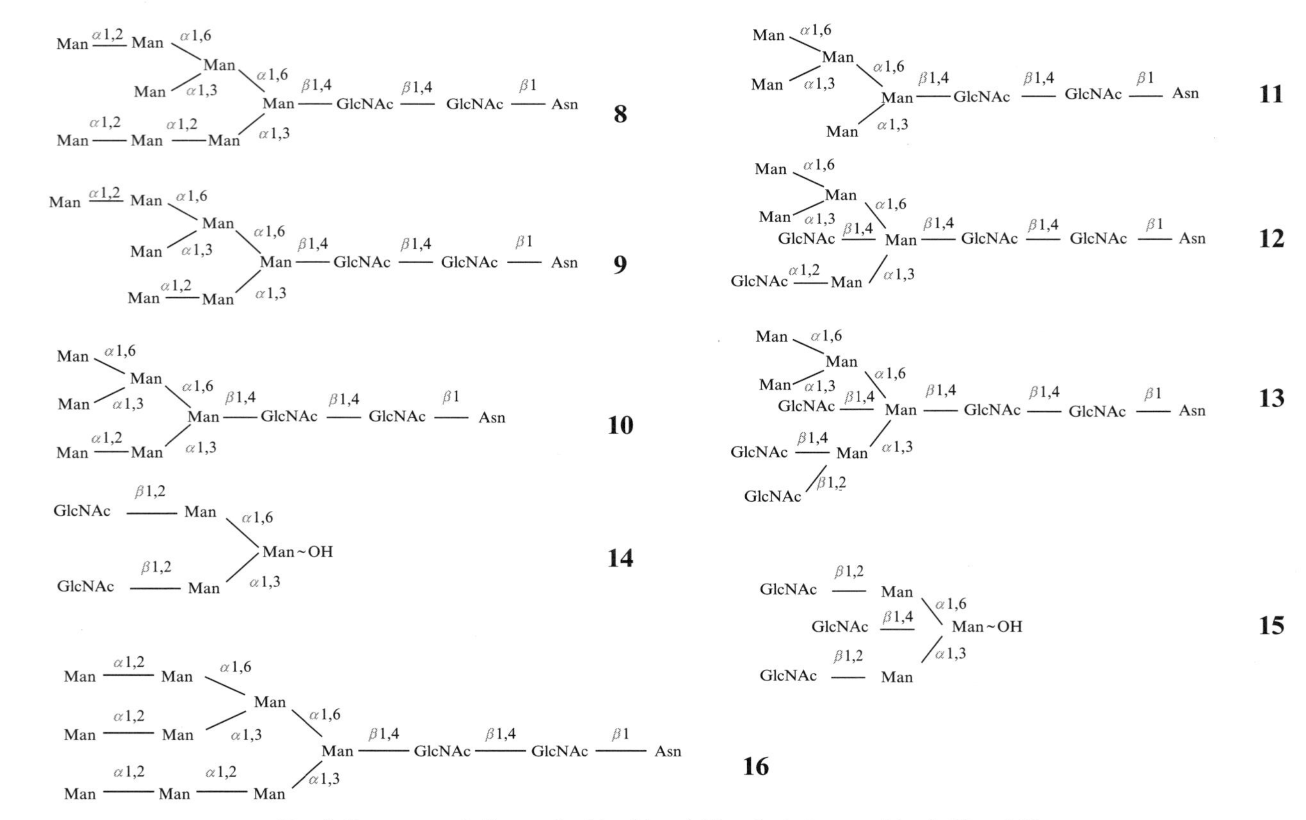

FIG. 2. Structures of oligosaccharides **14** and **15** and of glycopeptides **8–13** and **16**.

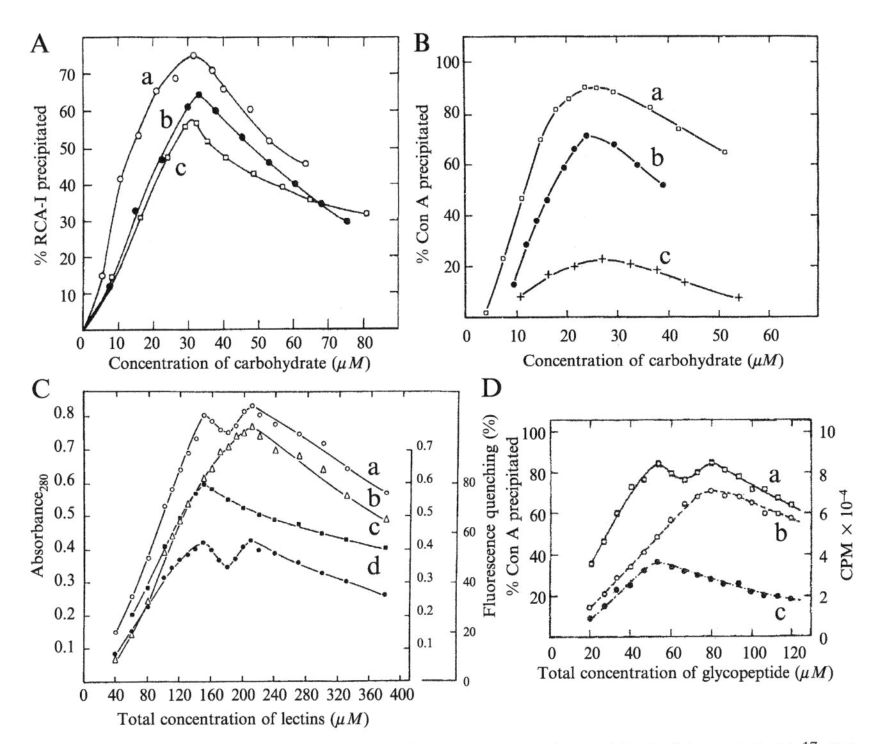

FIG. 3. (A) Curves showing precipitation of RCA-1 by **5** (a), **4** (b), and **3** (c).[17] (B) Precipitation of Con A by **10** (a), **12** (b), and **11** (c).[18] (C) Precipitation of an equimolar mixture of [^{14}C] MeEIL and RCA-I in the presence of [^{3}H] **5**: total protein (a), ^{14}C (b), fluorescence quenching (c), and ^{3}H (d) are shown.[20] (D) Precipitation profile of Con A (a) in the presence of a 1:1 mixture of [^{14}C] **11** and [^{3}H] **16**.[19] The counts per minute of **16** (b) and **11** (c) are also shown.

binding sites, but not the secondary sites (low affinity), which are located on their α(1–3) arms. Changes in the structures and affinities of both the primary and secondary sites of the glycopeptides influence their precipitation activities. These findings may provide insight into the relationship between the microheterogeneity of N-linked carbohydrates and their specificity as cell surface receptors associated with their cross-linking activities.[25] Bisected complex-type oligosaccharides such as **15** are also bivalent for Con A whereas the nonbisected analogs such as **14** are univalent for the lectin. NMRD and precipitation data indicate that nonbisected and

[25] L. Bhattacharyya, C. Ceccarini, P. Lorenzoni, and C. F. Brewer, *J. Biol. Chem.* **262,** 1288 (1987).

bisected complex-type carbohydrates bind with different mechanisms and conformations.[26] The presence of the bisecting GlcNAc residue in complex type carbohydrates has profound influence on their mode of binding to Con A, but the presence of the same in hybrid-type glycopeptides (such as **12**) has little effect on their interactions.

Cross-Linking of Lectins with Glycoproteins. Quantitative precipitation analysis of the interactions of Con A with the soybean agglutinin (SBA), which is a tetrameric glycoprotein possessing a single Man-9 oligomannose chain (**16**) per monomer, shows that SBA forms two different types of cross-linked complexes (1:1 and 2:1) with tetrameric Con A, depending on the relative ratio of the two molecules in solution. However, SBA forms only 1:1 cross-linked complexes with dimeric forms of Con A (Fig. 4A). Thus the total valency of the carbohydrate of SBA is a function of both the quaternary structure of Con A as well as the relative ratio of SBA to Con A. In addition, the individual Man-9 oligosaccharide, which as a glycopeptide is bivalent for binding to Con A, expresses univalency when present on the protein matrix of SBA.[21]

Galectin-1 from calf spleen and galactose-specific plant lectins such as EIL, *Erythrina cristagalli* lectin (ECL), and SBA form specific cross-linked complexes with asialofetuin (ASF), a 48-kDa monomeric glycoprotein that contains three N-linked oligosaccharides **7**. Formation of 1:9 and 1:3 stoichiometric cross-linked complexes (per monomer) of ASF to galectin-1, depending on their relative ratio in solution, has been documented.[27] Three triantennary N-linked complex-type oligosaccharide chains of ASF mediate the cross-linking interactions, and each chain expresses either trivalency in the 1:9 cross-linked complex or univalency in the 1:3 complex (Fig. 5B). The two dimeric *Erythrina* lectins also form 1:9 and 1:3 ASF:lectin cross-linked complexes. In the presence of tetrameric SBA, only a 1:3 ASF:lectin cross-linked complex is formed.[27]

The valency of the carbohydrate chains of ASF for a series of Gal-specific plant lectins and an animal lectin is determined by (1) the number and composition of the carbohydrate chains on ASF, (2) the quaternary structure and size of the lectins, and (3) the relative ratio of the glycoprotein to lectin in solution.

Formation of Unique Cross-Linked Complexes in Mixed Precipitation Systems. Evidence that specific lectins form homogeneous cross-linked lattices with multivalent carbohydrates can be obtained in mixed quantitative precipitation experiments. For example, the galactose-specific lectins from EIL and RCA-I show two separate peaks, respectively, in their

[26] L. Bhattacharyya, M. Haraldsson, and C. F. Brewer, *J. Biol. Chem.* **262,** 1294 (1987).
[27] D. K. Mandal and C. F. Brewer, *Biochemistry* **31,** 8465 (1992).

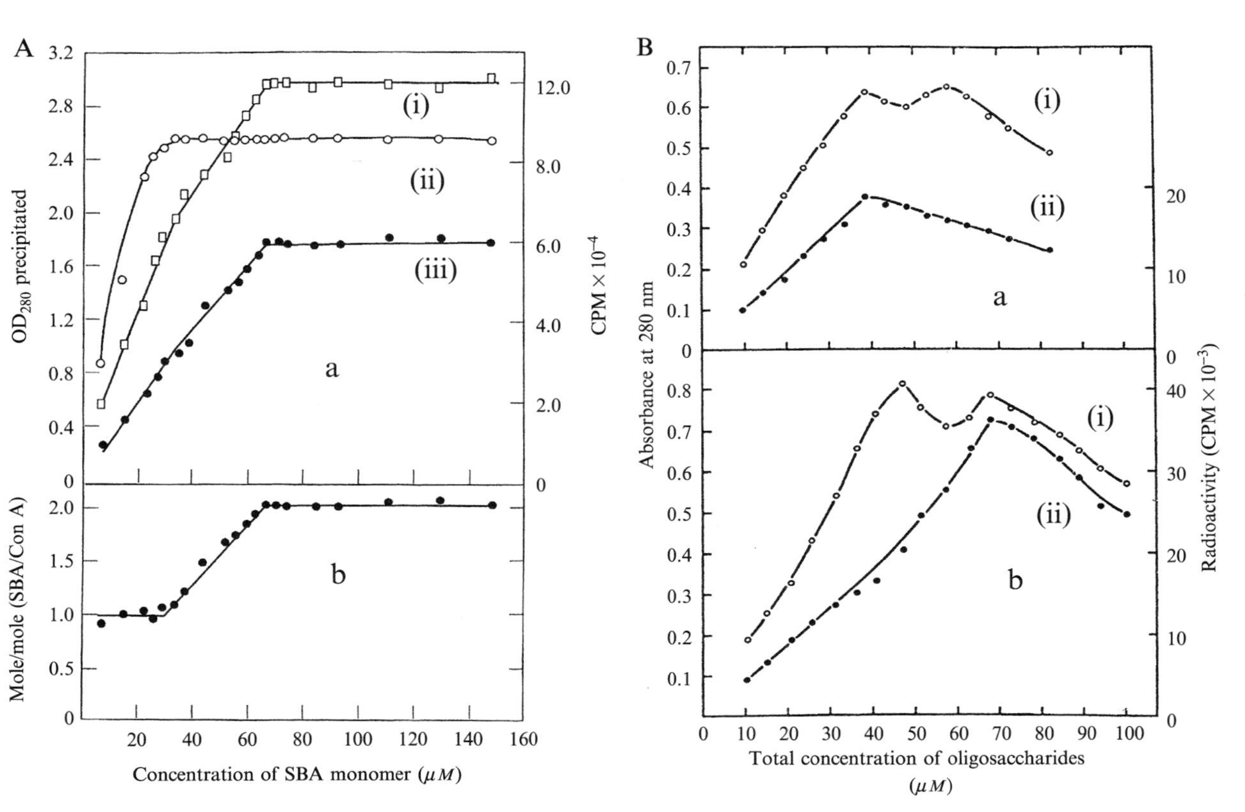

FIG. 4. (A) Quantitative precipitation of Con A by SBA: (a) profiles of total protein precipitated (i) and counts per minute of [^{3}H] Con A (ii) and [^{14}C] SBA (iii) in the precipitate; (b) ratio of moles of SBA precipitated per mole of Con A.[21] (B) Precipitation of EIL with equimolar mixtures of (a) [^{3}H] **4** and **5**, and (b) **4** and [^{3}H] **5**. Total protein (i) and counts per minute of ^{3}H (ii) are shown.[20]

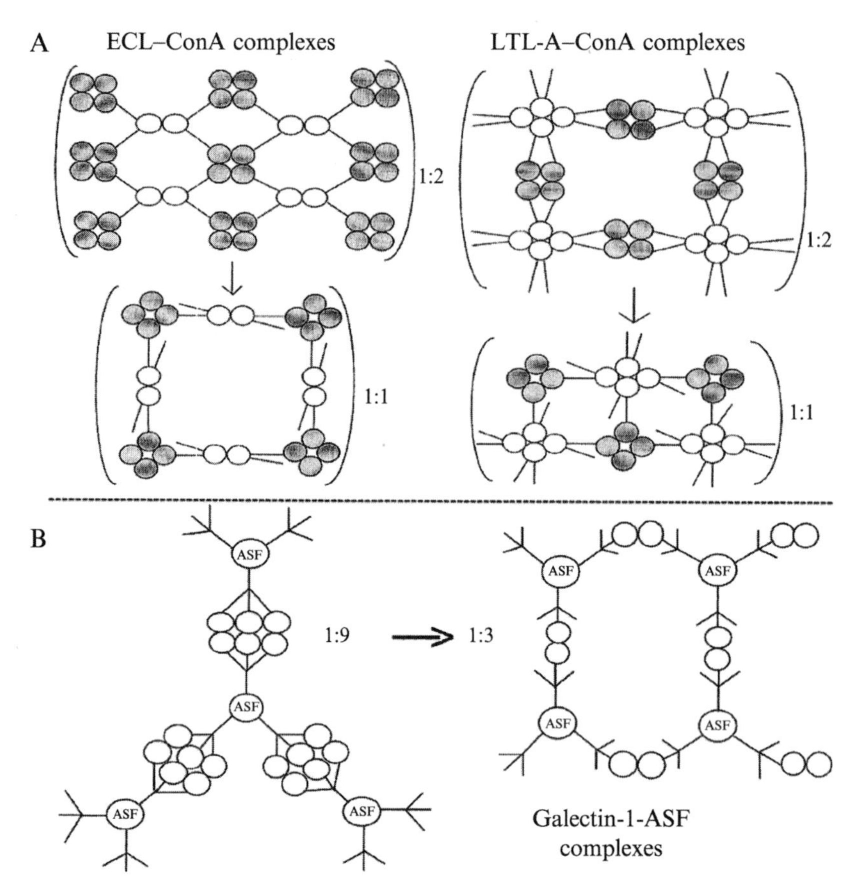

FIG. 5. Schematic representations of (A) 1:2 and 1:1 ECL–Con A complexes and of 1:2 and 1:1 LTL-A–Con A complexes. The arrows indicate transformation of one cross-linked complex to another complex. The hatched circles represent Con A tetramers and the open circles and connected line structures represent the protein and carbohydrate moieties, respectively, of different glycoprotein lectins.[22] (B) ASF:galectin-1 cross-linked complexes (1:9 and 1:3). A pair of circles (unmarked) represents the dimeric galectin-1.[29]

quantitative precipitation profiles in the presence of binary mixtures of **4** and **5** (Fig. 4B). Similar observations were made with SBA and **20–23**[28] and with Con A and glycopeptides **11** and **16** (Fig. 3D).[19] These results provide evidence for the formation of homogeneous cross-linked lattices for each lectin with each sugar ligand. Conversely, binary mixtures of EIL,

[28] D. Gupta, L. Bhattacharyya, J. Fant, F. Macaluso, S. Sabesan, and C. F. Brewer, *Biochemistry* **33,** 7495 (1994).

ECL, RCA-I, and SBA in the presence of a single multivalent oligosaccharide show evidence for the formation of separate cross-linked lattices between each lectin and the carbohydrate (Fig. 3C).[20]

The relative precipitation maxima of the oligosaccharides/glycopeplycopeptides is determined by mass action equilibria involving competitive binding of the two carbohydrates to the protein. These equilibria, in turn, are sensitive to the relative amounts and affinities of the carbohydrates at both their primary and secondary binding sites.

Mixed quantitative precipitation studies show that Con A forms the same unique stoichiometry cross-linked complexes with ovalbumin and glycoprotein lectins (LTL-A, ECL, EcorL, and SBA) in the presence of binary mixtures of the glycoproteins. These results provide evidence that each glycoprotein forms a unique homogeneous cross-linked lattice(s) with Con A in the presence of another glycoprotein (Fig. 5A).[22] Similarly, the dimeric galectin-1 forms homogeneous aggregated cross-linked complexes with ASF in the presence of other lectins with similar specificities and cross-linking activities.[29]

These findings indicate that specific lectin may bind to the carbohydrate moieties of glycoconjugate receptors (glycoproteins and glycolipids) and cross-link them into homoaggregates in the presence of other carbohydrates or lectins. Indeed, evidence indicates that galectin-1 forms homoaggregates of surface glycoproteins of T cells.[30] It is important to note that animal lectins in general possess multimeric structures[1] and are therefore potentially capable of forming cross-linked complexes with specific multivalent carbohydrate ligands, which may be important for their biological activities.

Kinetics of Precipitation

Methods

Kinetics of multivalent lectin–carbohydrate precipitation are determined by monitoring the time course of development of turbidity at 420 nm in Uvonic type 17Q cells (path length, 1 cm) in a Gilford 260 spectrophotometer coupled with a Cole-Palmer (Vernon Hills, IL) model 8373-10 recorder and an Endocal RTE-9B temperature control bath. The solution of lectin is quickly stirred on the addition of oligosaccharide and the absorbance is monitored continuously until it remains constant.[31]

[29] D. Gupta and C. F. Brewer, *Biochemistry* **33,** 5526 (1994).

[30] K. E. Pace, C. Lee, P. L. Stewart, and L. G. Baum, *J. Immunol.* **163,** 3801 (1999).

[31] L. Bhattacharyya, J. Fant, H. Lonn, and C. F. Brewer, *Biochemistry* **29,** 7523 (1990).

Results

Dependence of Cross-Linking Kinetics on Several Factors. The rate and extent of cross-linking and precipitation can be studied spectrophotometrically by light-scattering measurements. Structures of the carbohydrates and lectins, their affinities, and other physical factors profoundly influence the kinetics of precipitation. For example, the presence or absence of the galactose residues in **17** and **18** does not affect the affinities of the carbohydrates for isolectins such as *Lotus tetragonolobus* (LTL)-A and LTL-C, but has a strong influence on the temperature sensitivity and kinetics of the precipitation reactions (Fig. 6).[31]

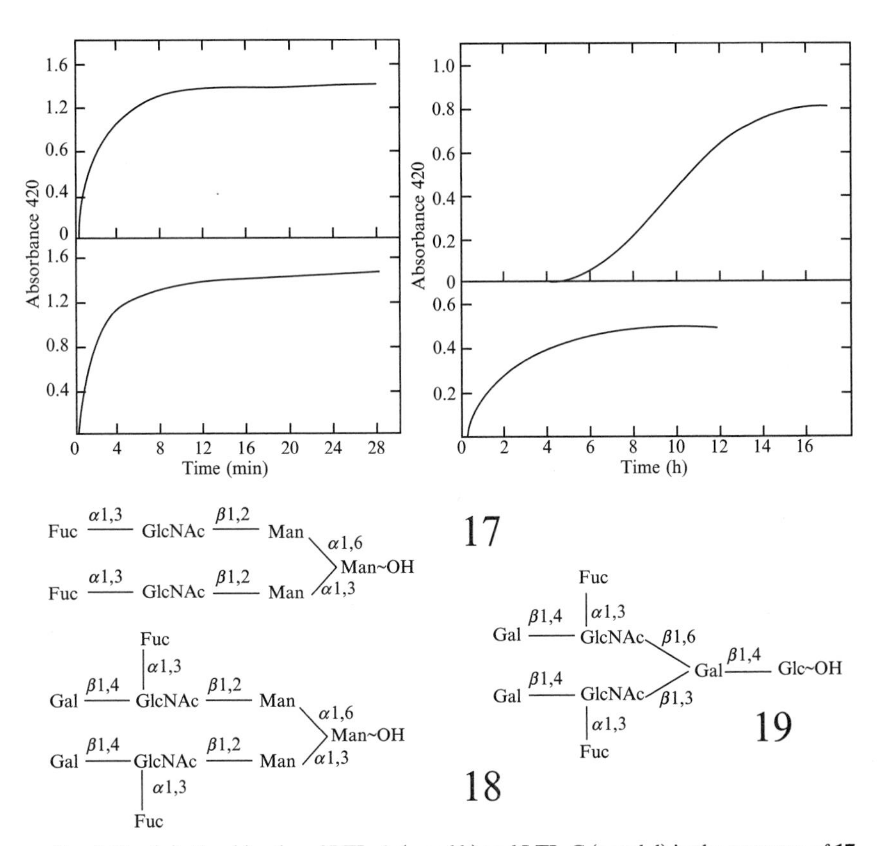

FIG. 6. Precipitation kinetics of LTL-A (a and b) and LTL-C (c and d) in the presence of **17** (a and c) and **18** (b and d).[31] *Bottom*: Structures of fucosyloligosaccharides **17–19** are shown.

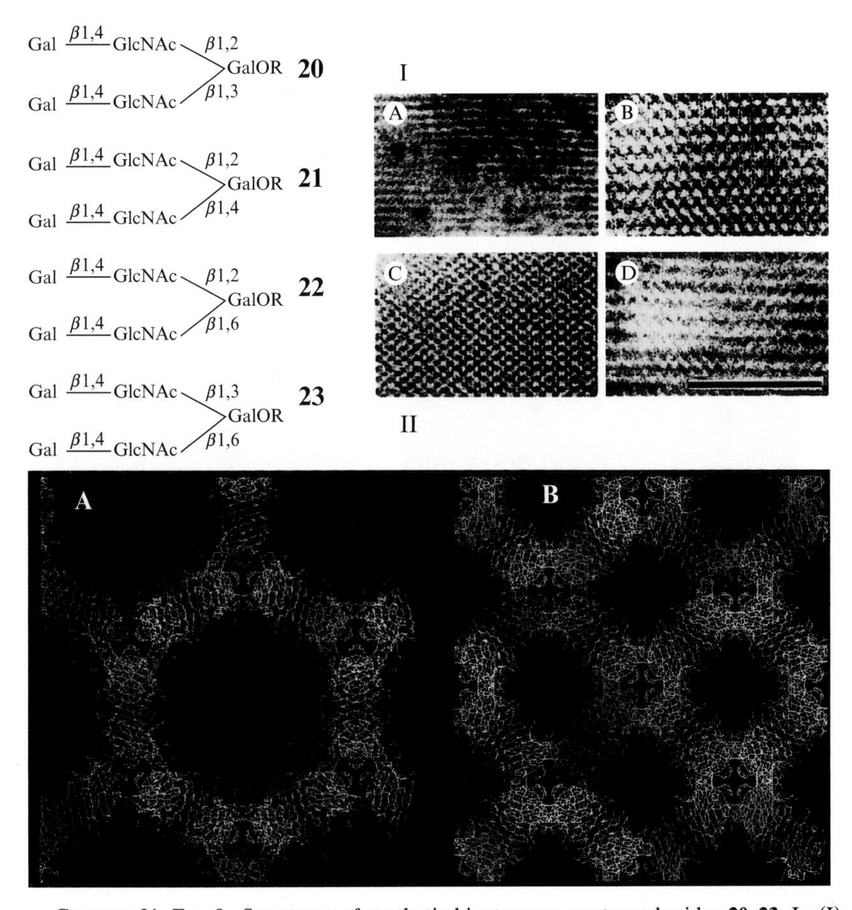

CHAPTER 31, FIG. 9. Structures of synthetic biantennary pentasaccharides **20–23**. In (I), negative stain electron micrographs of the precipitates of SBA with **20** (A), **21** (B), **22** (C), and **23** (D) are shown. Bar: (D) 0.4 m. Magnification in (A)–(D) is the same.[28] In (II), X-ray crystal structures of cross-linked complexes are shown. (A) The 6-fold axis of symmetry down the *c* axis is shown for the **23**–SBA cross-linked complex and (B) the 4-fold axis of symmetry down the *c* axis is shown for the **20**–SBA cross-linked complex. In these views, protein molecules are shown as colored trace models (monomers), with the cross-linking oligosaccharide not shown. The figures were generated with MOLPACK.[14]

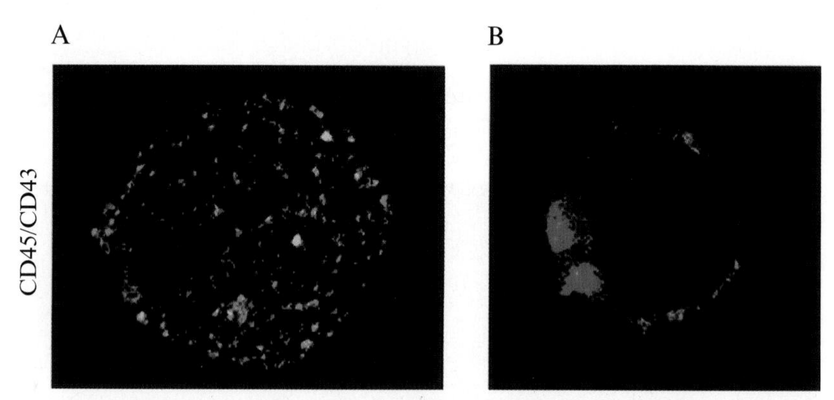

CHAPTER 31, FIG. 13. (A) Uniform distribution of CD45 (red) and CD43 (green) on the surface of MOLT-4 cells in the absence of galectin-1, with some areas of colocalization (yellow). *Top*: One 0.5-m slice from the top of a cell. *Bottom*: One 0.5-m slice through the center of a cell. (B) Segregation of CD45 (red) and CD43 (green) after galectin-1 treatment. *Top* and *bottom*: One 0.5-m slice each of two different cells.[30]

Electron Microscopic Studies of the Cross-Linked Complexes

Methods

Precipitates obtained from lectin–carbohydrate cross-linking interactions are negatively stained by placing the samples on 300 mesh carbon-coated Parlodion grids which have been freshly glow discharged, touched to filter paper, floated on a drop of 1% phosphotungstic acid, pH 7.0, and blotted immediately. Samples are observed at 80 kV in a 1200EX electron microscope (JEOL, Tokyo, Japan). For freeze–fracture studies, samples are placed in a gold double-replica device, frozen in liquid freon, and fractured in a Balzers BAF301 freeze–fracture unit (Bal-Tec, Balzers, Switzerland) at $-115°$. The fracture face is shadowed at a 45° angle with platinum and stabilized with carbon. Samples are observed as described.[31]

Results

Observation of Highly Ordered Cross-Linked Structures. Certain fucose-containing oligosaccharides, **17–19** (Fig. 6), are capable of cross-linking and precipitating with tetrameric isolectins LTL-A and LTL-C. EM of the precipitates of LTL-A shows a distinct lattice pattern for each oligosaccharide, indicating the presence of long-range order and well-defined geometry in each cross-linked complex (Fig. 8a–d). The negative stain electron micrographs of the precipitates of SBA with **1**, **2**, **5**, and **6** show similar ordered structures. Cross-linked complexes of SBA with **20–23** show distinct patterns by electron microscopy (Fig. 9).[28] X-ray crystallographic data establish that the cross-linked lattices for SBA with **20–23** are different from each other.[14]

The EM results with SBA demonstrate the existence of pseudo-2-fold axes of symmetry in naturally occurring and synthetic branch chain carbohydrates that correlate with observable EM patterns in their precipitates with SBA. The presence of such symmetry elements in both the carbohydrates and the lectin suggests that both types of molecules are designed to form unique homogeneous cross-linked complexes. Con A and *Dioclea grandiflora lectin* (DGL) also form unique EM-observable structures in complexes with mannose-containing synthetic bivalent sugars (T. K. Dam and C. F. Brewer, unpublished observations, 2002). Importantly, not all lectin–carbohydrate precipitates show observable EM patterns, even when their mixed quantitative precipitation profiles may show evidence for the formation of homogeneous cross-linking.

Observation of Distinct Cross-Linked Patterns in Mixed Precipitation Systems by EM. SBA and LTL-A form unique cross-linked complexes with distinct patterns with **2** and **17**, respectively, as observed by EM (Fig. 7A and B). Distinct patterns are observed for the two different

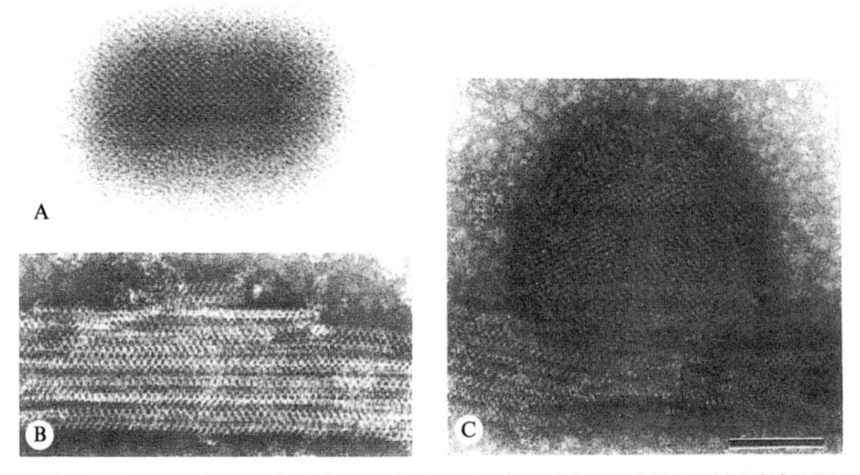

FIG. 7. Electron micrograph of the negatively stained precipitates of SBA with **2** (A), LTL-A with **17** (B), and a mixture of SBA (210 μM), LTL-A (100 μM), **2** (120 μM), and **17** (55 μM) (C). Bar: (C) 0.1 μM.[32]

complexes even when the precipitates form in a mixed precipitation system consisting of SBA, LTL-A, and oligosaccharides **2** and **17** (Fig. 7C).[32] Similarly, precipitates of SBA with a mixture of **1** and **6** show the coexistence of the individual patterns of both carbohydrates cross-linked complexes with the lectin.[28] In the mixed precipitation system the formation of each oligosaccharide–lectin lattice is mutually independent. The absence of any hybrid patterns in the mixed precipitates confirms unique molecular packing interactions that stabilize each lattice.

Structural Studies of Cross-Linked Complexes by X-Ray Powder Diffraction, Electron Micrograph Analysis, Simulation, and Model Building

X-Ray Diffraction

Methods. Pellets of the LTL-A/**19** precipitate are aspirated into clean, thin-walled glass capillaries (0.5 or 0.7 mm in diameter), and sealed with wax. X-ray diffraction patterns from the precipitates are obtained by radiation from a rotating anode X-ray source (RU200; Rigaku, Tokyo, Japan). A double-mirror camera system is used (CuK_a; $l = 1.54$ Å) to produce an X-ray beam with dimensions of approximately 0.3×0.3 mm at the film

[32] L. Bhattacharyya, M. I. Khan, J. Fant, and C. F. Brewer, *J. Biol. Chem.* **264,** 11543 (1989).

(diagnostic X-ray film; Eastman Kodak, Rochester, NY). Specimen-to-film distances are 70–130 mm for most experiments and the exposure times are usually 24–72 h.[33]

Results. All the observed reflections for the LTL-A/**19** precipitate can be indexed on an orthorhombic lattice with lattice dimensions 85.1 × 76.1 × 121.8 Å. The 76.1 × 121.8 Å dimensions correspond closely to those observed in electron micrographs of the lattice in projection. The 85.1-Å dimension corresponds to the thickness of the two-dimensional lattice. Only two "meridional" reflections (second and third order of the 85.1-Å repeat) are observed. This suggests a partial stacking of the lattices in the pellet. The remainder of the observed reflections correspond to "equatorial" reflections from the two-dimensional lattice.

Electron Micrograph Analysis

Methods. Electron micrographs of LTL-A/**19** cross-linked complexes are digitized with an EC850 digital imaging camera system (Eikonix). Optical densities of the electron micrographs in the range of 0–2 OD are coded into 8 bits (integers 0–255). At the setting used, each pixel corresponds to 2.07 Å in the structure. The areas selected from the micrographs for digital filtering contain ~260 unit cells.

To extract the periodic information and determine the symmetry of the lattice in projection, the electron micrographs are filtered to improve the signal-to-noise ratio. This procedure is implemented by computing the Fourier transform of the image, multiplying the transform by a set of Gaussian functions (one around each of the reciprocal lattice points), and then computing a back Fourier transform to obtain the filtered image.[33]

Results. The power spectrum (square of the Fourier transform) of the electron micrograph of the LTL-A/**19** precipitate in Fig. 8E (enlargement of Fig. 8D) is shown in Fig. 8J. Peaks fall on a lattice with dimensions of approximately $1/76 \times 1/122$ Å^{-1}. Peaks are observable to approximately 19-Å resolution in transforms from the most highly ordered specimens. Figure 8F is a filtered image of the electron micrograph of the lattice in Fig. 8E. In the filtered images of the LTL-A lattices, the stain-excluding material is concentrated at the vertices of the intersecting lines of stain-excluding material noted in the original micrographs. The LTL-A tetramer is centered at these points. The micrograph in Fig. 8F has mirror lines through these points, corresponding to in-plane 2-fold axes in the lattice. The Con A tetramer has 222 point group symmetry and it is likely that

[33] W. Cheng, E. Bullitt, L. Bhattacharrya, C. F. Brewer, and L. Makowsk, *J. Biol. Chem.* **273,** 35016 (1998).

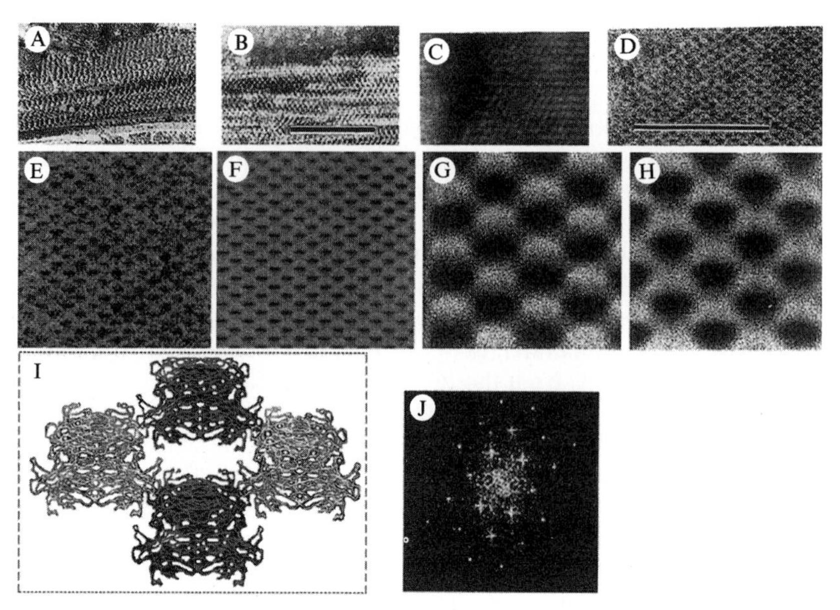

FIG. 8. (A) Freeze–fracture pattern of LTL-A with **17**; (B) negative stain pattern of the same. (C and D) Negative stain patterns of LTL-A with **18** and **19**, respectively. Bars: (B and D) 0.1 μm. Magnification in (A), (B), and (C) is the same.[31] (E) Enlargement of a portion of the electron micrograph shown in (D). The protein forms two sets of continuous, stain-excluding strands, rotated by 62–65° relative to one another. At each vertex, a pair of dimers, one on top of the other, form a tetramer. (F) Filtered image of electron micrograph from (E). Comparison of the simulated electron micrograph calculated from the molecular model of LTL-A (G) with the filtered micrograph (H). The simulated image was calculated with Con A dimer rotated 20° about the x axis to form the model for the LTL-A tetramer and the tetramer rotated 12° about the y axis. (I) Molecular model of the relative positions of the four LTL-A tetramers in the lattice. The molecular model of the Con A dimer as determined by X-ray crystallography has been used as a basis for predicting the molecular structure of LTL-A. (J) Power spectrum of the electron micrograph shown in (E).[33]

the LTL-A tetramer also exhibits this symmetry. One of the molecular 2-fold axes probably corresponds to the observed lattice 2-fold, greatly limiting the possible models for the molecular structure of the lattice. Modeling of the lattice structure requires the identification of the crystallographic 2-fold axis and the location of the two molecular 2-fold axes that are noncrystallographic.

Electron Micrograph Simulation

Methods. To investigate the orientation of the LTL-A tetramer in the lattice, a computer simulation of the filtered electron micrographs is employed.[33] Negative staining is simulated by computing the projected volume of the molecules, in essence to computationally surround the protein with "stain" of uniform density. No attempt is made to augment the density near charged groups (which would correspond to positive staining). The structure of the protein tetramer is modeled by using the three-dimensional atomic coordinates of the Con A tetramer[34] giving each atom a "weight" proportional to its van der Waals volume. Con A tetramers are rotated computationally into orientations consistent with the lattice symmetry and then projected onto the viewing plane. Calculations are made to determine the possible effect of multiple molecular layers on the images obtained. Further, the relative positions of the two dimers making up the tetramer of LTL-A are varied as suggested by differences that have been observed in the dimer–dimer interactions among homologous leguminous lectins. Mean phase differences are used as a criterion to evaluate the quality of simulations.

Results. Models of LTL-A molecules in the cross-linked lattice were generated by rotating LTL-A dimers up to 25.0° relative to one another. The best correspondence between calculated and observed micrographs and the lowest mean phase difference occurred for a relative rotation of the dimers of about 20° in a direction that brought them more nearly parallel to one another. Correspondence of the simulated micrograph with a filtered electron micrograph is demonstrated in Fig. 8G and B.

Molecular Model Building

Methods. Since the X-ray crystal structure of LTL-A is not known, the atomic coordinates of Con A, a related lectin, are used in model building. Using atomic coordinates for Con A[34] obtained from the Brookhaven Protein Data Bank, a fragment of the cross-linked lattice is constructed using the molecular software package FRODO on a Silicon Graphics IRIS workstation. Lattice dimensions are taken from the results of X-ray powder diffraction and the positions of the LTL-A dimers in the lattice from comparison of simulated and actual electron micrographs. Models for the lattice are examined to determine the presence of forbidden contacts and the relative positions of the carbohydrate-binding sites.[33]

Results. The model of the LTL-A/**19** lattice presented in Fig. 8I provides substantial information about the way in which LTL-A is cross-linked by **19**. The full 222 point group symmetry of the lectin is not reflected in the lattice symmetry, as only one molecular 2-fold axis is crystallographic. The structure of **19** is apparently not consistent with a lattice in which the point group symmetry of the LTL-A is optimally crystallographic. Rotation of the LTL-A tetramer by 12° about the *y* axis would result in a lattice that

[34] K. D. Hardman and C. F. Ainsworth, *Biochemistry* **11,** 4910 (1972).

would correspond to a single layer of a lattice with *C*222 symmetry. In this lattice, the cross-linking carbohydrate would span a 2-fold axis. **19** with its β(1–6) arm in the $\omega = 60^\circ$ rotamer conformation has a pseudo-2-fold axis of symmetry with respect to the outer Fuc residues, but not an exact 2-fold. The relative positions of the outer Fuc residues are consistent with the observed lattice symmetry. This can be compared with the conformation of a biantennary pentasaccharide possessing LacNAc arms linked β(1–6) and β(1–3) to a core Gal residue that was previously shown to bind with the β(1–6) arm in the $\omega = 60^\circ$ rotamer conformation in its cross-linked complex with SBA.[14] In this latter case, the pentasaccharide also possesses a pseudo-2-fold axis of symmetry with respect to its LacNAc moieties that is consistent with the lattice symmetry of the complex.

The structure of the LTL-A/**19** lattice can be compared with the structures of the cross-linked lattices formed between SBA and four biantennary pentasaccharides.[13,14] In those structures, each of the four three-dimensional lattices formed by SBA and the oligosaccharides is completely stabilized by carbohydrate cross-links that span the interprotein space. The presence of any protein–protein interactions at the tetramer–tetramer interface of the LTL-A/**19** lattice is difficult to assess on the basis of the model constructed here. Nevertheless, the LTL-A/**19** lattice is dissolved on addition of competing monovalent carbohydrate (i.e., Fuc), or prevented from forming in the presence of the monosaccharide. This indicates that the stability of the lattice is predominantly due to protein–carbohydrate interactions, as observed in the SBA/pentasaccharide lattices.[14]

LTL-A also forms cross-linked lattices with at least two other biantennary fucosyl oligosaccharides.[31] One is similar to the lattice described here, whereas the other forms a helical aggregate in which the topology of interactions is substantially different from the present lattice. The affinity of LTL-A for all three fucosyl oligosaccharides is similar, which suggests that their interactions with the protein are similar. This suggest that differences in the overall structures of the carbohydrates account for differences in the structures of the respective cross-linked lattices. Similar findings have been observed in the structures of cross-linked lattices formed between SBA and four analogs of the blood group I carbohydrate antigen.[14]

X-Ray Crystallography

Methods

Crystallization and Data Collection. The four SBA–pentasaccharide complexes are crystallized by the hanging drop vapor diffusion method, as previously described.[13] Single crystals of up to 3.0 mm on each edge can be grown by this method. Crystals of the 2,4- and 3,6-pentasaccharide complexes crystallize in space group $P6_422$, which is the same as for the 2,6-pentasaccharide complex previously described.[13] Crystals of the 2,3-pentasaccharide complex crystallize in space group $I4_122$. X-ray diffraction data are collected on a Siemens area detector system coupled to a Rigaku RU-200 rotating anode X-ray generator. The data are processed with the XENGEN software package (Siemens Analytical X-Ray Instruments, Madison, WI).[14]

Results

The structures of SBA complexed with four pentasaccharides, **20–23**, have been refined to approximately 2.4–2.8 Å. Noncovalent lattice formation in all four complexes is promoted uniquely by the bridging action of the two arms of each bivalent carbohydrate. Association between SBA tetramers involves binding of the terminal Gal residues of the pentasaccharides at identical sites in each monomer, with the sugar(s) cross-linking to a symmetry-related neighbor molecule. While the **21**, **22**, and **23** complexes possess a common $P6_422$ space group, their unit cell dimensions differ. The **20** cross-linked complex, on the other hand, possesses the space group $I4_122$. Thus, all four complexes are crystallographically distinct. Differences in the structure of the **20**–SBA lattice relative to the other three complexes can be observed in Fig. 9. The four cross-linking carbohydrates are in similar conformations, possessing a pseudo-2-fold axis of symmetry which lies on a crystallographic 2-fold axis of symmetry in each lattice. In the case of **22** and **23**, the symmetry of their cross-linked lattices requires different rotamer orientations about their β(1–6) glycosidic bonds. The results demonstrate that crystal packing interactions are the molecular basis for the formation of distinct cross-linked lattices between SBA and four isomeric pentasaccharides.[14]

The crystal structures of the four SBA–pentasaccharide complexes represent models for lectin–carbohydrate multidimensional clustering *in vivo*, and a common thermodynamic mechanism for selectively aggregating a dispersed population of multivalent receptors in biological systems. The crystallographic data together with previously described studies indicate that the carbohydrate moieties of specific glycoconjugate receptors can be cross-linked by a multivalent lectin into distinct homogeneous complexes.

Thermodynamic Studies by Isothermal Titration Calorimetry

Methods

ITC experiments with multivalent carbohydrates and lectins are performed with an MCS instrument from Microcal (Northampton, MA).

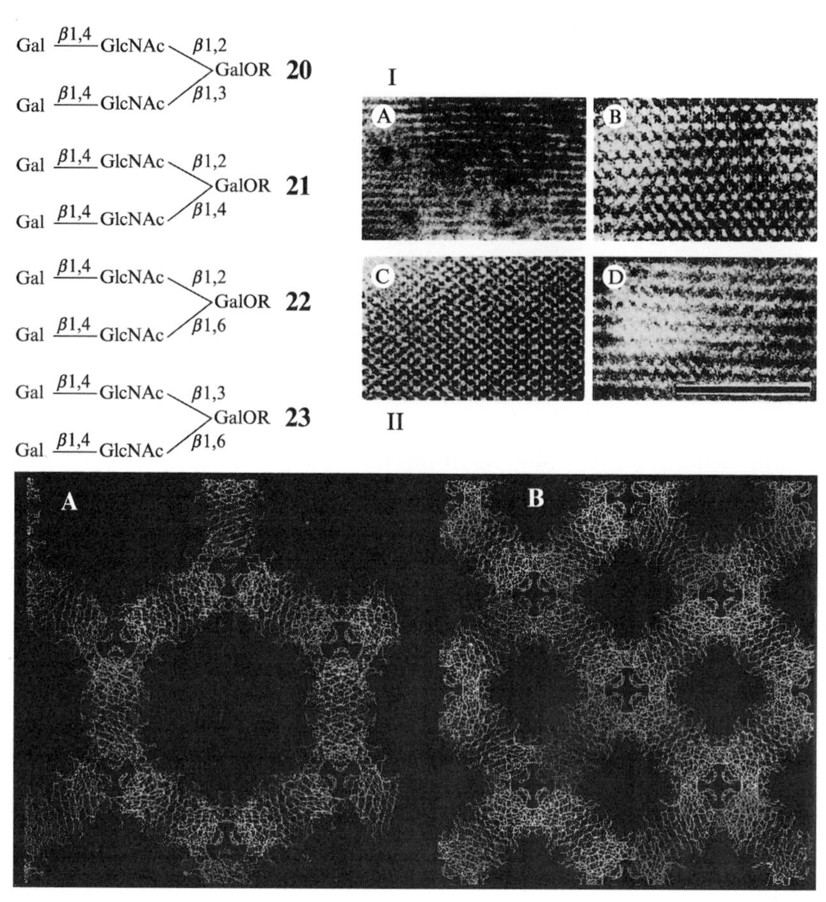

FIG. 9. Structures of synthetic biantennary pentasaccharides **20–23**. In (I), negative stain electron micrographs of the precipitates of SBA with **20** (A), **21** (B), **22** (C), and **23** (D) are shown. Bar: (D) 0.4 μm. Magnification in (A)–(D) is the same.[28] In (II), X-ray crystal structures of cross-linked complexes are shown. (A) The 6-fold axis of symmetry down the *c* axis is shown for the **23**–SBA cross-linked complex and (B) the 4-fold axis of symmetry down the *c* axis is shown for the **20**–SBA cross-linked complex. In these views, protein molecules are shown as colored trace models (monomers), with the cross-linking oligosaccharide not shown. The figures were generated with MOLPACK.[14] (See color insert.)

Injections of 4 μl of carbohydrate solution are added from a computer-controlled 250- or 100-μl microsyringe at intervals of 4 min into the sample solution of lectin (cell volume, 1.3424 ml) with stirring at 350 rpm. Control experiments performed by making identical injections of saccharide into a cell containing buffer without protein show insignificant heats of dilution. The experimental data are fitted to a theoretical titration curve, using software supplied by Microcal, with ΔH (enthalpy change in kcal/mol), K_a (association constant in M^{-1}), and n (number of binding sites per monomer) as adjustable parameters. Unlike some common practices, the n value is never set to a predecided value; rather, it is kept as a variable parameter. The quantity $c = K_a M_t(0)$, where $M_t(0)$ is the initial macromolecule concentration, is of importance in titration calorimetry.[35] All experiments are performed with c values greater than 1 but less than 200. Thermodynamic parameters are calculated from the equation

$$\triangle G = \triangle H - T\triangle S = -RT\ln K_a$$

where ΔG, ΔH, and ΔS are the changes in free energy, enthalpy, and entropy of binding, respectively. T is the absolute temperature and $R = 1.98$ cal mol^{-1} K^{-1}.[36]

Experimental Conditions for ITC Measurements with Multivalent Sugars. At pH 7.2 and NaCl concentrations greater than 0.15 M, where Con A and DGL are tetramers, multivalent carbohydrate analogs **25–27** are observed to bind and precipitate with both proteins at lectin concentrations between 25 and 60 μM and at nearly stoichimetric ratios of the sugars. Quantitative precipitation profiles of the lectins with the same sugars confirm their ability to precipitate the proteins. The precipitation reactions are inhibited by the presence of MeαMan or dissolved on addition of the monosaccharide. ITC measurements, however, require the presence of a nonprecipitating environment during the titration experiment. The dimer–tetramer equilibrium of Con A has been reported to be sensitive to pH[37,38] and NaCl (T. K. Dam and C. F. Brewer, unpublished observations, 2002). Therefore, ITC experiments are performed at low salt and pH (5.0), conditions under which the lectins are predominantly dimeric. A dimeric lectin does not form insoluble precipitates on binding a bivalent sugar, and the formation of precipitation with a tetravalent analog is considerably slower if the concentrations of the reactants are kept lower. By taking advantage of the higher affinities of the multivalent carbohydrates, such low-concentration titrations are performed. In particular, titrations with **25–27** are done with <20 μM lectins and 150–600 μM carbohydrates, significantly lower concentrations compared with those generally used in lectin–carbohydrate titrations. Time-dependent precipitation of the complex occurs with certain ligands after the run. Thus, the ITC experiments

[35] T. Wiseman, S. Williston, J. F. Brandt, and L.-N. Lin, *Anal. Biochem.* **179,** 131 (1989).
[36] T. K. Dam, R. Roy, S. K. Das, S. Oscarson, and C. F. Brewer, *J. Biol. Chem.* **275,** 14223 (2000).
[37] G. H. McKenzie, W. H. Sawyer, and L. W. Nichol, *Biochim. Biophys. Acta* **263,** 283 (1972).
[38] M. Huet, *Eur. J. Biochem.* **59,** 627 (1975).

with multivalent ligands are performed under conditions in which formation of insoluble complexes is arrested or considerably slowed and the quality of the run and fitting is excellent (Fig. 10).

Results

The driving force for the formation of homoaggregates of multimeric lectins with multivalent carbohydrates is solely thermodynamic. Multivalent ligands such as **1–5**, **8**, **9**, and **16** have been shown to possess higher affinities compared with those of their monovalent analogs.[17,39,40] Using several multivalent sugars (**25–27**) and multimeric lectins such as Con A and DGL, the thermodynamic basis of multivalent interactions has been investigated by ITC.[36]

As shown in Table II, the ΔH of a multivalent sugar binding to Con A (or DGL) is approximately its ITC-derived functional valency times the ΔH of its monovalent analog. The ITC data are consistent with the sequential binding of each arm (epitope) of a multivalent sugar to different lectin molecules. Therefore, the observed (macroscopic) ΔG is the average of the microscopic ΔG values of each arm of a multivalent carbohydrate. Because the microscopic ΔH is essentially identical at each arm, the increase in the observed K_a must be due to relatively more positive microscopic $T\Delta S$ at certain epitope(s) compared with the monovalent analog. On the basis of these results, the microscopic K_a value of the first epitope of a multivalent carbohydrate should be greater than those of the remaining epitopes because (1) the valency of a multivalent sugar will decrease with the progression of binding and (2) binding will become more and more difficult due to cross-linking effect (Fig. 12). These conclusions are confirmed by Scatchard and Hill plot analysis of the raw ITC binding data and by reverse ITC experiments (described below).[41,42]

Another important parameter derived from ITC analysis of multivalent carbohydrates binding to lectins is the n value, the number of binding sites per subunit of lectin. ITC results show that the n value is directly related to the functional valency of the carbohydrates (Table II), which may differ from its structural valency.[36]

[39] D. K. Mandal and C. F. Brewer, *Biochemistry* **32,** 5116 (1993).
[40] D. K. Mandal, N. Kishore, and C. F. Brewer, *Biochemistry* **33,** 1149 (1994).
[41] T. K. Dam, R. Roy, D. Pagé, and C. F. Brewer, *Biochemistry* **41,** 1351 (2002).
[42] T. K. Dam, R. Roy, D. Pagé, and C. F. Brewer, *Biochemistry* **41,** 1359 (2002).

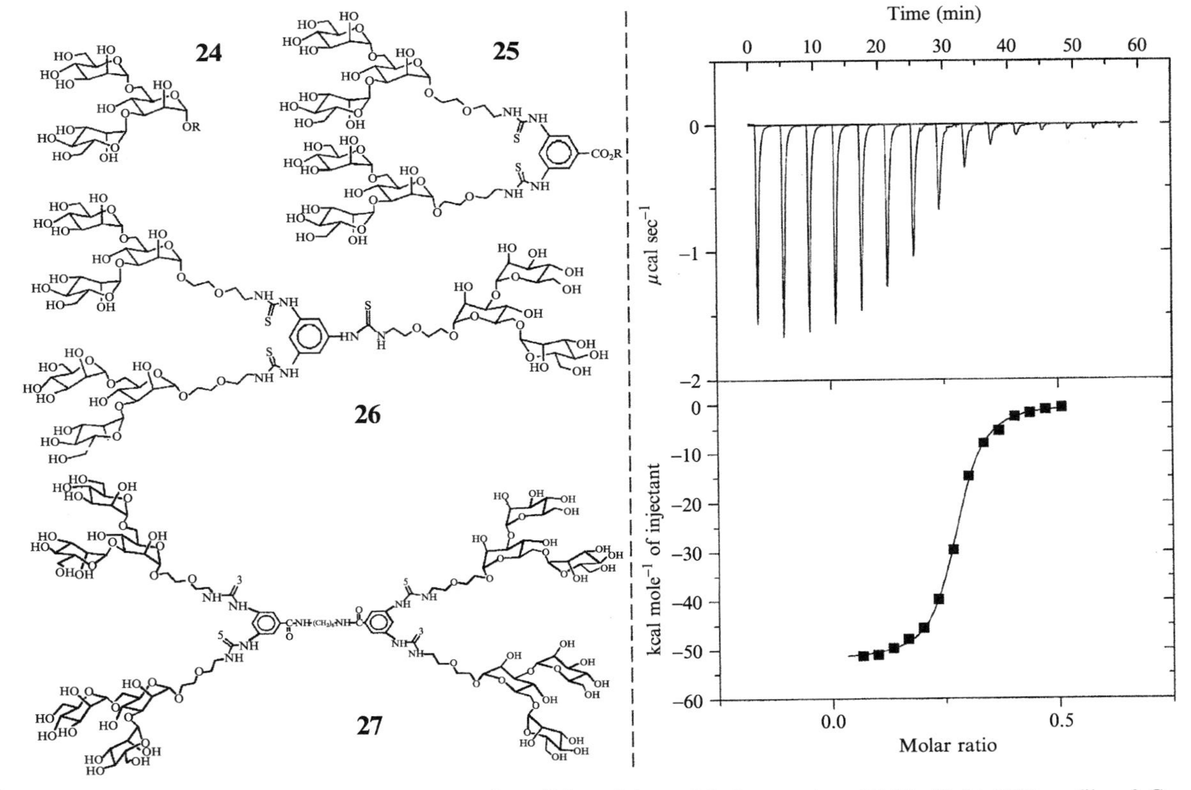

FIG. 10. *Left*: Structures of monovalent mannotriose (**24**) and its multivalent analogs **25**–**27**. *Right*: ITC profile of Con A (0.020 m*M*) with analog **27** (0.24 m*M*) at 27°.[36]

TABLE II
THERMODYNAMIC BINDING PARAMETERS FOR CON A AND DGL WITH MULTIVALENT SUGARS[a]

Compound	K_a ($M^{-1} \times 10^{-4}$)	$-\Delta G$ (kcal mol^{-1})	$-\Delta H$ (kcal mol^{-1})	$-T\Delta S$ (kcal mol^{-1})	n (no. sites/monomer)
Con A					
24	39	7.6	14.7	7.1	1.0
25	250	8.7	26.2	17.5	0.53
26	420	9.0	29.0	20.0	0.51
27	1350	9.7	53.0	43.3	0.26
DGL					
24	122	8.3	16.2	7.9	1.0
25	590	9.2	27.5	18.3	0.51
26	1000	9.6	32.2	22.6	0.40
27	6500	10.6	58.7	48.1	0.25

[a] At 27°. Errors in K_a range from 1 to 7% for Con A and from 7 to 10% for DGL; errors in ΔG are less than 1% for Con A and 1% for DGL; errors in ΔH are 1 to 4% for Con A and 1 to 7% for DGL; errors in $T\Delta S$ are 1 to 7% for Con A and 1 to 2% for DGL; errors in n are less than 2% for Con A and less than 1% for DGL. From Ref. 36.

Scatchard and Hill Plot Analysis of ITC Raw Data

Methods

The total concentration of ligand $X_t(i)$ as well as of lectin $M_t(i)$ after the ith injection and the heat evolved on the ith injection, $Q(i)$, are readily available from the ITC raw data file after each experiment. The concentration correction is automatically done by Origin software.

The concentration of bound ligand $X_b(i)$ after the ith injection is

$$X_b(i) = [Q(i)/(\Delta H \times V_0)] + X_b(i-1) \tag{1}$$

where $Q(i)$ (μcal) is the heat evolved on ith injection, ΔH (cal mol^{-1}) is the enthalpy change, V_0 (ml) is the active cell volume, and X_b (mM) is the concentration of bound ligand. X_b is equal to M_b, the concentration of bound protein, and, in the present study of multivalent ligands, the more general expression is $M_b = (X_b) \times$ (functional valency of ligand). The concentration of free ligand (X_f) after the ith injection is determined as follows:

$$X_f(i) = X_t(i) - X_b(i) \tag{2}$$

For Scatchard analysis, $r(i)$ is plotted against $r(i)/X_f(i)$, where $r(i)$ is $[X_b(i)] \times$ (functional valency of ligand)$/M_t(i)$, and Hill plots are constructed by plotting log $[Y(i)/1-Y(i)]$ versus $\log[X_f(i)]$, where $Y(i)$ is $[X_b(i)] \times$ (functional valency of ligand)$/M_t(i)$, which are modified versions of Scatchard and Hill plots (cf. Ref. 43) that take into account the functional valency of the ligand. Hill plots are disposed around the zero point on the ordinate as observed for monovalent mannotriose, only after multiplication with functional valencies of respective sugars.

A program has been created using Microsoft Excel for construction of Scatchard and Hill plots. Work sheet data are copied from the ITC raw data file and placed in appropriate columns of the program. After calculation the program shows the profiles of Hill and Scatchard plots. Delta Graph is used for further analysis of the plots.[41]

The validity of the information obtained from the Hill plot [$\log(Y/1-Y)$ versus $\log(X_f)$] is tested by directly fitting the binding data of monovalent mannotriose to the Hill equation. The Hill slope is observed to be the same by direct fitting or plotting of the Hill equation data. Attempts at directly fitting the ITC data for the multivalent analogs fail because the Hill slope values change throughout the binding process.

All the ITC raw data including those of control experiments are taken from titrations done at low and comparable concentration regimens. This has important implications for the quality of the data and subsequent explanations. Low concentration titrations inhibit precipitation and self-association of the molecules involved during the experiment and the use of identical experimental conditions and comparable concentrations of lectins and the carbohydrates in all experiments allow a valid comparison of the data and the profiles. As a consequence, fittings of the ITC data are excellent, which, in turn, significantly reduce the error margin in the raw data. If the raw data are taken from an ITC experiment with poor fitting, Scatchard and Hill plot profiles remain largely unreliable because of the significant error margin. The conclusions made from the present Scatchard and Hill plot analyses are experimentally confirmed by reverse ITC experiments (see below).

Results

Scatchard plots clearly indicate that multivalent carbohydrate binding to a lectin that results in cross-linking is associated with negative cooperativity, which is confirmed by Hill plot analysis.[41] A Scatchard plot of the ITC data for monovalent **24** is linear whereas that obtained with a multivalent analog such as **27** is curvilinear (Fig. 11A and B). The concave nature of the Scatchard plot in Fig. 11B suggests that multivalent analogs bind to both lectins with negative cooperativity.

[43] E. Di Cera, "Thermodynamic Theory of Site-Specific Binding Processes in Biological Macromolecules." Cambridge University Press, New York, 1995.

Hill plots of the binding of monovalent mannotriose to Con A and DGL are linear with slopes near 1.0, demonstrating a lack of binding cooperativity and allosteric transitions in the proteins. However, Hill plots for the binding of multivalent analogs (**25–27**) to both lectins are curvilinear with decreasing tangent slopes below 1.0, indicating increasing negative cooperativity on binding of the analogs to the lectins (Fig. 11C). The decreasing slope values of curvilinear Hill plots are consistent with decreasing affinity and functional valencies of the multivalent analogs on sequential binding of lectin molecules to the carbohydrate epitopes of the analogs. Figure 12 schematically presents a multivalent binding and shows how the microscopic binding constants decrease ($K_{a1} > K_{a2} > K_{a3} > K_{a4}$). Figure 12 is overly simplified because it does not take into account the other binding site of a lectin dimer. Involvement of both binding sites of a dimeric lectin molecule would lead to cross-linking. The degree of cross-linking will increase with the progression of binding. Therefore, cross-linking can be expected to contribute to the observed negative cooperativity.

Reverse Isothermal Titration Calorimetry

Methods

Reverse ITC experiments are performed to determine the microscopic binding parameters of individual epitopes, as described above, with the following exceptions. In individual titrations, injections of 4 μl of Con A solution are added from the computer-controlled 100-μl microsyringe at intervals of 4 min into a cell containing sugar solution (cell volume, 1.358 ml) dissolved in the same buffer as the lectin, while stirring at 350 rpm. Control experiments performed by making identical injections of Con A into a cell containing buffer with no sugar show insignificant heats of dilution.[42]

Results

Decreasing binding affinity on sequential binding was experimentally demonstrated by reverse ITC, which determined the thermodynamics of microscopic binding parameters of a lectin for the individual epitopes of multivalent analogs.[42] The reverse ITC measurements show an 18-fold greater microscopic affinity constant of Con A for the first epitope of the divalent analog **25** versus its second epitope, and a 53-fold greater microscopic affinity constant of Con A binding to the first epitope of the trivalent analog **26** versus its second epitope. The data also demonstrate that the microscopic enthalpies of binding of the two epitopes (ΔH_1 and ΔH_2) of **25** and **26** are essentially the same, and that differences in the

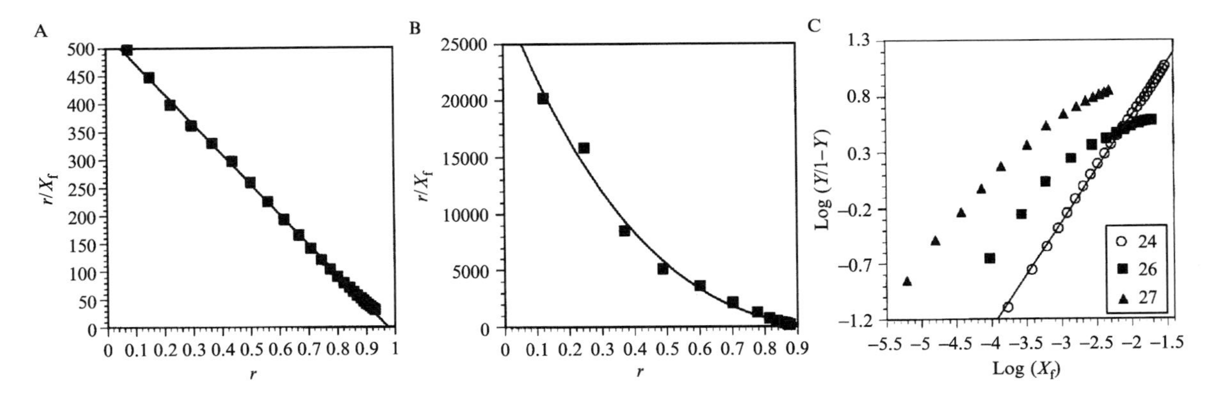

FIG. 11. Scatchard plot of the ITC raw data for monovalent **24** (0.8 m*M*) (A) and multivalent **27** (0.24 m*M*) (B) binding to Con A (0.020 m*M*). (C) Hill plots of ITC raw data for monovalent **24** (0.8 m*M*) and multivalent **26** (0.67 m*M*) and **27** (0.24 m*M*) binding to Con A (0.020 m*M*).[41]

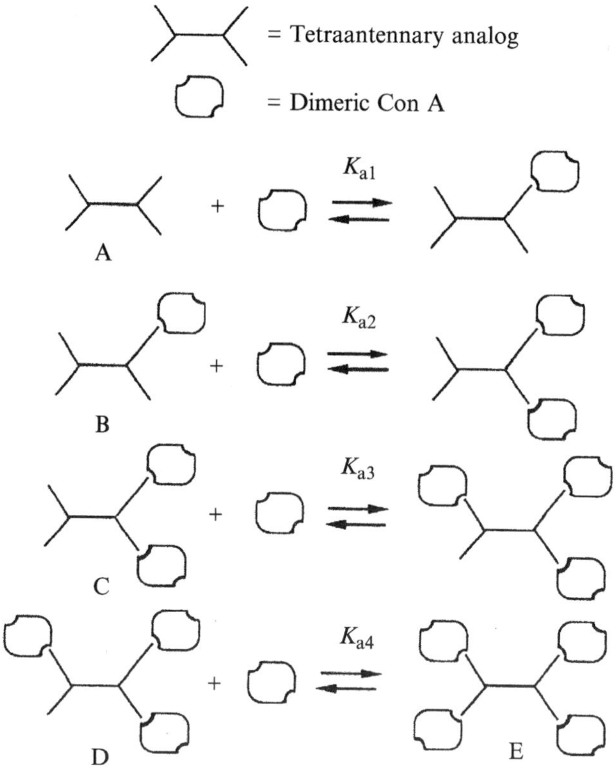

= Tetraantennary analog

= Dimeric Con A

$$K_{a1} > K_{a2} > K_{a3} > K_{a4}$$

FIG. 12. Four microequilibrium constants of the tetravalent analog **27** can be represented by K_{a1}, K_{a2}, K_{a3}, and K_{a4}, for binding of a dimeric Con A molecule to the first arm of **27** (species A), to the second arm of **27** (species B), etc. Hence, the observed (macroscopic) ΔG values of **27** (ΔG_{obs}) for Con A are the average of the four microscopic ΔG terms, or $\Delta G_{obs} = (\Delta G_1 + \Delta G_2 + \Delta G_3 + \Delta G_4)/4$. The relative values of ΔG_1, ΔG_2, ΔG_3, and ΔG_4 must decrease on the basis of the decreasing valencies of A, B, C, and D (which have the same valencies as tetra-, tri, bi-, and monovalent analogs). Thus, it is expected that $K_{a1} > K_{a2} > K_{a3} > K_{a4}$ for **27** binding to Con A as shown. The increasing level of cross-linking with the progression of binding will also contribute to the decreasing microscopic binding constants.[41]

microscopic K_a values of the epitopes are due to their different microscopic entropies of binding ($T\Delta S$) values (Table III). These findings are consistent with the increasing negative Hill coefficients obtained during Hill plot analysis. Compared with the binding of initial epitope(s) (in which the entropy is relatively more favorable and affinity is greater because of the higher possibilities of binding and recapture), increasing cross-linking

TABLE III

REVERSE ITC-DERIVED THERMODYNAMIC BINDING PARAMETERS FOR TRIMAN (24) AND MULTIVALENT SUGAR ANALOGS[a]

Compound	K_{a1} ($M^{-1} \times 10^{-5}$)	ΔK_{a2} ($M^{-1} \times 10^{-5}$)	$-\Delta G_1$ (kcal mole^{-1})	$-\Delta G_2$ (kcal mole^{-1})	n_1	n_2	ΔH_1 (kcal mole^{-1})	$-\Delta H_2$ (kcal mole^{-1})	$-T\Delta S_1$ (kcal mole^{-1})	$-T\Delta S_2$ (kcal mole^{-1})
24	6.2	—	7.9	—	0.99	—	13.1	—	5.2	—
25	161	8.8	9.8	8.1	0.97	0.94	12.5	12.3	2.7	4.2
26	460	8.6	10.4	8.1	1.05	1.09	13.3	12.2	2.9	4.1

[a] At 27°. Errors in K_a are less than 7%; errors in ΔG are less than 5%; errors in n are less than 4%; errors in ΔH are less than 4%; errors in $T\Delta S$ are less than 7%. From Ref. 42.

during multivalent binding makes the binding increasingly difficult, resulting in increasing unfavorable entropy and decreasing affinity.

Visualization of Receptor Segregation through Cross-Linking on Surface of Cell by Confocal Immunofluorescence Microscopy

Methods

MOLT-4 cells or human thymocytes (1×10^7) are incubated with 20 μM galectin-1 and 1.2 mM dithiothreitol (DTT), or 1.2 mM DTT alone as a control, for 10 min on ice, followed by 20 min in a 37° water bath to allow migration of counterreceptors on the cell surface.[30] Cells are cooled to 4°, and bound galectin-1 is dissociated with ice-cold 0.1 M β-lactose. Paraformaldehyde (2%) is added for 30 min on ice to fix the cells, and the reaction is quenched with 0.2 M glycine for 5 min on ice. The cells are stained by incubating them for 1.5 h at 25° with a 15-μg/ml concentration of each of the indicated combinations of two antibodies: 9.4 (CD45, mouse IgG_{2a}), DFT1 (CD43, mouse IgG_1), DFT1–biotin (CD43, mouse IgG_1–biotin conjugated), M-T701 (CD7, mouse IgG_1), LT7 (CD7, mouse IgG_{2a}), or UCHT1 (CD3, mouse IgG_1). The cells are washed and incubated at 25° for 1.5 h with the appropriate secondary reagents: fluorescein-conjugated goat anti-mouse IgG_1, fluorescein-conjugated goat anti-mouse IgG_{2a}, Texas Red-conjugated goat anti-mouse IgG_{2a}, fluorescein-conjugated streptavidin, Texas Red-conjugated streptavidin, or goat anti-mouse IgG_1. The cells are washed and, when required, a tertiary reagent, fluorescein-conjugated rabbit anti-goat IgG, is added for 1.5 h at 25°. The cells are washed, dropped on glass microscope slides, and mounted with ProLong Antifade mounting medium (Molecular Probes, Eugene, OR). The fluorescently labeled cells are viewed via the × 100 objective of a Leica (Deerfield, IL) confocal laser scanning microscope (CLSM). The cells are scanned by dual excitation of fluorescein (green) and Texas Red (red) fluorescence. Dual-emission fluorescent images are collected in separate channels as 0.5-μm optical slices. The images are processed on a Sun Workstation using AVS (Advanced Visualization Systems, Waltham, MA) image-processing software. Areas of red and green overlapping fluorescence are represented with a yellow signal. The confocal images are printed by a Fuji (Tokyo, Japan) Pictography 3000 printer.

Results

Receptor Segregation at Surface of Cell Leads to Apoptosis. Galectin-1 binding to susceptible T cells leads to cell death.[2] Galectin-1 binding also leads to a dramatic redistribution of surface glycoprotein receptors into

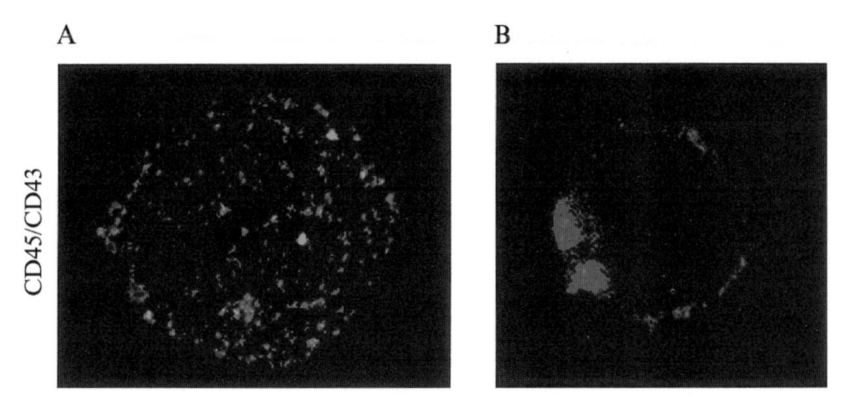

FIG. 13. (A) Uniform distribution of CD45 (red) and CD43 (green) on the surface of MOLT-4 cells in the absence of galectin-1, with some areas of colocalization (yellow). (B) Segregation of CD45 (red) and CD43 (green) after galectin-1 treatment.[30] (See color insert.)

segregated membrane microdomains on the cell surface of susceptible T cells. In the absence of galectin-1, counter receptors CD45 and CD43 are found randomly distributed on the surface of the cells. On galectin-1 binding these receptors segregate into distinct patches (Fig. 13). CD45 and CD3 colocalized on large islands on apoptotic blebs protruding from the cell surface. These islands also included externalized phosphatidylserine. In addition, the exposure of phosphatidylserine on the surface of galectin-1-treated cells occurred rapidly. CD7 and CD43 colocalized in small patches away from the membrane blebs, which excluded externalized phosphatidylserine. Receptor segregation was not seen on cells that did not die in response to galectin-1, including mature thymocytes, suggesting that spatial redistribution of receptors into specific microdomains is required to trigger apoptosis.[30]

Conclusions

Physical studies of multivalent lectin–carbohydrate cross-linking interactions described in the present article have provided insights into a new source of binding specificity, namely, the formation of homogeneous lectin–carbohydrate cross-linked lattices. The thermodynamic bases of such interactions have also been investigated. These findings have been applied toward understanding the role of galectin-1 in binding and segregating specific counter receptors on the surface of activated T cells leading to apoptosis.[30] Hence, studies of the physical interactions of multivalent lectins with multivalent carbohydrates and glycoproteins have provided a molecular basis for understanding their structure–function roles in biological systems.

Acknowledgment

This work was supported by Grant CA-16054 from the National Cancer Institute, Department of Health, Education, and Welfare, and by Core Grant P30 CA-13330 from the same agency (C.F.B.).

[32] Calorimetric Evaluation of Protein–Carbohydrate Affinities

By TRINE CHRISTENSEN and ERIC J. TOONE

Introduction

Protein–carbohydrate interactions control myriad biological recognition phenomena including fertilization, cell–cell recognition, immunological responses, and pathogen–host cell attachment. From a desire to control these events, significant effort has been expended toward the development of high-affinity ligands for various carbohydrate-binding proteins. On the other hand, the evaluation of protein–carbohydrate affinities is challenging. Carbohydrates lack the traditional photophysical properties typically used to measure binding phenomena. Protein–carbohydrate interactions are weak, typically in the millimolar to micromolar range, and many of the well-developed techniques for the study of ligand binding are inappropriate in this low-affinity domain. The saccharide ligands for many carbohydrate-binding proteins are scarce, and their availability is often limited by the elaborate and tedious synthetic protocols required for their preparation.

In response to these challenges, a variety of competitive assays have been developed for the analysis of protein–carbohydrate binding. The most widely used of this group is the inhibition of hemagglutination, or HIA, assay.[1,2] Because erythrocytes display various bound carbohydrates at high surface densities, many lectins cross-link, or agglutinate, red blood cells. In

[1] M. Heinrich, *Am. Lab.* **32,** 22 (2000).
[2] K. Landsteiner, "The Specificity of Serological Reactions." Dover, New York, 1962.

0076-6879/03 $35.00

a standard HIA assay, anticoagulated erythrocytes are treated with a solution of multivalent lectin. Soluble saccharide ligand is preincubated with lectin through a serial dilution, and the minimum concentration of ligand required to inhibit the hemagglutination assay is reported as a medias inhibitory concentration (IC_{50}). HIA offers the significant advantage of ease of use. On the other hand, the precision of the technique is low and the meaning of derived IC_{50} values is unclear. Because serial dilutions are used the precision of the technique is no more than ± one well—a factor of two in concentration. IC_{50} values are only tangentially related to true binding constants, in the traditional sense of the term, and many additional terms are reported in this value. Thus, although HIA is a useful tool for ordering a series of reasonably homologous ligands, it is not a useful technique for the evaluation of true binding constants.[3]

The more recently developed enzyme-linked lectin assay, or ELLA, circumvents some of the limitations of HIA.[4–6] In this methodology, an insoluble ligand is affixed to the wall of a microtiter plate, and the binding of an enzyme–lectin conjugate is evaluated after the well-known ELISA. Again, a competition assay is set up by the addition of a soluble ligand that, in turn, reduces the binding of the enzyme–lectin conjugate to immobilized saccharide. Readout is achieved by addition of a prodye enzyme substrate and evalution by ultraviolet–visible (UV–Vis) spectroscopy. This process greatly improves the precision of evaluation, because IC_{50} values are determined by curve fitting, rather than by visual inspection. In addition, the use of an enzyme–lectin conjugate seems to inhibit the aggregation of soluble lectin by multivalent ligands, rendering the technique less sensitive to aggregation artifacts. Still, the fundamental limitation—that IC_{50} values are not simple affinity constants—remains, and "affinities" determined by this method must be interpreted with some caution.

Many other biophysical techniques have been used to measure the affinity of protein–carbohydrate binding including surface plasmon resonance,[7,8] steady state and stopped-flow fluorescence spectroscopy,[9–11] and

[3] J. J. Lundquist and E. J. Toone, *Chem. Rev.* **102,** 555 (2002).
[4] D. Page and R. Roy, *Bioorg. Med. Chem. Lett.* **6,** 1765 (1996).
[5] D. R. Zanini and R. Roy, *J. Am. Chem. Soc.* **119,** 2088 (1997).
[6] J. P. McCoy, J. Varani, and I. J. Goldstein, *Cell Res.* **151,** 96 (1983).
[7] W. M. Mullet, E. P. Lai, and J. M. Yeung, *Methods* **22,** 77 (2000).
[8] W. Jager, "Carbohydrate Chemistry and Biology," Vol. 2. Wiley-VCH, Weinheim, Germany, 2000.
[9] T. Christensen, B. B. Stoffer, B. Svensson, and U. Christensen, *Eur. J. Biochem.* **250,** 638 (1997).
[10] K. Olsen, B. Svensson, and U. Christensen, *Eur. J. Biochem.* **209,** 777 (1992).
[11] K. Olsen, U. Christensen, M. R. Sierks, and B. Svensson, *Biochemistry*, **32,** 9686 (1993).

nuclear magnetic resonance (NMR).[12–14] Isothermal titration calorimetry (ITC) has emerged as a general and powerful technique for the study of a wide range of binding events.[15,16] ITC has been profitably employed for the study of protein–carbohydrate interaction, and has contributed in an important way to a fundamental knowledge of the molecular basis of carbohydrate affinity.[17–25] ITC is the only experimental technique that directly measures binding enthalpies. ITC is the only practical method of measuring changes in molar heat capacity accompanying binding; this parameter is of great value when considering the molecular basis of bulk thermodynamic binding parameters. Here, we describe the operational parameters, limitations, and pitfalls of ITC, especially as it pertains to the study of protein–carbohydrate interaction. The reader is referred to a series of important reviews and monographs that cover other aspects of ITC, including instrument design.[15,26–30]

Instrument Design and Basic Experimental Concerns

Isothermal titration calorimetry measures the heat evolved during the addition of ligand to its cognate binding partner as a function of ligand concentration.[15,27,30,31] This relationship depends on the concentrations of

[12] N. K. Sauer, M. D. Bednarski, B. A. Wurzburg, J. E. Hanson, G. M. Whitesides, and J. J. Skehel, *Biochemistry* **29,** 8388 (1989).
[13] K. A. Kronis and J. P. Carver, *Biochemistry* **24,** 834 (1985).
[14] S. H. Koenig, R. D. Brown, and C. F. Brewer, *Proc. Natl. Acad. Sci. USA* **70,** 475 (1973).
[15] E. Freire, O. L. Mayorga, and M. Straume, *Anal. Chem.* **62,** 950A (1990).
[16] G. P. Privalov and P. L. Privalov, *Methods Enzymol.* **323,** 31 (2000).
[17] J. J. Lundquist, S. D. Debenham, and E. J. Toone, *J. Org. Chem.* **65,** 8245 (2000).
[18] P. M. St. Hilaire, M. K. Boyd, and E. J. Toone, *Biochemistry* **33,** 14452 (1994).
[19] D. R. Bundle, R. Alibes, S. Nilar, A. Otter, M. Warwas, and P. Zhang, *J. Am. Chem. Soc.* **120,** 5317 (1998).
[20] N. Navarre, N. Amiot, A. van Oijen, A. Imberty, A. Poveda, J. Jimenez-Barbero, A. Cooper, M. A. Nutley, and G.-J. Boons, *Chem. Eur. J.* **5,** 2281 (1999).
[21] C. R. Berland, B. W. Sigurskjold, B. Stoffer, T. P. Frandsen, and B. Svensson, *Biochemistry* **34,** 1053 (1995).
[22] B. W. Sigurskjold, T. Christensen, N. Payre, S. Cottaz, H. Driguez, and B. Svensson, *Biochemistry* **37,** 10446 (1998).
[23] T. K. Dam and F. C. Brewer, *Chem. Rev.* **102,** 387 (2002).
[24] B. W. Sigurskjold and D. R. Bundle, *J. Biol. Chem.* **267,** 8371 (1992).
[25] B. W. Sigurskjold, E. Altman, and D. R. Bundle, *Eur. J. Biochem.* **197,** 239 (1991).
[26] L. Indyk and H. F. Fisher, *Methods Enzymol.* **295,** 350 (1998).
[27] S. Leavitt and E. Freire, *Curr. Opin. Struct. Biol.* **11,** 560 (2001).
[28] M. D. Doyle, *Curr. Opin. Biotechnol.* **8,** 31 (1997).
[29] D. R. Bundle and B. W. Sigurskjold, *Methods Enzymol.* **247,** 288 (1994).
[30] T. Wiseman, S. Williston, J. F. Brandts, and L. N. Lin, *Anal. Biochem.* **179,** 131 (1989).
[31] H. F. Fisher and N. Singh, *Methods Enzymol.* **259,** 194 (1995).

both binding partners, the association constant for complex formation, the enthalpy of complex formation, and the model (stoichiometry) of binding. Assuming sufficient data points exist to define unique solutions, a single titration furnishes an enthalpy of binding (ΔH°), the association constant (K), and the binding stoichiometry (n). The binding constant is related to the Gibbs free energy of binding (ΔG°) and the entropy of binding (ΔS°) can be calculated from known thermodynamic relationships, that is,

$$\Delta G^\circ = \Delta H^\circ - T\Delta S^\circ = -RT \ln K \quad (1)$$

where R is the universal gas constant and T is the absolute temperature.

Various calorimeter designs have been reported during the past 50 years. Today two instruments of similar design and with similar capabilities—those produced by Calorimetry Science (CSC, American Fork, UT) and MicroCal (Northampton, MA)—dominate the market. Figure 1 shows a schematic presentation of the CSC ITC. One of the CSC instruments utilizes removable cells and is suitable for use with solid samples; however, this model is not as sensitive as the other CSC instrument or the Microcal VP-ITC. The calorimeter consists of two almost identical, coin-shaped cells each with a volume of roughly 1.3 ml surrounded by an adiabatic jacket. The reference cell is loaded with buffer while the sample cell contains

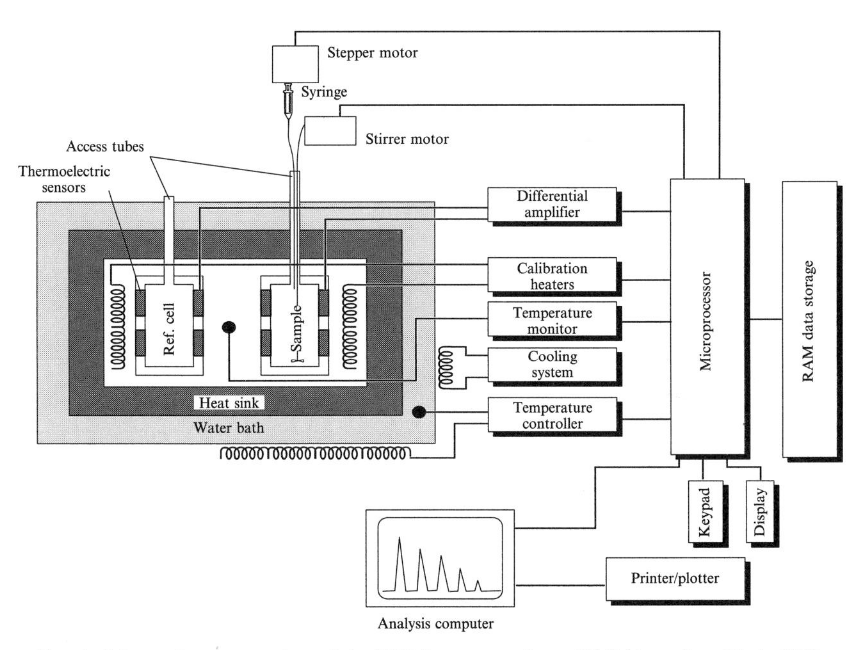

FIG. 1. Schematic presentation of the ITC instrument from CSC (American Fork, UT).

the protein of interest. An injection syringe with a paddle-shaped tip both adds ligand to and stirs the contents of the sample cell. The syringe is immersed into the sample cell and stirred at 300–400 rpm, ensuring complete mixing of ligand within a few seconds of addition. The time constant (detector response time) of both instruments is near 15 s^{-1}, and the kinetics of processes that occur on time scales slower than this can be measured by evaluating the evolution/absorption of heat as a function of time.[32]

Both instruments use a continuous power compensation protocol, rather than passive thermal conductivity, to measure the evolution or adsorption of heat. In this protocol, which takes advantage of the great precision with which temperature differences can be measured, a sample and reference cell are brought into thermal equilibrium and then heated at identical rates. The experiment is thus not strictly speaking isothermal, although changes in temperature typically amount to only a few tenths of a degree over the course of the titration. A compensating power is adjusted to keep the cells at identical temperatures.

The ITC experiment measures heat evolved as ligand is added to a solution of binding sites as a function of ligand concentration, yielding a curve equivalent in shape and, in at least some ways, meaning to a more familiar acid–base titration. Deconvolution searches for the inflection point on the curve; as a result, the curve shape must be amenable to numerical reduction. This shape is dependent on the concentration of binding sites in the cell and the binding constant. The unitless product of the concentration of binding sites and the binding constant, sometimes denoted as c, must range between 1 and 1000—more ideally between 10 and 100—to facilitate data reduction. Figure 2 shows curve shapes at various c values.

A consequence of this curve shape limitation is that the binding constant determines the concentration of binding sites required in the cell. Thus in dissociation units, a millimolar binding constant requires at least millimolar binding sites in the cell, whereas a nanomolar binding constant requires no more than micromolar binding sites in the cell. As a practical matter, the calorimetric experiment fails at the low-affinity limit for either solubility or material availability issues. At the high-affinity limit the experiment fails over sensitivity issues, and a typical high-affinity limit for direct titration is near $10^8\ M^{-1}$. Having determined the ideal concentration of binding sites, ligand concentrations are set at 10 to 20 times the concentration of binding sites. A minimum of 20 injections is required to properly define the curve for data reduction.

The calorimeter is calibrated using either a series of electrical pulses of known energy, or by measuring the enthalpy of a binding of known

[32] B. A. Williams and E. J. Toone, *J. Org. Chem.* **58,** 3507 (1993).

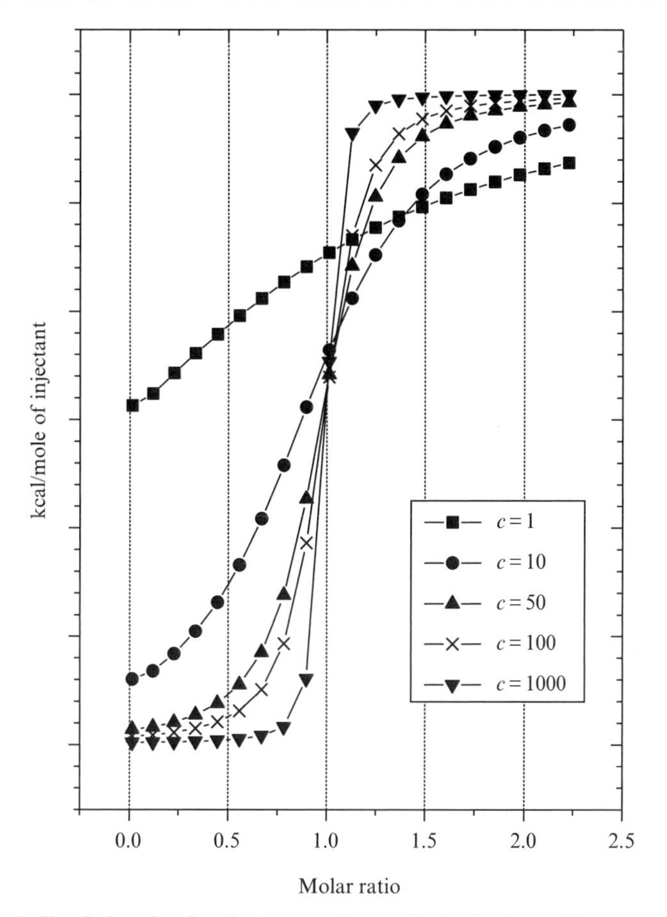

FIG. 2. Simulation showing the lower and upper limits for the unitless factor c.

enthalpy, often RNase–5′-CMP. In general, the electrical calibration is both more convenient and more accurate.

Data Reduction

The heat evolved on each injection is directly proportional to the amount of complex formed and the evolved heat decreases as saturation approaches. The raw data from a continuous power compensation instrument is power as a function time. Integration of these data with respect to time yields a plot of enthalpy as a function of ligand concentration; it is this latter relationship that is used to deduce both binding constants and enthalpies (Fig. 3).

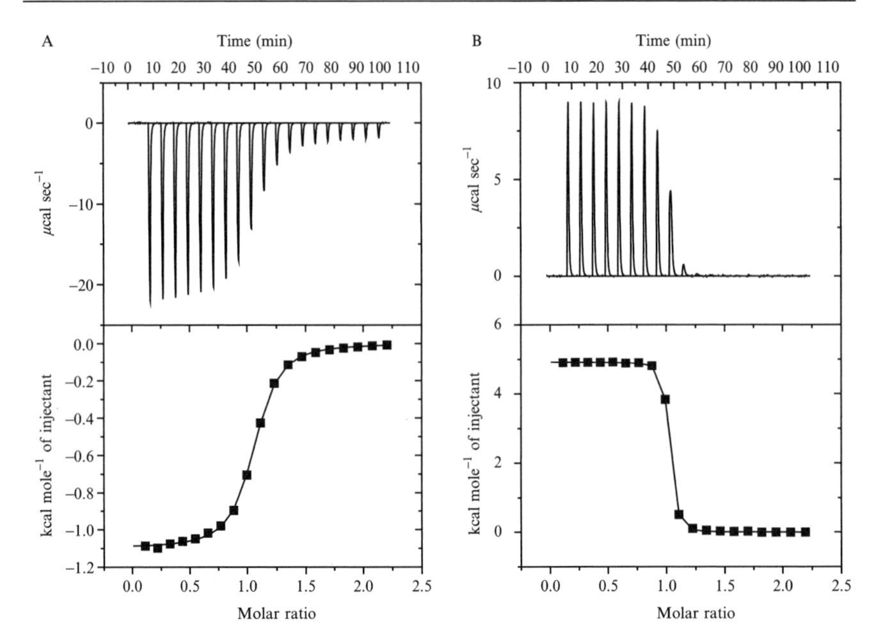

FIG. 3. (A and B) *Top*: Thermograms. *Bottom*: Binding isotherms. (A) Exothermic reaction between 4.96 mM nitrilotriacetic acid and 49.90 mM $CaCl_2$ in 20 mM HEPES, pH 8, at 45°. $K = (1.89 \pm 0.07) \times 10^4\ M^{-1}$, which gives a c value of 94 and an ideal titration, the enthalpy is -1.10 ± 0.01 kcal mol^{-1}, and the stoichiometry is 1.02 ± 0.01. (B) Endothermic reaction between 0.495 mM ethylenediaminetetraacetic acid and 4.95 mM $MgCl_2$ in 20 mM HEPES, pH 8, at 35°. In this case $c > 1000$, which means that the association constant cannot be determined; however, $K > 2 \times 10^6\ M^{-1}$, and it is possible to determine the enthalpy and stoichiometry from this tight binding experiment, which are $^{+}4.92 \pm 0.01$ kcal mol^{-1} and 0.98 $\pm$ 0.01, respectively. Both titrations were done on a MicroCal VP-ITC.

A binding event that involves n identical (i.e., noncooperative) binding sites is described by the reaction[15,26,29,30]

$$\mathrm{P} + n\mathrm{L} \overset{K}{\rightleftharpoons} \mathrm{PL}_n$$

where P and L represent the protein and the carbohydrate ligand, respectively. The heat absorbed or evolved during each injection (ΔQ) is proportional to the change in concentration of the formed complex (Δ[PL]), the number of binding sites (n), the molar enthalpy of binding (ΔH°), and the volume of the reaction cell (V_0):

$$\Delta Q = n\Delta[\mathrm{PL}]\Delta H^\circ V_0 \tag{2}$$

The incremental heat evolved as a function of ligand concentration is given by the expression

$$\frac{dQ}{d[\mathrm{L}]_0} = n\frac{d[\mathrm{PL}]}{d[\mathrm{L}]_0}\Delta H^\circ V_0 \tag{3}$$

Because the bound and free concentrations of both protein and ligand are unknown, the binding constant is expressed as a function of the total protein and ligand concentrations, $[\mathrm{P}]_0 = [\mathrm{P}] + [\mathrm{PL}]$ and $[\mathrm{L}]_0 = [\mathrm{L}] + n[\mathrm{PL}]$, respectively:

$$K = \frac{[\mathrm{PL}]}{[\mathrm{P}][\mathrm{L}]} = \frac{[\mathrm{PL}]}{([\mathrm{P}]_0 - [\mathrm{PL}])([\mathrm{L}]_0 - n[\mathrm{PL}])} \tag{4}$$

Isolating [PL] from Eq. (4) results in a quadratic equation for which only one real solution exists:

$$[\mathrm{PL}] = \frac{1}{2}\left\{\left([\mathrm{P}]_0 + \frac{[\mathrm{L}]_0}{n} + \frac{1}{nK}\right) - \sqrt{\left([\mathrm{P}]_0 + \frac{[\mathrm{L}]_0}{n} + \frac{1}{nK}\right)^2 - \frac{4[\mathrm{P}]_0[\mathrm{L}]_0}{n}}\right\} \tag{5}$$

The incremental change in [PL] with respect to addition of ligand can then be calculated as

$$\frac{d[\mathrm{PL}]}{d[\mathrm{L}]_0} = \frac{1}{2}\left[\frac{1}{n} - \frac{\dfrac{[\mathrm{L}]_0}{n[\mathrm{P}]_0} - 1 + \dfrac{1}{nK[\mathrm{P}]_0}}{\sqrt{\dfrac{[\mathrm{L}]_0^2}{[\mathrm{P}]_0^2} + \left(n + \dfrac{1}{K[\mathrm{P}]_0}\right)^2 - 2\dfrac{[\mathrm{L}]_0}{[\mathrm{P}]_0}\left(n - \dfrac{1}{K[\mathrm{P}]_0}\right)}}\right] \tag{6}$$

Substituting Eq. (6) into Eq. (3) leads to the following expression; Eq. (7) is used to fit the binding isotherm obtained from the ITC experiment:

$$\frac{dQ}{d[\mathrm{L}]_0} = \frac{1}{2}\Delta H^\circ V_0\left[1 - \frac{\dfrac{[\mathrm{L}]_0}{[\mathrm{P}]_0} - n + \dfrac{1}{K[\mathrm{P}]_0}}{\sqrt{\dfrac{[\mathrm{L}]_0^2}{[\mathrm{P}]_0^2} + \left(n + \dfrac{1}{K[\mathrm{P}]_0}\right)^2 - 2\dfrac{[\mathrm{L}]_0}{[\mathrm{P}]_0}\left(n - \dfrac{1}{K[\mathrm{P}]_0}\right)}}\right] \tag{7}$$

Equation (7) describes the relationship of evolved enthalpy as a function of ligand concentration for the formation of a 1:1 complex; similar equations can be derived for other stoichiometries. These derivations are straightforward in the simple case of equivalent noninteracting sites, and become progressively more complex for the eventuality of nonequivalent sites and interacting (cooperative) sites. In addition to more complex algebra, algebra that must typically be derived and entered by the operator, such systems require significantly more data for proper reduction than do less complex models.

Binding-Related Contributions to Net Enthalpies: Proton Transfer during Ligand Binding

Thermodynamic parameters are state, or path-independent, functions. An important corollary of this condition is that measured thermodynamic parameters contain contributions from all microscopic events that occur during the binding process. Because thermodynamic parameters are most often considered in the context of structural information in an attempt to derive the thermodynamic consequences of individual intermolecular interactions, it is vital to ensure that all contributing events are recognized and properly attributed. A common event during ligand binding is proton transfer, either to or from protein or ligand and buffer; such transfers occur as the pK_a values of various ionizable groups are altered by complex formation.

Proton exchange during binding is measured calorimetrically using the following expression:[33]

$$\Delta H^\circ = \Delta H^\circ_{\text{intrinsic}} + N\Delta H^\circ_{\text{ion}} \tag{8}$$

where ΔH° is the net calorimetric enthalpy, $\Delta H^\circ_{\text{intrinsic}}$ is the enthalpy of binding independent of proton exchange, N is number of protons exchanged with the buffer during binding ($N < 0$ if protons are released from the protein to the buffer and > 0 if protons are adsorbed by protein from buffer) and $\Delta H^\circ_{\text{ion}}$ is the buffer ionization enthalpy. Binding enthalpies are measured in several buffers of differing ionization enthalpy, and a plot of observed enthalpy of binding versus enthalpy of buffer ionization is created. The slope of this plot is equivalent to the number of protons transferred during binding, and the *y* intercept is the intrinsic enthalpy of binding. Ionization enthalpies of several common buffers are shown in Table I; additional values are available from several sources.[34–36] A typical plot evaluating the contribution of proton transfer to the enthalpy of ligand binding is shown in Fig. 4 for the binding of Ca^{2+} to 1,3-diaminopropane-*N*,*N*,*N*′,*N*′-tetraacetic acid.

Such an approach was used by Sigurskold *et al.* to parse binding enthalpies of both acarbose and 1-deoxynojirimycin to glucoamylase.[37] At pH 4.5, 0.5 ± 0.1 protons are released from glucoamylase during the binding of both 1-deoxynojirimycin and acarbose whereas at pH 7.5, 0.7 ± 0.1 protons are transferred from buffer to the protein. Glutamates at positions 179 and

[33] J. Gómez and E. Freire, *J. Mol. Biol.* **252,** 337 (1995).

[34] J. K. Grime, "Analytical Solution Calorimetry," Vol. 79. John Wiley & Sons, New York, 1985.

[35] T. Roig, P. Backman, and G. Olufsson, *Acta Chem. Scand.* **47,** 899 (1993).

[36] H. Fukuda and K. Takahashi, *Proteins Struct. Funct. Genet.* **33,** 159 (1998).

[37] B. W. Sigurskjold, C. R. Berland, and B. Svensson, *Biochemistry* **33,** 10191 (1994).

TABLE I
HEATS OF IONIZATION OF SOME WIDELY USED BUFFERS[a]

Buffer	ΔH_{ion} (kcal mol^{-1})
MES	3.03
Bis-Tris	6.99
ACES	7.20
ADA	2.75
MOPS	4.54
PIPES	2.08
BES	5.52
HEPES	3.92
TES	6.99
Ethyl glycinate	11.07
Glycinamide	10.70
HEPPS	4.29
Tricine	7.29
THAM	11.30
Glycylglycine	10.45
Bicine	6.27
TAPS	9.59
N,N-Dimethylglycine	7.53
CAPS	11.60

[a] From Ref. 34.

400 serve as the catalytic acid and base, respectively. The pK_a values of these key catalytic residues in the unbound state are 5.9 and 2.7, respectively; the observed proton transfer presumably reflects shifts in these values during ligand binding.

Evaluation of High- and Low-Affinity Systems

The calorimetric experiment can be deconvoluted to yield binding constants and enthalpies only when c ranges from 1 to 1000. Outside this range, derivation of a full set of thermodynamic parameters is not possible by direct calorimetric titration. Evaluation of binding enthalpies in systems where c exceeds 1000 is still straightforward, because beyond this value all ligand added is bound until saturation is reached. Integration of raw data thus yields enthalpies per injection that, when divided by the amount of ligand added per injection, provide molar enthalpies of binding. If a binding constant can be determined by some other technique, binding entropies are available by subtraction in the usual fashion. This methodology also

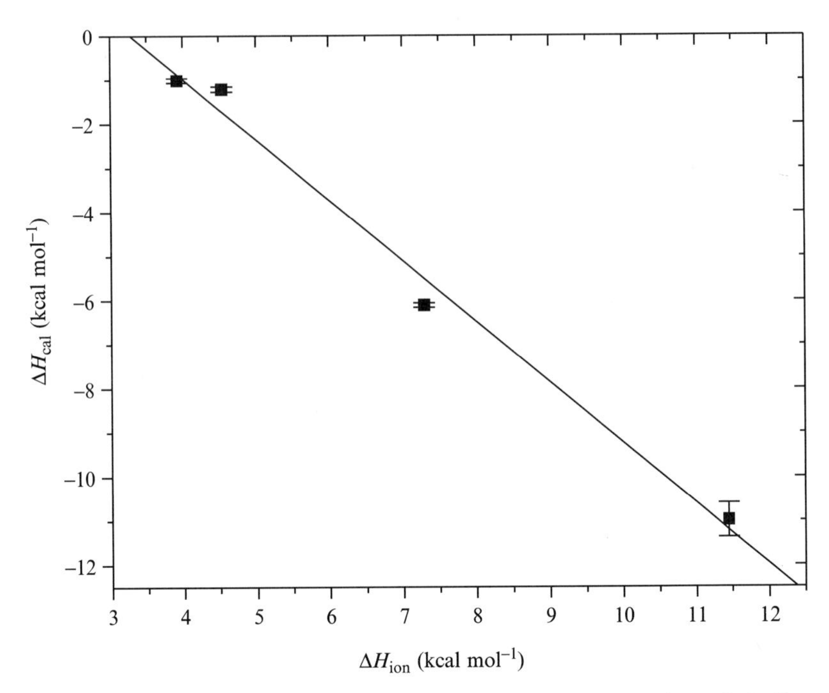

FIG. 4. Graph of the calorimetric enthalpy obtained from binding of $CaCl_2$ to 1,3-diaminopropane-*N,N,N′,N′*-tetraacetic acid as a function of the buffer ionization enthalpy at pH 8. Protons (1.37 ± 0.10) are released to the buffer during binding. The buffers used are HEPES, MOPS, Tricine, and Tris. The intrinsic enthalpy, the *y* intercept, is 4.51 ± 0.74 kcal mol^{-1}.

avoids the potential danger of correlated errors, a pitfall always present when two dependent variables are derived from a single measurement.

When it is not possible to perform titrations with *c* in an appropriate range binding constants can be determined by a competition, or displacement, titration. In this experiment, a "reference" ligand with known thermodynamic parameters is displaced by or displaces a tighter or weaker binding ligand, respectively (Fig. 5). Because thermodynamic parameters are state functions, the binding of the tight or weak binding ligand is the arithmetic sum of the observed thermodynamics of binding in the presence of the reference ligand and those of the reference ligand. This methodology has been most extensively developed by Sigurskjold and co-workers.[21,37–39]

[38] A. Solovicova, T. Christensen, E. Hostinova, J. Gasperik, J. Sevcik, and B. Svensson, *Eur. J. Biochem.* **264,** 756 (1999).

[39] B. W. Sigurskjold, *Anal. Biochem.* **277,** 260 (2000).

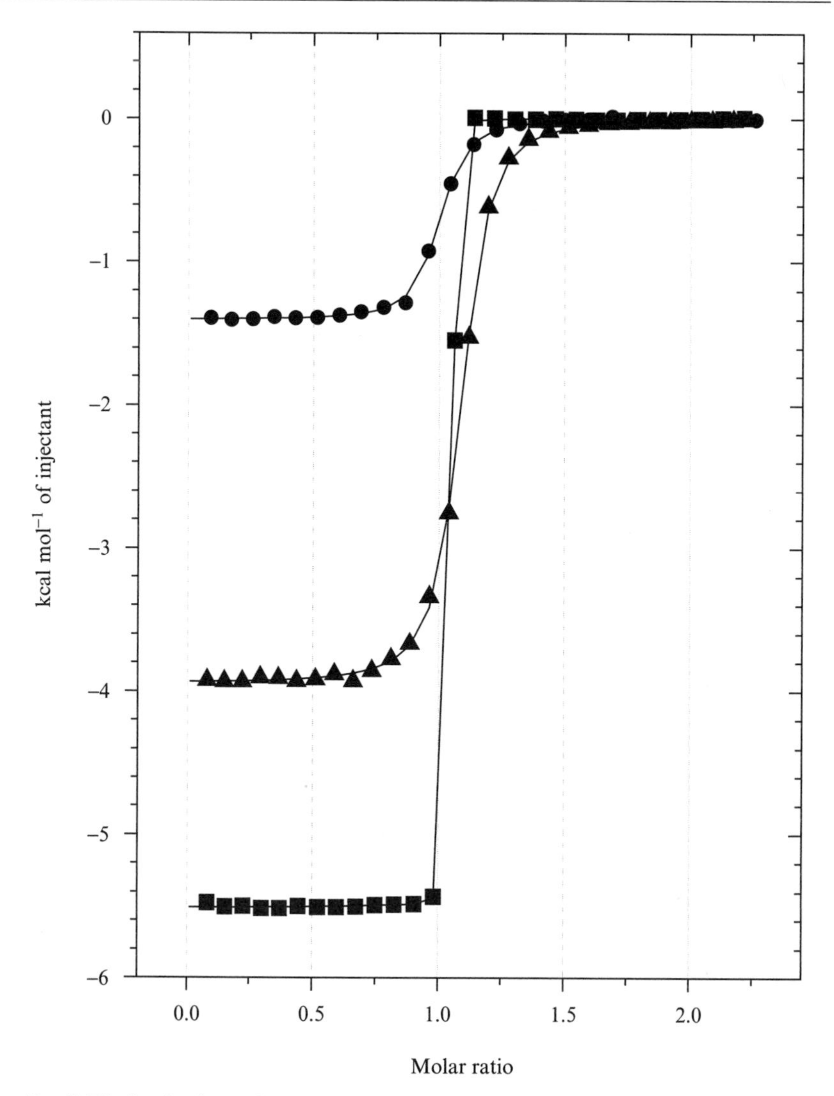

FIG. 5. Binding isotherms in 20 mM HEPES at pH 8, showing the displacement of Ba(II) by Ca(II) (triangles) from EDTA, and the individual titrations, Ba(II) + EDTA (circles) and Ca(II) + EDTA (squares). It is clear that (1) the association between EDTA and Ca(II) is too tight to measure directly and (2) the enthalpies for the two reactions are different and therefore the two ions are suitable for a displacement experiment. The obtained thermodynamic parameters are for (1) the Ba(II)–EDTA interaction, $K = (8.22 \pm 0.34) \times 10^5\ M^{-1}$ and $\Delta H = -3.75 \pm 0.05$ kcal/mol and (2) the Ca(II)–EDTA interaction, $K = (5.91 \pm 0.26) \times 10^8\ M^{-1}$ and $\Delta H = -5.25 \pm 0.12$ kcal/mol.

A displacement titration is described as follows:[39,40]

First titration:

$$P + L_1 \rightleftharpoons P \cdot L_1$$

Second titration:

$$P \cdot L_1 + L_2 \rightleftharpoons P \cdot L_2 + L_1$$

where P corresponds to the macromolecule, L_1 is the weaker binding ligand, and L_2 is the tighter binding ligand. Assuming $\Delta H_1^\circ \neq \Delta H_2^\circ$, $K_1/K_2 < 10^6$, and the initial concentration of the reference ligand is greater than the reciprocal of its binding constant (i.e., $[L_1]_0 > 1/K_1$) the displacement binding isotherm can be fit to Eq. (7). The apparent binding constants and enthalpies are related to the constituent microscopic values by Eqs. (9) and (10):[30]

$$K_{\mathrm{app}} = \frac{K_2}{1 + K_1[L_1]} \approx \frac{K_2}{1 + K_1[L_1]_0} \tag{9}$$

$$\Delta H_{\mathrm{app}}^\circ = \Delta H_2^\circ - \Delta H_1^\circ\left(\frac{K_1[L_1]}{1 + K_1[L_1]}\right) \approx \Delta H_2^\circ - \Delta H_1^\circ\left(\frac{K_1[L_1]_0}{1 + K_1[L_1]_0}\right) \tag{10}$$

where K_1 and K_2 are the association constant for the weak and tight binding ligands, respectively, $[L_1]$ is the concentration of free weak binding ligand, $[L_1]_0$ is the total concentration of the weak binding ligand, and ΔH_1° and ΔH_2° are the calorimetric enthalpies of the weak and tight binding ligands, respectively. If the concentration of the weak binding ligand is much larger than that of protein the apparent association constant and calorimetric enthalpy can be calculated approximately [Eqs. (9) and (10)].

During the first titration, the protein is diluted. The new concentration of protein and the concentration of ligand to be displaced in the cell can be calculated from Eqs. (11) and (12):

$$[P]_{\mathrm{new}} = [P]_{\mathrm{old}} \exp\left(-\frac{iV_i}{V_0}\right) \tag{11}$$

$$[L]_{\mathrm{new}} = [L]_{\mathrm{old}}\left[1 - \exp\left(-\frac{iV_i}{V_0}\right)\right] \tag{12}$$

where $[P]_{\mathrm{new}}$ and $[P]_{\mathrm{old}}$ are the concentrations of protein in the cell after and before the first titration, V_i is the volume of the ith injection, i is the

[40] Y.-L. Zhang and Z.-Y. Zhang, *Anal. Biochem.* **261,** 139 (1998).

number of injections, V_0 is the volume of the ITC cell, and $[L]_{new}$ and $[L]_{old}$ are the concentrations of ligand in the cell after the first titration and the concentration of ligand in the syringe in the first titration.

Interpretation of Calorimetric Data: Contributions from Solvophobic Effects

The goal of thermodynamic analysis of ligand binding is to reveal the energetic consequences of distinct intermolecular interactions. Again, a proper consideration of these values requires knowledge of all the contributing effects. We have previously considered the effect of proton transfer; a more significant and difficult to evaluate contribution to overall binding thermodynamics arises from solvent reorganization.

When two molecules form a complex in aqueous solution some fraction of each species is desolvated and the water of solvation is returned to bulk solution. Enthalpies of solvation of both neutral and charged species are enormous—on the order of hundreds of kcal mol^{-1}—and are expected to contribute strongly to observed thermodynamics of binding. Various parameterization schemes have been developed in an attempt to estimate the contribution of desolvation effects, particularly in the context of protein folding.[41–44] Although these formulas are useful empirical tools for systems close in structure to those from which they were developed, they are not general and their predictive utility diminishes rapidly as the structure under consideration diverges from those of the basis set. In addition, these techniques require information about the structure of the bound complex in order to calculate the polar and nonpolar surface area buried during binding.

It has long been recognized that changes in solvation are reflected in changes in the molar heat capacity of a solution. The constant pressure heat capacity of a solution is the temperature derivative of the enthalpy; the change in molar heat capacity during binding can therefore be obtained by measuring enthalpies of binding at different temperatures according to Kirchhoff's law:

$$C_p = \frac{\partial \Delta H^\circ}{\partial T} \tag{13}$$

[41] R. S. Spolar, J. R. Livingstone, and M. T. Record, *Biochemistry* **31,** 3947 (1992).
[42] K. P. Murphy, *Med. Res. Rev.* **19,** 333 (1999).
[43] E. Freire, *Methods Enzymol.* **240,** 502 (1994).
[44] S. P. Edgcomb and K. P. Murphy, *Curr. Opin. Biotechnol.* **11,** 62 (2000).

Change in molar heat capacity during dissolution is a property unique to water, and no other liquid, regardless of polarity, shows this trait.[45] In a seminal work, Sturtevant[46] suggested that the primary contributions to the heat capacity change during ligand binding arise from the hydrophobic effect and changes in internal vibrational modes during binding. Because Sturtevant was unable to distinguish the relative importance of these two terms, consideration of ΔC_p in terms of specific intermolecular interactions remained impossible.[46] A number of researchers, including Scheraga, Ben-Naim, Arnett, and Dahlberg,[47–54] explored the differential enthalpy of dissolution of various species in light and heavy water as a means to extract the contribution of changes in solvation to bulk thermodynamic parameters. Later, Toone and co-workers[55,56] used thermodynamic solvent isotope effects to consider the effect of solvent reorganization on ligand binding. The thermodynamic solvent isotope effect arises from the differential enthalpy of the H–O versus D–O hydrogen bond, and can be rationalized with the use of a Born–Haber cycle (Scheme 1).

Assuming equivalent structures in the isotopic liquids, the solvent isotope effect is proportional to the amount of nonpolar surface area transferred to and from bulk solution during binding. The difference between the observed enthalpies in protium and deuterium corresponds to the solvent reorganization and can be derived as follows:

$$\Delta J_i^\circ + \Delta J_{s,b}^\circ = \Delta J_{obs,X_2O}^\circ + \Delta J_{s,u}^\circ \tag{14}$$

The enthalpies can be measured in both H_2O and D_2O and Eq. (14) can be rearranged to show $\Delta\Delta H^\circ$:

$$\Delta\Delta H^\circ = \Delta H_{obs,H_2O}^\circ - \Delta H_{obs,D_2O}^\circ = \Delta H_{s,u}^\circ - \Delta H_{s,b}^\circ \tag{15}$$

A direct derivation of the contribution of solvent reorganization processes is feasible if the difference in enthalpy between the deuterium oxygen

[45] D. Mirejovsky and E. M. Arnett, *J. Am. Chem. Soc.* **105,** 1112 (1983).
[46] J. M. Sturtevant, *Proc. Natl. Acad. Sci. USA* **74,** 2236 (1977).
[47] D. B. Dahlberg, *J. Phys. Chem.* **76,** 2045 (1972).
[48] E. M. Arnett and D. R. McKelvey, *in* "Solute–Solvent Interactions" (J. F. Coetzee and C. D. Ritchie, eds.), p. 343. Marcel Dekker, New York, 1969.
[49] C. G. Swain and J. Thornton, *J. Am. Chem. Soc.* **84,** 822 (1962).
[50] N. Muller, *Acc. Chem. Res.* **23,** 23 (1990).
[51] A. Ben-Naim, *J. Chem. Phys.* **42,** 1512 (1965).
[52] Y. Marcus and J. Ben-Naim, *J. Chem. Phys.* **83,** 4744 (1985).
[53] G. Némethy and H. A. Scheraga, *J. Chem. Phys.* **36,** 3382 (1962).
[54] G. Némethy and H. A. Scheraga, *J. Chem. Phys.* **36,** 3401 (1962).
[55] M. C. Chervenak and E. J. Toone, *J. Am. Chem. Soc.* **116,** 10533 (1994).
[56] T. G. Oas and E. J. Toone, *Adv. Biophys. Chem.* **6,** 1 (1997).

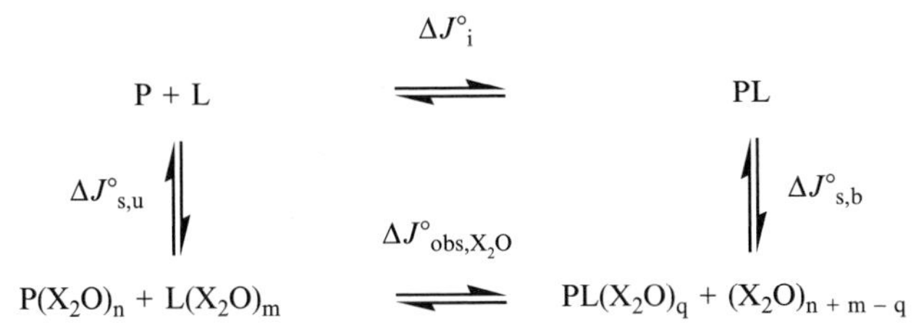

SCHEME 1. $\Delta J°$ represents any thermodynamic parameter; P, L, and PL represent, respectively, the desolvated form of the ligand, the protein, and the protein–ligand complex; X_2O represents protonated and deuterated water; and $P(X_2O)$, $L(X_2O)$, and $PL(X_2O)$ represent, respectively, the solvated forms of the protein, the ligand, and the complex. $\Delta J°_i$ and $\Delta J°_{obs}$ correspond to the thermodynamic parameters under desolvated and solvated conditions, respectively. $\Delta J°_{s,u}$ and $\Delta J°_{s,b}$ correspond to the solvation thermodynamic parameters of the unbounds species and the complex, respectively.

and protium oxygen hydrogen bond strengths is known. The D–O hydrogen bond is enthalpically favored over the corresponding H–O interaction by roughly 10%.[52,56,57] Unfortunately, the sign of the effect is indeterminate, and there exists no straightforward way to convert a $\Delta\Delta H°$ value into a solvation-associated enthalpy of binding.[56]

Toone and co-workers also noted a strong linear correlation between $\Delta\Delta H°$ and ΔC_p.[55,56] This correlation excludes a significant contribution of changes in low-frequency protein vibrational modes to overall changes in molar heat capacity, and ΔC_p can thus be considered exclusively in terms of solvent reorganization. Again, however, a simple correlation between ΔC_p and an enthalpy attributable to solvent reorganization does not exist.

Even without straightforward relationships between $\Delta\Delta H°$ or ΔC_p and a solvation contribution to the enthalpy of binding, evaluation of both terms provides a powerful tool for the consideration of solvation effects on ligand binding, and facilitates a more insightful evaluation of binding data than can be accomplished in its absence. Specifically, such data are of great utility during the comparison of the binding enthalpies of a series of related ligands. In such an exercise the response of the enthalpy or free energy of binding is evaluated as a function of incremental changes in ligand structure. Evaluation of either a thermodynamic solvent isotope effect or a constant pressure molar heat capacity allows assignation of changes in thermodynamic parameters to solute–solute or to solvent-associated events. Thus a change in the thermodynamics of binding in response to a change in ligand structure that is not accompanied by a change

[57] G. Némethy and H. A. Scheraga, *J. Phys. Chem.* **41,** 680 (1964).

in either $\Delta\Delta H^\circ$ or ΔC_p can confidently be ascribed to changes in solute–solute interactions. On the other hand, changes in ligand structure that produce changes in these solvation-associated thermodynamic terms presumably also result in a solvation-based contribution to changes in binding enthalpies.

Complications Arising from Multivalency: Cluster Glycoside Effect

Protein–carbohydrate interactions are uniformly weak, with dissociation constants ranging from millimolar to micromolar. Virtually all carbohydrate-binding proteins occur naturally as multisubunit assemblies; the suggestion has been made that such assemblies facilitate high affinity through multivalent interactions.[3,58–63] In this paradigm a multivalent ligand interacts with a multivalent lectin, forming a high-affinity complex. From this suggestion myriad multivalent carbohydrate ligands have been prepared, and many show significantly enhanced affinities in various assays of "affinity"; indeed, Rao and co-workers have reported an enhancement in affinity near 10^9 on a valence-corrected basis.[64] A molecular basis for this enhancement is unclear and currently the subject of some debate.[65,66] Regardless of the mechanism of enhancement, the evaluation of binding of multivalent ligands to multivalent lectins raises a host of issues that require special and careful attention.

The most serious of these involves ligand-induced aggregation. In several seminal papers Brewer and coworkers demonstrated the remarkable degree of specificity lectins display in the binding of multivalent ligands through the formation of highly ordered cross-linked lattices.[67,68] More recently, Toone and co-workers have suggested that many—if not most—enhancements in activity shown by multivalent glycosides in comparison with their monovalent counterparts arise from intermolecular binding and aggregation.[65,66] In this model a coupled equilibrium is driven by the

[58] J. M. Gardiner, *Expert Opin. Invest. Drugs* **7,** 405 (1998).
[59] K. Drickamer, *Nat. Struct. Biol.* **3,** 71 (1995).
[60] L. L. Kiessling and N. L. Pohl, *Chem. Biol.* **3,** 71 (1996).
[61] M. Mammen, S.-K. Choi, and G. M. Whitesides, *Angew. Chem. Int. Ed.* **37,** 2754 (1998).
[62] T. K. Lindhorst, *Top. Curr. Chem.* **218,** 201 (2002).
[63] R. T. Lee and Y. C. Lee, *Glycoconj. J.* **17,** 543 (2001).
[64] J. Rao, J. Lahiri, L. Isaacs, R. M. Weis, and G. M. Whitesides, *Science* **280,** 708 (1998).
[65] J. B. Corbell, J. J. Lundquist, and E. J. Toone, *Tetrahedron Asymmetry* **11,** 95 (2000).
[66] S. M. Dimick, S. C. Powell, S. A. McMahon, D. N. Moothoo, J. H. Naismith, and E. J. Toone, *J. Am. Chem. Soc.* **121,** 10286 (1999).
[67] D. Gupta and F. C. Brewer, *Biochemistry* **33,** 5526 (1994).
[68] D. Gupta, R. Arango, N. Sharon, and C. F. Brewer, *Biochemistry* **33,** 2503 (1994).

formation of protein–protein contacts and/or insolubility of macroscopic aggregates. Great care must therefore be taken while reducing and interpreting calorimetric data from the binding of multivalent ligands. Several observations are consistent with the formation of aggregates, including sharply diminished enthalpies of binding and the development of turbidity in solutions following titration.

The most straightforward method for eliminating aggregation as a source of affinity involves evaluation of binding constants as a function of protein concentration. In a simple 1:1 binding model, binding constants should be robust to changing protein concentration. On the other hand, the formation of macroscopic aggregates has some large molecularity in protein concentration, and apparent affinities are strong functions of protein concentration. When multivalency provides enhancements of even 10^2 or 10^3, varying protein concentrations through one or two orders of magnitude while maintaining a c value greater than 1 is typically feasible.

A second issue of some concern involves data reduction. A 1:1 model of binding could conceivably bind one carbohydrate ligand to one lectin-binding site; in this model the concentrations of carbohydrate epitopes and lectin-binding sites would be appropriate. A second 1:1 model of binding involves the binding of one multivalent ligand to one multivalent protein; in this instance the concentrations of ligand and lectin multimer would be appropriate. Both methods have been utilized in the literature.[65,69] As a practical matter, binding enthalpies and stoichiometries vary linearly as the concentration of epitope; enthalpy data from the two methods are thus related by a factor equivalent to the valency of the ligand. There seems little reason, then, to prefer one method over the other, so long as the data reduction protocols are clearly described.

Conclusions

A long-standing limitation in the field of protein–carbohydrate interaction arises from the lack of robust methodology for the evaluation of true binding constants. The last decade has been enormous strides toward rectifying this problem, and isothermal titration microcalorimetry is an important tool in this armamentarium. Calorimetry offers the significant advantage over other methodologies of directly measuring thermodynamic parameters. Care must be taken, however, to properly consider the contribution of each distinct event to an overall binding before interpretation of calorimetric data; consideration of binding in various buffers and the use of

[69] T. K. Dam, R. Roy, S. K. Das, S. Oscarson, and F. C. Brewer, *J. Biol. Chem.* **275,** 14223 (2000).

thermodynamic solvent isotope effects are important methodologies in this process. When properly considered and carefully conducted, ITC will undoubtedly continue to shed important light on the molecular basis of protein–carbohydrate affinity.

[33] Fluorescence Polarization to Study Galectin–Ligand Interactions

By PERNILLA SÖRME, BARBRO KAHL-KNUTSON, ULF WELLMAR, ULF J. NILSSON, and HAKON LEFFLER

Introduction

The galectin protein family is defined by a characteristic carbohydrate recognition domain (CRD) with affinity for β-galactosides and certain conserved amino acid sequence elements; it is an ancient protein family, with members found in mammals (14 galectins reported), other vertebrates (fish, birds, and amphibians), invertebrates (worms and insects), and even protists (sponge and fungus); galectin-related sequences are also found in plant genomes.[1] Galectins are typically soluble cytosolic proteins that can be secreted from cells by nonclassic pathways to interact with external glycoconjugates and have a variety of activities both extra- and intracellularly, with important implicated roles in immunity, inflammation, and cancer.[2–5] One obvious essential feature of galectins is their carbohydrate-binding specificity.

We began comparing the detailed specificity of various galectins (RL-14, RL-18, and RL-29, now known as galectin-1, -5, and -3, respectively) toward a panel of small to medium-size saccharides, using an assay involving inhibition of galectin binding to asialofetuin–Sepharose or lactosyl–Sepharose.[6] The advantage of this assay was that it was simple and rapid, and a small reaction volume (6–15 μl) made consumption of saccharides low. The disadvantage was that the galectins needed to be radioactively labeled, requiring different conditions for each galectin, and that the result

[1] D. N. W. Cooper, *Biochim. Biophys. Acta* **1572,** 209 (2002).
[2] C. F. Brewer, *Biochim. Biophys. Acta* **1572,** 255 (2002).
[3] F.-T. Liu, R. J. Patterson, and J. L. Wang, *Biochim. Biophys. Acta* **1572,** 263 (2002).
[4] G. A. Rabinovich, L. G. Baum, N. Tinari, R. Paganelli, C. Natoli, F.-T. Liu, and S. Iacobelli, *Trends Immun.* **23,** 313 (2002).
[5] A. Danguy, I. Camby, and R. Kiss, *Biochim. Biophys. Acta* **1572,** 285 (2002).
[6] H. Leffler and S. H. Barondes, *J. Biol. Chem.* **22,** 10119 (1986).

0076-6879/03 $35.00

obtained, the 50% inhibitory concentration (IC_{50}) of the test substance, is only an indirect estimate of K_d. Frontal affinity chromatography has been presented as an elegant, highly efficient way to compare the binding of a large library of fluorescently tagged saccharides to immobilized galectins.[7] One disadvantage of this technique is that each galectin needs to be immobilized, with possible confounding effects on interpreting binding data. A number of other techniques requiring immobilizing and/or labeling of either the galectin or ligand have been used by others, each with their advantages and disadvantages. Titration microcalorimetry has been applied to a few galectins and has given good information on solution-phase binding parameters, but requires high galectin and saccharide amounts.[8,9]

Here we describe fluorescence polarization (FP) as an alternative solution-phase binding assay for galectins. A fluorescent probe (in this case a fluorescein-tagged saccharide) is excited with plane-polarized light and the degree of polarization remaining in the emitted light is measured. This remaining polarization decreases in relationship to how much the fluorescent probe moves during the excited state life time (4 ns for fluorescein), that is, the method in principle measures the rotation speed of a fluorescent probe. When the probe is bound to protein (the galectin in this case) it rotates slower than when free, and the remaining fluorescence polarization is higher. Hence the amount of bound and free probe in solution can be measured without separation of the phases. A detailed description of the principles is given in Ref. 10.

Fluorescence polarization is an established technique,[10] and has also been used for studies of lectin–carbohydrate interactions.[11–15] However, instruments permitting the application of this method to small volumes in microtiter plate format have become available only more recently. The method is gaining increased popularity for various types of screening in

[7] J. Hirabayashi, T. Hashidate, Y. Arata, N. Nishi, T. Nakamura, M. Hirashima, T. Urashima, T. Oka, M. Futai, W. E. G. Muller, F. Yagi, and K.-i. Kasai, *Biochim. Biophys. Acta* **1572,** 232 (2002).

[8] D. Gupta, M. Cho, R. D. Cummings, and C. F. Brewer, *Biochemistry* **35,** 15236 (1996).

[9] K. Bachhawat-Sikder, C. J. Thomas, and A. Surolia, *FEBS Lett.* **500,** 75 (2001).

[10] J. R. Lakowicz, Ed., "Principles of Fluorescence Spectroscopy." Kluwer Academics/Ple-Plenum Press, New York, 1999.

[11] K. Kakehi, Y. Oda, and M. Kinoshita, *Anal. Biochem.* **297,** 111 (2001).

[12] M. I. Khan, M. K. Mathew, P. Balaram, and A. Surolia, *Biochem. J.* **191,** 395 (1980).

[13] M. I. Khan, N. Surolia, M. K. Mathew, P. Balaram, and A. Surolia, *Eur. J Biochem.* **115,** 49 (1981).

[14] Y. Oda, K. Nakayama, B. Abdul-Rahman, M. Kinoshita, O. Hashimoto, N. Kawasaki, T. Hayakawa, K. Kakehi, N. Tomiya, and Y. C. Lee, *J. Biol. Chem.* **275,** 26772 (2000).

[15] K. Kakehi, M. Kinoshita, Y. Oda, and B. Abdul-Rahman, *Methods Enzymol.* **362,** [34](2003) (this volume).

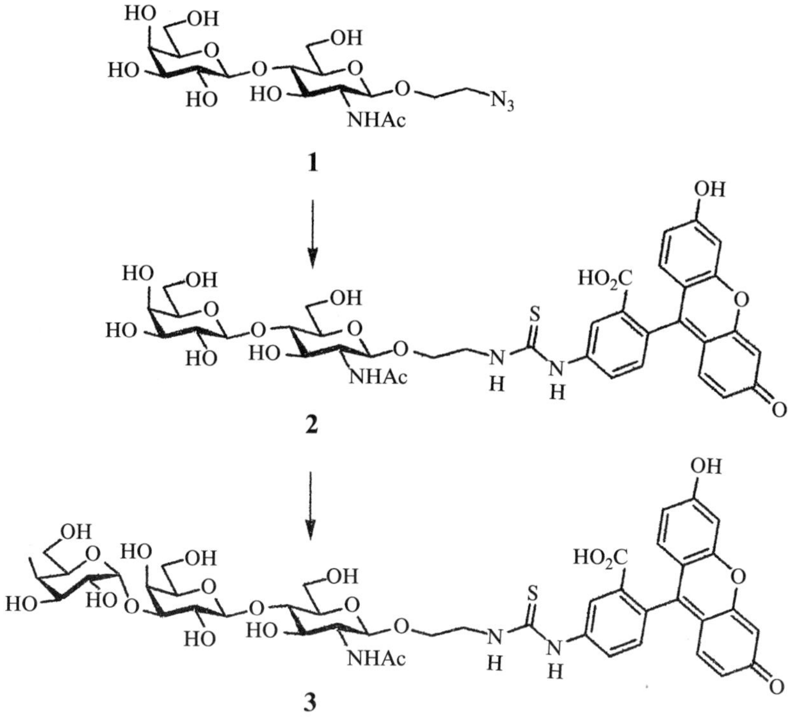

SCHEME 1. Summary of final synthetic steps for the two fluorescein-tagged saccharide probes **2** and **3**.

the pharmaceutical industry and for clinical drug measurements.[16] These are the reasons why we wanted to explore its usefulness for galectins.

Materials and Methods

Fluorescein-Tagged Saccharide Probes

All the compounds (see Scheme 1) are structurally identified by nuclear magnetic resonance (NMR) spectra (DRX-400 instrument; Bruker BioSpin, Billerica, MA) and high-resolution fast atom bombardment (FAB) mass spectra (SX-120 instrument; JEOL, Tokyo, Japan). Column chromatography is performed on SiO_2 (Matrex, 60 Å, 35–70 μm; Grace Amicon/Millipore, Bedford, MA) and thin-layer chromatography (TLC)

[16] M. S. Nasir and M. E. Jolley, *Combin. Chem. High throughput screen.* **2,** 177 (1999).

is carried out on SiO_2 60 F_{254} (Merck, Darmstadt, Germany) with detection under ultraviolet (UV) light and developed with aqueous sulfuric acid. UDP-α-D-galactose (disodium salt) and α-1,3-galactosyltransferase (porcine, recombinant *Escherichia coli*) are from Calbiochem-Novabiochem (San Diego, CA).

*2-(Fluorescein-5-thiourea)ethyl(β-D-galactopyranosyl)-(1→4)-2-acetamido-2-deoxy-β-D-glucopyranoside (**2**).* 2-Azidoethyl (β-D-galactopyranosyl)-(1→4)-2-acetamido-2-deoxy-β-D-glucopyranoside (**1**) is obtained as described elsewhere in this series.[17] This compound (6.6 mg, 0.0104 mmol) mmol) is dissolved in a mixture of ethanol and 1 *M* hydrochloric acid (110 μl, 0.11 mmol) and is hydrogenated (H_2, 10% Pd/C, 1 atm) for 3 h. The mixture is filtered through Celite and concentrated. The residue is dissolved in 0.1 *M* sodium bicarbonate buffer with 0.9% sodium chloride (pH 9.3, 2.5 ml). A solution of fluorescein 5-isothiocyanate (4.3 mg, 0.011 mmol) in *N,N*-dimethylformamide (200 μl) is added to the buffer solution and the mixture is kept at 4° overnight. The mixture is concentrated and chromatographed [SiO_2; CH_2Cl_2–methanol–H_2O, 65:35:5 (v/v/v)] to give **2** (4.2 mg, 87%).

*2-(Fluorescein-5-thiourea)ethyl(α-D-galactopyranosyl)-(1→3)-(β-D-galactopyranosyl)-(1→4)-2-acetamido-2-deoxy-β-D-glucopyranoside (**3**).* Compound **2** (1.5 mg) and UDP-D-galactose (4.8 mg, 7.9 μmol) are dissolved in a buffer (50 m*M* sodium cacodylate, 20 m*M* $MnCl_2$, pH 6.5, 1.0 ml). α-1,3-Galactosyltransferase is added and the reaction mixture is stirred at 37° overnight. The mixture is concentrated and purified by high-performance liquid chromatography (HPLC) to give **3** (0.9 mg, 54%).

Recombinant Galectin-3

Recombinant human galectin-3 is produced in *E. coli* BL21 (pET-3c vector) and purified by chromatography on lactosyl–Sepharose as described.[18] Before use lactose is removed by dialysis against phosphate-buffered saline (PBS: 118 m*M* NaCl, 67 m*M* sodium/potassium phosphate, pH 7.2) and the galectin is concentrated with CentriPrep concentrators (Millipore, Bedford, MA).

Fluorescence Polarization Binding Assay

To measure direct binding of fluorescent probe to galectin, FP is measured from above in 96-well microtiter plates (black polystyrene; Costar,

[17] P. Sörme, B. Kahl-Knulsson, U. Wellmar, B.-G. Magnusson, H. Leffler, and U. J. Nilsson, *Methods Enzymol.* **363** [12] (2003).

[18] S. M. Massa, D. N. Cooper, H. Leffler, and S. H. Barondes, *Biochemistry* **32,** 260 (1993).

Corning, NY) using a PolarStar instrument (BMG, Offenburg, Germany). The final sample volume in each well is 200 μl. For direct binding a series of different concentrations of galectin-3 is prepared in a microtiter plate (100 μl per well). Fluorescent probe (100 μl) is then added to a final concentration of 0.1, or 1 μM, and the plate is incubated under slow rotary shaking in the dark for 5 min. Control wells containing only fluorescent probe or fluorescein are included. All dilutions and measurements are done in PBS.

The FP is measured by two photomultipliers (PMT1 and PMT2) receiving light through two perpendicularly polarized channels. The values are given as anisotropy ($A = \text{PMT1} - \text{PMT2}/[\text{PMT1} + (2 \times \text{PMT2})]$) or polarization units $[P = \text{PMT1} - \text{PMT2}/(\text{PMT1} + \text{PMT2})]$.[10] The fluorescein-containing well is used to balance the gain of the photomultipliers (setting the K factor, according to instrument manual) assuming that free fluorescein should give a value of 35 mP. The total fluorescence intensity is measured as the sum of the two PMT channels, to detect quenching of the bound probe (difference between probe only or probe with maximum amount of galectin).

The temperature is set to 24, 30, 35, 40, or 45°, using the thermostat of the instrument. To measure at about 5° the plate is cooled on ice for 5 min before FP recording. The temperature in selected wells is measured before and after recording, using a TEMP 4 Acorn series thermistor temperature sensor (Oakton Instruments, Vernon Hills, IL).

To measure inhibitory potency of a nonfluorescent molecule, a range of concentrations of the molecule is included with a fixed concentration of galectin-3 and fluorescent probe. The galectin concentration and temperature are chosen to give an FP value of 100–150 mP, so as to give a span of about 50–100 mP units down to the value of free probe to study the inhibition.

Data plotting, regression analysis, and curve construction are done with Prism 3.0 (GraphPad Software, San Diego, CA).

Results and Discussion

The binding of galectin-3 to the two probes (Scheme 1) is shown in Fig. 1. An increasing anisotropy value is observed with increasing galectin concentration and a plateau is indicated. The maximum value approached represents the anisotropy with all probe bound to galectin. The intermediate values are the average of the amount of bound and free probe at each galectin concentration, if corrected for any significant fluorescence quenching of the bound probe. In this case fluorescence quenching is less than 10% even at the highest galectin concentration, so no correction is included.

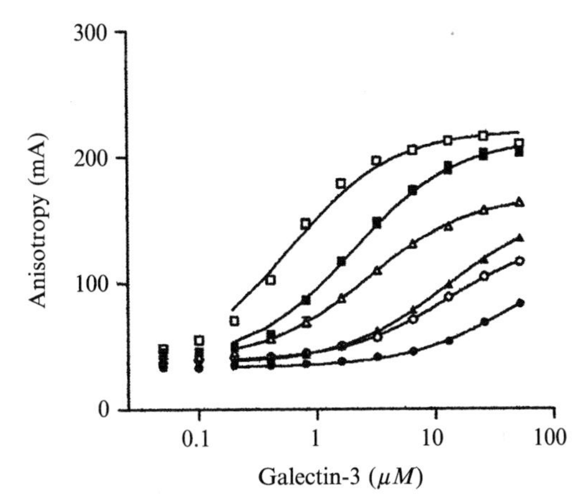

FIG. 1. Fluorescence polarization analysis of human galectin-3 binding to two fluorescein-tagged saccharides, LacNAc-Flu (**2** in Scheme **1**) and Galα1-3LacNAc-Flu (**3** in Scheme 1). Different concentrations of galectin (x axis) were added to a fixed concentration of probe (1 μM). The fluorescence anisotropy (y axis) was then measured at different temperatures. Each condition was prepared and analyzed in duplicate and each data point is plotted. However, for most conditions only one data point is observed because of excellent agreement between the two measurements. The curves were constructed by nonlinear regression, assuming the formula $Y = A_0 + (A_{max} - A_0)\ [X/(K_d + X)]$, where X is the galectin concentration (μM) and Y is the measured anisotropy (mA). The minimum (A_0) and maximum (A_{max}) anisotropy and the K_d were estimated by nonlinear regression.

TABLE I

ESTIMATED K_d AND A_{max} FOR PROBES **2** AND **3** BINDING TO HUMAN GALECTIN-3[a]

Probe	Temperature (°C)	A_0 (mA) (free probe)	A_{max} (mA) (estimated)	K_d (mA) (estimated)
LacNAc-Flu	5	45.8	213.6	2.1
	24	35.9	158.1	12.5
	45	32.1	122.9	42.0
Galα3LacNAc-Flu	5	52.6	220.3	0.6
	24	41.2	169.1	2.8
	45	37.6	135.9	12.7
LacNAc	7	[b]	[b]	115.7
	25	[b]	[b]	190.9
Galα3LacNAc	7	[b]	[b]	43.6
	25	[b]	[b]	53.6

[a] Based on data in Fig. 2. The values are compared with microcalorimetry data for the corresponding saccharides from Ref. 9.

[b] Microcalorimetry.[9]

The data as shown in Fig. 1 give three types of information (Table 1). First, the position of the curve along the x axis gives an estimate of the average K_d for the interaction between galectin and probe. Second, the A_{max} value give an estimate of the tumbling velocity of the bound probe. This depends on the size of the galectin–probe complex, temperature, and solution viscosity. However, it is also confounded (made lower) by propeller effects, that is, faster movement of the fluorescent moiety of the bound probe. Third, the shape of a fitted curve may give an indication of the complexity of the interaction between galectin and probe.

In the present case, as expected, a curve representing a simple one-to-one interaction can be fitted well to most data points by nonlinear regression (see Fig. 1). The fit of the lower part of the curve for Galα3LacNAc-Flu at 5° is not perfect because a requirement of this simplified formula is not fulfilled—the free galectin concentration cannot be approximated as the total galectin concentration because the bound fraction of the total is too large. Despite this fact, the higher points permit an estimate of K_d and A_{max} also in this case.

Varying the temperature has two main effects on the binding curve (Fig. 1). The curve moves along the x axis reflecting a change in K_d (lower at lower temperature). The A_{max} value is also higher at a lower temperature, reflecting slower tumbling of the probe–galectin complex due to the temperature itself and also to increased viscosity of the water.

The temperature variations could possibly be used to estimate thermodynamic parameters of the galectin–probe binding, using a van't Hoff plot

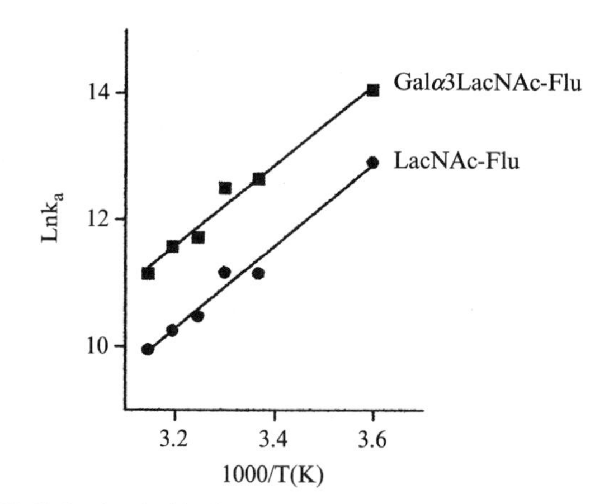

FIG. 2. van't Hoff plot for the binding of galectin-3 to fluorescein-tagged saccharide probes **2** and **3**. The K_d was estimated from the data presented in Fig. 1 (Table I), and in addition the corresponding data for measurements at 30, 35, and 40°.

(Fig. 2). This is especially valuable when comparing two probes differing only in the saccharide structure. The fact that the two lines in Fig. 2 are parallel indicates that the increased interaction with the trisaccharide probe, compared with the disaccharide probe, is mainly entropically driven. The slope of the curves indicate there is also a strong enthalpic component, about the same for both probes, which gives a relatively strong temperature dependence of the interaction. However, further work is required to ascertain that the van't Hoff plot is valid and represents the probe–galectin interaction in this assay. For now, it is more a preliminary summary of the binding data.

The estimated K_d values for the fluorescein-tagged saccharides are compared with those determined for the same free saccharides by microcalorimetry in Table I. It is clear that the binding of the fluorescein-tagged saccharides is stronger than the binding of the corresponding saccharide alone. In addition, the binding of the probes is much more temperature dependent, making it more than 50-fold stronger than the binding of the free saccharides at 5°. This indicates that the fluorescein moiety or ethylamine linker provides a stabilizing interaction with the galectin. On the other hand, the relative affinities of the two free saccharides, and of the two fluorescein-conjugated saccharides, are about the same, 3- to 4-fold higher for Galα3LacNAc.

As illustrated elsewhere in this series,[17] the FP assay can also be used to probe the potency of inhibitors. A range of concentrations of inhibitor to be tested is added to wells with a fixed concentration of galectin and probe. The concentrations and temperature are chosen so as to give a sufficient span between the uninhibited and completely inhibited situation to permit a reliable construction of an inhibition curve. We have found 100–150 mP for the uninhibited condition (50–100 mP above free probe) to be suitable, which requires 10 μM galectin-3 at room temperature or 1 μM at 5°.

There was no evidence of nonspecific interaction between the fluorescent tag and galectin-3. This is shown by the fact that the fluorescence polarization (or anisotropy) can be brought down to the base value of the free probe by inclusion of inhibitory saccharides such as lactose (not shown) or as shown in Ref. 17. Moreover, a fluorescent probe of the same structure as **2**, except that LacNAc had been replaced by Man, did not show any interaction with galectin-3 at the concentrations tested in Fig. 1 (not shown).

Background binding was also low when the FP assay was tested for detection of galectin in a complex mixture of other proteins. A total lysate of *Escherichia coli* expressing galectin-3 gave a clearly positive signal with LacNAc-Flu whereas a lysate of control *E. coli* not expressing recombinant protein did not (not shown).

Conclusions

The fluorescence polarization assay described above is simple and rapid. Two or three components, (for the inhibition assay), are mixed in the wells of a microtiter plate and incubated for 5 min, and then the whole plate (96 wells) can be read in 1–2 min. Although the method requires fluorescent probes, once a probe has been synthesized it lasts for many experiments (about 75,000 for 1 mg of the LacNAc-Flu probe used at 0.1 μM) because little is required per assay. Another advantage is that the galectin does not need to be labeled or immobilized and, thus, is native in the assay. Hence, the same conditions can be used to screen many different galectins, and possibly other lectins. A disadvantage is that relatively high protein concentrations are required. This can partially be overcome by doing the assay at lower temperatures. Here the strong temperature dependence of the interaction discussed above is of great practical use. The consumption of protein and probe could also be reduced by using 384- or 1536-well plates, which can also be read in the instrument, but would require a robot to load the samples reliably.

We are currently building a library of fluorescein-tagged probes with different saccharides and different linkers at the reducing end, as exemplified in Ref. 17. These will be of value for better defining the interactions with galectin-3 discussed above, and for defining the specificity of other galectins, galectin-like proteins, and lectins. The method is also useful for quality control of galectins, for example, after large-scale production, storage, and shipping for various biological experiments. The quantitation of galectin activity in bacterial lysates may help in selecting colonies and optimizing expression.

[34] Lectin from Bulbs of *Crocus sativus* Recognizing N-Linked Core Glycan: Isolation and Binding Studies Using Fluorescence Polarization

By KAZUAKI KAKEHI, MITSUHIRO KINOSHITA, YASUO ODA, and BADRULHISAM ABDUL-RAHMAN

Unlike many mannose-binding lectins, such as concanavalin A, which recognizes both mannose and glucose residues, *Crocus sativus* lectin (CSL) is truly mannose specific: its binding is inhibited only by mannooligosaccharides, and not by glucose or its oligomers or polymers.[1] Other mannose-specific lectins are also isolated from bulbs of *Tulipa gesneriana*,[2]

METHODS IN ENZYMOLOGY, VOL. 362
0076-6879/03 $35.00

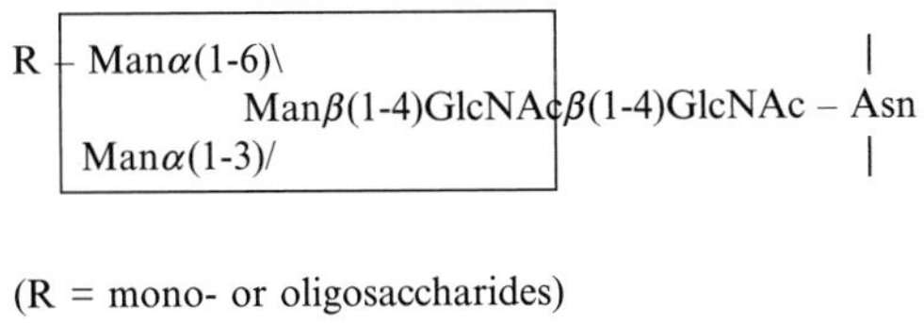

```
R ┤ Manα(1-6)\                                    |
               Manβ(1-4)GlcNAcβ(1-4)GlcNAc – Asn
    Manα(1-3)/                                    |
```

(R = mono- or oligosaccharides)

FIG. 1. Core oligosaccharide recognized by *Crocus sativus* lectin. R, Mono- or oligosaccharide.

Galanthus nivalis,[3] *Narcissus pseudonarcissus*, *Hippeastrum hybridum*,[4] and *Lycoris radiata*.[5] These mannose-binding lectins exhibit different binding specificities toward various yeast cells of different cell wall mannan structures, indicating that each lectin has its unique sugar-binding specificity.[6]

We also found that hen ovomucoid was a good inhibitor of CSL, but it did not inhibit other mannose-specific lectins from plant bulbs. We analyzed the binding specificity of CSL, using various carbohydrates as ligand, by fluorescence polarization (FP), flow injection analysis (FI), time-resolved fluorometry (TRF), and surface plasmon resonance (SPR), and found that CSL specifically recognizes the *N*-glycan core pentasaccharide (Fig. 1).

Isolation of CSL from Bulbs of *Crocus sativus*

Bulbs of *C. sativus* (autumn-flowering crocus, 20 bulbs, about 1 kg) are homogenized in distilled water (3 liters) with a blender. After centrifugation of the mixture at 10,000*g* for 10 min at 4°, solid ammonium sulfate is added to the supernatant. The precipitate formed between 40 and 60% ammonium sulfate saturation is collected by centrifugation and dialyzed against 20 m*M* Tris-HCl buffer (pH 7.5). The solution is applied to a mannan–Sepharose 4B column (2.5 × 25 cm) previously equilibrated with the same buffer. The column is washed with several bed volumes of the equilibrating buffer, and then with 1 *M* NaCl in the same buffer until the $A_{280\ nm}$ of the effluent is negligible. The lectin is eluted with 0.1 *M* acetic

[1] Y. Oda and Y. Tatsumi, *Biol. Pharm. Bull.* **16,** 978 (1993).

[2] Y. Oda and K. Minami, *Eur. J. Biochem.* **159,** 239 (1986).

[3] E. J. M. Van Damme, A. K. Allen, and W. J. Peumans, *FEBS Lett.* **215,** 140 (1987).

[4] H. Kaku, E. J. M. Van Damme, W. J. Peumans, and I. J. Goldstein, *Arch. Biochem. Biophys.* **279,** 298 (1990).

[5] Y. Oda, Y. Tatsumi, M. Kinoshita, S. Kurashimo, E. Honda, Y. Ohba, and K. Kakehi, *Bull. Pharm. Res. Technol. Inst.* **6,** 45 (1997).

[6] Y. Oda, M. Kinoshita, K. Nakayama, S. Ikeda, and K. Kakehi, *Anal. Biochem.* **269,** 230 (1999).

acid. Fractions containing the lectin are examined by agglutination assay toward yeast cells. The fractions showing agglutination activity are combined, neutralized with 2 *M* NaOH, dialyzed against distilled water, and lyophilized. The lyophilized material is dissolved in 20 m*M* Tris-HCl buffer containing 8 *M* urea and applied to a Sepharose CL-6B column (2.0×145 cm) equilibrated with the same buffer. The active fractions showing agglutination of yeast cells are pooled and further purified by passage through a DE-52 column (2.5×30 cm), equilibrated with 20 m*M* Tris-HCl buffer (pH 7.5). Elution is carried out with a linear gradient from 0 to 0.2 *M* NaCl in the equilibrating buffer after washing the column with buffer (500 ml). Four overlapped peaks are eluted between 0.01 and 0.1 *M* NaCl. These four active fractions show similar binding specificities toward mannan and are considered to be isolectins. The fractions are pooled, dialyzed against distilled water, and lyophilized. The lyophilized sample of CSL is kept in a refrigerator at -20°. The lectin is stable at least for 1 year without obvious loss of the activity.

Mannan–Sepharose 4B

Mannan (500 mg) isolated from baker's yeast is coupled with Sepharose 4B (80 ml), which has been activated with cyanogen bromide.[7] From the amount of the unbound mannan examined by phenol–sulfuric acid assay, about 2.5 mg of mannan is bound to 1 ml of the gel.

Agglutination Assay of CSL toward Yeast Cells

Agglutination activity is determined by serial 2-fold dilution of the fractions in a 96-well polyvinyl chloride plate. Each well contains 25 μl of yeast cell suspension (4×10^8 cells/ml), 25 μl of the lectin solution diluted with saline, and 50 μl of saline. Agglutination is observed after 30 min.

Molecular Mass of Lectin

The molecular mass of the lectin is determined with a TSK SW2000 column (7.8×300 mm) equilibrated with 50 m*M* Tris-HCl (pH 7.0) at 1.0 ml/min, using ferritin (molecular weight, 450,000), bovine serum albumin (molecular weight, 68,000), ovalbumin (molecular weight, 45,000), chymotrypsinogen A (molecular weight, 25,000), and cytochrome *c* (molecular weight, 12,500) as marker proteins. The molecular mass of the lectin is estimated to be approximately 48,000 Da on the basis of the elution volumes on the TSK SW2000 column. However, matrix-assisted laser

[7] P. Cuatrecases and C. B. Anfinsen, *Methods Enzymol.* **22,** 345 (1971).

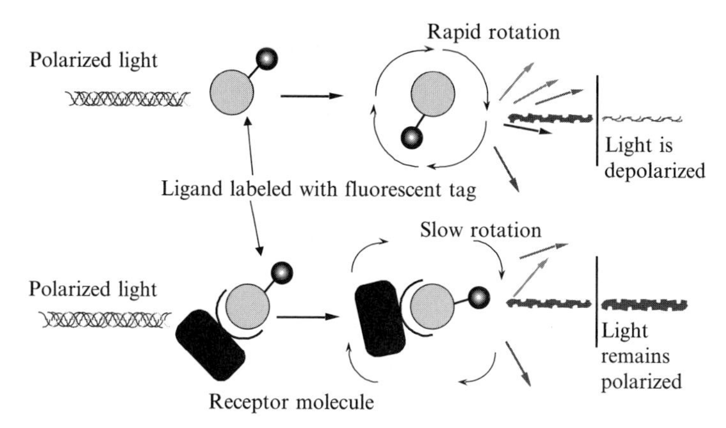

FIG. 2. Diagram showing the use of fluorescence polarization in the assay of a binding reaction.

desorption/ionization time of flight mass spectrometry (MALDI-TOF MS) of the lectin shows the molecular mass of the subunit to be 12,635 Da, when measured with sinapinic acid as the matrix and a Voyager DE PRO system (Applied Biosystems, Foster City, CA). These data indicate that CSL is a tetramer like many other plant lectins. A complete sequence of this lectin has been reported by Alvarez-Orti *et al.* (Swiss-Prot; accession number AAK29077). Van Damme *et al.* have also reported cloning of the lectin from *Crocus vernus* (spring-flowering crocus; same genus as *Crocus sativus*).[8] They have compared the polypeptide chains of both lectins and found that both lectins show similar amino acid sequences and that both have a tetrameric structure composed of different monomer molecules formed via posttranslational cleavage of the precursor molecule.

Principle of Fluorescence Polarization for Lectin-Binding Studies

FP gives a direct, nearly instantaneous measurement of the bound/free ratio of a tracer even in the presence of free tracer. Figure 2 illustrates how FP is employed in the measurement of a binding reaction. Small molecules in solution rotate and tumble rapidly, but the movement of larger molecules is slower. When fluorescently labeled small molecules in solution are excited with plane-polarized light (Fig. 2, top), the emitted light is depolarized because of the fast movement of the molecule. However, when the fast-moving fluorescently labeled small molecule is bound to a receptor

[8] E. J. M. Van Damme, C. H. Astoul, A. Barre, P. Rouge, and W. J. Peumans, *Eur. J. Biochem.* **267,** 5067 (2000).

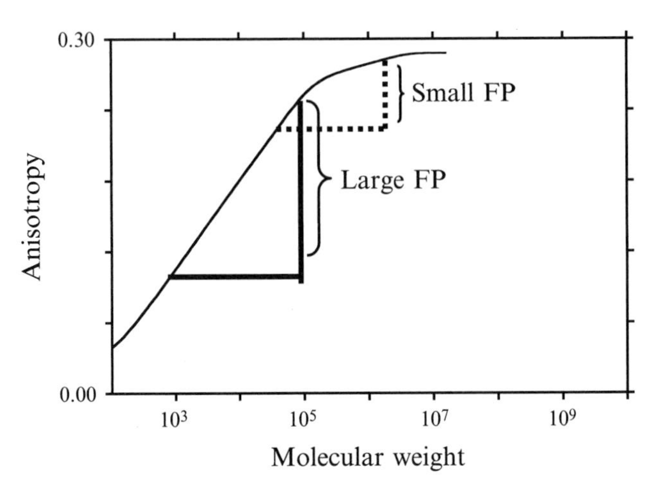

FIG. 3. Relationship between molecular weight and anisotropy.

of high molecular mass, the movement of the conjugate is restricted and becomes slow. When such a conjugate is irradiated with polarized light, the emitted light is obviously less depolarized (Fig. 2, bottom). The difference between these two states is dependent on the number of bound molecules and the binding constant or affinity constant. A binding isotherm can be easily constructed by titrating a binding protein into a solution of small, fluorescently tagged ligand molecules of fixed concentration. From such data, K_d and K_a can be obtained by nonlinear regression analysis.

The molecular mass (i.e., molecular size and shape) of the analyte molecule is an important factor in FP measurement because molecular size is considered to be inversely proportional to the molecular motion in solution. Figure 3 illustrates a theoretical model for the relationship between molecular size and anisotropy. Anisotropy, a parameter analogous to polarization, is also used, and is virtually interchangeable. They are mathematically closely related to each other, and most instruments can handle both polarization and anisotropy.[9]

When a flluorescently tagged molecule having a molecular mass of 1000 Da binds to a protein with a molecular mass of 100,000 Da, the FP (i.e., anisotropy) will be large as shown by the solid line in Fig. 3. However, if a fluorescent molecule of tens of kilodaltons binds to a macromolecular substance, the FP value becomes small (Fig. 3, dotted line). Therefore, we should select a fluorescently tagged ligand having low molecular mass for accurate FP analysis of the binding reaction.

[9] Y. C. Lee, *J. Biochem.* **121,** 818 (1997).

Sensitivity in FP measurement is usually high. When a fluorescent compound having high fluorescence quantum yield, such as fluorescein, is used as the labeling reagent, nanomole or even picomole-scale measurement is possible. Because of the physical properties of polarizing filters, fluorescent compounds having excitation in ultraviolet region (usually below 400 nm) are difficult to use in FP measurements.

Binding assay of CSL, Using FP Method

Preparation of N-Glycan Core Glycopeptide

Japanese quail ovomucoid is prepared according to the method reported by Fredericq and Deutsch.[10] Quail ovomucoid (1.58 g) is dissolved in 0.5 *M* Tris-HCl buffer (pH 8.5, 30 ml) containing 50 m*M* dithiothreitol (DTT), 6 *M* guanidine hydrochloride, and 5 m*M* EDTA, and the mixture is incubated for 1 h. Iodoacetamide (3.0 g) is added, and the mixture is kept in the dark at 37° for 30 min. The reaction mixture is dialyzed against water and lyophilized. The lyophilized material is dissolved in 50 m*M* Tris-HCl buffer (pH 7.5, 20 ml) and digested with Pronase (*Streptomyces griseus*; 1%, w/w) for 4 days at 55°, with daily addition of Pronase (0.5%, w/w). The digest is lyophilized and dissolved in 50 m*M* ammonium hydrogencarbonate and separated on a Sephadex G-50 column (5 × 180 cm) equilibrated and eluted with the same buffer, with 18-ml fractions collected. The fractions are analyzed with phenol–sulfuric acid[11] to identify the fractions containing glycopeptide. The glycopeptide-containing fractions are lyophilized, dissolved in 0.1 *M* acetic acid (5 ml), and further fractionated on a BioGel P4 column (2 × 138 cm), collecting 5-ml fractions. The fractions containing the trimannosyl core glycopeptide are pooled, and its identity is confirmed by sugar composition analysis by high-performance anion-exchange chromatography.[12] Reversed-phase high-performance liquid chromatography (RP-HPLC) [Hypercarb GCC column (Thermo Shandon, Pittsburgh, PA), using a gradient elution of acetonitrile in 10 m*M* NH_4OH buffer] indicates the presence of three glycopeptides. These glycopeptides have the same sugar composition and different peptide length,[13] and are used for the preparation of fluorescein-labeled glycopeptide.

[10] E. Fredericq and H. F. Deutsch, *J. Biol. Chem.* **181,** 499 (1949).

[11] M. Dubois, K. A. Gilles, J. K. Hamilton, P. A. Rebers, and F. Smith, *Anal. Chem.* **28,** 350 (1956).

[12] J.-Q. Fan, Y. Namiki, K. Matsuoka, and Y. C. Lee, *Anal. Biochem.* **219,** 375 (1984).

[13] Y. Oda, K. Nakayama, B. Abdul-Rahman, M. Kinoshita, O. Hashimoto, N. Kawasaki, T. Hayakawa, K. Kakehi, N. Tomiya, and Y. C. Lee, *J. Biol. Chem.* **275,** 26772 (2000).

Preparation of Fluorescein-Labeled Glycopeptide

$Man_3GlcNAc_2$-glycopeptide (700 nmol) prepared as described above is mixed with 20 μl of 0.2 *M* sodium borate and 30 μl of 45 m*M* 6-carboxyfluorescein succinimidyl ester in dimethyl sulfoxide. The mixture is allowed to react overnight at room temperature. The mixture is diluted to 200 μl and fractionated on a Sephadex G-10 column (0.5×30 cm), using water as eluent, with 1-ml fractions collected. Fluorescent fractions eluting earlier are pooled and stored at $-20°$ until use.

Inhibition Assay for Binding between CSL and Fluorescently Labeled $Man_3GlcNAc_2$-Glycopeptide, Using FP Method

FP measurement is carried out with a BEACON2000 (Panvera, Madison, WI), using fluorescein-labeled glycopeptide (see above) in phosphate-buffered saline (PBS) at a final volume of 100 μl at 25° and 492-nm (excitation) and 520-nm (emission) wavelengths. For the inhibition assay, final concentrations of 1 μM CSL, 100 p*M* fluorescent ligand, and various concentrations of each inhibitor oligosaccharide (Table I) are used. The mixture is mixed and allowed to reach equilibrium for 10 min at 25° before the measurement is made. The results are shown in Fig. 4.

Binding Specificity of CSL

As shown in Fig. 4, the best inhibitors share common structural elements (Fig. 1). Pyridylamination of the reducing end did not affect the affinity of the parent oligosaccharides (compare **1** and **2**). When GlcNAc is on the Manα(1-6) residue (**7**), it remains as potent as the unmodified parent compound (**1**). However, when GlcNAc is on the Manα(1-3) residue (**8**), the loss of potency is dramatic (10-fold change in IC_{50} value). Even more drastic is the loss of the Manα(1-3) residue, as shown by comparison of **2** and **3**. The influence of bisecting GlcNAc, judging by a comparison of the inhibitory effects of **5** and **9**, is not obvious.

In the inhibition assay based on a microtiter plate method, using Man_3-$GlcNAc_2$-glycopeptide labeled with europium, Manα(1-6)[Manα(1-3)]Manβ(1-4)GlcNAc showed a dramatic increase in affinity compared with Manα(1-6)[Manα(1-3)]Man. This indicates that the presence of GlcNAc linked to Manβ is important for the binding of CSL.

Comparison of K_d *Values Determined by Different Assay Methods*

Many assay methods are available for analysis of carbohydrate–lectin interactions. We compared the FP technique with some of the existing assay methods, using CSL. Table II shows K_d values for the binding

TABLE I

OLIGOSACCHARIDES USED IN INHIBITION STUDY

Compound	Structure
1	Manα(1-6) \ Manβ(1-4)GlcNAcβ(1-4)GlcNAc / Manα(1-3)
2	Manα(1-6) \ Manβ(1-4)GlcNAcβ(1-4)GlcNAc-PA / Manα(1-3)
3	Manα(1-6) \ Manβ(1-4)GlcNAcβ(1-4)GlcNAc-PA
4	GlcNAcβ(1-2)Manα(1-6) Fucα(1-6) \ \ Manβ(1-4)GlcNAcβ(1-4)GlcNAc-PA / GlcNAcβ(1-2)Manα(1-3)
5	GlcNAcβ(1-2)Manα(1-6) Fucα(1-6) \ \ GlcNAcβ(1-4)Manβ(1-4)GlcNAcβ(1-4)GlcNAc-PA / GlcNAcβ(1-2)Manα(1-3)
6	GlcNAcβ(1-2)Manα(1-6) \ Manβ(1-4)GlcNAcβ(1-4)GlcNAc-PA / GlcNAcβ(1-2)Manα(1-3)

(*continues*)

TABLE I (*continued*)

Compound	Structure
7	GlcNAcβ(1-2)Manα(1-6) \\ Manβ(1-4)GlcNAcβ(1-4)GlcNAc-PA / Manα(1-3)
8	Manα(1-6) \\ Manβ(1-4)GlcNAcβ(1-4)GlcNAc-PA / GlcNAcβ(1-2)Manα(1-3)
9	Galβ(1-4)GlcNAcβ(1-2)Manα(1-6) \\ Manβ(1-4)GlcNAcβ(1-4)GlcNAc-PA / Galβ(1-4)GlcNAcβ(1-2)Manα(1-3)

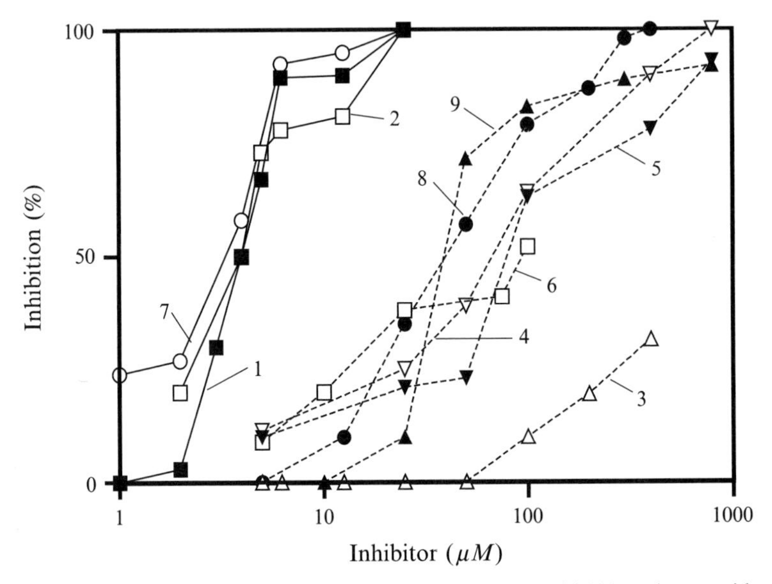

FIG. 4. Inhibition of CSL binding of fluorescein-labeled $Man_3GlcNAc_2$-glycopeptide.

TABLE II
ANALYSIS OF BINDING REACTION BETWEEN *Crocus sativus* LECTIN AND *N*-GLYCAN CORE PENTASACCHARIDE[a]

Method	Ligand	K_d (μM)
FP	Man3GlcNAc2-AMF	1.55 ± 0.22
	Man3GlcNAc2-GP-Flu	0.74 ± 0.17
FI	Yeast cells	0.30 ± 0.20
MP	Eu-Man3GlcNAc2-GP	0.54 ± 0.04
SPR	Taka-amylase	0.15 ± 0.02

[a] Using different assay methods. FP, Fluorescence polarization; FI, flow injection analysis; MP, microtiter plate method; SPR, surface plasmon resonance method; AMF, aminomethylfluorescein; GP, glycopeptide; Flu, fluorescein.

reaction between CSL and the *N*-glycan core pentasaccharide obtained by different assay methods.

In the FP assay, two fluorescent ligands were used. One is labeled with 5-aminomethylfluorescein (AMF) by reductive amination before the binding assay, and its GlcNAc residue at the reducing end becomes an acyclic aminoalditol. The other ligand for FP measurement is the glycopeptide described above. As shown in Table II, K_d values obtained with the two different ligands showed only a 2-fold difference. This is additional evidence that the lectin recognizes Manα(1-6)[Manα(1-3)]Manβ(1-4)GlcNAc. The authors developed an FI method using yeast cells as ligand.[6] This method is especially useful for binding studies of mannan-binding protein because yeast cell surfaces are covered with mannan and considered to be the best ligands for mannose-specific lectins. In the FI analysis, we labeled the lectin with hydroxycoumarin as a fluorescent tag and observed the inhibition of binding between the lectin and yeast cells by the core pentasaccharide. The K_d thus observed showed values similar to those observed in the FP assay. In the TRF assay based on a microtiter plate method, the lectin was coated onto microtiter wells by incubating the lectin solution overnight. The solution of europium-labeled trimannosyl core glycopeptide was added in a series of different concentrations. The microtiter plate method also showed values similar to those observed in the FP assay. Surface plasmon resonance analysis of the binding of CSL was also examined. Because of difficulty in immobilizing CSL that has only one amino group per subunit, Taka-amylase (α-amylase from *Aspergillus oryzae*) was coated onto a CM5 sensor chip after activation of the sensor with *N*-hydroxysuccinimide/carbodiimide in 10 mM sodium acetate (pH 3.6). The inhibition of binding between CSL and Taka-amylase by the core oligosaccharide was investigated by changing the concentrations of the core pentasaccharide. When Taka-amylase was used as the immobilized ligand for the binding study, K_d showed a smaller value of $(1.5 \pm 0.22) \times 10^{-7}$ *M*. It should be emphasized that, of all the methods examined, only the FP method is an in-solution equilibrium method; other methods use solid support.

In conclusion, CSL recognizes the structure depicted in Fig. 1. Its binding is influenced neither by the presence of bisecting GlcNAc nor by GlcNAc originally attached to Asn. Modification of the GlcNAc residue at the reducing terminal by pyridylamination or substitution by Fuc does not change the binding affinity. Masking of the terminal Man in the core structure is detrimental. Manα(1-3)Man is much more important than Manα(1-6)Man. Especially notable is the great enhancing effect by Manβ(1-4)-linked GlcNAc. This effect is not observed with either Con A[14] or *Dioclea grandiflora* lectin.[15] Binding by artocarpin, another Man-specific lectin, is also reported to be aided by β-Man-linked GlcNAc, but not as much as shown by CSL.[16] The binding specificity of CSL enables it to be a valuable tool for separation and structural analysis of *N*-glycans and related carbohydrates, such as for assay studies of β-*N*-acetylhexosaminidase in insect cells,[17] which cleaves the GlcNAc residue from 3-linked Man.

In the binding assay, it should be noted that the procedures for FP are performed in solution, and the method represents true in-solution equilibrium analysis. Washing is not necessary, and the overall procedures are simple. The time required to measure a series of sample solutions for kinetic analysis is usually less than 5 min. These characteristics are fundamentally different from those of surface plasmon resonance and microtiter plate-based, time-resolved fluorometry using europium ion. The sensitivity is comparable to that of an enzyme-linked immunosorbent assay.

[14] J. H. Naismith and R. A. Field, *J. Biol. Chem.* **271,** 972 (1996).

[15] D. A. Rozwarski, B. M. Swami, C. F. Brewer, and J. C. Sacchettini, *J. Biol. Chem.* **273,** 32818 (1998); T. K. Dam, S. Oscarson, and C. F. Brewer, *J. Biol. Chem.* **273,** 32812 (1998).

[16] S. Misquith, P. G. Rani, and A. Surolia, *J. Biol. Chem.* **269,** 30393 (1994).

[17] F. Altman, S. Schwihla, E. Staudacher, J. Gloessl, and L. Marz, *J. Biol. Chem.* **270,** 17344 (1995).

[35] Direct Fluorescence-Based Detection Methods for Multivalent Interactions

By XUEDONG SONG, JEANE SHI,* and BASIL I. SWANSON

Introduction

Multivalent interactions between ligands and receptors are common in biological systems.[1–7] There are three general scenarios of multivalent interactions. In the first scenario, each ligand has multiple binding sites (e.g., multiple subunits) for multiple receptors of one or more types. The fully bound form of the ligand has more than one receptor. In the second scenario, each ligand has only one binding site for one receptor. However, the ligand–receptor binding triggers an aggregation process of the ligand–receptor complexes to form an aggregated form (dimer or higher) and the aggregated form can be treated as an entity with multiple binding sites, equivalent to the multivalent complexes in the first scenario. This scenario commonly occurs in signal transduction processes and enzyme–substrate systems. In the third scenario, a ligand has multiple binding sites for different moieties of a branched receptor.

Here several fluorescence-based methods[8–14] for direct detection of multivalent ligand–receptor interactions of the first scenario are described. The methods should as well be applicable and useful for investigating ligand–receptor interactions of the second scenario.

* Deceased.

[1] W. L. Picking, H. Moon, H. Wu, and W. D. Picking, *Biochim. Biophys. Acta* **124,** 65 (1995).
[2] P. H. Fishman and E. E. Atikka, *J. Membr. Biol.* **54,** 51 (1980).
[3] M. E. Fraser, M. M. Chernaia, Y. V. Kozlow, and M. N. G. James, *Nat. Struct. Biol.* **1,** 59 (1994).
[4] P. M. Hilaire, M. K. Boyd, and E. J. Toone, *Biochemistry* **33,** 14452 (1994).
[5] K. Sandvig, S. Olsnes, J. E. Brown, O. W. Perterson, and B. Vandeurs, *J. Cell Biol.* **108,** 1331 (1989).
[6] E. A. Sudbeck, M. J. Jedrzejas, S. Singh, W. J. Brovillette, G. M. Air, W. G. Laver, Y. S. Babu, S. Bantia, P. Chand, N. Chu, J. A. Montgomery, D. A. Walsh, and M. Luo, *J. Mol. Biol.* **267,** 587 (1997).
[7] D. K. Takemoto, J. J. Skehel, and D. C. Wiley, *Virology* **217,** 452 (1996).
[8] X. Song, J. Nolan, and B. I. Swanson, *J. Am. Chem. Soc.* **120,** 4873 (1998).
[9] X. Song, J. Nolan, and B. I. Swanson, *J. Am. Chem. Soc.* **120,** 11514 (1998).
[10] X. Song and B. I. Swanson, *Langmuir* **15,** 4710 (1999).
[11] X. Song and B. I. Swanson, *Anal. Chem.* **71,** 2097 (1999).
[12] X. Song and B. I. Swanson, *Anal. Chim. Acta* **442**(1), 79-87 (2001).
[13] X. Song, J. Shi, J. Nolan, and B. I. Swanson, *Anal. Biochem.* **291,** 133 (2001).
[14] X. Song, J. Shi, and B. I. Swanson, *Anal. Biochem.* **284,** 35 (2000).

0076-6879/03 $35.00

Basic Principles

The principles of the direct detection methods for multivalent interactions are presented in Scheme 1 and 2. These methods are based on either distance-dependent fluorescence self-quenching or fluorescence resonant energy transfer (FRET). The methods mimic biological signal transduction mechanisms by directly triggering signal generation through binding events either in a homogeneous solution or on a biomimetic membrane surface. Self-assembled bilayers of phospholipids in the form of vesicles in aqueous solution, on supporting surfaces of silica microspheres (glass beads), or on the surface of planar optical waveguides are utilized to mimic cell membrane surfaces, which allow high lateral mobility of the fluorophore-labeled receptors on the surfaces.

In using fluorescence self-quenching as a signal generation mechanism (Scheme 1), the receptors covalently tagged with fluorescent probes are homogeneously anchored into a biomimetic membrane surface of lipids (or dissolved in aqueous solutions) with a surface density (or concentration) low enough to minimize any interaction between the fluorescent probes. In this case, the probes exhibit strong fluorescence. Binding of a ligand with multiple binding sites for the labeled receptors brings the fluorescent probes into close proximity to trigger a decrease in fluorescence intensity (as well as changes in other fluorescence properties such as fluorescence polarization, and lifetime) through a fluorescence self-quenching process. In using FRET, the receptors covalently attached to a fluorescent donor–acceptor pair are homogeneously decorated on the surface of biomimetic membrane (or dissolved in solutions). The donor and acceptor fluoresce independently to give strong fluorescence of the donor and weak fluorescence of the acceptor (ideally no acceptor fluorescence) if only the donor is excited. A multivalent ligand that induces aggregation of the receptors will bring the donor and acceptor into close proximity and trigger an FRET, resulting in an increase in acceptor fluorescence at the expense of donor fluorescence. A simultaneous increase in acceptor fluorescence and a decrease of donor fluorescence (as well as other properties such as fluorescence lifetime and polarization) can be achieved.

Use of organic fluorophores with small Stokes shifts as fluorescent probes often results in some nondesired background fluorescence of the acceptor in a simple FRET system (one-tiered FRET). In effect, it is not possible to excite the donor without also exciting the acceptor. To minimize or even eliminate the excitation of the acceptor when only excitation of the donor is desired, a two-tiered FRET scheme can be utilized. In such a method (Scheme 2), the fluorescence spectrum of the donor and absorption spectrum of the acceptor are not required to overlap, so that an acceptor

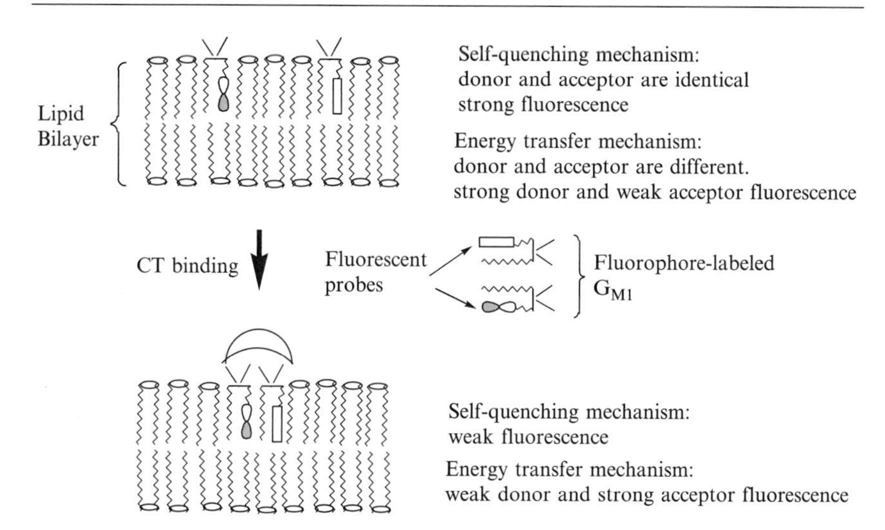

SCHEME 1. Schematic of signal generation mechanisms based on distance-dependent fluorescence self-quenching and one-tiered fluorescence resonant energy transfer.

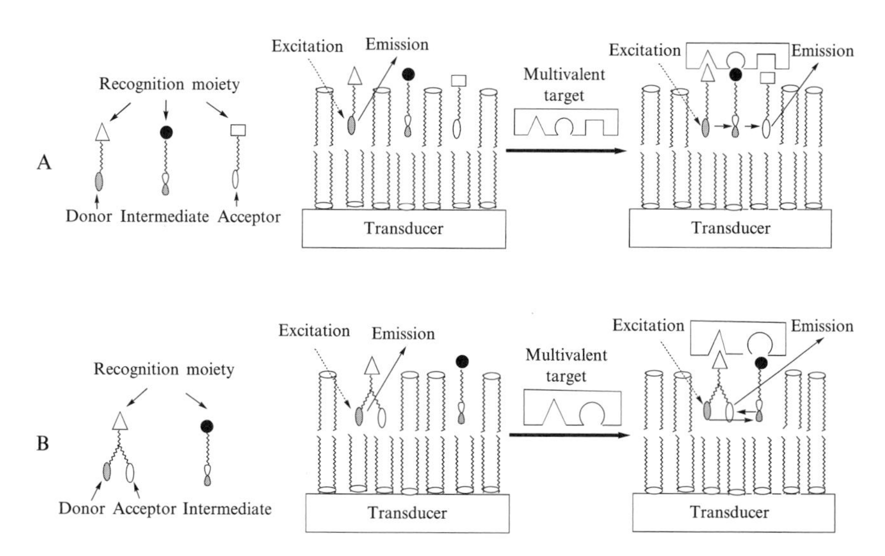

SCHEME 2. Schematic of aggregation-induced tow-tiered FRET for detection of multivalent interactions. (A) Fluorescent donor, intermediate, and acceptor are tagged separately to three individual receptors. (B) Fluorescent donor and acceptor are tagged to the same receptor and the intermediate to the other.

and a donor with largely separated absorption spectra can be selected to avoid simultaneous excitation of the donor and acceptor. For FRET still to occur between donor fluorescence and acceptor absorption, an intermediate fluorophore can be introduced to bridge the donor and acceptor. The absorption spectrum of the intermediate fluorophore should overlap with the fluorescence spectrum of the donor and its fluorescence spectrum should overlap with the excitation spectrum of the acceptor. In this case, a two-stage energy transfer from the donor to the intermediate, and then to the acceptor, can occur when the three fluorophores are brought into a close proximity. There are two possible configurations for such a scheme. In the first configuration (Scheme 2A), the donor, intermediate, and acceptor are separately attached covalently to receptors, and are then incorporated into a fluid biomimetic membrane surface (or dissolved in solutions). FRET from the donor to the acceptor can occur only if each ligand binds with the three labeled receptors together, and brings them into one complex. The second configuration (Scheme 2B) involves covalent attachment of both the donor and acceptor to the same receptor, and the intermediate to the other receptor. The ligand binding with the two labeled receptors can trigger FRET, resulting in a fluorescence change for the donor and acceptor, and providing a direct detection method to investigate the multivalent interaction with low background fluorescence of the acceptor.

Model System of Multivalent Interactions

The pentavalent cholera toxin (CT)–ganglioside M1(G_{M1}) interaction[15–17] was selected as a model system to demonstrate the detection methods. Cholera toxin, like most biological toxins such as ricin and Shiga toxin, is a protein with two subunits: A and B. The A subunit is a catalytic enzyme and is responsible for the cytotoxic activity, while the B subunit recognizes the target cell by binding receptor molecules (e.g., G_{M1} for CT) on its cell surface. In most cases, subunit B consists of multiple binding sites for its receptors with high specificity and affinity. CT has five identical binding sites for G_{M1} on the same side of the protein. To covalently tag G_{M1} with fluorescent probes without influencing its binding characteristics, lyso-G_{M1} with a free amino functional group in the nonbinding region is used for probe conjugation (see Scheme 3). In addition to the CT–G_{M1} system, the detection method using fluorescence self-quenching has also

[15] O. Livnah, E. A. Bayer, M. Wilchek, and J. L. Sussman, *Proc. Natl. Acad. Sci. USA* **90,** 5076 (1993).

[16] P. Cuatrecasas, *Biochemistry* **12,** 3547 (1973).

[17] P. H. Fisherman, J. Moss, and J. C. Osborne, *Biochemistry* **17,** 711 (1978).

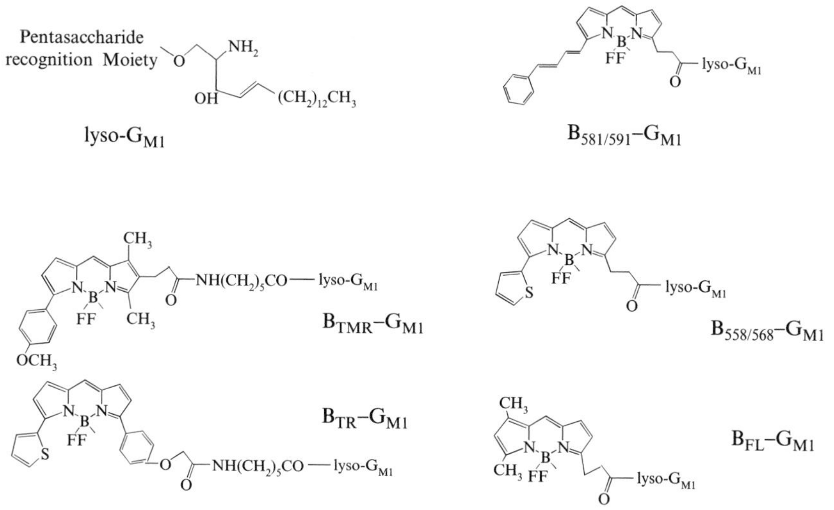

SCHEME 3. Molecular structures of lyso-G_{M1} and fluorescent probe-tagged G_{M1}.

been used to detect the tetrameric interaction between biotin and avidin, both in homogeneous solutions and on lipid bilayer surfaces.[12] Although the model system presented here is a multivalent interaction that involves an identical receptor, they are expected to be applicable for multivalent interactions with different binding sites (or subunits) for different binding receptors.

Selection of Fluorescent Probes

Selection of fluorescent probes depends on several factors, including the detection configuration of interest, signal generation mechanism, excitation source, and emission detector. For example, water-soluble probes (e.g., fluorescein) are preferred for studying multivalent interactions in homogeneous solutions, whereas hydrophobic and non-water-soluble fluorophores (e.g., BODIPY dyes) are often chosen to investigate multivalent interactions on biomimetic membrane surfaces. When the fluorescence self-quenching mechanism is used, fluorophores with high self-quenching efficiency (e.g., fluorescein, $BODIPY_{581/591}$) would be required, whereas an efficient FRET pair (e.g., fluorescein–rhodamine, rhodamine–Texas Red, $BODIPY_{TMR}$–$BODIPY_{TR}$) would be essential for the FRET method. Excitation light sources (e.g., laser) also play a role in choosing fluorescent probes, which should have acceptable absorbance at the wavelength of the excitation light. Criteria for a simple efficient FRET pair include a significant overlap between donor fluorescence and acceptor absorption spectra with a large critical FRET radius. It is much more challenging to screen a set of three fluorophores (e.g., $BODIPY_{FL}$, $BODIPY_{TMR}$, and $BODIPY_{TR}$) for two-tiered FRET because it is necessary to have significant overlap between donor fluorescence and intermediate absorption, and between intermediate fluorescence and the absorption spectra of the acceptor. Molecular Probes (Eugene, OR)[18] provides a wide variety of activated fluorescent probes for conjugation with functional groups (e.g., thiol and amino groups) of ligands. Several examples of fluorescent probes are shown in Scheme 3.

Sample Preparation

Preparation of Labeled G_{M1}

Lyso-G_{M1} (0.5 mg; purchased from Matreya, Pleasant Gap, PA) dissolved in 100 μl of 50 m*M* sodium bicarbonate buffer (pH 8.4) is added with 5 equivalents of the activated fluorescent probes [fluorophores activated as succiminidyl ester and dissolved in 200 μl of dimethylformamide (DMF) and the mixture is then vortexed for 3 h. Thin-layer chromatography (TLC) [CH_3Cl–methanol–H_2O, 100:60:10 (v/v/v) as a developing solvent] shows complete disappearance of lyso-G_{M1} and appearance of a new colored spot. The new colored spot is isolated and collected from the TLC plates. The compound is eluted from the silica gel through a micropipette, using CH_3Cl–methanol–water (100:60:10, v/v/v). The solvent is then removed by an N_2 stream, and the residue is dissolved in methanol. The methanol solution is then used to take an absorption spectrum to determine the concentration (or quantity) of the labeled G_{M1}, using the extinct coefficient constants of the fluorescent probes. The methanol is then removed by an N_2 stream and the residue is either stored in a freezer as a solid or dissolved in aqueous buffer. The buffer solution is filtered and stored in a freezer as a stock solution.

Preparation of Phospholipid Bilayer Vesicles with and without Labeled G_{M1}

Lipid Bilayer Vesicles. An appropriate stock solution of palmitoyl, 9-octadecenoylphosphatidylcholine (POPC), or other lipid that forms bilayer vesicles with a phase transition temperature below room temperature (to provide fluid membrane surfaces) in CH_2Cl_2 is transferred to a glass vial

[18] R. P. Haugland, "Handbook of Fluorescent Probes and Research Chemicals," 6th ed. Molecular Probes, Eugene, OR, 1997.

and the solvent is removed to form a thin film on the wall.[19] The lipid thin film is then put under vacuum for 2 h or more before an appropriate amount of buffer is added for hydration. Typically, Tris buffer (pH 7.0 to 8.5) or phosphate-buffered saline (PBS) buffer (pH = 7.4) is used for all sample preparations and measurements. After the mixture is hydrated at room temperature for 0.5 h, the mixture is probe-sonicated until a clear solution is obtained. The sonication time required to obtain a transparent solution depends on the sonication power and the volume of the samples. The transparent solution is then transferred to a centrifuge tube for centrifugation to remove the titanium particles. The supernatant is transferred to a new vial for direct use. The vesicle solution can also be stored at 4° for several weeks, and no significant difference in measurement results has been observed.

Lipid Bilayer Vesicles with Labeled Receptors on Outer Leaflet. Preformed lipid bilayer vesicles are incubated overnight with an appropriate amount of micellar solution of the labeled G_{M1} in a buffer (or comicelles of the labeled receptors with surfactants for non-water-soluble labeled receptors) at room temperature.[20] The labeled G_{M1} in the form of micelles is automatically diffused into the outer leaflet of the lipid vesicles.

Preparation of Glass Microspheres Coated with Lipid Bilayers and Labeled Ligands

Glass beads (5 μm in diameter, purchased from Bangs Laboratories, Fishers, IN) are incubated overnight with a preformed lipid bilayer vesicle solution (POPC, >0.2 m*M*) at room temperature to automatically form, by vesicle spreading, a supported lipid bilayer on the surface of the beads. The beads coated with POPC bilayers are separated from the rest of the solution by centrifugation and washed twice with buffer. The beads are then suspended in buffer for immediate use, or stored at 4° for future use. The lipid-coated beads should be kept constantly wet to prevent collapse of the supported lipid bilayers.

To incorporate labeled G_{M1} onto the outer layer of the lipid bilayers on the beads, an appropriate amount of labeled G_{M1} stock solution in buffer is added to lipid bilayer-coated beads and incubated at room temperature overnight with occasional shaking by hand. The beads are then separated from the rest of the solution by centrifugation and washed with buffer several times. The beads are then suspended in a proper buffer for direct use, or stored at 4° for future use. It is recommended that, to obtain consistent results, freshly lipid-coated beads be used.

[19] M. J. Hope, M. B. Bally, G. Webb, and P. R. Cullis, *Biophys. Acta* **55,** 812 (1985).

[20] P. L. Felgner, E. Freire, Y. Barenholz, and T. E. Thompson, *Biochemistry* **20,** 2168 (1981).

Measurements by Spectrofluorophotometer

Fluorescence Self-Quenching System

Titration. Stock solutions of labeled G_{M1} (either dissolved in homogeneous buffer solutions, or incorporated in surfaces of lipid vesicles or in lipid bilayer surfaces on glass beads) are dispensed into several identical vials (an equal amount of labeled G_{M1} in each vial) and buffer is added to the desired volume (typically 240 to 300 μl). Various amounts of CT stock solution in buffer (typically 150 n*M*) are then added to these vials and the samples are incubated at room temperature for about 30 to 60 min with occasional shaking by hand. Each sample is then transferred to a fluorescence cell with a magnetic stirring bar for fluorescence measurements. The excitation wavelength, slit width, and stirring speed for all the fluorescence measurements are kept constant. The fluorescence intensity at the fluorescence peak is obtained with a SPEX FluoroLog-2 spectrofluorophotometer (Jobin Yvon, Edison, NJ), and the percentage of fluorescence quenched is plotted as a function of CT concentration to obtain a titration curve. Figure 1 shows the titration curve of three sets of samples containing

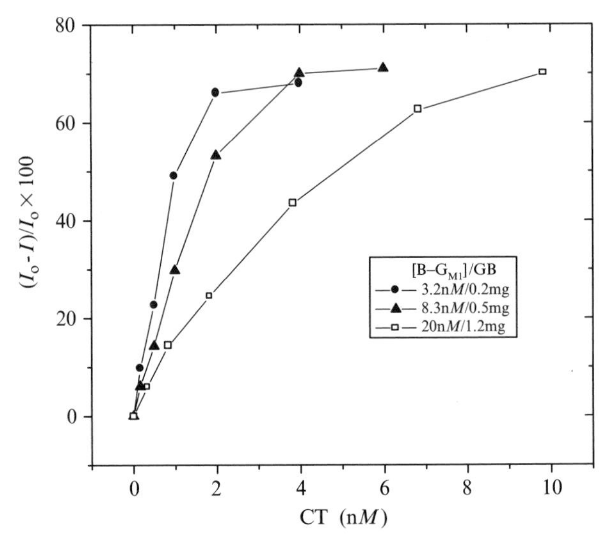

FIG. 1. Percentage of fluorescence quenched (excitation at 565 nm and emission at 600 nm) as a function of [CT] for three sets of $B_{581/591}$–G_{M1} (B–G_{M1}) in the outer leaflet of POPC bilayers coated on glass beads (GB)]. Each set of samples contains an equal amount of glass beads; different sets have different amounts of glass beads. Fluorescence spectra were collected 30 min after addition of CT under constant stirring. The standard deviation for three duplicates ranges from 4 to 7%. I_0, Fluorescence intensity without CT; I, fluorescence intensity with CT.

different amounts of $BODIPY_{581/591}$–G_{M1} incorporated into the outer layer of POPC bilayers coated on glass beads. The titration curve depends on the amount of POPC bilayers coated on glass beads and the amount of sample used for each set. To achieve maximal fluorescence quenching induced by the binding, the surface density of the labeled G_{M1} should be low enough ([labeled G_{M1}]/[POPC] $< 0.2\%$) to minimize prebinding fluorescence self-quenching. However, the surface density of the labeled G_{M1} should also be high enough ([labeled G_{M1}]/[POPC] $> 0.05\%$) so that the beads (or vesicles) have a sufficient number of labeled G_{M1} molecules for aggregation. The self-quenching method using binding-induced aggregation can be used for sensitive detection of both receptor (direction detection of ~100 p*M* CT) and ligand (competitive format-based detection for ~200 p*M* biotin). The high selectivity of the method has been demonstrated by the fact that no fluorescence change has been observed in the presence of much higher concentrations of other proteins such as albumin.

Kinetics. An appropriate amount of labeled G_{M1} stock solution (either dissolved in a homogeneous buffer solution, or incorporated in vesicle surfaces or in supported bilayer surfaces on glass beads) is dispensed into a fluorescence cell with a stirring bar and buffer is added to the desired volume (typically 240 to 300 μl). Fluorescence intensity at the peak wavelength is scanned as a function of time until the fluorescence intensity reaches a stable level. A small amount of CT stock solution is then injected, followed by an immediate fluorescence scan over time until no more fluorescence is noted. Figure 2 shows the fluorescence intensity of the fluorescent probes as a function of time on addition of various concentrations of CT.

FRET System

Titration. Sample preparation and measurement procedures for the titration of an FRET system are almost identical to those for the self-quenching system described above. Typically, equal amounts of stock solution (or a suspension in the case of glass beads) of labeled G_{M1} (either on the surfaces of lipid vesicles or incorporated in lipid bilayers coated on glass beads) are dispensed into a series of vials, and a different amount of CT is added to each vial. The samples are incubated for 30 to 60 min with occasional shaking by hand. After incubation, the fluorescence spectra for each sample are measured, and their ratios of acceptor fluorescence intensity to donor intensity at the fluorescence peaks are plotted versus [CT] to obtain a titration curve. Figure 3A shows the fluorescence spectra of B_{TMR}–G_{M1}/B_{TR}–G_{M1} in the outer leaflet of POPC bilayers coated on glass beads in the presence of various CT concentrations. The fluorescence intensity of the acceptor increases at the

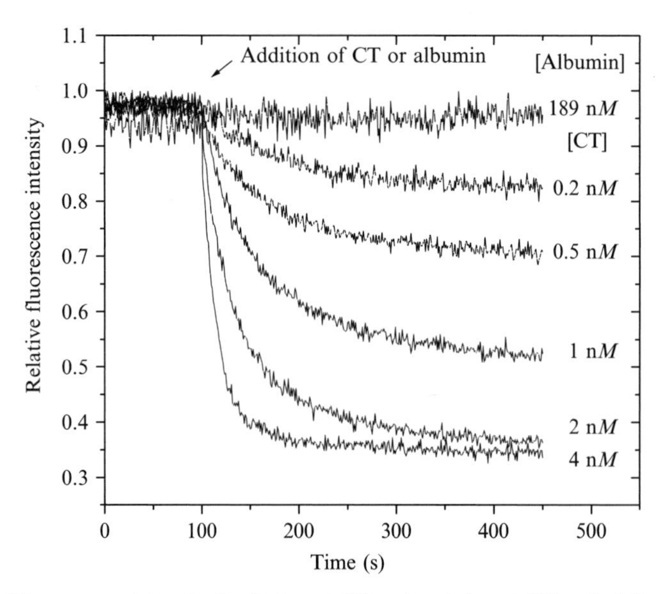

FIG. 2. Fluorescence intensity (excitation at 570 and emission at 600 nm) of $B_{581/591}$–G_{M1} (3.2 n*M*) in the outer leaflet of POPC bilayers coated on glass beads (0.2 mg) as a function of time on addition of CT or albumin. [POPC]/[$B_{581/591}$–G_{M1}] is estimated to be 600.

expense of donor fluorescence, resulting from CT–G_{M1} binding. As shown in Fig. 3B, the ratio of B_{TMR}–G_{M1}/B_{TR}–G_{M1} fluorescence intensities gives a reasonable linear relationship as a function of [CT] before leveling off caused by saturation of the labeled G_{M1}. The temperature effect (about room temperature) is negligible. The detection sensitivity depends on the total concentration of the labeled G_{M1} in the samples. High sensitivity (~50 p*M*) can be achieved by using sample with a low concentration of labeled G_{M1} ([B_{TMR}]/[B_{TR}–G_{M1}] = 1 n*M*, and [POPC] = 2 μM). High specificity for the system is demonstrated by the low level of interference produced by high concentrations of albumin. A large excess of CT (versus the availability of labeled G_{M1}) in a sample results in less change in fluorescence, which can be attributed to the formation of more monovalent or low-valent complexes, compared with the maximal fluorescence change in the presence of CT at approximately one-fifth the concentration of labeled G_{M1} in the sample. The decreased donor fluorescence and increased acceptor fluorescence induced by CT–G_{M1} bindings can be fully recovered by addition of excess G_{M1} to compete against the labeled G_{M1}. This competition process is slow and takes more than 12 h, as expected by the low off-rate that results from the multivalent nature of the CT binding for G_{M1}.

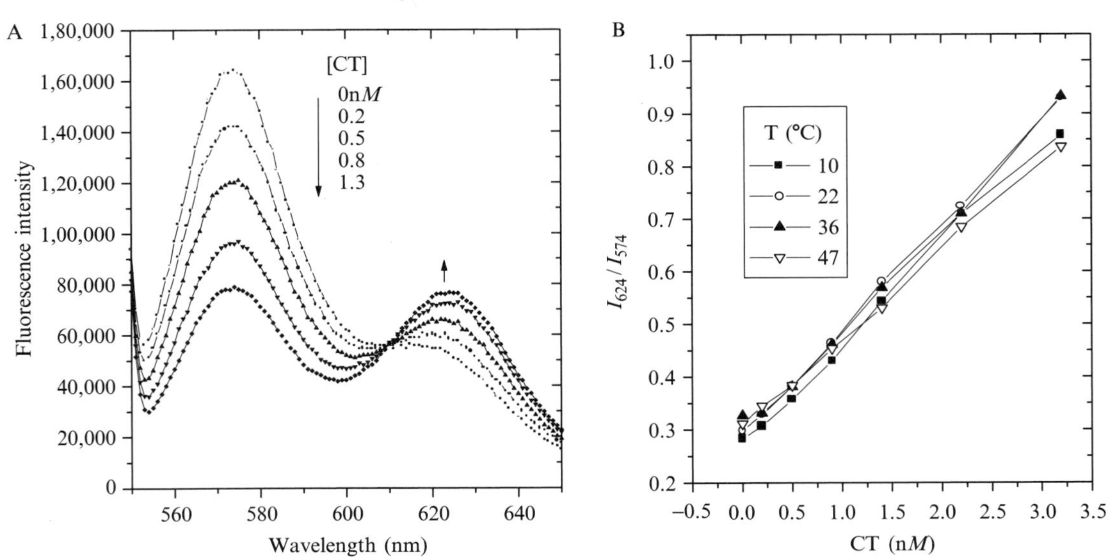

FIG. 3. (A) Fluorescence spectra (excitation at 530 nm) of B_{TMR}–G_{M1} and B_{TR}–G_{M1} in the outer leaflet of POPC bilayers supported on glass beads in the presence of various concentrations of CT. Sample preparation: 5 mg of glass beads coated with POPC bilayers was incubated in 31 μl of B_{TR}–G_{M1} (324 n*M*) and 28 μl of B_{TMR}–G_{M1} (356 n*M*) aqueous solution overnight. The beads were then washed with buffer three times and suspended in 1 ml of Tris buffer. The sample for each measurement has 120 μl of bead suspension diluted to 240 μl. (B) Plot of the fluorescence intensity ratio versus [CT] at various temperatures for four sets of samples. I_{624} and I_{574} are the fluorescence intensities at 624 and 574 nm, respectively.

An optimal surface density of labeled G_{M1} in the membrane surface for maximal signal change ($\Delta I_{624}/I_{574}$) has been found to range from 0.2 to 0.02% (molar) relative to [POPC]. Samples with a high density of labeled G_{M1} give a relatively small change in signal because of pre-binding FRET, whereas a relatively small signal change for samples with a low density of labeled G_{M1} is attributed to the formation of low-valent complexes. The optimal ratio of donor-labeled G_{M1} to acceptor-labeled G_{M1} for maximal fluorescence change has been found to be 1:1.

Kinetics. An appropriate amountof labeled G_{M1} (both donor and acceptor labeled) in a fluorescence cell with constant stirring is scanned using time-based scan mode at two wavelengths until the fluorescence intensities at these two wavelengths reach a stable level. A small amountof CT stock solution in buffer is then added, followed by immediate fluorescence scanning. Because the fluorescence intensities at two different wavelengths for the donor and acceptor are required to be measured simultaneously at a certain time point on addition of CT, a conventional fluorospectrometer with a single photomultiplier tube detector can be used only for slow binding systems because the monochromator is usually slow to switch from one wavelength to another. Therefore, it is difficult to obtain accurate data at a certain time point for rapid binding interactions. Nevertheless, it is still useful to obtain kinetic data for relatively slow multivalent binding interactions. Figure 4 shows the fluorescence change as a function of time on addition of CT for labeled G_{M1} on POPC bilayers supported on the surface of glass beads.

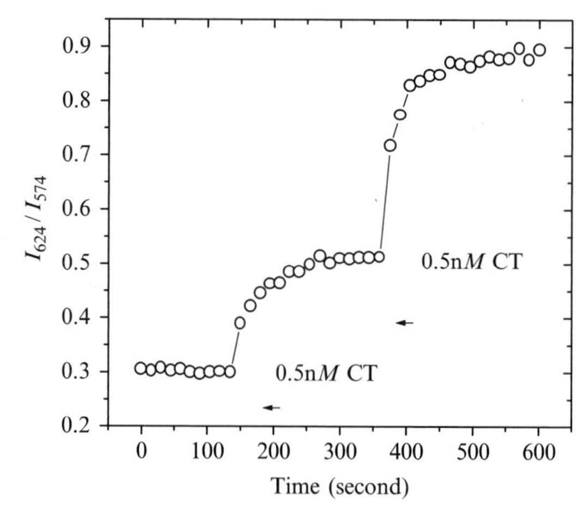

FIG. 4. Fluorescence intensity ratio of acceptor to donor (excitation at 530 nm) as a function of time on addition of CT for $[B_{TMR}–G_{M1}]/[B_{TR}–G_{M1}] = 2.5\ nM/2.5\ nM$ in the outer leaflet of POPC bilayers on glass beads (0.6 mg).

Several parameters can be tailored to obtain kinetic measurements. Those parameters include surface density of the labeled G_{M1}, ratio of donor-to acceptor-labeled G_{M1}, total concentration of labeled G_{M1}, temperature, and stirring speed.

Measurements by Flow Cytometer

Flow cytometry represents a powerful and versatile tool for a wide variety of applications in biological studies.[21,22] Its intrinsic capability to discriminate fluorescence molecules on particle surfaces from free fluorescent molecules in the volume surrounding the particles provides quantitative fluorescence measurements of particle-bound receptors without needing to separate the bound from the free receptors. The ability to construct homogeneous assays with continuous discrimination of free and bound ligands is particularly useful for kinetic studies. Flow cytometry also permits multiparameter measurements of individual cells or particles, which allows simultaneous identification of multiple targets in a mixture to provide a potential high-throughput screening technique.[23,24] The detection method described above can be readily adapted to flow cytometry for more accurate measurements.

Flow cytometric measurements of microsphere fluorescence are made on a FACSCalibur (BD Biosciences Immunocytometry Systems, San Jose, CA). The sample is illuminated at 488 nm (15 mW), and forward angle light scatter (FALS), 90° light scatter (side scatter, SSC), and fluorescence signals of the donor and acceptor are acquired through 530 (±30)-nm and 585 (±21)-nm bandpass filters, respectively (for a two-tiered FRET system, a long path filter at 625 nm is used). Analog detector signals are processed with a variable gain ratio module to give the ratio of acceptor to donor fluorescence on a particle-by-particle basis. For kinetic measurements, time is also measured by an internal clock in the data acquisition computer. Linear amplifiers are used for all measurements. Particles are gated on forward angle and 90° light scatter, and the mean fluorescence channel numbers are recorded.

[21] H. M. Shapiro, "Practical Flow Cytometry," Vol. 3. Wiley-Liss, New York, 1995.

[22] M. Melamed, T. Lindmo, and M. Mendelsohn, "Flow Cytometry and Sorting." Wiley-Liss, New York, 1990.

[23] H. M. Shapiro, *Cytometry* **3,** 227 (1983).

[24] J. R. Kettman, T. Davies, D. Chandler, K. G. Oliver, and R. J. Fulton, *Cytometry* **33,** 234 (1998).

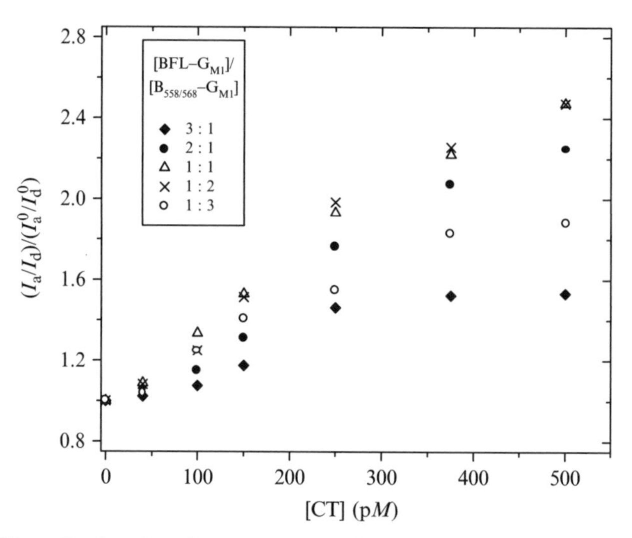

FIG. 5. Normalized ratio of mean acceptor/donor fluorescence measured by flow cytometer as a function of [CT] for a series of samples (2.5×10^5 beads in 500 μl of buffer). Each sample contains a constant concentration of the total labeled G_{M1} ([POPC]/[G_{M1}] = 1000) and a different molar ratio of B_{FL}–G_{M1}/$B_{558/568}$–G_{M1}. I_a and I_d are the mean fluorescence intensities of acceptor and donor with CT, respectively. I_a^0 and I_d^0 are the mean fluorescence intensities of acceptor and donor without CT, respectively.

Titration

Sample preparation for titration by flow cytometer is identical to that described above for the FRET system using a fluorospectrometer. Typical samples contain 2.5×10^5 glass beads. Receptor aggregation-induced RET is measured by flow cytometry. The titration results measured by flow cytometry for a $B_{FL}/B_{558/568}$–G_{M1} FRET system are similar to those obtained with a spectrofluorometer for glass microspheres. Detection sensitivity depends on the total concentration of labeled G_{M1} in the sample. Samples containing a lower concentration of labeled G_{M1} provide higher sensitivity. As a result of a higher signal-to-noise ratio for flow cytometry, the detection sensitivity improves compared with spectrofluorometry. Figure 5 shows titration data for a low range of CT concentrations in a series of samples containing the same amount of total labeled G_{M1} on the same number of glass beads (2.5×10^5), but different molar ratios of the donor- to acceptor-labeled G_{M1}. The titration data clearly show that the optimal ratio of donor-to acceptor-labeled G_{M1} is close to 1:1. The improvement in detection sensitivity also benefits from a strong excitation source and a sensitive fluorescence detection system built into the flow cytometer.

The fact that flow cytometry measures the fluorescence of individual particles rather than bulk averages allows the use of low bead concentration without loss of signal.

Kinetics

Because a flow cytometer can measure fluorescence intensities at several wavelengths simultaneously, using the proper filters for selection of detection wavelength, kinetic data of relatively fast multivalent binding can be accurately measured for the FRET system. Kinetic experiments are started by measuring substrate beads (coated with POPC bilayers and the labeled G_{M1} incorporated on the outer leaflet of the POPC bilayers) in 50 m*M* Tris buffer (pH 7.5) for 10 s to establish a baseline. The sample tube is removed from the tube holder and CT is added at 13 s. The tube is then vortexed and immediately reintroduced into the instrument. The time between mixing and data acquisition is about 10 s. The mean fluorescence channel number as a function of time is calculated with the IDLYK flow cytometry data analysis program (created at Los Alamos National Laboratory, Los Alamos, NM) and the ratio of B_{FL}–G_{M1} and $B_{558/568}$–G_{M1} fluorescence as a function of time is calculated. The time value for a given data point was the midpoint of the time window measured.

Figure 6 shows the ratio of mean acceptor to donor fluorescence as a function of time on addition of various concentrations of CT. As expected for a second-order process, a higher concentration of CT results in faster binding. For example, addition of 2.5 n*M* CT requires less than 1 min for completion of the binding, whereas it takes approximately 5 min for 0.5 n*M* CT. In addition to the concentration of CT added, the total concentration of labeled G_{M1} and its surface density also play important roles in determining the binding rate. As predicted, an increase in the total amount of labeled G_{M1} speeds up the binding rate. In the case of low [CT], the binding takes a relatively long time to complete, especially without stirring. The data collected after 5 min of measurement become relatively noisy, probably because of the formation of bead aggregates when the beads start to precipitate at the bottom of the sample tubes.

Two-Tiered FRET System

The two-tiered FRET is also useful for studying multivalent interactions. In this case, three fluorescent probes, donor, intermediate, and acceptor, are required. Sample preparations for the two-tiered FRET system are identical to those for the one-tiered FRET system. Data collection and processing from a commercial flow cytometer for the two-tiered energy transfer system are similar to those for the one-tiered FRET

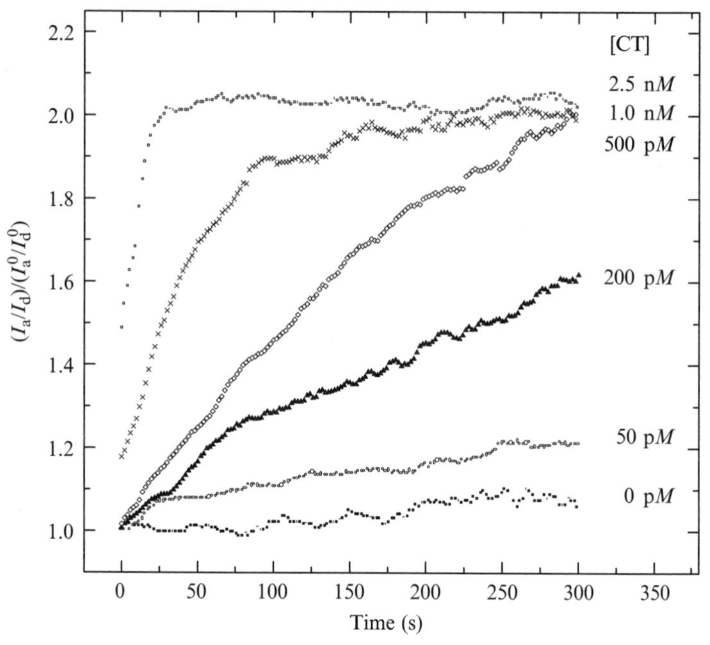

FIG. 6. Ratio of mean acceptor/donor fluorescence as a function of time on addition of different [CT] measured by flow cytometry. Each sample (500 μl) contains 2.5×10^5 beads coated with POPC/B_{FL}–G_{M1}/$B_{558/568}$–G_{M1} (1600:1:1, v/v/v) in POPC bilayers. I_a, I_d, I_a^0, and I_d^0 are defined in Fig. 5.

system, with the exception that a long path filter at 625 nm is used for the fluorescence spectrum of the acceptor.

Measurements by Fluorophotometer

Three BODIPY dyes, B_{FL}, B_{TMR}, and B_{TR} (see structures in Scheme 2), are selected as fluorescent donor, intermediate, and acceptor, respectively. For a POPC bilayer vesicle system in which the donor/intermediate/acceptor-tagged G_{M1} are asymmetrically incorporated into the outer leaflet of the bilayer vesicles, a strong fluorescence of B_{FL} is observed with little fluorescence of B_{TR} when the sample is excited at 488 nm. The fluorescence of B_{TR} increases at the expense of the fluorescence of B_{FL}–G_{M1} in the presence of CT (Fig. 7). Interestingly, the fluorescence (at 574 nm) of the intermediate B_{TMR} is found to change little, with or without CT. Compared with the system using one-stage energy transfer, the background fluorescence of the acceptor B_{TR} before CT binding has basically been eliminated for the two-tiered FRET system, and the overall fluorescence changes (the intensity ratio of acceptor to donor fluorescence; Fig. 8) are

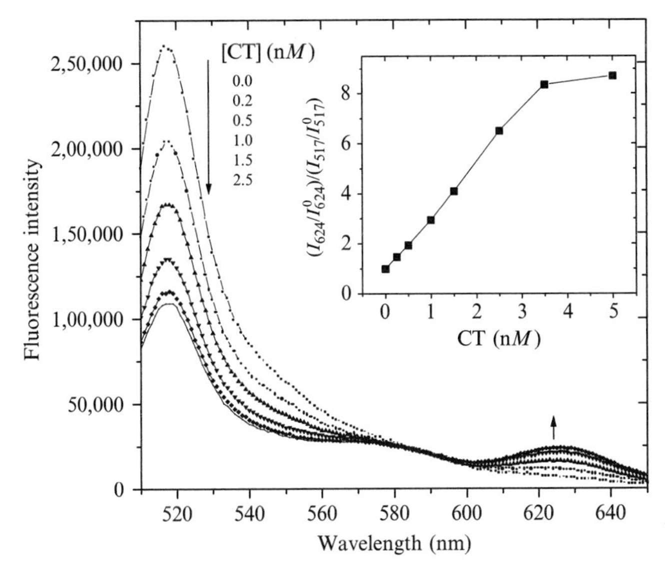

FIG. 7. CT-induced simultaneous increase in acceptor fluorescence and decrease in donor fluorescence for a sample containing $B_{FL}/B_{TMR}/B_{TR}$–G_{M1} [2:2:1 (v/v/v); the total concentration of labeled G_{M1} is 10 n*M*] in the outer leaflet of POPC vesicles ([labeled G_{M1}]/[POPC] = 0.1%). Excitation was at 488 nm. Each spectrum was taken 10 min after addition of CT under constant magnetic stirring. The inset is a zoom-out spectrum of acceptor fluorescence. *Inset*: Titration curve for the two-tiered energy transfer system. Each sample contains 10 μM POPC vesicles with B_{FL}–G_{M1} (3.3 n*M*), B_{TMR}–G_{M1} (3.3 n*M*), and B_{TR}–G_{M1} (3.3 n*M*) on the outer leaflet of the vesicles. I_{624}, I_{517}, and I^0_{624}, I^0_{517} represent the fluorescence intensities at 624 and 517 nm with and without CT, respectively. The error margins for the measurements are ~7%.

much larger. (The I_a/I_d ratio for the one-tiered FRET system increases up to ~2.5-fold by CT binding, whereas the I_a/I_d ratio for the two-tiered FRET system increases up to 8-fold; I_a and I_d are the fluorescence intensities of the acceptor and donor, respectively.)

As in the one-tiered FRET system, a surface density of total tagged G_{M1} (containing an equal molar amount of donor-, intermediate-, and acceptor-labeled G_{M1}) ranging from 0.2 to 0.05% (relative to the POPC concentration) is found to give optimal results. Samples with a donor:intermediate:acceptor ratio close to 1:1:1 are found to show the biggest fluorescence change in terms of the CT-induced fluorescence response.

Titration by Flow Cytometer

Figure 8 shows the titration data for the two-tiered energy transfer systems on the surface of POPC bilayers coated on glass microspheres.

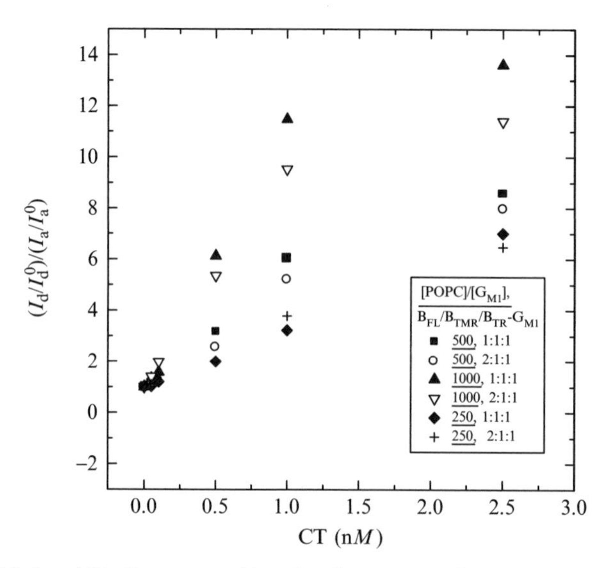

FIG. 8. CT-induced titration curves of two-tiered energy transfer systems measured by flow cytometer. Each sample contains 2.5×10^5 beads coated with POPC bilayers. The data were collected 1 h after the samples were incubated with CT at room temperature. I_a and I_d are the mean fluorescence intensities of acceptor and donor with CT, respectively. I_a^0 and I_d^0 are the mean fluorescence intensities of acceptor and donor without CT, respectively. The error margins for the measured data are ~9%.

The titration results depend not only on the surface density of the total tagged G_{M1}, but also on the ratio of donor-, intermediate-, and acceptor-tagged G_{M1} on the bilayer surfaces. Detailed studies show that samples with ~0.1% tagged G_{M1} relative to POPC and a donor:intermediate:acceptor ratio close to 1:1:1 provide the largest increase in I_a/I_d.

It is interesting to compare the titration results obtained from one-tiered and two-tiered energy transfer systems. On the one hand, the factors (e.g., surface density, ratio of the labeled G_{M1}, total concentration of the labeled G_{M1}) that influence detection performance (detection sensitivity and selectivity) are almost identical for both systems; on the other hand, the two-tiered ET system provides a much larger overall fluorescence change (increase in I_a/I_d) induced by the formation of multivalent G_{M1}–CT complexes than does the one-tiered FRET system. For the one-tiered FRET system, the maximal CT-induced I_a/I_d increase measured with a commercial flow cytometer (Becton Dickenson) is ~2.5-fold greater than background (without CT), whereas the I_a/I_d increase for the two-tiered FRET system can reach as high as 14-fold over background. The larger binding-induced fluorescence change and lower background fluorescence

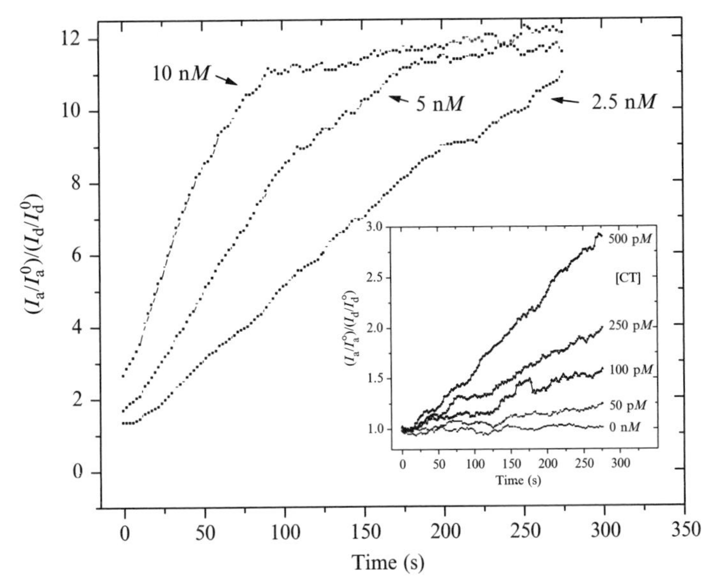

FIG. 9. CT-induced fluorescence response as a function of time for a two-tiered energy transfer system containing 2.5×10^5 beads with equal molar amounts of each labeled G_{M1} (0.1% relative to POPC). *Inset*: Response curves for lower CT concentration. Definitions for I_a, I_d, I_a^0, and I_d^0 are the same as in Fig. 8.

of the acceptor achieved with the two-tiered FRET system should provide better detection performance than the one-tiered FRET system, particularly for complex samples where interference from other components may be severe.

Kinetics by Flow Cytometry

Figure 9 shows the fluorescence response as a function of time on addition of various concentrations of CT for samples containing 2.5×10^5 POPC bilayer-coated beads with 0.1% total tagged G_{M1} (relative to POPC), and an equal molar amount of donor-, intermediate-, and acceptor-tagged G_{M1}. It clearly shows that I_a/I_d increases as a function of time on addition of CT. Without CT, I_a/I_d undergoes no change over time, whereas I_a/I_d increases ~50% over background 275 s after addition of 250 p*M* CT. In addition to the larger CT-induced fluorescence response for the two-tiered ET system, the I_a/I_d increase over time is slower than with the one-tiered FRET system. For example, for samples containing the same number of beads with the same surface density of labeled G_{M1} (0.1%) and an equal molar amount of donor- and acceptor-tagged G_{M1}

(for the one-tiered FRET system) or donor-, intermediate-, and acceptor-tagged G_{M1} (for the two-tiered FRET system), it took the two-tiered FRET system at least 275 s to reach equilibrium whereas the one-tiered FRET system took only ~30 s on addition of 2.5 n*M* CT.

The reason behind the observed slower fluorescence response to CT binding in the two-tiered FRET system is intriguing. Because the surface density and ratio of labeled G_{M1} in the samples used for both systems are similar, the lateral diffusion behaviors of labeled G_{M1} in the POPC bilayers coated on glass beads should be similar as well. The significant difference in fluorescence response over time may result from the fact that three fluorophores (donor, acceptor, and intermediate) of tagged G_{M1} are required to complex with the same CT to trigger a fluorescence change for the two-tiered FRET system, whereas only two fluorophores are required for the one-tiered FRET system. The formation of complexes of CT with at least three tagged G_{M1} molecules requires at least three steps: from free CT to monovalent complex, then to divalent complex, and then to trivalent complex, whereas the divalent CT–G_{M1} complex can be formed in only two steps to give a positive fluorescence response. The extra step may be attributed to the slower fluorescence response for the two-tiered FRET system, considering possible slow lateral diffusion of large G_{M1}–CT complexes. Such a hypothesis can be confirmed by comparing the results of the same G_{M1}–CT interactions, using configurations 2A and 2B in Scheme 2, in which the formation of a divalent G_{M1}–CT complex containing different tagged G_{M1} molecules can trigger a two-tiered FRET. Such a configuration can possess the benefits of the two-tiered FRET system while retaining a fast fluorescence response.

Acknowledgments

This research was supported by an LANL Laboratory Directed Research and Development grant. This article is dedicated to the memory of Jeane Shi, who was tragically killed in June 2001.

[36] Ultrasensitive Analysis of Sialic Acids and Oligo/Polysialic Acids by Fluorometric High-Performance Liquid Chromatography

By SADAKO INOUE and YASUO INOUE

Introduction

Sialic acids (Sia) occupy the outermost positions of animal glycan chains. Because of their location many molecules recognize Sia and bind or interact with them, and numerous examples show that such binding and interactions are important in animal life. In nature Sia occur in three parent structures: Neu5Ac, Neu5Gc, and KDN (2-keto-3-deoxy-D-*glycero*-D-*galacto*-nononic acid). KDN is different from the others in that it is a deaminoneuraminic acid whereas the other two are *N*-acylneuraminic acids. Modifications of these parent structures by esterification of hydroxyl groups at various positions by acetate (much less frequently lactate) and/or sulfate are also extensive. Thus, the structural diversity of the Sia family may correspond to a vast number of molecules recognizing them.[1,2]

Polysialic acids (polySia) are linear polymers of Sia and unique structural units expressed on various functional molecules. Because polySia frequently occur during the early developmental stages in animals, for examples, in embryonic tissues of vertebrates, and in eggs of fish and sea urchin, it has been anticipated that some important biological function may reside in their unique structures. PolySia chains expressed on the neural cell adhesion molecule (NCAM) regulate NCAM-mediated interaction of neural cells and have been extensively studied.[3] They are most abundantly expressed in embryonic animal brains whereas specific expression is observed in a limited part of the adult brain, where neurogenesis continues throughout the adult stage. PolySia are also reexpressed in several human tumors, and thus they serve as an oncodevelopmental antigen. In NCAM only one type of polySia structure, $\alpha2\rightarrow8$-linked polyNeu5Ac, is known to date. Monoclonal antibodies that recognize $\alpha2\rightarrow8$-linked oligo/polyNeu5Ac have been developed and used as sensitive detection and identification tools for this structure. Using such antibodies, the occurrence of $\alpha2\rightarrow8$-linked polyNeu5Ac structure has also been reported in molecules other than

[1] A. Varki, *Glycobiology* **2,** 25 (1992).
[2] Y. Inoue and S. Inoue, *Pure Appl. Chem.* **71,** 789 (1999).
[3] U. Rutishauser, *Curr. Opin. Cell Biol.* **8,** 679 (1996).

0076-6879/03 $35.00

NCAM.[4,5] However, many other oligo/polySia structures comprising different types of Sia, Neu5Gc, and KDN, and/or different types of interresidue linkages, occur in nature. The occurrence of $\alpha 2 \rightarrow 8$-linked polyNeu5Gc had been found in rainbow trout eggs[6] before the polyNeu5Ac structure was identified on NCAM, and a more divergent range of homo- and heteropolymers of Neu5Ac and Neu5Gc, that is, polyNeu5Ac, poly (Neu5Ac, Neu5Gc), and polyNeu5Gc, has been universally found in salmonid fish eggs.[7] We also reported $\alpha 2 \rightarrow 8$-linked oligoKDN[8] in the vitelline envelope of eggs and ovarian fluid of rainbow trout and $\alpha 2 \rightarrow 5$-$O_{glycolyl}$-linked polyNeu5Gc in sea urchin egg jelly[9] and egg surface glycoprotein.[10] Until now $\alpha 2 \rightarrow 9$-linked polyNeu5Ac has been known only in capsule polysaccharides of *Neisseria meningitidis* group C,[11] whereas polyNeu5Ac containing alternating $\alpha 2 \rightarrow 8$ and $\alpha 2 \rightarrow 9$ linkages occurs in *Escherichia coli* K92.[12] However, $\alpha 2 \rightarrow 9$-linked Neu5Ac dimer was reported in lactosaminoglycan glycopeptides derived from human PA1 embryonal carcinoma cells,[13] suggesting the possible occurrence of $\alpha 2 \rightarrow 9$-linked oligo/polyNeu5Ac in animals. Thus, although immunochemical methods are sensitive, different antibodies must be developed to detect all the different types of polySia.[14,15] Moreover, information about the degree of polymerization (DP) of polySia available by immunochemical methods is scant, although such information is considered to be important in understanding the biosynthesis and function of polySia. We have developed a highly sensitive analytical method that can be used not only for detection but also for DP analysis of any type of oligo/polySia structure. This method

[4] C. Zuber, P. M. Lackie, W. A. Catterall, and J. Roth, *J. Biol. Chem.* **267,** 9965 (1992).

[5] C. M. Martersteck, N. L. Kedersha, D. A. Drapp, T. G. Tsui, and K. J. Colley, *Glycobiology* **5,** 289 (1996).

[6] S. Inoue and M. Iwasaki, *Biochem. Biophys. Res. Commun.* **83,** 1018, (1978).

[7] S. Inoue and Y. Inoue, *in* "Glycoproteins" (J. Montreuil, J. F. G. Vliegenthart, and H. Schachter, eds.), p. 143. Elsevier, New York, 1997.

[8] A. Kanamori, S. Inoue, M. Iwasaki, K. Kitajima, G. Kawai, S. Yokoyama, and Y. Inoue, *J. Biol. Chem.* **265,** 21811 (1990).

[9] S. Kitazume, K. Kitajima, S. Inoue, F. A. Troy, J.-W. Cho, W. J. Lennarz, and Y. Inoue, *J. Biol. Chem.* **269,** 22712 (1994).

[10] S. Kitazume, K. Kitajima, S. Inoue, S. M. Haslam, H. R. Morris, A. Dell, W. J. Lennarz, and Y. Inoue, *J. Biol. Chem.* **271,** 6694 (1996).

[11] A. K. Bhattacharjee, H. J. Jennings, C. P. Kenny, A. Martin, and I. C. P. Smith, *J. Biol. Chem.* **250,** 1926 (1975).

[12] W. Egan, T.-Y. Liu, D. Dorow, J. S. Cohen, J. D. Robbins, E. C. Gotschlich, and J. B. Robbins, *Biochemistry* **16,** 3687 (1977).

[13] M. N. Fukuda, A. Dell, J. E. Oates, and M. Fukuda, *J. Biol. Chem.* **260,** 6623 (1985).

[14] A. Kanamori, K. Kitajima, S. Inoue, X. Zuo, C. Zuber, J. Roth, K. Kitajima, J. Ye, F. A. Troy, and Y. Inoue, *Histochemistry* **101,** 333 (1994).

[15] C. Sato, K. Kitajima, S. Inoue, and Y. Inoue, *J. Biol. Chem.* **273,** 2575 (1998).

used a simple fluorescence labeling reaction for sensitive detection and high-performance liquid chromatography for efficient resolution of oligo/polySia peaks up to a DP of about 90 with a detection threshold of 1.4 fmol per resolved peak.[16–18] This sensitivity is almost equivalent to that of immunochemical methods.[19] The method does not require antibodies and/or enzymes that are often commercially unavailable. Moreover, the whole procedure is less tedious than immunochemical methods, which involve many steps.

DMB Reaction of Oligo/polysialic Acid

The analysis described in this article is based on the highly sensitive fluorometric detection of the reaction products of 1,2-diamino-4,5-methylenedioxybenzene (DMB) and free oligo/polySia chains (Fig. 1) after efficient separation of them according to DP by anion-exchange high-performance liquid chromatography (HPLC).

DMB was originally developed as a fluorogenic reagent for α-keto acids (ulosonic acids) and applied to microdetermination of sialic acid (C_9-ulosonic acid, nonulosonic acid).[20] When free oligo/polySia chains are incubated with DMB, the reducing terminal residues of oligo/polySia are labeled with the fluorophore (quinoxalinone). Because the reaction occurs only in relatively strong acid, partial hydrolytic cleavage of interresidue sialyl linkages is unavoidable. In this method the differential temperature dependency of hydrolysis and the labeling reaction has been taken into account to optimize the conditions. Under the optimized conditions described below, DMB derivatives are obtained in sufficiently high yields, and hydrolytic cleavage of internal $\alpha 2 \rightarrow 8$-Neu5Ac linkages is kept at a low level. In acid-catalyzed hydrolysis, $\alpha 2 \rightarrow 3$-and/or $\alpha 2 \rightarrow 6$-sialyl linkages to the proximal sugar residues of core glycans are cleaved about 2.5-fold faster than $\alpha 2 \rightarrow 8$ linkages.[16,21] In any chromatographic method of oligo/polySia analysis, hydrolytic cleavage of oligo/polySia chains from the core glycans is necessary before separation and detection of oligo/polySia. lySia. Even under mild hydrolytic conditions, some interresidue sialyl linkages are cleaved. Typical ladder-like peaks (or bands) separated by

[16] S. Inoue and Y. Inoue, *J. Biol. Chem.* **276,** 31863 (2001).

[17] S. Inoue and Y. Inoue, *Biochimie* **83,** 605 (2001).

[18] S. Inoue, S.-L. Lin, Y. C. Lee, and Y. Inoue, *Glycobiology* **11,** 759 (2001).

[19] G. L. Poongodi, N. Surersh, S. C. B. Gopinath, T. Chang, S. Inoue, and Y. Inoue, *J. Biol. Chem.* **277,** 28200 (2002).

[20] S. Hara, Y. Takemori, M. Yamaguchi, M. Nakamura, and Y. Ohkura, *Anal. Biochem.* **164,** 138 (1987).

[21] S. Inoue, S.-L. Lin, and Y. Inoue, *J. Biol. Chem.* **275,** 29968 (2000).

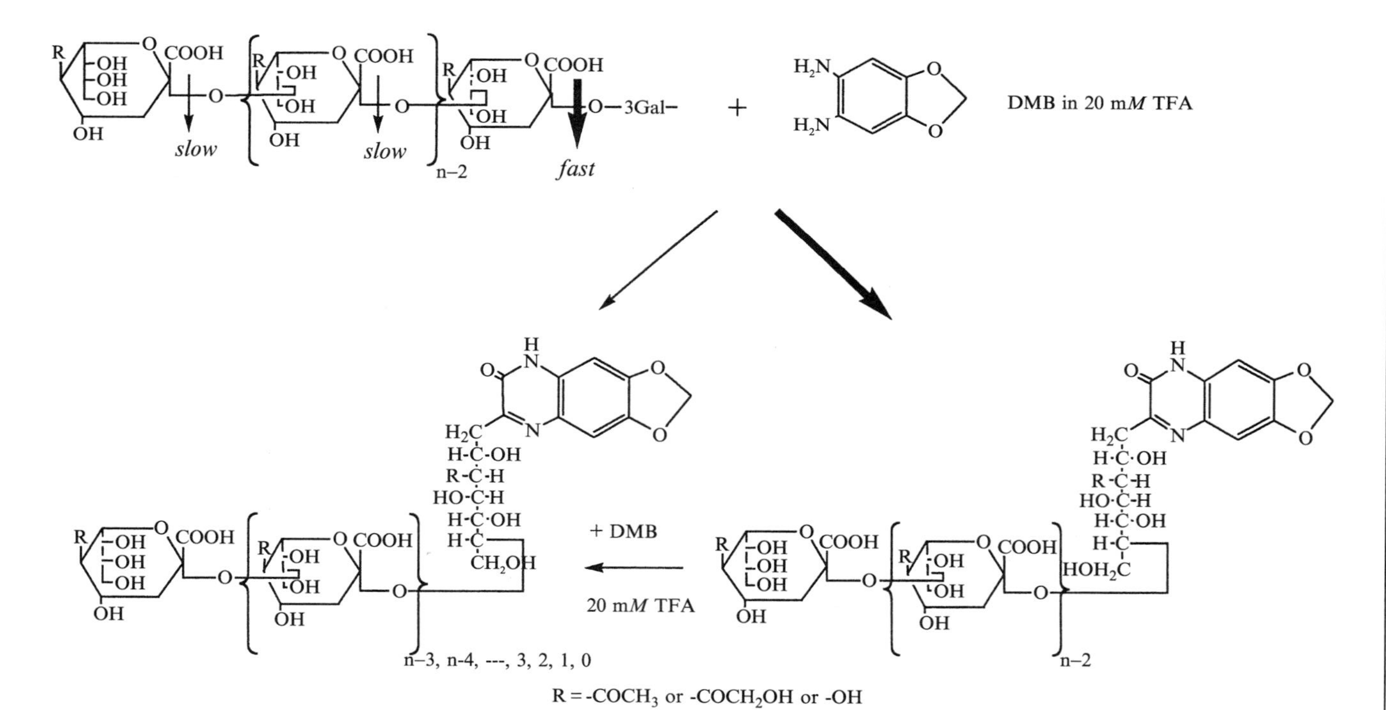

FIG. 1. DMB labeling of oligo/polySia chains linked to core glycans. In this example, $\alpha 2 \rightarrow 8$-linked oligo/polySia chain is linked $\alpha 2 \rightarrow 3$ to Gal residue of a core glycan. Preferential hydrolysis of proximal sialyl linkages and partial hydrolysis of internal sialyl linkages occur during incubation of the sample with DMB in 20 m*M* TFA, and liberated oligo/polySia chains react with DMB.

anion-exchange chromatography,[22,23] capillary electrophoresis,[24] and thin-layer chromatography[25] are indicative of the presence of oligo/polySia and help in the identification of the type of monomeric Sia residue by comparing the retention times (the migration rates) of all ladder-like peaks (bands) with the values for authentic oligo/polySia samples similarly processed (see below). The present method is superior to commonly used chromatographic methods of oligo/polySia analysis in that both hydrolytic cleavage of oligo/polySia chains from the core glycans and labeling with a highly sensitive fluorogenic reagent take place in one step under controlled conditions. Compared with highly sensitive oligo/polySia analysis by high-performance anion-exchange chromatography with pulsed amperometric detection (HPAEC-PAD),[22,23] the present method is almost 1000-fold more sensitive with an equivalent efficiency in the separation of high-DP polySia.

Reagents for DMB Labeling.

Stock solution of acidic reducing reagent containing 1 *M* 2-mercaptoethanol, 18 m*M* sodium hydrosulfite, and 40 m*M* trifluoroacetic acid (TFA): Dissolve 31.3 mg of sodium hydrosulfite ($Na_2S_2O_4$) in about 8 ml of water. Add 0.7 ml of 2-mercaptoethanol and 0.4 ml of 1 *M* TFA, and adjust the total volume to 10 ml. This solution can be kept at 4° for at least 1 year

DMB reagent: Dissolve 1.2 mg of 1,2-diamino-4,5-methylenedioxybenzene [DMB, $C_7H_8N_2O_2 \times 2HCl$, manufactured by Dojinbo (Kumamoto, Japan), and available from many suppliers] in 1 ml of the reducing reagent

This solution can be kept for weeks at 4°, and for at least 1 year at −20°. The concentration of DMB in the solution is 5.3 m*M*, which is lower than the 7 m*M* recommended for microanalysis of sialic acid by the original authors.[20] We found that 5.3 m*M* appeared to be the maximum convenient concentration of stock DMB in 40 m*M* TFA. Some DMB precipitates during storage and the solution cannot be clarified by mixing; the reagent is used without removing the precipitate. The same reagent can be used for labeling free Sia (liberated by hydrolysis), free oligo/polySia chains (produced by controlled hydrolysis), and oligo/polySia chains linked to core glycans.

[22] Y. Zhang, Y. Inoue, S. Inoue, and Y. C. Lee, *Anal. Biochem.* **250,** 245 (1997).

[23] S.-L. Lin, S. Inoue, and Y. Inoue, *Glycobiology* **9,** 807 (1999).

[24] M.-C. Cheng, S.-L. Lin, S.-H. Wu, S. Inoue, and Y. Inoue, *Anal. Biochem.* **260,** 154 (1998).

[25] S. Kitazume, K. Kitajima, S. Inoue, and Y. Inoue, *Anal. Biochem.* **202,** 25 (1992).

High-Performance Liquid Chromatography-Fluorescence Detection Analysis of Free and Glycan-Linked Oligo/PolySia Chains after Labeling with DMB: DMB/HPLC-FD

DMB Labeling

To a sample solution containing 100–1000 ng of polySia or 2–20 ng of oligoSia in 20 μl, add 20 μl of the DMB reagent. Incubate the mixture at 10° for 48 h. After the reaction, cool the reaction mixture on ice and add 10 μl of 0.5 *M* NaOH to bring the pH to about pH 13. For α2→8-linked high-DP polySia, this alkali treatment is necessary to hydrolyze lactones formed during the labeling reaction. Leave the alkalinized mixture for 30 min at room temperature or overnight at 4°. The pH of the final solution can be lowered to about pH 8 by adding an equal volume of 0.1 *M* acetic acid. This treatment is not usually necessary for polySia analysis. DMB-labeled samples can be kept for a few days at 4° or for more than 1 week at −20° if they are not analyzed immediately. The solution should be centrifuged on a microcentrifuge and an aliquot of the clear supernatant is injected. Commercial colominic acid is reacted with DMB under the same conditions and used as a reference in HPLC analysis.

HPLC Systems

A facility for gradient generation is necessary. For sensitive detection of high-DP peaks a fluorescence detector is operated at highest sensitivity [e.g., gain 100 for a Jasco (Tokyo, Japan) FP 920 or PMT gain 16 for a Hewlett-Packard (Palo Alto, CA) detector]. Wavelengths are set at 372 nm and 456 nm for excitation and emission, respectively.[26]

HPLC Columns

A DNAPac PA-100 column (Dionex, Sunnyvale, CA) gives excellent resolution of high-DP polySia. Using commercial high molecular weight colominic acid (molecular weight, 30,000; Nacalai Tesque, Kyoto, Japan), peaks of DP up to 90 are resolvable (Fig. 2). A CarboPac PA-100 (Dionex) can also be used for resolution of polySia peaks of DP up to 60. Higher salt concentrations are necessary to elute DMB-oligo/polySia from a CarboPac column than from a DNAPac column. When crude samples (e.g., cell or tissue extracts) are directly labeled and analyzed on a DNAPac column, overlapping peaks often interfere with peaks lower than DP 5. However, clean chromatograms are obtained for higher DP regions, and there is no problem in the determination of the highest

[26] S.-L. Lin, S. Inoue, and Y. Inoue, *Carbohydr. Res.* **329,** 447 (2000).

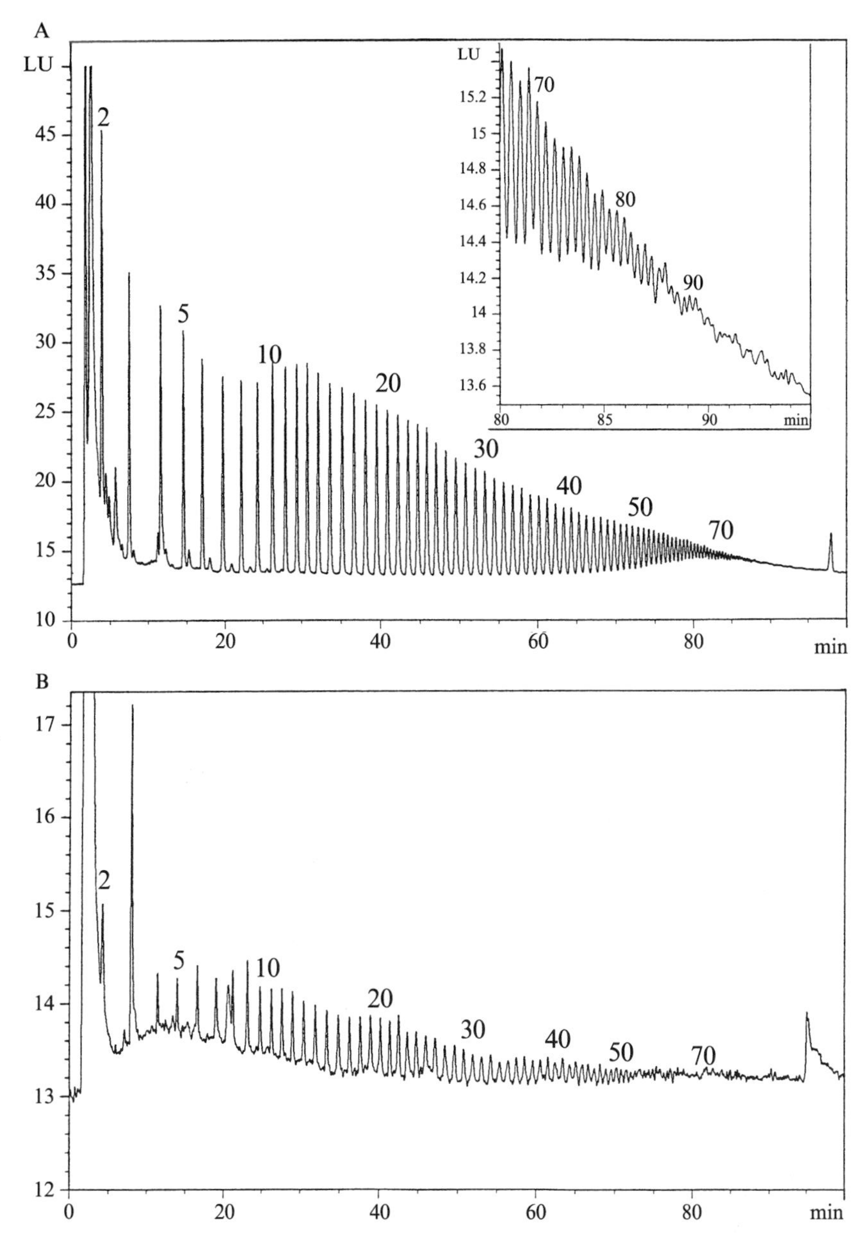

FIG. 2. Chromatographic profiles produced by a DNAPac PA-100 for commercial colominic acid, analyzed by the DMB/HPLC-FD method. One microgram of colominic acid was labeled with DMB under conditions described in text. Elution profiles produced with two different injection amounts of colominic acid, (A) 200 ng and (B) 10 ng, are depicted.

DP and the DP distribution analysis of polySia (DP $\geq$ 5) when crude samples are analyzed. The CarboPac column gives better resolution in the low-DP region (DP < 5). In analyzing low-DP oligoSia it is important to adjust the pH of the sample to pH 7–8 to avoid a change in the retention time.

Elution Conditions

Eluents: E1, H_2O; E2, 1 *M* $NaNO_3$; flow rate, 1 ml/min

A DNAPac PA-100 column is equilibrated with 0.02 *M* $NaNO_3$, and the concentration of $NaNO_3$ is increased up to 0.35 *M* by introducing E2 into E1. A typical gradient used for polySia analysis is as follows: E2, 2% for initial 5 min, 10% at 13 min, 20% at 27 min, 25% at 43 min, and 35% at 95 min. A CarboPac PA-100 column is equilibrated with 0.03 *M* $NaNO_3$. A typical gradient used for polySia analysis is as follows: E2, 3% for initial 5 min, 10% at 10 min, 20% at 27 min, 25% at 40 min, and 29% at 60 min. For oligoSia analysis only the initial sections of the gradient are used. Both CarboPac and DNAPac columns are washed with 0.4 *M* $NaNO_3$ for 2 min after every run.

Determination of DP and Analysis of DP Distribution in Samples Containing Oligo/PolySia with Various DP

As described above, cleavage of some inner sialyl linkages is unavoidable during the DMB-labeling reaction. However, partial cleavage of the inner linkages is small under labeling conditions. We have also shown that the stability of the inner linkages is not dependent on DP of oligo/polySia.[16] Therefore, the exact DP of polySia can be determined by the present method. A typical example for $\alpha 2 \rightarrow 8$-linked(Neu5Ac)$_n$ ($n = 26$, with a smaller amount at $n = 25$) is depicted in Fig. 3.

In the case of naturally occurring compounds such as NCAM, methods using anti-polySia antibodies do not provide much information about the DP of polySia chains expressed on them. Because hydrolytic cleavage of interresidue $\alpha 2 \rightarrow 8$ sialyl linkages is moderate whereas that of proximal $\alpha 2 \rightarrow 3$ sialyl linkage is extensive during DMB labeling (Fig. 1), peak distribution in the chromatographic profiles obtained by DMB/HPLC-FD reasonably reflects DP distribution of the parent compound. Shown in Fig. 4 is an example for a glycopeptide fraction derived from embryonic day 10 chicken brain. We can deduce the following conclusions from the elution profile: (1) the highest DP of polySia chains in this sample is about 60; (2) the proportion of DP > 50 is extremely low; and (3) the peak distribution is multiphasic, suggesting the original

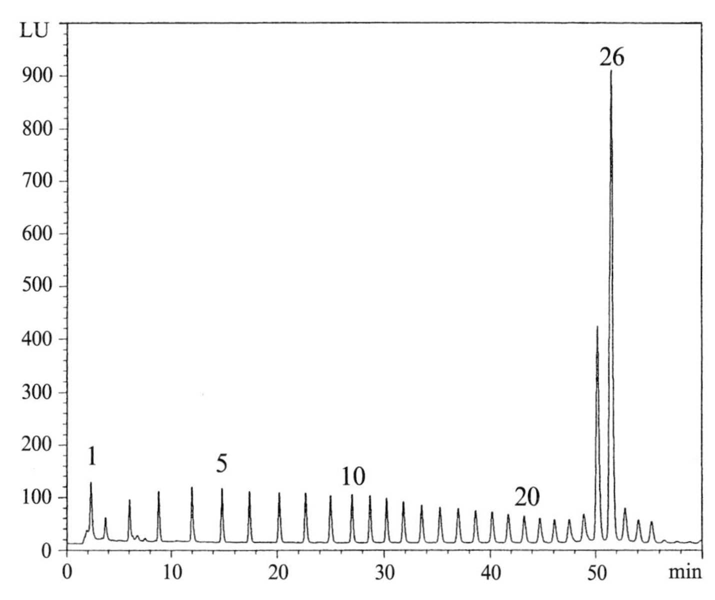

FIG. 3. DMB/HPLC analysis of polySia with a discrete DP value. A sample of commercial colominic acid was subjected to anion-exchange chromatography on a DNAPac PA-100 to fractionate oligo/polySia chains according to their DP. In this analysis, the material eluted under peak 26 (monitored by absorbance at 220 nm) was labeled with DMB and analyzed on a DNAPac PA-100 with fluorescence detection.

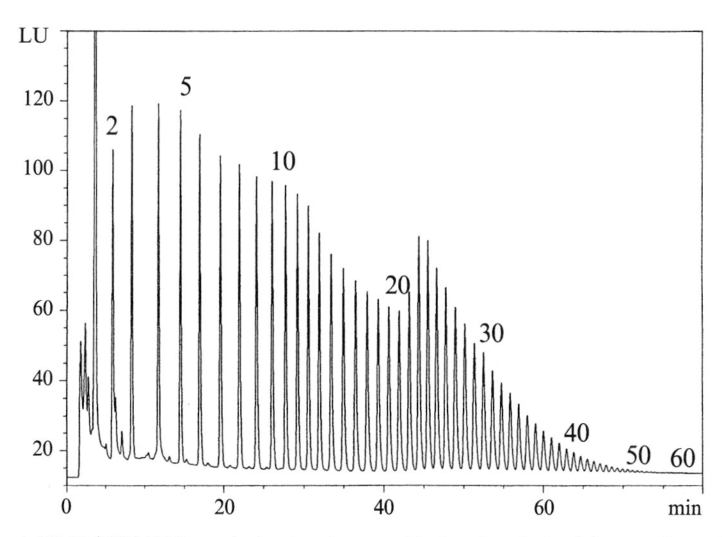

FIG. 4. DMB/HPLC-FD analysis of a glycopeptide fraction derived from embryonic day-10 chicken brain. A highly anionic glycopeptide fraction was separated after proteolysis of delipidated membrane fraction. A sample containing 650 ng of total Sia was labeled with DMB and 260 ng was injected into a DNAPac PA-100 column.

existence of two or three groups of oligo/polySia chains, that is, those below DP 5, those centered around DP 10, and those centered between DP 20 and 30.

Oligo/polySia Analysis in Detergent Extracts of Delipidated Membrane Fractions Derived from Minute Amounts of Cells and Tissues

Procedure

Homogenize cells or tissues (20–200 mg, wet weight) with a Polytron homogenizer (Kinematica, Litau, Switzerland) in 600 μl of chloroform–methanol–10 m*M* Tris-HCl (4:8:3, v/v/v), pH 8.0. Centrifuge the homogenate at 4000 rpm for 15 min. Suspend the residue in 600 μl of chloroform–methanol–water (4:8:3, v/v/v), and disperse the suspension by sonication. Collect the residue by centrifugation at 4000 rpm for 15 min, suspend it in 50% ethanol, and disperse the suspension by vortex mixing. Collect the residue containing delipidated membrane fraction and dissolve it with 100 μl of 0.5% Triton X-100 (Sigma, St. Louis, MQ) with sonication and incubation at 37° for 0.5–1 h. Remove insoluble material by centrifugation at 10,000 rpm for 10 min. In routine analysis 40 μl of the supernatant is incubated with 40 μl of the DMB reagent. When the starting material is available only in minute amounts, a suspension of membrane fraction in water can be labeled with an equal volume of the DMB reagent with vigorous shaking during the reaction. In this case the supernatant of the reaction mixture is used for analysis. As a typical example, polySia analysis in the pons tissue dissected from 7-day-old rat is shown in Fig. 5.

When extracts of cells and tissues are directly labeled with DMB, many interfering peaks appear during the initial 10 min of chromatography. However, the resolution of DMB–polySia (DP $\geq$ 5) is unaffected. The significance of this direct labeling by the DMB/HPLC-FD method is 2-fold. First, polySia is identified using minute amounts of cells and tissues, as used in immunochemical methods. Second, direct labeling avoids possible liberation and/or depolymerization of polySia chains during purification of samples, and thus can afford a reliable value of DP.

Quantification of polySia in Cell and Tissue Extracts

The polysia level in cell and tissue extracts can be estimated from the sum of the peak areas (DP $\geq$ 5) in the DMB/HPLC-FD chromatogram.[19] The value thus obtained represents the number of polySia chains in the analyzed sample.

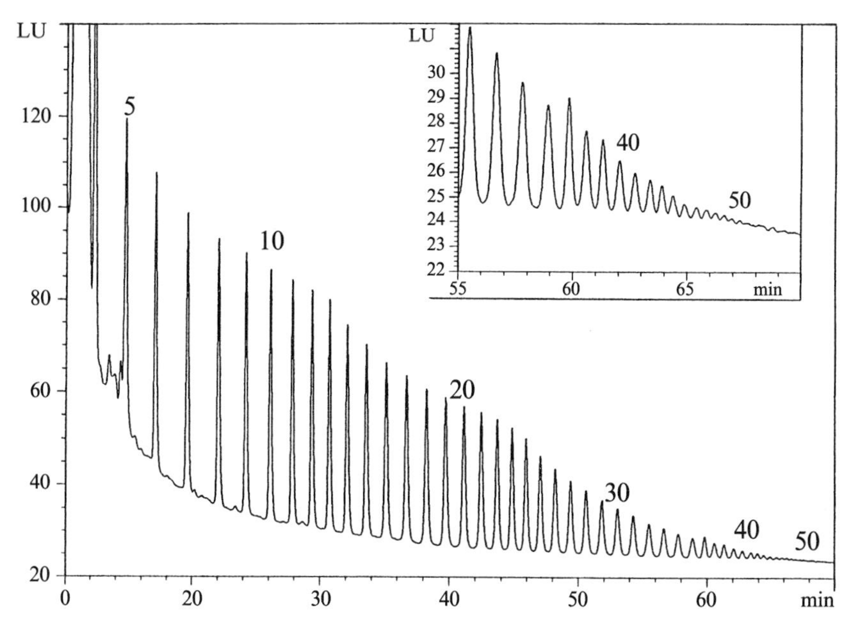

FIG. 5. DMB/HPLC-FD analysis of a detergent extract of pons dissected from 7-day-old rat. A delipidated membrane fraction obtained from about 20 mg of pons dissected from 7-day-old rat was solubilized with detergent and labeled with DMB. A portion of the reaction mixture, containing about 200 ng of total Neu5Ac, was injected into a DNAPac PA-100 column.

Evaluation of Sensitivity of the Method

The sensitivity of the method is comparable to that of immunochemical analysis using anti-α2→8-linked polyNeu5Ac antibodies. We usually apply 10 μg of protein (derived from 6–16 mg of wet cells) in one lane of a sodium dodecyl sulfate (SDS)–polyacrylamide gel that is subjected to Western blot analysis. In enzyme-linked immunosorbent assay (ELISA), sample added to a well is derived from 50 mg of wet cells, whereas sample injected into a column in DMB/HPLC-FD analysis is derived from 30 mg of wet cells (20–50 μg of protein, and 150–1000 ng of total Sia residues including those occurring in monomeric form) in HPLC analysis. For samples containing detectable amount of polySia by Western blot, peaks of polySia are always detectable in the HPLC chart, so that the DMB/HPLC-FD method can replace the immunochemical method.[17,19]

TABLE I
RETENTION TIMES OF VARIOUS TYPES OF OLIGO/POLYSIALIC ACIDS ON CARBOPAC PA-100 COLUMN[a]

n[b]	α2,8 (Neu5Ac)$_n$	α2,9 (Neu5Ac)$_n$	α2,8 (Neu5Gc)$_n$	α2,5-Oglycolyl (Neu5Gc)$_n$	α2,8 (KDN)$_n$
3	9.13	10.42	9.34	9.29	10.46
4	12.77	12.92	12.97	12.37	13.81
5	15.47	15.24	15.77	14.55	16.79
6	18.18	17.65	18.49	16.99	19.70
7	20.60	19.89	20.89	19.21	
8	22.65	21.86	22.90	21.17	
9	24.37	23.52	24.32	22.88	
10	25.86	25.00	25.59	24.36	

[a] For each type of oligo/polySia samples, average values for two or three independently labeled samples are given.

[b] Retention times of DMB-labeled monomer (no negative charge) and dimer (single negative charge) are influenced by the pH and concentration of the salt in injected samples.

Application of DMB/HPLC-FD to Various Types of Oligo/polySia

The DMB/HPLC-FD method may be used for detection and analysis of many types of naturally occurring oligo/polySia backbone structures including α2→9-linked polyNeu5Ac, α2→8-linked oligo/polyNeu5Gc, α2→5-$O_{glycolyl}$-linked oligo/polyNeu5Gc, and α2→8-linked oligoKDN. Among them monoclonal antibody is commercially available only for α2→8-linked oligo/polyNeu5Ac. We have developed monoclonal antibodies reactive to α2→8-linked oligoKDN,[14] and α2→8-linked oligo/polyNeu5Gc.[15] However, antibody against α2→5-$O_{glycolyl}$-linked Neu5Gc is not available. Differential analysis of different types of unlabeled oligo/polySia has been reported with the CarboPac PA-100 column (with elution in the presence of 0.1 *M* NaOH) and PAD.[22,23] DMB/HPLC-FD can also be used for this purpose using much smaller amount of samples. Retention times of different types of DMB-labeled oligo/polySia on a CarboPac PA-100 column are given in Table I.

Determination of Total Sialic Acid by a Modified DMB Method

Determination of total Sia is necessary to specify and quantify the sialic acid component of each sample for oligo/polySia analysis.

DMB Reagent

The same reagent used for oligo/polySia analysis is used. This DMB reagent appears more stable than that described by the original authors.[20]

Procedures

Acid Hydrolysis to Liberate Sialic Acid Monomers from Oligo/polySia Chains. To a 100-μl sample solution (10–100 ng of Sia) in a plastic centrifuge tube (400-ml capacity), add 100 μl of 0.2 *M* TFA. Cap tightly (a pushing cap is sufficiently tight) and mix. Incubate the mixture at 80° in a heating block for 2 h (oligoSia) or 4 h (polySia).

Using tightly capped small plastic tubes minimizes evaporation of medium during hydrolysis and degradation of liberated free sialic acid is small. Time of hydrolysis required for maximum liberation of free monomeric sialic acid depends on the DP of oligo/polySia,[21] the type of Sia residue,[27] the type of interresidue sialyl linkages.[28,29] When samples of α2→8-oligo/polyNeu5Ac are heated at 80° in 0.1 *M* TFA for 1 h, about 70% of total sialic acid is liberated from diSia whereas only 10–30% is liberated from polySia. Therefore, optimal hydrolysis time should be determined for each sample by examining a time course of sialic acid liberation. Hydrolysis time required for maximum liberation of free sialic acid in 0.1 *M* TFA at 80° is achieved in 2 h for α2→8-linked diSia and in 4 h for polySia. After hydrolysis, the acid is removed by evaporation on a centrifugal vacuum evaporator.

DMB-Labeling Reaction. Dissolve the dried residue of the hydrolyzed sample with 20 μl of water. Add 20 μl of the DMB reagent. Incubate the mixture at 55° for 2.5 h. After cooling, centrifuge the reaction mixture on a microcentrifuge to remove any precipitate.

Standard sialic acid should be reacted with the same DMB reagent used for unknowns. The fluorescence response determined by the peak area on the HPLC chromatogram is proportional to the injected amount of sialic acid over a wide range. However, it is desirable for accurate determination that the amounts of standard sialic acid be in the range of the unknown samples. In practice, it is convenient to prepare two kinds of standard solutions, one containing 100 ng and the other 10 ng of each

[27] D. Nadano, M. Iwasaki, S. Endo, K. Kitajima, S. Inoue, and Y. Inoue, *J. Biol. Chem.* **261,** 11550 (1986).

[28] T.-Y. Liu, E. C. Gotschlich, F. T. Dunne, and E. K. Jonssen, *J. Biol. Chem.* **246,** 4703 (1971).

[29] S. Kitazume, K. Kitajima, S. Inoue, F. A. Troy, W. J. Lennarz, and Y. Inoue, *Biochem. Biophys. Res. Commun.* **205,** 893 (1994).

sialic acid in 20 μl. Use of an internal standard is desirable for assignment of the peak; however, the retention time of KDO (3-deoxy-D-*manno*-octulosonic acid) is close to that of KDN. The DMB-labeled sialic acid can be kept for several days at 4° and for weeks at −20° if the response is compared with the standard sialic acid labeled on the same day and similarly stored.

Reversed-Phase HPLC Analysis of DMB-Labeled Sialic Acids

HPLC System. A single-pump HPLC system equipped with a fluorescence detector is used. For analyzing 1–10 ng of sialic acid, standard gain value (gain 1 for Jasco, and PMT gain 10 for Hewlett-Packard) is used. For samples containing a smaller amount (<1 ng), a higher gain (increasing sensitivity 10-fold) can be used without increasing noise. Wavelength is set at 373 nm for excitation and at 448 nm for emission.[20]

Column. An RP-18 column (4 × 250 mm; particle size, 5 μm) is used. We use a TSK-gel ODS 120T (Tosoh, Tokyo, Japan), a Microsorb C_{18} or Vydac 218TP54 (Rainin, Woburn, MA) column. Similar columns from other companies may give similar results. Use of a guard column is recommended for frequent analysis of crude biological materials.

Solvents. Use either of the following solvent systems, depending on the purpose. The flow rate is 1 ml/min.

Solvent 1: Acetonitrile–methanol–water (7:5:88, v/v/v): Use for identification and quantification of each KDN, Neu5Gc, and Neu5Ac

Solvent 2: Acetonitrile–methanol–water (8:15:77, v/v/v): Use for separation and quantification of Neu5Gc and Neu5Ac

Inject an aliquot (4–10 μl) of the reaction mixture, after DMB labeling, into the HPLC column. Peaks of DMB-labeled KDN, Neu5Gc, and Neu5Ac elute in this order when using either solvent 1 or solvent 2. Solvent 1 gives better resolution of KDN and Neu5Gc, but the retention time of the last eluting Neu5Ac is about 25 min. In solvent 2, Neu5Ac elutes at about 9 min, but KDN and Neu5Gc do not separate well when crude material is analyzed. Reequilibration requires 30 min, and 10 min is for elution by solvent 1 and solvent 2, respectively. Unless the fluorescence detector is used at the highest gain, the solvent can be recycled. We use the same 300–500 ml of solvent for 1 month. Baseline drift is negligible, the change in retention time is small, and there is practically no problem after 200 injections when using recycled solvent. The column life is also longer than that described by the original authors,[20] probably because sample size is small in our modification. Washing with acid and/or 50% methanol is not necessary if the column is washed with the elution solvent for 2 h after the last analysis. However, after frequent analysis of crude materials, the guard column must be replaced (because the pressure becomes too high).

Detection and Quantification of Internal Sialic Acid Residues in $\alpha 2 \rightarrow 8$-Linked Oligo/polyNeu5Acyl Chains after Periodate Oxidation of Nonreducing Terminal Sialic Acid

The internal *N*-acylneuraminic acid residues (Neu5Ac or Neu5Gc) in the $\alpha 2 \rightarrow 8$-linked oligo/polySia chain are resistant to periodate oxidation under conditions in which the side chain of the *N*-acylneuraminic acid residue at the nonreducing terminal position is oxidized. The oxidation product is then reduced with sodium borohydride to give 5-acylamide-3,5-dideoxy-L-*arabino*-2-heptulosonic acid (C_7) residue. The presence of C_9 in the hydrolysate of the oxidation/reduction product indicates the presence of internal 8-linked Sia residue in the original sample. If we can separately quantify C_9 and C_7, we can get information about the DP of oligo/polySia chains. HPAEC-PAD[30] is not the most sensitive and accurate method so far used for this purpose. In the method described here, C_7-Sia and C_9-Sia are labeled with DMB, separated, and quantified by reversed-phase HPLC with fluorescence detection. The following procedure is a modification of another method[31] and is applied only to oligo/polySia chains having no free reducing terminal residue. The reducing terminal residues of free oligo/polySia are extensively oxidized under these conditions unless they have been reduced to alcohols before samples are subjected to oxidation. Internal KDN in $\alpha 2 \rightarrow 8$-linked oligo/polyKDN, and internal Neu5Gc in $\alpha 2 \rightarrow 5$-O_{glycolyl}-linked oligo/polyNeu5Gc, are subjected to oxidation.

Mild Alkali Treatment of Samples

Incubate the sample in 0.1 *M* NaOH at 37° for 1–4 h depending on the extent of O-acylation. Then add one-fifth volume of 1 *M* acetic acid to adjust the pH to about pH 5. This alkali treatment is necessary when the presence of O-acylation in the side chain of *N*-acylneuraminic acid residues is anticipated. It should be noted that 9-O-acetylation extensively retards the rate of oxidation of the side chain.[32]

Reduction of Reducing Terminal Sialic Acid Residue

Reduction of the reducing terminal sialic acid residue[30] is necessary only for oligo/polySia having free reducing termini such as colominic acid and oligoSia prepared by partial hydrolysis of polySia. Incubate oligo/polySia

[30] G. Ashwell, W. K. Berlin, and O. Gabriel, *Anal. Biochem.* **222,** 495 (1994).
[31] C. Sato, S. Inoue, T. Matsuda, and K. Kitajima, *Anal. Biochem.* **261,** 191 (1998).
[32] M. R. Lifely, J. C. Lindon, J. M. Williams, and C. Moreno, *Carbohydr. Res.* **143,** 191 (1985).

sample (20 μg) with 0.04 *M* $NaBH_4$ in 100 μl of 0.01 *M* NaOH for 2 h at 37°. Add 10 μl of 2 *M* ammonium acetate to the ice-cold sample to stop the reaction and remove the salt by gel filtration using a column of Sephadex G-25 (1 × 10 cm).

Periodate Oxidation followed by Reduction with Borohydride

To 10 μl of sample solution (10–100 ng of Sia), add 10 μl each of 0.1 *M* sodium acetate (pH 5.5) and 0.015 *M* sodium metaperiodate and leave the mixture for 20 min at room temperature in the dark. Stop the reaction by adding 5 μl of 3% ethylene glycol or glycerol. Add 10 μl of freshly prepared 0.04 *M* sodium borohydride solution in 0.01 *M* NaOH and leave the mixture for 30 min at room temperature. Destroy excess borohydride by adding 10 μl of 0.2 *M* acetic acid. Adjust the volume with water to 100 μl.

Conditions of Periodate Oxidation Applied to α2→9-Linked polyNeu5Ac

Under the conditions described above, only a tiny proportion of internal residues in α2→9-linked polyNeu5Ac is oxidized. Incubation with periodate for 4 days at 4° is necessary to oxidize 90% of internal 9-linked Neu5Ac.[32]

Hydrolysis and DMB Labeling

Perform the reactions as described for the determination of total sialic acid.

HPLC Analysis of DMB-Labeled C_7 and C_9

Procedures are the same as those described for total sialic acid analysis. The retention time of DMB-C_7 is larger than that of DMB-C_9. The relative peak areas per mole of compounds are similar between C_7 and C_9. HPLC profiles obtained for reduced $(Neu5Ac)_3$ (expected C_7/C_9 ratio, 1), reduced α2→8-linked polyNeu5Ac (colominic acid), a glycopeptide sample from embryonic day-5 chicken brain, and alkali-treated and reduced α2→9-linked polyNeu5Ac are given in Fig. 6.

Application and Limitation

As little as 1 ng of internal *N*-acylneuraminic acid residues in α2→8-linked oligo/polySia is detected by this method. If carefully used, this method may be the most sensitive chemical method showing the possible presence of α2→8-linked oligo/polymers of *N*-acylneuraminic acid in a

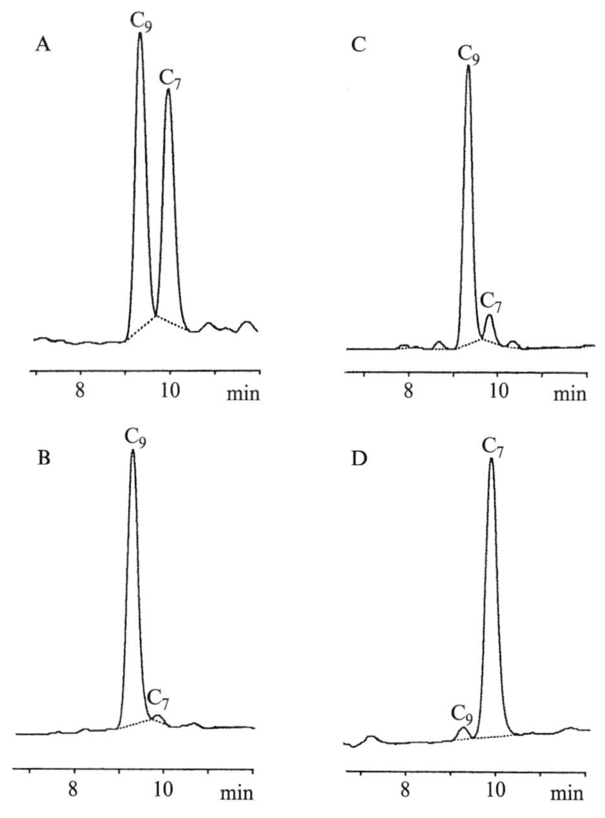

FIG. 6. HPLC analysis of DMB-labeled C_7 and C_9 produced from various oligo/polySia samples. (A) Reduced $(Neu5Ac)_3$, (B) reduced colominic acid, (C) a glycopeptide sample from embryonic day-5 chicken brain, and (D) alkali-treated and reduced $\alpha2\rightarrow9$-linked polySia. $\alpha2\rightarrow9$-linked polySia was a gift from Dr. Harold Jennings, National Research Council of Canada, Ottawa, Canada.

given sample. However, compared with oligo/polySia analysis by the DMB/HPLC-FD method, which detects typical ladder-like peaks derived from oligo/polySia, this method may lead to erroneous conclusions because any alkali-stable C_8 substitution in *N*-acylneuraminic acid makes it resistant to periodate oxidation. When Neu5Acyl residues in a sample occur only in $\alpha2\rightarrow8$-linked oligo/polySia chains, the mean DP values of oligo/polySia chains can be estimated from the C_9/C_7 ratio. In our hands, mean DP values of 12–55 were obtained for different batches of colominic acid samples. A wide range of DP values (10–40) was obtained for polySia-containing glycoprotein fractions partially purified from embryonic chicken brains at different developmental stages. Thus this method can be used for diagnostic purposes in detecting oligo/polySia.[33] However, purification of samples and additional structural information are prerequisites before evaluating the DP of oligo/polySia by this method. Most recently we discovered the ocurence of α 2 $\rightarrow$ 9 linked polyNeu5Ac in C-1300 murine neuroblastama cells (clone NB41A3) by employing the HPLC-based analysis described in the article.[34]

Acknowledgments

We thank Academia Sinica, Taiwan for supporting this research. This work was supported by the National Science Council, Taiwan, Grants NSC 89-2311-B-001-062, NSC 90-2311-B-001-155, and NSC 90-2311-B-001-102 (to S.I.), and NSC-89-2311-B-001-156, NSC 90-2311-B-001-096, and NSC 90-2311-B-001-140 (to Y.I.).

[33] C. Sato, H. Fukuoka, K. Ohta, T. Matsuda, R. Koshio, K. Kobayashi, F. A. Troy, and K. Kitajima, *J. Biol. Chem.* **275,** 15422 (2000).

[34] S. Inoue, G. L. Poongadi, N. Suresh, H. J. Jennings, and Y. Inoue, *J. Biol. Chem.* **278,** 8541 (2003).

[37] Identification of Lectins from Genomic Sequence Data

By KURT DRICKAMER and MAUREEN E. TAYLOR

Introduction

The sugar-binding activities of most lectins reside in discrete protein modules that are referred to as carbohydrate-recognition domains (CRDs).[1] Different lectin families are defined by the structures of the CRDs they contain. Three major classes of intracellular lectins are involved in intracellular protein trafficking: calnexin and calreticulin, L-type lectins that resemble lectins from legumes, and the mannose 6-phosphate receptors or P-type lectins. The galactose-binding galectins and lectins

[1] M. E. Taylor and K. Drickamer, "Introduction to Glycobiology." Oxford University Press, Oxford, 2002.

0076-6879/03 $35.00

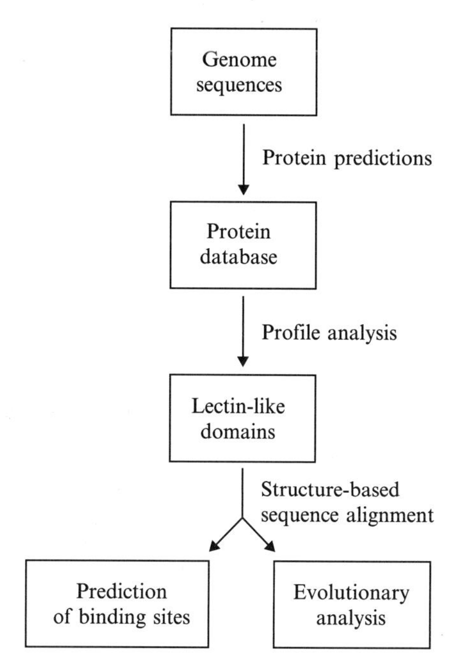

FIG. 1. Flow diagram of approaches to identification of potential carbohydrate-binding proteins.

containing R-type CRDs like those of the plant toxin ricin are found both inside cells and at the cell surface. In the extracellular space, cell–cell adhesion and signaling are mediated by I-type lectins, which are members of the immunoglobulin superfamily, and by Ca^{2+}-dependent C-type lectins. CRDs of a particular type usually form a subset of a larger family of protein domains that are homologous and have similar folds. The members of the larger family can be referred to as lectin-like domains, in order to avoid the unwarranted implication that all members of the larger family are likely to bind sugars.[2] Proteins that contain lectin-like domains are thus potential lectins. A genomic approach to identification and analysis of such proteins is summarized in Fig. 1.

A comprehensive database of many of the major known lectin families in humans and mice is currently available on the World Wide Web (Table I). The information is continuously updated as genome sequences are refined. This resource illustrates the type of information that can be derived from analysis of sequence data. In addition to consulting this site,

[2] W. I. Weis, M. E. Taylor, and K. Drickamer, *Immunol. Rev.* **163,** 19 (1998).

TABLE I
WORLD WIDE WEB SERVERS FOR ANALYSIS OF LECTIN SEQUENCES

Application	World Wide Web universal resource identifier[a]
Databases	
Database of known animal lectins	ctld.glycob.ox.ac.uk/ctld/lectins.html
Protein sequence database	www.ebi.ac.uk/swissprot/index.html
ProSite protein motifs	www.expasy.org/prosite
Pfam protein families motifs	www.sanger.ac.uk/Software/Pfam
InterPro protein motifs	www.ebi.ac.uk/interpro/index.html
Sequence retreval software (SRS)[b]	srs.ebi.ac.uk/
Servers	
Profile scanner for ProSite and Pfam	hits.isb-sib.ch/cgi-bin/PFSCAN
Profile scanner for InterPro	www.ebi.ac.uk/interpro/scan.html
Hydropathy analysis	www.expasy.ch/cgi-bin/protscale.pl
Coiled-coil identification	www.ch.embnet.org/software/COILS_form.html multicoil.lcs.mit.edu/cgi-bin/multicoil
Short motif identification	hits.isb-sib.ch/cgi-bin/PFSCAN using "Prosite patterns that match very frequently"
Multiple sequence alignment (ClustalW)	www.ebi.ac.uk/clustalw/index.html

[a] Mirror sites for many of these servers can be located by using the addresses shown.
[b] SRS provides a simple way to search most of the databases.

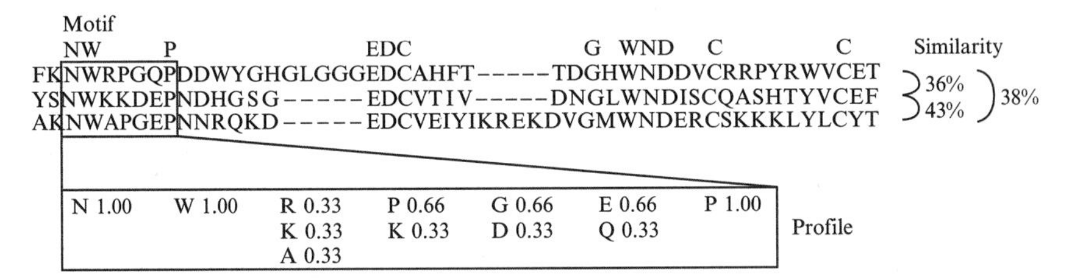

FIG. 2. Illustration of motifs, pairwise similarity, and profiles, using a small set of aligned sequences. Motifs identify absolutely conserved sequences at a small number of positions in all members of a family. Similarity scores reflect the overall number of conserved residues in any pair of sequences. Profiles encode both types of information, by indicating the frequency of each type of amino acid residue at each position in a sequence, as illustrated by a profile for the first seven residues of the sample sequences.

investigators may find it useful to undertake their own analysis of sequences of particular interest to them. For example, the screening protocols are broadly applicable to other organisms for which complete surveys have not been undertaken. The analysis can be applied to individual proteins or databases of proteins. In addition, cloned cDNAs may reveal errors in sequences and alternative splicing patterns that require further analysis. Such analysis will be particularly important when genes encoding lectin-like proteins are linked to inherited disease phenotypes or susceptibilities.

Identification of Lectin-Like Domains

Sugar-binding sites can be identified only in the context of a particular protein fold. Therefore, the first stage in screening for potential CRDs is to detect domains that have a fold that falls into one of the known CRD families. In the past, many lectin-like domains have been identified through pairwise sequence comparisons with known lectins or by looking for conserved sequence motifs found in all members of a particular lectin family. These methods have now been largely superseded by profile analysis.[3] The profile contains information about the probability of each amino acid occurring at each position in the domain (Fig. 2). It thus combines features of the pairwise comparisons, which look at the entire sequence and the motifs, which present information about all members of a domain family. As the number of sequences in a family grows, the number of absolutely invariant residues that can be used to define conserved sequence motifs becomes smaller, making profile scanning the preferred approach. Because they contain information about all members of a family, profiles are more powerful than pairwise sequence comparisons for detecting distantly related domains.

Several sets of profiles have been defined for the major lectin families (Table II). Profiles in the ProSite, Pfam (protein families), and InterPro formats give similar results in most screening procedures. For analysis of individual sequences, the sequences can be submitted directly to World Wide Web servers that perform screening for all the known profiles (Table I). All proteins entered into the SwissProt and National Center for Biotechnology Information (NCBI) databases of protein sequences are prescreened against these profiles and information about domains that are present are indicated in the annotations. Thus, it is possible to abstract all known proteins containing a particular type of domain by querying these databases with the profile identifier and the organism of interest as keywords. The Web sites associated with each set of profiles also contain lists of protein sequences that have been abstracted from the general protein sequence databases with each profile, but these lists may not be updated as frequently as the sequence databases.

[3] M. Gribskov, R. Lüthy, and D. Eisenberg, *Methods Enzymol.* **183,** 146 (1990).

TABLE II
PROFILES USED TO DETECT LECTIN-LIKE DOMAINS

Lectin group	ProSite profile	Pfam profile	InterPro profile	Notes
Calnexin		PF00262	IPR001580	ProSite entries do not include the CRD
L-type		PF03388		Entries for legume lectins do not detect animal L-type lectins
P-type		PF00878 PF02157	IPR000479 IPR000296	
C-type	PS50041 PS00615	PF00059	IPR001304	
Galectins	PS00309	PF00337	IPR001079	
I-type		PF00047	IPR003006	General motifs for immunoglobulin V-set domains
R-type		PF00652	IPR000772	

Prediction of Sugar-Binding Activity

Structure-based sequence alignments are used to examine lectin-like domains for potential sugar-binding sites. The structures of sugar-binding CRDs reveal the importance of particular residues in forming elements of secondary structure, turns, and hydrophobic packing interactions in the core of the protein. Conservation of these elements provides a basis for making adjustments to computer-generated alignments that are based simply on amino acid substitution matrices.

Structures of the lectins in complex with glycan ligands also highlight the roles of specific contact amino acids in the binding sites. For example, aromatic amino acids often pack against the hydrophobic face of galactose and the aromatic amino acid residues are conserved in galactose-binding sites such as those in the asialoglycoprotein receptors, the galectins, and ricin-type plant toxins.[4] Many residues can form hydrogen bonds with sugar hydroxyl groups and ionic interactions with charged sugars.[5] The binding sites of the C-type lectins are characterized by a constellation of amino acids that coordinate to a Ca^{2+} that in turn interacts with hydroxyl groups on the sugar. Binding sites in L-type lectins are stabilized by divalent cations, although these do not interact directly with the sugar ligand.[6]

[4] J. M. Rini, *Annu. Rev. Biophys. Biomol. Struct.* **24,** 551 (1995).
[5] W. I. Weis and K. Drickamer, *Annu. Rev. Biochem.* **65,** 441 (1996).
[6] H. Lis and N. Sharon, *Chem. Rev.* **98,** 637 (1998).

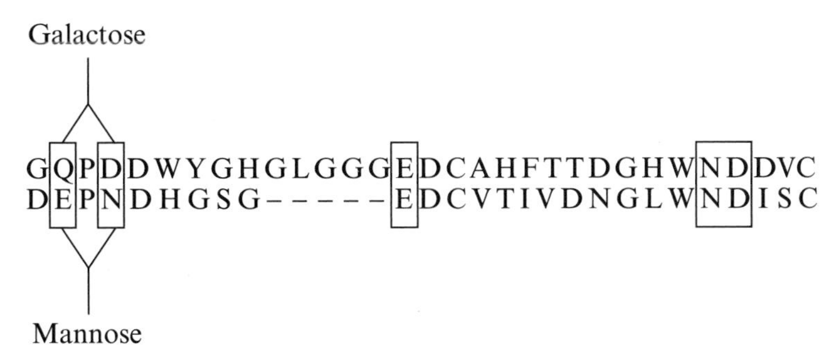

FIG. 3. Prediction of sugar-binding activity of C-type lectins from aligned amino acid sequences. The five residues shown in boxes form a Ca^{2+}-binding site that must be present to form a sugar-binding site. The specific types of sugars bound by CRDs are correlated with the residues found at two of the positions within this group.

Examination of aligned sequences reveals the presence or absence of the residues at appropriate positions to interact with sugar residues. Conservation of the binding site residues is not absolute, particularly in CRD families that bind a spectrum of different types of sugar ligands. Nevertheless, the presence of residues of a particular type, such as those that can make hydrophobic packing interactions or hydrogen bond donors and acceptors, is often preserved in binding sites. For example, the presence of at least a minimal set of residues that act as ligands for calcium ions seems to be essential for sugar binding to C-type CRDs.[5] For the I-type lectins, the profiles available identify all immunoglobulin superfamily members, so manual screening for the amino acid side chains that interact with sialic acid is essential to identify the small subset of domains that fall in the siglec (sialic acid-binding immunoglobulin-like lectin) family.[7]

Our knowledge of the sugar-binding sites in some lectin families highlights the importance of specific amino acids in discrimination between different potential ligands. Complete conservation of particular sets of residues can be a basis for making specific predictions about which ligands will bind to a particular domain (Fig. 3). Although these predictions have been reasonably accurate in some cases, they must be considered educated guesses and empirical evaluation of the predictions are essential.[8]

[7] A. P. May, R. C. Robinson, M. Vinson, P. R. Crocker, and E. Y. Jones, *Mol. Cell* **1,** 719 (1998).

[8] M. E. Taylor and K. Drickamer, *Methods Enzymol.* **363,** [1] In press (2003).

Biological Implications of Lectin Sequences

The context of a potential CRD within the sequence of a polypeptide can provide further insights into its possible biological functions. As many lectins are membrane bound, hydropathy plots can be used to identify signal sequences[9] and stop transfer sequences[10] and hence determine the probable organization of the CRD-containing polypeptides in a membrane (Table I). For example, the largest family of C-type lectins, including the asialoglycoprotein receptor and the DC-SIGN-related cell adhesion molecules, consists of type II transmembrane proteins, whereas the macrophage mannose receptor and the siglecs are examples of C-type and I-type lectins that are type I transmembrane proteins.

C-type lectins such as the type II membrane receptors and the soluble collectins form oligomers through coiled coils of α helices. Regions of polypeptide likely to serve as such protein–protein interaction domains can be identified as amphipathic helices formed by heptad repeats, in which every third or fourth amino acid is hydrophobic.[11] Further aspects of the domain organization of extracellular portions of the CRD-containing proteins often emerge from the profile analysis. In addition, screening of the intracellular domain for shorter sequence motifs can identify potential sites of phosphorylation and signals for endocytosis. These types of analyses can facilitate formation of hypotheses about the function of a protein containing a potential CRD. Such ideas require experimental validation, but they can be useful leads.

Further insights can also come from comparison of sequences of CRDs leading to development of dendrograms that reflect the likely pattern of evolution of the domains. For accurate comparisons, the CRDs must be of comparable lengths. Thus, the structure-based sequence comparisons should be used to establish the correct way to trim the ends of the sequences so they reflect the likely ends of the structural domains. Quantitative comparisons of sequence similarity between the trimmed domains are usually undertaken by means of a dynamic programming algorithm and cluster analysis is used to derive a dendrogram that reflects the relationships between the sequences. The ClustaiW program[12] is part of many sequence analysis packages and is also available on Web servers (Table II). The final dendrogram may be interpreted to reflect the likely order of divergence of lectin-like domains. The results of such analysis are often supported by comparisons of the overall domain

[9] G. von Heijne and L. Abrahmsén, *FEBS Lett.* **244,** 439 (1989).
[10] F. Jähnig, *Trends Biochem. Sci.* **15,** 93 (1990).
[11] A. Lupas, *Trends Biochem. Sci.* **21,** 375 (1996).
[12] D. G. Higgins and P. M. Sharp, *Gene* **73,** 237 (1988).

organization of the proteins containing the lectin-like domains, because the most similar lectin-like domains are usually found in proteins that have similar organization.[13]

The major limitation to computational approaches is that there is no systematic way to identity sugar-binding sites in novel structural contexts. Primary sugar-binding sites are shallow indentations on protein surfaces, so they can appear relatively easily in a variety of different types of protein modules.[4,5] The families discussed here are only a subset of the sugar-binding proteins found in animal cells. The approach described here can be extended to include new families when sufficient structural information becomes available, but there must first be experimental characterization of novel sugar-binding activity in a new type of module. Thus, it is important to be aware that the absence of a known type of lectin-like domain in a protein does not rule out the possibility that it may have sugar-binding activity. Nevertheless, the presence of a lectin-like domain with an appropriate binding-site configuration can provide a strong clue that such activity is worth investigating.

[13] K. Drickamer and R. B. Dodd, *Glycobiology* **9,** 1357 (1999).

[38] Exploring Enzyme Amplification to Characterize Specificities of Protein–Carbohydrate Recognition

By ANITA RAMDAS PATIL, SANDRA MISQUITH, TARUN KANTI DAM, VIVEK SHARMA, MILI KAPOOR, and AVADHESHA SUROLIA

Introduction

Carbohydrates play key roles in cellular recognition events. They are mostly conjugated to proteins and lipids, referred commonly as glycoconjugates.[1] A considerable amount of evidence has accumulated indicating that a large number of glycoconjugates become altered during normal and abnormal cellular development.[2] Understanding these diverse

[1] H. J. Allen and E. C. Kisailus, Eds., "Glycoconjugates: Composition, Structure, and Function." Marcel Dekker, New York, 1992.
[2] W. I. Weis and K. Drickamer, *Annu. Rev. Biochem.* **65,** 441 (1996).

0076-6879/03 $35.00

biological processes at the molecular level requires knowledge of the structure and biochemistry of cellular oligosaccharides.[3–5] Because of their exquisite carbohydrate specificity, ready availability, and diversity, lectins mostly from plants have become indispensable tools in biological research for the investigation of sugars of glycoconjugates.[6,7]

Several lectin–carbohydrate binding assays have been described in the literature that employ enzyme-linked solid-phase binding assays.[8–11] In principle, microtiter plate systems and associated photometric plate readers have become important tools for performing assays wherein appropriate lectin or glycoprotein species are immobilized onto the wells of microtiter plates.[12–19] These methods can be qualitative or quantitative, assayed by a radioactive, fluorescent, or enzyme label linked directly to a lectin or to a secondary anti-lectin antibody similar to classic indirect ELISA for antigen detection.[20–23] This article describes the principle and procedure employed for enzyme-linked lectinsorbent assay (ELLA) for measuring the carbohydrate-binding activity of lectins using a microtiter assay. The procedure is broadly divided into direct and indirect methods and is discussed here with representative lectin examples optimized in

[3] P. M. Rudd and R. A. Dwek, *Crit. Rev. Biochem. Mol. Biol.* **32,** 1 (1997).
[4] A. Kobata, *Eur. J. Biochem.* **209,** 483 (1992).
[5] T. W. Rademacher, R. B. Parekh, and R. A. Dwek, *Annu. Rev. Biochem.* **57,** 785 (1988).
[6] H. Lis and N. Sharon, *Chem. Rev.* **98,** 637 (1998).
[7] H. Lis and N. Sharon, *Annu. Rev. Biochem.* **55,** 35 (1986).
[8] J. Parkkinen and U. Oksanen, *Anal. Biochem.* **177,** 383 (1989).
[9] J. S. Haurum, S. Thiel, H. P. Haagsman, S. B. Laursen, B. Larsen, and J. C. Jensenius, *Biochem. J.* **293,** 873 (1993).
[10] M. Duk, E. Lisowska, J. H. Wu, and A. M. Wu, *Anal. Biochem.* **221,** 266 (1994).
[11] T. Hatakeyama, T. Fukuda, and N. Yamasaki, *Biosci. Biotechnol. Biochem.* **56,** 2072 (1992).
[12] M. A. Baker, G. J. Cerniglia, and A. Zaman, *Anal. Biochem.* **190,** 360 (1990).
[13] M. B. Ashour, S. J. Gee, and B. D. Hammock, *Anal. Biochem.* **166,** 353 (1987).
[14] W. Cockburn, G. C. Whitelam, S. P. Slocombe, and R. A. McKee, *Anal. Biochem.* **189,** 95 (1990).
[15] B. Kim, J. M. Buckwalter, and M. E. Meyerhoff, *Anal. Biochem.* **218,** 14 (1994).
[16] T. Hatakeyama, K. Murakami, Y. Miyamoto, and N. Yamasaki, *Anal. Biochem.* **237,** 188 (1996).
[17] I. A. van der Schall, T. J. Logman, C. L. Diaz, and J. W. Kijne, *Anal. Biochem.* **140,** 48 (1984).
[18] Y. Oda, K. Nakayama, B. Abdul-Rahman, M. Kinoshita, O. Hashimoto, N. Kawasaki, T. Hayakawa, K. Kakehi, N. Tomiya, and Y. C. Lee, *J. Biol. Chem.* **275,** 26772 (2000).
[19] M. M. Farajollahi, D. B. Cook, and C. H. Self, *Anal. Biochem.* **261,** 118 (1998).
[20] M. Duk, M. Ugorski, and E. Lisowska, *Anal. Biochem.* **253,** 98 (1997).
[21] N. Parker, C. A. Makin, C. K. Ching, D. Eccleston, O. M. Taylor, J. D. Milton, and J. M. Rhodes, *Cancer* **70,** 1062 (1992).
[22] A. M. Wu, S. C. Song, Y. Y. Chen, and N. Gilboa-Garber, *J. Biol. Chem.* **275,** 14017 (2000).
[23] M. T. Goodarzi, M. Rafiq, and G. Turner, *Biochem. Soc. Trans.* **23,** 168S (1995).

our laboratory. ELLA provides a sensitive biochemical assay with high specificity and broad flexibility to determine the sugar-binding affinities for different classes of lectins and has been applied to various protein–carbohydrate interaction studies.[24–30]

Experimental Design

Lectin Sources

Although most of the lectins are commercially available, they can be purified easily in the laboratory using routine chromatographic procedures. Affinity chromatography in particular has provided a powerful tool for the isolation of carbohydrate-binding proteins, particularly lectins from plant sources.[31] The widespread interest in lectins, coupled with the knowledge of their sugar specificity, has led to the development of efficient methods for isolation of the desired lectin from crude extracts of its natural source. The purification methods involve a single affinity chromatography step, and the purification to homogeneity is achieved with good yield in a reproducible manner. Usually the lectin is adsorbed on an affinity column made of an immobilized sugar derivative and is eluted either with the specific sugar or sometimes by lowering the pH of the eluant. The preferred elution system is a biospecific mono- or disaccharide solution that competes efficiently for binding to the lectin and facilitates the release of adsorbed lectin. It is therefore imperative to select lectins of well-defined specificities and employ easily available sugar competitors. Three general methods have been described, to obtain the affinity matrix, namely agarose activation and subsequent coupling of a ligand of choice,[32] direct reductive amination of the gel in the presence of the disaccharide,[33] and polysaccharide cross-linking.[34,35]

[24] J. P. McCoy, Jr., J. Varani, and I. J. Goldstein, *Anal. Biochem.* **130,** 437 (1983).
[25] A. L. Reddi, K. Sankaranarayanan, H. S. Arulraj, N. Devaraj, and H. Devaraj, *Cancer Left.* **149,** 207 (2000).
[26] V. Leriche, P. Sibille, and B. Carpentier, *Appl. Environ. Microbiol.* **66,** 1851 (2000).
[27] J. Keusch, Y. Levy, Y. Shoenfeld, and P. Youinou, *Clin. Chim. Acta* **252,** 147 (1996).
[28] C. K. Ching and J. M. Rhodes, *Br. J. Cancer* **59,** 949 (1989).
[29] D. Zhuang, S. Yousefi, and J. W. Dennis, *Cancer Biochem. Biophys.* **12,** 185 (1991).
[30] V. L. Thomas, B. A. Sanford, R. Moreno, and M. A. Ramsay, *Curr. Microbiol.* **35,** 249 (1997).
[31] B. B. Agrawal and I. J. Goldstein, *Biochim. Biophys. Acta* **133,** 376 (1967).
[32] I. Matsumoto, N. Seno, A. M. Golovtchenko-Matsumoto, and T. Osawa, *J. Biochem. (Tokyo)* **87,** 535 (1980).
[33] R. J. Baues and G. R. Gray, *J. Biol. Chem.* **252,** 57 (1977).
[34] G. R. Gray, *Arch. Biochem. Biophys.* **163,** 426 (1974).
[35] T. Majumdar and A. Surolia, *Prep. Biochem.* **8,** 119 (1978).

Carbohydrates and Glycoprotein Sources

The synthesis of carbohydrates up to a few oligomers has now become relatively routine and many monosaccharides, complex sugars, and glycoproteins are commercially available from sources such as Sigma (St. Louis, MO), Carbohydrate International (Arlöv, Sweden), Glyko (Novato, CA), and Dextra laboratories (London, UK). Also, a number of complex sugars from naturally occurring sources can be made in house; however, their preparation is not a trivial task.[36] The amount of sugar ligand remains a severe constraint for most of the assays and requires their judicious use. Sugars are generally stored at subzero temperatures in a powder form and are dissolved in buffer just before use. The solutions are stored frozen for repeated use over a few days, during experiments. Information about carbohydrate structures present on various glycoproteins is available in the literature. Also, immobilized lectins can be used for the separation of glycoproteins from their mixtures.

Enzyme Conjugates

Depending on the suitability for ELLA, various glycoproteins and lectin–enzyme and secondary antibody conjugates are available commercially from USB (Cleveland, OH), Vector Laboratories (Burlingame, CA), EY Laboratories (San Mateo, CA), Sigma, and Pierce (Rockford, IL). The use of enzymes as labels has made it possible to design the study of various glycoprotein–lectin and lectin–carbohydrate interactions that have been previously difficult owing to weak interactions or expensive ligands.[37,38] Incidentally, in the ELLA procedures discussed in this article no lectin–enzyme conjugate has been used. The direct methods have used the glycoprotein–enzyme for the assay whereas the indirect methods have used either the secondary antibody to the glycoprotein or to the lectin.

Setup of Assay

The wells of the polystyrene plates are sensitized by passive adsorption of the lectin or glycoprotein. The plates are then washed and can be stored for some time if required. The complementary molecule to the lectin or glycoprotein, at an optimized dilution, is incubated in the sensitized wells and the lectin–glycoprotein interaction is allowed to take place. Subsequent washing steps remove all unreacted components. The enzyme assay

[36] S. Misquith, P. G. Rani, and A. Surolia, *J. Biol. Chem.* **269,** 30393 (1994).

[37] G. T. Hermanson, "Bioconjugate Techniques." Academic Press, San Diego, CA, 1996.

[38] Patrik Englebienne, *in* "Immune and Receptor Assay in Theory and Practice." CRC Press, Boca Raton, FL, 2000.

is performed in a direct binding ELLA with the glycoprotein enzyme directly bound to the lectin. In the case of indirect ELLA, the IgG to the lectin or glycoprotein is added to the plate followed by subsequent washing and incubation with an enzyme-linked secondary antibody. The enzyme substrate, in solution, in both direct and indirect assays, is then incubated with the solid phase. The rate of degradation of the substrate, usually indicated by the color intensity of substrate solution, is proportional to the enzyme–lectin or enzyme–antibody conjugate concentration used. The reaction is stopped and the color intensity is estimated photometrically on an ELISA reader.

Coating of Solid Phase

The coating of the solid phase is mostly by simple adsorption. The rate and extent of adsorption depend on a number of factors such as the net charge of the protein molecule being adsorbed; size, molecular weight, diffusion coefficient, concentration, and the nature of coating buffer; its composition, molarity, and pH; the chemical nature of the solid phase; the shape of the wells; and the temperature and duration of adsorption. Typically the adsorption is conducted overnight at 4° or for 3 h at 37°. Phosphate-buffered saline (PBS, 10 m*M* sodium or potassium phosphate buffer, pH 7.2–7.8, containing 150 m*M* NaCl) is one of the most commonly used buffers for coating. Tris-buffered saline (TBS) is another commonly used buffer. The concentration of protein to be used in coating solution must be determined empirically. Usually, protein concentrations of 1–10 μg/ml have been used. The coated plates can be stored wet with buffer at 4° for about 1 week. After thorough coating, the plates are washed with buffer containing nonionic detergents such as Tween 20 for effective removal of unbound protein.

Blocking and Ligand Incubation

The remaining unoccupied sites on the plate are saturated with an unrelated, noninterfering protein to prevent nonspecific adsorption. Bovine serum albumin (BSA, 1–5%), gelatin (0.02–0.1%), or casein (2–5%) is commonly used. The most suitable blocking agent depends on the individual system, that is, on the coated protein, the detecting reagent, and the nature of the sample. The time and temperature of ligand incubation are also to be optimized with respect to each ligand.

Enzyme Assay

The assay needs to be simple, sensitive, and precise. Several characteristics of the enzyme to be conjugated need to be considered in choosing the enzyme. These include the turnover number, specific activity, stability, and presence of reactive residues on the enzyme available for cross-linking. The conditions for maximum activity of enzyme conjugate need to be optimized in the assay. In practice, among the enzymes commonly used, horseradish peroxidase (HRP) is by far the most commonly used enzyme in lectin and glycoprotein conjugates. Because of its small size (40 kDa), robust nature, and high stability, HRP is quite suitable for obtaining a moderate sized protein–enzyme conjugate. HRP is stable for years in a freeze–dried state. HRP catalyzes the reaction of hydrogen peroxide with certain organic, electron-donating substrates to yield highly colored products.

Enzyme Substrates

Many chromogenic substrates are used with HRP. The common ones, in the order of decreasing absorbance intensity, are 3,3′,5,5′-tetramethylbenzidine (TMB) and *o*-phenylenediamine (OPD). TMB and OPD are most commonly used as a 0.40-mg/ml solution in 0.1 *M* citrate buffer, pH 4–5. The optimum substrate concentration (hydrogen peroxide) is between 0.01 and 0.0005%, depending on the chromogen used. The concentration must be optimized because too high a concentration exerts an inhibitory effect on the system. Before reading the absorbance, the enzyme reaction must be stopped in order to ensure identical kinetics between the samples. This is usually accomplished by the addition of 0.1 to 0.5 volume of 1 *M* sulfuric acid or hydrochloric acid. The optimum wavelengths for reading are 450 nm for acidified TMB and 490 nm for acidified OPD. For alkaline phosphatase, *p*-nitrophenyl phosphate is the most popular substrate and is used at a concentration of 1 mg/ml in 0.1 *M* Tris-HCl buffer, pH 8.1 (bacterial enzyme) or 10% diethanolamine, pH 9.8, both containing 0.01% magnesium chloride required for the activity of the enzyme. Glucose oxidase is usually coupled to peroxidase for the detection of hydrogen peroxide generated by it.

Enzyme Inhibition

The enzymes used in the assay are inhibited by various metal ions. For example, horseradish peroxidase is inhibited by fluoride and azide, alkaline phosphatase is inhibited by Cu^{2+}, Ag^{+}, Hg^{2+}, and EDTA, and glucose oxidase is inhibited by heavy metals and hydroxylamine. Care should be taken to avoid the presence of these inhibitors during the assays. Sodium azide is added to the plant lectin samples during storage and hence must be dialyzed away completely before the assay. Also, PBS solutions should be prepared as stocks without azide for the assay wash solutions.

Data Analysis

The most common modes of expression of ELLAs are untransformed absorbance readings at a single dilution. For quantification, the data can be expressed in a variety of ways such as semilogarithmic plots and log–log plots. The responses obtained with the standards (glycoprotein sugars or oligosaccharides) are used to generate a calibration curve. Absorbance from the competition ELLAs using different monosaccharides, disaccharides, and oligosaccharides is interpolated through this curve to obtain the concentration of the inhibiting saccharide for 50% activity, which is then used to obtain the binding affinity for a given lectin–carbohydrate interaction. When reference curves are not available the high positive values are used as an internal references and are given a highest unit value. The rest of the values are then derived as the fraction or percentage values with respect to the highest value.

Enzyme-Linked Lectinsorbent Assay: Representative Procedures

Materials

Preparation of Reagents for Assay. TBS (10 m*M* Tris-HCl buffer, pH 7.5, containing 150 m*M* NaCl), PBS (10 m*M* phosphate-buffered saline, pH 7.5), and 10 m*M* citrate phosphate buffer, pH 6.2, are the commonly used buffers in ELLA. All the reagents should be of analytical grade or better. Also, the solutions should be prepared with double-distilled deionized water. Filtering the buffer solutions before the assay reduces the background and variation among duplicate wells. Artocarpin, garlic lectin, chicken IgG, and bovine liver calreticulin are purified in the laboratory by standard procedures.[36,39–41] Rabbit anti-chicken IgG–HRP conjugate, *o*-phenylenediamine (OPD), 3,3′,5,5′-tetramethylbenzidine (TMB), bovine serum albumin, and invertase are obtained from Sigma. Glucose oxidase is from Roche Molecular Biochemicals (Indianapolis, IN). Skimmed milk powder is procured from local stores. The milk solution is freshly prepared in boiling water and filtered through Whatman (Clefton, NJ) No. 1 filter paper to remove the presence of insoluble particles, which again cause background noise in the assay plates.

Accuracy and reproducibility remain the primary goal for performing a ligand-binding assay. The physical state and temperature of individual assay reagents and specimens awaiting analysis must be chosen to preserve

[39] T. K. Dam, K. Bacchawat, P. G. Rani, and A. Surolia, *J. Biol. Chem.* **273,** 5528 (1998).
[40] V. Sharma, M. Vijayan, and A. Surolia, *J. Biol. Chem.* **271,** 21209 (1996).
[41] A. R. Patil, C. J. Thomas, and A. Surolia, *J. Biol. Chem.* **275,** 24348 (2000).

maximal stability during short- and long-term storage. Recommended storage temperatures will often require refrigeration or freezing of all or part of the assay reagents.

Equipment. Multiwell 96-well polystyrene plates with flat bottom [Corning (Corning, NY) or Nunc (Raskilde, Denmark)], micropipettes, and an ELISA plate reader are required.

Direct Methods

The absorbed lectin directly binds a glycoprotein enzyme as its ligand, as well as the assay molecule, which is then used to develop the product on substrate addition. The enzyme assay can be direct or coupled as illustrated by the ensuing examples.

Procedure 1: Carbohydrate Binding Specificity of Artocarpin.

PRINCIPLE. Artocarpin belongs to the class of mannose-specific plant lectins and investigations of its carbohydrate binding have revealed a high degree of specificity toward complex N-linked fucosylated glycan found in some plant glycoproteins such as peroxidase. Misquith *et al.*[36] have studied carbohydrate binding of artocarpin, wherein horseradish peroxidase, a mannose-containing glycoprotein serves both as a ligand and assay enzyme for the detection of binding as described below (Fig. 1A). This perhaps represents the best example of ELLA.

METHOD

1. Artocarpin is purified from jackfruit seeds, using mannose–Sepharose affinity matrix.[36]

2. Fifty microliters of the artocapin solution (0.1–10 μg/ml) in TBS is added to the wells of the titer plate and left overnight at 4°. The plates are subsequently washed three times with TBS and then blocked with 3% bovine serum albumin in TBS for 1 h at room temperature. The binding of horseradish peroxidase to various amounts of artocarpin coated in the wells of microtiter plates is studied initially. In another set of experiments, the amount of lectin is fixed and horseradish peroxidase (HRP) is varied. On the basis of these initial experiments, 250 ng of artocarpin and 50 ng of horseradish peroxidase are found to be optimal for all subsequent sugar inhibition studies.

3. After washing the plates three times with the blocking buffer, different sugars are added at various concentrations and allowed to interact with the lectin for 1 h.

4. A fixed concentration of horseradish peroxidase (50 ng) is then added to each well and the plate is incubated at room temperature for 2 h.

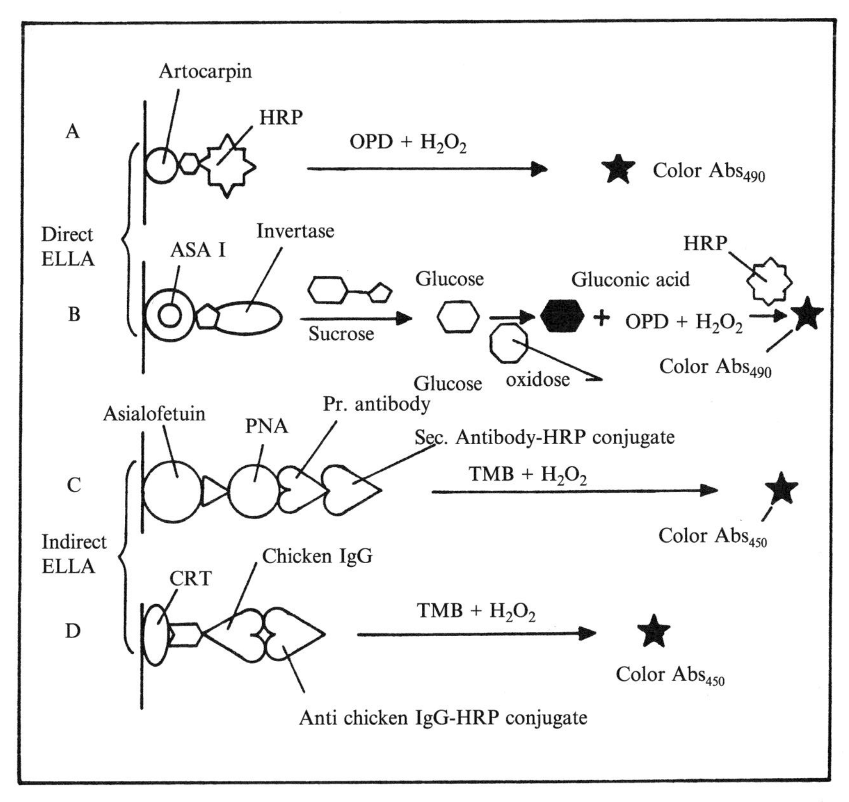

FIG. 1. Diagrammatic representation of various ELLA schemes. Abs, Absorbance; ASA I, *Allium sativum* agglutinin I; CRT, calreticulin; PNA, peanut (*Arachis hypogaea*) agglutinin; IgG, immunoglobulin G; HRP, horseradish peroxidase; OPD, *o*-phenylenediamine; TMB, 3,3′,5,5′-tetramethylbenzidine.

5. Subsequently, the plates are washed three times with TBS and the amount of HRP bound is estimated using *o*-phenylenediamine as the substrate (OPD at 0.5 mg/ml in 100 m*M* citrate buffer, pH 5.0, containing 5 μl of H_2O_2). The reaction is stopped by addition of 2 *N* HCl and the absorbance is recorded by an ELISA reader at 490 nm.

COMMENTS ON PROCEDURE. Inhibition of the lectin binding has been studied with various monosaccharide, oligosaccharide, and glycoprotein ligands. The structures of various oligosaccharides (**1–20**) used in ELLA are shown in Fig. 2. Relative affinities of the lectins for different sugars have been determined from the concentration of sugars required for 50% inhibition of the binding of glycoprotein enzyme used in this study. Using this type of approach, the binding affinities of artocarpin for various mono-,

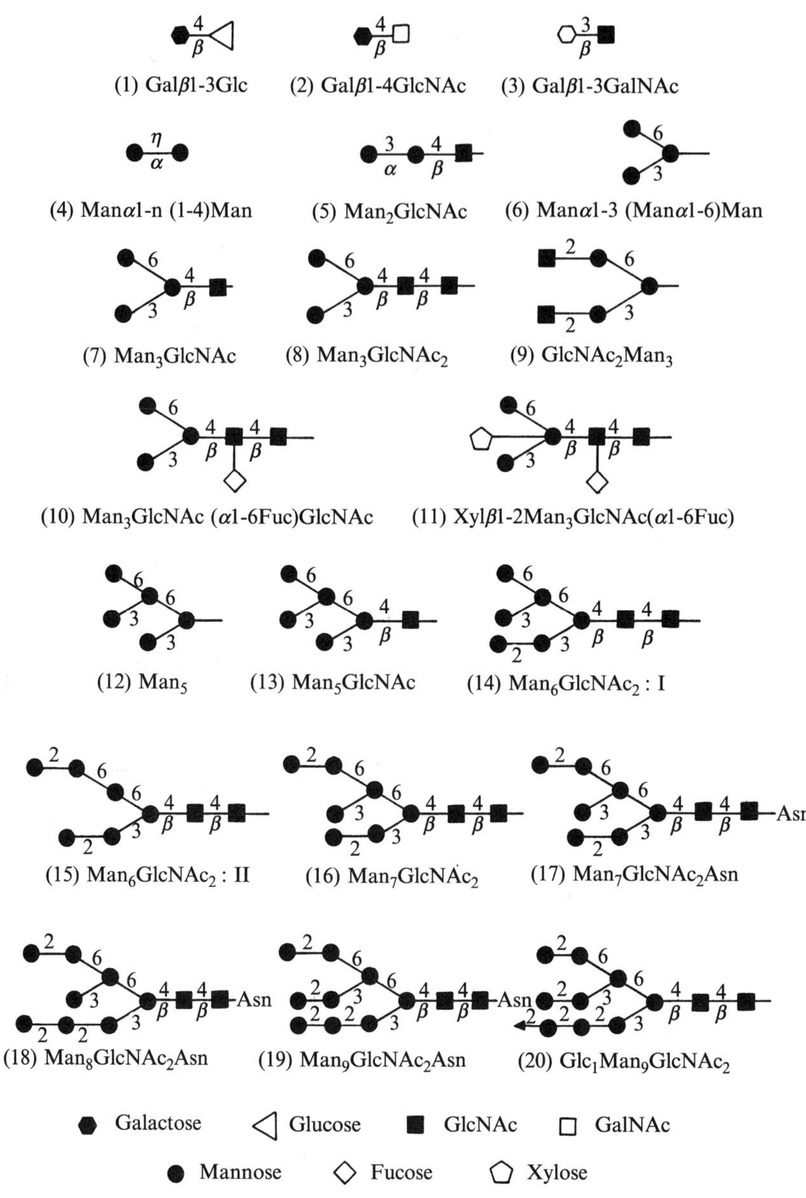

FIG. 2. Structures of various oligosaccharides used in ELLA.

TABLE I
INHIBITION OF BINDING OF HORSERADISH PEROXIDASE TO ARTOCARPIN BY OLIGOSACCHARIDES[a]

Oligosaccharide	Concentration for 50% inhibition (m*M*)
Mannose	0.98 (1)
Manα1-3[Manα1-6]Man or (Man_3)	0.074 (13)
Man_2GlcNAc	0.015 (65)
Man_3GlcNAc	0.14 (7)
$GlcNAc_2Man_3$	2.3 (0.43)
Man_5 or (Mannopentaose)	0.096 (10)
Man_5GlcNac	0.0045 (218)
Man_6GlcNac; I	0.048 (2)
Man_6GlcNac; II	NI up to 0.34 m*M*
$Man_7GlcNac_2$Asn	NI up to 0.2 m*M*
$Man_8GlcNac_2$Asn	NI up to 0.2 m*M*
Man_3GlcNAc(α1-6 Fuc)GlcNAc	0.0018 (544)
$GlcNAc_2Man_3$GlcNAc(α1-6 Fuc)GlcNAc	0.01 (98)
$Gal_2GlcNAc_2Man_3$GlcNAc(α1-6 Fuc)GlcNAc	0.011 (89)
Man_3GlcNAc(α1-6Fuc)GlcNAc	0.0030 (327)
Xylβ 1-2Man_3GlcNAc(α1-6Fuc)GlcNAc (horseradish peroxidase oligosaccharide)	0.0006 (1633)

[a] With their relative inhibitory potencies. Values in parentheses indicate the relative potencies to mannose. NI, No inhibition. Reproduced with permission from S. Misquith, P. G. Rani, and A. Surolia, *J. Biol. Chem.* **269,** 30393 (1994) and the publisher, ASBMB.

di-, and oligosaccharides, and glycoprotein substrates, have been elucidated as shown in Table I. These studies have clearly demonstrated that mannose is preferred over glucose and mannopentaose and mannotriose are more potent as ligands as compared with Manα1-3Man, whereas, Manα1-2Man and Manα1-4Man are barely inhibitory. Finally, the lectin shows strongest affinity for the xylose-containing heptasaccharide, with fucosylation (α1-3) in the core region, present in horseradish peroxidase. Further studies by isothermal titration calorimetry (ITC) to determine the binding affinity of the lectin to the trimannoside and its various deoxy derivatives confirmed α1-3Man to be the primary contributor to its binding affinity.[42]

Procedure 2. Carbohydrate Binding Specificity of Garlic Lectin

PRINCIPLE. The garlic lectin (*Allium sativum* agglutinin, ASA) belongs to monocot mannose-binding lectins and displays extraordinary avidity

[42] P. G. Rani, K. Bachhawat, G. B. Reddy, S. Oscarson, and A. Surolia, *Biochemistry* **39,** 10755 (2000).

for invertase, a high-mannose glycoprotein. This has been determined by hemagglutination inhibition studies using a series of simple sugars and several glycoproteins. Of all the glycoproteins used, invertase was found to be the strongest inhibitor. This finding has been used to design a sensitive enzyme-based system to investigate the binding specificities of various mannose derivatives, oligosaccharides and glycoproteins.[39] Garlic bulbs have two mannose-binding lectins, ASA I and ASA II. In this study ASA refers to ASA I only. The lectin-bound invertase is assayed by its activity on sucrose and the amount of glucose released determined by a coupled assay using glucose oxidase and horseradish peroxidase with OPD as the substrate and the absorbance recorded at 490 nm in an ELISA reader (Fig. 1B).

METHOD

1. The lectin ASA I is purified by affinity chromatography on mannose–Sepharose matrix followed by gel filtration on a Sephadex G-100 (or Bio-Gel P-100) column as described by Dam *et al.*[39]

2. The lectin solution (50 μl, 0.1–10 μg/ml) in PBS is coated onto ELISA plates and left overnight at 4°. After washing three times with 20 mM PBS, the wells are blocked with 3% bovine serum albumin for 1 h at room temperature. The binding of invertase to varying amounts of lectins is checked in order to optimize the concentrations of these components. In another set of experiments, the amounts of horseradish peroxidase and glucose oxidase are varied, respectively, keeping the amounts of lectin and invertase fixed. By these experiments, 0.35 μg of lectin, 20 ng of invertase, and 3 units of glucose oxidase have been found optimal for the rest of the studies.

3. The adsorbed lectin in the wells of the titer plate is then incubated with a 50-μl solution of sugars for 1 h at room temperature. A fixed concentration of invertase is then added to each well and incubated for 30 min. The wells are washed three times with blocking buffer and once with PBS.

4. An equal volume of 50 mM sucrose solution in 100 mM acetate buffer, pH 5.0, is added to each well. After incubation for 1 h, glucose oxidase in the same buffer is then added to the solution to produce H_2O_2 from the liberated glucose. H_2O_2 thus generated is assayed using HRP and OPD (0.05 mg/well in 100 mM citrate buffer). The reaction is stopped by 2 N HCl and the absorbance recorded by an ELISA plate reader at 490 nm.

COMMENTS ON PROCEDURE. The relative affinities of the various ligands for ASA I have been determined from the concentration of sugars required for 50% inhibition of binding, as described above. The potencies of the ligands for ASA I increase in the following order; mannobiose, mannotriose $\approx$ mannopentaose $<<$ Man_9-oligosaccharide (Table II). The best mannooligosaccharide ligand is $Man_9GlcNAc_2Asn$. The strongest inhibitors are the high mannose-containing glycoproteins, which carry large glycan chains. Not surprisingly, therefore, invertase exhibited the highest

TABLE II
INHIBITION OF ASA-INVERTASE BINDING BY OLIGOSACCHARIDES[a]

Oligosaccharide	Concentration for 50% inhibition (ASA I) (m*M*)
Mannose	72 (1)
α-Methylmannopyranoside	10.5 (6.8)
4-Methylumbelliferyl α-D-mannopyranoside	8 (9)
4-Methylumbelliferyl β-D-mannopyranoside	NI*
p-Nitrophenyl-α-mannopyranoside	97 (0.74)
N-Acetylmannosamine	NI
Manα1-2Man	8.3 (8.6)
Manα1-3Man	6.2 (11.6)
Manα1-4Man	NI
Manα1-6Man	9.8 (7.3)
Man_2GlcNAc	4.8 (15)
Manα1-3[Manα1-6]Man or (Man_3)	2.5 (29)
$Man_3GlcNAc_2$	0.51 (141)
$GlcNAc_2Man_3$	1.9 (37.8)
Man_5 or (Mannopentaose)	2.5 (28.8)
Man_3GlcNAc(α1-6 Fuc)GlcNAc	0.46 (157)
Man_3GlcNAc(α1-3 Fuc)GlcNAc	0.87 (82.7)
Xylβ1-2Man_3GlcNAc(α1-3 Fuc)GlcNAc	2.8 (25.7)
Xylβ1-2$Man_3GlcNAc_2$	1.3 (55.4)
$Man_7GlcNac_2$Asn	0.0110 (6545)
$Man_8GlcNac_2$Asn	0.01 (7200)
$Man_9GlcNac_2$Asn	0.0051 (14120)

[a] Values in parentheses indicate the relative potencies to mannose. NI, No inhibition; NI*, no inhibition up to 100 m*M*. Reproduced with permission from T. K. Dam, K. Bachhawat, P. G. Rani, and A. Surolia, *J. Biol. Chem.* **273,** 5528 (1998) and the publisher, ASBMB.

binding affinity. Subsequent studies[43] on binding of various mannooligosaccharides to ASA I by biosensor corroborates the ELLA results about the special preference of ASA I for terminal α-1,2-linked mannose residues. An increase in binding propensity could be directly correlated to the addition of α-1,2-linked mannose to the mannooligosaccharide at its nonreducing end. Mannonaose glycopeptide ($Man_9GlcNAc_2$Asn), the highest oligomer studied, exhibited the greatest binding affinity. ITC studies of the lectin with the core trimannoside also showed similar binding affinities. Thus, the sensitivity of ELLA compares fairly well with other, more direct binding methods, namely surface plasmon resonance (SPR) and ITC.

[43] K. Bachhawat, C. J. Thomas, B. Amutha, M. V. Krishnasastry, M. I. Khan, and A. Surolia, *J. Biol. Chem.* **276,** 5541 (2001).

Indirect Methods

In this variation of ELLA the lectin–glycoprotein conjugate is detected with an anti-lectin or anti-glycoprotein antibody. In a manner similar to classical indirect ELISA, Sharma *et al.*[40,44] used lectin capture method where the glycoprotein (asialofetuin)-bound peanut agglutinin (PNA) was quantified by the double antibody method using anti-PNA IgG as the primary antibody and goat anti-rabbit IgG conjugated to horseradish peroxidase as the secondary antibody to detect sugar specificities of the wild-type and mutant peanut agglutinins (Fig. 1C). Various amounts of sugar inhibitors were used in the solid-phase microtiter plate assay and the binding affinities of wild-type PNA was compared with two mutants, L212N and L212A (denoting the mutation of a critical leucine residue at position 212, to asparagine and alanine, respectively). The results showed that the mutations did not alter the affinity for monosaccharide ligand, galactose, but unlike the wild-type protein, the mutant L212N did not bind the sugar Galβ1-4GlcNAc even at high concentrations. This study underlines the subtle differences in the binding of lactose and T-antigen (Galβ1-3GlcNAc) with PNA. Also, the mutant L212N displays exquisite T-antigen specificity, whereas L212A interacted extremely poorly with T-antigen and recognized LacNAc with greater efficacy.

In another variation of ELLA (Patil *et al.*[41]) the carbohydrate binding of endoplasmic reticulum lectin chaperone calreticulin was studied by its binding to glycoprotein IgG from chicken serum followed by detection with using rabbit anti-chicken IgG linked to horseradish peroxidase as the secondary antibody. An inhibition assay using various concentrations of oligosaccharide $Glc_1Man_9GlcNAc_2$ was performed. In both studies the substrate TMB (3,3′,5,5′-tetramethylbenzidine dihydrochloride)–hydrogen peroxide was used to read the absorbance at 450 nm (Fig. 1D).

Procedure 3: Binding of Calreticulin to the Glycoprotein, Chicken Immunoglobulin G.

PRINCIPLE. Calreticulin is specific for the complex heterooligosaccharide $Glc_1Man_9GlcNAc_2$ and knowledge of its carbohydrate specificity has come from a series of elegant experiments with the dolichol phosphate precursor present in endoplasmic reticulum (ER) membrane, radiolabeled substrates, thin-layer chromatography, and gel electrophoresis experiments with the newly synthesized glycoproteins.[45,46] Among the various naturally

[44] V. Sharma, V. R. Srinivas, P. Adhikari, M. Vijayan, and A. Surolia, *Glycobiology* **8,** 1007 (1998).

[45] A. J. Parodi, *Annu. Rev. Biochem.* **69,** 69 (2000).

[46] R. G. Spiro, *J. Biol. Chem.* **275,** 35657 (2000).

occurring glycoproteins, chicken IgG is one of the few to retain the hetero-oligosaccharide $Glc_1Man_9GlcNAc_2$ on the heavy chain of the protein. This observation by Raju *et al.*[47] proved critical in developing on ELLA for calreticulin. We first established the purification of bovine liver calreticulin and then studied its interaction with chicken IgG by ELLA.[41] Another advantage of using chicken IgG as glycoprotein substrate is the fact that secondary antibodies against this IgG are commercially available as HRP conjugates, which can be assayed with TMB as substrate.

METHOD

1. Calreticulin used in this study is purified by ion-exchange chromatography, adsorption chromatography on hydroxylapatite, and gel-filtration chromatography by established protocols.[41]

2. Fifty microliters of calreticulin dissolved in 10 m*M* PBS buffer, pH 7.5, is dispensed into each well of a polystyrene microtiter plate (Corning) and incubated overnight at 37° in an incubator or moist chamber. The incubation time is standardized from 1 h to overnight incubation at 37°. Overnight incubation is found to be most efficient for this procedure. Calreticulin-coated plates are stable for almost 1–2 weeks at 4°. The commonly used blocking solutions such as BSA, ovalbumin, and gelatin show high background color and are substituted with skimmed milk powder as blocking reagent.

3. After discarding the sensitizing CRT solutions, 200 μl of 5% skimmed milk powder in PBS is dispensed and incubated for 2 h at 37° in a moist chamber to block any remaining active sites on the plastic surface.

4. The minimum amount of CRT required to observe maximum IgG binding is first optimized and then kept constant against varying IgG concentrations. The IgG-binding time period is also standardized and 4 h at 25° has been found to be optimal.

5. Chicken IgG is added at various concentrations (0.1–120 μg) in PBS containing skimmed milk powder, which prevents nonspecific IgG binding. Appropriate calreticulin and IgG controls are also kept with either the lectin or the IgG not added in the reaction steps.

6. After discarding the supernatants and washing three times with PBS, 50 μl/well of an appropriate dilution (1:10,000) of rabbit anti-chicken IgG conjugated to horseradish peroxidase in PBS is incubated at 25° for 30 min in a moist chamber.

7. Subsequently plates are washed four times with PBS before adding the substrate (200 μl/well). The substrate TMB is freshly prepared in citrate phosphate buffer (0.6% in 10 m*M* citrate phosphate buffer, pH 6.2, containing 1 μl/ml of H_2O_2). The plates are kept in the dark for 5–10 min

[47] T. S. Raju, J. B. Briggs, S. M. Borge, and A. J. Jones, *Glycobiology* **10,** 477 (2000).

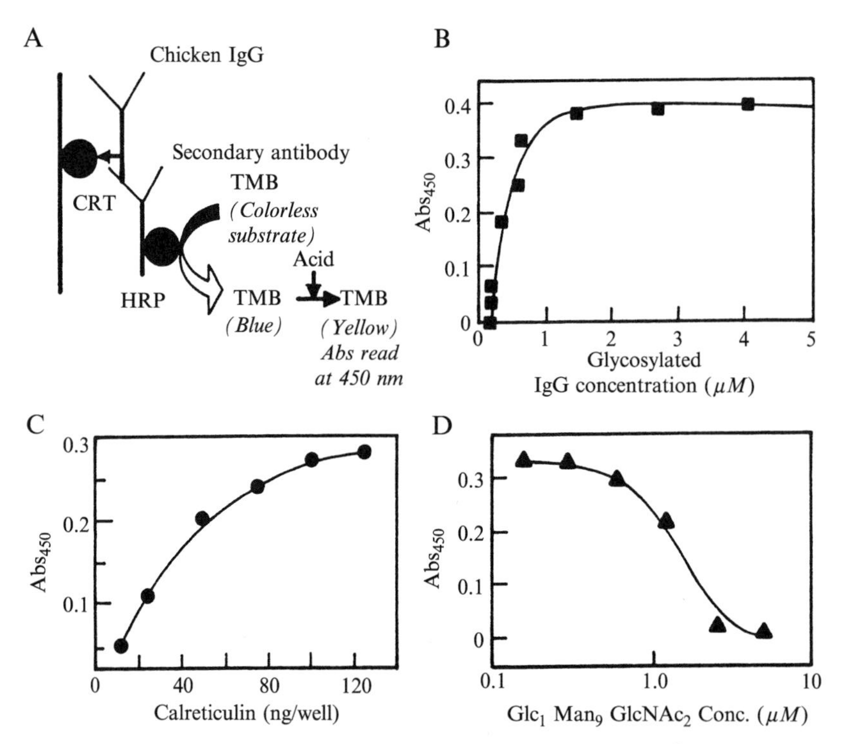

FIG. 3. Binding profiles of calreticulin to IgG by ELISA at 25°. (A) Schematic representation of ELISA for calreticulin. (B) Except for calreticulin, all other parameters, namely 40 μg/well per 50 μl of IgG, 50 μl/well of a 1:10,000 dilution of secondary antibody, were constant. (C) Constant calreticulin concentration (125 ng/well in 50 μl) and 50 μl/well of a 1:10,000 dilution of secondary antibody. (D) Inhibition curve for inhibition of IgG binding by oligosacaccharide, $Glc_1Man_9GlcNac_2$ (calreticulin, 125 ng/well in 50 μl, IgG concentration of 10 μg/well in 50 μl and 50 μl/well of 1:10,000 dilution of secondary antibody). Reproduced with permission from A. R. Patil, C. J. Thomas, and A. Surolia, *J. Biol. Chem.* **275,** 24348 (2000) and the publisher, the American Society for Biochemistry and Molecular Biology.

min and the development of the color is stopped by adding 50 μl of 1 *N* sulfuric acid before the background becomes too dark. The plates are read at 450 nm in an ELISA reader against an air blank and then subtracted from the IgG and CRT blanks after averaging. All the IgG dilutions are performed in triplicate to minimize errors in the observations.

COMMENTS ON PROCEDURE. These ELLA studies using chicken IgG as the glycoprotein substrate have provided a tool to assay calreticulin activity *in vitro*. Using this assay, immobilization of 125 ng of calreticulin on the plates leads to maximum binding of IgG as observed at 4 μM, showing a sigmoidal binding curve (Fig. 3). The association binding constant (K_a) as derived from this curve is $4.9 \times 10^6\ M^{-1}$. The IgG binding could be inhibited to 50% of the maximum, at 1.45 μM oligosaccharide ($Glc_1Man_9GlcNAc_2$), thereby indicating the affinity of the interaction to be $6.9 \times 10^5\ M^{-1}$. These values are in good agreement with the K_a values determined by SPR.[42] These studies have been instrumental in stimulating further ELLA and Biacore studies of various ER chaperone–substrate systems.[48,49]

Assay Specificity and Sources of Error. Despite automation and convenient solid-phase antibodies and conjugates, enzyme-linked assays still require a high level of care and technical skill to achieve optimal performance. Assay optimization can be regarded as an exercise in identifying the sources of error and determining the appropriate reagents that minimize their influence. Two main causes of nonspecificities in ELLA procedures are the presence of cross-reactants in the blocking solutions of BSA, gelatin, or skimmed milk powder; and antibodies or the presence of certain salt conditions that influence the avidity of lectin–glycoprotein or lectin–antibody reaction. Fundamental to evaluation is a thorough understanding of biochemistry and analytical principles, which govern the assay. Understanding how the various assay components contribute to sensitivity leads to more effective assay optimization and troubleshooting. The signal-to-noise ratio must be high for the assay to be highly sensitive. Increased incubation time, optimal temperature, improved mixing, and washing can be predicted to increase ligand binding and precision in the assays.

Acknowledgments

This work was supported by grants from the Department of Science and Technology and the Department of Biotechnology, Government of India to A.S.

[48] E. M. Frickel, R. Riek, I. Jelesarov, A. Helenius, K. Wuthrich, and L. Ellgaard, *Proc. Natl. Acad. Sci. USA* **99,** 1954 (2002).

[49] J. M. Holaska, B. E. Black, D. C. Love, J. A. Hanover, J. Leszyk, and B. M. Paschal, *J. Cell Biol.* **152,** 127 (2001).

Author Index

Numbers in parentheses are footnote reference numbers and indicate that an author's work is referred to although the name is not cited in the text.

A

C

D

E

F

H

I

J

L

M

N

O

P

S

T

W

Z

Subject Index

A

B

C

D

E

F

G

H

I

K

L

O

T

ISBN 0-12-182265-6

90000

9 780121 822651